15th International Conference "Intelligent Systems" (INTELS'22)

15th International Conference "Intelligent Systems" (INTELS'22)

Editors

Askhat Diveev
Ivan Zelinka
Arutun Avetisyan
Alexander Ilin

MDPI

Basel • Beijing • Wuhan • Barcelona • Belgrade • Novi Sad • Cluj • Manchester

Editors

Askhat Diveev
Federal Research Center "Computer Science and Control", Russian Academy of Sciences
Moscow, Russia

Ivan Zelinka
Faculty of Electrical Engineering and Computer Science, VSB-Technical University of Ostrava
Ostrava, Czech Republic

Arutun Avetisyan
Ivannikov Institute of System Programming, Russian Academy of Sciences
Moscow, Russia

Alexander Ilin
Faculty of Computational Mathematics and Cybernetics, Lomonosov Moscow State University
Moscow, Russia

Editorial Office
MDPI
St. Alban-Anlage 66
4052 Basel, Switzerland

This is a reprint of articles from the Proceedings published online in the open access journal *Engineering Proceedings* (ISSN 2673-4591) (available at: https://www.mdpi.com/2673-4591/33/1).

For citation purposes, cite each article independently as indicated on the article page online and as indicated below:

Lastname, A.A.; Lastname, B.B. Article Title. *Journal Name* **Year**, *Volume Number*, Page Range.

ISBN 978-3-0365-9288-6 (Hbk)
ISBN 978-3-0365-9289-3 (PDF)
doi.org/10.3390/books978-3-0365-9289-3

Contents

engineering proceedings

Editorial

Statement of Peer Review †

Askhat Diveev [1,*], Ivan Zelinka [2], Arutun Avetisyan [3] and Alexander Ilin [4]

1 Federal Research Center "Computer Science and Control", Russian Academy of Sciences, Moscow 119333, Russia
2 Faculty of Electrical Engineering and Computer Science, VSB-Technical University of Ostrava, 70833 Ostrava, Czech Republic; ivan.zelinka@vsb.cz
3 Ivannikov Institute of System Programming of the Russian Academy of Sciences, Moscow 109004, Russia; arut@ispras.ru
4 Faculty of Computational Mathematics and Cybernetics, Lomonosov Moscow State University, Moscow 119991, Russia; iline@cs.msu.ru
* Correspondence: aidiveev@mail.ru
† Presented at the 15th International Conference "Intelligent Systems" (INTELS'22), Moscow, Russia, 14–16 December 2022.

In submitting conference proceedings to *Engineering Proceedings*, the volume editors of the proceedings certify to the publisher that all papers published in this volume have been subjected to peer review that was administered by the volume editors. Reviews were conducted by expert referees to the professional and scientific standards expected of a proceedings journal.

- Type of peer review: single-blind; double-blind; triple-blind; open; other (please describe): single-blind
- Conference submission management system: https://intels.confreg.org/
- Number of submissions sent for review: 110
- Number of submissions accepted: 80
- Acceptance rate (number of submissions accepted/number of submissions received): 73%
- Average number of reviews per paper: 1
- Total number of reviewers involved: 14

Conflicts of Interest: The authors declare no conflict of interest.

Disclaimer/Publisher's Note: The statements, opinions and data contained in all publications are solely those of the individual author(s) and contributor(s) and not of MDPI and/or the editor(s). MDPI and/or the editor(s) disclaim responsibility for any injury to people or property resulting from any ideas, methods, instructions or products referred to in the content.

Citation: Diveev, A.; Zelinka, I.; Avetisyan, A.; Ilin, A. Statement of Peer Review. *Eng. Proc.* **2023**, *33*, 1. https://doi.org/10.3390/engproc2023033001

Published: 8 May 2023

engineering proceedings

MDPI

Proceeding Paper

Stabilization of Movement along an Optimal Trajectory and Its Solution †

Askhat Diveev [1,*], Elena Sofronova [1], Nurbek Konyrbaev [2] and Ainur Bexeitova [2]

[1] Federal Research Center "Computer Science and Control" of the Russian Academy of Sciences, Vavilova Str., 44/2, 119333 Moscow, Russia
[2] Institute of Engineering and Technology, Korkyt Ata Kyzylorda University, Aiteke bi Str. 29A, Kyzylorda 120014, Kazakhstan
* Correspondence: aidiveev@mail.ru; Tel.: +7-905-711-4427
† Presented at the 15th International Conference "Intelligent Systems" (INTELS'22), Moscow, Russia, 14–16 December 2022.

Abstract: The extended optimal control problem is considered. It is necessary to find an optimal control function, that not only provides the achievement of terminal state with optimal value of the given quality criterion, but also is implemented in the control system of a real object. It means, that the control function should depend on the state space vector, and the optimal solution should keep optimality property at small perturbations of the found solution. To solve this problem machine learning control by symbolic regression is used. In the extended optimal control problem, the problem statement of stabilization system synthesis for movement along the optimal trajectory is included. Synthesis problem is solved by the network operator method. In the synthesis problem a domain of initial conditions is considered instead of one point of initial state. It provides less sensitivity of found solution to perturbations of initial states. An example of solving the extended optimal control problem with complex phase constraints in the form of bottleneck for four quadcopters is presented.

Keywords: optimal control; stabilization system; control synthesis; symbolic regression; evolutionary computations

Citation: Diveev, A.; Sofronova, E.; Konyrbaev, N.; Bexeitova, A. Stabilization of Movement along an Optimal Trajectory and Its Solution. *Eng. Proc.* **2023**, *33*, 12. https://doi.org/10.3390/engproc2023033012

Academic Editors: Ivan Zelinka, Arutun Avetisyan and Alexander Ilin

Published: 8 June 2023

1. Introduction

Experts in the field of control are aware that the solution of the optimal control problem in the classical statement [1] cannot be implemented directly in a real control object, even in the presence of a sufficiently accurate mathematical model of the control object. The main reason for it is that the solution to the classical optimal control problem is a time function independent from coordinates of the state space vector. Therefore, the use of the obtained solution in the control object results in an open-loop control system that is sensitive even to small disturbances. The solution of the optimal control problem is only needed to obtain the optimal program trajectory in the state space. To provide a movement of a control object along this optimal trajectory, it is necessary to develop an additional control system with feedback control. It should be noted here that the development of an additional control system changes the mathematical model of the object that was used in the optimal control problem. Thus, the optimal trajectory, obtained for a control object, when solving an optimal control problem, is no longer optimal for an object with a control system that provides movement along an optimal trajectory.

There are several ways to solve the optimal control problem and provide the movement of an object along the optimal trajectory in the state space. In practice, engineers first make these robots stable relative to some point in space, and then control the movement of the robots by repositioning these stable points [2,3]. Usually, they set these points along a given trajectory to allow robots to move along that trajectory. They do not solve the optimal problem, because the control object stops near a stable point, and the task is to follow these points subsequently. To ensure stability in the state space, engineers usually apply linear P-, PI- or PID-regulators in a feedback loop.

To provide practical feasibility of the solution, it is possible to solve the general control synthesis problem instead of the optimal control problem. For this purpose, it is necessary to change one initial state into a domain of initial states and to look for a control function of the state space vector. The search for a structure of function of many arguments is a difficult task. Here, it is possible to use methods of symbolic regression [4,5]. This perspective direction allows the search of mathematical expressions in coded form by means of a special evolutionary genetic algorithm. As a result, we obtain rather complex mathematical expressions, but they can be directly applied in a real control object.

Another way to solve the optimal control problem that can be implemented in a real object is to use a synthesized control specially developed for this purpose [6]. According to this approach, firstly, the control synthesis problem is solved, and a control object becomes stable in the state space relative to some equilibrium point. After that, the positions of this stable equilibrium point are defined by the quality criterion of the optimal control problem so that, if these positions are switched over equal time intervals, a control object every time trends to the stable equilibrium point and moves along the optimal trajectory. As a result, it achieves the given terminal state with the optimal value of the given quality criterion.

In this work, the third way of solution is considered, when after solving the optimal control problem a stabilization system for motion along the optimal trajectory is obtained. This problem is a control system synthesis for tracking the given trajectory. According to this approach, a control object is included in the control system as a reference model for the trajectory generation needed for tracking. Unlike some known works on tracking [7–10], here the optimal trajectories obtained by solving the optimal control problem are considered and a tracking system synthesis problem is solved by means of symbolic regression. In this work, mathematical statements of the extended optimal control problem [11] and the tracking system synthesis problem are presented. In the experimental part, the optimal control problem for a group of four quadcopters is presented.

2. The Extended Optimal Control Problem

Consider the optimal control problem with additional requirements to the optimal trajectory.

The mathematical model of control object is given

$$\dot{\mathbf{x}} = \mathbf{f}(\mathbf{x}, \mathbf{u}), \tag{1}$$

where $\mathbf{x}$ is a state-space vector of control object, $\mathbf{x} \in \mathbb{R}^n$, $\mathbf{u}$ is a control vector, $\mathbf{u} \in \mathrm{U} \subseteq \mathbb{R}^m$, U is a compact set.

A compact set U determines restrictions on the control and is often presented in the form of algebraic inequalities

$$u_i^- \le u_i \le u_i^+,\ i = 1, \ldots, m, \tag{2}$$

where u_i is a component of the control vector, $\mathbf{u} = [u_1 \ldots u_m]^T$, u_i^-, u_i^+ are the given values, $i = 1, \ldots, m$.

The initial state is given:

$$\mathbf{x}(0) = \mathbf{x}^0 \in \mathbb{R}^n. \tag{3}$$

The terminal state is given:

$$\mathbf{x}(t_f) = \mathbf{x}^f \in \mathbb{R}^n, \tag{4}$$

where t_f is a terminal time not given but limited, $t_f \le t^+$, t_f is defined on achievement of the terminal state, t^+ is the given limit time of control process. If the control object does not reach the terminal state in t^+, then it is considered that the control object will never reach the terminal state (4).

The quality criterion is given

$$J = \int_0^{t_f} f_0(\mathbf{x}, \mathbf{u})dt \to \min_{\mathbf{u} \in \mathrm{U}}. \tag{5}$$

It is necessary to find a control function in the following form

$$\mathbf{u} = \mathbf{v}(t),\ t \in (0; t_f). \tag{6}$$

If we replace a control vector $\mathbf{u}$ by the found control function $\mathbf{v}(t)$ in the right part of ODE system (1), then the obtained ODE system

$$\dot{\mathbf{x}} = \mathbf{f}(\mathbf{x}, \mathbf{v}(t)) \tag{7}$$

will have a particular solution that reaches the given terminal state (4) from the given initial state (3) with the optimal value of the quality criterion (5).

Let $\mathbf{v}^*(t)$ be the optimal control function. In the second stage, consider the following ODE system

$$\begin{aligned} \dot{\mathbf{x}}^* &= \mathbf{f}(\mathbf{x}^*, \mathbf{v}^*(t)), \\ \dot{\mathbf{x}} &= \mathbf{f}(\mathbf{x}, \mathbf{u}). \end{aligned} \tag{8}$$

In (8), the first subsystem is the reference model that generates optimal trajectory. For the reference model, the initial state (3) is given.

For the second subsystem, a domain of initial states is given

$$\mathrm{X}_0 \subseteq \mathbb{R}^n, \tag{9}$$

where X_0 is a compact set.

It is necessary to find a control function in the following form

$$\mathbf{u} = \mathbf{h}(\mathbf{x}^* - \mathbf{x}) \in \mathrm{U}, \tag{10}$$

where x^* is the given optimal trajectory or the particular solution of the reference system in (8) from the given initial state (3).

The control function should minimize the following quality criterion

$$J_1 = \int \cdots \int_{\mathrm{X}_0} \max_t \|\mathbf{x}^*(t) - \mathbf{x}(t, \mathbf{y})\| dt \to \min_{\mathbf{h}(\mathbf{x}^* - \mathbf{x}) \in \mathrm{U}}, \tag{11}$$

where $\mathbf{y} \in \mathrm{X}_0$.

3. Symbolic Regression

To solve the synthesis problem and find the control function (10), symbolic regression is used.

Symbolic regression is a perspective computational method that allows to find mathematical expressions fully automatically. If someday humanity manages to create artificial intelligence, that is, write a program that will reflect and form the goals of the calculation, then this will be performed not manually but automatically. Symbolic regression is such an approach. It allows finding the structure and parameters of mathematical expressions automatically in a coded form. Previously, when the problem of finding a mathematical expression arose, for example, for deriving the physical laws, humans wrote mathematical expressions with accuracy to values of some constant parameters. Only for finding these parameters on experimental data, numerical methods were used, for example, the least-squares method. Note, that any artificial neural network is a function with a given structure and a big number of unknown parameters. The search of these parameters by optimization algorithms is called machine learning.

In this work, a network operator method is used to solve the tracking system synthesis problem and to find the control function as a function of a state space vector.

To code mathematical expressions, symbolic regression uses an alphabet of elementary functions. The network operator uses an alphabet of unary and binary functions and codes a mathematical expression in the form of an oriented computational graph. Consider an example of coding a mathematical expression by the network operator.

Let us be given a mathematical expression

$$y = q_1 \sin(x_1 + q_2 \cos(x_1 - x_2)). \tag{12}$$

The expression has two parameters q_1, q_2 and two variables x_1, x_2. The alphabet of the following elementary functions is enough to present this mathematical expression.

A set of unary functions

$$\mathrm{F}_1 = \{f_{1,1}(z) = z, f_{1,2}(z) = -z, f_{1,3}(z) = \sin(z), f_{1,4}(z) = \cos(z)\}, \tag{13}$$

A set of binary functions

$$\mathrm{F}_2 = \{f_{2,1}(z_1, z_2) = z_1 + z_2, f_{2,2}(z_1, z_2) = z_1 z_2\}. \tag{14}$$

Elementary function designations contain two indexes. The first index is the number of arguments of the function, and the second index is its number in the set. Using elementary functions, expression (12) can be written as

$$y = f_{2,2}(q_1, f_{1,3}(f_{2,1}(x_1, f_{2,2}(q_2, f_{1,4}(f_{2,1}(x_1, f_{1,2}(x_2))))))). \tag{15}$$

In the network operator method, functions with two arguments $f_{2,i}(z_1, z_2)$ must be commutative, associative and have a unit element, e_i. If one of the arguments is equal to a unit element, then the result of the calculations of the function is equal to the second argument, $f_{2,i}(e_i, z_2) = z_2$. For the addition function, the unit element is zero, $e_1 = 0$, and for multiplication, the unit element is one, $e_2 = 1$.

Figure 1 presents the network operator for the mathematical expression (15). Arguments of the mathematical expression are placed in the source nodes of the graph, the numbers of the binary functions are placed in other nodes of the graph and the numbers of the unary functions are located over arcs. To present the network operator in the PC memory, all nodes must be numerated so that the node number from which the arc outputs is less than the node number where the arc inputs. It can be done if the graph has no loops. Node numbers are pointed in the upper parts of the nodes.

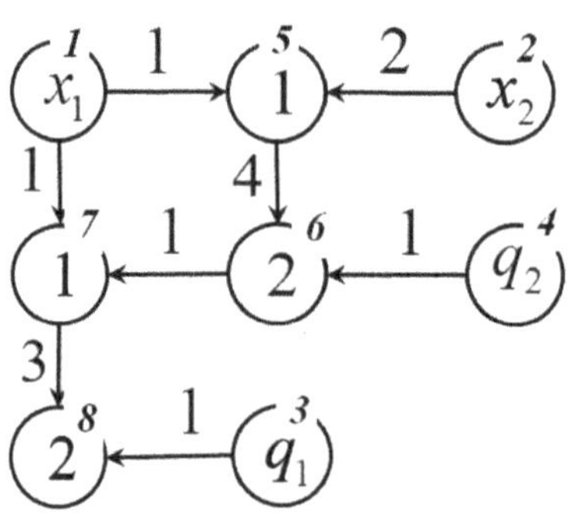

Figure 1. The network operator of the mathematical expression.

The network operator is presented in the PC memory in the form of an upper triangular integer network operator matrix. The network operator matrix for the considered mathematical expression is

$$\mathbf{\Psi} = \begin{bmatrix} 0 & 0 & 0 & 0 & 1 & 0 & 1 & 0 \\ 0 & 0 & 0 & 0 & 2 & 0 & 0 & 0 \\ 0 & 0 & 0 & 0 & 0 & 0 & 0 & 1 \\ 0 & 0 & 0 & 0 & 0 & 1 & 0 & 0 \\ 0 & 0 & 0 & 0 & 1 & 4 & 0 & 0 \\ 0 & 0 & 0 & 0 & 0 & 2 & 1 & 0 \\ 0 & 0 & 0 & 0 & 0 & 0 & 1 & 3 \\ 0 & 0 & 0 & 0 & 0 & 0 & 0 & 2 \end{bmatrix} \tag{16}$$

In the network operator matrix, the zero elements on the main diagonal point to the lines corresponding to the source nodes. The non-zero elements of the main diagonal are the binary function numbers. Non-zero elements above the main diagonal are the unary function numbers.

In order to calculate the mathematical expression by the network operator matrix, it is necessary to set a nodes vector of a dimension equal to the number of nodes of the network operator graph

$$\mathbf{z} = [z_1 \dots z_L]^T, \tag{17}$$

where $L \times L$ is a dimension of the network operator matrix.

Firstly, it is necessary to initialize the nodes vector. The components of the nodes vector corresponding to the source nodes are equal to the arguments of the mathematical expression, and the remaining components are equal to the unit elements of the corresponding binary functions.

Further, the values of the components of the nodes vector change after passing each line of the matrix and finding a non-zero non-diagonal element in it, as

$$z_j^{(i)} \leftarrow \begin{cases} f_{2,\psi_{j,j}}(z_j^{(i-1)}, f_{1,\psi_{i,j}})(z_i^{(i-1)}), \text{ if } \psi_{i,j} \neq 0 \\ z_j^{(i-1)} \end{cases} ,i = 1,\dots,L-1,\ j = i+1,\dots,L, \tag{18}$$

where $\psi_{i,j}$ is an element of the network operator matrix $\mathbf{\Psi} = [\psi_{i,j}]$, $i,j = 1,\dots,L$.

For the network operator matrix (16), a nodes vector has the following changes

$$\begin{aligned} \mathbf{z}^{(0)} &= [x_1\ x_2\ q_1\ q_2\ 0\ 1\ 0\ 1]^T, \\ \mathbf{z}^{(1)} &= [x_1\ x_2\ q_1\ q_2\ x_1\ 1\ x_1\ 1]^T, \\ \mathbf{z}^{(2)} &= [x_1\ x_2\ q_1\ q_2\ x_1 - x_2\ 1\ x_1\ 1]^T, \\ \mathbf{z}^{(3)} &= [x_1\ x_2\ q_1\ q_2\ x_1 - x_2\ 1\ x_1\ q_1]^T, \\ \mathbf{z}^{(4)} &= [x_1\ x_2\ q_1\ q_2\ x_1 - x_2\ q_2\ x_1\ q_1]^T, \\ \mathbf{z}^{(5)} &= [z_1^{(4)} \dots z_5^{(4)}\ q_2 \cos(x_1 - x_2)\ x_1\ q_1]^T, \\ \mathbf{z}^{(6)} &= [z_1^{(5)} \dots z_6^{(5)}\ \sin(x_1 + q_2 \cos(x_1 - x_2))\ q_1]^T, \\ \mathbf{z}^{(7)} &= [z_1^{(6)} \dots z_7^{(6)}\ q_1 \sin(x_1 + q_2 \cos(x_1 - x_2))]^T. \end{aligned} \tag{19}$$

For more details on the network operator and the genetic algorithm that searches for the mathematical expression using the network operator, see the monographs [4,5].

4. Computational Experiment

Consider the optimal control problem for spatial movement of four similar quadcopters.

The mathematical model of control object is

$$\begin{aligned} \dot{x}_1^j &= x_4^j, \\ \dot{x}_2^j &= x_5^j, \\ \dot{x}_3^j &= x_6^j, \\ \dot{x}_4^j &= u_4^j(\sin(u_3^j)\cos(u_2^j)\cos(u_1^j) + \sin(u_1^j)\sin(u_2^j)), \\ \dot{x}_5^j &= u_4^j\cos(u_3^j)\cos(u_1^j) - g_c, \\ \dot{x}_6^j &= u_4^j(\cos(u_2^j)\sin(u_1^j) - \cos(u_1^j)\sin(u_2^j)\sin(u_3^j)), \end{aligned} \tag{20}$$

where $j = 1, \ldots, 4$, $g_c = 9.80665$.

The control is constrained

$$\begin{aligned} -\pi/12 &\leqslant u_1^j \leqslant \pi/12, \\ -\pi &\leqslant u_2^j \leqslant \pi, \\ -\pi/12 &\leqslant u_3^j \leqslant \pi/12, \\ 0 &\leqslant u_4^j \leqslant 12, \quad j = 1, \ldots, 4. \end{aligned} \tag{21}$$

For the system (20), the initial states are given:

$$\begin{aligned} \mathbf{x}^{1,0} &= [0\ 5\ 0\ 0\ 0\ 0]^T, & \mathbf{x}^{2,0} &= [10\ 5\ 0\ 0\ 0\ 0]^T, \\ \mathbf{x}^{3,0} &= [0\ 5\ 10\ 0\ 0\ 0]^T, & \mathbf{x}^{1,0} &= [10\ 5\ 10\ 0\ 0\ 0]^T. \end{aligned} \tag{22}$$

The terminal states are given

$$\begin{aligned} \mathbf{x}^{1,f} &= [10\ 5\ 10\ 0\ 0\ 0]^T, & \mathbf{x}^{2,f} &= [0\ 5\ 10\ 0\ 0\ 0]^T, \\ \mathbf{x}^{3,f} &= [10\ 5\ 0\ 0\ 0\ 0]^T, & \mathbf{x}^{1,f} &= [0\ 5\ 0\ 0\ 0\ 0]^T. \end{aligned} \tag{23}$$

The static phase constraints are given

$$\varphi_i(\mathbf{x}^j) = r_i - \sqrt{(x_i - x_1^j)^2 + (z_i - x_3^j)^2} \leqslant 0, \tag{24}$$

where $i = 1, \ldots, 4$, $j = 1, \ldots, 4$, $r_1 = r_2 = r_3 = r_4 = 1.5$, $x_1 = 1.5$, $z_1 = 2.5$, $x_2 = 1.5$, $z_2 = 7.5$, $x_3 = 8.5$, $z_3 = 2.5$, $x_4 = 8.5$, $z_4 = 7.5$.

The dynamic phase constraints are given

$$\chi(\mathbf{x}^{j_1}, \mathbf{x}^{j_2}) = s - \sqrt{\sum_{i=1}^{3}(x_i^{j_1} - x_i^{j_2})^2} \leqslant 0, \tag{25}$$

where $j_1, j_2 \in \{1, 2, 3, 4\}$, $j_1 \neq j_2$, $s = 1.5$.

Initially, the optimal control problem is solved by the direct approach. For this purpose, the time axis is divided into equal time intervals. In each interval, a control function is found as a piecewise-linear time function. It is necessary to find the values of the constant parameters on the boundaries of the time intervals. Taking into account the constraints on control, the control function is

$$u_i^j = \begin{cases} u_i^{j,+}, \text{ if } \tilde{u}_i^j > u_i^{j,+} \\ u_i^{j,-}, \text{ if } \tilde{u}_i^j < u_i^{j,-} \\ \tilde{u}_i^j, \text{ otherwise} \end{cases}, \tag{26}$$

where

$$\tilde{u}_i^j = q_{i+(k-1)m} + (q_{i+km} - q_{i+(k-1)m})\frac{t-(k-1)\Delta t}{\Delta t}, \quad (k-1)\Delta t \leqslant t \leqslant k\Delta t, \tag{27}$$

$i = 1, \ldots, m$, m is a dimension of the control vector, $m = 4$, $k = 1, \ldots, N$, N is the number of intervals.

It is necessary to find the control functions for each control object as (26). The time limit of the control process is $t^+ = 5.6$. The time interval is $\Delta t = 0.4$. To solve the optimal control problem, it is necessary to find 240 parameters, $4 \cdot m \cdot (N+1) = 4 \cdot 4 \cdot \left(\left\lfloor \frac{5.6}{0.4} \right\rfloor + 1\right) = 4 \cdot 4 \cdot 15 = 240$.

To solve the optimal control problem by direct approach, the evolutionary hybrid algorithm is used [6,12]. The algorithm includes evolutionary transformations of three evolutionary algorithms, the genetic algorithm (GA) [13], particle swarm optimization (PSO) algorithm [14] and grey wolf optimizer (GWO) algorithm [15].

Figure 2 shows projections of the found optimal trajectories on the horizontal plane. Here, the solid line is for the trajectory of the first robot, dashed line—the second robot, dot-dashed line—the third robot, dots—the forth robot, and circles are the static phase constraints.

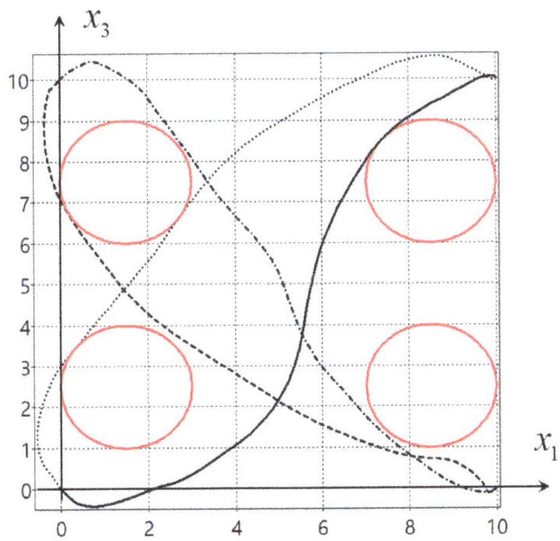

Figure 2. Optimal trajectories on the horizontal plane.

At the second stage, the tracking system synthesis problem is solved. To solve the synthesis problem, we used the network operator method.

The domain of initial states (9) was replaced by the set of points

$$x_i(0) = x_i^{j,0} \pm 0.2, \quad i = 1,2,3, \quad j = 1, \ldots, 4. \tag{28}$$

In the quality criterion (11), an integral over domain was replaced by the sum of all $3^3 = 27$ initial states.

The following solution was found by the network operator method

$$u_i^j = \begin{cases} u_i^{j,+}, \text{if } \hat{u}_i^j > u_i^{j,+} \\ u_i^{j,-}, \text{if } \hat{u}_i^j < u_i^{j,-} \\ \hat{u}_i^j, \text{otherwise} \end{cases} \tag{29}$$

where

$$\hat{u}_1^j = \mu(A), \tag{30}$$

$$\hat{u}_2^j = \mu(A) - \mu^3(A), \tag{31}$$

$$\hat{u}_3^j = \rho_{17}(\hat{u}_2^j) + \rho_{19}(B + \mu(A)) + \rho_{17}(C), \tag{32}$$

$$\begin{aligned} \hat{u}_4^j \;=\; & \hat{u}_3^j + \ln(\hat{u}_2^j) + \mathrm{sgn}(B + \mu(A))\sqrt{|B + \mu(A)|} + \rho_{19}(B) + \\ & \arctan(D) + \mathrm{sgn}(E) + \arctan(F) + \exp(q_2(x_2^{j,*} - x_2^j)) + \sqrt{q_1}, \end{aligned} \tag{33}$$

$$A = q_3(x_3^{j,*} - x_3^j) + q_6(x_6^{j,*} - x_6^j),$$

$$B = G + \tanh(E + \sin(C)\sqrt[3]{F}) + \exp(H) - C,$$

$$C = q_1(x_1^{j,*} - x_1^j) + q_4(x_4^{j,*} - x_4^j)$$

$$D = E + \sin(C)\sqrt[3]{F} + \tanh(E) + \rho_{18}(V),$$

$$E = F + \arctan(A) + \arctan(H) - V,$$

$$F = H + \mathrm{sgn}(x_5^{j,*} - x_5^j) + (x_2^{j,*} - x_2^j)^3,$$

$$G = \exp(D) + \cos(q_6(x_6^{j,*} - x_6^j)) + \mathrm{sgn}(E + \sin(C)\sqrt[3]{F})\sqrt{|E + \sin(C)\sqrt[3]{F}|} + E^{-1} - q_6,$$

$$H = \rho_{17}(A) + V^3 + C + \vartheta(q_5(x_5^{j,*} - x_5^j)) + x_5^{j,*} - x_5^j,$$

$$V = \sin(q_6(x_6^{j,*} - x_6^j)) + q_5(x_5^{j,*} - x_5^j) + \cos(q_1) + \vartheta(x_2^{j,*} - x_2^j),$$

$$\vartheta(\alpha) = \begin{cases} 1, \text{ if } \alpha > 0 \\ 0, \text{ otherwise} \end{cases}, \quad \mu(\alpha) = \begin{cases} \alpha, \text{ if } |\alpha| < 1 \\ \mathrm{sgn}(\alpha), \text{ otherwise} \end{cases},$$

$\rho_{17}(\alpha) = \mathrm{sgn}(\alpha)\ln(|\alpha| + 1)$, $\rho_{18}(\alpha) = \mathrm{sgn}(\alpha)(\exp(|\alpha|) - 1)$, $\rho_{19}(\alpha) = \mathrm{sgn}(\alpha)\exp(-|\alpha|)$, $q_1 = 11.52295$, $q_2 = 7.54395$, $q_3 = 15.45825$, $q_4 = 12.39771$, $q_5 = 15.94922$, $q_6 = 10.84229$, $\mathbf{x}^{j,*} = [x_1^{j,*} \ldots x_6^{j,*}]^T$ is a program trajectory, generated by the reference model.

Figure 3 shows the trajectories from eight random initial states of given domain (28). Here, blue lines mark trajectories from (3) and black lines mark trajectories from (28). As it can be seen from the experiment, the obtained control system (29) for optimal trajectory tracking guarantees the presence of an attracting neighborhood for the optimal trajectory in the extended optimal control problem.

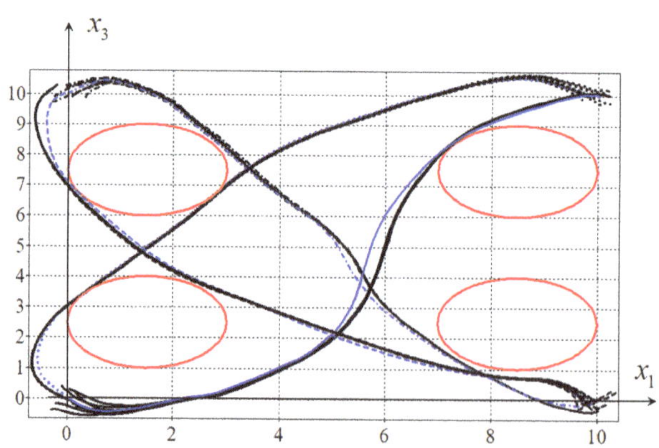

Figure 3. The optimal trajectory and disturbances trajectories from eight initial states.

5. Results

The paper presents the statements of the optimal control problem and the stabilization system synthesis with respect to the optimal program trajectory. To solve the synthesis problem, machine learning by the network operator method was used. A brief description of the network operator method is presented. A computational example is given for a group of four quadcopters. To analyze the quality of the obtained solutions, we simulated the synthesized stabilization system under perturbations of the initial state. The experiments showed the high quality of the stabilization system.

6. Discussion

The implementation of the control system is related to the solution of the control synthesis problem. As a result of solving the synthesis problem, we obtain a feasible feedback control. It is a complex problem that does not have universal computational methods. In this paper, we applied an approach based on the machine learning by the symbolic regression method. The method is universal, but requires lots of calculations. It is necessary to continue the study of this approach for other complex control problems.

Author Contributions: Conceptualization, A.D.; methodology, A.D.; software, A.D. and E.S.; validation, A.D., E.S., N.K. and A.B.; investigation, A.D. and E.S.; writing—original draft preparation, A.D., E.S. and N.K.; writing—review and editing, E.S.; visualization, A.B.; supervision, N.K.; project administration, N.K.; funding acquisition, N.K. All authors have read and agreed to the published version of the manuscript.

Funding: This research was funded by the Science Committee of the Ministry of Education and Science of the Republic of Kazakhstan (Grant No. AP14869851).

Institutional Review Board Statement: Not applicable.

Informed Consent Statement: Not applicable.

Data Availability Statement: Experimental data is available at: https://cloud.mail.ru/public/L51J/oDPtP3Xhc (accessed on 7 June 2023).

Conflicts of Interest: The authors declare no conflict of interest.

References

1. Pontryagin, L.S.; Boltyanskii, V.G.; Gamkrelidze, R.V.; Mishchenko, E.F. *The Mathematical Theory of Optimal Process*; Gordon and Breach Science Publishers: New York, NY, USA; London, UK; Paris, France; Montreux, Switzerland; Tokyo, Japan, 1985; Volume 4, p. 360.
2. Egerstedt, M.; Hu, X.; Stotsky, A. Control of a Car-Like Robot Using a Virtual Vehicle Approach. *IEEE Trans. Robot. Autom.* **2001**, *17*, 947–951. [CrossRef]
3. Walsh, G.; Tilbury, D.; Sastry, S.; Murray, R.; Laumond, J.P. Stabilization of Trajectories for Systems with Nonholonomic Constraints. *IEEE Trans. Autom. Control* **1994**, *1*, 216–222. [CrossRef]
4. Diveev, A.I.; Sofronova, E.A. The Network Operator Method for Search of the Most Suitable Mathematical Equation. In *Bio-Inspired Computational Algorithms and Their Applications*; Gao, C., Ed.; Intech: Rijeka, Croatia, 2013; pp. 19–42.
5. Diveev, A.I.; Shmalko, E.Y. *Machine Learning Control by Symbolic Regression*; Springer: Cham, Switzerleand, 2021; p. 155.
6. Diveev, A.; Sofronova, E. Synthesized Control for Optimal Control Problem of Motion Along the Program Trajectory. In Proceedings of the 8th International Conference on Control, Decision and Information Technologies (CoDIT), Istanbul, Turkey, 17–20 May 2022.
7. Udin, N. Trajectory Tracking Control System Design For Autonomous Two-Wheeled Robot. *Inform. Telecommun. Electron.* **2018**, *3*, 90–97. [CrossRef]
8. Samir, A.; Hammad, A.; Hafez, A.; Mansour, H. Quadcopter Trajectory Tracking Control using State-Feedback Control with Integral Action. *Int. J. Comput. Appl.* **2017**, *9*, 1–7. [CrossRef]
9. Siti, I.; Mjahed, M.; Ayad, H.; El Kari, A. New Trajectory Tracking Approach for a Quadcopter Using Genetic Algorithm and Reference Model Methods. *Appl. Sci.* **2019**, *9*, 1780. [CrossRef]
10. Dong, W.; Gu, G.-Y.; Zhu, X.; Ding, H. High-performance trajectory tracking control of a quadrotor with disturbance observer. *Sens. Actuators A Phys.* **2014**, *1*, 67–77. [CrossRef]
11. Diveev, A.I. Refined Optimal Control Problem and Its Solution Using Symbolic Regression. In *Lecture Notes in Networks and Systems*; Springer: Cham, Switzerleand, 2022; Volume 507, pp. 294–305.
12. Diveev, A. Hybrid evolutionary algorithm for optimal control problem. In *Intelligent Systems and Applications*; Arai, K., Ed.; IntelliSys: Amsterdam, The Netherlands, 2022; Volume 543, pp. 726–738.

13. Davis, L. *Handbook of Genetic Algorithms*; Van Nostrand Reinhold: New York, NY, USA, 1991.
14. Kennedy, J.; Eberhart, R. Particle swarm optimization. In Proceedings of the IEEE International Conference on Neural Networks, Perth, Australia, 27 November–1 December 1995; Volume IV, pp. 1942–1948.
15. Mirjalili, S.; Mirjalili, S.M.; Lewis, A. Grey Wolf Optimizer. *Adv. Eng. Softw.* **2014**, *69*, 46–61. [CrossRef]

engineering proceedings

MDPI

Proceeding Paper

Quality of Labeled Data in Machine Learning: Common Sense and the Controversial Effect for User Behavior Models †

Maxim Bakaev * and Vladimir Khvorostov

Faculty of Automation and Computer Engineering, Novosibirsk State Technical University, Pr. K. Marksa 20, 630073 Novosibirsk, Russia; xvorostov@corp.nstu.ru

* Correspondence: bakaev@corp.nstu.ru

† Presented at the 15th International Conference "Intelligent Systems" (INTELS'22), Moscow, Russia, 14–16 December 2022.

Abstract: Intelligent systems today are increasingly required to predict or imitate human perception and behavior. In this, feature-based Machine Learning (ML) models are still common, since collecting appropriate training data from human subjects for the data-hungry Deep Learning models is costly. Considerable effort is put into ensuring data quality, particularly in crowd-annotation platforms (e.g., Amazon MTurk), where fees of top workers can be several times higher than the median. The common knowledge is that quality of input data is beneficial for the end quality of ML models, though quantitative estimations of the effect are rare. In our study, we investigate how labeled data quality affects the accuracy of models that predict users' subjective impressions—per the scales of Complexity, Aesthetics and Orderliness assessed by 70 subjects. The material, about 500 web page screenshots, was also labeled by 11 workers of varying diligence, whose work quality was validated by another 20 verifiers. Unexpectedly, we found significant *negative* correlations between the workers' precision and R^2s of the models, for two out of the three scales ($r_{11} = -0.768$ for Aesthetics, $r_{11} = -0.644$ for Orderliness). We speculate that the controversial effect might be explained by a bias in the indiligent labelers' output that corresponds to subjectivity in human perception of visual objects.

Keywords: web interfaces; intelligent systems; machine learning; image recognition

Citation: Bakaev, M.; Khvorostov, V. Quality of Labeled Data in Machine Learning: Common Sense and the Controversial Effect for User Behavior Models. *Eng. Proc.* **2023**, *33*, 3. https://doi.org/10.3390/engproc2023033003

Academic Editors: Askhat Diveev, Ivan Zelinka, Arutun Avetisyan and Alexander Ilin

Published: 9 May 2023

1. Introduction

One of the implicit assumptions in Machine Learning (ML) is that the data that get through the preliminary screenings and tweaks to the model training stage are appropriate. As for ML models that seek to predict or simulate human behavior, such as user behavior models (UBMs) in the field of Human–Computer Interaction (HCI), the situation is rather more sophisticated. The actual interaction-related data, which are generally the input of the predictive UBMs [1], arguably cannot be "bad", as long as they reflect the human "imperfection". However, there are also increasingly important subjective dimensions, from perceptional "how pleasant is our website design" in HCI to "how likely is it that you would recommend our service to a friend" in marketing. By definition, the subjective impressions are usually directly provisioned by human subjects—although indirect methods do exist, e.g., facial emotion recognition. Correspondingly, Deep Learning is slow to take off in this field, and an ample share of the models are feature based and rely on labeled data and the subjective assessments.

There is a general consensus that inaccurately annotated data are a hindrance and that the labeled data quality does not come for free. In micro-task platforms, such as Amazon Mechanical Turk (MTurk), filtering of crowdworkers can be carried out by a reputation that is principally based on the Approval Rate supplied by task requesters [2]. The fees charged by higher-paid workers are about four times above the *median* ones in MTurk [3], even though it has been shown that even top workers can be indiligent [4].

Reputation might have seemed an easy solution to crowd-labeled data quality a decade ago [2], but the arsenal of methods and tools has been rapidly expanding since then [5], as we subsequently outline in Section 2.1. The currently mainstream data quality control methods are *majority/group consensus* and *ground truth*, which necessarily imply redundancy (several workers performing the same task), wasting up to 33% of the output.

Even if data labeling work is carried out by volunteers and is technically free, their limited effort should be used efficiently too. Although volunteers generally have higher motivation than crowdworkers, redundancy might still be necessary to reach the certainty thresholds [6]. Setting the latter is actually a major problem for a requester, which we believe is not adequately covered in existing studies. Similarly to software debugging, more is always better, and there is no hard threshold to improving the quality of the data, only the one advised by practicability. Many developments to improve input data for UBMs, e.g., the enhanced version of the robust *Aalto Interface Metrics* (https://github.com/aalto-ui/aim, accessed on 1 June 2022) [7], are underway with the best intentions. Unfortunately, estimating the concrete "return on investment" in data quality remains problematic, as quantitative studies of its end effect in ML are scarce.

In our paper, we explore the relation between the completeness and precision of the input data produced by 11 human labelers and the quality of the ensuing 33 user behavior models built for 487 web page screenshots assessed by another 70 participants. Rather unexpectedly, we find that the significant correlation between the labelers' precision and the quality of the models constructed for the subjective scales of aesthetics and orderliness is **negative**. We attribute this preposterous result to the bias in indiligent labelers that brings their output closer to some subjective dimensions of human visual perception. We did not find any significant correlations for the labeling of completeness—even for complexity, which is known to be affected by the number of visual elements. Our results question the traditional data quality measures' applicability for human-related data, although further research is necessary.

The outcome has been preliminary reported and discussed at the *2021 Fall Conference of Ergonomic Society of Korea (ESK)*. In the current paper, we present the extended version of our results, referencing some of our previous related publications, such as [8,9]. In Section 2, we briefly review the research relevant to human behavior data quality in ML and describe our experiment. In Section 3, we construct the models and analyze the effects of the input data on their quality. In the final section, we discuss the findings and their possible causes, and outline directions for further research.

2. Method and Related Work

2.1. Data Quality Control in ML

As noted by philosophers long ago, the concept of *good* is very subjective. In relation to ML, it was recently demonstrated that the understanding of "good data" varies considerably for different stakeholders [10]. The concept of *quality*, though more objective and operational, is domain specific [11] and multi-dimensional [12]. With respect to data, it commonly involves the aspects of completeness, consistency, lack of duplicates, accuracy, timeliness for the purpose, and so on—some researchers identify as many as 20 dimensions. Since ML is predominantly concerned with *precision* and *recall* of the models, it associates data quality for the most part with *completeness* and *accuracy*.

The importance of these two dimensions in data quality was well recognized even before the ML era, and the related methodologies were classified as the ones helping "selection, customization, and application of data quality assessment and improvement techniques" [13]. Currently, the data quality control incorporates techniques for data collection planning, cleaning, profiling, evaluation, monitoring, etc. About a decade ago, strong focus in the field was established concerning crowd data, due to rapid advancement of crowdworking platforms such as Amazon MTurk (2005), microworkers.com (2009), Yandex.Toloka (2014), etc. The whole family of related meta-tools dedicated to data quality control emerged, such as CDAS (2012), Crowd Truth (2014), iCrowd (2015), DOCS for

AMT (2015), and others [9]. A comprehensive review of quality control in crowdsourcing can be found in [5], where the methods are organized into three major groups: individual, group and computation based. The former two generally imply involvement of humans in the assessment of the annotators or of the tasks' output.

It should be noted though that there has been a certain decrease in research enthusiasm towards crowd data since then, as the involved disadvantages had been acknowledged [4]. ML and Intelligent Systems came to rely more on unstructured and uncontrolled data sources [14], see Big Data [11,15] and data scrapped from the web [16]. A recent related publication carefully catalogs the software tools for data quality measurement and monitoring, listing a whopping 667 of them [17]. All in all, the quantitative engineering of data quality is better developed in the fields where data generation is easier to control. A recent example of such a field is IoT (see review in [18]), while the most established one is industrial data, where datasets are well structured and plentiful. Researchers in the field of industrial data quality already formulate it as a *dataset selection problem* and propose, e.g., the criteria of *estimated relative return improvement* and *estimated action stochasticity* [19]. However, those working with human-related data more often than not have no luxury of choosing between several datasets relevant to their specific problem.

2.2. Human Factor in Data Quality

The comparative rarity of reusing human behavior-related data outside of reproducibility and meta-analysis studies (e.g., [20] using the dataset from [8]) is partially due to its high value. The latter mainly comes from costly human time needed to generate or label the data, but its potential economic value may be involved too—think of social networks users' behavior data. Another reason that decreases the chance for finding appropriate data for a specific problem is that human data are a task too and context-dependent, and there is never a perfect match of factors and conditions. Moreover, their quality is arguably less formalizable on the scale from "good" to "bad", and the emerging concept of "fitness-for-use" [21] might prove to be more appropriate than "quality".

In the dawn of the AI/ML era, the human factor in data was rather considered a nuisance (cf. user needs in the era of mainframes). For instance, *10 reasons for bad data quality* comprehensively listed by Lee et al. in 2006, include "subjective judgments during data generation" [22]. Lately though, there is more recognition that human-related data are special, and specific quality dimensions are introduced, such as ethical ones [23]. The latter are arguably a response to the recently highlighted "inappropriate" behavior of trained AIs, who started to demonstrate "racist", "sexist", or "offensive" behavior [24]—just in accordance with the patterns they found in human-generated training data.

Still, the urgency of ML methods to describe human behavior is widely recognized, and UBMs that both incorporate domain knowledge and are trained on practical data is a popular implementation. The models' output are certain key performance characteristics—the examples in HCI field are success rate, time to complete a task, dimensions of subjective satisfaction, etc. The corresponding input data generally would need to specify the characteristics of the target users and the parameters of a candidate UI [1]. The techniques for parameterizing the UI can rely on manual labeling, on automated design mining algorithms (see in [25]), or on their combination [9]. Indeed, the algorithms for calculating the features are already numerous, and the developers put in a considerable effort in improving them [7]). However, the data quality studies in the field rather focus on adherence to "best practices" [26] and the reasons leading to "bad" models [27]. Quantitative studies of the data effect are rare, if any.

So, we undertook the following experimental study to relate the measured dimensions of the input data quality and the quality parameters for some simple UBMs.

2.3. *The Experiment Description*

2.3.1. Material

The material in our experiment was screenshots of website homepages belonging to universities and colleges from all over the world (but only their English versions). First, we automatically collected 10,639 screenshots in PNG format using a dedicated Python script crawling through various catalogs, DBPedia, etc. Then, we manually selected 497 of them for the experiment (see [8] for more detail)—hereafter, references as the UIs. To ensure better diversity of UI elements, the screenshots were made for full web pages, not just of the part above the fold or of a fixed size.

2.3.2. Procedure

UI Assessment

In a dedicated online survey (see details in [8]), the participants provided the subjective assessments of their impressions for each UI, per the three visual perception scales that we employed:

- How visually complex is the UI: *Complexity;*
- How aesthetically pleasant is the UI: *Aesthetics;*
- How orderly is the UI: *Orderliness.*

Complexity and aesthetics were elected as arguably the most popular dimensions in studies of subjective visual perception [20]. Orderliness was added mostly for the purpose of validity control of the assessments, as most studies in HCI are uniform about the positive correlation of UI regularity with aesthetics and the negative one with complexity. For each of three the scales, Likert ratings were used (1—lowest, 7—highest). The participants were instructed to provide their honest subjective assessments and were told that there are no right or wrong answers. The screenshots were randomly assigned to each participant successively, and the completeness of the assessment for all the 3 scales per UI was mandatory and controlled by the survey software.

UI Labeling

The labelers used LabelImg (Version 1.8.1, from https://github.com/tzutalin/labelImg, accessed on 1 June 2022), a third-party dedicated software tool that saves the output as XML files in PASCAL VOC format. They were asked to draw bounding rectangles around UI elements in the screenshots, as precisely as possible, and to choose one of the 20 pre-defined classes for the element: *image, background image, text, textinput, link, button, etc.* (see the complete list in [9]). The participants were provided with the written instruction on UI labeling and on technical usage of LabelImg and asked to process as many UI elements in each UI as possible. The screenshots were distributed among them near evenly, but no random assignment was performed.

The Labeling Verification

For each UI element in each screenshot, the verifiers could specify the labeling as *correct* or *incorrect.* In addition, for each UI the they were asked to subjectively assess completeness, i.e., if all the visible UI elements had been labeled, on the scale from 1 (very few elements) to 100 (all elements). The verifiers had the written instruction with recommendations for making the *correct/incorrect* decision, based on the UI elements' bounding box precision and the correct specification of its class. To support the verification procedure, we have developed a custom web-based software. The previously labeled UIs were distributed among the verifiers near evenly, but without a random assignment.

2.3.3. Subjects

There were 3 groups of human participants, mostly students of Novosibirsk State Technical University (NSTU), who performed the aforementioned activities:

1. *The UI assessment* was performed by 70 participants (43 females, 27 males), whose age ranged from 18 to 29 (mean 20.86, SD = 1.75).
2. *The UI labeling* was performed a few months later by another 11 participants (6 male, 5 female), with the age ranging from 20 to 24 (mean = 20.5, SD = 0.74).
3. *The verification* of the labelers' output was performed a few months later by yet another 20 participants (10 male, 10 female), whose ages ranged from 20 to 22 (mean = 21.1, SD = 0.45).

All the participants took part in the study voluntarily, and informed consent was obtained. They had normal or corrected to normal vision and reasonably high experience in the general usage of IT.

2.3.4. Design

The mean UI assessment ratings per the screenshots on the three scales of complexity, aesthetics, and orderliness (ScaleC, ScaleA, and ScaleO, respectively) were used as the *output variables* for the 3 × 11 = 33 user behavior models that we would construct for each scale and each labeler.

The *input data* for the models were 8 factors, whose values we automatically calculated for each UI from the labeling data, using our dedicated Python script:

1. number of all UI elements,
2. number of text elements,
3. share of the text elements' area in the screenshot,
4. number of image elements,
5. share of the image elements' area,
6. number of background image elements,
7. share of the background image elements' area,
8. share of whitespace (the screenshot area minus all the other labeled elements).

From the 20 labeled classes, we deliberately chose the most visually prominent ones: *text*, *image* and *background image*, since our experiment implied visual perception of the material, but no interaction with the UIs—hence, no *link*, *radiobutton*, *selectbox*, *textinput*, and so on.

So, our experiment had between-subject design. The main independent variables were *subjective completeness (SC)* and *Precision*, averaged for each of the 11 labelers:

$$Precision = \frac{correct}{correct + incorrect}. \quad (1)$$

The (derived) dependent variables were the quality parameters (R^2s) of the user behavior models. We also controlled for another derived variable, the number of screenshots processed by each labeler (UI).

Our **hypothesis** was that higher SC and precision, corresponding to better quality of the labeling data, should result in the better quality of the models (R^2s).

3. Results

3.1. Descriptive Statistics

In total, we collected 12705 assessments for the 497 UIs. Further, the 11 labelers specified 42,716 elements in 495 UIs (see [Table 1] in [9]), and the quality of their work was evaluated by 20 verifiers. Some UIs had technical problems or incomplete evaluations, so, we remained with 487 valid UIs (98.0%), for which the descriptive statistics are presented in Table 1. The first and second names of the labelers are abbreviated in the IDs.

Table 1. The descriptive statistics per the labelers (M ± SD).

	UI Labeling		UI Assessment		
ID	**UIs**	**Elements**	**ScaleC**	**ScaleA**	**ScaleO**
AA	54	4802	3.67 ± 0.55	4.20 ± 0.86	4.43 ± 0.63
GD	44	3520	3.55 ± 0.56	4.07 ± 0.74	4.47 ± 0.53
KK	44	3927	3.33 ± 0.59	4.32 ± 0.71	4.59 ± 0.57
MA	44	5349	3.60 ± 0.63	4.02 ± 0.76	4.36 ± 0.56
NE	44	4994	3.57 ± 0.65	3.97 ± 0.90	4.34 ± 0.64
PV	43	4544	3.69 ± 0.74	4.34 ± 0.74	4.66 ± 0.58
PE	42	2569	3.69 ± 0.64	3.79 ± 1.07	4.16 ± 0.80
SV	43	3737	3.54 ± 0.63	4.22 ± 0.90	4.46 ± 0.68
ShM	41	1675	3.55 ± 0.71	4.05 ± 0.88	4.43 ± 0.56
SoM	45	3266	3.62 ± 0.73	4.25 ± 0.91	4.44 ± 0.68
VY	43	3630	3.47 ± 0.61	4.07 ± 0.83	4.52 ± 0.67
Total	**487**	**42,013**	**3.57 ± 0.64**	**4.12 ± 0.86**	**4.44 ± 0.64**

To check for the homogeneity of the UI assessments per the 11 labelers, we ran ANOVA tests for all three scales. We found a barely significant effect of ID only on ScaleO ($F_{10,476} = 1.87$, $p = 0.047$), but not on ScaleC ($F_{10,476} = 1.21$, $p = 0.284$) or ScaleA ($F_{10,476} = 1.63$, $p = 0.096$). The post-hoc test for ScaleO (Tukey HSD, since there were many levels of the independent variables) found significant difference (at $\alpha = 0.05$) only between labelers PV and PE ($p = 0.012$). The variances were not different ($p = 0.372$), so the ANOVA assumptions were met. Pearson correlations for the assessments per UIs were highly significant between ScaleA and ScaleO ($r_{487} = 0.771$, $p < 0.001$), as well as between ScaleC and ScaleO ($r_{487} = -0.145$, $p = 0.001$), but not between ScaleC and ScaleA.

In the verification, 37,053 labeled elements were specified as correct and 4967 as incorrect, and the mean Precision per labelers was 88.7%, which indicates a reasonably good work quality. The Pearson correlation between Precision and SC per labelers was not significant ($p = 0.727$), which suggests that these two aspects of UI labeling quality are distinct. The correlation between SC and the average number of correct objects was significant ($r_{11} = 0.622$, $p = 0.041$), unlike for the number of all labeled objects ($r_{11} = 0.170$, $p = 0.618$), which reinforces the meaningfulness of the verification.

3.2. The Effect of the Input Data Quality in the Models

To construct the UBMs, we relied on simple linear regression, since we only had a limited number of data samples (41–54) for each labeler. So, we built 33 models, each having the same 8 factors calculated from each labeler's output. The R^2s obtained for the models are presented in Table 2, together with the mean labelers' quality parameters obtained from the UI's verifications.

Since the number of screenshots processed by each labeler (UI) was not exactly the same (see in Table 1), we checked its correlations with R^2s for each of the three scales. We found that neither of the Pearson correlations was significant at $\alpha = 0.05$, so treating all labelers' models universally is justified.

The subsequent Pearson correlations analysis revealed that the SC did not have a significant correlation (at $\alpha = 0.05$) with the models' quality parameter (R^2) for either of the scales. Even for ScaleC, the correlation was $r_{11} = -0.062$ ($p = 0.856$), whereas the visual complexity of a user interfaces is known to be influenced by the number of elements [25]. For the sake of checking the conceptual validity of our SC variable, we also checked the association between the factual *average number of elements per UI* for each labeler and the R^2s. Again, neither of the Pearson correlations were significant (at $\alpha = 0.05$), the correlation for ScaleC being $r_{11} = 0.274$ ($p = 0.415$).

Table 2. The labelers' and the models' quality.

UI Labeling Quality			Models' Quality (R^2s)		
ID	SC	Precision	ScaleC	ScaleA	ScaleO
AA	73.0%	89.0%	0.108	0.149	0.114
GD	84.3%	89.9%	0.261	0.345	0.222
KK	82.5%	95.5%	0.261	0.252	0.152
MA	75.1%	72.0%	0.362	0.486	0.295
NE	78.3%	85.1%	0.316	0.488	0.416
PV	81.7%	91.6%	0.363	0.289	0.199
PE	72.0%	77.9%	0.165	0.568	0.611
SV	80.4%	97.4%	0.277	0.176	0.213
ShM	77.5%	89.5%	0.337	0.324	0.215
SoM	56.0%	95.9%	0.304	0.309	0.198
VY	95.5%	92.8%	0.204	0.110	0.169
Avg.	**77.8%**	**88.7%**	**0.269**	**0.318**	**0.255**

For precision, we found significant **negative** correlations with the R^2s for ScaleA ($r_{11} = -0.768, p = 0.006$) and ScaleO ($r_{11} = -0.644$, $p = 0.032$), but not for ScaleC ($r_{11} = -0.051$, $p = 0.883$). Recognizing the possible inaccuracy of our quality measures, we tried treating R^2 and the precision as ordinal variables—this is rather practical, since task requesters are often interested in only accepting the output from the best labelers. However, the results did not change very much for Kendall's tau-b correlation measure: $\tau_{11} = -0.491, p = 0.036$ for ScaleA and $\tau_{11} = -0.418, p = 0.073$ for ScaleO.

4. Discussion and Conclusions

Seeking to explore the effect of input data quality, we undertook an experimental study with 101 human participants and 497 web UIs. Our assumption was that better quality of the UI labeling should result in better quality of UBMs.

Contrary to our expectations, we found significant *negative* correlations between the labeling quality parameters and the resulting models' quality (see Table 2) for the subjective impression dimensions of aesthetics ($r_{11} = -0.768$) and orderliness ($r_{11} = -0.644$). Before deciding to report the negative research results in the current paper, we revisited the possible biases. However, the following considerations re-enforce the validity of our findings:

1. *Invalid UI assessment:* there was almost no significant difference in the distribution of the ratings per the labelers.
2. *Invalid UI labeling:* dimensions of precision (88.7%) and SC (77.8%) indicated high work quality and were distinct.
3. *Invalid Verification:* SC was correlated ($r_{11} = 0.622$) with the number of correct objects, but not with the number of all objects.
4. *Invalid subjective impressions scales:* as expected, ScaleA and ScaleO had significant positive correlation ($r_{487} = 0.771$), while ScaleC and ScaleO had significant negative correlation ($r_{487} = -0.145$). The relation between ScaleA and ScaleC was more controversial, as known from the literature [20], and we did not find a significant correlation.
5. *Imperfection in quality measurement:* we tried the objective measure for SC (elements per UI) and ordinal scale correlation (Kendall's tau-b) for Precision, but there were no major changes in the outcomes.
6. *Uncontrolled differences in the models:* the sample sizes varied from 41 to 45 (and even to 54 for one of the labelers), but there was no correlation between UI and the models' R^2.

The discovered negative correlations between the labelers' precision and the quality of the resulting models are not entirely clear to us, and we do not yet have a convincing explanation. We would like to note that the effect was found for the scale of aesthetics and the related scale of orderliness, but not for the less subjective scale of complexity. It is believed that aesthetics judgements for visual objects are rather high level, involving the factors of layout, visual hierarchy, colors, etc. Individual elements are grouped according to Gestalt principles, and imprecisions and omissions might even contribute to that—think of an Impressionist painting. Correspondingly, we might speculate that the indilligent workers would have a bias towards picking the UI elements and labeling them in a way matching the actual human perception. However, a much closer look at their output would be required before making any justified conclusions.

Among the limitations of our study, we see the relative minimalism of the linear regression UBMs. We only employed eight factors, and often they would not even be significant in the models. Correspondingly, the absolute quality levels of some models were rather modest, while the average R^2 per the 33 models turned out to be 0.281. The latter is arguably acceptable for our small-scale study that deliberately incorporated the potentially low-quality input data. For instance, in our another study with the same set of university websites screenshots, R^2s ranged from 0.105 to 0.248 (similarly, aesthetics had the highest R^2 of the three scales) [8]. However, recognizably, there the number of factors was smaller, and the number of samples was higher. In any case, in the current study we were interested in the relative values and never intended to use the models in production.

Our further research prospects involve experimentation with more labelers and a more diverse set of the web UIs. Having collected more data, we plan to employ artificial neural network (ANN) models, instead of the simple linear regression ones. ANNs are known as universal approximators and can naturally handle systematic bias in data.

Author Contributions: Conceptualization, M.B.; methodology, M.B.; software, V.K.; validation, M.B. and V.K.; formal analysis, M.B.; investigation, M.B.; resources, M.B. and V.K.; data curation, M.B. and V.K.; writing—original draft preparation, M.B.; writing—review and editing, M.B.; visualization, M.B.; supervision, M.B.; project administration, M.B.; funding acquisition, M.B. All authors have read and agreed to the published version of the manuscript.

Funding: This research was funded by RFBR, grant number 19-29-01017.

Institutional Review Board Statement: The study was conducted according to the guidelines of the Declaration of Helsinki and approved by the Ethics Committee of Faculty of Humanities of Novosibirsk State Technical University (protocol code 7_02_2019).

Informed Consent Statement: Informed consent was obtained from all subjects involved in the study.

Data Availability Statement: The data presented in this study are available on request from M.B. (the first author). The data are not publicly available due to privacy reasons.

Acknowledgments: We would like to thank those who contributed to the project. Sebastian Heil, Martin Gaedke, Anna Stepanova and Galina Hamgushkeeva; as well as Alyona Bakaeva, who provided inspiration for this work.

Conflicts of Interest: The authors declare no conflicts of interest.

References

1. Oulasvirta, A. User interface design with combinatorial optimization. *Computer* **2017**, *50*, 40–47. [CrossRef]
2. Peer, E.; Vosgerau, J.; Acquisti, A. Reputation as a sufficient condition for data quality on Amazon Mechanical Turk. *Behav. Res. Methods* **2014**, *46*, 1023–1031. [CrossRef] [PubMed]
3. Hara, K.; Adams, A.; Milland, K.; Savage, S.; Callison-Burch, Ch.; Bigham, J.P. A data-driven analysis of workers' earnings on Amazon Mechanical Turk. In Proceedings of the 2018 CHI Conference on Human Factors in Computing Systems, Montreal, QC, Canada, 21–26 April 2018; pp. 1–14.
4. Saravanos, A.; Zervoudakis, S.; Zheng, D.; Stott, N.; Hawryluk, B.; Delfino, D. The hidden cost of using Amazon Mechanical Turk for research. In *International Conference on Human-Computer Interaction*; Springer International Publishing: New York, NY, USA, 2021; pp. 147–164.
5. Daniel, F.; Kucherbaev, P.; Cappiello, C.; Benatallah, B.; Allahbakhsh, M. Quality control in crowdsourcing: A survey of quality attributes, assessment techniques, and assurance actions. *ACM Comput. Surv. (CSUR)* **2018**, *51*, 1–40. [CrossRef]
6. Salk, C.; Moltchanova, E.; See, L.; Sturn, T.; McCallum, I.; Fritz, S. How many people need to classify the same image? A method for optimizing volunteer contributions in binary geographical classifications. *PLoS ONE* **2022**, *17*, e0267114. [CrossRef] [PubMed]

7. Oulasvirta, A.; De Pascale, S.; Koch, J.; Langerak, T.; Jokinen, J.; Todi, K.; Laine, M.; Kristhombuge, M.; Zhu, Y.; Miniukovich, A.; et al. Aalto Interface Metrics (AIM): A service and codebase for computational GUI evaluation. In Proceedings of the 31st Annual ACM Symposium on User Interface Software and Technology Adjunct Proceedings, Berlin, Germany, 14–17 October 2018; pp. 16–19.
8. Boychuk, E.; Bakaev, M. Entropy and compression based analysis of web user interfaces. In *International Conference on Web Engineering*; Springer International Publishing: New York, NY, USA, 2019; pp. 253–261.
9. Heil, S.; Bakaev, M.; Gaedke, M. Assessing completeness in training data for image-based analysis of web user interfaces. *CEUR Workshop Proc.* **2019**, *2500*, 17.
10. Thakkar, D.; Ismail, A.; Kumar, P.; Hanna, A.; Sambasivan, N.; Kumar, N. When is Machine Learning Data Good? Valuing in Public Health Datafication. In Proceedings of the CHI Conference on Human Factors in Computing Systems, New Orleans, LA, USA, 29 April–5 May 2022; pp. 1–16.
11. Gudivada, V.; Apon, A.; Ding, J. Data quality considerations for big data and machine learning: Going beyond data cleaning and transformations. *Int. J. Adv. Softw.* **2017**, *10*, 1–20.
12. Wiemer, H.; Dementyev, A.; Ihlenfeldt, S. A Holistic Quality Assurance Approach for Machine Learning Applications in Cyber-Physical Production Systems. *Appl. Sci.* **2021**, *11*, 9590. [CrossRef]
13. Batini, C.; Cappiello, C.; Francalanci, C.; Maurino, A. Methodologies for data quality assessment and improvement. *ACM Comput. Surv. (CSUR)* **2009**, *41*, 1–52. [CrossRef]
14. Bakaev, M.; Avdeenko, T. Intelligent information system to support decision-making based on unstructured web data. *ICIC Express Lett.* **2015**, *9*, 1017–1023.
15. Taleb, I.; Serhani, M.A.; Dssouli, R. Big data quality: A survey. In Proceedings of the IEEE International Congress on Big Data (BigData Congress), San Francisco, CA, USA, 2–7 July 2018; pp. 166–173.
16. Bakaev, M.; Khvorostov, V.; Heil, S.; Gaedke, M. Web intelligence linked open data for website design reuse. In *International Conference on Web Engineering*; Springer International Publishing: New York, NY, USA, 2017; pp. 370–377.
17. Ehrlinger, L.; Wöß, W. A survey of data quality measurement and monitoring tools. *Front. Big Data* **2022**, *5*, 850611. [CrossRef] [PubMed]
18. Alwan, A.A.; Ciupala, M.A.; Brimicombe, A.J.; Ghorashi, S.A.; Baravalle, A.; Falcarin, P. Data quality challenges in large-scale cyber-physical systems: A systematic review. *Inf. Syst.* **2022**, *105*, 101951. [CrossRef]
19. Swazinna, P.; Udluft, S.; Runkler, T. Measuring Data Quality for Dataset Selection in Offline Reinforcement Learning. In Proceedings of the IEEE Symposium Series on Computational Intelligence (SSCI), Orlando, FL, USA, 5–7 December 2021; pp. 1–8.
20. Miniukovich, A.; Marchese, M. Relationship between visual complexity and aesthetics of webpages. In Proceedings of the 2020 CHI Conference on Human Factors in Computing Systems, Honolulu, HI, USA, 25–30 April 2020; pp. 1–13.
21. Jonietz, D. A concept for fitness-for-use evaluation in Machine Learning pipelines. Presented at the 35th Conference on Neural Information Processing Systems (NeurIPS 2021), Sydney, Australia, 6–14 December 2021.
22. Lee, Y.W.; Pipino, L.L.; Funk, J.D.; Wang, R.Y. *Journey to Data Quality*; The MIT Press: Cambridge, MA, USA, 2006.
23. Hagendorff, T. Linking Human And Machine Behavior: A New Approach to Evaluate Training Data Quality for Beneficial Machine Learning. *Minds Mach.* **2021**, *31*, 563–593. [CrossRef] [PubMed]
24. Ciarochi, J. Racist robots: Eradicating algorithmic bias. *Triplebyte Compil. Blog.* **2020**. Available online: https://triplebyte.com/blog/racist-robots-detecting-bias-in-ai-systems (accessed on 1 June 2022).
25. Bakaev, M.; Heil, S.; Khvorostov, V.; Gaedke, M. Auto-extraction and integration of metrics for web user interfaces. *J. Web Eng.* **2018**, *17*, 561–590. [CrossRef]
26. Geiger, R.S.; Cope, D.; Ip, J.; Lotosh, M.; Shah, A.; Weng, J.; Tang, R. "Garbage in, garbage out" revisited: What do machine learning application papers report about human-labeled training data? *Quant. Sci. Stud.* **2021**, *2*, 795–827. [CrossRef]
27. Sambasivan, N.; Kapania, S.; Highfill, H.; Akrong, D.; Paritosh, P.; Aroyo, L.M. "Everyone wants to do the model work, not the data work": Data Cascades in High-Stakes AI. In Proceedings of the 2021 CHI Conference on Human Factors in Computing Systems, Yokohama, Japan, 8–13 May 2021; pp. 1–15.

engineering proceedings

MDPI

Proceeding Paper

Predictive Diagnosis of Breast Cancer Based on Cytokine Profile †

Marina Barulina [1,2,3,*,‡], Yuliya Gergenreter [4,‡], Natalia Zakharova [4,‡], Vladimir Maslyakov [3,4,‡], Vladimir Fedorov [4,‡] and Ivan Ulitin [1,2,‡]

1 Institute of Precision Mechanics and Control, ul Rabochaya, 24, 410028 Saratov, Russia
2 Faculty of Computer Science and Information Technology, Saratov National Research State University Named after N.G. Chernyshevsky, St. Astrakhanskaya, 83, 410012 Saratov, Russia
3 Department of Natural Science Disciplines, Private Medical University Reaviz, ul Verkhny Rynok 10, 410012 Saratov, Russia
4 Department of Clinical Laboratory Diagnostics, Saratov State Medical University Named after V. I. Razumovsky, ul. B. Kazachia, 112, 410012 Saratov, Russia
* Correspondence: barulina@iptmuran.ru
† Presented at the 15th International Conference "Intelligent Systems" (INTELS'22), Moscow, Russia, 14–16 December 2022.
‡ These authors contributed equally to this work.

Abstract: A predictive model for the early diagnosis of breast cancer based on the concentration of some cytokines in the tumor microenvironment in the blood was built in this paper. In the work, the influence of the following cytokines was studied: monocytic chemoattractant protein-1, vascular endothelial growth factor, tumor necrosis factor-alpha, interferon gamma, transforming growth factor-beta1, granulocyte colony stimulating factor, and granulocyte-macrophage colony stimulating factor. As a result of preliminary statistical analysis, some combinations of these cytokines that allowed for almost reliable detection of the presence or absence of breast cancer were identified. Based on the identified combinations, new features were constructed. A machine learning model was trained using gradient boosting for its classification method. The built model has an accuracy equal to 1.0 at this stage, so the authors find it reasonable to carry out additional tests of the model for more patients. However, even at this stage, it can be concluded that the concentration of cytokines in the blood serum is applicable for the early diagnosis of breast cancer.

Keywords: tumor microenvironment; cytokines; machine leaning; breast cancer; blood analysis; predictive analysis

Citation: Barulina, M.; Gergenreter, Y.; Zakharova, N.; Maslyakov, V.; Fedorov, V.; Ulitin, I. Predictive Diagnosis of Breast Cancer Based on Cytokine Profile. *Eng. Proc.* **2023**, *33*, 4. https://doi.org/10.3390/engproc2023033004

Academic Editors: Askhat Diveev, Ivan Zelinka, Arutun Avetisyan and Alexander Ilin

Published: 9 May 2023

1. Introduction

Breast cancer (BC) is the most common type of cancer in women in the world. BC accounts for about 12 percent of newly diagnosed cancers and 25 percent of all cancers in women [1]. According to the statistics of the World Health Organization [1], more than 2.3 million cases of BC were diagnosed worldwide in 2020, which accounted for approximately 24.5% of all cancer cases, and more than 680 thousand patients with diagnosed BC died. The early diagnosis of BC significantly reduces the risk of death. The survival rate for early BC diagnosis approaches 99.1%, and the mortality rate reaches 70% for late diagnosis [2] (Figure 1). At the same time, the early diagnosis of breast cancer cannot be considered sufficient, since the localized stage accounts for only 64% of newly diagnosed cases of breast cancer (Figure 2) [2].

The incidence and mortality of breast cancer increased rapidly in developing countries [3–5]. Therefore, along with early diagnostic methods such as mammograms, clinical breast exams, and breast self-exams, it is of interest to develop new methods of additional minimally invasive clinical diagnostics that could improve early detection and prognosis in breast cancer. Such new approaches can be created using methods and algorithms of artificial intelligence, thereby creating a predictive system for the early diagnosis of breast cancer based on a number of biophysiological parameters of the patient. This paper is devoted to the creation of a predictive system for the early diagnosis of breast cancer by analyzing the tumor microenvironment and its cytokines. The tumor microenvironment (TME) is a microenvironment that is formed as a result of the interaction between tumor tissue and the cells of the immune system that infiltrate it. The concentration of TME cytokines in the blood varies depending on the activity of the tumor and the presence of a metastatic process. As shown in [6], the mediator imbalance of the MCO, its special characteristics, can be used for diagnosing and predicting tumor development. In the previous work of [6], the authors analyzed changes in the concentration of cytokines in the TME for breast cancer for two cases—for the incubation of tumor biopsies and for biopsies of unchanged breast tissue with the determination of spontaneous cytokine production for analyzing the cytokine concentrations in blood serum. This work describes the experience of creating a predictive system based on the cytokine concentrations in blood serum, since this method seems to be more convenient for patients for early BC diagnosis.

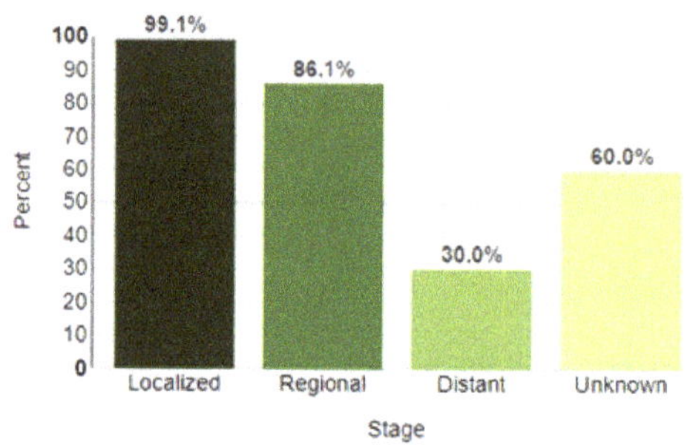

Figure 1. 5-Year relative survival.

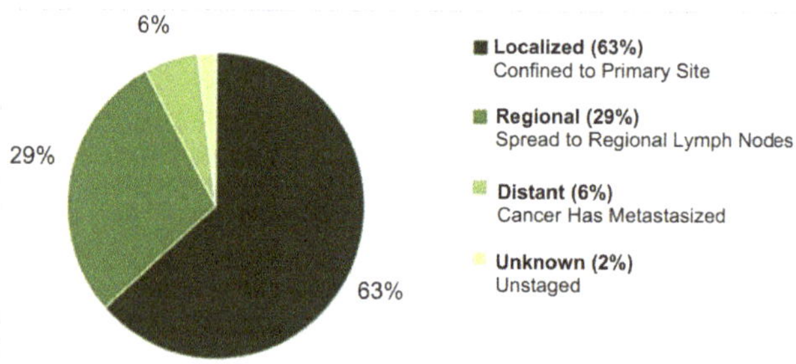

Figure 2. Percent of cases by stage.

The purpose of this work is to develop a predictive system for early BC diagnosis by the cytokine concentration in the blood serum.

2. Materials and Methods

Clinical research was carried out in the Regional Clinical Oncological Dispensary, Saratov, Russia. A total of 80 patients (50–69 years) with diagnosed breast cancer were involved in this clinical research.

The inclusion criteria were as follows: 1. age 50–69 years, 2. the I–IV stage of BC, 3. the signed consent to informed participation in the study.

The exclusion criteria were as follows: 1. malignant tumors of other localizations, 2. acute infectious diseases, 3. chronic diseases in an acute phase (bronchitis, bronchial asthma, pyelonephritis, etc.), 4. concomitant background in the form of somatic diseases in the sub- and decompensation stage (cardiovascular insufficiency, insufficiency circulatory, hepatic, renal failure, and diabetes mellitus).

The comparison group consisted of 26 almost healthy women with ages ranging from 41 to 62 years old.

The study was approved by the local ethics committee of the Medical University "Reaviz", Saratov branch (protocol No. 7 dated 21 July 2017).

Healthy women of the comparison group and patients were examined in accordance with the Clinical Guidelines of the Russian Federation. The patients were classified by the BC stages according to the TNM classification, revision 7 (2012). The patients were divided into three groups. The zero study group included healthy women without BC. The first study group (local cancer, 37 (46%) patients) included patients with the first and second stages of BC; the second study group (spread cancer, 43 (54%) patients) included patients with the third and fourth stages.

Samples of seven blood serum cytokines were taken from women of the comparison group (n = 26) and from the patients with diagnosed BC (n = 80) before therapy. The following serum cytokines were obtained: MCP–1 (monocyte chemoattractant protein-1), VEGF (vascular endothelial growth factor) , TNF–α (tumor necrosis factor-alpha), IFN–γ (interferon gamma), TGF–β1 (transforming growth factor-beta1), G–CSF (granulocyte colony stimulating factor), and GM–CSF (granulocyte-macrophage colony-stimulating factor). The levels of MCP–1, VEGF, TNF–α, and IFN–γ were determined by solid-phase ELISA (a set of reagents of Vector Best JSC, Novosibirsk). For the study of TGF–β1, G–CSF, and GM–CSF, a "sandwich" ELISA variant (three-stage) with mono- and polyclonal antibodies to cytokines (R&D Systems, UK) was used. To activate the inactive form of TGF–β1, blood serum was kept for 60 min before ELISA in an environment with pH 1–2.0, followed by neutralization to pH 7.2–7.6 [6].

3. Statistical Analysis

At the first stage, a statistical analysis was carried out to determine the parameters that had the greatest impact on the stage of cancer. Figure 3 shows the matrix of paired correlation coefficients for the studied parameters for three groups of patients (no BC—stage 0, local cancer—stage 1, spread cancer—stage 2), and for two groups of patients, divided according to the principle—breast cancer is absent (stage 0) and breast cancer is diagnosed in any (local or spread) form (stage 1).

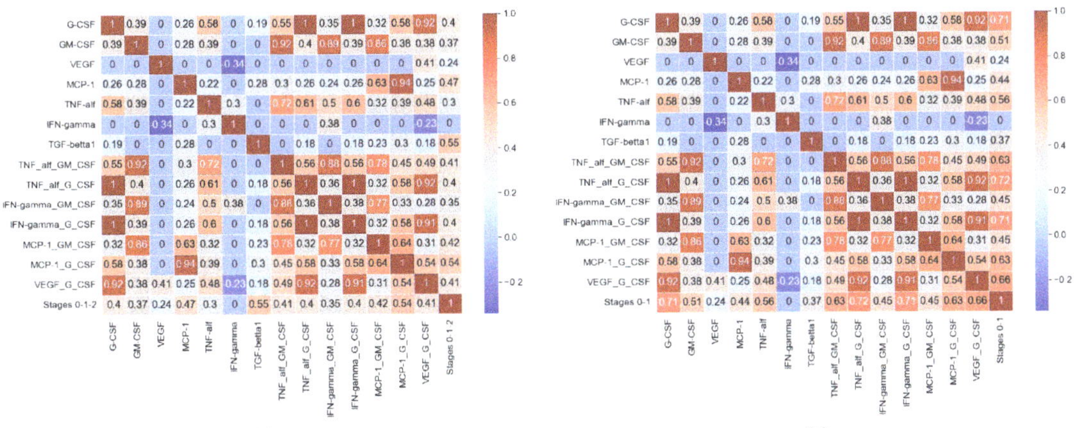

(**a**) (**b**)

Figure 3. Correlation matrices. (**a**) three groups (0—no BC, 1—local BC, 2—spread BC); (**b**) two groups (0—no BC, 1—local or spread BC).

As can been seen in Figure 3a, there was a positive correlation between the zero stage of BC (no BC), the local, and the spread BC stages and the cytokine concentrations in the blood serum: TGF–β1 (R = 0.58; $p < 0.05$), MCP–1 (R = 0.5; $p < 0.05$), and G–CSF (R = 0.43;

$p < 0.05$) showed positive correlations. meaning retained For GM–CSF (R = 0.41; $p > 0.05$), TNF–α (R = 0.32; $p > 0.05$), and VEGF (R = 0.24; $p > 0.05$) the correlation with the BC stage was absent or had a non-linear nature. A completely different situation occurred when patients were divided into two groups according to the absence of BC and its presence. The results of the *t*-test of the statistical significance of differences between these groups for the considered cytokines are shown in Table 1.

Table 1. Comparison of the baseline cytokines characteristics for BC patients and healthy women (N total= 107).

Cytokine	No BC Mean	±SD	Yes BC Mean	±SD	t-Value	t Critical	*p*
G–CSF	7.81	2.90	158.25	73.27	10.43	1.98	<0.01
GM–CSF	0.98	0.21	10.04	6.05	6.67		<0.01
VEGF	440.78	239.98	535.54	132.15	2.56		<0.05
MCP–1	186.70	90.75	404.07	212.44	5.06		<0.01
TNF–α	2.15	0.98	6.97	3.47	6.97		<0.01
IFN–γ	20.39	3.36	18.93	5.54	1.22		>0.05
TGF–β1	28,159.62	15,728.94	43,275.01	16,419.02	4.11		<0.01

From Table 1, it follows that the concentrations of all considered cytokines, except for IFN–γ, were statistically significant and different for patients with and without BC. To visualize the obtained results, scatter diagrams were constructed for all pairwise combinations of cytokines. Some of these diagrams are shown in Figure 4. As follows from the scatter diagrams, although IFN–γ did not correlate in any way with the presence or absence of BC, combinations of IFN–γ and some other cytokines made it possible to quite clearly identify the intervals for these cytokines values in which not a single case of BC was observed.

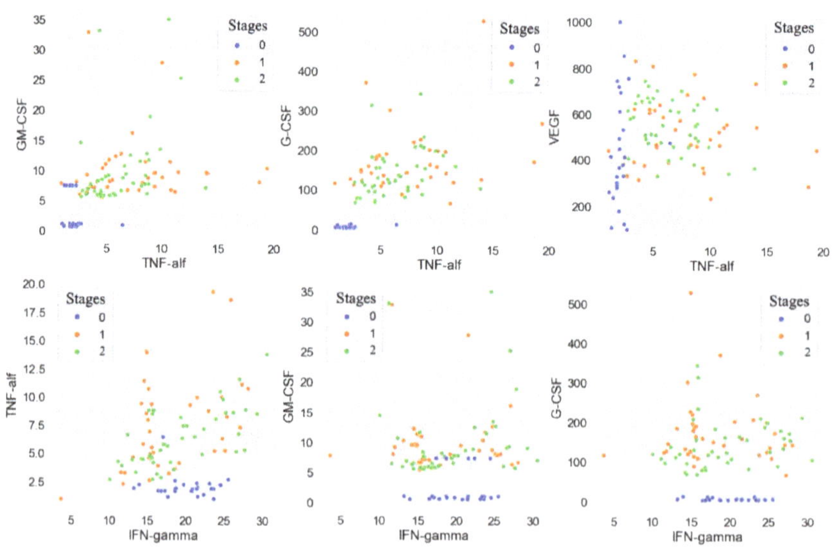

Figure 4. Scatterplots of some pairwise combinations of cytokines. Blue dots are a group of healthy women, orange dots are a group of local BC, and green dots are a group with spread BC.

Based on the results of a preliminary statistical analysis, it was decided that it was necessary to construct new combinations of features for training a predictive machine learning model.

4. Machine Leaning and Predictive Algorithm

The following combinations of cytokines were taken as new features for training the predictive classification model: TNF–α and GM–CSF, TNF–α and G–CSF, IFN–γ and GM–CSF, IFN–γ and G–CSF, MCP–1 and GM–CSF, MCP–1 and G–CSF, VEGF, and G–CSF.

An importance diagram is shown in Figure 5. As can be seen in Figure 5, the vast majority of constructed parameters had a more significant effect on the target variable than the cytokines. Therefore, the model was trained on the constructed parameters and on the G–CSF cytokine.

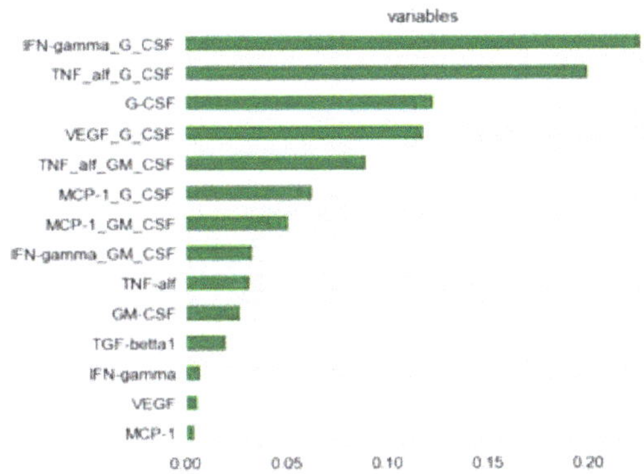

Figure 5. Percent of Cases by Stage.

The formulated problem was reduced to solving the classification problem when we assigned the patient to one of two categories (0—no BC and 1—yes BC) based on the values of the parameters. To solve the problem, the Gradient Boosting for Classification algorithm and supervised learning were used. Gradient Boosting is a machine learning technique that is widely used to solve regression and classification problems. In fact, it represents the decision as a set of weak prediction models, which are typically decision trees [7,8]. This method works well on samples with heterogeneous data and is able to effectively find non-linear relationships in data of various natures. The main set of patients was randomly divided into training and test sets, but in such a way that the resulting sets were statistically the same with a reliability coefficient no worse than 0.05. The best parameters for the Gradient Boosting for Classification algorithm were selected using a Randomized Search on Hyperparameters (RandomizedSearchCV), which allows for the quick exploration of a wide ranges of parameters of predictive algorithms and their impact on the accuracy of the generated forecast. For the case under consideration, the following parameters were recognized as the best parameters for the Gradient Boosting for Classification: n_estimators = 1500, min_samples_split = 8, min_samples_leaf = 3, max_features = 5, max_depth = 4, and learning_rate = 0.001, where n_estimators is the number of gradient boost steps; learning_rate is the learning rate; max_features is the number of features to consider when looking for the best split; and max_depth is the maximum depth of individual regression scores, which limits the node quantities in the tree.

Calculating the ROC AUC metrics score on the test set gave the following results: an accuracy (overall correct predictions) of 1.0 and an AUC of 1.0. A confusion matrix is shown in Figure 6. As can be seen in Figure 6, all predicted (Pred) values of the target variable coincided with true (True) for the test set.

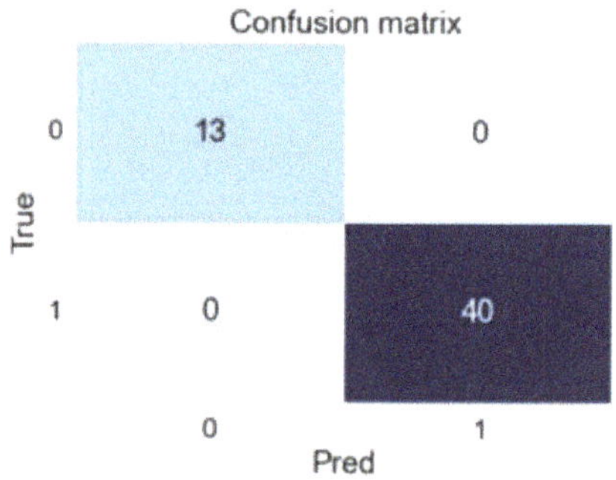

Figure 6. Percent of cases by stage.

5. Conclusions

In this work, the predictive system for the early diagnosis of breast cancer was built based on machine learning methods. The concentration of cytokines in blood serum and their combinations were used for a prognosis. The accuracy of the prediction value being equal to 1 can be explained by the following reasons. Firstly, the model was trained on constructed features, which were pairs of cytokines. Furthermore, these pairs were defined at the stage of preliminary statistical analysis as combinations that were clearly different for patients who have BC and those who do not. Therefore, it was possible to divide the patients into two categories based on these pairs of cytokines. Secondly, the training and test sets were obtained randomly from the initial sample but in such a way that they were not statistically distinguishable. Third, the result obtained can be explained by the insufficient number of patients in the initial set. Therefore, as the next step, the further testing of the constructed predictive system for a larger quantity of patients has to be carried out. However, even at this stage, it can be concluded that the concentration of cytokines in the blood serum is applicable for the early diagnosis of breast cancer.

Author Contributions: Conceptualization, M.B. and Y.G.; methodology, M.B. and N.Z.; software, I.U.; validation, Y.G., N.Z. and V.M.; formal analysis, V.F. and V.M.; investigation, Y.G.; resources, M.B.; data curation, N.Z.; writing—original draft preparation, V.F. and I.U.; writing—review and editing, M.B., Y.G. and V.M.; visualization, Y.G.; supervision, N.Z.; project administration, M.B.; funding acquisition, V.M. All authors have read and agreed to the published version of the manuscript.

Funding: This research received no external funding.

Institutional Review Board Statement: Not applicable.

Informed Consent Statement: Not applicable.

Data Availability Statement: Not applicable.

Conflicts of Interest: The authors declare no conflict of interest.

References

1. World Health Organization. Breast Cancer. 2021. Available online: Https://www.who.int/news-room/fact-sheets/detail/breast-cancer (accessed on 16 June 2022).
2. National Cancer Institute. The Surveillance, Epidemiology, and End Results (SEER) Program. Cancer Stat Facts: Female Breast Cancer. 2022. Available online: Https://seer.cancer.gov/statfacts/html/breast.html (accessed on 16 June 2022).
3. Lei, S.; Zheng, R.; Zhang, S.; Wang, S.; Chen, R.; Sun, K.; Wei, W. Global patterns of breast cancer incidence and mortality: A population-based cancer registry data analysis from 2000 to 2020. *Cancer Commun.* **2021**, *41*, 1183–1194. [CrossRef] [PubMed]
4. Heer, E.; Harper, A.; Escandor, N.; Sung, H.; McCormack, V.; Fidler-Benaoudia, M.M. Global burden and trends in premenopausal and postmenopausal breast cancer: A population-based study. *Lancet Glob. Health* **2020**, *8*, e1027–e1037. [CrossRef] [PubMed]
5. Xia, C.; Dong, X.; Li, H.; Cao, M.; Sun, D.; He, S.; Yang, F.; Yan, X.; Zhang, S.; Li, N.; et al. Cancer statistics in China and United States, 2022: Profiles, trends, and determinants. *Chin. Med. J.* **2022**, *135*, 584–590. [CrossRef] [PubMed]

6. Gergenreter, Y.S.; Zakharova, N.B.; Barulina, M.A.; Maslyakov, V.V.; Fedorov, V.E. Analysis of the cytokine profile of blood serum and tumor supernatants in breast cancer. *Acta Biomed. Sci.* **2022**, *7*, 134–146.
7. Piryonesi, S.M.; El-Diraby, T.E. Data Analytics in Asset Management: Cost-Effective Prediction of the Pavement Condition Index. *J. Infrastruct. Syst.* **2020**, *26*, 04019036. [CrossRef]
8. Hastie, T.; Tibshirani, R.; Friedman, J.H. Boosting and Additive Trees. In *The Elements of Statistical Learning*, 2nd ed.; Springer: New York, NY, USA, 2009; pp. 337–384.

engineering proceedings

MDPI

Proceeding Paper

Using the STEGO Neural Network for Scintigraphic Image Analysis †

Ivan Ulitin [1,2,*,‡], Marina Barulina [1,2,‡] and Marina Velikanova [3,‡]

1 Institute of Precision Mechanics and Control, ul Rabochaya, 24, Saratov 410028, Russia
2 Faculty of Computer Science and Information Technology, Saratov National Research State University Named after N.G. Chernyshevsky, St. Astrakhanskaya, 83, Saratov 410012, Russia
3 Laboratory of Radioisotope Diagnostics, Saratov State Medical University Named after V. I. Razumovsky, ul. B. Kazachia, 112, Saratov 410012, Russia
* Correspondence: ulitin@iptmuran.ru
† Presented at the 15th International Conference "Intelligent Systems" (INTELS'22), Moscow, Russia, 14–16 December 2022.
‡ These authors contributed equally to this work.

Abstract: Currently, neural networks are being widely implemented for the diagnosis of various diseases, including cancer of various localizations and stages. The vast majority of such solutions use supervised or unsupervised convolutional neural networks, which require a great deal of training data. Using unsupervised image segmentation algorithms can be considered the preferred trend since their use significantly reduces the complexity of neural network training. So, developing unsupervised image segmentation algorithms is one of the topical tasks of machine learning. This year, a team of developers from Google, MIT, and Cornell University developed the STEGO algorithm, which is an unsupervised and non-convolutional neural network. As its author stated, the STEGO algorithm performs well at image segmentation problems compared with other machine learning models. And this algorithm does not need a large amount of training data, unlike convolutional neural networks, which are widely used for medical image analysis. So, the aim of this work is to check the possibility of using this neural network for scintigraphy image segmentation by testing whether the STEGO algorithm is relevant when applied to a scintigraphy dataset. To achieve this goal, the intersection over union metric (IoU) was chosen for evaluating the correctness of the detection of the location of metastases. The training dataset consists of scintigraphic images of patients with various types of cancer and various metastasis appearances. Another version of this metric (mIoU, mean intersection over union) was also used by the creators of STEGO to assess the quality of the model to segment images with different kinds of content. Since the calculated metrics are not good enough, the use of this algorithm for scintigraphic image analysis is not possible or requires the development of a special methodology for this.

Keywords: artificial neural networks; machine learning in medicine; deep learning; image segmentation

Citation: Ulitin, I.; Barulina, M.; Velikanova, M. Using the STEGO Neural Network for Scintigraphic Image Analysis. *Eng. Proc.* **2023**, *33*, 5. https://doi.org/10.3390/engproc2023033005

Academic Editors: Askhat Diveev, Ivan Zelinka, Arutun Avetisyan and Alexander Ilin

Published: 9 May 2023

1. Introduction

1.1. Artificial Intelligence and It's Applications

Artificial intelligence (AI) in the world is one of the most promising information technology industries, and there are many examples of improving the living conditions of people through the implementation of machine learning algorithms into their everyday activities [1]. However, despite the already quite widespread use of AI in the economy and other human routines, there are areas of human life where the introduction of neural networks is quite specific, and medicine is among them. Using methods for processing data about a person by implementing a neural network to make a diagnosis, recommend medicines, and carry out a course of treatment entails a high level of responsibility for the lives of people to whom these methods will be applied. However, artificial intelligence in medicine is the use of machine learning models to search medical data and uncover insights to help improve health outcomes and patient experiences. Thanks to recent advances in computer science and informatics, artificial intelligence (AI) is quickly becoming an integral part of modern healthcare. AI algorithms and other applications powered by AI are being used to support medical professionals in clinical settings and in ongoing research.

1.2. Scintigraphy

Scintigraphy is a modern method of radionuclide visualization based on the registration of irradiation generated by the radioactive substance inside a patient. Scintigraphy is used for the evaluation of the functioning of various organs and tissues. Diagnostic methods such as X-ray, ultrasound, CT, or MRI are focused on detecting structural changes in body tissues, and are not always able to distinguish the disease in its early stages, when deviations manifest themselves at the level of biochemical changes in tissues. Radionuclide research methods are indicated for diseases of the heart, the brain, the kidneys, and the liver and look indispensable in identifying and assessing the prevalence of oncological processes. Currently, scintigraphy is widely used in the USA, Europe, and a number of other countries. So more than 17 million radionuclide studies on more than 15 million people were conducted in 2007 in the USA. In Europe, in the same year, over 12 million studies were conducted [2].

1.3. Machine Learning Usage in Scintigraphy

It was worth expecting the appearance of attempts to apply deep learning methods to radionuclide diagnostics. Recent research has been completed on these areas that has resulted in an AI model that demonstrates high-quality diagnostics of cancer bone metastasis [3,4]. The area under the curve (AUC) of receiver operating characteristic (ROC) was 0.988 for breast cancer, 0.955 for prostate cancer, 0.957 for lung cancer, and 0.971 for other cancers [3]. This model, after retraining, demonstrated performance comparable to that of physicians individually classifying bone metastasis. Further AI-consulted interpretation also improved diagnostic sensitivity and accuracy. In total, this AI model performed a valuable benefit for nuclear medicine physicians in the timely and accurate evaluation of cancer bone metastasis [3]. In the work of Papandrianos A. et al. [5], several convolutional neural network algorithms, such as esNet50, VGG16, MobileNet, and DenseNet, were compared with each other in terms of efficiency for bone metastasis diagnosis by means of the analysis of scintigraphy images in RGB and gray scales. As has been shown, some of these neural networks can achieve high accuracy, up to 92.50%, with the specific values of the models. However, convolutional neural networks have great disadvantages. For example, convolutional neural networks need a lot of training data and can't recognize the orientation and position of objects [6]. Therefore, the authors [5] could not create any classification model suitable for practical use since they had only 408 images. A brief overview of some image processing algorithms regarding the task of scintigraphic image analysis were discussed in ref. [7]. In this work [7], the authors proposed a new machine learning approach in which the problem is rephrased to the multi-label learning problem and each bone will be labeled with a non-empty subset of all possible labels. The articles above do not exhaust the methods and algorithms of deep learning for scintigraphic images [8,9]. Thus, the problem of developing deep learning algorithms for the analysis of scintigraphic images is a task of current interest. Of special interest is the use of completely new architectures of neural networks and deep learning algorithms not originally intended for the analysis of medical images such as scintigraphy. One of such neural network architectures is the self-supervised transformer with energy-based graph optimization (STEGO) framework, which was published recently in [10]. This framework distils unsupervised features into high-quality, discrete semantic labels .

The purpose of this work is to evaluate the applicability of STEGO to the scintigraphic image segmentation problem and to compare it to other machine and deep learning methods that can be used for scintigraphic image analysis.

2. Dataset

Scintigraphy images were obtained during clinical research that was carried out at Saratov State Medical University named after V. I. Razumovsky. The inclusion criteria were the presence of kidney or prostate cancer with metabone metastases. The data for this research was prepared by the scientific department of Saratov State Medical University and contains information on 57 patients. The dataset includes data of 54 men and 3 women, patients ages are normally distributed (Figure 1). Patient ages normally distributed except two outliers (Figure 1). 17 patients have prostate cancer (about 30%), and patients with another type or combination of types of cancer account for 40 (70%).

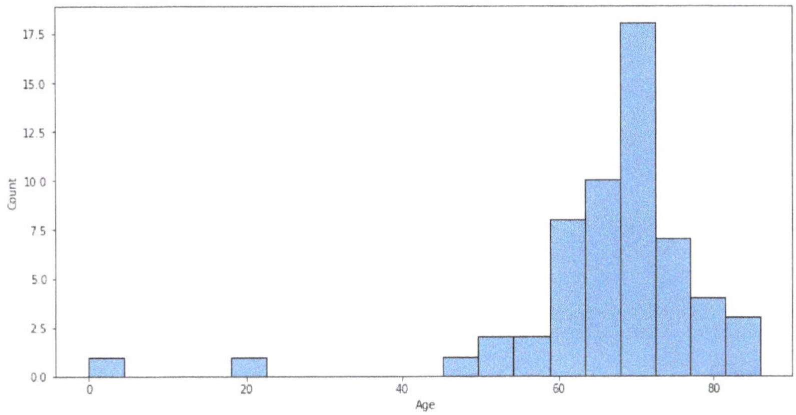

Figure 1. Normal distribution of patient ages with two outliers.

Patient data was in DICOM format. It is a special medical data format that may contain a set of images of human bones and metastases on them for each patient on both the front and dorsal sides. The images can be represented by different color palettes without decreasing in quality due to DICOM format specifics (Figure 2). The dataset contains three different types of images (like in Figure 2) for the front and dorsal views of the body for each person in scope. This is raw scintigraphy images.

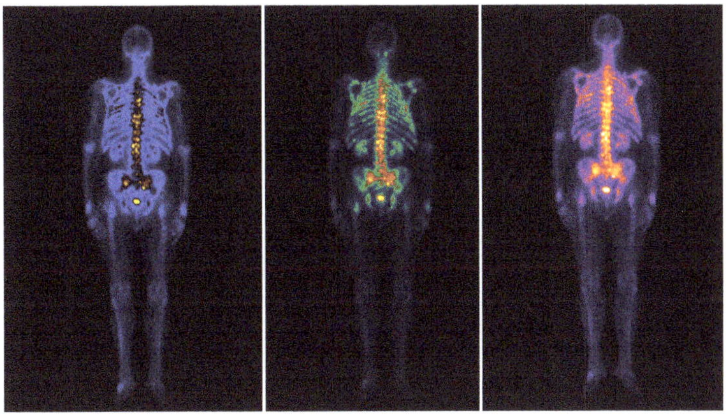

Figure 2. A sample of scintigraphy image represented by different colour palettes (Blue/Yellow, Blue/Green/Red/Yellow, Warm Metal).

Since patient data contains images of the front and dorsal sides of the body, we can analyze the spread of metastases more precisely because metastasis can look varied on different points of view.

Metastasis means that cancer spreads to a different body part from where it started. Metastases most commonly develop when cancer cells break away from the main tumor and enter the bloodstream or lymphatic system. These systems carry fluids around the body. This means that the cancer cells can travel far from the original tumor and form new tumors when they settle and grow in a different part of the body. The main task of scintigraphy is to show these affected body parts by inserting the radioactive substance inside a patient.

The principle of color indication on scintigraphic images is as follows (Figure 2). For example, the color gamma of the left image in Figure 2 is blue/yellow, where the intensity of the yellow color indicates the size and quantity of bone metastases. That is, an area containing many metastases will be more intensely yellow than an area containing fewer metastases. The less metastasis an area contains, the bluer it looks. For the other two images in Figure 2, the principle of colour indication is the same, and yellow colour also shows the quantity of metastases.

3. Implementation of STEGO for Scintigraphic Image Analysis

The STEGO algorithm (hereinafter-STEGO) first time described in [10] by scientists from MIT's Computer Science and Artificial Intelligence Laboratory (CSAIL), Microsoft, and Cornell University. As it was stated, STEGO can jointly discover and segment objects without any human labels at all, down to the pixel. The input to the STEGO is an image. The STEGO retrieves global image information by pooling spatial data. The neural network is made up of a frozen backbone that serves as a source of learning feedback and as an input to the segmentation head for predicting distilled features. This segmentation head is a simple feed forward network with ReLU activations. Thus, STEGO looks quite promising for medical image analysis.

For example, when it is used in an ensemble as an image preprocessing neural network, the output of the work can later be presented as an input to a classifier neural network that will predict the level and area of the lesion, diagnosis based on this area, and so on. The creation of ensembles for solving the problem of image segmentation of medical data has previously been discussed in various papers, such as ref. [11]. In this work, Robert Gordon University's members considered the approach of creating a two-layer ensemble, which was validated using the Hausdorff and Dice metric. The algorithm proposed in ref. [11] showed the best results and has a 0.892 value of Dice Measure and a 48.831 value of Hausdorff measure out of a few other two-layer ensembles that contain quite popular neural networks such as VGG-16, ResNet-34, UNet, and others. Training of this network was carried out on Kvasir-SEG dataset (which consists of 1000 gastrointestinal polyp images) and the CAMUS dataset (the Cardiac Acquisitions for Multi-structure Ultrasound Segmentation).

In this work, raw scintigraphic images were taken as input data for STEGO. As a result, good low-resolution images 246 × 246 px were obtained, which almost ideally divide the input pre-scaled scintigraphic image into segments.

An example of STEGO output can be seen in Figures 3 and 4. Both Figures 3 and 4 show three images. The first image is the input image of STEGO, the second one is processed image by cluster prediction, and the last one is the image after linear probe prediction. As can be seen in Figures 3 and 4, the resulting images of cluster predictions look almost uniformly colored, so the cluster prediction algorithm is not good enough for the scintigraphic images' analysis. The resulting images of linear probe predictions look more promising for further analysis since areas with and without metastases are well separated there.

The patient in Figure 3a has metastases on the ribs since we see yellow colored areas on the initial image (the left image), and the body areas without metastases are colored in various shades of blue on the initial image. In the output image of the STEGO algorithm (linear probe predictions), the yellow blurry areas are replaced by red areas with clear boundaries. The same situation exists for the image of the patient with metastasis on the spine (Figure 3b). Thus, the areas of metastasis are clearly identified, and the resulting images of the linear probe prediction algorithm can be used further for building high-level models, for example, classification models.

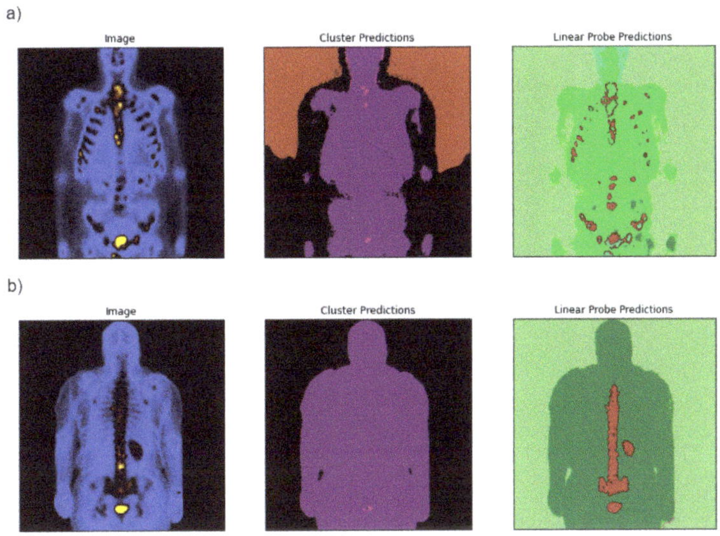

Figure 3. The initial scintigraphy image and STEGO outputs for patients with metastasis on the ribs (**a**) and on the spine (**b**).

It should be noted that the result of the STEGO of processing scintigraphic images is sensitive to the color scheme of these images. So the wrong segments can be highlighted (Figures 3 and 4).

In Figure 4, the input image shows metastases located on the spine, but linear probe predictions of the STEGO algorithm split the image into two segments, green and red, and we can clearly see a large area with metastasis (left red area in the right image) between the arm and thigh of the patient, which is an obvious mistake.

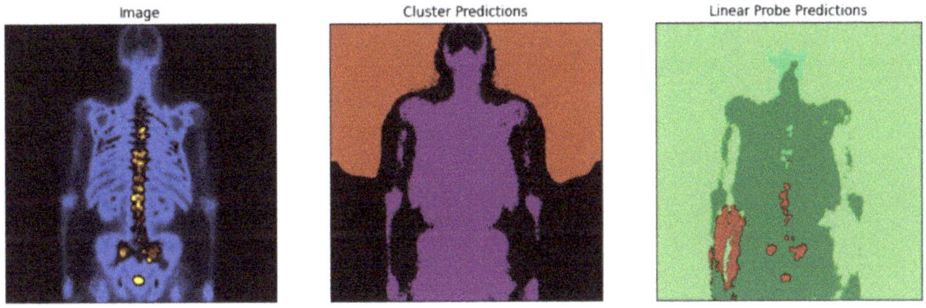

Figure 4. The initial scintigraphy image and incorrect STEGO outputs for a patient with metastasis on the spine.

If the color gamma changes from blue/yellow to rainbow colors, the performance of dividing into segments by STEGO can get worse (Figure 5). Thus, not all possible color palettes allow for high-quality of image segmentations.

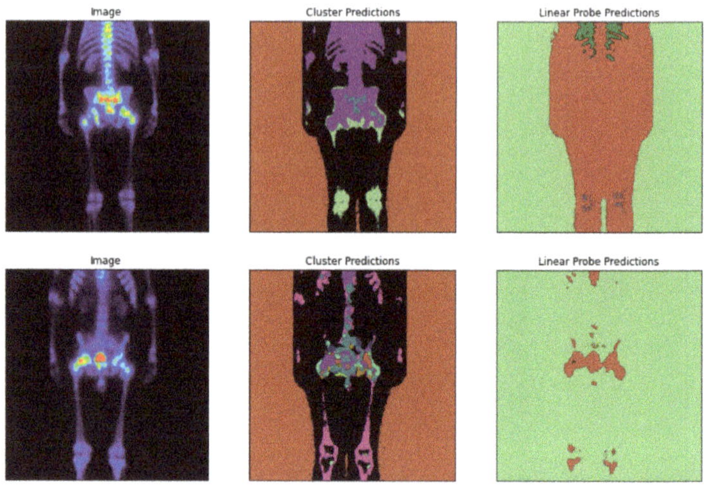

Figure 5. The initial scintigraphy image and incorrect STEGO outputs.

4. Quality Metrics

To check the quality of STEGO performance on a scintigraphic image dataset, an IoU (intersection over union) was chosen as a standard metric for the problem of image segmentation. The IoU is the area of overlap between the predicted segmentation and the ground truth divided by the area of union between the predicted segmentation and the ground truth. To apply this metric, it is necessary to create a mask for each patient, which will determine the true position of metastases on the body. Since the size of the output image from STEGO is 246 × 246 px, after applying the algorithm, it is needed to increase the image size to the mask size (690 × 940 px) for the correct calculation of the metric.

Due to the fact that linear probe prediction in STEGO makes segments of images by coloring them into different color palettes, the metric is calculated as follows. STEGO output setting to grayscale, then metric applied over the mask and over every gray shade, thereby giving the list of results (intersection over union for the mask and each gray shade). Among the obtained results, the one that has the highest value determines the value of this metric relative to this particular image. The metric is applied like this to all images, and thus forming a list of metric values for each of the images in the dataset. The mIoU can be calculated by finding the average value of the elements of the resulting list.

According to the steps and actions described above, and also taking into account the above dataset and its characteristics, the value of the metric is about 0.05.

5. Conclusions

In the work, the possibility and efficiency of using the unsupervised algorithm STEGO for scintigraphic image analysis were studied.

It was shown that cluster predictions of the STEGO algorithm look almost uniformly colored, so the cluster prediction algorithm is not good enough for the scintigraphic image analysis. Whereas, the linear probe predictions of the STEGO algorithm look more promising for further analysis, but the quality of image segmentation highly depends on the original color scheme of the initial images.

Two metrics were used for quality checking. The intersection over union metric (IoU) was chosen for evaluating the correctness of detection of the location of metastases. The training dataset consists of scintigraphic images of patients with various types of cancer and various metastasis appearances. Another version of this metric (mIoU, mean intersection over union) was also used by the creators of STEGO to assess the quality of the model to segment images with different kinds of content.

Since the calculated metrics are not good enough, the use of the STEGO algorithm for scintigraphic image analysis is not possible, or requires the development of a special methodology for this.

Author Contributions: Conceptualization, M.B. and I.U.; methodology, M.B.; software, I.U.; validation, M.V.; formal analysis, I.U. and M.B.; investigation, M.B.; resources, M.V.; data curation, M.V.; writing—original draft preparation, I.U.; writing—review and editing, M.B.; visualization, I.U.; supervision, I.U.; project administration, M.B.; funding acquisition, M.B. All authors have read and agreed to the published version of the manuscript.

Funding: This research received no external funding.

Institutional Review Board Statement: Not applicable.

Informed Consent Statement: Not applicable.

Data Availability Statement: Not applicable.

Acknowledgments: The authors are grateful to the following employees of the Saratov State Medical University named after N.N. V. I. Razumovsky for administrative and technical support: Tatyana Kalyuta, Alexander Fedonnikov, Alexander Korolev and Igor Glushakov.

Conflicts of Interest: The authors declare no conflict of interest.

References

1. West, D.M.; Allen, J.R. How Artificial Intelligence Is Transforming the World. Report. Available online: Https://www.brookings.edu/research/how-artificial-intelligence-is-transforming-the-world/ (accessed on 14 July 2022).
2. Notghi, A.; Low, C. Myocardial perfusion scintigraphy: Past, present and future. *Br. J. Radiol.* **2011**, *84*, S229–S236. [CrossRef] [PubMed]
3. Zhao, Z.; Pi, Y.; Jiang, L.; Xiang, Y.; Wei, J.; Yang, P.; Zhang, W.; Zhong, X.; Zhou, K.; Li, Y.; et al. Deep neural network based artificial intelligence assisted diagnosis of bone scintigraphy for cancer bone metastasis. *Sci. Rep.* **2020**, *10*, 17046. [CrossRef] [PubMed]
4. Aoki, Y.; Nakayama, M.; Nomura, K.; Tomita, Y.; Nakajima, K.; Yamashina, M.; Okizaki, A. The utility of a deep learning-based algorithm for bone scintigraphy in patient with prostate cancer. *Ann. Nucl.* **2020**, *34*, 926–931. [CrossRef] [PubMed]
5. Papandrianos, N.; Papageorgiou, E.; Anagnostis, A.; Feleki, A. A Deep-Learning Approach for Diagnosis of Metastatic Breast Cancer in Bones from Whole-Body Scans. *Appl. Sci.* **2020**, *10*, 997. [CrossRef]
6. Islam, M.A.; Kowal, M.; Jia, S.; Derpanis, K.G.; Bruce, N.D.B. Position, padding and predictions: A deeper look at position information in cnns. *arXiv* **2021**, arXiv:2101.12322.
7. Lukašajn, L.L.; Kononenko, I. *Image Segmentation and Parameterization for Automatic Diagnostics of Whole-Body Scintigrams: Basic Concepts*; Chapman & Hall: London, UK, 2008.
8. Chiu, J.S.; Wang, Y.F.; Su, Y.C.; Wei, L.H.; Liao, J.G.; Li, Y.C. Artificial neural network to predict skeletal metastasis in patients with prostate cancer. *J. Med. Syst.* **2009**, *33*, 91–100. [CrossRef] [PubMed]
9. Moustakidis, S.; Siouras, A.; Papandrianos, N.; Ntakolia, C.; Papageorgiou, E. Deep Learning for Bone Metastasis Localisation in Nuclear Imaging data of Breast Cancer Patients. In Proceedings of the IISA 2021—12th International Conference on Information, Intelligence, Systems and Applications, Chania Crete, Greece, 12–14 July 2021.
10. Hamilton, M.; Zhang, Z.; Hariharan, B.; Snavely, N.; Freeman, W.T. Unsupervised Semantic Segmentation by Distilling Feature Correspondences. *arXiv* **2022**, arXiv:2203.08414.
11. Dang, T.; Nguyen, T.T.; McCall, J.; Elyan, E.; Moreno-García, C.F. Two layer Ensemble of Deep Learning Models for Medical Image Segmentation. *arXiv* **2021**, arXiv:2104.04809.

engineering proceedings

MDPI

Proceeding Paper

Synthesis of a Feedback Controller by the Network Operator Method for a Mobile Robot Rosbot in Gazebo Environment †

Elizaveta Shmalko *,‡ and Yuri Rumyantsev ‡

Federal Research Center "Computer Science and Control" of the Russian Academy of Sciences, 119333 Moscow, Russia; urock@fastsense.tech
* Correspondence: e.shmalko@gmail.com
† Presented at the 15th International Conference "Intelligent Systems" (INTELS'22), Moscow, Russia, 14–16 December 2022.
‡ These authors contributed equally to this work.

Abstract: The article presents an approach based on machine learning with symbolic regression for the synthesis of a stabilization system for a mobile robot. This approach is universal and allows you to numerically solve the synthesis problem in a general setting without the need to form a training sample, instead relying only on the value of the functional. The synthesis is implemented to stabilize the mobile robot Rosbot in the Gazebo simulation environment. The feedback stabilization system is received by the network operator method. The advantage of the method is that it can be applied to a control object of any complexity and linearity.

Keywords: control synthesis; stabilization; machine learning; symbolic regression; Gazebo mobile robot

Citation: Shmalko, E.; Rumyantsev, Y. Synthesis of a Feedback Controller by the Network Operator Method for a Mobile Robot Rosbot in Gazebo Environment. *Eng. Proc.* **2023**, *33*, 6. https://doi.org/10.3390/engproc2023033006

Academic Editors: Askhat Diveev, Ivan Zelinka, Arutun Avetisyan and Alexander Ilin

Published: 16 May 2023

1. Introduction

Synthesis of a feedback stabilization system is one of the key tasks in applied robotics. Feedback is needed to level out the differences between the model and the real object, as well as other possible uncertainties and noise. From a mathematical point of view, this problem belongs to the class of problems for the synthesis of optimal control systems, where it is necessary to find a control vector function that depends on the state vector of the object and delivers a minimum to the quality functional (see Figure 1).

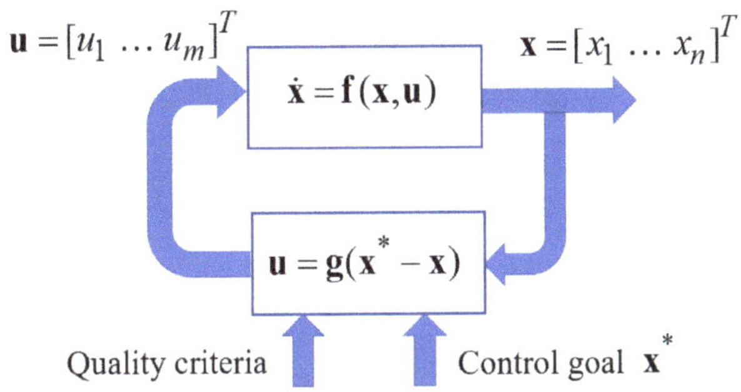

Figure 1. The problem of synthesis of the stabilization system.

There are two main strategies for the synthesis of a control system: parametric and structural-parametric.

Parametric synthesis includes all methods in which the control structure is specified, and only the parameters are optimally tuned from the point of view of the functional. This synthesis strategy is by far the most common. It also includes the most popular various PID controllers [1,2], other controllers based on neural networks [3,4], and fuzzy logic [5,6], etc. In all these approaches, some preliminary knowledge of the developer about the object is required in order to set the controller structure as correctly as possible; nevertheless, there is still no reason to consider the chosen structure as optimal, as only the parameters are adjusted according to the optimality criterion.

With the structural-parametric approach, not only are the parameters optimized, but the optimal structure of the feedback control function is also sought. Among the analytical methods for solving the problem of stabilization system synthesis, the most popular are methods based on solving the Riccati equation [7], but, for linear systems only, as well as more modern analytical approaches of backstepping [8] and analytical design of aggregated controllers [9,10], they also depend on the types of right-hand sides of nonlinear differential equations that describe the control object model. This challenging task for nonlinear systems can be addressed by a dynamic programming (DP) algorithm assisted by the Hamilton–Jacobi–Bellman (HJB) equation. However, it always results in difficulty due to the well-known curse of dimensionality. Recently, modern techniques such as adaptive DP (ADP) [11–14] and reinforcement learning (RL) [15–17] have received increasing attention as powerful machine learning and optimization strategies for addressing the control problems of nonlinear systems regarding the use of neural networks for numerical approximating solutions of the HJB equation. However, these approaches also have many computational difficulties, primarily related to the definition and training of the neural networks used.

Thus, the application of machine learning methods opens up broad prospects, but it is necessary to develop novel control methods addressing general mathematical statements of the control synthesis problem in order to satisfy the requirement of optimal performance in control synthesis tasks. This motivates our research.

In this paper, we apply symbolic regression methods to solve the problem of synthesis of a stabilization system. These methods also belong to the class of machine learning methods, but unlike neural networks, they allow us to search for not only for parameters but also for the optimal structure of the control function. These methods use evolutionary optimization algorithms for the structural-parametric search for the control function, basing it directly on the value of the quality functional.

A wheeled mobile robot is considered as a control object.

The paper develops an applied software implementation of the robot control system. To create it, the most popular robotic operating system ROS is used today. It provides the ability to work with all aspects of the control system, including hardware abstraction, low-level control, message passing between processes, and package management. The developed software systems were tested in the Gazebo simulation environment, which is one of the most popular robotic simulators due to its compatibility with ROS. Gazebo is a 3D simulator that aims to model a robot in a way that gives you a close substitute for how the robot would behave in a real physical environment. Gazebo has a fairly reliable simulation of physics and various physical phenomena, takes into account the influence of forces, and also has a large number of plug-ins for simulating the operation of sensors, such as lidars or cameras. Due to these facts, most developers of control systems for robotic systems around the world use this ROS/Gazebo bundle as a standard for testing the developed control algorithms [18,19].

In this work, we used the ready-made Gazebo model ROSbot 2.0 integrated into ROS [20]. As a position source, a plugin was used that gives the true coordinates of the robot. The robot model in the simulator is shown in Figure 2.

For the selected object, the problem of synthesizing the stabilization system was successfully solved by machine learning based on symbolic regression via the network operator method [21].

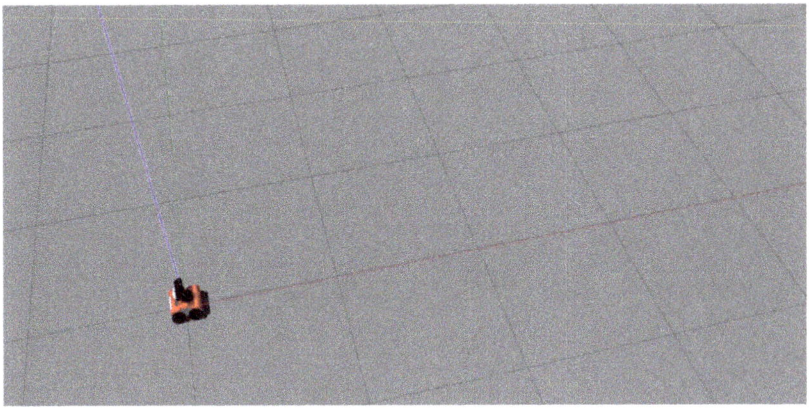

Figure 2. ROSbot in Gazebo simulator.

2. Problem Statement of the Stabilization System Synthesis

The main goal of the introduction of the stabilization system is to provide a stability property for the object in some domain $X_0 \subseteq \mathbb{R}^n$.

Let us be given a mathematical model of a control object. This model can be derived from physical laws or identified by some machine learning technique [22]. Generally, this model is described by a system of ordinary differential equations with a free control vector on the right hand side

$$\dot{\mathbf{x}} = \mathbf{f}(\mathbf{x}, \mathbf{u}), \tag{1}$$

where the state of the object is described by $\mathbf{x} \in \mathbb{R}^n$, and control by $\mathbf{u} \in \mathrm{U} \subseteq \mathbb{R}^m$; U is a compact set, $m \leq n$,

$$\begin{aligned} \mathbf{x} &= [x_1 \dots x_n]^T, \\ \mathbf{u} &= [u_1 \dots u_m]^T,\ m \leq n, \\ \mathbf{f}(\mathbf{x}, \mathbf{u}) &= [f_1(\mathbf{x}, \mathbf{u}) \dots f_n(\mathbf{x}, \mathbf{u})]^T. \end{aligned} \tag{2}$$

An area of initial conditions is given

$$X_0 \subseteq \mathbb{R}^n. \tag{3}$$

It is necessary to find a control function in the form

$$\mathbf{u} = \mathbf{h}(\mathbf{x}^* - \mathbf{x}), \tag{4}$$

where $\mathbf{x}^*$ is a fixed point in the state space, which becomes an equilibrium point of the differential equation

$$\dot{\mathbf{x}} = \mathbf{f}(\mathbf{x}, \mathbf{h}(\mathbf{x}^* - \mathbf{x})), \tag{5}$$

where control function $\mathbf{h}(\mathbf{x}^*, \mathbf{x}) = [h_1(\mathbf{x}^* - \mathbf{x}) \dots h_m(\mathbf{x}^* - \mathbf{x})]^T : \mathbb{R}^n \times \mathbb{R}^n \to \mathbb{R}^m$ has the following properties:

$$\begin{aligned} \mathbf{h}(\mathbf{x}^*, \mathbf{x}) \in \mathrm{U} \subseteq \mathbb{R}^n, &\quad \forall \quad \mathbf{x}^* \in X_0,\ \exists \tilde{\mathbf{x}}(\mathbf{x}^*) \quad \text{such that} \\ \mathbf{f}(\tilde{\mathbf{x}}(\mathbf{x}^*), \mathbf{h}(\mathbf{x}^*, \tilde{\mathbf{x}}(\mathbf{x}^*))) &= 0, \\ \det(\mathbf{A} - \lambda \mathbf{E}) = (-1)^n(\lambda - \lambda_1) \cdot \ldots \cdot (\lambda - \lambda_n) &= \textstyle\prod_{j=1}^n (\lambda - \lambda_j) = 0, \\ \lambda_j = \alpha_j + i\beta_j,\ j &= 1, \dots, n, \\ \alpha_j < 0,\ j = 1, \dots, n,\ i &= \sqrt{-1}, \\ \mathbf{A} = \tfrac{\partial \mathbf{f}(\tilde{\mathbf{x}}(\mathbf{x}^*), \mathbf{h}(\mathbf{x}^*, \tilde{\mathbf{x}}(\mathbf{x}^*)))}{\partial \mathbf{x}}&, \\ \mathbf{E} = \operatorname{diag}(\underbrace{1, \dots, 1}_{n})&. \end{aligned} \tag{6}$$

The properties (6) indicate that $\forall \mathbf{x}^* \in X_0$ for the system $\dot{\mathbf{x}} = \mathbf{f}(\mathbf{x}, \mathbf{h}(\mathbf{x}^*, \mathbf{x}))$ means that there is always a stable equilibrium point $\tilde{\mathbf{x}}(\mathbf{x}^*) \in \mathbb{R}^n$. Additionally, the equilibrium point possesses attractor properties, since near this point all solutions converge.

Computationally, to provide a stability property to the equilibrium point $\tilde{\mathbf{x}}$, the synthesis problem (1)–(4) is solved with the terminal point $\mathbf{x}^f = \tilde{\mathbf{x}}$, the initial domain $X_0 \subset X$, and the quality criterion

$$J = \max\{t_{f,1}, \ldots, t_{f,K}\} + a_1 \sum_{i=1}^{K} \Delta_{f,i} \to \min, \tag{7}$$

where a_1 is a weight coefficient,

$$\Delta_{f,i} = \left\| \mathbf{x}^f - \mathbf{x}(t_{f,i}, \mathbf{x}^{0,i}) \right\|, \tag{8}$$

$t_{f,i}$ is a time of achievement of the terminal position from the initial condition $\mathbf{x}^{0,i}$ of the set of initial conditions $X_0 = \{\mathbf{x}^{0,1}, \ldots, \mathbf{x}^{0,K}\}$, $i \in \{1, \ldots, K\}$,

$$t_{f,i} = \begin{cases} t, & \text{if } t < t^+ \text{ and } \Delta_{f,i} \le \varepsilon \\ t^+ & \text{otherwise} \end{cases}, \tag{9}$$

t^+ and ε are given positive values, and $\mathbf{x}(t, \mathbf{x}^{0,i})$ is a partial solution of the system

$$\dot{\mathbf{x}} = \mathbf{f}(\mathbf{x}, \mathbf{h}(\mathbf{x}^f - \mathbf{x})), \tag{10}$$

for initial conditions $\mathbf{x}(t_0) = \mathbf{x}^{0,i}$, $i \in \{1, \ldots, K\}$,

$$\left\| \mathbf{x}^f - \mathbf{x} \right\| = \sqrt{\sum_{i=1}^{n} (x_i^f - x_i)^2}. \tag{11}$$

Since we are solving the problem of synthesizing a stabilization system using machine learning, we need machine confirmation of the achievement of the desired properties. Let us introduce the following definition of a machine criterion for a differential equation system to obtain some property. To define the property of the whole system (1), it is enough to set a quantity K of partial solutions that obtain this property.

Definition 1. *If D experiments are carried out, and in every i experiment K_i partial solutions of the differential equation perform the required property from any $M_i \geq K_i$ randomly selected initial conditions from the initial domain,*

$$\lim_{D \to \infty} \sum_{i=1}^{D} \frac{K_i}{M_i} \to 1, \tag{12}$$

and so the existence of this property for the differential equation in this domain is proven by a machine.

In other words, as the number of experiments increases, the probability of a "bad" event, when the system does not have the desired property, tends to zero. From a mathematical point of view, this means that all private solutions for a domain of initial conditions have this property except for solutions for a subset of a zero measure.

Based on the proposed formulation of the problem statement, let us consider in the next section a solution of the stabilization system synthesis problem via a machine learning approach of symbolic regression for a mobile robot in a Gazebo simulation environment.

3. Stabilization System Synthesis for ROSbot in Gazebo

We consider the ROSbot 2.0 virtual robot implemented in the Gazebo physical simulation environment. The robot is a platform on four non-rotating wheels around the vertical axis. An electric motor is attached to each wheel. A differential control scheme is used: the robot moves forward and backward by applying the same voltage to all four electric motors; turning the robot to the right or left is carried out by supplying more voltage to the electric motors of the left or right wheels, respectively.

The robot motion model is described by the following system of differential equations [23]

$$\begin{aligned} \dot{x}_1 &= 0.5(u_1+u_2)\cos(x_3), \\ \dot{x}_2 &= 0.5(u_1+u_2)\sin(x_3), \\ \dot{x}_3 &= 0.5(u_1-u_2), \end{aligned} \tag{13}$$

where $\mathbf{x} = [x_1\ x_2\ x_3]^T$ is a vector of state, $\mathbf{u} = [u_1\ u_2]^T$ is a control vector.

Physical control of the robot in Gazebo is implemented using two signals: u^v — the desired linear speed; and u^ω — the desired angular velocity. As we are aiming to stabilize the system, it can be assumed that control signals are completely directly transmitted to the system $v = u^v$, $\omega = u^\omega$. In this case, the (13) equations are converted to the following:

$$\begin{cases} \dot{x}_1 = u^v\cos(x_3), \\ \dot{x}_2 = u^v\sin(x_3), \\ \dot{x}_3 = u^\omega. \end{cases} \tag{14}$$

The stabilization system is synthesized in advance and then is programmed into an onboard computer. For the solution, a machine learning approach based on symbolic regression was chosen. Symbolic regression allows you to search for a solution to the problem without training data, simply according to the formal statement, relying in the search process on the criterion of minimizing the functional. Moreover, this approach is universal and can be equally applicable to models of any kind, including non-linear models or models in the form of neural networks.

The network operator method [21] was used in the calculations. This symbolic regression method is good because it uses the principle of variation of the basic solution, which significantly speeds up the process of finding a solution that is close to optimal. The method encodes possible solutions as a square upper triangular matrix. It is in the form of a matrix that describes the sequence of calculation of the control function that the resulting stabilization system is placed in the on-board computer.

In the calculations, the following parameters were set.

The control values were constrained $-10 \le u_i \le 10$, $i = 1, 2$.

The initial domain was defined by 26 elements:

$$\begin{aligned} \tilde{X}_0 = \{&[-2.5\ -2.5\ -5\pi/12]^T, [-2.5\ -2.5\ 0]^T, [-2.5\ -2.5\ 5\pi/12]^T, \\ &[-2.5\ 0\ -5\pi/12]^T, [-2.5\ 0\ 0]^T, [-2.5\ 0\ 5\pi/12]^T, [-2.5\ 2.5\ -5\pi/12]^T, \\ &[-2.5\ 2.5\ 0]^T, [-2.5\ 2.5\ 5\pi/12]^T, [0\ -2.5\ -5\pi/12]^T, [0\ -2.5\ 0]^T, \\ &[0\ -2.5\ 5\pi/12]^T, [0\ 0\ -5\pi/12]^T, [0\ 0\ 5\pi/12]^T, [0\ 2.5\ -5\pi/12]^T, \\ &[0\ 2.5\ 0]^T, [0\ 2.5\ 5\pi/12]^T, [2.5\ -2.5\ -5\pi/12]^T, [2.5\ -2.5\ 0]^T, \\ &[2.5\ -2.5\ 5\pi/12]^T, [2.5\ 0\ -5\pi/12]^T, [2.5\ 0\ 0]^T, [2.5\ 0\ 5\pi/12]^T, \\ &[2.5\ 2.5\ -5\pi/12]^T, [2.5\ 2.5\ 0]^T, [2.5\ 2.5\ 5\pi/12]^T\}. \end{aligned} \tag{15}$$

The stabilization point was chosen as

$$\mathbf{x}^* = [x_1^*\ x_2^*\ x_3^*]^T = [0\ 0\ 0]^T. \tag{16}$$

It is necessary to find a control function in the form

$$u_i = h_i(x_1^* - x_1, x_2^* - x_2, x_3^* - x_3, r_1, r_2, r_3), \tag{17}$$

where r_1, r_2, r_3 are constant parameters, $i = 1, 2$, such that a robot from all 26 initial conditions (15) reaches the stabilization point (16) with minimal time and highest accuracy.

A time consuming computational experiment has been carried out, and as the result of synthesis the network operator method found a solution in the form of a network operator matrix. The dimension of the matrix in this experiment was $L \times L$, where $L = 24$.

This matrix encodes a rather sophisticated symbolic formula. Each non-zero element represents one of either unary (such as $sin(x)$, $e^x \ldots$) or binary operation (such as $x + y$, $x \times y$). To be able to calculate control values, the matrix has to be decoded online on an onboard computer in real time. The computational complexity of such a process is limited by $O(L^2)$ but is often reasonably low because a lot of matrix elements are zero.

$$C_{NOP} = \begin{bmatrix}
0\ 0\ 0\ 0\ 0\ 0\ 1\ 10\ 0\ 0\ 12\ 1\ 0\ 0\ 0\ 0\ 0\ 0\ 0\ 0\ 0\ 0\ 10 \\
0\ 0\ 0\ 0\ 0\ 0\ 0\ 1\ 0\ 0\ 0\ 0\ 0\ 0\ 0\ 0\ 0\ 0\ 12\ 0\ 0\ 0\ 0 \\
0\ 0\ 0\ 0\ 0\ 0\ 0\ 0\ 1\ 0\ 0\ 0\ 0\ 2\ 9\ 0\ 0\ 0\ 0\ 10\ 0\ 0\ 0 \\
0\ 0\ 0\ 0\ 0\ 0\ 1\ 0\ 0\ 0\ 0\ 0\ 0\ 0\ 13\ 0\ 0\ 0\ 0\ 0\ 0\ 0\ 0 \\
0\ 0\ 0\ 0\ 0\ 0\ 0\ 1\ 0\ 0\ 0\ 0\ 0\ 0\ 0\ 0\ 0\ 0\ 1\ 0\ 0\ 0\ 0\ 0 \\
0\ 0\ 0\ 0\ 0\ 0\ 0\ 0\ 1\ 0\ 0\ 0\ 0\ 0\ 0\ 0\ 0\ 0\ 0\ 0\ 0\ 0\ 0\ 19 \\
0\ 0\ 0\ 0\ 0\ 0\ 2\ 0\ 0\ 8\ 0\ 5\ 0\ 4\ 13\ 10\ 0\ 0\ 0\ 14\ 15\ 0\ 0\ 0 \\
0\ 0\ 0\ 0\ 0\ 0\ 0\ 2\ 0\ 1\ 10\ 9\ 0\ 0\ 0\ 0\ 0\ 0\ 0\ 0\ 0\ 0\ 0\ 0 \\
0\ 0\ 0\ 0\ 0\ 0\ 0\ 0\ 2\ 1\ 0\ 0\ 8\ 0\ 0\ 0\ 12\ 0\ 0\ 0\ 19\ 0\ 0\ 0 \\
0\ 0\ 0\ 0\ 0\ 0\ 0\ 0\ 0\ 1\ 1\ 8\ 0\ 0\ 0\ 1\ 8\ 0\ 0\ 0\ 14\ 12\ 0\ 0 \\
0\ 0\ 0\ 0\ 0\ 0\ 0\ 0\ 0\ 0\ 1\ 1\ 0\ 5\ 4\ 23\ 1\ 0\ 0\ 0\ 15\ 0\ 0\ 23 \\
0\ 0\ 0\ 0\ 0\ 0\ 0\ 0\ 0\ 0\ 0\ 1\ 17\ 10\ 10\ 0\ 0\ 0\ 0\ 16\ 0\ 16\ 0\ 16 \\
0\ 0\ 0\ 0\ 0\ 0\ 0\ 0\ 0\ 0\ 0\ 0\ 1\ 0\ 15\ 0\ 14\ 0\ 0\ 0\ 0\ 0\ 0\ 0 \\
0\ 0\ 0\ 0\ 0\ 0\ 0\ 0\ 0\ 0\ 0\ 0\ 0\ 1\ 9\ 0\ 0\ 0\ 0\ 0\ 0\ 0\ 0\ 0 \\
0\ 0\ 0\ 0\ 0\ 0\ 0\ 0\ 0\ 0\ 0\ 0\ 0\ 0\ 1\ 1\ 10\ 0\ 0\ 0\ 0\ 12\ 0\ 13 \\
0\ 0\ 0\ 0\ 0\ 0\ 0\ 0\ 0\ 0\ 0\ 0\ 0\ 0\ 0\ 1\ 8\ 0\ 0\ 16\ 0\ 0\ 0\ 0 \\
0\ 0\ 0\ 0\ 0\ 0\ 0\ 0\ 0\ 0\ 0\ 0\ 0\ 0\ 0\ 0\ 1\ 1\ 0\ 0\ 0\ 0\ 0\ 0 \\
0\ 0\ 0\ 0\ 0\ 0\ 0\ 0\ 0\ 0\ 0\ 0\ 0\ 0\ 0\ 0\ 0\ 2\ 1\ 0\ 15\ 0\ 0\ 0 \\
0\ 0\ 0\ 0\ 0\ 0\ 0\ 0\ 0\ 0\ 0\ 0\ 0\ 0\ 0\ 0\ 0\ 0\ 2\ 0\ 17\ 0\ 0\ 0 \\
0\ 0\ 0\ 0\ 0\ 0\ 0\ 0\ 0\ 0\ 0\ 0\ 0\ 0\ 0\ 0\ 0\ 0\ 0\ 1\ 5\ 0\ 0\ 17 \\
0\ 0\ 0\ 0\ 0\ 0\ 0\ 0\ 0\ 0\ 0\ 0\ 0\ 0\ 0\ 0\ 0\ 0\ 0\ 0\ 1\ 1\ 0\ 13 \\
0\ 0\ 0\ 0\ 0\ 0\ 0\ 0\ 0\ 0\ 0\ 0\ 0\ 0\ 0\ 0\ 0\ 0\ 0\ 0\ 0\ 1\ 1\ 17 \\
0\ 0\ 0\ 0\ 0\ 0\ 0\ 0\ 0\ 0\ 0\ 0\ 0\ 0\ 0\ 0\ 0\ 0\ 0\ 0\ 0\ 0\ 1\ 4 \\
0\ 0\ 0\ 0\ 0\ 0\ 0\ 0\ 0\ 0\ 0\ 0\ 0\ 0\ 0\ 0\ 0\ 0\ 0\ 0\ 0\ 0\ 0\ 1
\end{bmatrix} \tag{18}$$

4. Verification in Gazebo

For a real time onboard computation network, the operator decode function has been implemented in C++ [24] and later incorporated into the ROS node Rosbot controller. Rosbot controller node accepted robot ground truth coordinates and movement goal as inputs and generated control signals u^v, u^ω every 100 ms as outputs.

The robot was directed to reach several goals. After reaching its first goal, the robot started heading to the next and so on, as shown in Figure 3.

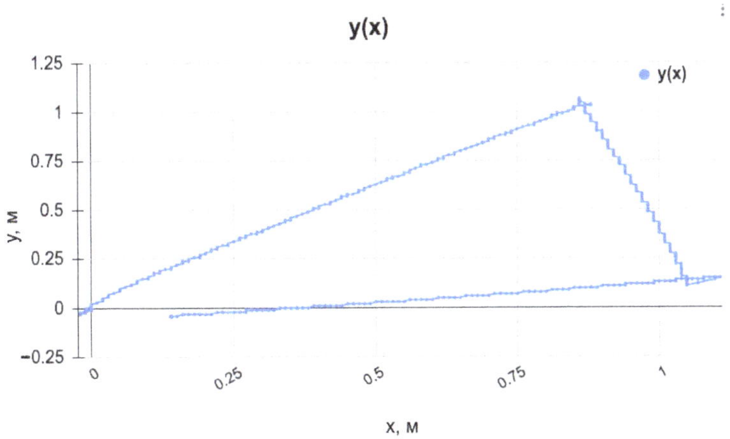

Figure 3. Rosbot trajectory on XY plane under network operator control.

In general we noticed that the robot's movements were stable and predictable.

5. Discussion

In this paper, a Rosbot controller synthesized by a computer algorithm using the network operator method has been verified to work in a Gazebo simulated environment. The synthesis algorithm required only a mathematical model of the robot as an input. The mathematical structure of the control function has been found automatically without any human input. The synthesized function has been verified to be able to control Rosbot in a stable and predictable way.

The presented numerical approach to the synthesis of a stabilization system by a robot, implemented on the Gazebo simulator robot, is a universal machine learning approach for the synthesis of control systems, and opens up broad prospects for its use in various technical problems. The main advantage of the approach is that it is not tied to the type of control object model, and allows you to find the feedback control function in automatic mode using the symbolic regression algorithm.

Author Contributions: Conceptualization, E.S.; methodology, E.S.; software, Y.R.; validation, Y.R.; formal analysis, E.S. and Y.R.; investigation, Y.R.; data curation, Y.R.; writing—original draft preparation, E.S.; writing—review and editing, E.S.; visualization, Y.R. All authors have read and agreed to the published version of the manuscript.

Funding: This research received no external funding.

Institutional Review Board Statement: Not applicable.

Informed Consent Statement: Not applicable.

Data Availability Statement: Not applicable.

Conflicts of Interest: The authors declare no conflicts of interest.

References

1. Qi, Z.; Shi, Q.; Zhang, H. Tuning of Digital PID Controllers Using Particle Swarm Optimization Algorithm for a CAN-Based DC Motor Subject to Stochastic Delays. *IEEE Trans. Ind. Electron.* **2020**, *67*, 5637–5646. [CrossRef]
2. Castillo-Zamora, J.J.; Camarillo-GóMez, K.A.; PéRez-Soto, G.I.; RodríGuez-ReséNdiz, J. Comparison of PD, PID and Sliding-Mode Position Controllers for V–Tail Quadcopter Stability. *IEEE Access* **2018**, *6*, 38086–38096. [CrossRef]
3. Hu, F.; Zeng, C.; Zhu, G.; Li, S. Adaptive Neural Network Stabilization Control of Underactuated Unmanned Surface Vessels with State Constraints. *IEEE Access* **2020**, *8*, 20931–20941. [CrossRef]
4. Li, S.; Ahn, C.K.; Guo, J.; Xiang, Z. Neural Network-Based Sampled-Data Control for Switched Uncertain Nonlinear Systems. *IEEE Trans. Syst. Man Cybern. Syst.* **2021**, *51*, 5437–5445. [CrossRef]

5. Wang, T.-Y.; Chang, C.-D. Hybrid Fuzzy PID Controller Design for a Mobile Robot. In Proceedings of the 2018 IEEE International Conference on Applied System Invention (ICASI), Chiba, Japan, 13–17 April 2018; pp. 650–653. [CrossRef]
6. Chwa, D.; Boo, J. Adaptive Fuzzy Output Feedback Simultaneous Posture Stabilization and Tracking Control of Wheeled Mobile Robots with Kinematic and Dynamic Disturbances. *IEEE Access* **2020**, *8*, 228863–228878.[CrossRef]
7. Kolmanovskii, V.B.; Lafay, J.-F.; Richard, J.-P. Riccati equations in stability theory of difference equations with memory. In Proceedings of the 1999 European Control Conference (ECC), Karlsruhe, Germany, 31 August–3 September 1999; pp. 3630–3636. [CrossRef]
8. Khalil, H.K. *Nonlinear Systems*; Prentice Hall: Hoboken, NJ, USA, 2002.
9. Kolesnikov, A.A.; Kuz'menko, A.A. Backstepping and ADAR Method in the Problems of Synthesis of the Nonlinear Control Systems. *Mekhatronika Avtom. Upr.* **2016**, *17*, 435–445. [CrossRef]
10. Podvalny, S.L.; Vasiljev, E.M. Analytical synthesis of aggregated regulators for unmanned aerial vehicles. *J. Math. Sci.* **2019**, *239*, 135–145. [CrossRef]
11. Bellman, R.E. *Dynamic Programming*; Princeton University Press: Princeton, NJ, USA, 1957.
12. Werbos, P.J. Approximate dynamic programming for real-time control and neural modeling, In *Handbook of Intelligent Control: Neural, Fuzzy, and Adaptive Approaches*; White, D.A., Sofge, D.A., Eds.; Van Nostrand Reinhold: New York, NY, USA, 1992; Chapter 13.
13. Liu, D.; Wei, Q.; Wang, D.; Yang, X.; Li, H. *Adaptive Dynamic Programming with Applications in Optimal Control*; Springer: Cham, Switzerland, 2017.
14. Al-Tamimi, A.; Lewis, F.L.; Abu-Khalaf, M. Discrete-time nonlinear HJB solution using approximate dynamic programming: Convergence proof. *IEEE Trans. Syst. Man Cybern. B Cybern.* **2008**, *38*, 943–949. [CrossRef]
15. Si, J.; Wang, Y.-T. Online learning control by association and reinforcement. *IEEE Trans. Neural Netw.* **2001**, *12*, 264–276. [CrossRef] [PubMed]
16. Kaelbling, L.P.; Littman, M.L.; Moore, A.W. Reinforcement learning: A survey. *J. Artif. Intell. Res.* **1996**, *4*, 237–285. [CrossRef]
17. Modares, H.; Lewis, F.L. Optimal tracking control of nonlinear partially-unknown constrained-input systems using integral reinforcement learning. *Automatica* **2014**, *50*, 1780–1792. [CrossRef]
18. Lavrenov, R.; Magid, E.; Matsuno, F.; Svinin, M.; Suthakorn, J. Development and Implementation of Spline-based Path Planning Algorithm in ROS/Gazebo Environment. *SPIIRAS Proc.* **2019**, *18*, 57–84. [CrossRef]
19. Zhang, B.; Liu, P. Control and benchmarking of a 7-DOF robotic arm using Gazebo and ROS. *Peer J. Comput. Sci.* **2021**, *7*, e383. [CrossRef] [PubMed]
20. Rosbot_description. Available online: https://github.com/husarion/rosbot_description (accessed on 14 May 2023).
21. Diveev, A.; Shmalko, E. Symbolic Regression Methods. In *Machine Learning Control by Symbolic Regression*; Springer: Cham, Switzerland, 2021. [CrossRef]
22. Shmalko, E.; Rumyantsev, Y.; Baynazarov, R.; Yamshanov, K. Identification of Neural Network Model of Robot to Solve the Optimal Control Problem. *Inform. Autom.* **2021**, *20*, 1254–1278. [CrossRef]
23. Diveev, A.I.; Shmalko, E.Y. Machine-Made Synthesis of Stabilization System by Modified Cartesian Genetic Programming. *IEEE Trans. Cybern.* **2022**, *52*, 6627–6637. [CrossRef] [PubMed]
24. Network Operator. Available online: https://github.com/urock/network_operator (accessed on 14 May 2023).

engineering proceedings

MDPI

Proceeding Paper

Additional Requirement in the Formulation of the Optimal Control Problem for Applied Technical Systems †

Elizaveta Shmalko *,‡ and Askhat Diveev ‡

Federal Research Center "Computer Science and Control" of the Russian Academy of Sciences, Moscow 119333, Russia; aidiveev@mail.ru

* Correspondence: e.shmalko@gmail.com

† Presented at the 15th International Conference "Intelligent Systems" (INTELS'22), Moscow, Russia, 14–16 December 2022.

‡ These authors contributed equally to this work.

Abstract: This paper considers the difficulties that arise in the implementation of solutions to the optimal control problem. When implemented in real systems, as a rule, the object is subject to some perturbations, and the control obtained as a function of time as a result of solving the optimal control problem does not take into account these factors, which leads to a significant change in the trajectory and deviation of the object from the terminal goal. This paper proposes to supplement the formulation of the optimal control problem. Additional requirements are introduced for the optimal trajectory. The fulfillment of these requirements ensures that the trajectory remains close to the optimal one under perturbations and reaches the vicinity of the terminal state. To solve the problem, it is proposed to use numerical methods of machine learning based on symbolic regression. A computational experiment is presented in which the solutions of the optimal control problem in the classical formulation and with the introduced additional requirement are compared.

Keywords: optimal control; stability; control synthesis; feasibility of control; synthesized control

1. Introduction

The main disadvantage of the optimal control problem [1] is that its solution is an optimal control as a function of time, and it cannot be implemented in practice since its implementation leads to an open control system that is insensitive to model disturbances.

Consider a well-known optimal control problem

$$\begin{aligned} \dot{x}_1 &= x_2, \\ \dot{x}_2 &= u, \end{aligned} \tag{1}$$

where $\mathbf{x} = [x_1\ x_2]^T$ is a state vector, u is a control signal.

The control values are limited

$$-1 \leq u \leq 1. \tag{2}$$

It is necessary to find a control that will move the object (1) from the initial state

$$\mathbf{x}(0) = [1\ 1]^T, \tag{3}$$

to the given terminal position

$$\mathbf{x}^f = [0\ 0]^T \tag{4}$$

as fast as possible

$$J = t_f \to \min. \tag{5}$$

Citation: Shmalko, E.; Diveev, A. Additional Requirement in the Formulation of the Optimal Control Problem for Applied Technical Systems. *Eng. Proc.* **2023**, *33*, 7. https://doi.org/10.3390/engproc2023033007

Academic Editors: Ivan Zelinka, Arutun Avetisyan and Alexander Ilin

Published: 16 May 2023

The analytical solution of the stated problem was presented in [1]. According to the maximum principle, the optimal control takes only limit values (2) and has no more than one switch. According to (3), initially, the control has the value $u = -1$. Then, when a certain state is reached, the control switches to the value $u = 1$.

$$u = \begin{cases} -1, \text{ if } t < t^* \\ 1, \text{ otherwise} \end{cases} \quad (6)$$

where t^* is the moment of control switching.

Let us find the moment of switching. A particular solution of the system (1) from the initial state (3) has the following form

$$\begin{aligned} x_1 &= -0.5t^2 + t + 1, \\ x_2 &= -t + 1. \end{aligned} \quad (7)$$

The general solution (1) for the control $u = +1$ is the following

$$\begin{aligned} x_1 &= 0.5t^2 + x_{2,0}t + x_{1,0}, \\ x_2 &= t + x_{2,0}, \end{aligned} \quad (8)$$

where $x_{1,0}$, $x_{2,0}$ are the coordinates of the switching point.

Let us express in (8) x_1 as a function of x_2

$$x_1 = x_2^2 - 0.5x_{2,0}^2 + x_{1,0}, \quad (9)$$

The relation for the switching point follows from the terminal conditions (4)

$$x_{1,0} = 0.5x_{2,0}^2. \quad (10)$$

Let us now find the moment of time for a particular solution (7) that satisfies the relation (10).

$$\begin{aligned} -0.5t^2 + t + 1 &= 0.5(-t+1)^2; \\ t^2 - 2t - 0.5 &= 0; \\ t^* = 1 + \sqrt{1.5} &= 2.22474487. \end{aligned} \quad (11)$$

The switching time (11) is the solution to this optimal control problem. To determine the value of the functional (5), we calculate the coordinates of the switching point. Substituting (11) into the particular solution (7), we obtain

$$x_1 = -0.75, \ x_2 = \sqrt{1.5}. \quad (12)$$

From the second equation in (8), we obtain the optimal time of reaching the terminal state

$$\tilde{t} = t^* + \sqrt{1.5} = 3.44948974. \quad (13)$$

Now, we introduce perturbations into the initial conditions (3)

$$x_1(0) = 1 + \delta_1, \ x_2(0) = 1 + \delta_2, \quad (14)$$

where δ_1, δ_2 are random variables from a limited range.

During the time t^*, the object does not get to the switching point (12), and after switching accordingly, does not get into the terminal state. Based on the optimal value of the functional (13), let us limit the control time to $t^+ = 3.5$ and determine the state of the object at the moment t^+, taking into account the switching of the control at the moment (11)

$$
\begin{aligned}
\hat{x}_{1,0} &= -0.5t^{*2} + (1+\delta_2)t^* + 1 + \delta_1, \\
\hat{x}_{2,0} &= -t^* + 1 + \delta_2, \\
x_1(t^+) &= 0.5(t^+ + t^*)^2 + \hat{x}_{2,0}(t^+ - t^*) + \hat{x}_{1,0} = \\
& \quad 0.00127565 - (1.27525513 + t^*)\delta_2 + \delta_1, \\
x_2(t^+) &= t^+ - t^* + \hat{x}_{2,0} = 0.05051026 + \delta_2.
\end{aligned}
\tag{15}
$$

Figure 1 shows trajectories of eight randomly perturbed solutions of the problem (1)–(5) from the range

$$x_1(0) = 1 \pm 0.25,\ x_2(0) = 1 \pm 0.25. \tag{16}$$

In Figure 1, the blue curve represents the optimal unperturbed solution.

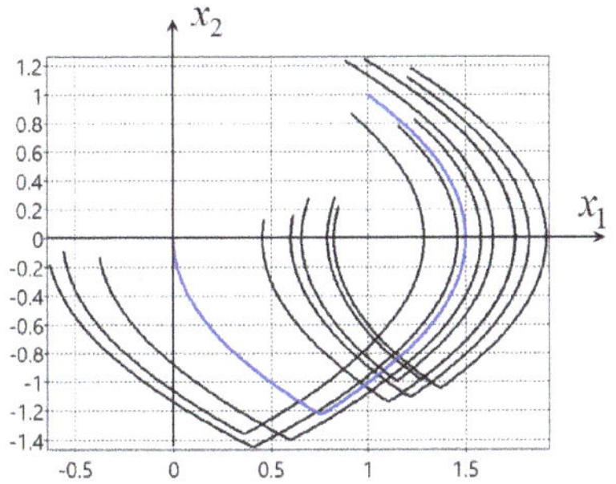

Figure 1. Optimal and perturbed solutions with control (6).

All perturbed solutions do not reach the terminal state. It is obvious from the plots that the solution of the optimal control problem as a function of time (6) cannot be implemented in practice since according to the model (1) with control (6) due to disturbances, it is impossible to assess the state of the control object.

In this regard, it is necessary to introduce additional requirements into the formulation of the optimal control problem so that the resulting controls can be directly implemented on a real plant.

2. Optimal Control Problem Statement with Additional Requirement

Consider the formulation of the optimal control problem, the solution of which can be directly implemented on the plant.

Given the mathematical model of the control object

$$\dot{\mathbf{x}} = \mathbf{f}(\mathbf{x}, \mathbf{u}), \tag{17}$$

where $\mathbf{x} \in \mathbb{R}^n$, $\mathbf{u} \in \mathrm{U} \subseteq \mathbb{R}^m$, $\mathbf{f} = [f_1(\mathbf{x}, \mathbf{u}) \ldots f_n(\mathbf{x}, \mathbf{u})]^T$.

Given the initial

$$\mathbf{x}(0) = \mathbf{x}^0 \in \mathbb{R}^n. \tag{18}$$

and terminal conditions

$$\mathbf{x}(t_f) = \mathbf{x}^f \in \mathbb{R}^n, \tag{19}$$

where t_f is the time to reach the terminal conditions, not specified, but limited, $t_f \leq t^+$, t^+ is the specified limit time of the control process.

A quality criterion is set. It may include conditions for fulfilling phase constraints

$$J_1 = \int_0^{t_f} f_0(\mathbf{u}, \mathbf{x})dt \to \min_{\mathbf{u} \in \mathrm{U}}. \tag{20}$$

We need to find a control in the form

$$\mathbf{u} = \mathbf{g}(\mathbf{x}, t) \in \mathrm{U}. \tag{21}$$

The found control must be such that the particular solution $\mathbf{x}(t, \mathbf{x}^0)$ of the system

$$\dot{\mathbf{x}} = \mathbf{f}(\mathbf{x}, \mathbf{g}(\mathbf{x}, t)) \tag{22}$$

from the initial state (18) reaches the terminal state (19) with the optimal value of the quality criterion (20). Moreover, the optimal particular solution $\mathbf{x}(t, \mathbf{x}^0)$ of the system (22) would have a neighborhood $\Delta(t) > 0$ such that if for any other particular solution $\mathbf{x}(t, \mathbf{y})$ of the system (22) from another initial state $\mathbf{y} \in \mathbb{R}^n$ at time t', $0 \le t' \le t^+$

$$\|\mathbf{x}(t', \mathbf{y}) - \mathbf{x}(t', \mathbf{x}^0)\| \le \Delta(t'), \tag{23}$$

then $\forall t$, $t' \le t \le t_f$, this particular solution does not leave the neighborhood of the optimal

$$\|\mathbf{x}(t, \mathbf{y}) - \mathbf{x}(t, \mathbf{x}^0)\| \le \Delta(t),\ t' \le t \le t_f. \tag{24}$$

The neighborhood $\Delta(t)$ shrinks near the terminal state. This means that for any particular solution from the neighborhood of the optimal one for which the conditions (23) are satisfied $\exists t'' < t^+$ such that

$$\|\mathbf{x}(t'', \mathbf{y}) - \mathbf{x}(t'', \mathbf{x}^0)\| \le \varepsilon, \tag{25}$$

where ε is a given small positive value.

The existence of a neighborhood of the optimal solution in many cases can worsen the value of the functional. For example, in a problem with phase constraints, which are obstacles on the path of movement of the control object to the terminal state, the optimal trajectory often passes along the boundary of the obstacle. Such a trajectory will not have a neighborhood, so in this case, it is necessary to find another trajectory that will not be optimal according to the classical formulation of the optimal control problem but allows variations in the initial values with a small change in the value of the functional.

3. Overview of Methods for Solving the Extended Optimal Control Problem with Additional Requirement

Therefore, an additional requirement has been put forward in the formulation of the optimal control problem, which makes it possible to implement the obtained controls on real objects. Consider the existing methods for solving the problem in the presented extended formulation.

The solution to the problem of general control synthesis for a certain region of initial conditions leads to the fact that each particular solution from this region of initial conditions will be optimal. In this case, each particular solution has a neighborhood containing other optimal solutions. The neighborhood will be open, but will also shrink near the terminal state.

For the model (1), there is a solution to the problem of general control synthesis, in which a control is found that ensures the optimal achievement of the terminal state from any initial condition.

$$u^* = \begin{cases} -1, \text{if } h(x_1, x_2) \ge 0 \\ 1, \text{otherwise} \end{cases}, \tag{26}$$

where

$$h(x_1, x_2) = \begin{cases} x_1 + 0.5x_2^2, \text{if } x_1 < 0 \\ x_1 - 0.5x_2^2, \text{otherwise} \end{cases}. \tag{27}$$

Plots of eight perturbed solutions for the object model (1) with control (26) are shown in Figure 2.

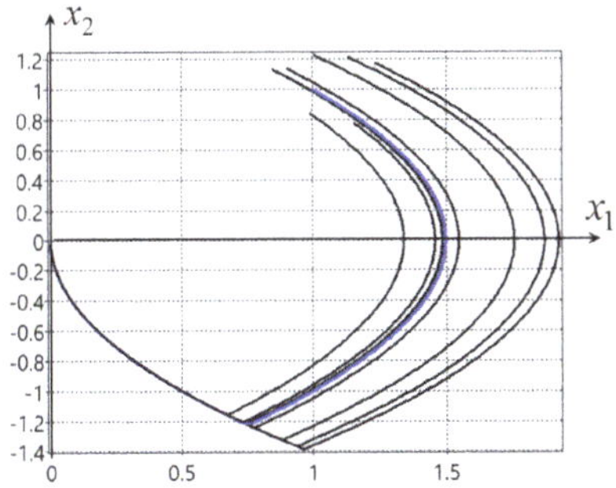

Figure 2. Optimal and perturbed solutions with control (26).

As can be seen from Figure 2, all perturbed solutions have reached the terminal state. This control (26) is practically feasible.

However, the problem of general control synthesis is a complex mathematical problem for which there is no universal numerical solution method. In this case, this problem was solved because the plant model is simple, the optimal control takes only two limit values, and for both of these values, general solutions of the differential equations of the model (1) are obtained.

Another approach to solving the optimal control problem and fulfilling additional requirements is stabilizing motion along the trajectory based on the theory of stability of A.M. Lyapunov [2]. As a result of constructing the stabilization system, the optimal trajectory should become asymptotically stable. The construction of such a stabilization system is not always possible; in particular, in the problem under consideration (1), the control resources are exhausted to obtain the optimal trajectory and there are no more control resources for the stabilization system.

Another approach that also allows solving the optimal control problem in the presented extended formulation with additional requirements is the synthesized control method. It consists of two stages [3,4]. Initially, the problem of control synthesis is solved in order to ensure the stability of the control object relative to some point in the state space. In the second stage, the problem of optimal control is solved, while the coordinates of the stability points of the control object are used as control. It should be noted that the solution of the control synthesis problem at the first stage will significantly change the mathematical model of the control object. When solving the control synthesis problem, the functional of the optimal control problem is not used to ensure stability; therefore, various methods for solving the control synthesis problem will lead to different mathematical models of a closed control system and to different solutions to the optimal control problem. The presence of a neighborhood with attraction properties for the optimal solution requires the choice of such a position of the stability points in the state space so that particular solutions from a certain region of initial states, being attracted to these stability points, are close to each other, moving to the terminal state.

Consider the application of the synthesized optimal control to problems (1)–(5). For stabilization system synthesis different methods can be used from traditional regulators [5] to modern machine learning techniques [6–8]. As far as the considered object is rather simple (1), it is enough to use a proportional regulator. Taking into account the limits on control, it has the following form

$$u = \begin{cases} \text{sgn}(\tilde{u}), \text{ if } |\tilde{u}| > 1 \\ \tilde{u} - \text{otherwise} \end{cases}, \tag{28}$$

where

$$\tilde{u} = k_1(x_1^* - x_1) + k_2(x_2^* - x_2). \quad (29)$$

The object is stable if $k_1 > 0$, $k_2 > 0$. A stable equilibrium point exists if $|\tilde{u}| < 1$. The coordinates of the equilibrium point for the given k_1, k_2 depend on the values of x_1^*, x_2^*

$$\tilde{x}_1 = \frac{k_2 x_2^* + k_1 x_1^*}{k_1}, \ \tilde{x}_2 = 0. \quad (30)$$

As a result, we obtain the following control object model

$$\begin{array}{rcl} \dot{x}_1 & = & x_2 \\ \dot{x}_2 & = & k_1(x_1^* - x_1) + k_2(x_2^* - x_2) \end{array}. \quad (31)$$

The control in the model is the vector $\mathbf{x}^* = [x_1^* \ x_2^*]^T$, whose values are limited by the following inequalities

$$-1 \leq k_1(x_1^* - x_1) + k_2(x_2^* - x_2) \leq 1. \quad (32)$$

To solve the problem of optimal control, we include in the quality criterion the accuracy of hitting the terminal state $\mathbf{x}^f = [0 \ 0]^T$

$$J_5 = t_f + p_1 \|\mathbf{x}^f - \mathbf{x}\|, \quad (33)$$

where t_f is a terminal time, p_1 is a weight coefficient, $p_1 = 1$.

When solving the problem, we divide the control time t^+ into intervals Δt, and on each interval, we look for the values x_1^*, x_2^*, taking into account the constraints (32). The following parameter values were used: $k_1 = 2$, $k_2 = 2$, $p_1 = 1$, $t^+ = 3.5$, $\Delta_t = 0.5$, $\varepsilon_1 = 0.001$. An evolutionary hybrid algorithm [9] was used for the solution. The optimal solution gave the value of the functional (33) $J_5 = 3.6343$.

Figure 3 shows particular solutions of the system (31) with random perturbations of the initial values in the range (16). The blue curve in the figure shows the unperturbed optimal solution. The black dots show the positions of the found $\mathbf{x}^*$ control points. As we see from the experimental results, the perturbed solutions stabilize in the vicinity of the optimal one. Compared to Figure 1, it is obvious that the resulting model (31) is feasible.

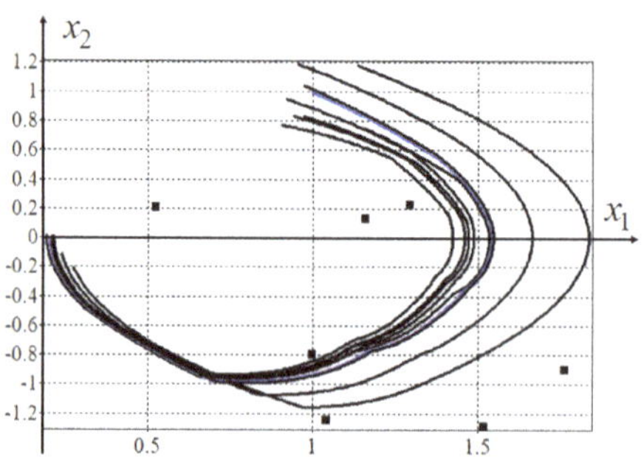

Figure 3. Perturbed and unperturbed solutions under synthesized optimal control.

4. Discussion

This paper raises the problem of the feasibility of optimal controls obtained as a result of solving the classical formulation of the optimal control problem. It is shown that when disturbances appear, the solutions turn out to be unsatisfactory. In the paper, an additional requirement for the desired control function is introduced. The introduced requirement ensures the stability of solutions near the optimal solution. Possible approaches to solving the proposed extended optimal control problem are considered. The best solution for problems

(1)–(5) is the solution to the general control synthesis problem. However, for more complex objects, it is not always possible to solve the problem of general synthesis. This paper considers the method of synthesized optimal control, which satisfies the introduced requirement of the feasibility of optimal control, and at the same time finds solutions that are close to optimal through the use of machine learning methods and evolutionary algorithms.

Author Contributions: Conceptualization, A.D. and E.S.; methodology, A.D.; software, E.S. and A.D.; validation, E.S. and A.D.; formal analysis, A.D.; investigation, A.D. and E.S.; data curation, E.S.; writing—original draft preparation, E.S.; writing—review and editing, E.S. All authors have read and agreed to the published version of the manuscript.

Funding: This research was performed with partial support of the Russian Science Foundation grant number 23-29-00339.

Institutional Review Board Statement: Not applicable.

Informed Consent Statement: Not applicable.

Data Availability Statement: Not applicable.

Conflicts of Interest: The authors declare no conflict of interest.

References

1. Pontryagin, L.S.; Boltyanskii, V.G.; Gamkreidze, R.V.; Mishchenko, E.F. *The Mathematical Theory of Optimal Processes*; Division of John Wiley and Sons, Inc.: New York, NY, USA, 1962.
2. Krasovskii, N.N. *Stability of Motion*; Stanford University Press: Redwood City, CA, USA, 1963.
3. Diveev, A.; Shmalko, E.; Serebrenny, V.; Zentay, P. Fundamentals of Synthesized Optimal Control. *Mathematics* **2021**, *9*, 21. [CrossRef]
4. Shmalko E. Feasibility of Synthesized Optimal Control Approach on Model of Robotic System with Uncertainties. In *Electromechanics and Robotics*; Ronzhin, A., Shishlakov, V., Eds.; Smart Innovation, Systems and Technologies; Springer: Singapore, 2022; Volume 232, pp. 131–143.
5. Ang, K.H.; Chong, G.; Li, Y. PID control system analysis, design, and technology. *IEEE Trans. Control Syst. Technol.* **2005**, *13*, 559–576. [CrossRef]
6. Hu, F.; Zeng, C.; Zhu, G.; Li, S. Adaptive Neural Network Stabilization Control of Underactuated Unmanned Surface Vessels With State Constraints. *IEEE Access* **2020**, *8*, 20931–20941. [CrossRef]
7. Yu, H.; Park, S.; Bayen, A.; Moura, S.; Krstic, M. Reinforcement Learning Versus PDE Backstepping and PI Control for Congested Freeway Traffic. *IEEE Trans. Control Syst. Technol.* **2022**, *30*, 1595–1611. [CrossRef]
8. Diveev, A.; Shmalko, E. *Machine Learning Control by Symbolic Regression*; Springer International Publishing: Cham, Switzerland, 2021.
9. Diveev, A. Hybrid Evolutionary Algorithm for Optimal Control Problem. In *Intelligent Systems and Applications*; Arai, K., Ed.; Lecture Notes in Networks and Systems Series; Springer: Cham, Switzerland, 2023; Volume 544.

engineering proceedings

MDPI

Proceeding Paper

Autonomous Navigation of Mobile Robot Assisted by Its Identified Neural Network Model †

Igor Prokopiev ‡, Elizaveta Shmalko *,‡ and Askhat Diveev ‡

Federal Research Center "Computer Science and Control" of the Russian Academy of Sciences, 119333 Moscow, Russia; fvi2014@list.ru (I.P.); aidiveev@mail.ru (A.D.)
* Correspondence: e.shmalko@gmail.com
† Presented at the 15th International Conference "Intelligent Systems" (INTELS'22), Moscow, Russia, 14–16 December 2022.
‡ These authors contributed equally to this work.

Abstract: Autonomous navigation is one of the key tasks in the development of control systems for real autonomous mobile objects. This paper presents the developed technology for accurately determining the position of a mobile robot in an autonomous operating mode without an external positioning system. The approach involves using a high-precision model of a real robot identified by a neural network. The robot adjusts its position, determined using odometry and video camera, according to the position of the robot, obtained using an accurate model. To train the neural network, a training set is used that takes into account the features of the movement of a wheeled robot, including wheel slip. In the experimental part, the problem of autonomous movement of a mobile robot along a given trajectory is considered.

Keywords: identification; neural network; machine learning; mobile robot; localization

Citation: Prokopiev, I.; Shmalko, E.; Diveev, A. Autonomous Navigation of Mobile Robot Assisted by Its Identified Neural Network Model. *Eng. Proc.* **2023**, *33*, 8. http://doi.org/10.3390/engproc2023033008

Academic Editors: Ivan Zelinka, Arutun Avetisyan and Alexander Ilin

Published: 17 May 2023

1. Introduction

Currently, there are many types of navigation of autonomous unmanned robots. Satellite navigation is certainly a common way. However, it also has a number of applicability limitations. For example, indoors, the satellite signal usually does not reach devices through concrete and metal structures. In this case, due to the fact that the space inside buildings is often limited to relatively small areas, it is possible to use navigation tools such as triangulation [1], navigation through various markers such as QR codes [2,3], barcodes [4], ArUco markers for position correction [5–7], and SLAM navigation [8,9], as well as combinations of the above methods.

In unexplored spaces, the most promising of the mentioned technologies is the SLAM navigation method [10]. This method can be used in an unprepared and unknown to the robot room to create a map and then use it. At present, this concept is widely used in the field of robotics for solving problems of autonomous motion [11–13]. According to SLAM, the robot needs to know its location at every moment of time, and also gradually scans the surrounding space with the help of sensors, thus compiling a map of the area. The map is built gradually as the robot explores new areas.

The main source of information about the location of the robot is odometry obtained in one way or another (wheels, computer vision, IMU, or a combination of both). However, any sensory system has its weak points. Therefore, for example, cameras can fail when lighting changes, or wheel slip can occur, causing an odometry error. In this regard, it becomes necessary to support the navigation system in some way independent of the readings of the sensors. This explains the relevance of this study.

The paper proposes the use of model-based navigation as a parallel system. In this case, one system can correct its values relative to the other.

Receiving a mathematical model for the control object is often time-consuming and sometimes even impossible using traditional methods. To build a mathematical model of a robot, the laws of theoretical mechanics and the Lagrange equations of the second kind for generalized coordinates are used. It should be noted that the analytical derivation of the robot dynamics equations in most cases will not accurately determine some parameters of the model, for example, friction coefficients of the robot moving parts or moments of inertia of rotating structures. Thus, the analytical output of mechanics equations allows one to build models with accuracy to the values of some parameters.

In view of the increasing diversity and extremely complex nature of control objects, including the variety of modern robotic systems, the identification problem is becoming increasingly important, which allows one to build a mathematical model of the control object, having input and output data about the system. Conducting experiments in order to identify the correspondence of the mathematical model to a real object is the necessary stage of solving the identification problem.

The identification of a nonlinear system is of particular interest, since most real systems have nonlinear dynamics, and if, earlier, the identification of a system model consisted of the selection of optimal parameters for the selected structure, then the emergence of modern machine learning methods opens up broader prospects and allows one to automate the identification process itself. Thus, this process should also be automated and, therefore, must be considered as the task for machine learning.

A modern technique to identify a mathematical model of a robot is the use of artificial neural networks. To perform this, it is necessary, depending on the assumed complexity of the robot model, to choose the structure and type of the neural network, then, depending on the problem to be solved, select a set of controls and build a training sample. Next, it is necessary to train the neural network so that the resulting neural network reacts to control actions in the same way as a real object. It is this approach that is presented in this paper and applied in the described experiments for a mobile robot.

Further, the paper presents a mathematical formulation of the model identification problem. A universal approach for solving it based on modern methods of machine learning using neural networks is proposed. As an example, a solution to the navigation problem for a mobile wheeled robot based on the proposed approach using its identified model is presented.

2. Problem Statement of the Control Object Identification by an Artificial Neural Network

In the identification problem, the mathematical model of the control object is not fully or partially known, but the researcher has a real control object or its physical simulator. In this case, the real control object or physical simulator layout is an unknown function. The space of the input vectors of this function is the space of admissible controls for this object.

Let some control functions be set as functions of time

$$\mathbf{u}^j(\cdot) = \mathbf{v}^j(t),\ t \in [0; t^+], \tag{1}$$

where $\mathbf{u}^j \in \mathrm{U} \subseteq \mathbb{R}^m$, $j = 1, \ldots, K$, U is a compact set that takes into account control constraints, and t^+ is a limit value of time.

The control functions $\mathbf{v}^j(t)$, $j = 1, \ldots, K$, should be selected so that it is possible to determine dynamic abilities and speed limits in straight lines and at turns by the movement of the control object.

To obtain a training sample, it is necessary to conduct natural experiments. In the experiment, the control object is subjected to a given control $\mathbf{u}^j(\cdot)$, $j = 1, \ldots, K$ and at fixed times, $t_i = i\Delta t$, the state of the control object is stored, $\mathbf{x}(t_i)$, $i = 0, \ldots, D_j$, where D_j is a number of time intervals for experiment with control $\mathbf{u}^j(\cdot)$, $j = 1, \ldots, K$.

A training sample for an artificial neural network is a set of stored values of the vectors of the control and the state in determined moments of time.

$$\mathrm{T}_j = (T_i^j = \{\mathbf{u}^j(t_i), \mathbf{x}^j(t_i), t_i = i\Delta t\}_i : i = 0, \ldots, D_j). \tag{2}$$

The input vector for the artificial neural network is an element of the training sample $\mathbf{y}(t_i) = T_i^j, i = 1, \ldots, D_j - 1, j \in \{1; K\}$.

The output vector for the artificial neural network $\mathbf{z}(t_i)$ must coincide with the state space vector for the next moment of time $\mathbf{x}^j(t_{i+1}), i = 0 \ldots, D_j - 1, j \in \{1; K\}$:

$$\mathbf{z}(t_i) = \mathbf{f}_{NN}(\mathbf{y}(t_i)),\ i = 0, \ldots, D_i - 1,\ j \in \{1; K\}. \tag{3}$$

where $\mathbf{f}_{NN}(\mathbf{y}(t_i))$ is a function described by the artificial neural network.

A target functional of optimization in the machine learning process of the artificial neural network is

$$J_0 = \sum_{j=1}^{K} \sum_{i=0}^{D_j-1} \|\mathbf{x}^j(t_{i+1}) - \mathbf{z}(t_i)\| \to \min. \tag{4}$$

In this case, the trained neural network is an approximation of the mathematical model of the control object in the form of a system of finite-difference recurrent equations.

If the time interval Δt was selected to be quite small, then it is possible to use the artificial neural network as approximation of right parts of ordinary differential equation system in the Cauchy form. For this purpose, it is necessary to determine a rate of change of state variables for each time moment:

$$\dot{\mathbf{x}}^j(t_i) = \frac{\mathbf{x}^j(t_{i+1}) - \mathbf{x}^j(t_i)}{\Delta t},\ i = 0, \ldots, D_j - 1,\ j = 1, \ldots, K. \tag{5}$$

If the rate vector (5) is used as the output vector when training the neural network, then the mathematical model of the control object has the following form:

$$\mathbf{x}^i(t_{i+1}) = \mathbf{x}^i(t_i) + \Delta t \mathbf{f}_{NN}(\mathbf{x}^j(t_i), \mathbf{u}^j(t_i), t_i),\ i = 1, \ldots, D_j - 1,\ j = 1, \ldots, K. \tag{6}$$

An important step in system identification is to determine the type of model used, since the overwhelming majority of methods implement the so-called parametric identification [14]. First, a certain model structure is selected, which is considered suitable for describing a given object. Next, an identification experiment is carried out in which the input and output signals are measured, and then the identification method implements the tuning of the model parameters in accordance with some adaptive laws, so that the response of the model to the input signal can approximately correspond to the response of the real system to the same input action. Most often, object identification is used using linear systems [15], since it is easy to determine the effect of various input signals on the output for them. Although linear models are attractive for many reasons, they have their limitations, especially since all physical systems are nonlinear to some extent, and in many cases linear models are not suitable for representing these systems. In this regard, there is currently a significant interest in methods for identifying nonlinear systems, especially using machine learning methods based on neural networks [16–18].

Neural networks, being a universal approximator, provide a powerful tool for identifying nonlinear systems. In view of the wide distribution and availability of software, neural networks have gained immense popularity. Combined approaches have also been developed, based, for example, on the reference model [19]. This article presents the implementation of a combined approach to the identification of a control object based on a neural network [20].

3. Application Object Description

A wheeled mobile robot shown in Figure 1 is taken as a control object.

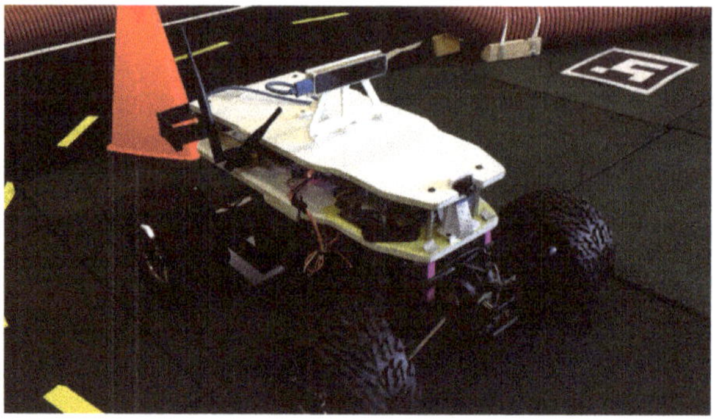

Figure 1. A control object.

The work uses a mixed approach to identification, when part of the object model is known based on physical laws, and the other part is determined by a neural network trained on the basis of a specially formed training sample.

The robot motion model is described by a system of differential equations:

$$\begin{cases} \dot{x} = v\cos(\theta), \\ \dot{y} = v\sin(\theta), \\ \dot{\theta} = \omega. \end{cases} \tag{7}$$

where x, y, $and\theta$ define the object state, and v, ω are control signals.

Robot control is implemented using two signals: the desired linear velocity u^v and the desired angular velocity u^ω. In the model (7), it is assumed that control signals are completely directly transmitted to the system $v = u^v$, $\omega = u^\omega$. In fact, a completely different picture is observed: firstly, the speeds cannot change infinitely quickly, and the system always has a certain dynamics, and secondly, as a rule, there is some regulator at a lower level that directly controls the voltage supplied to the motors. Therefore, in order to adequately describe the robot motion model, we introduced two additional equations into the (7) system that describe an object's dynamics.

$$\begin{cases} \dot{x} = v\cos(\theta), \\ \dot{y} = v\sin(\theta), \\ \dot{\theta} = \omega, \\ \dot{v} = f^v(v, \omega, u^v, u^\omega), \\ \dot{\omega} = f^\omega(v, \omega, u^v, u^\omega), \end{cases} \tag{8}$$

where f^v and f^ω are unknown functions to be identified.

Factors such as friction of the wheels on the surface, inertia, and uneven distribution of the robot's mass do not allow writing the functions f^v, f^ω in an explicit form.

For the numerical implementation of the identification problem, Equation (8) is presented in the finite-difference form with the time step Δt:

$$\begin{cases} x_{k+1} = x_k + \Delta t \cdot v_k cos(\theta_k), \\ y_{k+1} = y_k + \Delta t \cdot v_k sin(\theta_k), \\ \theta_{k+1} = \theta_k + \Delta t \cdot \omega_k, \\ v_{k+1} = F^v(v_k, \omega_k, u_k^v, u_k^\omega, \Delta t), \\ \omega_{k+1} = F^\omega(v_k, \omega_k, u_k^v, u_k^\omega, \Delta t). \end{cases} \tag{9}$$

To identify the system, we used the approximation of the function F by the parametric function F_ϕ. The learning function with parameters ϕ takes the current state of the robot, the control vector, and the time step as input, and outputs the state of the robot at the next moment of time:

$$[v_{k+1}, \omega_{k+1}] = F_{\phi}(v_k, \omega_k, u_k^v, u_k^{\omega}, \Delta t). \quad (10)$$

Parameters ϕ are selected in such a way that the trajectory of the movement of the object described by the system of Equation (9) with the model (10) repeats the real trajectory of the robot as closely as possible.

4. ANN Model Identification

For training, the principle of supervised learning was chosen, in which the neural network is trained using the training dataset "input–reference output", and then checked using a set of validation and test data that did not fall into the training sample.

To collect data, multiple runs of the robot were implemented under various types of control. Figures 2 and 3 show the example trajectories and sample control data.

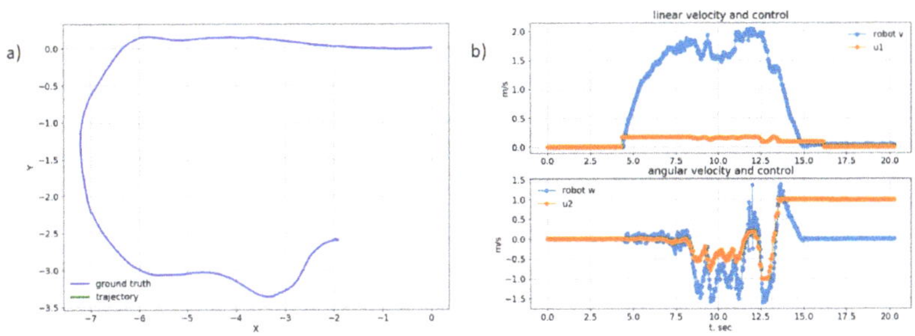

Figure 2. Sample 1: (**a**) trajectory; (**b**) control.

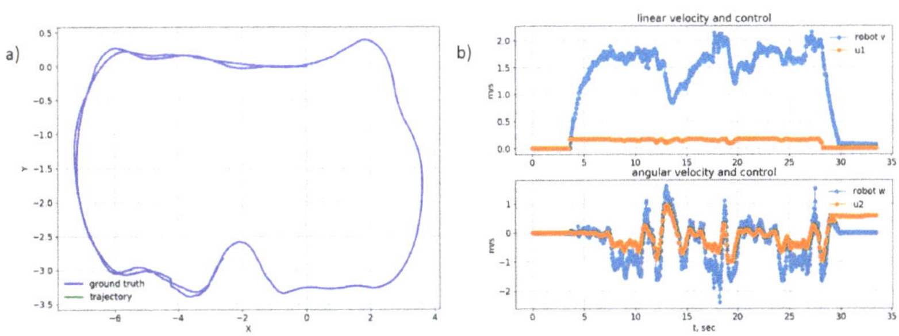

Figure 3. Sample 2: (**a**) trajectory; (**b**) control.

An artificial neural network with the multilayer perceptron architecture was chosen as the training model (10). PyTorch was used to train the neural network. A neural network with three layers was chosen, with 128 neurons on all hidden layers, an RELU activation function, and Adam optimization algorithm.

To assess the quality of the model, the accuracy of predicting the trajectory of movement by the model was evaluated on passages that were not included in the training sample. The neural network receives as input the initial state of the system and the control sequence for the test drive. As in the training phase, the trajectory of the robot's movement is predicted. To compare the predicted and actual trajectories, the ATE metric was used—the absolute translation error:

$$J_{ATE} = \frac{1}{N}\sum_{t=0}^{N}\sqrt{(x_t - \hat{x}_t)^2 + (y_t - \hat{y}_t)^2}, \quad (11)$$

where N is the number of points in the trajectory; x_t and y_t are actual trajectory point coordinates; $\hat{x}_t$ and $\hat{y}_t$ are coordinates of the predicted trajectory point.

To assess the quality of the predicted angle, the MAE (mean absolute error) metric was used.

$$J_{MAE} = \frac{1}{N}\sum_{t=0}^{N} |yaw_t - \hat{yaw}_t|, \tag{12}$$

where yaw_t is the actual value of the yaw angle, and $\hat{yaw}_t$ is the predicted value of the yaw angle.

The results of simulations of the trained neural network model are shown in Figures 4 and 5. As can be seen from the presented plots, the robot copes with the task of following the given trajectory even in the complex environment with strong surface slopes.

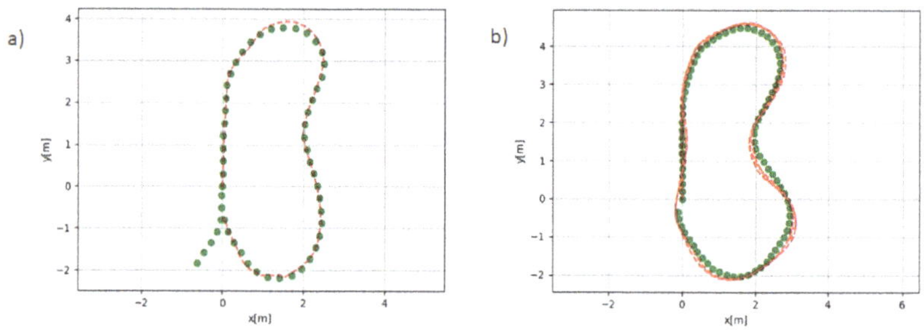

Figure 4. The trajectory to follow is given by green points at the distance of (**a**) 30 cm ; (**b**) 20 cm. The robots's trajectory is a red line.

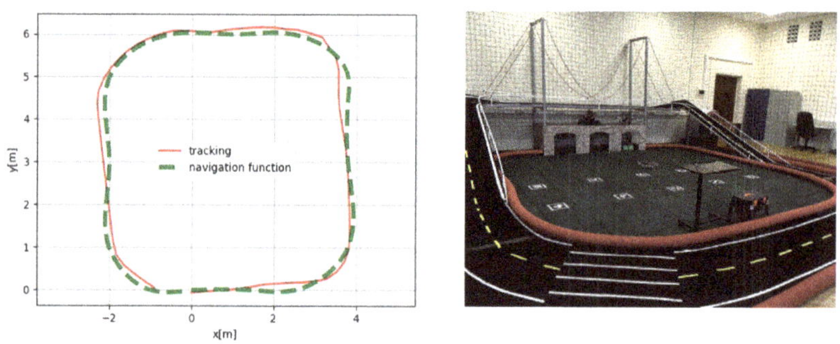

Figure 5. The robot's trajectory (red line) in a complex environment of the Robotic Center of FRC CSC RAS.

5. Discussion

This paper presents an approach to the localization of a mobile robot based on the identified neural network model. The presented results show how the robot copes quite accurately with the task of following various trajectories. The resulting neural network model can be used as an independent method of localization or as an auxiliary algorithm for correcting the position in case of failure of sensors and cameras.

Author Contributions: Conceptualization, I.P. and E.S.; methodology, A.D. and E.S.; software, I.P.; validation, E.S.; formal analysis, A.D.; investigation, I.P. and E.S.; data curation, E.S.; writing—original draft preparation, E.S.; writing—review and editing, E.S. All authors have read and agreed to the published version of the manuscript.

Funding: This research received no external funding.

Institutional Review Board Statement: Not applicable.

Informed Consent Statement: Not applicable.

Data Availability Statement: Not applicable.

Conflicts of Interest: The authors declare no conflicts of interest.

References

1. Sharp, N.; Soliman, Y.; Crane, K. Navigating intrinsic triangulations. *ACM Trans. Graph.* **2021**, *38*, 55. [CrossRef]
2. Li, Z.; Huang, J. Study on the use of Q-R codes as landmarks for indoor positioning: Preliminary results. In Proceedings of the 2018 IEEE/ION Position, Location and Navigation Symposium (PLANS), Monterey, CA, USA, 23–26 April 2018; pp. 1270–1276. [CrossRef]
3. Novoselov, S.; Sychova, O.; Tesliuk, S. Development of the Method Local Navigation of Mobile Robot a Based on the Tags with QR Code and Wireless Sensor Network. In Proceedings of the 2019 IEEE XVth International Conference on the Perspective Technologies and Methods in MEMS Design (MEMSTECH), Polyana, Ukraine, 22–26 May 2019; pp. 46–51. [CrossRef]
4. Zhou, C.; Liu, X. The Study of Applying the AGV Navigation System Based on Two Dimensional Bar Code. In Proceedings of the 2016 International Conference on Industrial Informatics—Computing Technology, Intelligent Technology, Industrial Information Integration (ICIICII), Wuhan, China, 3–4 December 2016; pp. 206–209. [CrossRef]
5. Sani, M.F.; Karimian, G. Automatic navigation and landing of an indoor AR. drone quadrotor using ArUco marker and inertial sensors. In Proceedings of the 2017 International Conference on Computer and Drone Applications (IConDA), Kuching, Malaysia, 9–11 November 2017; pp. 102–107. [CrossRef]
6. Marut, A.; Wojtowicz, K.; Falkowski, K. ArUco markers pose estimation in UAV landing aid system. In Proceedings of the 2019 IEEE 5th International Workshop on Metrology for AeroSpace (MetroAeroSpace), Turin, Italy, 19–21 June 2019; pp. 261–266. [CrossRef]
7. Tian, W.; Chen, D.; Yang, Z.; Yin, H. The application of navigation technology for the medical assistive devices based on Aruco recognition technology. In Proceedings of the 2020 IEEE/RSJ International Conference on Intelligent Robots and Systems (IROS), Las Vegas, NV, USA, 24 October 2020–24 January 2021; pp. 2894–2899. [CrossRef]
8. Geng, K.; Chulin, N.A. UAV Navigation Algorithm Based on Improved Algorithm of Simultaneous Localization and Mapping with Adaptive Local Range of Observations. *Her. Bauman Mosc. State Tech. Univ. Ser. Nat. Sci.* **2017**, *3*, 114. [CrossRef]
9. Won, D.H.; Chun, S.; Sung, S.; Kang, T.; Lee, Y.J. Improving mobile robot navigation performance using vision based SLAM and distributed filters. In Proceedings of the 2008 International Conference on Control, Automation and Systems, Seoul, Republic of Korea, 14–17 October 2008; pp. 186–191. [CrossRef]
10. Cheeseman, P.; Smith, R.; Self, M. A stochastic map for uncertain spatial relationships. In Proceedings of the 4th International Symposium on Robotics Research, Santa Cruz, CA, USA, 9–14 August 1987; pp. 467–474.
11. Lu, F.; Milios, E. Globally Consistent Range Scan Alignment for Environment Mapping. *Auton. Robot.* **1997**, *4*, 333–349. [CrossRef]
12. Biswas, J.; Veloso, M. Depth camera based indoor mobile robot localization and navigation. In Proceedings of the IEEE International Conference on Robotics and Automation, St. Paul, MN, USA, 14–18 May 2012; pp. 1697–1702. [CrossRef]
13. Tu, Y.; Huang, Z.; Zhang, X.; Yu, W.; Xu, Y.; Chen, B. The Mobile Robot SLAM Based on Depth and Visual Sensing in Structured Environment. In *Robot Intelligence Technology and Applications 3*; Kim, J.H., Yang, W., Jo, J., Sincak, P., Myung, H., Eds.; Advances in Intelligent Systems and Computing Series; Springer: Cham, Switzerland, 2015; Volume 345.
14. Zeng, R.; Kang, Y.; Yang, J.; Zhang, W.; Wu, Q. Terrain Parameters Identification of Kinematic and Dynamic Models for a Tracked Mobile Robot. In Proceedings of the 2018 2nd IEEE Advanced Information Management, Communicates, Electronic and Automation Control Conference (IMCEC), Xi'an, China, 25–27 May 2018; pp. 575–582. [CrossRef]
15. Cox, P.; Toth, R. Linear parameter-varying subspace identification: A unified framework. *Automatica* **2021**, *123*, 109296. [CrossRef]
16. De León, C.L.; Vergara-Limón, S.; Vargas-Treviño, M.A.; López-Gómez, J.; Gonzalez-Calleros, J.M.; González-Arriaga, D.M.; Vargas-Treviño, M. Parameter Identification of a Robot Arm Manipulator Based on a Convolutional Neural Network. *IEEE Access* **2022**, *10*, 55002–55019. [CrossRef]
17. Ge, W.; Wang, B.; Mu, H. Dynamic Parameter Identification for Reconfigurable Robot Using Adaline Neural Network. In Proceedings of the 2019 IEEE International Conference on Mechatronics and Automation (ICMA), Tianjin, China, 4–7 August 2019; pp. 319–324. [CrossRef]
18. Liu, G.P. *Nonlinear Identification and Control: A Neural Network Approach*; Springer Science & Business Media: Berlin/Heidelberg, Germany, 2012.
19. Williams, G.; Drews, P.; Goldfain, B.; Rehg, J.M.; Theodorou, E.A. Aggressive driving with model predictive path integral control. In Proceedings of the 2016 IEEE International Conference on Robotics and Automation (ICRA), Stockholm, Sweden, 16–21 May 2016; pp. 1433–1440.
20. Shmalko, E.; Rumyantsev, Y.; Baynazarov, R.; Yamshanov, K. Identification of Neural Network Model of Robot to Solve the Optimal Control Problem. *Inform. Autom.* **2021**, *20*, 1254–1278. [CrossRef]

engineering proceedings

MDPI

Proceeding Paper

The System Architecture and Methods for Efficient Resource-Saving Scheduling in the Fog †

Anna Klimenko

Institute of IT and Security Technologies, Kirovogradskaya St. 25-2, 117534 Moscow, Russia; anna_klimenko@mail.ru

† Presented at the 15th International Conference "Intelligent Systems" (INTELS'22), Moscow, Russia, 14–16 December 2022.

Abstract: The problem of resource-saving scheduling in a fog environment is considered in this paper. The objective function of the problem in question presupposes the fog nodes' reliability function maximizing. Therefore, to create a schedule, the following is required: the history of the fog devices' state changes and the search space, which consists of preselected nodes of the cloud-fog broker neighbourhood. The obvious approach to providing the scheduler with this information is to poll the fog nodes, yet this can consume the unacceptable time because of the QoS requirements. In this paper, the system architecture and general methods for efficient resource-saving scheduling is presented. The system is based on distributed ledger element usage, which provides the nodes with the proper awareness about the surroundings. The usage of the distributed ledger allows not only for the creation of the resource-saving schedule but also the reduction of the scheduling problem-solving time, which frees addition time that can be used for the solving of user tasks. The latter also affects the overall resource-saving via reliability. The novelty of this paper consists in the development of the hybrid ledger-based system, which integrates and arranges the elements of various ledger types to solve the newly formulated problem.

Keywords: scheduling problem; optimization; fog computing; distributed ledger

Citation: Klimenko, A. The System Architecture and Methods for Efficient Resource-Saving Scheduling in the Fog. *Eng. Proc.* **2023**, *33*, 9. https://doi.org/10.3390/engproc2023033009

Academic Editors: Askhat Diveev, Ivan Zelinka, Arutun Avetisyan and Alexander Ilin

Published: 17 May 2023

1. Introduction

The scheduling problem is known as a problem of high importance in the field of distributed computing, including scheduling in the fog- and edge- computing environments. Resource-saving scheduling is a problem with a particular objective function, that is to maximize the reliability function values of the nodes at the end of the user operation. Moreover, the problem solution must not consume much time due to the quality of service (QoS) and quality of experience (QoE) requirements for the fog or edge systems.

Fog computing presupposes the cloud-fog broker (CFB) functioning in the fog layer of the network. The general tasks of the CFB are to receive the computational problem and data from the edge device; to schedule this problem, assigning the subtasks to the fog devices nearby—or if there are no devices with appropriate amount of resources in the fog,to send the problem and data to the cloud—and to return a result to the user's device [1].

There are some issues in this scheme that created by the peculiarities of the fog layer and of the resource-saving scheduling problem:

- There can be an uncertainty about the fog nodes in the CFB neighbourhood. The main reason for this is the dynamic nature of the fog layer [2,3];
- The distances between fog nodes affect the time of data exchange between the fog nodes. Time consumption leads to the impossibility of meeting the QoS requirements and also worsens the reliability functions of the fog nodes because of possible overloading;

- The scheduling problem itself must be solved as soon as possible because of the need to maintain the QoE and QoS requirements;
- Finally, the forming of a resource-saving schedule requires the nodes' workload history, at least when the log of the device resource states changes.

The issues listed above generate the following tasks which must be performed to generate the schedule:

- The search space for the optimization problem solving must be formed on the basis of available fog nodes;
- The workload history must be used for the schedule formation.

The most obvious solution to implementing the scheduling process under the listed conditions is to request all the needed data from the fog nodes in the CFB neighbourhood, and, if sufficient resources are there, to assign the tasks to the node. Yet, distances between nodes can produce unacceptable delays for the formation of scheduling problem search space (as well as for workload information retrieval), while the increase of the scheduling formation time leads to the decrease of the time for the computational task solution. Therefore, the general goal of this paper is to design the architecture and basic functioning methods of the system for efficient resource-saving scheduling in the fog.

The following tasks are considered to achieve the general goal:

- Resource-saving scheduling problem formalization and analysis;
- A state-of-the-art analysis in the field of scheduling in the fog, resource saving, and distributed ledger;
- A comparison of the approaches to providing CFB with the appropriate data;
- System architecture and basic functional method development.

The main contribution of this paper is the development of the architecture and basic functional methods of the system for efficient resource-saving scheduling in the fog. The novelty of this paper consists in the development of the hybrid ledger-based system, which integrates and arranges the elements of various ledger types to solve the newly formulated problem.

2. The Resource-Saving Scheduling Problem and Its Analysis

The scheduling problem in the fog environment takes place when the CFB distributes the computational tasks within the set of nearby fog nodes [1].

The process of task scheduling in general involves np-hard problem-solving, which can be implemented via various methods, including up-to-date heuristics and methaheuristics (genetic algorithms, ant colony algorithms, simulated annealing, etc.). These methods are quite efficient; however, no discrete optimization can be performed without the formation of a search space [4], and a schedule-forming procedure has its time restrictions: with the increase of scheduling time the time for functional tasks implementation reduces. Moreover, the resource-scheduling problem requires the data on the workload history of the nodes because this is the key parameter for the workload distribution.

Consider the CFB, which receives the user task the and data to solve. The problem is to schedule the task within the formed fog node community, so as $P_0(\tau) \rightarrow max,\ where\ P_0(\tau)$ is the overall reliability function value of the fog nodes community, including CFB,τ is the moment of the user operation completion.

Consider the network graph $G = \ < V, U >$ where V is a set of computational nodes, and R is a set of ribs. $V = v_i \ = \ < i, p_i, R_i(t_0), L_i >$, where i is the node identifier, p_i is the node performance, $R_i(t_0)$ is the computational resource of the node at the moment of scheduling problem solution, and L_i—is the workload of the node at the moment of t_0. $U = u_i$, where u_i is the data transmission rate of the network rib j. The user operation is described as an acyclic graph, whose vertexes are assigned to tasks and whose ribs are assigned to information connections between them. $O \ = \ < T, C >$, where T is the set of subtasks, and C is the set of information connections. $T = t_j \ = \ < j, w_j, d_j >$, where j is the subtask identifier, w_j is the computational complexity of the subtask, and d_j is the data volume transferred to the network. The problem solution is the following tasks assignment: $A = \begin{bmatrix} t_{11} & \dots & t_{1m} \\ \dots & \dots & \dots \\ \dots & t_{ij} & t_{nm} \end{bmatrix}$ such as $P_0(\tau) \rightarrow max.$ where $P_0(\tau) = \prod_i P_i(\tau)$, $P_i(\tau) = \exp\left(-\lambda_0 \tau 2^{(kD_j/10)}\right)$.

where $P_j(\tau)$ is the reliability function value,

D is the node workload;

k is the coefficient of node temperature increase depending on the current workload, and

t_{i_j} is the moment of assignment of task j to the node i.

The constraint for this problem is as follows: $\tau < t_c onst$; that is, the user operation completion time must be less than the declared time for this operation. One can see that the formal presentation of the resource-saving scheduling problem is quite common for the scheduling ones and can be solved by one of the existing and described methods (e.g., simulated annealing or some greedy approaches).

However, to solve this optimization problem in the fog, the following data must be provided:

- The search space, which is the set of nodes that are available for the task assignment;
- The workload story for the fog nodes, which is needed to estimate the reliability function values, as it is the integral part of the overall objective function.

Therefore, to solve the resource-saving scheduling problem in the fog, some means must be developed to provide the CFBs with the information required.

3. State-of-the-Art Analysis

The basic research area of this paper covers the following problem fields:

- The computational resources, node failure rate and reliability function connected to the nodes workload;
- The particular scheduling problem-solving under the time constraints and the uncertainty of the resources;
- The distributed ledger technologies and their application to the node information provisioning.

We consider the field of the computational resources, failure rate, and reliability. In the current paper the term "computational resource" refers to the reliability function of the computational node. The reliability function value depends on the failure rate, while the failure rate is related to the device temperature and workload:

$$\lambda = \lambda_0 * 2^{(\Delta T/10)}; \quad (1)$$

where λ is a resulting failure rate, λ_0 is the failure rate under conditions of the unloaded device, and ΔT is the temperature difference between the temperature of the unloaded device and the temperature of the loaded one.

In the study [5], the coefficient is determined, which connects the node temperature and the workload. Consequently, the reliability function is determined as follows:

$$P_i(\tau) = \exp(-\lambda_0 \tau 2^{(kD/10)}). \quad (2)$$

Therefore, by varying the node workload, the reliability function values can be improved at the particular time moment. This approach to system reliability improvement has been used in other studies [6,7], with a variation of objective function forms. However, no attention has been paid to the workload distribution under the uncertainty of the search space, which is highly related to the fog environment.

The scheduling problem is considered carefully in a wide range of publications because the problem itself is not new [8]. Additionally, some studies have examined resource planning and scheduling in the fog but have not considered the questions of reliability, resource-saving, or scheduling under the search space uncertainty.

Notably, in [9], the method of the online formation of the search space in the dynamic fog environment with the usage of onthologies to speed up the search space formation is considered. However, this approach presupposes the formation of the search space online by means of node polling, which may be unacceptable in conditions with time and the QoS restrictions.

Distributed ledger technologies, which first emerged as cryptocurrency systems, have been applied to the particular areas of fog and edge computing, with some examples listed below:

- Security issues in the fog [10];
- Data management [11];
- Consensus mechanisms [12].

Regarding the current paper, distributed ledger technologies can provide synchronized local data storing as the source of available resources, node states, and workload history data. It must be mentioned that as there is no report of ledger usage aimed at improving node reliability,except for in [13]. In this study, the distributed ledger is used to store the information about the device's workload, yet there are no methods providing the data integrity and system functioning in general.

Therefore, despite the rare efforts, there are still no basic approaches to the design and development of particular systems for efficient problem-solving of resource-saving scheduling in the fog.

4. A Comparison of the Approaches to CFB Data Provisioning and the Approach Choice for System Design

As mentioned in the previous section, to solve the resource-saving scheduling problem the following is required:

- A search space for the optimization of problem-solving;
- Information about the workload history of the fog nodes.

All these requirements can be met by means of particular data storing on the fog nodes.

Estimate the time t_0 as the time when all the needed data will be gathered from the nearby fog bodes. The minimum possible value of t_0 is the minimum of t_i, which is the time of information gathering from the nearest fog node.Thus, the lower estimation is as follows: $t_0 \geq t_{\min}$. The estimations of the time needed for the information gathering are as follows, with D_k as the distance between CFB and the fog device:

$$t_{\min} \leq t_0 \leq \sum_{i=1}^{n} t_{\min} D_k \tag{3}$$

Consider a case in which all the needed information is stored on the fog node locally. There is no polling, and the time of information gathering for the search space forming and for the node workload history includes the time of local storage analysis.
Therefore, the estimation of the time for the local data processing and analysis can be as follows:

$$t_0 = \xi(V_{ledger}), \tag{4}$$

where V_{ledger} is the volume of the local data storage, and $\xi(V_{ledger})$ is the time of ledger analysis. Thus, the polling strategy implementation depends on the network diameter and nodes as well as the data transmission channels state, while the local storage of the appropriate information makes it possible to form the search space and analyse the workload history of the nodes much more efficiently. The comparison between the approaches to the information gathering is shown in Figure 1

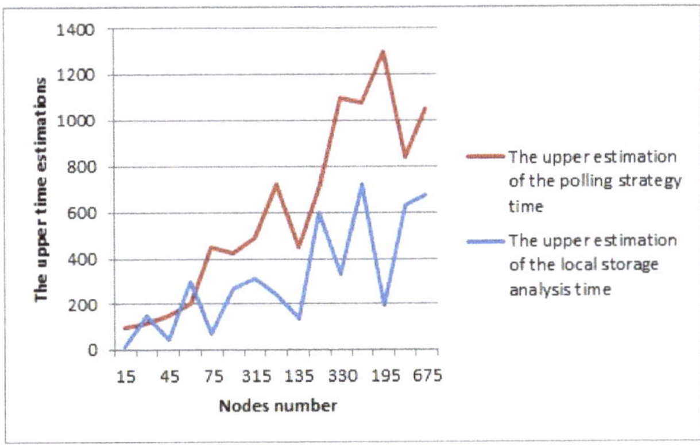

Figure 1. The time estimations for the polling strategy and the local data storage.

The following must be highlighted: the time decrease of the resource-saving scheduling problem-solving increases the resources of the fog nodes due to the possibility of decreasing the workload.

It is well-known that there is a wide range of distributed ledger types, each which use different data storage and consensus methods. For example, there are many consensus methods for blockchain-based ledgers, including proof of work, proof of stake, proof of authority (the competitive consensus), and PBFT (cooperative). Some consensus methods have been developed for ledgers with other storage types, e.g., Nano (block lattice), Swirlds Hashgraph consensus algorithm (Hedera Hashgraph), and others [14].

To provide the CFBs with the appropriate data, the choice of data storage method and the method of consensus on data must be formed on the basis of the general functional requirements of the resource-saving scheduling problem-solving system architecture:

- There are three node types in the system: user (edge) nodes, fog nodes, and CFB nodes. User nodes are the sources of the tasks and data,fog nodes are supposed to participate in the tasks and data processing, and CFB nodes produce schedules and distribute computational workload through the fog nodes and cloud.
- The need to deal with the local copies of the workload history and resource states forms the main requirement to the distributed ledger functioning: the data on the resource state change of the fog node must be placed into the ledger in an order of events.
- The additional requirement for the fog node information collecting and disseminating is that information about the node resource state change must be disseminated through the network as soon as possible.
- In the case of assigning the fog nodes to the CFB, there is a need to detect CFB failure and to restore the information provisioning as soon as possible.

block lattice/Nano technology of data storing and synchronization is the most appropriate for the storing of the changing resource and workload states of the fog nodes. Yet, it is insufficient to synchronize the data only: there must be a mechanism to detect the CFB failure and to recover the system of fog brokers. The view stamped replication concept has potential in this regard.As we have the fully replicated data, there is no need for an exchange with the leader elections procedure. Thus, the time needed for leader election is acceptable.

5. Development of the System Architecture and Basic Functional Methods

Consider the atomic transaction as follows: $Q = \; < timestamp, Load, node_{id}, t, broker_{id} >$, where $transaction_{id}$ is the transaction identifier, $round_{id}$ is the epoch number, $Load$ is the workload estimation of the node, $Node_{id}$ is the unique fog node identifier, t is the moment the device exploitation begins, and $broker_{id}$ is fog broker identifier. Transactions of the described structure are stored in the block lattice—type ledger. Each fog node is assigned to its own blockchain within the block lattice. Each blockchain implements the storage of the nodes' states. Every time the node changes its state (e.g., workload change), this fact is placed into its assigned blockchain and disseminates through the fog broker nodes, which are the owners of the block lattice storage.

The basic architecture of the system is presented in the Figure 2.

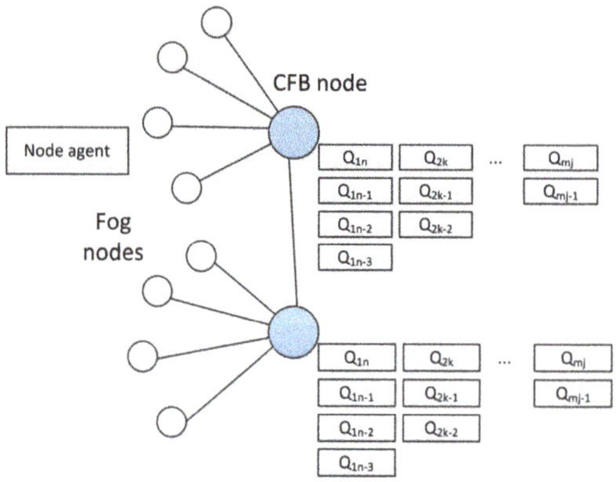

Figure 2. The basic architecture of the system.

The general system architecture implements the following functionality. The fog device software performs the following:

- Provides the node state registration;
- Provides the storage of the device-broker assignment information;
- Provides the fog node–broker interaction;
- Provides the procedures of receiving and processing the computational tasks as planned by a fog broker.

The CFB software performs the following:

- Interaction with other brokers;
- Interaction with the fog nodes;
- Interaction with the ledger replicas and the providing of its functioning;
- Scheduling problem-solving.

Figure 3 contains the software architecture of the fog and broker node.

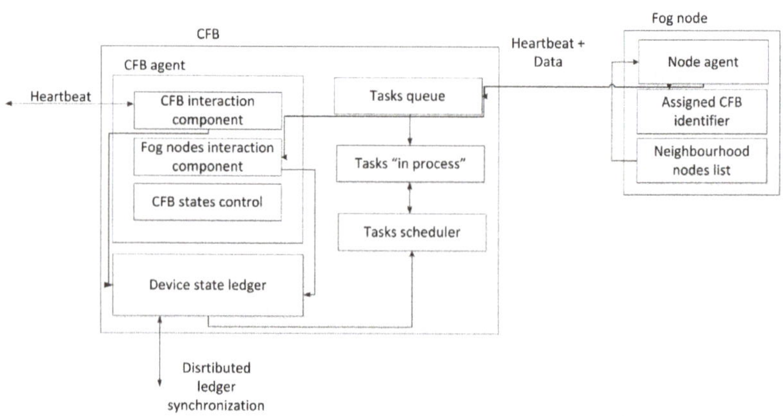

Figure 3. The structure of the CFB component and its interaction with the fog device.

The basic functions of the system are implemented by means of the following methods.

5.1. *Adding a New Broker Node to the Broker's Network*

1. The broker agent initializes.
2. Broker agent sends a request to the neighbour nodes about their participation in the process of information delivery for the resource-saving scheduling problem-solving. If some nodes are found, then the following occurs:
    - Request of the list of fog brokers from the neighbour fog node.
    - Sending of a request of the ledger copy to the nearest fog broker node.
    - Addition of the new own blockchain to the block lattice, which is assigned to the new device in the fog layer.
3. If there is no devices with ledger copies, then, the current fog broker is the first ledger copy.
4. Creation the first ledger copy with the blockchain assigned to the broker node.
5. Creation of the new list of brokers, each with its identifier being added to the list.
6. Implementation of the search through the network cluster to gather the information about the fog nodes' current states.
7. Addition of new blockchains to the ledger and their being assigned with the new-found nodes.
8. Sending of the broker ID to the active fog nodes.

5.2. *Broker Failure*

The failure of the fog broker does not make it necessary to remove its own blockchain from the ledger. However, the broker failure can become the cause errors in the scheduling problem-solving when the neighbourhood fog nodes send their state descriptions to the failed broker. In the case of broker failure, the fog nodes do not have the command to send their data somewhere else. This problem can be solved in various ways:

- The fog node sends the state data to some of the nearest fog brokers. In the case of broker failure, no data are lost, yet the question of data duplication emerges.
- The fog node sends state data to the nearest fog broker.

In the case of broker failure, some state information will be lost, yet this situation can be detected within the fog broker network, and the interaction between the fog nodes and brokers can be recovered. To detect the broker failures, the broker community is formed. The brokers' awareness about their neighbourhood is implemented by means of message exchange. In case of distributed leader usage (viewstamped replication concept), broker nodes send the "heartbeat" messages to the leader and receive the same messages from it. One can see that in the distributed leader approach, the information exchange within the broker network will be quite acceptable.

The main stages of the system functioning based on the distributed leader approach are outlined below.

5.3. *Functioning Stage*

1. The leader sends "heartbeat" messages to its followers.
2. The follower sends the proof of the message's receipt.

5.4. *Follower Failure*

1. If the current leader has not received the heartbeat message, then;
2. The leader searches its ledger copy for the fog nodes assigned to the failed broker;
3. The leader searches the fog broker, which is nearest to the fog nodes assigned to the failed broker;
4. The request of the new fog broker setup is sent to the fog nodes;
5. The new assigned broker sends the request for fog node states to the fog nodes;
6. States are put into the ledger copy;

7. The state of the failed fog broker is put into the ledger as a state with the full utilization (no resources available);
8. New data are replicated through all the brokers.

5.5. Leader Failure

1. Followers wait for the "heartbeat" from the leader node;
2. If there is no "heartbeat", the new leader is elected, for instance, by means of some simple procedure (round robin);
3. The search of the new broker, which is near the nodes of the failed broker, is conducted;
4. The request of the new fog broker setup is sent to the fog nodes;
5. The new fog broker sends the state request to the new fog nodes;
6. The received states are put into the ledger copy;
7. The state of the failed broker is put to the ledger as the state of full utilization (the lack of available resources);
8. New data are replicated to the ledger copies.

5.6. New Fog Device Addition

1. Within the new fog node emergence, the fog node agent sends its identifier into the network.
2. The first broker receives this message, sends the confirmation to the new node, and blocks it for the other brokers. From this moment the fog node is presupposed to be assigned to the current broker.
3. The assigned broker requests the node state information and puts it into the ledger.

5.7. Fog Node Failure

In the case of fog node failure, there can be a situation in which the assigned tasks can not be performed. In this case, the fog broker can reschedule the problem, taking into account that fact that the time for user computations has decreased. The failed fog node state is put into the ledger as "no resource available".

6. Conclusions

In the current paper, the architecture and some functional methods are proposed for decreasing the resource-saving scheduling problem time.

The distributed ledger technology, block lattice + Nano concept elements were chosen for the data provisioning implementation, while viewstamped replication protocol elements were chosen for the control and recovery of the CFB network. The architecture and methods proposed provide the CFB nodes with the data needed for the following:

- Search space formation;
- Resource-saving scheduling problem-solving.

Furthermore, the possibility to process the local data without the node polling allows the scheduling problem-solving time to be shortened, thus increasing the time for functional user task solution. This leads to a decrease in fog node's workload decrease, and in this way, the provision of additional time is achieved.

Funding: This study received no external funding.

Institutional Review Board Statement: Not applicable.

Informed Consent Statement: Not applicable.

Data Availability Statement: Not applicable.

Conflicts of Interest: The author declares no conflict of interest.

References

1. Xu, J.; Hao, X.; Zhang, R.; Sun, X. A Method Based on the Combination of Laxity and Ant Colony System for Cloud-Fog Task Scheduling. *IEEE Access* **2019**, *7*, 116218–116226. [CrossRef]
2. Alenizi, F.; Rana, O. Fog Computing: Towards Dynamically Controlling the Offloading Threshold and Managing Fog Resources in Online Dynamic Systems. *Preprints.org* **2021**. [CrossRef]
3. Giang, N.; Lea, R. Developing applications in large scale, dynamic fog computing: A case study. *Softw. Pract. Exp.* **2020**, *50*, 519–532. [CrossRef]
4. Zhiglyavsky, A.; Zhilinskas, A. *A Global Extremum Search Methods*; Nauka: Moscow, Russia, 1991; p. 247.
5. Klimenko, A.; Melnik, E. Information and Control Systems with Distributed Ledger Usage: A Reliability Issue. In *Artificial Intelligence in Intelligent Systems: Proceedings of 10th Computer Science On-Line Conference 2021*; Springer International Publishing: New York, NY, USA, 2021. [CrossRef]
6. Klimenko, A.; Kalyaev, I. A Technique to Provide an Efficient System Recovery in the Fog- and Edge-Environments of Robotic Systems. In Proceedings of the Interactive Collaborative Robotics: 6th International Conference, St. Petersburg, Russia, 27–30 September 2021. [CrossRef]
7. Korovin, I.; Melnik, E.; Klimenko, A. The Fog-Computing Based Reliability Enhancement in the Robot Swarm. In Proceedings of the Interactive Collaborative Robotics: 4th International Conference, ICR 2019, Istanbul, Turkey, 20–25 August 2019. [CrossRef]
8. Blazewicz, J.; Moseley, B.; Pesch, E.; Trystram, D.; Zhang, G. New Perspectives in Scheduling Theory. *J. Sched.* **2021**, *24*, 161–169. [CrossRef]
9. Klimenko, A.; Safronenkova, I. An Ontology-Based Approach to the Workload Distribution Problem Solving in Fog-Computing Environment. In *Artificial Intelligence Methods in Intelligent Algorithms: Proceedings of 8th Computer Science On-line Conference 2019*; Springer International Publishing: New York, NY, USA, 2019. [CrossRef]
10. Wu, D.; Ansari, N. A Cooperative Computing Strategy for Blockchain-Secured Fog Computing. *IEEE Internet Things J.* **2020**, *7*, 6603–6609. [CrossRef]
11. Alam, T. IoT-Fog-Blockchain Framework: Opportunities and Challenges. *Int. Fog Comput.* **2020**, *3*, 1–20.[CrossRef]
12. Son, B.; Lee, J.; Jang, H. A Scalable IoT Protocol via an Efficient DAG-Based Distributed Ledger Consensus. *Sustainability* **2020**, *12*, 1529. [CrossRef]
13. Kalyaev, I.; Melnik, E.; Klimenko, A. Distributed Ledger Based Workload Logging in the Robot Swarm. In Proceedings of the Interactive Collaborative Robotics: 4th International Conference, ICR 2019, Istanbul, Turkey, 20–25 August 2019. [CrossRef]
14. Masood, F.; Faridi, A. An Overview of Distributed Ledger Technology and its Applications. *Int. J. Comput. Sci. Eng.* **2018**, *6*. 422–427. [CrossRef]

engineering proceedings

Proceeding Paper

Monitoring the Condition of a Patient with Parkinson's Disease †

Yulia Shichkina *,‡, Roza Fatkieva ‡ and Nikita Isaenko ‡

Department of Computer Science and Engineering, Saint Petersburg Electrotechnical University LETI, St. Petersburg 197022, Russia; rikki2@yandex.ru (R.F.); isaenko.nikita.7305@gmail.com (N.I.)
* Correspondence: strange.y@mail.ru
† Presented at the 15th International Conference "Intelligent Systems" (INTELS'22), Moscow, Russia, 14–16 December 2022.
‡ These authors contributed equally to this work.

Abstract: Currently, interest in the development of Internet-of-Things technologies is increasingly penetrating the field of clinical medicine. This paper provides an overview of the use of Internet-of-Things technologies in medical practice using the Scopus database of publications. The classification of publications on the topic of research directions with promising development trends has been performed. With this in mind, the concept of the architecture of a system for monitoring the condition of a patient with Parkinson's disease is presented. The necessary hardware and software solutions have been developed, taking into account the needs in order to more effectively adjust treatment and monitor the course of the disease. To more accurately determine the state of progression of the patient's disease, tests have been designed and developed to assess overall emotional and physical well-being. For the most effective correction and control of drug therapy, a prototype of a drug administration device with four compartments for a drug has been designed, each compartment of which is individually controlled by a specially developed Bluetooth data transmission protocol. Access to each compartment is individual, which ensures the versatility of the device for taking several medications throughout the day. The practical application of the solution has also shown the relevance of its use in the task of studying the course of Parkinson's disease as well as for monitoring the condition of a patient with existing concomitant diseases and the degree of their influence on the course of the underlying disease.

Keywords: Parkinson's disease; monitoring; IoT; styling; insert

Citation: Shichkina, Y.; Fatkieva, R.; Isaenko, N. Monitoring the Condition of a Patient with Parkinson's Disease. *Eng. Proc.* **2023**, *33*, 10. https://doi.org/10.3390/engproc2023033010

Academic Editors: Askhat Diveev, Ivan Zelinka, Arutun Avetisyan and Alexander Ilin

Published: 23 May 2023

1. Introduction

According to the World Health Organization, 35 per 100,000 people suffer from Parkinson's disease in the world. According to experts, this figure may increase and represents an acute social and economic problem due to the high cost of treatment, disability and the decrease in the quality of life. On the other hand, the use of E-health technology is becoming more common in the field of healthcare and clinical research due to the convergence of mobile and wireless technologies, which allows continuous monitoring of chronically ill patients, better care and feedback, shorter hospitalization time, increased medical capacity and reduced cost of medical services.

We conducted a search for the publication activity of research aimed at E-health in the Scopus database and analyzed a representative sample of publications based on 10,668 keywords and 499 publications over the period of 2016–2021 using VOSviewer software, which allowed us to identify the main areas of IoT in medicine.

Direction Health care: allows you to trace the following associative links relevant to the protection of public health.

Direction Internet in medicine (number of entries—643 publications): concerns the cycle of organizational activity and interaction for research, storage and transfer of medical data. The following sections were highlighted in publications related to the Internet:

Direction Telemedicine: the use of computer and telecommunication technologies for the exchange of medical information. It is one of the fastest growing segments of healthcare in the world. The following sections are highlighted:

The main directions of publication activity and analysis of relevant works: Some studies were grouped according to a specific feature, e.g., wearable devices, reviews in the field of IoT development and technological innovations in medicine, possible solutions and implemented projects and systems.

Consider research related to wearable devices [1–3]: The paper [1] discusses consumer trends in wearable electronics, commercial and new devices, as well as production methods. In [2], it is proposed to look at skin-like electronics by reviewing several recent reports on various strategies for using materials and methodologies for integrating stretchable conductive and semiconductor nanomaterials that are used as electrodes and active layers in stretchable sensors, transistors, multiplexed matrices and integrated circuits. The first part of the review [3] briefly discusses issues related to the use of smart wearable devices, including technologies, users, technology-related activities and the consequences of using technologies. The second part of this review is devoted to the risks of using smart wearable devices.

Works that address the areas of IT development and technological innovation [4–7]: The review [4] describes four main areas of interest for respiratory care: pulse oximetry, pulmonary ventilation, activity tracking and air quality assessment. While some issues still need to be addressed, smart wearable technologies provide unique opportunities for the future of personalized respiratory medicine. While some challenges remain to be solved, intelligent wearable technologies offer unique opportunities for the future of personalized respiratory medicine. The article [5] analyzes the state of the Internet of Things in the medical environment, demonstrating an expanded range of healthcare applications based on IoT. Considerations are presented regarding data-mining applications, such as forecasting, classification and clustering of risks, which are considered fundamental problems for ensuring the accuracy of assistance processes. Digital innovations are changing medicine, and hemodynamic monitoring will not be an exception. The study [6] describes technological and digital innovations that are likely to change hemodynamic monitoring over the next 5–10 years. The paper [8] provides an overview of the common problems faced by wearable technology in the transition from a new device to an effective, functioning and reliable clinical tool for modern medicine. The publication [7] presents the results of a study focused on the analysis of modern definitions of the digital double (DT), a study of the main characteristics that DT should possess, and the study of areas in which DT applications are currently used.

Research in the field of home health care [9,10]: The article [9] examines home healthcare technologies that are currently being used to improve this situation by reminding users about medication schedules, remotely monitoring and updating new patient medical data, which can be performed by a doctor via the Internet. The conclusion of the work [10] shows that the use of a wearable monitoring system and remote monitoring of health status in everyday life improved diabetes control in middle-aged employees of the company. Research aimed at solutions or at finding solutions to improve IoT systems in medicine includes that addressing architecture [11,12], hardware [13,14], projects and systems [15,16].

Publications describing architectural solutions: In [11], the structure of a cloud healthcare system based on digital-twin healthcare (CloudDTH) is proposed. CloudDTH aims to achieve interaction and convergence between medical physical and virtual spaces. Accordingly, a new concept of digital-twin healthcare (DTH) is proposed and discussed. An energy-efficient architecture of the Internet of Medical Things "from fog to cloud" is proposed in [12] to optimize energy consumption. The proposed architecture uses Bluetooth-enabled biosensors since Bluetooth technology is energy efficient and also helps to enable sleep and wake modes.

Works describing hardware components for the implementation of systems: In [13], an attempt is made to study the hardware components necessary for the development of wearable devices that are used in the developing context of the Internet of Medical Things (IoMT). This means that they can be used to monitor diseases by capturing biosignals. From a practical point of view, the medical health monitoring system in hospital wards was developed using a CC2430 microcontroller, a human information sensor, as well as microelectronics and modern wireless communication technologies. The results of the experiment confirmed the reliability of the network node and the accuracy of data transmission. It is concluded that the medical monitoring system basically meets the design requirements set out in [14].

Research aimed at describing the development and implementation of systems: One paper [10] describes in detail the development and implementation of a system that improves commercial CGMS by adding Internet-of-Things (IoT) capabilities to them, which allows remote monitoring of patients and, thus, can warn users about potentially dangerous situations. The proposed system uses smartphones to collect blood-glucose values from CGM and then sends them to the remote monitoring system. A practical example presented in [16] concerns the design of a WSN (Wireless Sensor Network) in the concept of the Internet of Things in terms of the service life of the system. A hierarchical routing methodology was approved and showed energy efficiency. Simulation results showed that the chosen method prolongs the service life of the WSN in comparison with other clustering schemes studied.

The presented review shows that the patient's interaction with electronic resources and timely response to deviations from indicators makes it possible to promptly adjust the prescribed treatment. However, these works do not solve the problem of operational control over the condition of a patient with Parkinson's disease with the possibility of therapy correction and require the introduction of a software and hardware platform for calculating the indicators necessary for remotely monitoring the dynamics and prognosis of the patient's condition, which makes the development relevant.

2. System Architecture Concept

Means of remotely monitoring involve assessment of the patient's condition outside the medical institution; therefore, there is a need to endow the software and hardware complex with the following functions: collection, systematization and storage of individual indicators of the patient's condition; analysis and prediction of the patient's condition based on the data obtained; and correction of the received therapy.

Using this approach, it is possible to present the stages of parameter measurement and assessment of the patient's current condition in the form of the following steps:

1. Determination of the necessary minimum of indicators of the functioning of the body systems from the set $X = (x_1, x_2, ..., x_n)$ critical for each disease;
2. Evaluation of the current indicator $x_i(t_i)$ of the patient's condition through direct measurements;
3. Determination of dynamic $x_i(t_{i-1}, t_t)$ changes in the indicators for a given period of time;
4. Trend calculation T_t indicator $x_i(t_i)$ to eliminate high-frequency noise $\xi(t_i)$ and random variations $e(t_i)$;
5. Verification of criteria for deviation from the physiological state by evaluating deviations from the trend $T(t_{i-1}, t_t)$ and making a decision about the presence of an exacerbation by comparing the obtained indicators with the threshold ones according to formal criteria while checking the correctness of the obtained values;
6. Formation of a control effect according to changes in a group of parameters by correcting the therapy.

Observation of changes in parameter values over time allows us to obtain time series reflecting the dynamics of the patient's conditions. However, it should be borne in mind that the formation of parameters for each indicator and the construction of a forecast requires large time resources and is complicated by the need to take into account deviations of the indicator from the average value.

The modern development of remote-monitoring technologies should be based on the individual characteristics of the patient (personalized medicine), assessment of the quality of the patient's registration of the obtained indicators and methods of evaluating the results obtained. To implement this approach, a hardware and software complex is proposed, consisting of a device for measuring indicators and an information system that provides storage, processing and analysis of indicators, developed in accordance with the concept of a service-oriented architecture and implemented using services. The architecture of the platform consists of the following levels:

Client level: allows systematization of collection and provides measurement of patient indicators on a sensor or mobile device or centrally on a server.

Corporate level: allows authentication of the user and evaluation of his current state.

The level of integration of the system processes: interaction between corporate levels of various systems is provided.

3. Practical Implementation for Monitoring the Condition of a Patient with Parkinson's Disease

Parkinson's disease is a disease of the nervous system, consisting mainly in tremors—involuntary and uncontrolled movements of the musculoskeletal system. The disease is common: from 1% of people aged 60 years and up to 4% of people over the age of 85 years are affected [8]. Parkinson's disease is considered incurable, so all means are aimed at relieving symptoms. An urgent and important task is to monitor the patient's condition, which includes: assessment of general well-being, marking of medication intake, manifestation of dyskinesia and performing tests characterizing the dynamics of the disease. This allows doctors to adjust the course of treatment and medication intake; to investigate the behavior and nature of the course of the disease, it is advisable to purposefully collect these data using a mobile application that implements the functionality necessary for data collection.

The use of the mobile application allows users to generate a report on the patient's condition, which opens by default upon successful identification or when the application is launched in the case of completed identification, as shown in Figure 1.

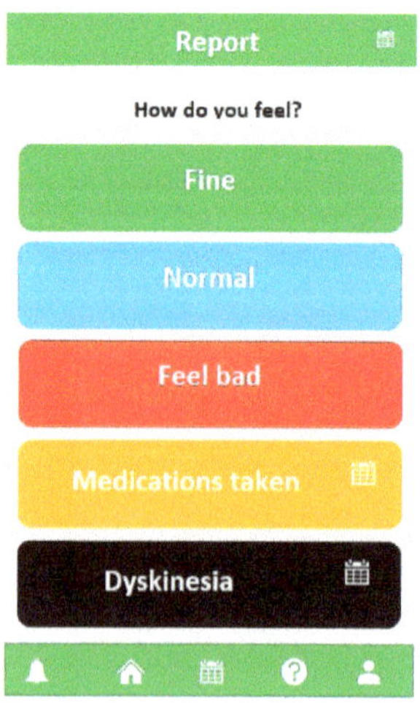

Figure 1. Status report screen.

The user can make a report of his general well-being using the buttons "Good", "Normal" and "Bad", as well as send verification of taking medications or manifestations of dyskinesia by clicking on the buttons "Medications taken" and "Dyskinesia", respectively.

In the dialog that appears by clicking, the user can specify whether the action was performed at that moment or not. In the case of refusal, the user is asked to choose the exact time. Both actions work identically to each other, so Figures 2 and 3 show the result of an attempt to send verification of taking medications.

The user forms a description of his own well-being in as much detail as possible in the form of various tests presented in Figure 4.

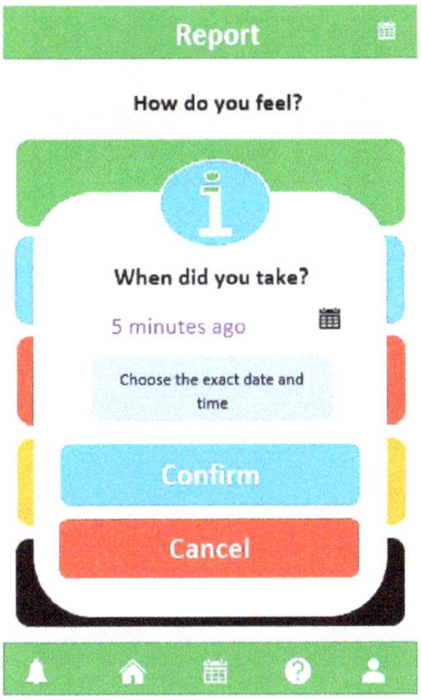

Figure 2. Sending verification of taking medications.

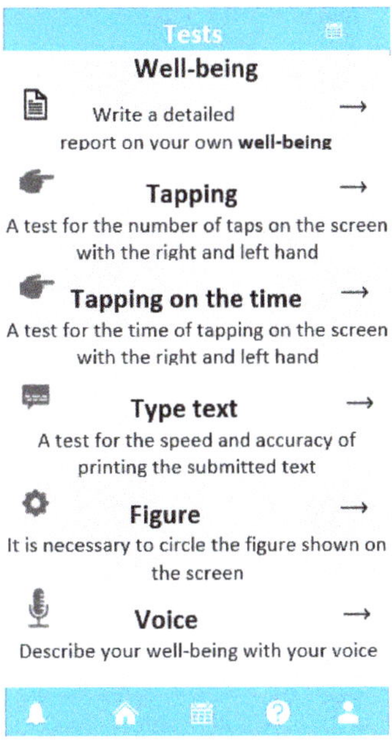

Figure 3. Test screen.

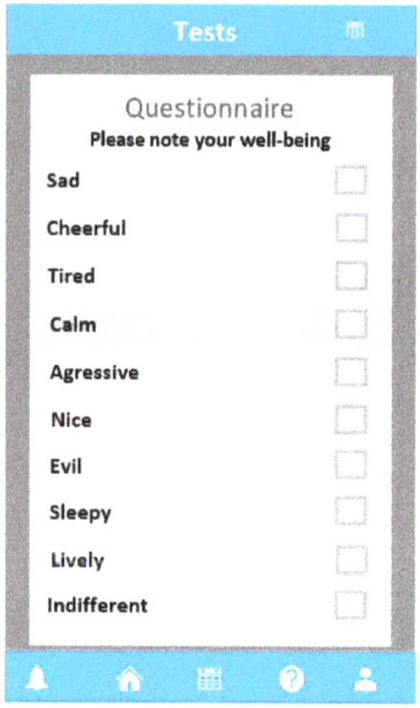

Figure 4. Description of well-being.

According to the parameters studied during testing, it is possible to determine the accuracy of typing on the phone's keyboard keys, the speed of hand work and the most frequent typos, as well as the accuracy, time and number of clicks. The collected data allow us to assess the degree and severity of hand tremors. However, if necessary, the user can describe his own well-being by voice. When passing the test, audio samples are recorded in PCM format (pulse width modulation, i.e., raw data from the microphone), which are then converted to WAV (WAVE) format using a specially written algorithm by adding a special header to the beginning of the file with a description of the audio track parameters. According to the audio file, the emotional state of a person and his well-being and mood are further determined.

Events are time-configurable notifications on the user's mobile phone that arrive at a strictly specified time. Note the following types of events: taking medications (daily), checking the condition (daily), performing tests (daily) and visiting a doctor (once). The "Doctor's visit" event requires the choice of the date and time of the visit and also has the ability to send data about the course of the disease to the doctor's email.

Competent and timely medication intake helps to improve the overall effectiveness of the course of medication. Taking medications requires entering the time of taking the medication and the duration of taking the medication (Figures 5 and 6). In addition, the user, together with the hardware component of the complex, can configure the module compartment with the medicine.

To monitor the use of medicines, a device has been developed that can store medicines in itself and communicate via Bluetooth with a mobile phone to configure and control physical access to the medicine, as well as notify of successful medication intake. The device is built on the basis of the Arduino platform. An ATmega32u4 microcontroller is used on an Arduino Micro board. The device uses a real-time RTC module for non-volatile recording of the current time, servos to restrict user access to compartments with medicines and Hall sensors to determine the position of each compartment (open or closed). The device is powered by a 9V battery connected to the VCC and GND power pins.

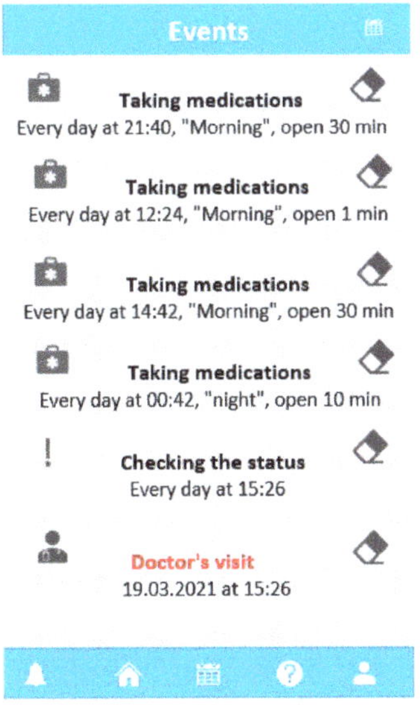

Figure 5. Event screen.

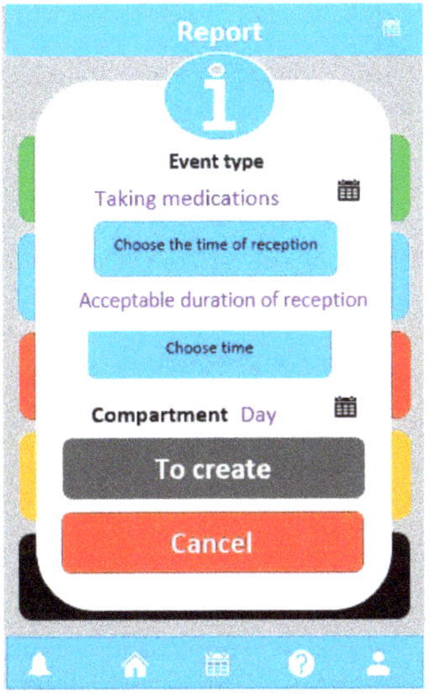

Figure 6. Description of well-being.

Physically, the device is a pillbox with four compartments for medicines. Each compartment has a hook that hooks the servo rotor in closed mode. There is a neodymium magnet on the hook, which acts with its magnetic field on the Hall sensor located in the compartment. To open the compartment, voltage needs to be applied to the corresponding servo, which will move its rotor so that the hook does not cling to it, after which the compartment can be physically opened. When the compartment is opened, the hook moves upwards, and the magnetic field of the magnet attached to it is removed from the sensitivity zone of the Hall sensor (Figure 7).

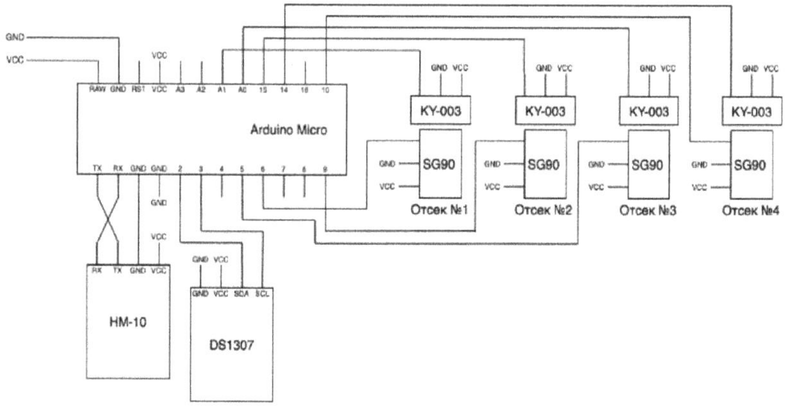

Figure 7. Device diagram.

The microcontroller is programmed in the Arduino IDE environment to work with compartments, the opening of which is perceived as events. An event is an abstract unit of work that occurs at a strictly defined time and contains the time of occurrence, its type (open or close a compartment) and the number of the compartment. In the loop, the program compares the current time with the time of each event. When an event occurs, it is processed, i.e., opening or closing of the corresponding compartment, after which the event is overwritten in memory with a time shift, depending on the type of event, either on the next day or for the duration of the reception, while its type changes. Bluetooth protocol is used to transmit event data to a mobile device. Connection to the HM-10 module takes place through the FFE0 service, which has the characteristic of reading and writing FFE1—it is used for data transmission. In the process of event processing, a certain set of data is recorded, which is read by a microcontroller (or a mobile phone) for its subsequent processing. When Bluetooth is turned on and the application is opened, devices are searched for a few seconds to detect the HM-10. Upon successful connection, the application sends a time update command to the module to correct possible inaccuracies of the real-time module (as a rule, such modules, depending on the cost, may lag by 1–2 s every few days). In case of successful execution of the command, the module sends a command to the phone to request events that the application stores in the phone's memory.

On the event screen (Figure 6), the user can create an event of the type "Taking medications". If there is a Bluetooth connection, such events are generated into a command that is sent to the device. Any change in the list of events of this type leads to the re-formation and sending of the command. With the Bluetooth module connected, medication intake is monitored and taken into account. For example, when opening and then closing one of the compartments, the device sends the appropriate command to the application, which is perceived as a successful medication intake and is indicated by a similar manual click on the "Medications taken" button. Thus, the drug intake control system automates and monitors the intake of medications, allowing the patient not to mark the intake manually.

4. Conclusions

The practical application of the complex has also shown the relevance of its use in the task of studying the course of Parkinson's disease, as well as for monitoring the condition of a patient with existing concomitant diseases and the degree of their influence on the course of the underlying disease. It is advisable to include the development of methods for constructing a digital double of the patient and the formation of methods for predicting the course of the disease for further research.

Author Contributions: Conceptualization, methodology and formal analysis, Y.S.; project administration and writing—review and editing, R.F.; software, validation, formal analysis, original draft preparation and visualization, N.I. All authors have read and agreed to the published version of the manuscript.

Funding: This research was funded by Russian Science Foundation (RSF) project 22-21-00686.

Institutional Review Board Statement: Not applicable.

Informed Consent Statement: Not applicable.

Data Availability Statement: Not applicable.

Conflicts of Interest: The authors declare no conflict of interest.

References

1. Yetisen, A.K.; Martinez-Hurtado, J.L.; Ünal, B.; Khademhosseini, A.; Butt, H. Wearables in Medicine. *Adv. Mater.* **2018**, *30*, 17069102. [CrossRef] [PubMed]
2. Son, D.; Bao, Z. Nanomaterials in Skin-Inspired Electronics: Toward Soft and Robust Skin-like Electronic. *Nanosyst. ACS Nano* **2018**, *12*, 11731–11739. [CrossRef] [PubMed]
3. Afzal, M.; Riazul Islam, S.M.; Hussain, M.; Lee, S. Precision Medicine Informatics: Principles, Prospects, and Challenges. *IEEE Access* **2020**, *8*, 13593–13612. [CrossRef]
4. Aliverti, A. Wearable technology: Role in respiratory health and disease. *Breathe* **2017**, *13*, e27–e36. [CrossRef] [PubMed]
5. Scarpato, N.; Pieroni, A.; Di Nunzio, L.; Fallucchi, F. E-health-IoT universe: A review. *Int. J. Adv. Sci. Eng. Inf. Technol.* **2017**, *7*, 2328–2336. [CrossRef]
6. Michard, F. Hemodynamic monitoring in the era of digital health. *Ann. Intensive Care* **2016**, *6*, 15. [CrossRef] [PubMed]
7. Fernández-Caramés, T.M.; Froiz-Míguez, I.; Blanco-Novoa, O.; Fraga-Lamas, P. Enabling the internet of mobile crowdsourcing health things: A mobile fog computing, blockchain and iot based continuous glucose monitoring system for diabetes mellitus research and care. *Sensors* **2019**, *19*, 3319. [CrossRef] [PubMed]
8. Godfrey, A.; Hetherington, V.; Shum, H.; Bonato, P.; Lovell, N.H.; Stuart, S. From A to Z: Wearable technology explained. *Maturitas* **2018**, *113*, 40–47. [CrossRef] [PubMed]
9. Zanjal, S.V.; Talmale, G.R. Medicine Reminder and Monitoring System for Secure Health Using IOT. *Procedia Comput. Sci.* **2016**, *78*, 471–476. [CrossRef]
10. Behmanesh, A.; Sayfouri, N.; Sadoughi, F. Technological Features of Internet of Things in Medicine: A Systematic Mapping Study. *Wirel. Commun. Mob. Comput.* **2020**, *2020*, 9238614. [CrossRef]
11. Liu, Y.; Zhang, L.; Yang, Y.; Zhou, L.; Ren, L.; Wang, F.; Liu, R.; Pang, Z.; Deen, M.J. A Novel Cloud-Based Framework for the Elderly Healthcare Services Using Digital Twin. *IEEE Access* **2019**, *7*, 49088–49101. [CrossRef]
12. Dziak, D.; Jachimczyk, B.; Kulesza, W.J. Wirelessly interfacing objects and subjects of healthcare system—IoT approach. *Elektronika ir Elektrotechnika* **2016**, *22*, 66–73. [CrossRef]
13. Dash, S.; Shakyawar, S.K.; Sharma, M.; Kaushik, S. Big data in healthcare: Management, analysis and future prospects. *J. Big Data* **2019**, *6*, 54. [CrossRef]
14. Kamel Boulos, M.N.; Peng, G.; Vopham, T. An overview of GeoAI applications in health and healthcare. *Int. J. Health Geogr.* **2019**, *18*, 7. [CrossRef] [PubMed]
15. Qureshi, F.; Klishnan, S. Wearable hardware design for the internet of medical things (IoMT). *Sensors* **2018**, *18*, 3812. [CrossRef] [PubMed]
16. Pramanik, P.K.D.; Solanki, A.; Debnath, A.; Nayyar, A.; El-Sappagh, S.; Kwak, K.-S. Advancing Modern Healthcare with Nanotechnology, Nanobiosensors, and Internet of Nano Things: Taxonomies, Applications, Architecture, and Challenges. *IEEE Access* **2020**, *8*, 65230–65266. [CrossRef]

engineering proceedings

MDPI

Proceeding Paper

Mathematical Model of Information Exchange in the Autonomous Underwater Vehicle Network †

Elizaveta G. Litunenko *,‡, Alexander M. Gruzlikov ‡, Nikolai V. Kolesov ‡ and Iurii M. Skorodumov ‡

State Research Center of the Russian Federation—Concern CSRI Elektropribor, JSC, 30, Malaya Posadskaya Str., Saint Petersburg 197046, Russia
* Correspondence: lisa.litunenko@gmail.com
† Presented at the 15th International Conference "Intelligent Systems" (INTELS'22), Moscow, Russia, 14–16 December 2022.
‡ These authors contributed equally to this work.

Abstract: Nowadays, a lot of attention is given to autonomous underwater exploration. To reduce the time of such explorations groups of networked underwater vehicles are used. The implementation of such communication is rather problematic because of low message transfer speed and other factors related to the underwater environment. In such networks, messages often pass through a chain of relay nodes to reach their destination point. Moreover, vehicles can form queues of messages to be transmitted. The order of messages in the queue can influence the total transmission time. In complex information exchange scenarios, it is crucial to order the messages that should be delivered to ensure optimal management of the process. The proposed article discusses an optimization of messaging in a network of autonomous underwater vehicles. The authors propose algorithms for scheduling information exchanges in an underwater network. These algorithms are suboptimal. Algorithms that satisfy the criteria of the minimum upper bounds for either the total or average delivery time have been proposed. The authors considered situations when messages are unordered, as well as when messages are partially pre-ordered.

Keywords: mathematical model; scheduling; delivery time; autonomous underwater vehicles; underwater network

Citation: Litunenko, E.G.; Gruzlikov, A.M.; Kolesov, N.V.; Skorodumov, I.M. Mathematical Model of Information Exchange in the Autonomous Underwater Vehicle Network. *Eng. Proc.* **2023**, *33*, 11. https://doi.org/10.3390/engproc2023033011

Academic Editors: Askhat Diveev, Ivan Zelinka, Arutun Avetisyan and Alexander Ilin

Published: 30 May 2023

1. Introduction

Problems and advances in underwater robotics have been in the focus of specialists' attention in recent decades [1,2]. These issues are also widely discussed in the modern scientific and technical literature. Significant interest in this field is explained by the fact that nowadays autonomous underwater underwater vehicles (AUVs) are used for a multitude of applications, among which are geological exploration, prospecting, and oceanographic and other kinds of works and research that are often conducted in extreme conditions. To reduce the time, such works are often carried out by a group of vehicles based on a certain multi-agent concept, which implies information exchange between the agents to coordinate joint actions. In this case, AUVs are considered as a network, in which the vehicles exchange information using underwater sonar communication [3–7]. The implementation of such communication is rather problematic in view of a low speed of sound propagation in water, the frequency dependence of the attenuation coefficient of a hydroacoustic signal, multipath propagation, and signal detection under conditions of a priori uncertainty in the noise–signal environment. As a consequence, these problems lead to a significant limitation of both the rate (kilobits per second) and radius of information exchange between the AUVs. As a result, in the general case, instead of reaching the destination node directly, a message passes via a chain of relay nodes. In addition, due to the AUV mobility, the message delivery route may change following the changes in the AUV network topology. These restrictions significantly increase the requirements for organizing information exchanges. Therefore, it necessary to optimize the routes and sequences of the messages to be transmitted.

This article addresses one of the theoretical aspects of the practical developments in AUV networking carried out at the Concern CSRI Elektropribor, JSC, namely, the organization of information exchanges in an AUV network. Ways of optimizing the sequence of message transmissions, i.e., scheduling information exchanges, are proposed and discussed. The problem statement is given in Section 1. The proposed scheduling algorithms are considered in Sections 2 and 3.

2. Problem Statement

For scheduling algorithms development, we will assume that the considered network has the following properties:

- The network is an undirected and connected graph;
- The network is stationary [8];
- The nodes are identic and use the same operating algorithm;
- Every node has information about all another nodes location;
- The routing algorithm uses routing tables that are assumed to be known [9];
- The nodes emit messages periodically.

The property of stationarity provides a constant value of the intensity of the flow of information exchanges in the network. The coordinates of all network nodes are contained in routing tables, which update in the process of information interaction between network nodes. We assume that during the information exchange the devices do not move significantly, so the messages are transmitted along the planned routes and always reach the addressees. The sequence of messages transmitted by nodes includes messages generated by the node itself, as well as messages relayed from other nodes.

Generally, the order of messages transmitted by a node affects on the effectiveness of message delivery. In what follows, the criterion of efficiency is represented either by the total delivery time Δ_s of all messages from the sequence being transmitted, or the average delivery time $\overline{\Delta}$ for messages. In this case, the delivery time Δ of a message at the k-th position in the queue of the node involved is taken to mean the sum of the waiting time $e^{w}_{[k]}$ in the queue of the transmitting node and the message transmission time $e^{t}_{[k]}$ (from the moment the message transmission starts to the end of its reception) between the transmitting and destination nodes:

$$\Delta_{[k]} = e^{w}_{[k]} + e^{t}_{[k]}.$$

In a general case, the message does not reach the destination node directly, but through a chain of relay nodes; therefore, the message transmission route is multi-hop. Then, the message delivery time is calculated by the formula:

$$\Delta_{[k]} = \sum_{i=1}^{r_{[k]}} (e^{w}_{[k],i} + e^{t}_{[k],i}).$$

where $r_{[k]}$ is the number of route steps for the message at the k-th position in the queue, and i is the number of the route step.

This expression can be represented in the following form:

$$\Delta_{[k]} = e^{w}_{[k]} + e^{t}_{[k]} + \sum_{i=1}^{r_{[k]}} (e^{w}_{[k],i} + e^{t}_{[k],i}).$$

While planning a multi-hop route, it is impossible to estimate in advance the size of the message queue at each of the relay nodes. Therefore, it is assumed that the waiting time in the queue at each relay node will have a maximum or minimum value. This article considers algorithms under conditions when each relay node considers the waiting time in the queue to be maximum, then instead of the message delivery time $\Delta_{[k]}$, the upper bound on the delivery time $\hat{\Delta}_{[k]}$ is estimated:

$$\hat{\Delta}_{[k]} = e^{w}_{[k]} + e^{t}_{[k]} + (r_{[k]} - 1)\overline{n}E + \sum_{i=2}^{r_{[k]}} e^{t}_{[k],i}, \quad (1)$$

where $\overline{n}$ is the maximum of the queue length and E is maximum message duration.

To solve the problem of optimizing information exchanges in the network of underwater vehicles in terms of reducing the message time delivery, we use the known scheduling algorithms [10–14] and develop new ones, which are presented below.

3. Scheduling Unordered Messages

In the simplest case, messages to be transmitted are not pre-ordered. The scheduling algorithm for such messages uses an upper bound on the total delivery time of all messages $\hat{\Delta}_s$ as an optimality criterion. The expression for it is given below:

(1):

$$\hat{\Delta}_s = \sum_{k=1}^{n} \hat{\Delta}_{[k]} = \sum_{k=1}^{n} [e^{w}_{[k]} + e^{t}_{[k]} + (r_{[k]} - 1)\overline{n}E + \sum_{i=2}^{r_{[k]}} e^{t}_{[k],i}].$$

Then, we can prove the optimality of the following scheduling algorithms.

Algorithm 1 can be described by the following expression:

$$e_{[1]} \leq e_{[2]} \leq ... \leq e_{[n]}.$$

Algorithm 1 Scheduling unordered messages

1: $i := 1$.
2: Determine a message of minimum duration on the set of messages, which are not placed in the queue already.
3: Place the message determined at step 2 in the queue at the i-th position.
4: Exclude the message determined at step 2 from the set of unplaced ones, $i := i + 1$. If $i > n$, then finish; otherwise, go to step 2.

If some priorities need to be set on a sequence of messages, the optimality criterion may be the upper bound for the total weighted message delivery time $\hat{\Delta}^{w}_{s}$:

$$\hat{\Delta}^{w}_{s} = \sum_{k=1}^{n} w_{[k]} \hat{\Delta}_{[k]}.$$

where $w_{[k]}$ is the weight of a message at the k-th position in the queue.

Then, the optimal Algorithm 2 has the following form.

Algorithm 2 Scheduling unordered prioritised messages

1: Determine for each message the ratio of duration to weight.
2: $i := 1$.
3: Determine the message with the minimum duration-to-weight ratio on the set of unscheduled messages.
4: Place the message determined at step 3 in the queue at the i-th position.
5: Exclude the message determined at step 3 from the set of unplaced ones, $i := i + 1$. If $i > n$, then finish; otherwise, go to step 3.

This algorithm can be described by the following expression:

$$\frac{e_{[1]}}{w_1} \leq \frac{e_{[2]}}{w_2} \leq ... \leq \frac{e_{[n]}}{w_n}.$$

4. Scheduling Partially Ordered Messages

Consider the situation when the messages to be transmitted are divided into p disjoint groups with a strict order of messages inside them. The size of i-th group is $n_i, i = \overline{1, p}$. In this case, interruptions in a groups are prohibited, so the task of message scheduling is reduced to the scheduling groups of pre-ordered messages.

The optimal scheduling algorithm for these conditions with a $\hat{\Delta}_s$ criterion is presented below (Algorithm 3).

Algorithm 3 Scheduling partially ordered messages with a $\hat{\Delta}_s$ criterion (interruptions in a groups are prohibited)

1: Calculate for each i-th group of messages the total duration $e_i^{'}$:

$$e_i^{'} = \sum_{j=1}^{n_i} e_{i,j}, j = \overline{1, n_i}.$$

2: $i := 1$.
3: Determine the group with the minimum duration on the set of unscheduled message groups.
4: Place the group determined at step 3 in the queue at the i-th position.
5: Exclude the group determined at step 3 from the set of unplaced ones, $i := i + 1$. If $i > p$, then finish; otherwise, go to step 3.

This algorithm can be described by the following expression:

$$e_{[1]}^{'} \le e_{[2]}^{'} \le \ldots \le e_{[p]}^{'}.$$

If some priorities need to be set on a sequence of message groups, and if $\hat{\Delta}_{g[k]}$ is the upper bound on the delivery time for the group at the k-th position, the optimality criterion should be minimized:

$$\hat{\Delta}_g^w = \sum_{k=1}^{p} w_{[k]} \hat{\Delta}_{g[k]}. \quad (2)$$

where $\hat{\Delta}_{g[k]}$ is the upper bound for the message group delivery time.

Algorithm 3 is transformed as follows (Algorithm 4).

Algorithm 4 Scheduling prioritised message groups with a $\hat{\Delta}_g^w$ criterion (interruptions in a groups are prohibited)

1: Calculate for each i-th group of messages the total duration $e_i^{'}$, and determine for each group the ratio of duration to weight.
2: $i := 1$.
3: Determine the group with the minimum duration-to-weight ratio on the set of unscheduled groups.
4: Place the group determined at step 3 in the queue at the i-th position.
5: Exclude the group determined at step 3 from the set of unplaced ones, $i := i + 1$. If $i > p$, then finish; otherwise, go to step 3.

This algorithm can be described by the following expression:

$$\frac{e_{[1]}^{'}}{w_1} \le \frac{e_{[2]}^{'}}{w_2} \le \ldots \le \frac{e_{[p]}^{'}}{w_p}.$$

Obviously, all the results described above are also valid if the optimality criterion is represented by an upper bound of message-averaged delivery time $\overline{\hat{\Delta}}$ (Algorithms 1 and 2) or the upper bound of message group-averaged delivery time $\overline{\hat{\Delta}}_g$ (Algorithm 3), as well as (2) rather than the upper bound of the total delivery time.

In some cases, it is required to use the upper bound of the message-averaged delivery time $\bar{\hat{\Delta}}$ as an optimization criterion. Then, the following optimal Algorithm 5 is true.

Algorithm 5 Scheduling partially ordered messages with a $\bar{\hat{\Delta}}$ criterion (interruptions in a groups are prohibited)

1: Calculate for each i-th group of messages the total duration $e_i^{'}$, and determine for each group the ratio of duration to group size $n_{[k]}$.
2: $i := 1$.
3: Determine the group with the minimum duration-to-weight ratio on the set of unscheduled groups.
4: Place the group determined at step 3 in the queue at the i-th position.
5: Exclude the group determined at step 3 from the set of unplaced ones, $i := i + 1$. If $i > p$, then finish; otherwise, go to step 3.

This algorithm can be described by the following expression:

$$\frac{e_{[1]}^{'}}{n_{[1]}} \leq \frac{e_{[2]}^{'}}{n_{[2]}} \leq \ldots \leq \frac{e_{[p]}^{'}}{n_{[p]}}. \tag{3}$$

Consider the situation where interrupts of message groups are allowed. For such a case, Algorithm 6 is valid.

Algorithm 6 Scheduling partially ordered messages with a $\bar{\hat{\Delta}}$ criterion (interruptions in a groups are allowed)

1: The conditional upper bound of the average delivery time is calculated for each message j in the i-th group (the upper bound of the average delivery time, provided that the subgroup of messages preceding the (j + 1)-th message in the i-th group is at the beginning of the queue being scheduled):

$$\bar{\hat{\Delta}}_{i,j} = \frac{\sum_{h=1}^{j} \hat{\Delta}_{i,h}}{j}.$$

2: $\bar{\hat{\Delta}}_{i,h_i}$ is calculated for each i-th group: $\bar{\hat{\Delta}}_{i,h_i} = min(\bar{\hat{\Delta}}_{i,1}, \bar{\hat{\Delta}}_{i,2}, \ldots, \bar{\hat{\Delta}}_{i,n_i})$.
3: Group i^* is chosen on the condition that $i^* = arg(min_i \bar{\hat{\Delta}}_{i,h_i})$, so that the first h_{i^*} messages are at the beginning of the queue.
4: The values of $\bar{\hat{\Delta}}_{i,h_i}$ are calculated without the messages that are already placed in a queue.
5: Steps 3 and 4 are repeated until all the messages are scheduled.

Example 1. *Let us form queues for groups of messages* $S_1 = \{s_{11}, s_{12}, s_{13}\}, S_2 = \{s_{21}, s_{22}\}$, $S_3 = \{s_{31}, s_{32}, s_{33}, s_{34}\}$. *Upper bounds for message delivery times are shown in Table 1.*

Table 1. The upper bound of the message delivery time.

Message Group	$\hat{\Delta}_{i1}$	$\hat{\Delta}_{i2}$	$\hat{\Delta}_{i3}$	$\hat{\Delta}_{i4}$
S_1	5	15	4	-
S_2	3	14	-	-
S_3	10	2	5	7

While interrupts are prohibited, in accordance to Algorithm 5, we will compute $\hat{\Delta}$ *for each group:*
$\hat{\Delta}_1 = 8, \hat{\Delta}_2 = 8.5, \hat{\Delta}_3 = 6$.
The optimal queue: $Q_1 = S_3 S_1 S_2 = s_{31}, s_{32}, s_{33}, s_{34}, s_{11}, s_{12}, s_{13}, s_{21}, s_{22}$.
While interrupts are allowed, we will use Algorithm 6.

We order messages in the queue in eight steps, which are shown in Table 2.

Table 2. The upper bound of the message delivery time.

Message Group	$\tilde{\Delta}_1$				$\tilde{\Delta}_2$				$\tilde{\Delta}_3$				$\tilde{\Delta}_4$		$\tilde{\Delta}_5$		$\tilde{\Delta}_6$		$\tilde{\Delta}_7$	$\tilde{\Delta}_8$
S_1	5	10	8	-	5	10	8	-	15	9.5	-	-	15	9.5	15	9.5	15	9.5	4	-
S_2	3	8.5	-	-	14	-	-	-	14	-	-	-	14	-	14	-	14	-	14	14
S_3	10	6	5.6	6	10	6	5.6	5	10	6	5.6	6	5	6	7	-	-	-	-	-

The optimal queue: $Q_2 = s_{21}, s_{11}, s_{31}, s_{32}, s_{33}, s_{34}, s_{12}, s_{13}, s_{22}$.

5. Conclusions

This paper is devoted to the scheduling of information exchanges in AUV networks. Algorithms for scheduling exchanges, which are suboptimal for the current network topology and satisfy the criteria of minimum of upper bounds for either the total or average delivery time, have been proposed.

Author Contributions: Conceptualization, A.M.G., N.V.K., E.G.L. and I.M.S.; formal analysis, A.M.G. and N.V.K.; investigation, I.M.S.; methodology, A.M.G. and N.V.K.; software, E.G.L.; writing—original draft, N.V.K.; writing—review and editing, E.G.L. All authors have read and agreed to the published version of the manuscript.

Funding: This research was funded by the Russian Science Foundation, project no. 22-29-00339.

Institutional Review Board Statement: Not applicable.

Informed Consent Statement: Not applicable.

Data Availability Statement: The data presented in this study are available on request from the corresponding author.

Conflicts of Interest: The authors declare no conflict of interest.

References

1. Inzartsev, A.V.; Kiselev, L.V.; Kostenko, V.V.; Matvienko, Y.V.; Pavin, A.M.; Shcherbatyuk, A.F. *Underwater Robotic Systems: Systems, Technologies, Applications*; Dal'nauka: Vladivostok, Russia, 2018.
2. Giger, G.; Kandemir, M.; Dzielski, J. Graphical Mission Specification and Partitioning for Unmanned Underwater Vehicles. *J. Softw. (JSW)* **2008**, *3*, 42–54. [CrossRef]
3. Fedosov, V.P.; Tarasov, S.P.; Pivnev, P.P.; Voronin, V.V.; Kucheryavenko, S.V.; Legin, A.A.; Lomakina, A.V.; Frants, V.A. *Communication Networks for Underwater Autonomous Robotic Complexes*; YuFU: Taganrog, Russia, 2018. (In Russian)
4. Pankratov, F.S.; Malakhov, I.M. Actual and perspective methods of constructing wireless access networks in underwater acoustic communication. *Large-Scale Syst. Control* **2021**, *91*, 120–143.
5. Hamilton, A.; Holdcroft, S.; Fenucci, D.; Mitchell, P.; Morozs, N.; Munafò, A.; Sitbon, J. Adaptable Underwater Networks: The Relation between Autonomy and Communications. *Remote Sens.* **2020**, *12*, 3290. [CrossRef]
6. Tuphanov, I.; Scherbatyuk, A. Some marine trial results of centralized control system for marine robot group. *Large-Scale Syst. Control* **2020**, *59*, 233–246.
7. Kebkal, K.G.; Mashoshin, A.I.; Morozs, N.V. Solutions for underwater communication and positioning network development. *Gyroscopy Navig.* **2019**, *10*, 161–179. [CrossRef]
8. Kleinrock, L. *Queueing Systems. Volume 1: Theory*; Wiley: Hoboken, NJ, USA, 1979.
9. Tel, G. *Introduction to Distributed Algorithms*; Cambridge University Press: New York, NY, USA, 1994.
10. Conway, R.W.; Maxwell, W.L.; Miller, L.W. *Theory of Scheduling*; Dover Publications: Mineola, NY, USA, 1967.
11. Malashenko, Y.E.; Nazarova, I.A.; Novikova, N.M. Analysis of two-layer resource supply flow networks. *J. Comput. Syst. Sci. Int.* **2020**, *59*, 387–399. [CrossRef]
12. Lazarev, A.A.; Gafarov, E.R. *Schedule Theory. Tasks and Algorithms*; Lomonosov Moscow State University: Moscow, Russia, 2011.
13. Liu, J.W.S. *Real-Time Systems*; Prentice Hall: Hoboken, NJ, USA, 2000.
14. Cottet, F.; Delacroix, J.; Mammeri, Z.; Kaiser, C. *Scheduling in Real-Time Systems*; Wiley: Hoboken, NJ, USA, 2002.

engineering proceedings

MDPI

Proceeding Paper

Optimality Conditions for the Principle of Trajectory Division †

Valentin Bereznev

Federal Research Center Computer Science and Control of RAS, Vavilov Str. 44, build.2, Moscow 119333, Russia; va_bereznev@mail.ru; Tel.: +7-915-167-0295

† Presented at the 15th International Conference "Intelligent Systems" (INTELS'22), Moscow, Russia, 14–16 December 2022.

Abstract: This paper considers the problem of controlling a mobile robot in the presence of circular obstacles. To solve this problem, it is proposed to use the previously suggested principle of dividing permissible trajectories into a sequence of rectilinear sections and arcs of circles that are the boundaries of circular obstacles. The conditions for the solution based on this principle of optimality are obtained.

Keywords: controlling a mobile robot; principle of dividing the feasible trajectory; optimality conditions

Citation: Bereznev, V. Optimality Conditions for the Principle of Trajectory Division. *Eng. Proc.* **2023**, *33*, 2. https://doi.org/10.3390/engproc2023033002

Academic Editors: Askhat Diveev, Ivan Zelinka, Arutun Avetisyan and Alexander Ilin

Published: 9 May 2023

Suppose that the problem of controlling a mobile robot is solved using the principle of dividing feasible trajectories [1]. In the case of the phase space $\mathbb{R}_2$, this principle is an alternative to the classical optimal control theory based on the maximum principle [2]. According to this principle, a connected graph is first constructed, then the shortest path on the graph is searched. Taking into account that it is not the distance between the vertices that is used as the edges of the graph, but rather the minimum time rewuired for the robot to travel along the edge, we can assume that the time-optimal problem has been solved. However, this is true only under the essential condition that the robot, in bypassing a circular obstacle, always moves along the border of the obstacle, i.e., that the minimum allowable radius of curvature of the route is selected on the turn. Nevertheless, there is doubt that such a choice is always optimal in terms of performance. This is clearly seen in the example of auto racing, when, in seeking to overtake an opponent, the driver tries to avoid heavy braking on a curve, instead choosing a larger radius of curvature. If the driver chooses too high a speed at this point, slippage may occur, i.e., loss of grip with the surface of the track, resulting in displacement in the direction of the centrifugal force of inertia or even in the vehicle overturning.

On this basis, let us consider the main factors influencing the reasonable choice of the radius of curvature for the trajectory of the device when bypassing a circular obstacle. As already noted, an object moving along the arc of a circle is affected by the so-called centrifugal force of inertia F, which acts along the radius from the center of the circle and is calculated by the formula

$$F = m\omega^2 r,$$

where m is the mass of the object, ω is its angular speed, and r is the radius of the circle. This force is opposed by the force G of adhesion to the surface of the highway acting in the opposite direction, which is determined by weather conditions, the condition of the road surface, and the wheels of the car or robot.

Let us consider a concrete example (see Figure 1) in which a controlled device must move from point A to point B while bypassing a circular obstacle of radius r_1 (a circle of dark color) or an obstacle of radius $r_2 > r_1$ (a circle of light color).

For simplicity, assume that the distances from points A and B to the center of the circles are the same and are equal to S. The points C, D, M, and N are the tangent points of the lines drawn from A and B to the small and large circles. Let us define the conditions under which, from the point of view of time consumption, it is preferable to bypass a circular obstacle along an arc of a small radius or along an arc of a large one, provided that $m_1 = m_2 = m$ and

$$F_i = m\omega_i^2 r_i = G, \quad i = 1,2, \tag{1}$$

i.e., the passage of the turn is guaranteed against slippage.

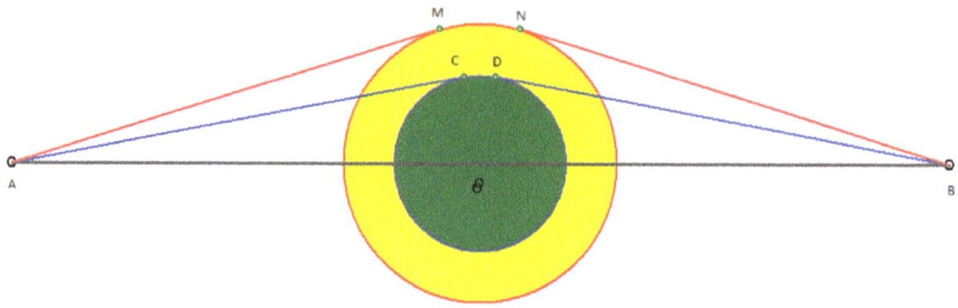

Figure 1. Passage of a vmrage with a different radius.

Let $r_2 = \gamma r_1, \gamma > 1$. Note that the coefficient γ cannot be arbitrarily large, as at $\gamma > S/r_1$ point A already falls inside the circle and the task loses its meaning. Due to the symmetry of the tracks in both cases with respect to a straight line passing through the center of the circles parallel to the ordinate axis, we estimate the time of passage of the left half of the trajectories. Let us first determine the time spent on moving along the arcs of circles. Suppose that the linear speed of motion along a small arc, corresponding to the fulfillment of the condition (3) at $i = 1$, is equal to $v_1 = v^-$ and determines the feasible linear speed v_2. It follows from (3) that $\omega_2 = \omega_1\sqrt{1/\gamma}$, i.e., $\omega_2 < \omega_1$. As is well known, $v_i = \omega_i r_i, i = 1,2$; therefore, $v_2 = \omega_2\gamma r_1 = \omega_1 r_1\sqrt{\gamma} = v_1\sqrt{\gamma} > v_1 = v^-$.

It is easy to see that the angles based on the half-arcs of the circles are equal to $\alpha_1 = \arcsin(r_1/S)$ and $\alpha_2 = \arcsin(\gamma r_1/s)$, respectively. Then, the lengths of the arcs themselves are $s_1 = r_1\alpha_1$ and $s_2 = \gamma r_1\alpha_2$. Due to the uniform motion along these arcs with speeds v_1 and $v_2 = v_1\sqrt{\gamma}$, we obtain expressions for time $t_{11} = r_1\alpha_1/v_1$ and $t_{12} = r_1\sqrt{\gamma}\alpha_2/v_1$. Thus,

$$\Delta T_1 = t_{12} - t_{11} = \frac{r_1(\sqrt{\gamma}\alpha_2 - \alpha_1)}{v_1} > 0, \tag{2}$$

as the expression in parentheses is obviously positive.

Now, let us estimate the time for the device to pass straight-line line break sections of the route to the points of contact C and M. Recall that, according to the assumptions, the robot starts moving at point A at a speed of v^+, and should arrive at the touch point at a speed of $v_i, i = 1,2$. Let us first determine the time spent in both cases on braking, which we denote by $t_{2i}, \quad i = 1,2$. We have $t_{2i} = (v^+ - v_i)/|u^-|$, i.e.,

$$\Delta T_2 = t_{22} - t_{21} = \frac{v^+ - \sqrt{\gamma}v_1}{|u^-|} - \frac{v^+ - v_1}{|u^-|} = \frac{(1-\sqrt{\gamma})v_1}{|u^-|}, \tag{3}$$

meaning that in this case the distance traveled by the robot during braking is equal to $d_i = (v^+ - v_i)^2/(2|y^-|)$. Thus, with the maximum speed v^+, the robot travels the distance $S\cos\alpha_i - d_i$, with the time spent on this task being

$$t_{3i} = \frac{1}{v^+}(S\cos\alpha_i - d_i) = \frac{1}{v^+}\left(S\cos\alpha_i - \frac{(v^+ - v_i)^2}{2|u^-|}\right).$$

Hence ,

$$\Delta T_3 = t_{32} - t_{31} = \frac{S(\cos\alpha_2 - \cos\alpha_1)}{v^+} - \frac{(v^+ - \sqrt{\gamma}v_1)^2}{2v^+|u^-|} + \frac{(v^+ - v_1)^2}{2v^+|u^-|}. \tag{4}$$

Thus, it follows from (4)–(6) that in both cases the difference in the time taken by the device to pass half of the trajectory is expressed by

$$\begin{aligned}\Delta T &= \frac{r_1(\sqrt{\gamma}\alpha_2 - \alpha_1)}{v_1} + \frac{(1-\sqrt{\gamma})v_1}{|u^-|} + \\ &+ \frac{S(\cos\alpha_2 - \cos\alpha_1)}{v^+} - \frac{(v^+ - \sqrt{\gamma}v_1)^2}{2v^+|u^-|} + \frac{(v^+ - v_1)^2}{2v^+|u^-|} = \\ &= \frac{r_1(\sqrt{\gamma}\alpha_2 - \alpha_1)}{v_1} + \frac{S(\cos\alpha_2 - \cos\alpha_1)}{v^+} + \frac{(1-\gamma)v_1^2}{2v^+|u^-|}.\end{aligned} \tag{5}$$

Let us now evaluate the second term in this expression. Taking into account that $\sin x < x$ for $x > 0$, and assuming that $\alpha_1 + \alpha_2 < \pi$, we have

$$\begin{aligned}&\frac{S(\cos\alpha_2 - \cos\alpha_1)}{v^+} = -\frac{2S}{v^+}\sin\frac{\alpha_2+\alpha_1}{2}\sin\frac{\alpha_2-\alpha_1}{2} > \\ &> -\frac{S}{v^+}\sin\left(\arcsin\frac{r_1}{S}\right)(\alpha_2 - \alpha_1) = -\frac{r_1}{v^+}(\alpha_2 - \alpha_1).\end{aligned}$$

Therefore,

$$\begin{aligned}\Delta T &> \frac{r_1(\sqrt{\gamma}\alpha_2 - \alpha_1)}{v_1} - \frac{r_1}{v^+}(\alpha_2 - \alpha_1) + \frac{(1-\gamma)v_1^2}{2v^+|u^-|} > \\ &> \frac{r_1(\alpha_2 - \alpha_1)}{v_1} - \frac{r_1}{v^+}(\alpha_2 - \alpha_1) + \frac{(1-\gamma)v_1^2}{2v^+|u^-|} = \\ &= r_1(\alpha_2 - \alpha_1)\left(\frac{1}{v_1} - \frac{1}{v^+}\right) + \frac{(1-\gamma)v_1^2}{2v^+|u^-|}.\end{aligned} \tag{6}$$

Let $v^+ = \theta v_1, \quad \theta > 1$. Then, the last expression in (8) takes the form

$$\Delta T > \frac{r_1(\alpha_2 - \alpha_1)(\theta - 1)}{\theta v_1} + \frac{(1-\gamma)v_1}{2\theta|u^-|};$$

moreover, the right side of this inequality takes a positive value when

$$\gamma < 1 + \frac{2r_1(\alpha_2 - \alpha_1)|u^-|(\theta - 1)}{v_1^2}. \tag{7}$$

Obviously, the estimate for the second half of the route takes the same form (9) with the replacement of $|u^-|$ by u^+. Thus, the following statement is true.

Theorem 1. *Assume that the coupling force G and the maximum speed v_1 when passing a turn along the minimum allowable radius r_1 are known. Then, $\Delta T > 0$ for any γ satisfying the inequality*

$$1 < \gamma < 1 + \frac{2r_1(\alpha_2 - \alpha_1)\tilde{u}(\theta - 1)}{v_1^2},$$

where $\tilde{u} = \min\{|u^-|, u^+\}$.

Thus, with a sufficient natural condition for the parameter γ, the solution of the time-optimal problem based on the principle of trajectory division is optimal. This is confirmed by computational experiments.

Funding: This research received no external funding.

Institutional Review Board Statement: Not applicable.

Informed Consent Statement: Not applicable.

Data Availability Statement: Not applicable.

Conflicts of Interest: The author declares no conflict of interest.

References

1. Bereznev, V.A. The principle of dividing feasible trajectories in a robot control problem. *Procedia Comput. Sci.* **2021**, *186*, 456–459. [CrossRef]
2. Pontryagin, L.; Boltyanskii, V.; Gamkrelidze, R.; Mishchenko, E.L. *The Mathematical Theory of Optimal Processes*; Publishing House: Interscience, NY, USA, 1962.

engineering proceedings

MDPI

Proceeding Paper

Improved Social Network User Recommendation System—The Machine Learning Approach †

Yana A. Bekeneva *,‡ and Titus U. Eze ‡

Department of Computer Science and Engineering, Saint Petersburg Electrotechnical University "LETI", ul. Professora Popova 5, 197022 St. Petersburg, Russia; titusugochukwu28@gmail.com

* Correspondence: yabekeneva@etu.ru

† Presented at the 15th International Conference "Intelligent Systems" (INTELS'22), Moscow, Russia, 14–16 December 2022.

‡ These authors contributed equally to this work.

Abstract: In this paper, we propose a recommendation system for social media users which makes recommendations on the basis of the user profile information and the contents posted by users-the bio-aware algorithm. Text mining techniques are used to pre-process the words before feeding it to an LDA model which handles the topic and feature extraction. In this way, we determine the similarity between users based on their interests. We further trained a machine learning model which is able to identify and score the top interests of a particular social media user. Other users who share similar scores are shown as recommendations to each other.

Keywords: social network; people interaction; information searching; information extraction; bio-aware approach

Citation: Bekeneva, Y.A.; Eze, T.U. Improved Social Network User Recommendation System—The Machine Learning Approach. *Eng. Proc.* **2023**, *33*, 13. https://doi.org/10.3390/engproc2023033013

Academic Editors: Askhat Diveev, Ivan Zelinka, Arutun Avetisyan and Alexander Ilin

Published: 9 June 2023

1. Introduction

In 2020, an estimated 46% of the world population is active in at least one social network [1]. When they register on these networks, they are asked to follow some people to see their content. Users then interact with each other's content, send messages to each other and so on. To further improve this interaction, there is a need to develop more effective recommendation systems as users are more likely to spend more time online and engage with the content of people they identify better with. Users are often recommended to their fellow classmates, people who live in the same city or friends of their friends, the list goes on. Familiarity relationships are simulated through repeated exposure to profiles [2].

There have been previous attempts which include both content-based and graph-based approaches. The first focuses on measuring the topic similarity among social media users, while the second infers the relationship between users based on a graph. While it takes a lot of data to develop algorithms that would produce these recommendations, this paper explores a peer-to-peer recommendation system with a bio-aware algorithm which measures the similarity interests among users based on their profile biography and their last post content to offer suggestions.

Recommender systems are generally tools that help in filtering and sorting information as it is relevant and interesting to the user. This is as a result of the huge data problem that came with the adoption of the world wide web.

Netflix's recommendation system also plays an important role in the success of the platform. Many times, people come to the site without knowing exactly what to watch but completely believe that the site will recommend what fits in to their taste. When creating an account for the first time, Netflix asks a user to choose about five different movies they like. From this list, Netflix is able to populate their feed with similar movies that match their taste using its recommendation system.

Social media is no different. The basics of social interaction are such that there is a variety of topics to talk about, and even better when the discussions are with people who share similar views with you. When someone registers on a social network such as Twitter, they are asked to follow some people. These are usually celebrities, government officials or new agencies in their country. From this, Twitter is able to decide what kind of users and topics the user is interested in and then populate their feed accordingly. Previously, you could accurately understand what a post on Twitter was about from the hashtags it has. Recently, Twitter introduced topics. People can follow certain topics ranging from sports, celebrity news, particular fields such as cryptography, space and so on. With this, they can accurately suggest relevant posts on the user's timeline.

Traditional approaches to building recommender systems are listed as follows: content filtering, collaborative filtering and hybrid filtering. Other modern approaches include location-content-aware, context-aware, semantic based, cross-lingual and peer-to-peer.

Even though a lot of research has gone into improving the state of recommender systems using machine learning approaches, there are still many challenges that affect these systems to date. Some of them are described below.

Cold start problems occur when there is not enough data. It could be a cold start of items in which case we do not have enough information about the items or users. For example, it would be difficult to recommend movies to a user if they have not watched any movies at all or given any information that will help in profiling them.

Large datasets are often required to develop a commercial recommender system. This leads to the usage of a large and spare user-item matrix for filtering which in turn affects the performance of the recommendation process. The cold-start problem is caused by data sparsity.

As the number of customers and items increases, it poses a scalability problem to traditional collaborative filtering algorithms since the complexity will be too large. Most resources are used for the purpose of determining the similarity in interests between users, and items with similar attributes [3].

This is one of the most commonly encountered challenges, especially with content-based filtering, since it always would recommend items that users are already familiar with. One of the properties of a good system is that it should be diverse.

2. Related Works

Obtaining user profile information can be useful when grouping users and building recommendation systems. Since unlabelled data are considerably more frequent than labeled data in many social network datasets, inferring hidden users attributes and using graph-based semi-supervised learning algorithms are more suitable for this case [4,5].

In order to obtain a good recommendation from a system, there is a need for clear distinctive information on a user's preferences [6]. This is generally achieved by recording feedback from the user; the posts they like, or using sentiment analysis to score their comments on posts [7].

Gurini et al. [8] emphasizes the use of implicit sentiment analysis which further improves the performance of recommendation systems. In this approach, they were able to build more complete user profiles than traditional content-based approaches by defining a novel weighting function that takes into account sentiment and size in relation to the user's interests.

N. R. Vajjhala et al. [9] proposed a solution to identify user interests using IBM Watson platform which makes use of Natural language processing for advanced text analysis. Their solution however does not cover for cases with multiple preferences.

Arru et al. [10] proposed a system that is based on a novel user model, termed bag-of-signal. Guy et al. [11,12] recommend people based on articulated social network information by combining different sources to derive factors that might influence the similarity measure.

Recent studies, e.g., [13], suggest a user's interests are always changing and there is a need to capture that. Their solution combines the user interactions, social features and post history to capture their interests.

Hannon et al.'s proposal [14] recommends users through a hybrid system of both content based and collaborative filtering. In the content-based method, users correlation scores are a function of their followees' posts, their followers' posts or a combination of both. In this case, users with similar posts will be recommended to each other. For the collaborative filtering approach, users are represented by the IDs of their followees, followers or a combination of them. Then tf-idf weighting scheme is used to find users similar followers/followees and recommend them to the target user.

A recurring problem for many of these solutions is that they largely suffer from the cold-start problem. We need information about the user's followers, followees and their posts in order to develop the system. A user needs to have followed a substantial number of users in order to effectively generate relevant suggestions.

3. Approach Description

The bio-aware algorithm proposed in this paper seeks to eliminate this and also solve the data sparsity and scalability problems. It also reduces the computing time and resources involved in working with datasets comprising many features.

This is based on the assumption that most social media users include details about themselves, who they are and what they like or talk about in their biography. When combined with their most recent posts we are able to generate a list of topics that the user is interested in.

To use social media platforms such as Twitter, a registered user needs to describe who they are, their name, short biography, location, etc. A user can follow other users and see information about them, posts they have made, their followers and the people they follow termed 'followees'.

One determining factor as to why a user follows another user can sometimes depend on how their interests align. Interests could be, for example, if they both like football, or they are fans of a celebrity. Studies have shown that people with the most followers are usually new agencies, celebrities or famous organizations. It was also discovered that the lesser the number of followers a user has, the less likely they are to make a tweet while the more followers they have, the less likely they are willing to reciprocate the following.

Our purpose is to present a user with a set of recommendations which they will most likely identify with. If a user already deliberately followed other users, we could assume they already find these people interesting either based on the recent tweets they made, their description of what they do and from how popular they are (Figure 1).

Since we are dealing with a lot of raw user generated data, which are mostly unstructured in nature, we need to pre-process them firstly using about different techniques namely, tokenization, stemming, removing stop words and noise removal.

There are different feature extraction techniques which include bag of words, TF-IDF, word embedding, and Natural Language Processing "NLP". Studies have shown that for machine learning algorithms using text classification.

To find how similar user's interests are to each other, we compare the similarity based on their LDA. We would define the similarity measure as the dot product between the corresponding vectors representing the contributions of each word to a topic.

Merge fields: since we need to feed the LDA model a string text, there is a need to combine the relevant fields which we will be using for our topic extraction.

Preprocessing: in (Figure 1), we have listed the preprocessing techniques and steps that will be used in this work namely; tokenization, lemmatization, removal of stop words, removal of numbers, emails, URLs, punctuations and white spaces.

Topic extraction: after preprocessing, we perform topic extraction using the LDA model to identify the top ten words in each topic. The topics here are not labeled so we use wordcloud to visualize the data.

Topic weight score: here, we obtain the weight score for a user's top topics and assign it as a new corresponding field named 'topic'.

Get recommendations: we retrieve a user's list of recommendations by checking other users with similar weight scores.

To obtain a model that is able to predict a user's topic weight scores, we explore the following steps: split the dataset into train and test data; using random forest classifier to train the topic column; predict the topic weight score for the test data; obtain recommendations.

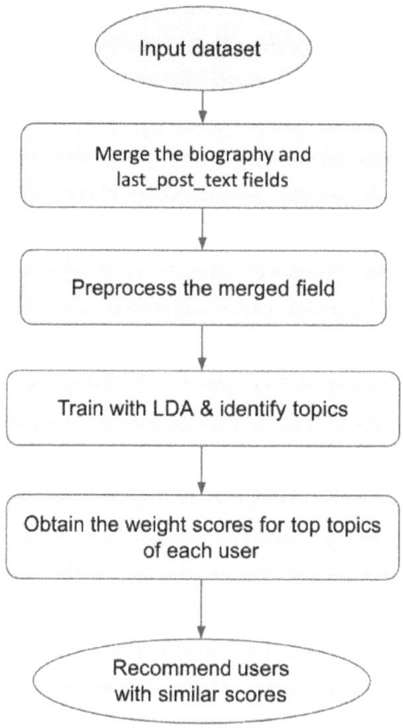

Figure 1. Proposed algorithm design for building our classification model.

4. Experiment Results

For the sake of experiment, Twitter was chosen as the social media platform as a case study since it was easier to obtain useful data than other platforms.

We used popular processing tools like numpy for array manipulations, pandas for dataframes, sklearn for feature extraction, re for formatting the string data, nltk library for preprocessing, keras for building models, matplotlib and wordcloud for data visualization.

Using the wordcloud library, we are able to visualize the results of our LDA training model. We can infer that from the most popular words, this topic has to do with world sports since it captures words such as league, game, player, baseball and so on.

We can deduce from Figure 2 that this topic relates more to startups and businesses. People who talk about or are interested in investment, finance, funding, web development and the rest. Figure 3 can be summarized as relating to people who are interested in skill acquisition or building networks.

We built a learning model that can predict the topic weight score of new datasets so as to reduce the build time and save resources. To do this, we split our dataset into training and testing subsets. Our testing dataset was 20% of the entire dataset.

Figure 2. Topic visualization map for posts related to business

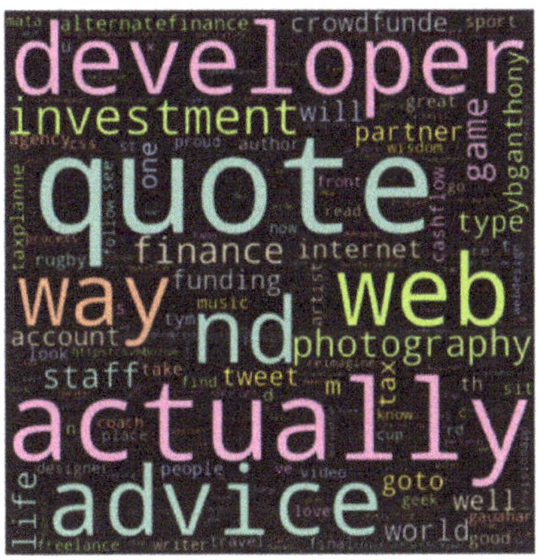

Figure 3. Topic visualization map for posts related to network technologies

5. Comparison of with Other Approaches

When we consider both traditional approaches, e.g., content-based filtering, collaborative filtering and hybrid filtering with our approach, we see a clear improvement (Table 1).

Table 1. Results of comparison of different approaches.

Approaches	Cold Start	Privacy	Multiple Languages	Sparsity
Content-filtering	-	-	-	+-
Collaborative filtering	+-	+	-	-
Hybrid filtering	+	-	-	+
Cross-lingual	-	+	+	-
Bio-aware	+	+	-	+

The bio-aware algorithm proposed in this paper seeks to eliminate most of the challenges listed above such as data sparsity, scalability, cold-start and privacy problems. It also reduces the computing time and resources involved in working with datasets comprising many features since we also develop a model that is able to automatically output the topic weight score for an already preprocessed dataset.

The recommendation system developed with this algorithm is independent of a particular platform and can be used on any social media platform of choice. Twitter was chosen as the use case here because of the availability of relevant dataset.

6. Conclusions

It is difficult to verify the effectiveness of our algorithm without comparing it to the user's actual followee's list, which is not provided in the dataset used for this project. It's also important to acknowledge the limits of the approach used in this work.

From the experiment results, it is obvious that the preprocessing techniques and steps used have a great influence on how the classification performs. Removing mentions from the text could possibly increase the effectiveness. For the purpose of this work, we only chose to process English words which is not the case in the real world. Posts and user bio can be in a user's native language. As much as we tried to remove some English stop words, it did not capture the everyday English words which appeared so many times but do not add to the relevance of our analysis.

Even though this approach tries to eliminate the cold-start problem, there are still few cases where a user does not add anything in their bio or have not made any posts at all. For this case, we suggest that only the popular users be shown as recommendations here. There might also be cases where a user does not capture what they are interested in on their bio or on their last post. To solve this, we propose that only the last five posts with more than 80 characters be considered in the case that there are more.

It is also important to note that our approach does not capture topic classification on the subcategory basis. If someone mentions that they like baseball, we should only offer them suggestions of people who also like baseball and not people who like sports generally. To do this, we might have to increase the number of topics that the LDA model should classify into. A proposed solution to this might be to have a known list of possible topics that is already labeled against each user in the dataset. These days, platforms such as Twitter and Instagram already allow you to choose topics or hashtags based on your preferences and content is recommended around these topics.

Author Contributions: Conceptualization, Y.A.B.; Methodology, Y.A.B., T.U.E.; Software, T.U.E.; Supervision, Y.A.B.; Validation, Y.A.B., T.U.E.; Visualization, T.U.E.; Writing—original draft, T.U.E., Y.A.B.; Writing—review and editing, Y.A.B. All authors have read and agreed to the published version of the manuscript.

Funding: This research received no external funding.

Institutional Review Board Statement: Not applicable.

Informed Consent Statement: Not applicable.

Data Availability Statement: Not applicable.

Conflicts of Interest: The authors declare no conflict of interest.

References

1. Statista Research Department. Number of Social Network Users Worldwide from 2017 to 2025. 2021. Available online: https://www.statista.com/statistics/278414/number-of-worldwide-social-network-users/ (accessed on 1 July 2022).
2. Guy, I.; Inbal, R. Do You Know? Recommending People to Invite into Your Social Network. In Proceedings of the 14th International Conference on Intelligent User Interfaces, IUI '09, Sanibel Island, FL, USA, 8–11 February 2009; pp. 77–86.
3. Jain, S.; Grover, A.; Thakur, P.S.; Choudhary, S.K. Trends, Problems And Solutions of Recommender System. In Proceedings of the International Conference on Computing, Communication and Automation (ICCCA2015), Greater Noida, India, 15–16 May 2015; pp. 955–958.
4. Ding, Y.; Yan, S.; Zhang, Y.; Dai, W.; Dong, L. Predicting the attributes of social network users using a graph-based machine learning method. *Comput. Commun.* **2015**, *73*, 9. [CrossRef]
5. Zheleva, E. To join or not to join: The illusion of privacy in social networks with mixed public and private user profiles. In Proceedings of the 19th Century International Conference of the World Wide Web, Madrid, Spain, 20–24 April 2009; pp. 531–540.
6. Yue, L.; Chen, W.; Li, X.; Zuo, W.; Yin, M. A survey of sentiment analysis in social media. *Knowl. Inf. Syst.* **2019**, *60*, 617–663. [CrossRef]

7. Javed, U.; Shaukat, K.; Hameed, I.A.; Iqbal, F.; Alam, T.M.; Luo, S. A review of content-based and context-based recommendation systems. *Int. J. Emerg. Technol. Learn. (IJET)* **2021**, *16*, 274–306. [CrossRef]
8. Gurini, D.F.; Gasparetti, F.; Micarelli, A.; Sansonetti, G. A Sentiment-Based Approach to Twitter User Recommendation. *RSWeb@ RecSys* **2013**, *1066*, 1–4.
9. Vajjhala, N.R.; Rakshit, S.; Oshogbunu, M.; Salisu, S. Novel user preference recommender system based on Twitter profile analysis. *Soft Comput. Tech. Appl.* **2021**, *1248*, 85–93.
10. Arru, G.; Feltoni Gurini, D.; Gasparetti, F.; Micarelli, A.; Sansonetti, G. Signal-based user recommendation on twitter. In Proceedings of the 22nd International Conference on World Wide Web, Rio de Janeiro, Brazil, 13–17 May 2013; pp. 941–944.
11. Palomares, I.; Porcel, C.; Pizzato, L.; Guy, I.; Herrera-Viedma, E. Reciprocal recommender systems: Analysis of state-of-art literature, challenges and opportunities towards social recommendation. *Inf. Fusion* **2021**, *69*, 103–127. [CrossRef]
12. Gurini, D.F.; Gasparetti, F.; Micarelli, A.; Sansonetti, G. Temporal people-to-people recommendation on social networks with sentiment-based matrix factorization. *Future Gener. Comput. Syst.* **2018**, *78*, 430–439. [CrossRef]
13. Tsou, M.H.; Zhang, H.; Park, J.; Nara, A.; Jung, C.T. Spatial Distribution Patterns of Geo-tagged Twitter Data Created by Social Media Bots and Recommended Data Wrangling Procedures. In *Empowering Human Dynamics Research with Social Media and Geospatial Data Analytics*; Springer: Cham, Switzerland, 2021; pp. 257–273.
14. Sardianos, C.; Ballas Papadatos, G.; Varlamis, I. Optimizing parallel collaborative filtering approaches for improving recommendation systems performance. *Information* **2019**, *10*, 155. [CrossRef]

engineering proceedings

MDPI

Proceeding Paper

Use of Dynamic Models in Cognitive Cyber-Physical Systems †

Michael Chervontsev [1], Alexey Subbotin [1,*], Alexander Vodyaho [1] and Nataly Zhukova [1,2]

[1] Department of Computer Science and Engineering, Saint-Petersburg State Electrotechnical University, St. Professor Popov, 5, St. Petersburg 197376, Russia

[2] Saint-Petersburg Federal Research Centre of the Russian Academy of Sciences, 14th Line V.O., 39, St. Petersburg 199178, Russia

* Correspondence: ansubbotin@stud.etu.ru; Tel.: +7-921-365-6535; Fax: +7-812-440-7701

† Presented at the 15th International Conference "Intelligent Systems" (INTELS'22), Moscow, Russia, 14–16 December 2022.

Abstract: This article discusses an approach to the development of large-scale cognitive cyber-physical systems characterized by a high level of structural, functional and architectural dynamics. The main idea of the proposed approach is the use of several modern paradigms to build cyber-physical systems, such as continuous architecture, agile architecture, digital twins and digital threads. A three-level model of a cognitive cyber-physical system is proposed. In order to ensure the required level of flexibility of the system, such possibility should be laid at earlier stages of the life cycle, i.e., at the development stage. At the upper level, the system is described in terms of a continuous architecture; at the middle level, the system is represented as a system with agile architecture, which is described as a multi-level relatively finite automaton, and at the lower level, structural and functional models are used that illustrate the system in the process of functioning. This article provides examples of using the proposed approach.

Keywords: cognitive cyber-physical systems; dynamic models; continuous architecture; agile architecture; digital twins; digital threads

Citation: Chervontsev, M.; Subbotin, A.; Vodyaho, A.; Zhukova, N. Use of Dynamic Models in Cognitive Cyber-Physical Systems. *Eng. Proc.* **2023**, *33*, 14. https://doi.org/10.3390/engproc2023033014

Academic Editors: Askhat Diveev, Ivan Zelinka, Arutun Avetisyan and Alexander Ilin

Published: 9 June 2023

1. Introduction

One of the distinctive features of the modern stage of development of society is very rapid changes in people's lives and production methods. Modern companies and organizations adapt their strategy, business models, products and services, as well as business processes and information systems, to increase the level of their digitalization with the help of intelligent services and products with digital improvement. The potential of the Internet and related digital technologies, such as the Internet of Things, artificial intelligence, data analyses, cloud computing, mobile systems and cyber-physical systems (CPS), involves strategic factors contributing to the development of digital platforms with rapidly developing ecosystems of intelligent services for digital products [1,2]. Currently, the CPS paradigm is widely used for building real-life systems. One can define CPS as systems based on the integration of entities of a different physical nature when separate subsystems of a different physical nature function as a single whole [3].

Nowadays, another point of interest for IT developers is cognitive systems. In the IT industry, the concept of "cognitiveness" came from cognitive psychology, which studies the processes occurring in the human brain during information processing. This concept later formed the basis of one of the branches of artificial intelligence, which is engaged in the development of artificial cognitive systems taking into account advances in neurophysiology and cognitive psychology. As the simplest cognitive system, we can consider a system that can have certain knowledge about itself and the external world, the functioning of which is realized in the form of achieving goals. We can say that a cognitive system should have at least a model of itself and models of the external world (context) presented in the form of knowledge. The following definitions can be proposed: cognitive systems can be attributed to systems that are able to perceive information about the state of the external world and their own state, working with a model of themselves and a model of the external world; systems that do not use these models can be defined as intelligent systems.

One of the main distinctive features of modern CPS is the variability of their structure and behavior [4]. Thus, we can say that modern CPS may have the ability of evolution, i.e., they can be considered as developing systems that may have such typical features as (i) a spontaneous change of the system state; (ii) counteraction (reaction) to the influence of the environment (other systems), leading to a change in the initial state of the environment; and (iii) permanent changes in the structure and behavior of the system. If a developing system evolves at the expense of its own resources, then such systems are called self-developing (self-sufficient developing).

Practically all self-developing cognitive CPS are Software Intensive Systems (SWIS) [5], which use mechanisms of working with knowledge.

2. Background

The CPS conceptual model consists of five levels: physical, network, data storage, processing and analytics and an application layer. Large-scale CPS are most often implemented on fog and edge computing platforms [6].

The mainstream of the CPS evolution is associated with the permanent complication of the implemented functionality, and nowadays many modern CPS can be considered as ambient intelligence systems. A significant part of modern CPS can be attributed to cognitive systems.

In relation to the IT domain, the term "cognitiveness" is most often associated with the derived concepts of cognitive modeling, cognitive systems and cognitive architectures and systems that implement cognitive behavior.

Cognitive modeling can be defined as a decision-making support process taking into account the mutual influence of various events and the relationship between them. Cognitive analyses, using the apparatus of cognitive models and technologies, promptly solve such tasks as building a model of the situation, assessing the impact of external and internal factors on possible scenarios of the situation, and identify trends in the development of situations. The cognitive modeling technology is usually used to solve management problems under unforeseen circumstances.

Classical systems with cognitive architectures use human-inspired heuristic algorithms, which are found by psychologists and biologists who are often part of the development team. The history of the creation of this kind of cognitive system comprises at least 40 years. For the most part, these are experimental systems. When constructing them, concepts such as patterns of situations and fusion are most often used [2]. For the most part, these are also decision-making support systems. As an example of using this approach, we can consider the concept of the Cognitive Internet of Things [7].

The main idea of this paradigm is that the modern Internet of Things is just a collection of a wide variety of devices equipped with status sensors and actuators. To manage and maintain such a complex infrastructure in a proper condition, it is necessary to endow it with intelligence, i.e., the ability to perceive information about its own state and, if necessary, issue control actions for the purpose of reconfiguring the structure. The cognitive system here is the Internet of Things management system, which can be built as a distributed intelligence system. Within the framework of this paradigm, no specific solutions are given; it is determined only what should be done, but not how the mechanisms of cognition should be implemented.

3. Problem Statement

A significant part of modern self-developing cognitive CPS are big and complex distributed systems with a high level of not only structural and functional but also architectural variability. One of the key problems associated with the construction of self-developing cognitive CPS is the problem of ensuring their manageability at all levels. To solve this problem, in turn, it requires having sufficient information about the current state of the system, which is usually presented as a set of models.

Designing self-developing cognitive CPS is also quite a difficult problem. Very often, the use of traditional SWIS design approaches [8] does not give the desired results. The development of new approaches to the design of self-developing cognitive CPS is required.

4. The proposed Approach

The main idea of the suggested approach is that it is proposed to use several modern paradigms for the construction of CPS, such as continuous architecture [9], agile architecture [10], digital twins [11] and digital threads [12]. The proposed approach can be considered as an adaptation of the digital threads' paradigm in relation to cognitive CPS.

The implementation is supposed to be based on a Model-Based Approach [13] and assumes the use of a three-level CPS life cycle model (LC).

A three-level model of self-developing cognitive CPS. In order to provide the required level of CPS capabilities such as flexibility, this opportunity should be laid at earlier stages of the CPS LC, i.e., at the development stage. This model can be used as the basis of a development viewpoint or encapsulated into a separate architectural viewpoint [8].

The proposed model describes the LC of continuous agile architecture at all five LC stages on three levels.

The structure of the model is shown in Figure 1.

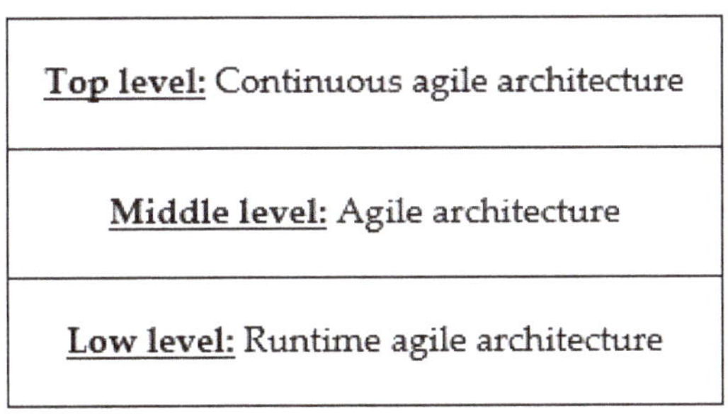

Figure 1. Three-level model.

At the top level, CPS are described in terms of a continuous architecture [9].

At the middle level, the system is represented as an agile architecture and is described as a multilevel relatively finite state automaton [14].

At the low level, structural and functional models describing the system in the process of operation are used.

These models describe CPS at different LC stages. The top level describes all LC stages. The middle level refers mainly to the stages of operation and modernization. The low level relates mainly to the stage of operation.

Top level. At this level, the process of transformation of the architectural representations of the CPS during the transition between the individual stages of the LC cycle is described (Figure 2).

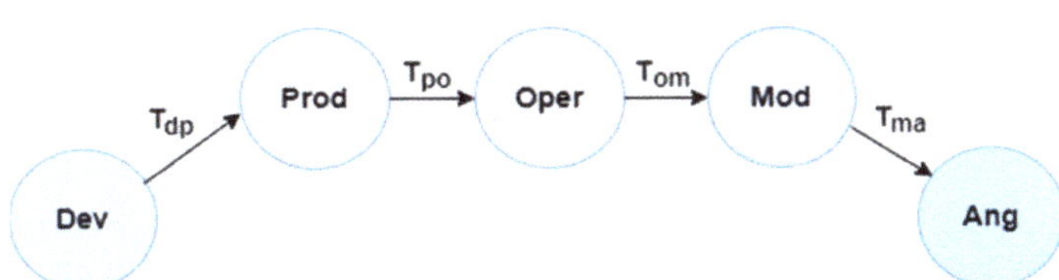

Figure 2. Transformation of artifacts.

In general, five phases of the life cycle can be distinguished: Development (Dev), Production (Prod), Operation (Oper), Modernization (Mod) and Annihilation (Ang). The chain of transformations can be represented as

Art Dev--> Tdp_--> Art Prod--> Tpo--> Art Oper--> Tmo--> Art Mod--> Tma-->--> Tmo--> Art Ang.

At each phase of the LC, a certain set of artifacts (Art) is formed. Artifacts can belong to the following classes:

Art classes: = <Model I Code I Physical Entity>. Each of the phases uses its own set of artifacts.

One can define four basic types of transformations:

Mi--> Mi-1, M--> Code, M--> PhE, Mi1--> Mi2, where M are models and PhE are physical entities.

In general, at each stage of the LC, its own set of entities is used, which are interconnected by transformation mechanisms.

At the top level, one can discuss a single model. At the middle level, each phase uses its own artifacts. Each phase uses its own agile architectures (AA). Each of the AA architectures is matched with a set of runtime architectures (RTAA).

Each phase uses its own artifact systems, but some common features can also be distinguished.

Artifacts, the classification of which is shown in Figure 3, can be divided into three groups: entities, models and metrics.

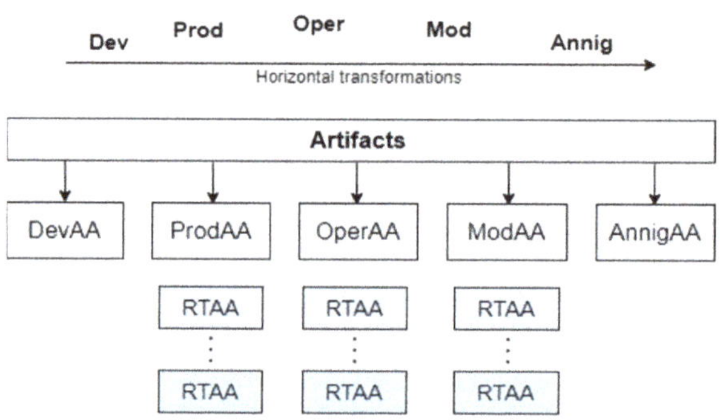

Figure 3. Artifact system structures.

Entities can be physical (PE) or virtual (VE). Physical entities can be either physical elements of the CPS, or their physical models. Both CPS elements and CPS models can act as virtual ones. A virtual entity can concede as a code or as a business process.

Currently, models are becoming one of the key elements of CPS. Models can be both physical and virtual (PM and VM). In this paper, physical models are not considered. A complete classification of the models can be found, e.g., in [13]. From the point of view of the subject of research, models can be divided into models describing a class of systems (VCM) and models describing instances of a class (VIM).

The model and the entity being modeled can be in the following relations: (i) model implementation M--> E; (ii) model building E--> M; and (iii) synchronization of the model and the entity state M<-->E.

Middle level. At this level, the mechanisms of structural, functional and architectural reconfiguration are supported. This level describes the operational phase.

At the middle level, the system can be described as a multilevel relatively finite state automaton, the vertices of which correspond to architectural states (As), and the transitions are illustrated by the fitness functions (FF) [10].

This model can be used both on exploitation and the modernization phase.

In the first case, we are discussing maintaining agility mechanisms in the runtime mode, and in the second case, about maintaining these mechanisms in the process of CPS modernization.

Low level. At the low level, structural and functional mining mechanisms are supported, i.e., tracking the current structural and functional (architectural) state of the observed CPS.

The architectural state of the observed CPS can be described using different models depending on the goals.

Approaches to implementation. The implementation of self-organizing cognitive CPS on the fog and edge computing platforms is associated with solving a number of problems, the central of which are the following: (i) it is obvious that cognitive CPS should be focused on working with knowledge; however, working with ontologies and knowledge graphs requires sufficiently powerful computing resources that are not always available even at a fog level, and really full-scale support of mechanisms for working with knowledge can only be organized at the cloud level; (ii) neural networks can be considered as an effective tool for implementing cognitive functions, but their training can really be organized only at the cloud level.

The generalized policy related to the implementation of the cognitive functions of the CPS can be formulated as follows. Functions related to managing slow business processes, such as self-organization management, can be implemented in terms of ontologies and knowledge graphs mainly in a cloud environment; neural networks and federated learning algorithms can be used to implement functions with higher performance requirements.

The maturity models. The proposed model approach is not always possible and appropriate to be applied at full scale. In some cases, it is more appropriate to apply it partially or only at the design stage. In this context, by analogy with the well-known methods of assessing the maturity of CMMI [15], one can discuss the maturity levels of the application of the suggested model approach. Five maturity levels can be distinguished: (i) level 0—the model approach is not applied; (ii) level 1—the model approach is applied only at separate stages, e.g., at the development stage; (iii) level 2—the model approach is used to a limited extent, both at the development and runtime, e.g., it can be used only to control structural dynamics or only to control dynamic business processes; (iv) level 3—the model approach is used to solve both structural dynamic management problems and dynamic business process monitoring, but the possibilities of building models are limited, e.g., one can only tune the properties of individual model elements, or one can add only elements of certain types to the model; and (v) level 4—all models can be synthesized in the runtime, no restrictions are imposed on the complexity and structure of models and model transformation mechanisms are used.

5. Case Study

The proposed approach was used by the authors in the construction of a number of real systems. It should be noted that in order to solve real problems, it was not required to use it at full scale.

Below is a brief description of real projects in which the proposed approach was used.

Case 1. A cable TV network maintenance system. This project solves the problem of maintenance of a complex distributed system. The main task was to create cognitive CPS based on the existing cable TV network, with the abilities of self-repair and self-optimization, i.e., implementing elements of cognitive behavior. When solving this problem, it turned out to be possible to limit ourselves only to structural dynamic models that are focused on supporting the operation phase [16].

Case 2. Production system. The purpose of the development was to modernize the production system with mobile entities at the sensory level, built on the platform of the Industrial Internet of Things. There are two key subtasks in this project: collecting data at the sensory level in conditions of a very high level of electromagnetic interference and the task of building a corporate knowledge graph. This system is focused mainly on work in the runtime and uses structural and functional models operating at a low level. The use of top-level models allows for increasing flexibility and simplifying the process of integration into the corporate information system. The presence of a dynamic model illustrating the production system in the form of a knowledge graph opens up possibilities for the construction of cognitive production systems. This example can be considered as the use of the proposed approach for the construction of dynamic digital twins [17,18].

Case 3. In the third case, the model approach was used to solve the problem of building an educational trajectory muster level curriculum in the IT domain. This is the task of synthesizing business processes based on the model. This task is also interesting because educational systems can be considered as one of the varieties of socio-cybernetic systems where it is required to work with competence models. This indicates the possibility, with certain improvements related to the adaptation of the model, of using this approach to solve data processing problems in socio-cybernetic systems, for example, social networks.

Case 4. Digital twin of an electric piano. This example illustrates the possibility of constructing a dynamic digital twin. The idea is as follows. The piano often breaks down and there is a special connector for connecting to a computer for diagnostics. Modern electric pianos have great capabilities, can change the timbre of the voice through a connected microphone, have good sound-reproducing speakers that are not inferior to professional equipment for artists of the musical genre, have additional accompaniment functions and have a training mode for beginners to learn to play the piano. Many electric pianos have a special diagnostic input, for which the authors of the article developed a digital twin (Figure 4).

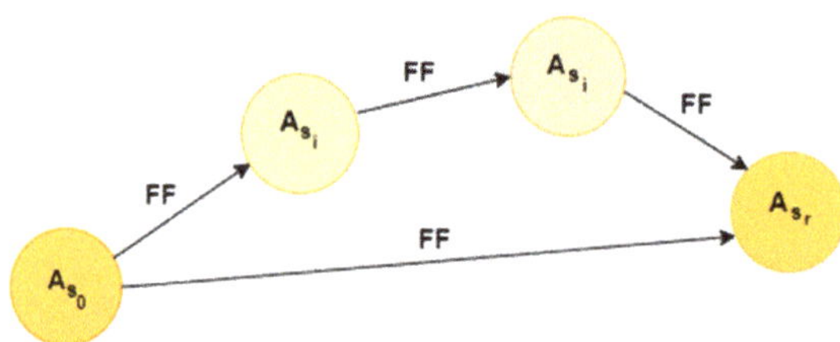

Figure 4. Middle level model.

The program has a menu through which it is possible to download and save the model (File) and screenshots for subsequent sending to the statistics server and various settings (Figure 5). On the left of the form, there is a static image of the object (electric piano), which can be hidden by pressing the "Hide" button. On the right is a schematic image of the object, which can be moved horizontally and vertically through the blue circle. It is possible to zoom in front of the magnifying glass image (plus and minus buttons), manually set the size by pressing the "Set" button and reset to the original dimensions, which are presented by default, by pressing the "Reset" button. On the left, you can select the event levels, Easily, Danger and Seriously, and see the current diagnostic status "Good"; meaning everything is fine and the components are in good working order. Below is the configuration of sensors and signal sources, Keys, Pedals, Speakers, Functions and Connectors, with the possibility of additional configuration using the "More..." button. The log with additional functions is displayed on the right: clear, save to a file, open a file with a log and open it on the site in a table with a search using the "Web" button. Only the last 20 lines of the log are displayed on the form element to avoid memory overflow errors.

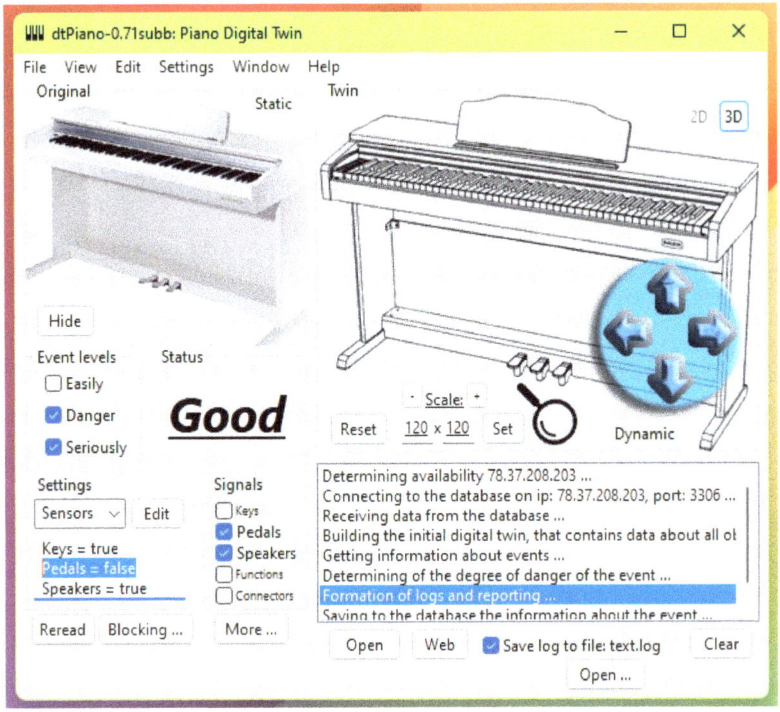

Figure 5. Digital Twin for Troubleshooting.

6. Conclusions

The above-described approach can be considered, first of all, as an approach to the development of large-scale cognitive heterogeneous CPS with a high level of structural, functional and architectural dynamics, built on fog computing platforms. The key idea of the developed approach is to use a system of dynamic models of the CPS, which is kept up-to-date throughout the entire system LC.

Implementation of the suggested approach allows for solving a number of important problems, such as increasing the level of CPS intelligence and increasing the level of service availability, to reach a new level of complexity of the created CPS. The presence of model knowledge about the past states of the system allows for determining the root causes of events and predicting future states of the system to implement proactive management. In addition, the availability of knowledge about past states allows for using learning mechanisms.

A necessary and sufficient condition for using the proposed model approach is a sufficiently high complexity of the structure and behavior of CPS that have high structural dynamics and adaptive behavior.

Currently, work on the further development of the suggested approach is being carried out in three directions: (i) creating new models; (ii) expanding the scope of the approach, in particular, for its application for the analysis of social networks; and (iii) development of new algorithms of model synthesis and model transformations.

Author Contributions: Conceptualization, A.V. and N.Z.; methodology, A.V.; software, A.S.; validation, M.C., A.V. and N.Z.; formal analysis, A.V.; investigation, N.Z.; resources, A.S.; data curation, A.S.; writing—original draft preparation, A.S.; writing—review and editing, A.V.; visualization, A.S.; supervision, N.Z.; project administration, N.Z.; funding acquisition, N.Z. All authors have read and agreed to the published version of the manuscript.

Funding: This research received no external funding.

Institutional Review Board Statement: Not applicable.

Informed Consent Statement: Not applicable.

Data Availability Statement: Not applicable.

Conflicts of Interest: The authors declare no conflict of interest.

References

1. Sanfelice, R.G. Analysis and Design of Cyber-Physical Systems. In *A Hybrid Control Systems Approach//Cyber-Physical Systems: From Theory to Practice*; Rawat, D., Rodrigues, J., Stojmenovic, I., Eds.; CRC Press: Boca Raton, FL, USA, 2016; ISBN 978-1-4822-6333-6.
2. Kotseruba, I.; Tsotsos, J.K. 40 years of cognitive architectures: Core cognitive abilities and practical applications. *Artif. Intell. Rev.* **2018**, *53*, 17–94. [CrossRef]
3. Calinescu, R.C.; Camara Moreno, J.; Paterson, C. Socio-Cyber-Physical Systems: Models, Opportunities, Open Challenges. In Proceedings of the 5th International Workshop on Software Engineering for Smart Cyber-Physical Systems, Montreal, QC, Canada, 28 May 2019.
4. Capilla, R.; Bosch, J.; Kyo-Chul, K. (Eds.) *Systems and Software Variability Management*; Springer: Berlin/Heidelberg, Germany, 2013.
5. Anthony, J.L. *Architecting Software Intensive Systems. Practitioner's Guide*; Taylor & Francis Group, LLC.: Abingdon, UK, 2009; 453p.
6. Buyya, R.; Srirama, S.N. (Eds.) *Fog and Edge Computing Principles and Paradigms*; John Wiley & Sons, Inc.: Hoboken, NJ, USA, 2019; 490p.
7. Wu, Q.; Ding, G.; Xu, Y.; Feng, S.; Du, Z.; Wang, J.; Long, K. Cognitive Internet of Things. A New Paradigm beyond Connection. *IEEE Internet Things J.* **2014**, *1*, 129–143. [CrossRef]
8. Bass, L. *Software Architecture in Practice*, 3rd ed.; Bass, L., Clements, P., Kazman, R., Eds.; Addison-Wesley: Upper Saddle River, NJ, USA, 2013; 661p.
9. Ford, N.; Parsons, R.; Kua, P. *Building Evolutionary Architectures*; O'Reilly Media: Sebastopol, CA, USA, 2017; 272p.
10. Bloomberg, J. *The Agile Architecture Revolution: How Cloud Computing, REST-Based SOA, and Mobile Computing Are Changing Enterprise IT*; Wiley & Sons, Inc.: Hoboken, NJ, USA, 2013; 278p.
11. van der Valk, H.; Hunker, J.; Rabe, M.; Otto, B. Digital twins in simulative applications: A taxonomy. In Proceedings of the 2020 Winter Simulation Conference, Orlando, FL, USA, 14–18 December 2020. [CrossRef]
12. Digital Thread. Available online: https://searcherp.techtarget.com/definition/digital-thread (accessed on 1 September 2022).
13. Weilkiens, T.; Lamm, J.; Roth, S.; Walker, M. *Model-Based System Architecture*; John Wiley & Sons, Inc.: Hoboken, NJ, USA, 2016; 375p.
14. Osipov, V.; Stankova, E.; Vodyaho, A.; Lushnov, M.; Shichkina, Y.; Zhukova, N. Automatic Synthesis of Multilevel Automata Models of Biological Objects. In Proceedings of the Computational Science and Its Applications–ICCSA 2019: 19th International Conference, Saint Petersburg, Russia, 1–4 July 2019; Springer: Cham, Switzerland, 2019; Volume 11620.
15. CMMI. Available online: http://www.sei.cmu.edu/ (accessed on 1 September 2022).
16. Osipov, V.; Zhukova, N.; Vodyaho, A. About one approach to multilevel behavioral program synthesis for television devices. *Int. J. Comput. Commun.* **2017**, *11*, 17–34.
17. Vodyaho, A.; Abbas, S.; Zhukova, N.; Chervoncev, M. Model Based Approach to Cyber–Physical Systems Status Monitoring. *Computers* **2020**, *9*, 47. [CrossRef]
18. Vodyaho, A.; Osipov, V.; Zhukova, N.; Chernokulsky, V. Data Collection Technology for Ambient Intelligence Systems in Internet of Things. *Electronics* **2020**, *9*, 1846. [CrossRef]

engineering proceedings

MDPI

Proceeding Paper

Particle Swarm Optimization for Target Encirclement by a UAV Formation †

Tagir Muslimov

Ufa State Aviation Technical University, Ufa 450077, Russia; tagir.muslimov@gmail.com

† Presented at the 15th International Conference "Intelligent Systems" (INTELS'22), Moscow, Russia, 14–16 December 2022.

Abstract: This paper presents an idea of using particle swarm optimization (PSO) to tune the control system of a decentralized unmanned aerial vehicle (UAV) formation. Simulations were run on a consensus-based decentralized UAV formation. Vector field guidance was used to control the formation. A fitness function is proposed that is based not only on the error of distance to the circular path, but also on the relative inter-UAV distance error. To demonstrate the effectiveness of the proposed method, the obtained results of such tuning are compared to those obtainable by the conventional trial and error method.

Keywords: UAV formation flight; collective circumnavigation; target tracking; vector field guidance; drone flocking

1. Introduction

Use of autonomous robots in groups is a promising area of research in today's mobile robotics, and it receives much attention. Decentralized control of autonomous robots is one of the more complex yet effective approaches. Thus, many papers cover decentralized control applications in ground-based robots [1–3] and autonomous unmanned aerial vehicles (UAVs) alike [4,5]. Most papers cover decentralized control of rotary-wing UAV groups, mainly quadcopter formations [6,7].

Currently, control of decentralized swarms of autonomous robots [8–10] and unmanned aerial vehicles (UAVs) [11,12] is a promising area of research. However, beside control and trajectory-planning algorithms, these formations require optimizing their transient trajectories. Since autonomous robot formations, including UAV formations, are complex nonlinear interconnected systems, soft computing could be effective for such optimization.

Control optimization research focuses on evolutionary algorithms [13] including genetic algorithms [14,15] and particle swarm optimization [16,17]. Some papers show implementation of particle swarm optimization for single-quadcopter controllers [18–20], UAV formations [21–23], UAV trajectory optimization [24], UAV movement planning [25] and UAV formations [26] in an uncertain environment. However, swarm optimization for vector field-controlled UAV formations remains under-researched. Thus, the goal hereof is to test the feasibility of applying particle swarm optimization to a decentralized consensus-based UAV formation controlled via vector field guidance.

Citation: Muslimov, T. Particle Swarm Optimization for Target Encirclement by a UAV Formation. *Eng. Proc.* **2023**, *33*, 15. https://doi.org/10.3390/engproc2023033015

Academic Editors: Askhat Diveev, Ivan Zelinka, Arutun Avetisyan and Alexander Ilin

Published: 9 June 2023

2. Preliminary Remarks and Statement of Problem

The assumptions here are the same as in [12,27]: no wind, inter-UAV communication enabled and sufficiently accurate computation of the relative inter-UAV distance in the group.

The dynamics of a decentralized UAV formation can be tested via full models as well as via high-level models that approximate the movement of the formation provided that the UAVs are equipped with fine-tuned onboard autopilots. Full models are preferable, e.g., for final testing of control algorithms or for testing formation-wide stability. High-level models, also referred to in the literature as guidance models, are more suitable for simulating spatial movement planning algorithms as well as for trajectory optimization. However, full models are still useful when the trial and error method is used to find the initial values for the formation controller coefficients that are further to be used for trajectory optimization.

UAV formation trajectory optimization is a subtask of cooperative target tracking. This task is sometimes referred to as collective circumnavigation or target encirclement. The idea is to maintain a certain preset distance not only between the UAVs (through specified angular values) but also to the target encirclement orbit, which is a moving path. Formal statement of the problem can be found in Section 4. We covered a similar problem in [28], where we used a genetic algorithm to solve it.

3. Consensus-Based UAV Formation Control Algorithm

This paper uses a decentralized neighbor-to-neighbor interaction topology, where UAVs only receive data from the neighboring aircraft. The topology can be defined as a graph referred to as the "interaction graph" hereinafter. Paper [12] shows a mathematical model of interaction in a consensus-based UAV formation. The same model is described later in this section. Let $\mathcal{N}$ be the set of all UAV agents.

Let $\mathbf{e}_\theta \in \mathbb{R}^{N\times 1}$ be this vector, where $\mathbb{R}^{N\times 1}$ is a space of $N \times 1$-dimensional matrices with components from $\mathbb{R}$. To find this vector, use some elements of the vector of all possible relative phase shift angle errors $\bar{\mathbf{e}}_\theta = \left(\hat{e}_{i,j}\right) \in \mathbb{R}^{N(N-1)\times 1}$, where $\hat{e}_{i,j}$ is the value of error for the directly interacting ith and jth agents. The choice is dictated by the interaction architecture; in this research, the control action vector is set as such for open-chain interaction in the same manner as described in [12,27]:

$$\mathbf{e}_\theta = \begin{bmatrix} e_1 \\ \vdots \\ e_k \\ \vdots \\ e_N \end{bmatrix} = \begin{bmatrix} \hat{e}_{12} \\ \vdots \\ -\hat{e}_{k-1,k} + \hat{e}_{k,k+1} \\ \vdots \\ -\hat{e}_{N-1,N} \end{bmatrix} = \hat{\mathbf{M}}_\theta \breve{\mathbf{e}}_\theta + \mathbf{D}, \tag{1}$$

where $\mathbf{D} = -\mathbf{M}_\theta \mathbf{H}_\theta^{-1} \left(\left(\mathbf{P}_\theta^d\right)^{\mathrm{T}}, \hat{P}_\theta \right)^{\mathrm{T}}$ is a system control vector in the space of relative distances (an $(N-1)$-dimensional space generated by the interaction graph incidence matrix columns), and $\mathbf{H}_\theta$ is a matrix that specifies the agents for agent-to-agent distance measurements, defined as follows:

$$\mathbf{H}_\theta = \begin{bmatrix} \mathbf{q}_1 \\ \mathbf{q}_2 \\ \vdots \\ \mathbf{q}_N \end{bmatrix}, \quad \mathbf{q}_i = \begin{bmatrix} \vdots \\ 1 \\ \vdots \\ -1 \\ \vdots \end{bmatrix}^{\mathrm{T}}, \quad i < N, \ \mathbf{q}_N = \begin{bmatrix} 1 \\ 1 \\ \vdots \\ 1 \end{bmatrix}^{\mathrm{T}},$$

where $\mathbf{H}_\theta \in \mathbb{R}^{N\times N}$, $\mathbf{q}_i \in \mathbb{R}^{1\times N}$ and the positions of "1" and "-1" in $\mathbf{q}_i$ are determined according to the structure of the interaction graph.

$\mathbf{P}_\theta^d \in \mathbb{R}^{(N-1)\times 1}$ is the vector of the desired inter-UAV phase shift angles and $\hat{P}_\theta = \sum\limits_{k=1}^{N} \varphi_k$ is the total of the current UAV phase angles in an inertial coordinate system;

$\breve{\mathbf{e}}_\theta = \left(\breve{e}_{i,i+1}\right)_{i=\overline{1,N-1}} \in \mathbb{R}^{(N-1)\times 1}$ is the vector of current phase shift angles for directly co-engaged agents, calculated by the triple scalar product, e.g., when the final movement is directed clockwise, the following applies:

If $\mathbf{n}\cdot(\mathbf{d}_i\times\mathbf{d}_{i+1})\geq 0$, then $\breve{e}_{i,i+1}=\beta=\arccos\frac{(\mathbf{d}_i,\,\mathbf{d}_{i+1})}{\|\mathbf{d}_i\|\|\mathbf{d}_{i+1}\|}$ and $\breve{e}_{i,i+1}=2\pi-\beta$ in other cases, where $\mathbf{d}_k$, $k\in\mathcal{N}$ is the vector of aircraft-to-moving-target distance at a given time, $\mathbf{n}=(0,\ 0,\ 1)^{\mathrm{T}}$;

$\mathbf{M}_\theta\in\mathbb{R}^{N\times N}$ is an interaction matrix that in cases of decentralized neighbor–neighbor interactions as herein is as follows:

$$\mathbf{M}_\theta=\begin{bmatrix}-1 & 1 & 0 & \cdots & 0\\ 1 & -2 & \ddots & \ddots & \vdots\\ 0 & \ddots & \ddots & 1 & 0\\ \vdots & \ddots & 1 & -2 & 1\\ 0 & \cdots & 0 & 1 & -1\end{bmatrix};$$

$\hat{\mathbf{M}}_\theta\in\mathbb{R}^{N\times(N-1)}$ is a matrix derived from the matrix $\mathbf{M}_\theta\mathbf{H}_\theta^{-1}$ by removing the Nth column.

For collective target encirclement, this paper uses the same control laws as in [12,27]. For control based on angular errors, it uses the speeds of the UAVs in the formation. The following control command vector $\mathbf{v}^c$ is set for UAV speeds:

$$\mathbf{v}^c=\begin{bmatrix}v_1 & v_2 & \ldots & v_N\end{bmatrix}^{\mathrm{T}}=v\mathbf{1}_N+\mathbf{L},\tag{2}$$

where $\mathbf{1}_N=\begin{bmatrix}1 & 1 & \ldots & 1\end{bmatrix}^{\mathrm{T}}\in\mathbb{R}^{N\times 1}$ and the vector $\mathbf{L}=\left(v_f(2/\pi)\arctan\{k_\theta e_i+k_{\dot\theta}\dot{e}_i\}\right)_{i=\overline{1,N}}\in\mathbb{R}^{N\times 1}$ is found given (1), k_θ is the positive tuning coefficient, $k_{\dot\theta}$ is the positive tuning coefficient for the derivative signal, v_f is the maximum norm of the additional velocity vector that is to be adjusted for the constraints of the real-UAV dynamics, and v is the ultimate linear cruise speed of the UAVs provided that the target is stationary.

Path error-based control relies on the heading angles of the UAVs in the formation. The following control law from [12,27] is applied to the heading-angle command vector $\boldsymbol{\varnothing}^c$ with slightly modified coefficients:

$$\boldsymbol{\varnothing}^c=\left(\varphi_i+\lambda\left[\frac{\pi}{2}+\arctan\left(k_o^i(d_i-\rho)+k_{\dot o}^i\dot{d}_i\right)\right]\right)_{i=\overline{1,N}}\in\mathbb{R}^{N\times 1},\tag{3}$$

where d_i is the ith UAV-to-target distance, $\dot{d}_i$ is the corresponding derivative signal, k_o^i is the tuning coefficient for the distance-to-circular-path signal for the ith UAV, $k_{\dot o}^i$ is the tuning coefficient for the distance-to-circular-path derivative signal for the ith UAV, ρ is the radius of the circular path that the UAV follows whilst encircling the target, φ_i is the phase angle of circumvolution around the target for the ith UAV.

4. Implementation of Particle Swarm Optimization

The standard `particleswarm` solver from MATLAB2015b with default parameters was used for particle swarm optimization. A four-UAV formation was tested for this paper, hence eight tuning parameters. These parameters are coefficients in the control law (3):

$$\mathbf{K}=\begin{bmatrix}k_o^1 & k_{\dot o}^1 & \ldots & k_o^4 & k_{\dot o}^4\end{bmatrix}\in\mathbb{R}^{1\times 8}.$$

We also added upper and lower bounds as follows:

$$\begin{bmatrix}0.1 & 0.1 & \ldots & 0.1 & 0.1\end{bmatrix}\leq\mathbf{K}\leq\begin{bmatrix}30 & 30 & \ldots & 30 & 30\end{bmatrix}$$

These constraints were chosen in order to preserve the UAV formation stability. Stability can be lost if the control law coefficients go beyond certain limits in the absence of adaptive control.

The initial guess for the vector **K** was as follows:

$$\mathbf{K}_{initial} = \begin{bmatrix} 1 & 2 & \ldots & 1 & 2 \end{bmatrix} \in \mathbb{R}^{1\times 8}.$$

For the fitness function, we chose the following:

$$F_{fitness} \triangleq \int_0^{t_n} t \sum_{i=1,\ldots,4} (|d_i - \rho| + |e_i|)\,dt, \quad (4)$$

where t_n is the particle swarm optimization time; the remaining parameters are defined in Equations (2) and (3). Thus, this solution optimizes not only for the error of each UAV's distance to the ultimate orbit of target encirclement but also for the relative neighbor-to-neighbor distance errors.

The formal statement of the goal would be as follows:

$$\begin{aligned} &\text{minimize} \quad F_{fitness} \triangleq \int_0^{t_n} t \sum_{i=1,\ldots,4} (|d_i - \rho| + |e_i|)\,dt, \\ &\text{s.t.} \quad \begin{bmatrix} 0.1 & 0.1 & \ldots & 0.1 & 0.1 \end{bmatrix} \le \mathbf{K} \le \begin{bmatrix} 30 & 30 & \ldots & 30 & 30 \end{bmatrix} \end{aligned}$$

5. Simulation Results

5.1. Simulation Parameters

For simulation, we ran a high-level UAV model from [29]. The formation consisted of four UAVs of the same type. To make an initial guess, we also ran full UAV models of this formation. This allowed us to find, by trial and error, a controller coefficient that would keep the entire formation system stable. Control laws (2) and (3) were used in the simulation. The simulation parameters are shown in Table 1.

Table 1. Simulation parameters.

Parameter	Symbol	Values
Initial UAV heading vector (rad)	$\boldsymbol{\varnothing}(0)$	$\begin{bmatrix} 0 & 0 & 0 & 0 \end{bmatrix}^T$
Initial UAV speed vector (m/s)	$\mathbf{v}(0)$	$\begin{bmatrix} 13 & 13 & 13 & 13 \end{bmatrix}^T$
Circular path radius (m)	ρ	200
Initial target coordinates (m)	$\mathbf{c} = \begin{bmatrix} c_e(0) & c_n(0) & h(0) \end{bmatrix}^T$	$\begin{bmatrix} 500 & 500 & 100 \end{bmatrix}^T$
Ultimate cruise speed of the formation in case of a stationary target (m/s)	v	13
Target speed (m/s)	v_{target}	2
Target course angle (rad)	χ_{target}	$\pi/4$
Vector of initial UAV coordinates in the ICS (m)	$\mathbf{p}_1(0)$, $\mathbf{p}_2(0)$, $\mathbf{p}_3(0)$, $\mathbf{p}_4(0)$	$\begin{bmatrix} 180 & 910 & -100 \end{bmatrix}^T$, $\begin{bmatrix} 245 & 247 & -100 \end{bmatrix}^T$, $\begin{bmatrix} 710 & 135 & -100 \end{bmatrix}^T$, $\begin{bmatrix} 915 & 760 & -100 \end{bmatrix}^T$
Desired inter-UAV phase shift angles (°)	$\mathbf{P}_\theta^d$	$\begin{bmatrix} 270 & 260 & 290 \end{bmatrix}^T$
Initial tuning coefficients	$k_\theta, k_{\dot{\theta}}, k_o^i, k_{\dot{o}}^i$	35, 15, 1, 2

5.2. Simulation Results and Discussion

Figure 1 shows how the fitness function changed during optimization. Apparently, the function's value stopped declining drastically after 25 iterations. However, a significantly increasing number of iterations would be required to further reduce the value.

Running the optimization algorithm returned the following $\mathbf{K} = \mathbf{K}_{opt}$ values that were used in the simulation:

$$\mathbf{K}_{opt} = \begin{bmatrix} 4.7426, & 25.1807, & 2.6214, & 23.8265, & 20.6034, & 29.4922, & 18.7241, & 26.8984 \end{bmatrix}^T$$

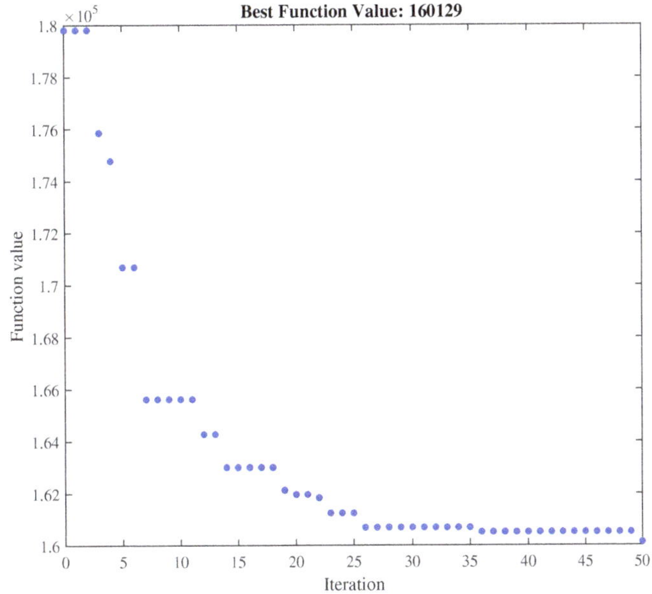

Figure 1. Fitness function value change during simulation.

Figure 2 shows UAV formation angular errors before and after optimization. As can be seen in the graphs, optimization enabled the formation to reach the pre-specified relative angular positions somewhat faster. Figure 3 shows how path errors changed in the UAV formation. As can be seen from the graphs, UAV1 and UAV2 showed the most drastic changes. Apparently, transient trajectories before and after optimization are different (Figure 4). Even though the trajectories look similar in the figure, they are still different. That is especially noticeable in the trajectories for UAV1 and UAV2.

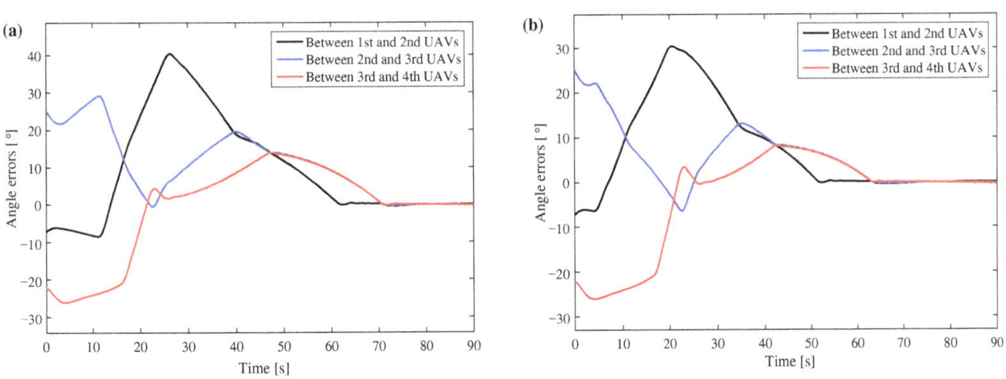

Figure 2. UAV angular errors at time $t = 90$ s. (**a**) Angular errors before particle swarm optimization; (**b**) angular errors after particle swarm optimization.

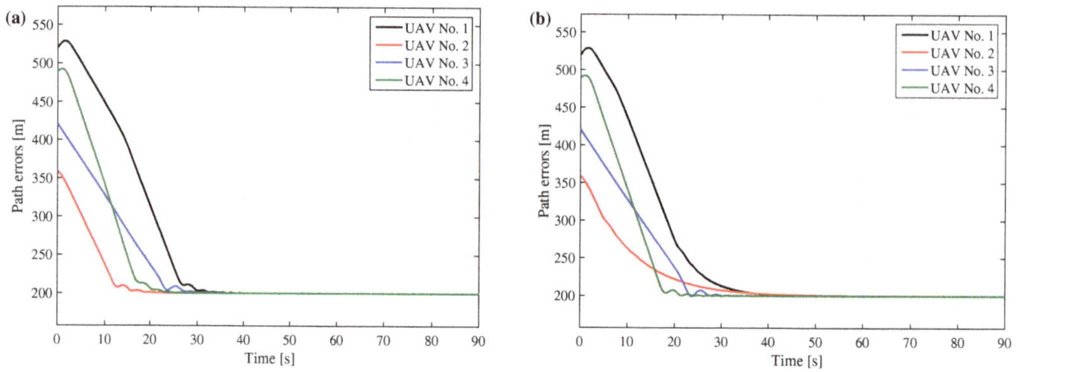

Figure 3. UAV path errors at time $t = 90$ s. (**a**) Path errors before particle swarm optimization; (**b**) path errors after particle swarm optimization.

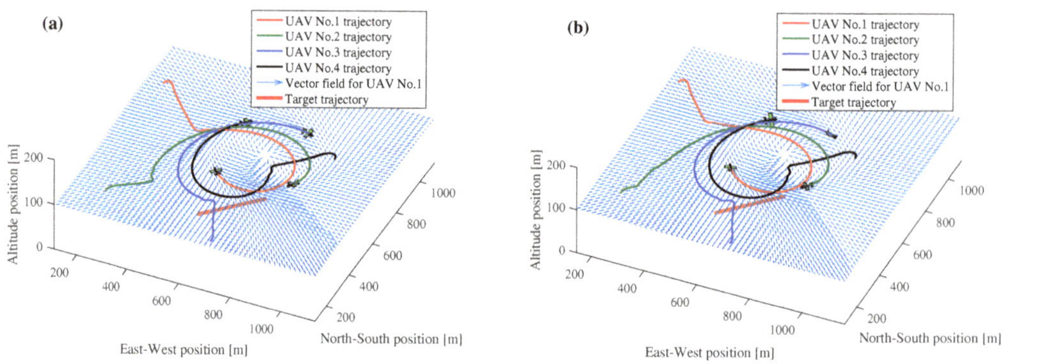

Figure 4. UAV formation trajectories at time $t = 90$ s. (**a**) Trajectories before particle swarm optimization; (**b**) trajectories after particle swarm optimization.

Notably, although the controller coefficients were tuned only for the control law (3), the fitness function included the total angular error of the formation (4). The reason for this was that control by path errors is tied to control by angular errors in a decentralized UAV formation system. This connection can be seen, among other things, in the simulation results in Figure 2.

6. Conclusions

The paper demonstrates a successful use of particle swarm optimization for target encirclement and tracking by a UAV formation. The formation itself was a vector field-controlled decentralized formation. Simulations showed a reduction in the proposed fitness function as well as a change in the pattern of transient trajectories. A close connection was found between optimizing the path error controller optimization and the quality of transient trajectories for angular errors in the UAV formation.

Funding: This work was supported by the Ministry of Science and Higher Education of the Russian Federation (Agreement No. 075-15-2021-1016).

Institutional Review Board Statement: Not applicable.

Informed Consent Statement: Not applicable.

Data Availability Statement: Not applicable.

Conflicts of Interest: The authors declare no conflict of interest.

References

1. Darintsev, O.V.; Yudintsev, B.S.; Alekseev, A.Y.; Bogdanov, D.R.; Migranov, A.B. Methods of a Heterogeneous Multi-agent Robotic System Group Control. *Procedia Comput. Sci.* **2019**, *150*, 687–694. [CrossRef]
2. Ivanov, D.Y. Distribution of roles in groups of robots with limited communications based on the swarm interaction. *Procedia Comput. Sci.* **2019**, *150*, 518–523. [CrossRef]
3. Veselov, G.; Sklyrov, A.; Mushenko, A.; Sklyrov, S. Synergetic Control of a Mobile Robot Group. In Proceedings of the 2014 2nd International Conference on Artificial Intelligence, Modelling and Simulation, Madrid, Spain, 18–20 November 2014; pp. 155–160. [CrossRef]
4. Bennet, D.J.; McInnes, C.R.; Suzuki, M.; Uchiyama, K. Autonomous Three-Dimensional Formation Flight for a Swarm of Unmanned Aerial Vehicles. *J. Guid. Control Dyn.* **2011**, *34*, 1899–1908. [CrossRef]
5. Milyakov, D.A.; Merkulov, V.I. The Approach to Managing a Group of UAVs as a System with Distributed Parameters. *Procedia Comput. Sci.* **2019**, *150*, 39–45. [CrossRef]
6. Diveev, A.I.; Shmalko, E.Y.; Hussein, O. Synthesized Optimal Control of Group Interaction of Quadrocopters Based on Multi-Point Stabilization. *Instrum. Eng.* **2020**, *133*, 114–133. [CrossRef]
7. Titkov, I.P.; Karpunin, A. Collision-aware formation assignment of quadrotors. *Procedia Comput. Sci.* **2021**, *186*, 727–735. [CrossRef]
8. Pavlovskii, V.E.; Pavlovskii, V.V. A mathematical model of a 2D homogeneous swarm of robots. *Sci. Tech. Inf. Process.* **2016**, *43*, 306–314. [CrossRef]
9. Senotov, V.D.; Aliseychik, A.P.; Pavlovsky, E.V.; Podoprosvetov, A.V.; Orlov, I.A. Algorithms for swarm decentralized motion control of group of robots with a differential drive. *Keldysh Inst. Prepr.* **2020**, *123*, 1–39. [CrossRef]
10. Zakiev, A.; Tsoy, T.; Magid, E. Swarm robotics: Remarks on terminology and classification. In Proceedings of the Interactive Collaborative Robotics: Third International Conference, ICR 2018, Leipzig, Germany, 18–22 September 2018; pp. 291–300.
11. Ollervides-Vazquez, E.J.; Rojo-Rodriguez, E.G.; Garcia-Salazar, O.; Amezquita-Brooks, L.; Castillo, P.; Santibañez, V. A sectorial fuzzy consensus algorithm for the formation flight of multiple quadrotor unmanned aerial vehicles. *Int. J. Micro Air Veh.* **2020**, *12*, 1756829320973579. [CrossRef]
12. Muslimov, T.Z.; Munasypov, R.A. Multi-UAV cooperative target tracking via consensus-based guidance vector fields and fuzzy MRAC. *Aircr. Eng. Aerosp. Technol.* **2021**, *93*, 1204–1212. [CrossRef]
13. Diveev, A.; Shmalko, E. Hybrid evolutionary algorithm for synthesized optimal control problem for group of interacting robots. In Proceedings of the 2019 6th International Conference on Control, Decision and Information Technologies (CoDIT), Paris, France, 23–26 April 2019; pp. 876–881. [CrossRef]
14. Bożko, A.; Ambroziak, L.; Pawluszewicz, E. Genetic Algorithm for Parameters Tuning of Two Stage Switching Controller for UAV Autonomous Formation Flight. *Adv. Intell. Syst. Comput.* **2021**, *1390*, 154–165.
15. Kim, M. Error Dynamics-Based Guidance Law for Target Observation using Multiple UAVs with Phase Angle Constraints via Evolutionary Algorithms. *J. Control Autom. Electr. Syst.* **2021**, *32*, 1510–1520. [CrossRef]
16. Ali, Z.A.; Zhangang, H. Multi-unmanned aerial vehicle swarm formation control using hybrid strategy. *Trans. Inst. Meas. Control* **2021**, *43*, 2689–2701. [CrossRef]
17. Ali, Z.A.; Han, Z.; Masood, R.J. Collective Motion and Self-Organization of a Swarm of UAVs: A Cluster-Based Architecture. *Sensors* **2021**, *21*, 3820. [CrossRef] [PubMed]
18. Derrouaoui, S.H.; Bouzid, Y.; Guiatni, M. PSO Based Optimal Gain Scheduling Backstepping Flight Controller Design for a Transformable Quadrotor. *J. Intell. Robot. Syst.* **2021**, *102*, 67. [CrossRef]
19. Kiyak, E. Tuning of controller for an aircraft flight control system based on particle swarm optimization. *Aircr. Eng. Aerosp. Technol.* **2016**, *88*, 799–809. [CrossRef]
20. Saribas, H.; Kahvecioglu, S. PSO and GA tuned conventional and fractional order PID controllers for quadrotor control. *Aircr. Eng. Aerosp. Technol.* **2021**, *93*, 1243–1253. [CrossRef]
21. Biantoro, N.; Halim, M.; Nazaruddin, Y.Y.; Juliastuti, E. PSO-based Optimization of Formation Control and Obstacle Avoidance for Multiple Quadrotors. In Proceedings of the 2021 International Conference on Instrumentation, Control and Automation (ICA), Bandung, Indonesia, 25–27 August 2021; pp. 133–137. [CrossRef]
22. Hoang, V.T.; Phung, M.D.; Dinh, T.H.; Zhu, Q.; Ha, Q.P. Reconfigurable Multi-UAV Formation Using Angle-Encoded PSO. In Proceedings of the 2019 IEEE 15th International Conference on Automation Science and Engineering (CASE), Vancouver, BC, Canada, 22–26 August 2019; pp. 1670–1675. [CrossRef]
23. Sruthy, A.N.; Jacob, J.; Ramch, R. PSO Based Integral Backtepping Control for Leader-Follower Quadrotors. In Proceedings of the 2020 Fourth International Conference on Inventive Systems and Control (ICISC), Coimbatore, India, 8–10 January 2020; pp. 466–471. [CrossRef]
24. Vijayakumari, D.M.; Kim, S.; Suk, J.; Mo, H. *Receding-Horizon Trajectory Planning for Multiple UAVs Using Particle Swarm Optimization*; AIAA Scitech 2019 Forum; American Institute of Aeronautics and Astronautics: Reston, VA, USA, 2019. [CrossRef]
25. Patley, A.; Bhatt, A.; Maity, A.; Das, K.; Ranjan Kumar, S. *Modified Particle Swarm Optimization Based Path Planning for Multi-Uav Formation*; AIAA Scitech 2019 Forum; American Institute of Aeronautics and Astronautics: Reston, VA, USA, 2019. [CrossRef]

26. Skrzypecki, S.; Tarapata, Z.; Pierzchała, D. Combined PSO Methods for UAVs Swarm Modelling and Simulation. In Proceedings of the Modelling and Simulation for Autonomous Systems: 6th International Conference, MESAS 2019, Palermo, Italy, 29–31 October 2019; pp. 11–25.
27. Muslimov, T.Z.; Munasypov, R.A. Coordinated UAV Standoff Tracking of Moving Target Based on Lyapunov Vector Fields. In Proceedings of the 2020 International Conference Nonlinearity, Information and Robotics (NIR), Innopolis, Russia, 3–6 December 2020; pp. 1–5. [CrossRef]
28. Muslimov, T. Application of Genetic Algorithm for Vector Field Guidance Optimization in a UAV Collective Circumnavigation Scenario. In *Robotics in Natural Settings: CLAWAR 2022*; Springer International Publishing: Cham, Switzerland, 2022.
29. Beard, R.W.; McLain, T.W. *Small Unmanned Aircraft: Theory and Practice*; Princeton University Press: Princeton, NJ, USA, 2012.

engineering proceedings

MDPI

Proceeding Paper

Active Simultaneous Localization and Mapping Method Based on Model Prediction †

Anna N. Daryina *,‡ **and Igor V. Prokopiev** ‡

Federal Research Centre "Computer Science and Control" of Russia Academy of Sciences, Vavilov Str., 44, 2, 119333 Moscow, Russia; fvi2014@list.ru
* Correspondence: daryina@ccas.ru
† Presented at the 15th International Conference "Intelligent Systems" (INTELS'22), Moscow, Russia, 14–16 December 2022.
‡ These authors contributed equally to this work.

Abstract: In the process of controlling an unmanned vehicle, it is practically important that under conditions of rapidly changing dynamic constraints, control laws be developed that would be optimal with respect to a given quality functional or a multicriteria functional. When static and dynamic constraints do not allow the optimal movement to be chosen to a given quality functional, the authors consider the transition to another quality functional using the predictive integral path model and the method of active simultaneous localization and mapping. In this case, the strategy for choosing the state space is more efficient than the strategy for choosing the control space. The practical question is how to achieve this. The paper presents a method and experiments using an unmanned vehicle platform at a test site in the form of a complex environment, showing the feasibility of the method.

Keywords: method of active simultaneous localization and mapping; model predictive path integral; mobile robot; nonlinear problem; optimal trajectory; optimal control

Citation: Daryina, A.N.; Prokopiev, I.V. Active Simultaneous Localization and Mapping Method Based on Model Prediction. *Eng. Proc.* **2023**, *33*, 16. https://doi.org/10.3390/engproc2023033016

Academic Editors: Askhat Diveev, Ivan Zelinka, Arutun Avetisyan and Alexander Ilin

Published: 9 June 2023

1. Introduction

Currently, much attention is being paid to solving the problems of controlling mobile robots. When they are controlled, the requirements for autonomy, the accuracy of determining the state, and the ability to determine and overcome static and dynamic constraints increase. In this regard, the problems of studying the trajectories of a mobile robot and calculating control are very relevant. The automation of mobile robots is the subject of many studies. However, despite fundamental and applied research, the problems of managing them have not been fully resolved. The main problem is the accuracy of determining its local state, and the calculation of the optimal trajectory in complex environments with static and dynamic constraints requires assessing the entire space of possible states and finding the best solution.

We investigated the analysis of the external environment and the robot's localization in it using the method of simultaneous localization and mapping (SLAM). The efficiency and reliability of optimization in problems was developed in [1–4]. All these approaches formulate SLAM as a maximum a posteriori estimation problem, and often use factor graphs [5] to establish the interdependence between variables.

The simultaneous assessment of the state of a robot equipped with onboard sensors and the construction of a model (map) of the environment has been developing for the last 35 years [6]. The map represents the position of landmarks and obstacles, including dynamic ones, with a description of the environment in which the robot operates.

Therefore, the improvement of SLAM methods is still relevant. In this article, we present a new method of active SLAM based on MPPI [7] (ASLAM-MPPI), which can calculate quasi-optimal control to achieve a terminal state, taking into account the bypass of dynamic constraints in real time based on the forecast period.

The article deals with the problem of optimizing the control of a mobile robot in real time based on the SLAM method.

One of the main causes of algorithm failures is data aggregation. For example, in feature-based visual SLAM, each visual function is associated with a specific breakpoint. Perceptual smoothing is a phenomenon in which different sensory inputs result in the same sensor signature. This makes this task very difficult. In the presence of a data perception alias, the association establishes erroneous matches of the measurement states, which, in turn, leads to incorrect estimates. Standard approaches to solving the problem are based on optical flow or descriptor matching [8] and provide reliable tracking. At a high frame rate, the angle of view of the sensor (laser, camera) does not change significantly, so the appearance and all features at time $t+1$ remain close to those observed at time t. The other method is long-term data association in an interface, which includes loop closure, discovery, and validation.

We propose a solution to this problem using the ASLAM-MPPI algorithm. We present the first predictive control method with active SLAM for mobile robots that combines control and planning in relation to action goals and perception.

The model prediction method (MPC) has long been used in robotic systems. A more flexible method of MPC—prediction of the integral path model (model predictive path integral—MPPI)—a sample-based algorithm that can be optimized according to general cost criteria, convex and nonconvex, was first implemented in [9].

We have extended this method by integrating it with active SLAM so that it is applicable to a larger class of stochastic robotic systems with dynamic constraints. Active SLAM includes integration with object recognition. At the same time, the structure optimizes the perception goals for reliable sensing of key points for closing the loop by sampling the trajectory for maximum testing of points of interest. Loop closure is one of the most important contributions of SLAM [10]. Reliable loop closure allows one to change the card globally. When the cycle is closed, the constructed map and the calculated trajectory of the robot are more accurate.

The keyframe must satisfy a certain test, which usually contains three movements, three rotations, and one scaling parameter. When the candidate has enough checks, we are sure that the loop is found.

Given both the goals of perception and actions for motion planning, the choice of a trajectory is difficult due to possible conflicts arising from their respective requirements. Our predictive model, taking into account the recognition of significant landmarks for the perception goal, will require choosing such a movement to make the landmark as noticeable as possible for testing key frames. To carry this out, a condition for switching to the goals of perception is introduced into the cost function.

2. Active SLAM method based on MPPI

The problem of controlling the robot's movement in order to minimize the uncertainty of its display on the map and localization is usually called active SLAM. This definition comes from the well-known active perception [8]. Theoretical management approaches for active SLAM include the use of a predictive management model [9,11].

In our work, we developed a method based on MPPI [12–14] and active SLAM.

The ASLAM-MPPI algorithm allows for real-time processing of the complex nonlinear dynamics of an unmanned vehicle and the environment. However, like most classical algorithms, it suffers from instability when simulating dynamics that are different from the true dynamics of a car. One way to solve this problem is to define the parameters of the object model and take into account the costs of understanding the scene using SLAM in real time.

The ASLAM-MPPI algorithm consists of the following steps:

- Initialization (localization using the SLAM method) of the state vector x_0;
- Control calculation.

To implement the second step c, we consider a sequence of controls V_m, where m is a realization of a random trajectory from previous iterations, consisting of (thousands) trajectories.

Control costs are collected for each trajectory and mapped to its weights

$$\Omega(V_m) = e^{\left(-\frac{1}{\lambda}\left(S(V_m) - \sum_{t=0}^{T-1}(v_t)^T \Sigma^{-1} v_t - \rho\right)\right)},$$

where ρ sets the minimum value of the cost among all the selected trajectories intended to exclude the algorithm; $S(V_m) = q_1(x_T^m) + \sum_{t=0}^{T-1} q_2(x_t^m)$ is part of the cost, depending on the state.

The quasi-optimal control at each time step is calculated as

$$U(T,K) = U(t,K-1) + \frac{\sum_{t=1}^{T-1}[\Omega(V_m)\xi^m]}{\sum_{T=1}^{m}\Omega(V_m)},$$

where $\xi^m \sim N(0, \Sigma)$ is Gaussian noise with zero mean.

Figure 1 shows the discrepancies between the state of the unmanned vehicle in the model and the state obtained using SLAM at the Gaussian noise level, with the same control without feedback. In Figure 1, the red line shows the trajectory described by the robot in the real state without feedback. The blue line indicates the trajectory described by the robot according to the model after identification. The green line indicates the trajectory obtained during the second run on the model.

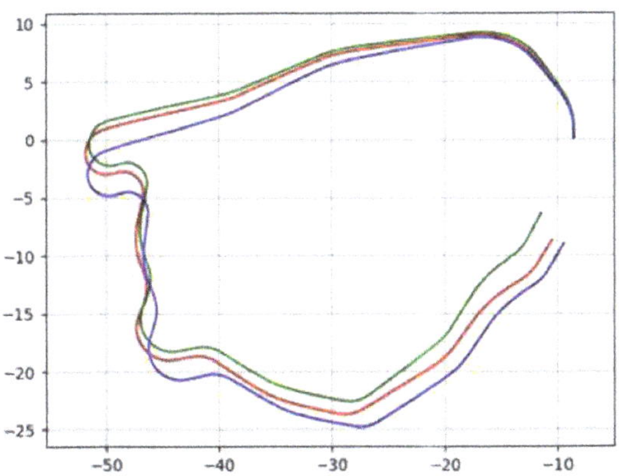

Figure 1. Discrepancies in the state of the unmanned vehicle in the model with the state obtained using SLAM at the Gaussian noise level, with the same control without feedback.

In global planning, it is desirable that significant landmarks (obtaining unambiguous functions by the robot) appear in the area of the navigation trajectory, but are not an obstacle. At the same time, the global trajectory can be optimal in the sense of fault safety.

Optimal trajectory planning for autonomous robots has long been studied. Rosenblatt [15] developed a jet navigation system to enable multiple targets. Each goal is considered a behavior, and their task is to weigh a set of discrete motion controls based on its expected utility. The benefits from each behavior are also weighted by the central arbiter depending on the current mission of the system.

To be able to control the robot in real time, we need to update the map quickly enough [16]. This is very important for robots with fast dynamics, such as Figure 2 (the picture on the left) unmanned vehicle. To meet these requirements, you need to use accelerations, for example, using the CUDA GPU.

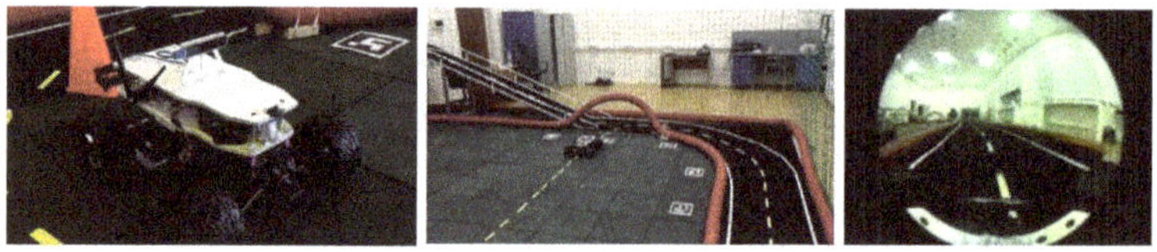

Figure 2. Landmarks in the form of yellow markings, red cones, and radio frequency markers.

Our simple solution to refine and adjust the position and orientation of the mobile robot is to use a marker when initializing the unmanned vehicle and close loop. The disadvantage of the visual–inertial SLAM is its sensitivity to vibrations and in constantly changing light conditions. Between adjustments using radio frequency markers, the position and orientation of the mobile robot is correlated with the dynamic model of the mobile robot, and if the threshold is exceeded, the system uses the position and orientation of the model until it converges with the SLAM.

For dynamic obstacles in previous works [17], we used a trajectory generator and precalculated trajectories, which were selected by the cost function. The ASLAM-MPPI algorithm allows us to generate thousands of trajectories in real time and choose the one that takes into account the dynamic constraint (Figure 3). In Figure 3, we can see red, blue, and yellow fields; this is cost map. The light blue line is a trajectory of the ASLAM-MPPI method, visualized on RVIZ. The optimal trajectory has a violet color.

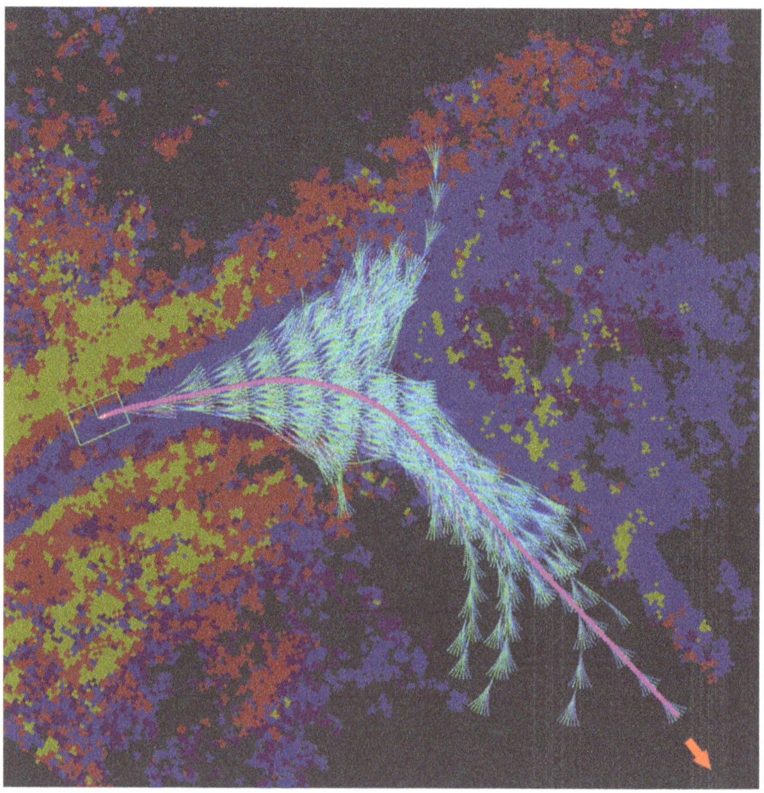

Figure 3. The set of obtained trajectories, where the arrow indicates the direction of the optimal trajectory.

As a result, the ASLAM-MPPI algorithm works as follows:

1. Using ASLAM algorithms, we obtain the local state of the unmanned vehicle (UV) and the employment map, taking into account dynamic obstacles;
2. The A-star [18] algorithm calculates the global trajectory. Active SLAM includes terminal states in the circle of significant landmarks, for example, radio frequency markers;
3. The ASLAM-MPI algorithm models predict, from the initial state, thousands of trajectories, taking into account the UV dynamics in the direction of the global trajectory. The prediction window depends on the calculated speed of dynamic obstacles;
4. Using the cost function, the optimal trajectory is calculated according to a given criterion;
5. The control corresponding to the optimal trajectory is applied;
6. Then, the algorithm is repeated;
7. If the UV localization exceeds the threshold of discrepancy with the forecast of the local state of the ASLAM-MPI algorithm, the localization system is corrected.

3. An example of using the method

Consider the movement of a robot with an Ackermann geometry chassis [19]:

$$\begin{cases} \dot{x} = u_1 \cos\theta, \\ \dot{y} = u_1 \sin\theta, \\ \dot{\theta} = \frac{u_1}{L} \tan u_2, \end{cases}$$

where x, y are the coordinates of the center of the rear axis of the mobile robot; u_1 is its linear speed; θ is an angle of rotation around the axis x; u_2 is the rotation angle (positive counterclockwise); L is the distance between the front and rear axles of the wheels of the mobile robot.

The technology described above was tested in the robotics center of the Russian Academy of Sciences, where a polygon simulating an industrial facility with complex structures, ramps in the form of a steep hill, a bridge, and other restrictions is deployed. In Figure 4, we can see the part of the track with the form of bridge and a mobile robot with an Ackermann geometry chassis, which we use in our experiments.

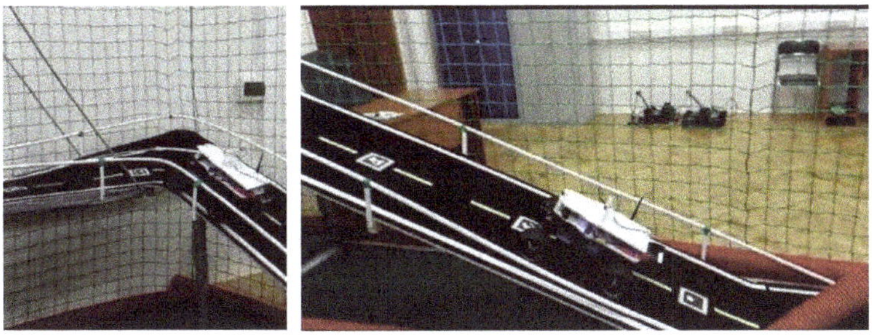

Figure 4. The track.

Static phase restrictions are given:

$$\varphi(x,y) = r^2 - (x_i^* - x)^2 - (y_i^* - y)^2 \leq 0, \tag{1}$$

where r is the obstacle dimensions; x_i^*, y_i^* are the coordinates of the center of the obstacle.

If $f(x) < 0$, where $f(x)$ is the condition for recognizing labels in the algorithm, then the algorithm switches to another cost function, such as

$$G(Z_k) = F(Z_k) + (1 - \vartheta(f(x)))y,$$

where

- $F(Z_k)$ is from [17];

- $y = \sin(x + \Delta x, y + \Delta y)$, where (x, y) stands for current coordinates and $(\Delta x, \Delta y)$ is distance to the recognized label;
- ϑ is a Heaviside function:

$$\vartheta(a) = \begin{cases} 0, & a < 0, \\ 1, & a \geq 0. \end{cases}$$

The dynamics model of the robot shown in Figure 2 (the picture on the left) was implemented on a neural network model structure having the following mathematical form:

$$g_i(\varphi(k, \mathbf{q}), \mathbf{q}) = \hat{y}_i(k|\mathbf{q}) = \hat{y}_i(k|w, W) = = F_i\left(\sum_{j=1}^{n_h} W_{ij} f_j\left(\sum_{l=1}^{n_\varphi} w_{jl}\varphi_l + w_{j0},\right) + W_{i0}\right)$$

where $\mathbf{q}$ represents the configurable parameters of the neural network, including weight coefficients and offsets (w_{jl}, W_{ij}); $F_i(x) = ax$ is the activation function of neurons in the output layer $a = const$; n_h is the number of neurons in the hidden layer; $f_i(x) = \tanh(x)$ is the activation function of neurons of the hidden layer; n_φ is the number of inputs.

Our neural network for the ASLAM-MPPI method includes two hidden layers with two nonlinearities, which means that the overall network configuration is 6-32-32-4. It takes as input four state variables (roll, longitudinal speed, transverse speed, course), as well as controlled steering and speed control, and they output the time derivative of the state variables. This network was trained on real data collected from the real polygon (Figure 4) in manual UV control mode. Stochastic gradient descent with ADAM was used for training.

As an embedded computing platform, NVIDIA Jetson TX2 was used with 256 CUDA cores and a four-core ARM Cortex-A57 processor, which made it possible to parallelize the computer network.

Experiments were conducted with ROS (melodic version) on a Linux computer with an Intel Core i7 processor, on which the MPPI algorithm was deployed. The control commands were transmitted via WiFi to the unmanned vehicle shown in Figure 2.

Closed trajectories were defined in the form of generated points in space, located on the dotted central track marking.

The absolute trajectory error (ATE) measurement showed the same error on the slide and on the horizontal plane, and was, on average, 0.99 percent of the length of the path traveled. The length of the path passing along the slide was 23.3 m. A deviation error of more than 0.03 m leads to accidents on the turns of the slide. Thus, the quality criterion of SLAM and control is the number of tracks passing through the slide.

In complex conditions, when spatial constraints severely constrict the space of acceptable movements, the strategy of selecting the state space for optimization is more efficient than sampling in the control space.

Thus, the quality criterion of the ASLAM-MPI algorithm is the number of tracks passing through the slide.

The experiment showed that the optimal passage of the slide firstly passes through the specified states, the exact definitions of which depend on the localization system of the robot.

We used the Intel Realsense T265, a standalone 6-degree-of-freedom tracking sensor that runs a visually-inertial SLAM algorithm that accurately evaluates movement with 6 degrees of freedom and obtains the positions and orientation of the tracking camera at 200 Hz, which is 10 times quicker than localization with the best-known ORB-SLAM2 algorithm [20].

Active SLAM shows an average of seven laps on a track with a slide without failures, which also depend on the control settings. A SLAM based on the Intel Realsense T265 tracking camera allows you to make only one lap, and on the second one, there is a failure due to an accumulated error, where the accumulation of errors leads to the inability to overcome turns on the slide. The slope of the slide is 45 degrees, and represents a serious limitation for the mobile robot. Figure 5 shows a track without a slide, because a SLAM method based on an Intel Realsense T265 security camera allows one to make 2–4 laps (depends on speed), after which a failure occurs due to an accumulated error.

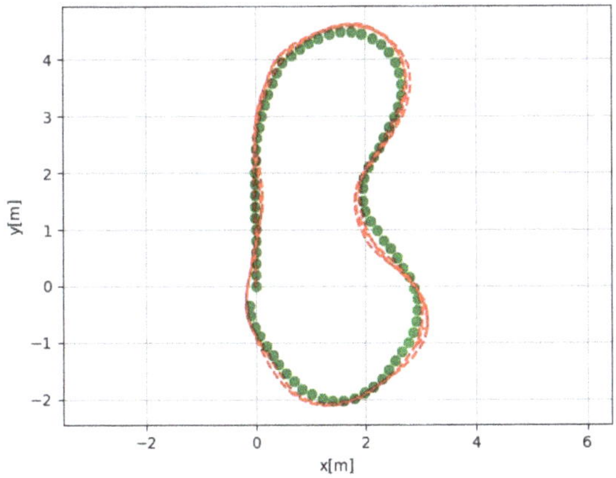

Figure 5. An accumulated error on a track without a slide.

In Figure 6, we can see the polygon and its employment map.

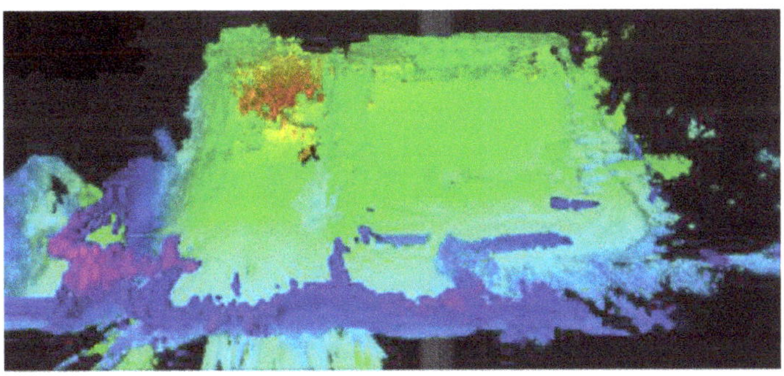

Figure 6. Polygon employment map.

We see a further development of the method in accelerating the study of the robot's map (like in Figure 6) with the help of autoencoders. We are also continuing to work on the automatic configuration of the cost function $F(Z_K)$ [17]. It has also been supposed to conduct experiments and compare the proposed method with previously known ones.

4. Conclusions

The result of the presented work is a new method for controlling an unmanned vehicle based on the choice of control in the state space that satisfies a given quality functional and is resistant to uncertain restrictions and interferences arising in a wide class for the SLAM method. The use of active SLAM allows, firstly, for a more reliable data association in the interface, including loop closures, which is important for correcting the orientation and posture of an unmanned vehicle, as well as detecting key points for odometry. Secondly, the method allows for controls to be developed based on navigation in unknown spaces. In addition, the method allows for the use of a variety of sensors for difficult conditions. At the same time, the method can use known cartographic information.

Classical control schemes are unable to predict costs and the corresponding trajectories (for example, PID controllers, LQR, iLQR). In addition, the formulation of the aggregate perception and action, as an optimization problem, has advantages in the reliability of the control system.

Our approach can be compared with the method [21] based on a predictive control model (MPC) with deep convolutional neural networks (CNN) for real-time scene understanding. A fully convolutional network is able to study the complex out-of-plane transformation required to project the observed image pixels onto the ground plane and predict the map of controlled objects in the area in front of the car. The solution is to train a deep neural network to convert visual inputs from a single monocular camera into a cost functions in a local coordinate system oriented to the robot is excellent. This cost map can then be directly introduced into the predictive model control algorithm. However, the disadvantage of the method may be nonvisited areas, where the training sample may not have enough reliable cost maps. The predictive controller (MPC) model uses only the information that it can see at the output of the neural network, and plans in advance for 1.5 s to issue a control signal. This 1.5-second time horizon leads to extremely timid behavior, because the available forward view has a short distance.

The practical result is the implementation of the method on a transport platform (see Figure 6) and the solution of the problem of following a global trajectory along a complex trajectory, including steep ascents and descents with spatial constraints that significantly narrow the choice of trajectory.

MPPI with deep convolutional neural networks cannot recognize ascents and descents from the slide without additional information. Thus, the cost map created by this method cannot be directly introduced into the predictive model control algorithm. Classical control schemes (for example, PID, LQR, and iLQR) can reliably track the global trajectory, but are not stable in cases of failures in the localization system.

As a result of conducted field experiments on a real UV, the ASLAM-MPPI method shows significant performance (algorithm speed, fault tolerance) compared to, for example, the ROS navigation stack [22].

Author Contributions: Conceptualization, A.N.D. and I.V.P.; methodology, A.N.D.; software, I.V.P.; formal analysis, A.N.D.; investigation, I.V.P.; resources, I.V.P.; data curation, A.N.D.; writing—original draft preparation, I.V.P.; writing—review and editing, A.N.D. and I.V.P. All authors have read and agreed to the published version of the manuscript.

Funding: This research received no external funding.

Institutional Review Board Statement: Not applicable.

Informed Consent Statement: Not applicable.

Data Availability Statement: Not applicable.

Conflicts of Interest: The authors declare no conflict of interest.

References

1. Dellaert, F.; Kaess, V. Square Root SAM: Simultaneous Localization and Mapping via Square Root Information Smoothing. *Int. J. Robot. Res.* **2006**, *25*, 1181–1203. [CrossRef]
2. Kaess, M.; Johannsson, H.; Roberts, R.; Ila, V.; Leonard, J.J.; Dellaert, F. iSAM2: Incremental Smoothing and Mapping Using the Bayes Tree. *Int. J. Robot. Res.* **2012**, *31*, 217–236. [CrossRef]
3. Thrun, S.; Montemerlo, M. The GraphSLAM Algorithm With Applications to Large-Scale Mapping of Urban Structures. *Int. J. Robot. Res.* **2005**, *25*, 403–430. [CrossRef]
4. Montemerlo, M.; Thrun, S.; Koller, D.; Wegbreit, B. FastSLAM: A factored solution to the simultaneous localization and mapping problem. In Proceedings of the AAAI National Conference on Artificial Intelligence, Edmonton, AL, Canada, 28 July–1 August 2002; pp. 593–598.
5. Kschischang, F. R.; Frey, B. J.; Loeliger, H. A. Factor Graphs and the Sum-Product Algorithm. *IEEE Trans. Inf. Theory* **2001**, *47*, 498–519. [CrossRef]
6. Cadena, C.; Carlone, L.; Carrillo, H.; Latif, Y.; Scaramuzza, D.; Neira, J.; Reid, I.; Leonard J.J. Past, Present, and Future of Simultaneous Localization And Mapping: Towards the Robust-Perception Age. *IEEE Trans. Robot.* **2016**, *32*, 1309–1332. [CrossRef]
7. Williams, G.; Goldfain, B.; Drews, P.; Rehg, J.M.; Theodorou E.A. Autonomous Racing with AutoRally Vehicles and Differential Games. *arXiv* **2017**, arXiv:1707.04540.

8. Bajcsy, R. Active Perception. *Proc. IEEE* **1988**, *76*, 966–1005. [CrossRef]
9. Leung, C.; Huang, S.; Dissanayake, G. Active SLAM using Model Predictive Control and Attractor Based Exploration. In Proceedings of the IEEE/RSJ International Conference on Intelligent Robots and Systems, Beijing, China, 9–15 October 2006; pp. 5026–5031.
10. Ardón, P.; Kushibar, K.; Peng, S. A Hybrid SLAM and Object Recognition System for Pepper Robot. *arXiv* **2019**, arXiv:1903.00675. Available online: https://github.com/PaolaArdon/Salt-Pepper (accessed on 10 September 2022).
11. Leung, C.; Huang, S.; Kwok, N.; Dissanayake, G. Planning Under Uncertainty using Model Predictive Control for Information Gathering. *Robot. Auton. Syst.* **2006**, *54*, 898–910. [CrossRef]
12. Williams, G.; Drews, P.; Goldfain, B.; Rehg, J.M.; Theodorou, E.A. Aggres-sive driving with model predictive path integral control. In Proceedings of the IEEE International Conference on Robotics and Automation, Jeju Island, Republic of Korea, 27–31 August 2016; pp. 1433–1440.
13. Williams, G.; Aldrich, A.; Theodorou, E.A. Model predictive path integral control: From theory to parallel computation. *J. Guid. Control. Dyn.* **2017**, *40*, 1–14. [CrossRef]
14. AutoRally. Available online: https://github.com/AutoRally/autorally (accessed on 12 July 2022).
15. Rosenblatt, J. K. Optimal selection of uncertain actions by maximizing expected utility. In Proceedings of the IEEE International Symposium on Computational Intelligence in Robotics and Automation, Monterey, CA, USA, 8–9 November 1999; pp. 95–100.
16. Guizilini, V.; Ramos F. Towards real-time 3d continuous occupancy mapping using hilbert maps. *Int. J. Robot. Res.* **2018**, *37*, 566–584. [CrossRef]
17. Daryina, A.N.; Prokopiev, I.V. Unmanned vehicle's control real-time method based on neural network and selection function. *Procedia Comput. Sci.* **2021**, *186*, 217–226. [CrossRef]
18. Hart, P. E.; Nilsson, N. J.; Raphael, B. Correction to A Formal Basis for the Heuristic Determination of Minimum Cost Paths. *Sigart Newsl.* **1972**, *37*, 28–29. [CrossRef]
19. Kuwata, Y.; Teo, J.; Karaman, S.; Fiore, G.; Frazzoli, E.; How, J.P. Motion planning in complex environments using closed-loop prediction. Presented at the AIAA Guidance, Navigation and Control Conference and Exhibit, Honolulu, HI, USA, 18–21 August 2008.
20. Mur-Artal, R.; Tardós, J. D. ORB-SLAM2: an open-source SLAM system for monocular, stereo, and RGB-D cameras. *IEEE Trans. Robot.* **2017**, *33*, 1255–1262. [CrossRef]
21. Drews, P.; Williams, G.; Goldfain, B.; Theodorou, E.A.; Rehg, J.M. Aggressive Deep Driving: Combining Convolutional Neural Networks and Model Predictive Control. *arXiv* **2017**, arXiv:1707.05303.
22. ROS. Available online: https://wiki.ros.org/navigation (accessed on 24 August 2022).

engineering proceedings

MDPI

Proceeding Paper

Real-Time Hardware Identification of Complex Dynamical Plant by Artificial Neural Network Based on Experimentally Processed Data by Smart Technologies †

Valerii I. Kruzhkov [1,2,*], Yuri V. Mitrishkin [1,2] and Eugenia A. Pavlova [1]

1 V.A. Trapeznikov Institute of Control Science of RAS, Moscow 117997, Russia
2 Faculty of Physics, M. V. Lomonosov Moscow State University, Moscow 119991, Russia
* Correspondence: kruzhkov.vi14@physics.msu.ru; Tel.: +7-916-567-2375
† Presented at the 15th International Conference "Intelligent Systems" (INTELS'22), Moscow, Russia, 14–16 December 2022.

Abstract: Artificial neural networks with different structures are used for identification of complex dynamic plant with distributed parameters. The plant is a high-temperature plasma in the spherical Globus-M2 tokamak. Experimental data from it were processed by plasma reconstruction code based on Picard iterations, namely, the Flux-Current Distribution Identification (FCDI) code. This represents smart technology employed to obtain distributed plasma parameters by minimizing the difference between measured and reconstructed signals. An artificial neural network was then applied to identify the data obtained by the FCDI code on the hardware as a real-time testbed realized on a Speedgoat computer. The aim of this repeated identification is to increase the operational response speed in real time in the closed-loop control system of the plasma shape.

Keywords: neural networks; recurrent neural networks; plasma reconstructing

Citation: Kruzhkov, V.I.; Mitrishkin, Y.V.; Pavlova, E.A. Real-Time Hardware Identification of Complex Dynamical Plant by Artificial Neural Network Based on Experimentally Processed Data by Smart Technologies. *Eng. Proc.* **2023**, *33*, 17. https://doi.org/10.3390/engproc2023033017

Academic Editors: Askhat Diveev, Ivan Zelinka, Arutun Avetisyan and Alexander Ilin

Published: 13 June 2023

1. Introduction

An artificial neural network is a powerful and universal algorithm of identification and approximation. It can be used to predict parameters of a complex dynamical plants under control. The advantages of neural network models include low computational complexity and wide possibilies of hardware implementation. From this perspective, it is especially beneficial to employ neural networks with a small number of neurons.

In tokamak experiments, plasma parameters cannot be directly measured because of the high temperature plasma that is able to disable diagnostics. The simplest and most widely used method to detect the boundary and integral parameters of the plasma is the use of magnetic sensors which measure the changes in magnetic field and flux outside the plasma, but within the region of the magnetic field produced by the plasma. The plasma shape is to be identified from the signals of the tokamak diagnostic system. This inverse problem is known as the plasma equilibrium reconstruction problem [1,2]. The FCDI algorithm was developed for plasma equilibrium reconstruction in the Globus-M2 tokamak and uses the Picard iteration approach [3]. The FCDI code is able to obtain the plasma shape offline after cessation of experiment and cannot be used online because of high computational complexity and high task execution time. For real-time use, it is necessary to achieve identification of the FCDI code via a fast model.

There are different approximation methods of a plasma reconstruction algorithm [4]. Neural networks were used for plasma reconstruction in COMPASS tokamak (UK) [5], ASDEX-Upgrade (Germany) [6], D-IIID (USA) [7], and models of ITER [8]. The EFIT reconstruction code was approximated through neural networks by researchers in [9]. In these works, fully connected neural networks with different layer numbers, activation functions, layers sizes were employed.

An artificial neural network was used for identification of a reconstruction algorithm for Globus-M2 in [4,10]. In [10], a neural network was used for identification of moving filaments of the plasma reconstruction approach. In [4], several methods of FCDI algorithm identification were considered including artificial neural networks. The present work offers further development in the use of artificial neural networks for plasma shape prediction in the Globus-M2 tokamak. Neural network structure, optimization of hyperparameters and real-time hardware implementation are considered.

A real-time imitation platform was constructed for working out algorithms of tokamaks plasma control [11]. It is composed of two Speedgoat Performance computers including digital-to-analog and analog-to-digital converters connected by analog feedback. The first computer models the controller with a plasma reconstruction algorithm; the second computer models the tokamak with plasma and actuators. Such an approach provides several opportunities for digital control systems development. In particular, the imitation platform allows us to measure the working time required of the plasma reconstruction algorithm for plasma shape calculation. Calculation of TET of the Globus-M2 plasma shape control system with a plasma reconstruction algorithm on a Speedgaot computer gives less than 100 μs. Consequently, the time for plasma shape calculation should be about 10 μs.

2. Experimental Signals

This paper deals with the spherical Globus-M2 tokamak (Ioffe Institute, Saint-Petersburg, Russia) [12]. The basic parameters of this device are as follows: major radius R = 0.36 m, minor radius a = 0.24 m, aspect ratio R/a = 1.5, toroidal magnetic field, B t max = 1 T, plasma current 0.5 MA, impulse duration less than 0.7 s. Plasma confinement and control are provided by 8 poloidal field coils: horizontal field coil, vertical field coil, central solenoid, 4 poloidal field coils and a correcting coil. Plasma diagnostics consists of measurements of plasma current and 24 magnetic fluxes loops.

A plasma discharge consists of several phases. The phase when plasma is present inside the vacuum vessel and is not in contact with the inner walls is known as the diverter phase. In the Globus-M2 tokamak, a duration of the diverter phase is about 30 ms. For improving plasma confinement and control, it is necessary to use feedback control of the plasma shape. The plasma shape is determined at 6 points on a separatrix. The separatrix is the largest closed equal line of magnetic flux in Figure 1. In this paper, time series from 51 Globus-M2 discharges were processed. The FCDI algorithm was used to obtain the plasma shape in these discharges on the diverter phase. An example from the experimental data can be seen in Figure 2. The time step of the experimental data is 1 ms.

(**a**) (**b**)

Figure 1. (**a**) Globus-M2 tokamak (**b**) Vertical cross-section with poloidal system, diagnostic probes (red dots) and points on separatrix (g1–g6).

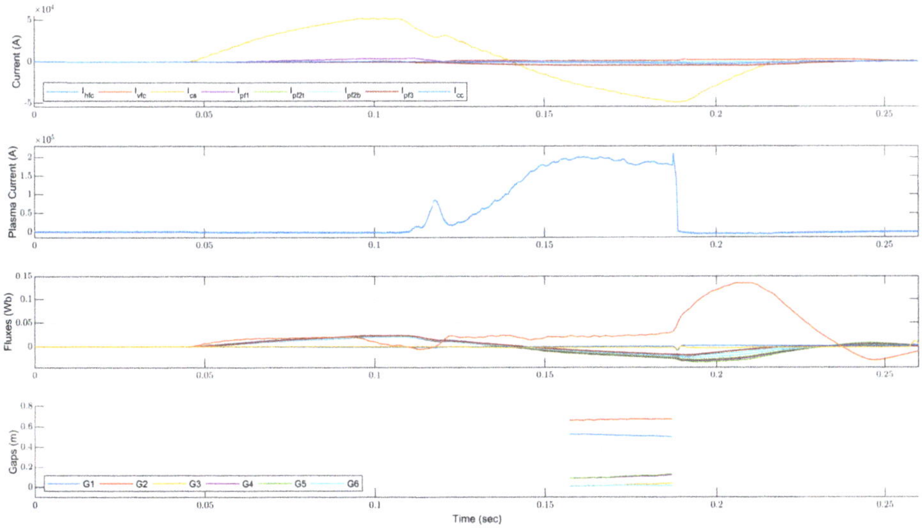

Figure 2. Experimental signals from shot #37255 and Gaps reconstructed by the FCDI algorithm at a divertor phase of the plasma discharge.

3. Artificial Neural Networks

The simplest type of artificial neural network is a single layer perceptron [13]. The signal from inputs multiplied by weights, summarized with bias, and converted by a nonlinear activation function is written as follows:

$$output = activation(W \cdot input + b), \tag{1}$$

3.1. Fully Connected Feedforward Neural Networks

An artificial neural network consisting of a set of fully connected layers is called a fully connected feedforward neural network [14]. In these neural networks, the layers are connected consequentially without loops (Figure 3). Values of x_i are referred to as hidden layers.

$$\begin{aligned} x_1 &= activation_1(W_1 \cdot input + b_1), \\ x_2 &= activation_2(W_2 \cdot x_1 + b_2), \\ &\dots \\ x_{n-1} &= activation_{n-1}(W_{n-1} \cdot x_{n-2} + b_{n-1}), \\ output_{NN} &= W_n \cdot x_{n-1} + b_n; \end{aligned} \tag{2}$$

Regarding the identification of the FCDI algorithm, the neural network has 33 input signals (8 coil currents, plasma current and 24 magnetic fluxes) and 6 output signals (2 strike point coordinates and 4 gaps between the plasma separatrix and the vacuum vessel). In general cases, the sizes the of hidden layers differ, and in the case of the FCDI code identification neural networks with hidden layers of the same size s were considered

$$W_1 \in \mathbb{R}^{33 \times s}, W_2, \dots, W_{n-1} \in \mathbb{R}^{s \times s}, W_n \in \mathbb{R}^{s \times 6}, \tag{3}$$

$$b_1, \dots, b_{n-1} \in \mathbb{R}^{s}, b_n \in \mathbb{R}^{6}. \tag{4}$$

An activation function is a nonlinear function that is needed for separating different hidden layers. In general, activation functions in various layers can differ. In the case of FCDI identification, a neural network with different activation functions but the same in all layers was considered. These are ReLU, LeakyReLU, ELU, Sigmoid, LogSigmoid, Tanh.

Values of n and s and type of the activation function are called hyperparameters of the neural network (Figure 4). It is necessary to obtain optimal values of hyperparameters.

$$
\begin{gathered}
ReLU(x) = \begin{cases} 0 & ,x<0 \\ x & ,x \geqslant 0 \end{cases}, \quad LeakyReLU(x) = \begin{cases} 0.01x & ,x<0 \\ x & ,x \geqslant 0 \end{cases}, \\
ELU(x) = \begin{cases} e^x - 1 & ,x<0 \\ x & ,x \geqslant 0 \end{cases}, \quad Sigmoid(x) = \frac{1}{1+e^{-x}}, \\
LogSigmoid(x) = \log \frac{1}{1+e^{-x}}, \quad Tanh(x) = tanh(x) = \frac{e^x - e^{-x}}{e^x + e^{-x}}.
\end{gathered} \tag{5}
$$

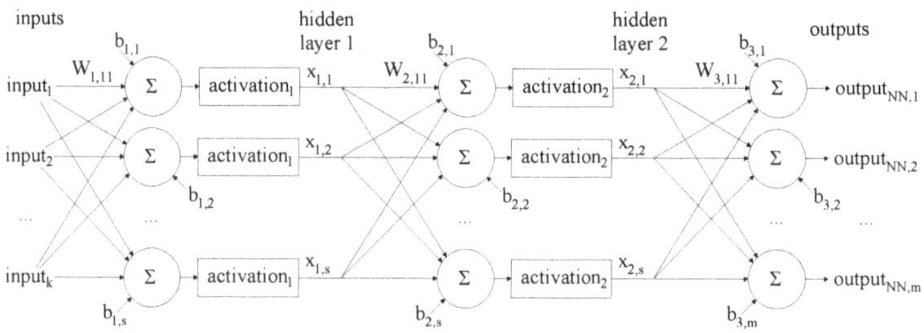

Figure 3. Structure of the fully connected feedforward neural network containing two hidden layers at a size of s. Input size is k, output size is m.

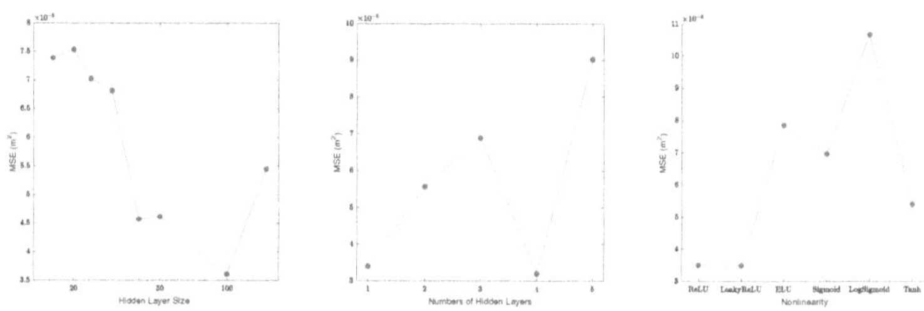

Figure 4. Dependency of accuracy on hyperparameter values of the fully connected feedforward neural network. Accuracy is measured on test data.

Training of the neural network involves finding values of weights W_i and biases b_i to obtain minimum of the loss-function between a real output signal and a signal predicted by the neural network. In case of FCDI identification, the MSE (Mean Square Error) loss is used. This is a multi-parameter non-linear optimization problem. The most widely used algorithm for training artificial neural networks is backpropagation with a gradient method [15]. The main idea rests in calculation of the gradient of the loss function with respect to weights from output to input. A gradient descent with backpropagation does not guarantee finding the global minimum of the loss function, but only a local minimum. For minimizing this problem, a stochastic gradient method with momentum is used. Furthermore, normalization of signals was made to improve performance of the backpropagation algorithm. Training of the neural network for the FCDI identification was achieved using the Adam [16] algorithm in PyTorch. The experimental data were divided into two parts: 46 training shots and 5 testing shots. Thus, the size of training data is 1795 points and the size of the test data is 145 points. Results are shown in Figure 5.

$$loss(output_{NN}, output_{FCDI}) \xrightarrow[W_i, b_i]{} \min, \tag{6}$$

$$loss(x,y) = MSE(x,y) = \sum_{i=1}^{N} (x_i - y_i)^2 / N. \tag{7}$$

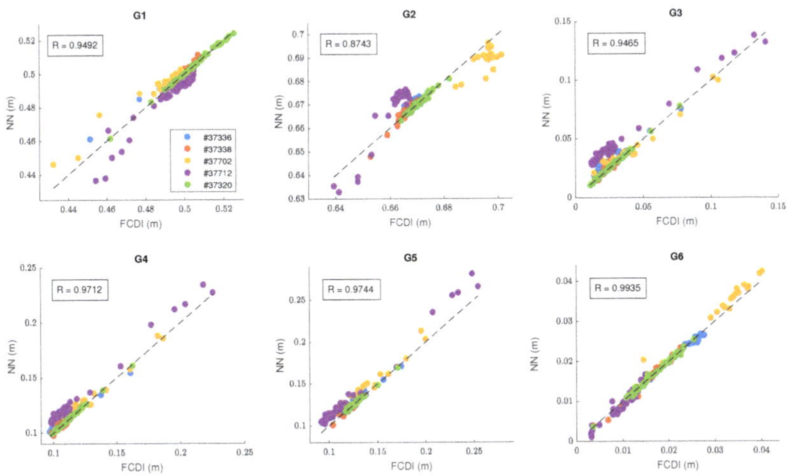

Figure 5. Results of the FCDI identification by the fully connected feedforward neural network with one hidden layer of the size of 100 neurons and the LeakyReLU activation function. In total, 5 test signals from Globus-M2 shots #37336, #37338, #37702, #37712 and #37320 are shown.

3.2. Recurrent Neural Networks

The feedforward neural network only uses a current input signal for calculation of the output signal. In the case of dynamic system identification, it could be useful to apply values from previous time moments. Recurrent neural networks use an additional hidden layer h. The values of the hidden layer h_{t-1} on the time step $t-1$ uses for calculation next step h_t. Accuracy of RNN in dependence of hyperparameters is shown in Figure 6.

$$
\begin{gathered}
h_t = activation(W_1 \cdot input + W_h \cdot h_{t-1} + b_1), \\
output = W_2 \cdot h_t + b_2; \\
W_1 \in \mathbb{R}^{33 \times s}, W_2 \in \mathbb{R}^{s \times 6}, W_h \in \mathbb{R}^{s \times s}, \\
b_1 \in \mathbb{R}^{s}, b_2 \in \mathbb{R}^{6}.
\end{gathered} \tag{8}
$$

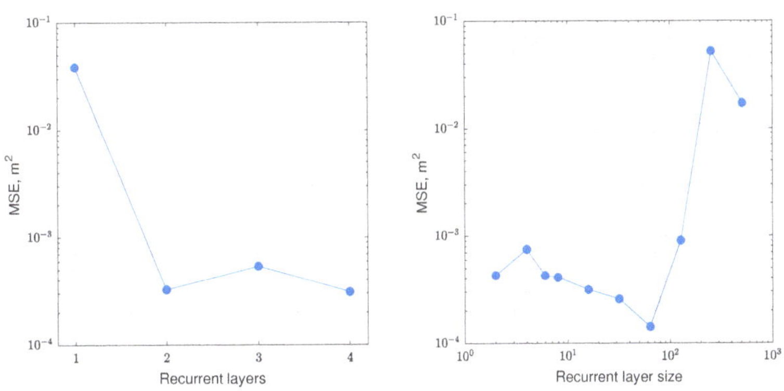

Figure 6. Dependency of accuracy on hyperparameter values of the recurrent neural network. Accuracy is measured on test data.

3.3. Implementation of Neural Network Models in Real-Time Hardware

The imitation platform with Speedgoat computers is based on a SimulinkRT operation system. Real-time test bed for plasma control in tokamaks consists of two Speedgoat performance real-time target machines that are connected in feedback: one computer plays the role of the controlled plant model and the other one acts as the MIMO controller. PyTorch models were transferred to Matlab with the help of SciPy library. The neural networks with different structures were testified in real-time regime (Figure 7). Each neuron is connected with all neurons of adjacent layers. Because of this, the computation complexity of one layer of the neural network is proportional to the square of the layer size.

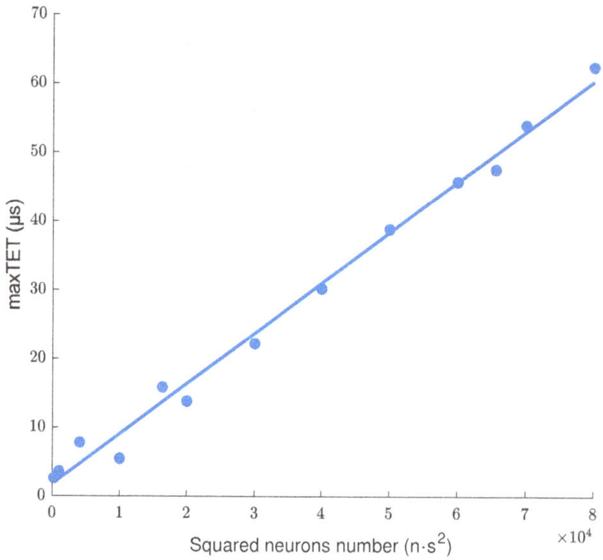

Figure 7. Dependency of Task Execution Time (TET) on number of neurons.

$$TET(n,s) = a \cdot n \cdot s^2 + b,\ a = 7.33 \times 10^{-10}\ \text{s},\ b = 1.79 \times 10^{-6}\ \text{s} \tag{9}$$

Plasma shape control in Globus-M2 requires plasma-reconstruction works in about 10 μs. Consequently, for real-time usage neural networks containing 100–150 neurons are applicable.

4. Conclusions

The artificial neural network model of the FCDI plasma reconstruction code was obtained along with an optimal structure and hyperparameters. The optimal model for FCDI algorithm identification is a fully connected feedforward neural network with one hidden layer and LeakyReLU activation function. Modeling on the imitation platform showed the possibility of using the neural network model for real-time high-performance control systems, with results demonstrating a Task Execution Time of 4.4 μs and a Mean Square Error of 3.4×10^5 m^2.

Author Contributions: Conceptualization, Y.V.M.; methodology, Y.V.M.; funding acquisition, Y.V.M.; software, V.I.K.; supervision, Y.V.M.; validation, V.I.K.; visualization, V.I.K. and E.A.P.; writing—original draft preparation, V.I.K., Y.V.M. and E.A.P.; writing—review and editing, Y.V.M. and E.A.P. All authors have read and agreed to the published version of the manuscript.

Funding: This research was sponsored by the Russian Science Foundation (RSF), Grant No. 21-79-20180.

Institutional Review Board Statement: Not applicable.

Informed Consent Statement: Not applicable.

Data Availability Statement: The data used in this paper are not currently publicly accessible.

Conflicts of Interest: The authors declare no conflict of interest.

References

1. Mitrishkin, Y.V.; Korenev, P.S.; Prokhorov, A.A.; Kartsev, N.M.; Patrov, M.I. Plasma control in tokamaks. Part 1. Controlled thermonuclear fusion problem. Tokamaks. Components of control systems. *Adv. Syst. Sci. Appl.* **2018**, *18*, 26–52. [CrossRef]
2. Mitrishkin, Y.V.; Kartsev, N.M.; Kuznetsov, E.A.; Korostelev, A.Y. *Metodi i Sistemi Magnitnogo Upravleniya Plazmoi vs. Tokamakakh [Methods and Systems of Plasma Magnetic Control in Tokamaks]*; KRASAND: Moscow, Russia, 2020. (In Russian)
3. Korenev, P.S.; Mitrishkin, Y.V.; Patrov, M.I. Rekonstruktsiya ravnovesnogo raspredeleniya parametrov plasmi tokamaka po vneshnim magnitnim izmereniyam i postroyeniye lineynikh plazmennikh modeley [Reconstruction of equilibrium distribution of plasma parameters on the base of external magnetic measurements and construction of plasma linear models]. *Mechatron. Autom. Control* **2016**, *17*, 254–265. (In Russian)
4. Mitrishkin Yuri, V.; Korenev Pavel, S.; Konkov Artem, E.; Kruzhkov Valerii, I.; Ovsyannikov Nicolai, E. New Identification Approach and Methods for Plasma Equilibrium Reconstruction in D-Shaped Tokamaks. *Mathematics* **2022**, *10*, 40. [CrossRef]
5. Bishop, C.M.; Haynes, P.S.; Smith, M.E.U.; Todd, T.N.; Trotman, D.L. Real-Time Control of a Tokamak Plasma Using Neural Networks. *Neural Comput.* **1995**, *7*, 206–217. [CrossRef]
6. Coccorese, E.; Morabito, C.; Martone, R. Identification of noncircular plasma equilibria using a neural network approach. *Nucl. Fusion* **1994**, *34*, 1349–1363. [CrossRef]
7. Lister, J.B.; Schnurrenberger, H. Fast non-linear extraction of plasma equilibrium parameters using a neural network mapping. *Nucl. Fusion* **1991**, *31*, 1291–1300. [CrossRef]
8. Albanese, R.; Coccorese, E.; Gruber, O.; Martone, R.; McCarthy, P.; Morabito, F.C. Identification of Plasma Equilibria in ITER from Magnetic Measurements Via Functional Parameterization and Neural Networks. *Fusion Technol.* **1996**, *30*, 219–236. [CrossRef]
9. Joung, S.; Kim, J.; Kwak, S.; Bak, J.G.; Lee, S.; Han, H.; Ghim, Y. Deep neural network Grad-Shafranov solver constrained with measured magnetic signals. *Nucl. Fusion* **2019**, *60*, 016034. [CrossRef]
10. Prokhorov, A.; Mitrishkin, Y.; Korenev, P.; Patrov M. *The Plasma Shape Control System in the Tokamak with the Neural Network as a Plasma Equilibrium Reconstruction Algorithm*; Elsevier Ltd.: London, UK, 2020; pp. 857–862. [CrossRef]
11. Mitrishkin, Y.V. Plasma magnetic control systems in D-shaped tokamaks and imitation digital computer platform in real time for controlling plasma current and shape. *Adv. Syst. Sci. Appl. Int. Inst. Gen. Syst. Stud.* **2022**, *22*, 1–15.
12. Minaev, V. B.; Gusev, V.K.; Sakharov, N.V.; Varfolomeev, V.I.; Bakharev, N.N.; Belyakov, V.A.; Bondarchuk, E.N.; Brunkov, P.N.; Chernyshev, F.V.; Davydenko, V.I.; et al. Spherical tokamak Globus-M2: Design, integration, construction. *Nucl. Fusion* **2017**, *57*, 066047. [CrossRef]
13. Rosenblatt, F. The perceptron: A probabilistic model for information storage and organization in the brain. *Psychol. Rev.* **1958**, *65*, 386. [CrossRef]
14. Schmidhuber, J. Deep learning in neural networks: An overview. *Neural Netw. Off. J. Int. Neural Netw. Soc.* **2015**, *61*, 85–117. [CrossRef] [PubMed]
15. Goodfellow, I.; Bengio, Y.; Courville, A. Deep learning. *Genet. Program. Evolvable Mach.* **2018**, *19*, 305–307. [CrossRef]
16. Kingma, D.; Ba, J. Adam: A Method for Stochastic Optimization. In Proceedings of the International Conference on Learning Representations, Banff, AB, Canada, 14–16 April 2014.

engineering proceedings

MDPI

Proceeding Paper

Improving Collaborative Robotic Complex Efficiency: An Approach to the Intellectualization of the Control System †

Mikhail Gorkavyy, Yuri Ivanov, Sergey Sukhorukov, Sergey Zhiganov, Makrel Melnichenko, Alexander Gorkavyy and Daniil Grabar *

Komsomolsk-na-Amure State University, Komsomolsk-na-Amur 681013, Russia
* Correspondence: gorbat308@yandex.ru; Tel.: +7-4217-241-139; Fax: +7-4217-536-150

† Presented at the 15th International Conference "Intelligent Systems" (INTELS'22), Moscow, Russia, 14–16 December 2022.

Abstract: This paper proposes an original approach to improving the efficiency of technological processes based on collaborative robots, which differs from the existing ones by the possibility of intensifying the process of modeling the human factor when forming a control law. A structural and functional diagram of the model of a standard cobot control system in the basic industrial configuration is presented. The shortcomings of a standard solution for the formation of laws for controlling the movement of a cobot in a nondeterministic environment in the same workspace with a person are demonstrated. Structural and functional solutions are proposed to outline a strategy for increasing the degree of synergistic effect of human–machine interaction. The effect can be achieved through the introduction of an extended system of sensors and analytics and an intelligent module robot trajectory movement formation and optimization under disturbing influences. The results of the comparison between the standard control system of the cobot and the prototype of the intelligent system are presented. As an example, the operation of the implementation of collision avoidance that occurs due to the appearance of a stationary object in the working area is given. The results obtained demonstrate a significant time- and energy-saving effect (from 15% to 182% depending on the operation) in the case of using an intelligent control system. A feature of the proposed approach is to strengthen the integration links of intelligent analysis and optimization modules, which allow real-time multimodal processing of sensory data and environmental modeling to predict human actions and form cobot reactions.

Keywords: collaborative robots; approach; intelligent control system; energy efficiency; optimization; image analysis; decision making

Citation: Gorkavyy, M.; Ivanov, Y.; Sukhorukov, S.; Zhiganov, S.; Melnichenko, M.; Gorkavyy, A.; Grabar, D. Improving Collaborative Robotic Complex Efficiency: An Approach to the Intellectualization of the Control System. *Eng. Proc.* **2023**, 33, 18. https://doi.org/10.3390/engproc2023033018

Academic Editors: Askhat Diveev, Ivan Zelinka, Arutun Avetisyan and Alexander Ilin

Published: 13 June 2023

1. Introduction

In the context of the evolutionary transition from the concept of Industry 4.0 to Industry 5.0 in modern industrial enterprises, it becomes necessary to intensify the processes of organizing production on the basis of collaborative robots [1], which have proven themselves in solving not only general industrial problems [2–5] but also in the processes of assisting a person in the piece and small-scale production of innovative products [6] in various industries: mechanical assembly, medicine, electronics production [7], etc.

There is a sufficient number of models of collaborative robots from the world's leading manufacturers, such as KUKA, ABB, FANUC, Kawasaki, and Yaskawa [8], that can be integrated into enterprise processes. However, in the basic configuration, collaborative robots supplied to enterprises cannot unleash the potential of a possible synergistic effect of interaction between a person and a robot, and, as a rule, the collaborative property is reduced only to ensuring the safety of a person in the same workspace with a robot inside a cyberphysical system. The synthesis of control systems for a collaborative process only on the basis of a robot equipped with an internal sensor system (a complex of force–torque sensors), which makes it possible to implement mechanisms for organizational and technical avoidance of collisions between the robot and scene objects, is difficult due to the limitations of these means of measuring and evaluating the internal environment in the working space of the robot. The resulting solutions will not be highly efficient due to the impossibility of carrying out control according to the forecast, while control by mistake will entail additional time and energy costs and will not guarantee the achievement of the desired result in principle (bringing the tool center point (TCP) of the robot to the position required by the technological operation). Improving the efficiency of control systems for cyberphysical systems (CPS) based on cobots is possible due to the expansion of sensory tools (in particular vision systems) [9] and intelligent modules that plan and optimize the trajectory movements of a robot taking into account changes in the external environment [1,10].

In this paper, the authors propose an approach to the formation of structural and functional models of a collaborative robotic complex equipped with an intelligent control system with the possibility of self-learning and additional training in order to develop multimodal adaptive algorithms and methods for controlling the behavior of collaborative robotic systems, taking into account emergency situations and extreme conditions in a non-deterministic environment, as well as an approximate assessment of the economic potential of solutions for expanding the control system of the KUKA LBR iiwa 7 collaborative robot, using the example of a typical technological operation in the event of a collision.

In carrying out the study, the authors used the basic approaches of control theory and elements of mathematical methods of the vector–matrix description of control systems and methods of system analysis, including methods of structural and functional decomposition; elements of the proposed solutions are confirmed on the basis of the results of an experiment performed on industrial equipment from KUKA Robotics.

2. An Approach to the Intellectualization of the Control System

Figure 1 shows a block diagram of the control of a cyberphysical system (CPS) consisting of a decision maker (HS), a collaborative robot (CS) and a library of behavior models (BMs) inside a standard robot control system.

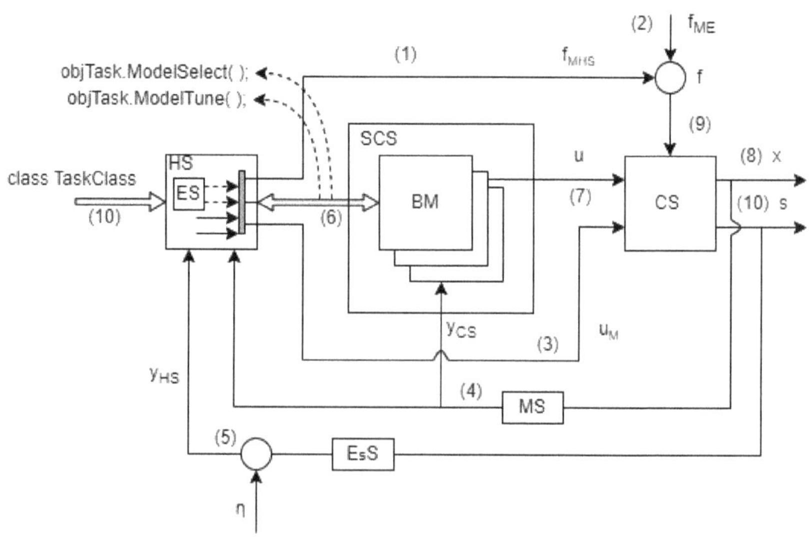

Figure 1. Structural and functional diagram of the CPS control system.

A typical CPS operates in two main modes: implementation of a technological process and debugging of a technological process (performing individual operations, generating a program code, setting up a strategy for responding to data from the internal sensor system (CR) of a collaborative robot).

The vector–matrix description of the linear part [11] of the cyberphysical system control system can be presented below (for some cases, it can also include additional methods such as [12,13]).

$$\begin{cases} \dot{\boldsymbol{x}}(t) = \boldsymbol{A} \cdot \boldsymbol{x}(t) + \boldsymbol{B}_{u_1} \cdot \boldsymbol{u}(t) + \boldsymbol{B}_{u_2} \cdot \boldsymbol{u}_M(t) + \boldsymbol{B}_{M_1} \cdot \boldsymbol{f}_{MHS}(t) + \boldsymbol{B}_{M_2} \cdot \boldsymbol{f}_{ME}(t) \\ \boldsymbol{y}_{CS}(t) = \boldsymbol{C}_{MS} \cdot \boldsymbol{x}(t) \\ \boldsymbol{y}_{HS}(t) = \boldsymbol{C}_{EsS} \cdot \boldsymbol{s}(t) + \boldsymbol{\eta}(t) \end{cases} \quad (1)$$

$$Procedure : TaskClass => < objTask, objTaskModelTune(\ldots), objTaskModelSelect(\ldots) > \quad (2)$$

where $\boldsymbol{A}$—functional matrix of the state of the object (power block);
$\boldsymbol{B}_{u_1}, \boldsymbol{B}_{u_2}$—control matrix;
$\boldsymbol{B}_{M_1}, \boldsymbol{B}_{M_2}$—disturbance matrices;
$\boldsymbol{\eta}(t)$—noise vector;
$\boldsymbol{y}_{CS}(t), \boldsymbol{y}_{HS}(t)$—measure vectors.

$$\boldsymbol{x}(t) = [x_1(t), x_2(t)\ldots x_7(t)]^T, \text{ where } \boldsymbol{x}_i(t) = \begin{bmatrix} \varphi_i \\ \omega_i \\ i_i \\ e_i \\ \vdots \end{bmatrix},$$

$\boldsymbol{u}(t) = [u_1(t), u_2(t)\ldots u_7(t)]^T$—control law;
$\boldsymbol{u}_M(t) = [u_{M_1}(t), u_{M_2}(t)\ldots u_{M_7}(t)]^T$—torque;
$\boldsymbol{f}_{MHS}(t) = [f_{MHS_1}(t), f_{MHS_2}(t)\ldots f_{MHS_7}(t)]^T$—torque;
$\boldsymbol{f}_{ME}(t) = [f_{ME_1}(t), f_{ME_2}(t)\ldots f_{ME_7}(t)]^T$—torque;
$\boldsymbol{s}_{CS}(t) = [s_{CS_1}(t), s_{CS_2}(t)\ldots s_{CS_n}(t)]^T$,

where $n \to \infty$ (all points of the robot surface) $s_{CSi}(t) = \begin{bmatrix} X_{CS_i} \\ Y_{CS_i} \\ Z_{CS_i} \end{bmatrix}$.

In accordance with the task, which is a set of documentation describing the technological process, HS selects a preinstalled M_i model, if any, or generates a new one via channel (6). The model includes a set of movement trajectories CS, speed modes, types of movement, levels of activation of the (internal) sensory system and algorithms for responding to activation in the event of a signal (9). CS implements movements according to u (channel (7)) generated by BM. The values of the state variables x (channel (8)) of the collaborative system are available to both the SCS and HS (channel (4)) measurement system (MS). In addition, some of the state variable CSs are approximately estimated using the natural human senses (vision, touch and hearing) via the channel (5) estimate system (EsS). This feedback option, in addition to monitoring the general state of the process, is used when forming/correcting a model from BMs by physical impact on the robotic arm (channel (3)) and moving it to the desired point while maintaining its coordinates in the model. The internal sensor system and algorithms for responding to sensors in BMs allow CSs to "compensate" for disturbing influences $f = f_{ME} + f_{MHS}$ within a limited range. With the standard configuration of the CS, the response to disturbance occurs "by mistake", and the built-in sensor system in the online mode does not allow forming an idea of an obstacle in statics, or, moreover, in dynamics. These facts do not allow the CPS control system to guarantee the restoration of the motion algorithm after the appearance of a disturbance. In addition, such a system, in which the role of the control unit is performed by a person, is difficult to optimize, especially according to a complex system of criteria, for example, including speed and energy efficiency.

Moreover, an important factor that reduces the efficiency of the CPS (Figure 1) is the possibility of generating a disturbance of the HS itself (channel (1)), both in the process debugging mode and in the execution mode.

The presented problems and limitations inherent in the standard CPS, which are presented in most of the volume of collaborative systems of industrial enterprises, do not allow building a full-fledged synergistic system [6] or organizing effective human–machine interaction in an innovative technological process, taking into account emergency situations in a nondeterministic environment.

The most promising solution for improving the CPS, according to [1,14], is to add to the typical CPS system (Figure 1) the means of "feeling" the collaborative machine: an extended (multimodal) sensor system [9] and advanced intelligent algorithms as part of distributed system of analysis and control [1].

This research by the authors is aimed at the synthesis of control laws within the CPS and aimed at changing the values of state variables not only of CSs but also of HS, through the formation of advice or instructions. At the same time, it is proposed to associate the interface of an intelligent control system with CS, thereby creating the illusion of sensing and animating a collaborative machine [15] to bot.

The approach to control is proposed to be implemented according to the structural and functional diagram compiled by the authors (Figure 2), which expands the diagram of Figure 1 with an intelligent control system as part of a subsystem for collecting, recognizing and primary data analysis, a subsystem for generating an intelligent logical inference of a control law and a subsystem for organizing an impact on HS, including interface.

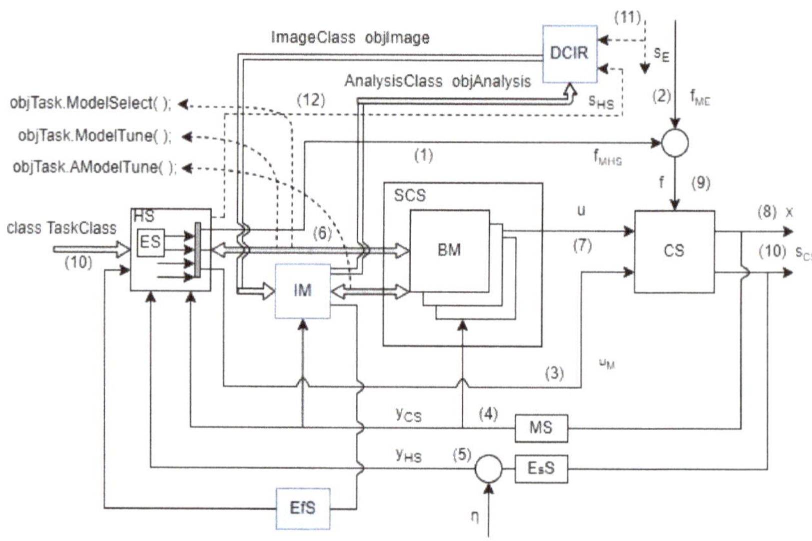

Figure 2. Structural and functional diagram of an intelligent control system for a cyberphysical system.

When forming the control law, the intelligent system relies on the optimization mechanisms of the intelligent module (IM), among which the key ones are minimizing the length of movement trajectories, minimizing the execution time of a technological operation, minimizing the number of avoidance of collisions, minimizing energy consumption [16–18] and others, as well as their combinations within a weighted system of criteria.

A description of the functions of the presented blocks as part of an intelligent control system for a cyberphysical system is presented below:

(1) IM—Dynamic synthesis of the CPS performance criteria system based on the selected BM prototype and data from the Data Collect and Image recognizing (DCIR) block (environmental changes), prediction of the CPS state change, BM adjustment/synthesis, resulting change u, DCIR control and formation of prescriptions and scenarios for an emotional system (ES).

(2) DCIR—Data collection of video and audio streams, recognition of key points of the operator (HS) and other objects in the 3D space of the working area and prediction of the direction vector (displacement) of key points.

(3) Effector system (EsS)—Implementation of the impact on HS through sound, graphic, tactile or a combination of these effects in the physical and virtual interfaces (VR/AR); the impact is implemented both in the mode of information and in the mode of recommendations for actions, for example, in the AR instruction format. There should also be a regime for archiving data and providing it to a higher level of management in order to control the behavior of HS.

Thanks to the introduction of an intelligent system with external sensors, two channels become available for collecting and analyzing information—(11) and (12)—which determines the geometry of physical objects (an operator from HS and any external objects) that cause or may create disturbances in the future f_{MHS} and fme, respectively. The DCIR module provides the IM with arrays of key points of objects, as well as a prediction of their displacement vector, thereby allowing the intelligent module, taking into account the knowledge of the CS behavior model and its current state, to quickly correct the behavior model or switch it. Therefore, the system implements control by prediction or in a hybrid mode: a combination of a series of control actions by error and by prediction (if collisions cannot be avoided).

In addition, due to the simultaneous availability of channel (12) with the archive of HS action patterns in the IM, it seems possible to detect anomalies in human behavior that are counterproductive, generating (causing) f_{MHS} disturbance through channel (1), caused by the emotional component of the ES system HS. Thus, the DCIR–IM–EFS interaction can be eliminated, or the influence of f_{MHS} significantly reduced, by eliminating the perturbation generator itself in HS via channel (12) or by promptly correcting the model in SCS and forming a new control law (channel (7)), correcting the state CS even before the triggering of its standard sensor system.

In addition to solving the problems of avoiding collisions, the IM must take into account the efficiency criteria for CS operation established in BMs, including models for increasing energy efficiency and models for minimizing the time of execution of a technological operation.

In order to justify the technical and economic feasibility of promising work, an experiment was conducted on the basis of a draft prototype of an intelligent system for planning the trajectory movements of an industrial robot equipped with force–torque sensors. The experiment was aimed at identifying differences between the values of the integral indicators of the technological process (the possibility of avoiding a collision, the total duration of the operation, taking into account the time of avoiding a collision and the energy consumed in this case) of positioning the tool by a robot under the control of a system built according to the scheme in Figure 1—option 1, as well as the system corresponding to the scheme in Figure 2—option 2.

Figure 3a shows the trajectory of the tool movement by the robot in space (option 1) in the absence of an obstacle on the path of movement; the values of the execution time and energy consumed obtained in this case are presented in the first column of Table 1. Figure 3b shows the trajectory of the tool movement by the robot in space (option 1) under the conditions of the occurrence of a static obstacle of an unknown shape (disturbance f_{ME} (Figure 2)) and the system working off the collision by the "probing" method; the values of the execution time and energy consumed are presented in column 2 of the Table 1. Figure 3c shows the trajectory of the tool movement by the robot in space (option 2) under the conditions of the occurrence of a static obstacle of unknown shape (disturbance f_{ME}

(Figure 2)), recognized by the DCIR module and working out by the collision avoidance system based on the planned movement trajectory by the IM; the resulting execution times and consumed energy are presented in columns 3–7 in Table 1. It should be noted that the trajectories were obtained using only one criterion—collision avoidance. It can be seen from the data that the operation execution time and the energy consumed at the same time are different for the five implementations in option 2, which demonstrates the possibility and expediency of connecting to the problem of avoiding collisions, criteria for ensuring reduction in operation time, energy efficiency, or their weighted combinations. It is also worth noting that option 1 cannot always provide avoidance of collisions and may cause (in the case of standard behavior algorithms of the standard system) the activation of the standby mode, which will increase the execution time of the operation for an indefinite period, or an emergency stop (according to signals from current sensors).

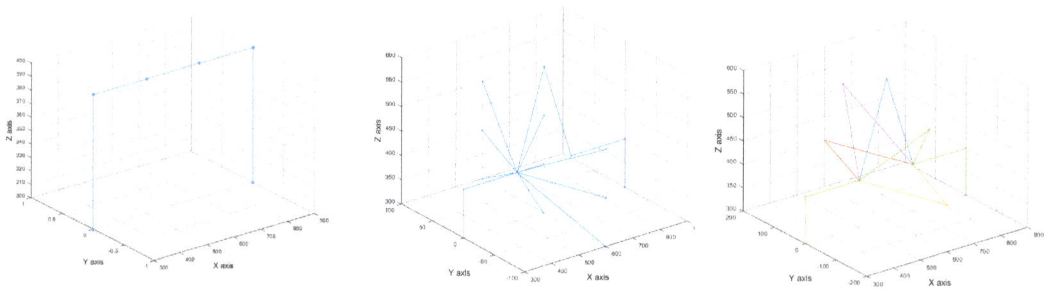

(**a**) the operator sets how to get around the obstacle (**b**) automatic search for the passage of obstacles (**c**) using computer vision to pass an obstacle

Figure 3. Trajectories of robot TCP movement.

Table 1. Experimental data.

	Trajectory Complex 1	**Trajectory Complex 2**	**Trajectory Complex 3**	**Trajectory Complex 4**	**Trajectory Complex 5**	**Trajectory Complex 6**	**Trajectory Complex 7**
Characteristics of the technological operation							
Control system variant number	1	1	2	2	2	2	2
The presence of a disturbance	not existing	existing	existing	existing	existing	existing	existing
Operation execution time, s	17.09	48.055	20.907	20.834	20.836	20.919	20.918
Consumed energy, W·h	1.251	3.532	1.524	1.513	1.524	1.501	1.530

The obvious (Figure 4) difference between the results obtained during the operation of a standard control system and an expanded one (more than twice as good results in energy and time) allows us to conclude that it is expedient to expand standard collaborative robots with external intelligent systems, which will positively affect the efficiency of the production process.

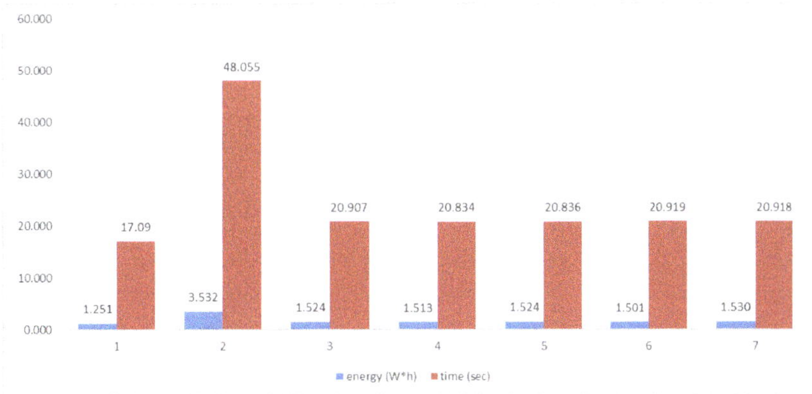

Figure 4. The results of measurements of the time of passage by the robot from point A to point B.

3. Conclusions

The approach proposed by the authors to the intellectualization of industrial cyber-physical systems conceptually embeds the main promising areas of research of the authors: the development of algorithms for multimodal analysis of the scene (space of the working area) in real time, including the detection, classification and prediction of the behavior of the operator and objects—the functionality of the DCIR block; development of methods for the formation of CS movement trajectories and their operational correction based on the results of the analysis of external disturbances (learning algorithms)—the functionality of the IM block—and development of a dialogue and effector system that increases the overall efficiency of CPS by influencing HS—the functionality of the EFS block—into the overall CPS control system, presented in the form of a structural–functional diagram (Figure 2).

The results of a simplified experiment based on a simulation prototype of an intelligent system demonstrate significant physical and economic effects (from 15% to 182% depending on the operation) that can be obtained under the conditions of production processes when standard collaborative solutions are completed with an extended sensor system and intelligent modules for optimizing trajectory movements, according to the proposed approach, which determines the possibility of assuming the feasibility of promising research in the field of improving intelligent scene methods and optimizing the movement of collaborative robots in a nondeterministic environment.

In the future, it is planned to produce a seminatural experiment on the industrial collaborative system KUKA LBR iiwa 7 R800 using moving objects that have a disturbing effect. A promising area of research is the implementation of the DCIR module using neural network algorithms for pattern recognition.

Author Contributions: Conceptualization, M.G. and Y.I.; investigation, M.M.; methodology, M.G. and Y.I.; software, D.G.; validation, S.S., S.Z. and M.M.; writing—original draft preparation, M.G.; writing—review and editing, A.G. and D.G. All authors have read and agreed to the published version of the manuscript.

Funding: The work was supported by the Russian Science Foundation (project No. 22-71-10093).

Institutional Review Board Statement: Not applicable.

Informed Consent Statement: Informed consent was obtained from all subjects involved in the study.

Data Availability Statement: Not applicable.

Conflicts of Interest: The authors declare no conflict of interest.

References

1. Giberti, H.; Abbattista, T.; Carnevale, M.; Giagu, L.; Cristini, F. A Methodology for Flexible Implementation of Collaborative Robots in Smart Manufacturing Systems. *Robotics* **2022**, *11*, 9. [CrossRef]
2. Gorkavyi, M.A.; Gorkavyi, A.I.; Gudim, A.S.; Melnichenko, M.A.; Egorova, V.P. Intelligent robotic systems implementation in the shipbuilding enterprise processes production. *Mar. Intell. Technol.* **2021**, *2*, 97–104.
3. Frolov, A.V. The experience of the portal machine control system modernizing for automatic arc welding and surfacing in a shielded gas. *Lect. Notes Netw. Syst.* **2021**, *200*, 622–631.
4. Frolov, A.V. Installation of automatic welding of elements of ship pipelines. *Mar. Intellect. Technol.* **2021**, *2*, 91–96.
5. Liu, Z.; Liu, Q.; Xu, W.; Wang, L.; Zhou, Z. Robot learning towards smart robotic manufacturing: A Review. *Robot. Comput. Integr. Manuf.* **2022**, *77*, 102360. [CrossRef]
6. Babkin, A.V.; Fedorov, A.A.; Liberman, I.V.; Klachek, P.M. Industry 5.0: Concept, formation and development. *Econ. Ind.* **2021**, *14*, 375–395. [CrossRef]
7. Grigore, L.S.; Priescu, I.; Joita, D.; Oncioiu, I. The integration of collaborative robot systems and their environmental impacts. *Processes* **2020**, *8*, 494. [CrossRef]
8. Soida, M.; Żak, J.; Bydoń, S. Structural Analysis of 6R Robotic Arm. Comparison of Different Complexity Models. *Adv. Intell. Syst. Comput.* **2021**, *1336*, 269–278.
9. Cheng, Y.; Sun, L.; Liu, C.; Tomizuka, M. Towards efficient human-robot collaboration with robust plan recognition and trajectory prediction. *IEEE Robot. Autom. Lett.* **2020**, *5*, 2602–2609. [CrossRef]
10. Sun, X.; Zhang, R.; Liu, S.; Lv, Q.; Bao, J.; Li, J. A digital twin-driven human–robot collaborative assembly-commissioning method for complex products. *Int. J. Adv. Manuf. Technol.* **2022**, *118*, 3389–3402. [CrossRef]
11. Gorkavyy, A.I.; Gorkavyy, M.A.; Melnichenko, M.A.; Berkh, A.V. Improving the functioning of the cybernetic control system of the mechatronic module of the robotic complex. *Lect. Notes Netw. Syst.* **2021**, *3*, 702–709.
12. Cherniy, S.P.; Buzikayeva, A.V.; Gudim, A.S. A model of multi-cascade fuzzy logic controller implemented using different variations of inference algorithms. In Proceedings of the International Multi-Conference on Industrial Engineering and Modern Technologies, FarEastCon 2019, Vladivostok, Russia, 1–4 October 2019; pp. 91–96.
13. Savelyev, D.O.; Gudim, A.S.; Solovev, D.B. Stabilizing the transients in the objects and systems controlling the compensation of nonlinear acs (automatic control system) elements. In Proceedings of the International Science and Technology Conference EastConf, Vladivostok, Russia, 1–2 March 2019; pp. 576–579.
14. Calvo, R.; Gil, P. Evaluation of Collaborative Robot Sustainable Integration in Manufacturing Assembly by Using Process Time Savings. *Materials* **2022**, *15*, 611. [CrossRef] [PubMed]
15. Reyes, M.E.; Meza, I.V.; Pineda, L.A. Robotics facial expression of anger in collaborative human–robot interaction. *Int. J. Adv. Robot. Syst.* **2019**, *16*, 1–13. [CrossRef]
16. Yin, S.; Ji, W.; Wang, L. A machine learning based energy efficient trajectory planning approach for industrial robots. *Procedia CIRP* **2019**, *81*, 429–434. [CrossRef]
17. Palomba, I.; Wehrle, E.; Carabin, G.; Vidoni, R. Minimization of the energy consumption in industrial robots through regenerative drives and optimally designed compliant elements. *Appl. Sci.* **2020**, *10*, 7475. [CrossRef]
18. Soriano, L.A.; Rubio, J.J.; Orozco, E.; Cordova, D.A.; Ochoa, G.; Balcazar, R.; Cruz, D.R.; Meda-Campaña, J.A.; Zacarias, A.; Gutierrez, G.J. Optimization of sliding mode control to save energy in a SCARA robot. *Mathematics* **2021**, *9*, 3160. [CrossRef]

engineering proceedings

MDPI

Proceeding Paper

Using an Ensemble of Deep Neural Networks to Detect Human Keypoints in the Workspace of a Collaborative Robotic System †

Yuri Ivanov *, Sergey Zhiganov, Mikhail Gorkavyy, Sergey Sukhorukov and Daniil Grabar

Department of Energy and Management, Komsomolsk-na-Amure State University, 681013 Komsomolsk-na-Amur, Russia

* Correspondence: ivanov_ys@icloud.com; Tel.: +7-4217-241-139; Fax: +7-4217-536-150

† Presented at the 15th International Conference "Intelligent Systems" (INTELS'22), Moscow, Russia, 14–16 December 2022.

Abstract: It is suggested that the use an ensemble of deep neural networks can determine the spatial position of the operator using keypoints with a multicamera sensor system. The advantage of the algorithm is the use of a multicamera system that allows keypoints to be linked to the local coordinate system of an industrial robotic complex. The testing of this work was made on the basis of modern embedded computing hardware and software. The effectiveness of the proposed approach is demonstrated even when only a subset of key points is found in the frame, as well as when they partially overlap. A software module in Python has been developed for detecting and localizing key points of the operator and industrial manipulator. The proposed approach will make it possible to plan the robot's trajectories for the safe execution of joint operations in one workspace. The developed algorithm will be used to predict the operator's actions in the workspace and detect abnormal situations and possible intersections in the trajectories of the collaborative robot.

Keywords: detection; recognition; classification; human pose estimation; deep neural network; video stream; keypoint detection

Citation: Ivanov, Y.; Zhiganov, S.; Gorkavyy, M.; Sukhorukov, S.; Grabar, D. Using an Ensemble of Deep Neural Networks to Detect Human Keypoints in the Workspace of a Collaborative Robotic System. *Eng. Proc.* **2023**, *33*, 19. https://doi.org/10.3390/engproc2023033019

Academic Editors: Askhat Diveev, Ivan Zelinka, Arutun Avetisyan and Alexander Ilin

Published: 13 June 2023

1. Introduction

The human pose estimation task is one of the most interesting areas of research. It has applications in various fields, such as games, healthcare, augmented reality and sports [1]. The operator position estimation is of great importance in collaborative robotics, since the solution of this task will increase the efficiency of robotics and expand the possibilities of its application. A real-time response is required not only in security applications [2,3], where human actions must be detected in time, but also in industrial applications where human movement is predicted, to prevent collisions with robots in shared workspaces. For these reasons, research on motion-capture systems without markers and wearable sensors has been especially in demand in recent years.

When solving the problem using computer vision methods, it is necessary to find the coordinates of each joint (arm, head, torso, etc.), called keypoints, in the video frame, and form a skeletal representation of the human body. Thus, the detection of keypoints includes the simultaneous detection of people and the localization of their keypoints.

This problem is particularly difficult due to the heterogeneity of objects that have various and potentially complex shapes, as well as the difficulties arising from background noise and partial overlaps between objects (occlusions).

Thus, the most crucial problems can be pointed out, which are [4]:

- variability of lighting conditions;
- partial occlusions and layering of objects on the scene video;
- the complexity of the human skeleton structure;
- loss of three-dimensional information that occurs as a result of observation from one point, etc.

There are several approaches to the human body modeling:

- model based on the skeleton;
- contour model;
- three-dimensional model.

Though the posture estimation has been studied for many years, this task still remains very complex and largely unsolved. There is still no universal approach that can give satisfactory results in general non-laboratory conditions.

The paper [5] considers a linear Tensor-on-Tensor regression model to predict human behavior. However, in this work, only reference points and connections of the human upper limbs are constructed, i.e. only hand movements in the working area are analyzed.

An approach to detect hazardous operator behavior was proposed in [6]. The authors simulate dangerous behavior through time series analysis to detect hazards. Unfortunately, the authors did not give examples of real testing of the proposed algorithms, and all tests were carried out only in simulation.

Industrial cobots must be able to detect the human presence, identify and determine intentions based on hand gestures, actions, etc. There is a promising tendency here towards marker-less recognition [7]. High-level motion planning is usually combined with the ability of a robot to recognize the intentions of a human partner [8–10].

In [11] the authors applied the promising technology ViT (Visual Transformers) and obtained the maximum score in the Pose Estimation on MPII Human Pose benchmark [12].

The approaches that use multiple sensors or camera networks appear to be promising, which increase reliability in the case of occlusion and visual field limitations.

The paper [13] proposed an approach to estimate 3D poses of multiple persons in calibrated RGB-Depth camera networks. Each single-view outcome was computed by using a CNN for 2D pose estimation and extending the resulting skeletons to 3D by means of the sensor depth. The authors presented their solution in the form of an open-source library OpenPTrack [14].

The study [15] proposed a convolutional neural network (CNN) approach for estimating human body pose using a small number of cameras, including outdoors scenes. The authors [16] proposed a purely geometric approach to infer a multiview pose from a synchronous set of 2D skeletons.

The paper [17] proposed an approach to estimate a 3D human pose from multiview video recordings, taken with unsynchronized and uncalibrated cameras. A unique approach to self-calibrate the system using the detected keypoints was employed in the research.

It should be noted that despite the high accuracy of some approaches, they are rather resource intensive and involve the transmission of a video stream and calculations using server graphics processing units (GPUs). However, in industrial systems, the necessity to use embedded computing modules based on GPUs of lower performance as on-board computers for robots arises. The fact imposes significant restrictions on the complexity of algorithms due to their low power consumption and small size.

It is important to emphasize once again that the application possibilities of cobots will be significantly increased thanks to timely recognition, analysis and prediction of human actions based on data from a multimodal sensory system and the development of control actions, taking into account emergency situations and extreme conditions in real time. Thus, the technological solutions, including the development of software and algorithmic support for the implementation of joint work of various types of robots with a person, is a promising task.

From our point of view, the most promising is the approach that combines a multicamera system, as well as additional sensor devices, thereby providing multimodal data processing based on neural network approaches. For this reason, this article proposes to use an ensemble of deep neural networks (NN) to determine the spatial operator posed using keypoints and to link them to the world coordinate system of the robot.

The scientific novelty of the project is in the proposed set of methods, approaches and algorithms aimed at ensuring effective interaction between the components of the operator-cobot system.

2. Problem Statement

The task of increasing the efficiency of cobot-person interactions is formulated as follows. Judging by the incoming video stream from surveillance cameras, microphones, and other sensors installed in the working area, it is necessary to recognize, localize in space and build a predictive dynamic model of the operator's behavior, which should be further synchronized and adapted to the collaborative control system of the manipulation robot with a different number of degrees of freedom to perform joint, diverse and previously unknown scenarios.

The research deals with the problem of detecting and localizing human keypoints. Thus, the particular task solved in this article can be formulated as follows.

By the incoming video stream from several surveillance cameras, it is necessary to recognize and localize keypoints of the operator in three-dimensional space. To preliminarily calibrate the system, it is necessary to apply calibration methods "by template" and to apply computer vision methods based on deep neural networks in order to localize keypoints.

One needs to develop software in the Python environment and conduct a full-scale experiment with video scenes which come from surveillance cameras installed at a production facility.

Mathematical Formulation of the Keypoint Detection Task

The task of detecting human keypoints in an image can be formulated as a regression task.

Let there be: a set of images $\omega \in \Omega$, defined using features $x_i, i = 1, n$, the totality of which for the image ω is represented by vector descriptions $\mathbf{\Phi}(\omega) = (x_1(\omega), \ldots x_n(\omega)) = \mathbf{x}$ and a set of values of the dependent variable $\mathbf{y}$, corresponding to them, each of which is a vector of values $\mathbf{y} = (y_1, y_2, \ldots y_n)$, and $\mathbf{y}_i \in \mathbb{R}$.

A priori information is represented with a training set (dataset) $\mathbb{D} = ((\mathbf{x}^j, \mathbf{y}^j)), j = \overline{1, L}$, defined using a table, each row j, contains a vector image description $\mathbf{\Psi}(\omega)$ and the value of the target variable. Note that the training set characterizes the unknown mapping $^*\mathbf{F} : \Omega \to \mathbb{Y}$.

We specify the regression task to the keypoints detection task. Let there be a video stream frame $\mathbf{I}^t$, where t - is the number of the current frame. Judging by the available frames $\mathbf{I}^t$ of a continuous video stream $\mathbf{V} = (\mathbf{I}^1 \ldots, \mathbf{I}^t, \ldots, \mathbf{I}^\tau)$ and a priori information given by the training set $\mathbb{D} = ((\mathbf{x}^j, \mathbf{y}^j)), j = \overline{1, L}$ for deep learning of a supervised NN, it is required to solve the problem of image recognition: To detect the key points of an object on a video stream frame. Key points can be represented as a vector $\mathbf{y} = (y_1, y_2, \ldots y_n)$, containing object coordinates.

Generally accepted metrics are used to evaluate the task of detecting the keypoints of a human as a performance criterion [18].

Average precision (AP), which is averaged over various Object Keypoint Similarity (OKS) thresholds is set to $0,50 : 0,05 : 0,95$.

OKS is calculated based on the distance between predicted points and ground truth points normalized to the human scale. The scale and constant of the keypoint is needed to equalize the importance of each keypoint: the location of the neck position is more precise than the position of the hips [19]: OKS $= exp(-\frac{d_i^2}{2s^2k_i^2})$, where d_i - Euclidean distance between true keypoint and predicted keypoint; s–scale. The square root of the object segment area; k–is a constant for the keypoint that controls the decline; each keypoint has its own coefficient (the circles for the shoulders and knees can be larger than for the nose or eyes). The OKS metrics only shows how close the predicted keypoint is to the true keypoint (value between 0 and 1). Perfect predictions will have OKS $= 1$, while the predictions for which all keypoints differ by more than a few standard deviations $s \cdot k_i$, will have OKS ≈ 0. Mean Average precision is also calculated using OKS with thresholds of 0.50 and 0.75 (AP50 and AP75).

Percentage of Detected Joints–PDJ. A detected joint is considered correct if the distance between a predicted keypoint and a true one is within a certain fraction of the diagonal of the bounding box (diameter of the torso).

The use of the PDJ metrics implies that the accuracy of all joints is estimated using the same error threshold.

$$\mathrm{PDJ} = \frac{\sum_{i=1}^{n} bool(d_i < 0.05) * diagonal}{n},$$

where d_i–is the Euclidean distance between the true keypoint and the predicted keypoint; $bool(condition)$–a function that returns 1 if the condition is true, 0, if it is false; n–the number of keypoints in the image.

Percentage of Correct Key-points PCK [20] considers a body part to be correctly located if the estimated endpoints of the body segments are within 50% of the segment length of their true location. PCKh-0.5 [21]. PCK modification using 50% head segment length matching threshold.

3. Problem Solution

The algorithm which is aimed to detect and localize keypoints in the world coordinate system is divided into a separate subtasks solution (Figure 1):

1. Preliminary calibration of the multi-camera system and its binding to the cobot coordinate system is performed.
2. Human keypoints are detected on each of the cameras.
3. Aggregation of detected keypoints is performed and their coordinates in the world system are compared.

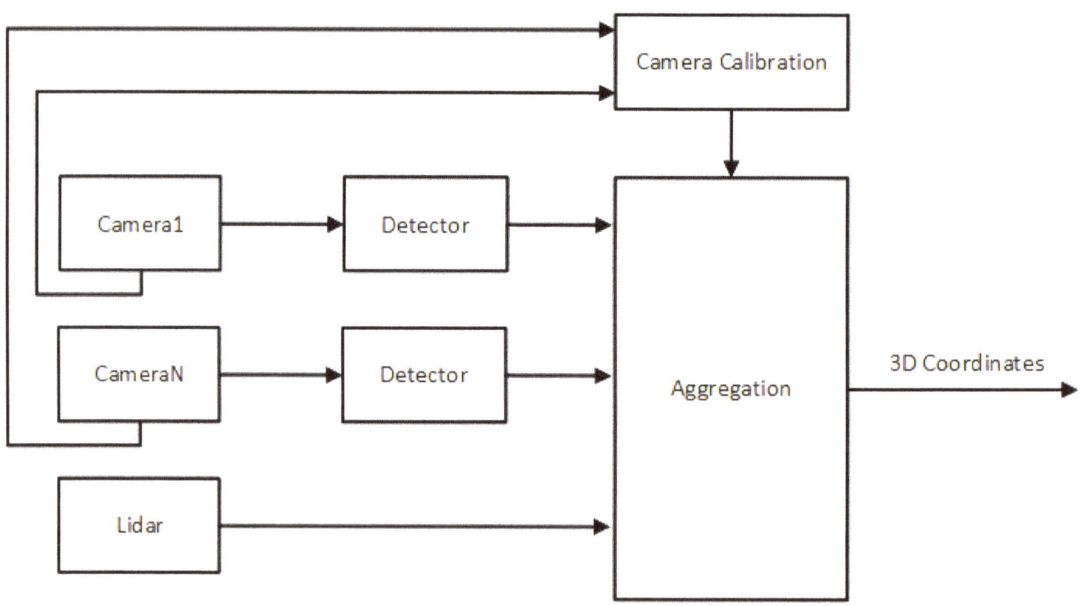

Figure 1. Algorithm for detecting and localizing keypoints in the world coordinate system

3.1. Calibration of the Multi-Camera System

Let there be a multi-camera system consisting of two cameras observing one scene. The configuration and location of the system is shown in Figure 2. The camera registers a scene containing N reference points. The task is to use the three-dimensional coordinates of the reference points (X_i^p, Y_i^p, Z_i^p) and the coordinates of their projection in the camera image plane (c_{x_i}, c_{y_i}), where $i = 1...N$ evaluation of matrix elements **A**. As a rule, a calibration object in the form of a checkerboard is formed for camera calibration, since the use of alternating black and white squares has a sharp gradient in two directions. The intersections of the checkerboard lines are used as corners.

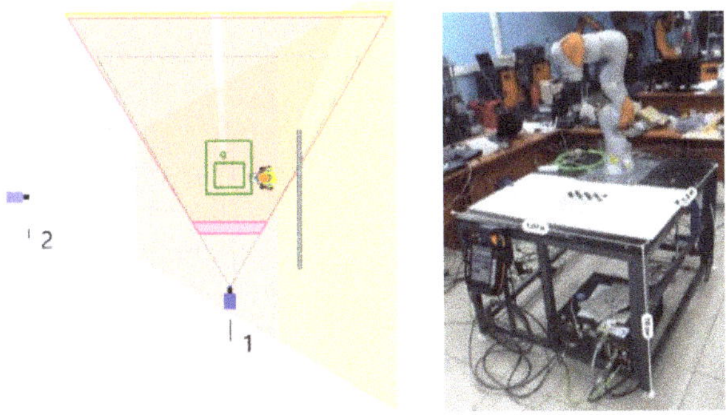

Camera 1: installation height – 1910 mm., distance to the table – 4700 mm., focal length - 4.7 mm.,

Camera 2: installation height – 1510 mm., distance to the table – 2210 mm., focal length – 1.5 mm.

Figure 2. Observed scene

K images for each camera are generated with a calibration object depending on the number of rotation angles N. To calculate four internal parameters $(f/w, f/h, c_x, c_y)$ where (c_x, c_y) are the natural coordinates of the point, f is the distance from the optical center, w, h are the dimensions along the axes ox, oy and six external parameters such as (ψ, ϕ, θ) rotation angles and (T_x, T_y, T_z) transfer parameters. The number of frames and the number of corners is calculated using the following expression: $2 \cdot N \cdot K \geq 6 + 4$. To obtain the coordinates of reference points (X_i^p, Y_i^p, Z_i^p) and the coordinates of their projection in the camera image plane (c_x, c_y) the *findChessboardCorners* function for finding corners on the chessboard is used, provided in the OpenCV library (Figure 3). Using the method [22] the matrix element of camera internal parameters is evaluated:

$$\mathbf{A} = \begin{bmatrix} f/w & 0 & c_{x0} \\ 0 & f/h & c_{y0} \\ 0 & 0 & 1 \end{bmatrix}.$$

The result of the function is a matrix of the camera internal characteristics **A**, distortion vector **r**, rotation vector **m** and transfer vectors **q**.

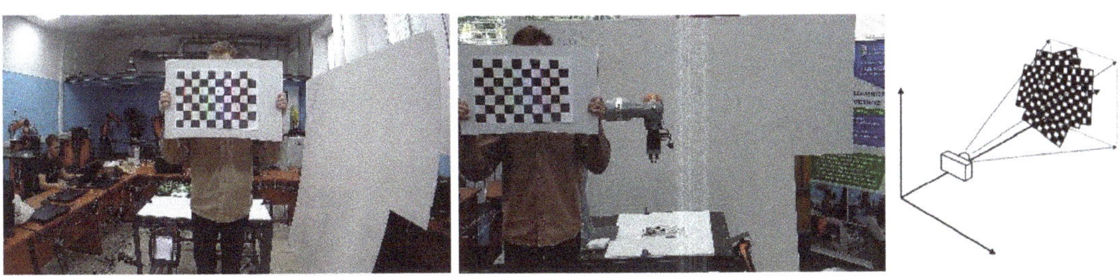

Figure 3. Camera calibration

3.2. Human KeyPoint Detection

The following approach worked out by the authors [23] based on the OpenPifPaf project [24] was used as an algorithm to detect the operator keypoints. The authors [23] proposed a number of modifications that improve the quality of keypoints detection. The underlying OpenPifPaf is a bottom-up detector based on composite fields. The architecture of the model is shown in Figure 4. The input gets an $h \times w$ image with three color channels.

The neural network encoder generates PIF and PAF fields 17 × 5 and 19 × 7 channels. The decoder converts the PIF and PAF fields into pose estimates containing 17 joints each. Each connection is represented by x and y coordinates and a confidence score. ResNet is used as encoders [25] or ShuffleNetV2 [26].

The Part Intensity Field (PIF) and Part Association Field (PAF) blocks are 1 × 1 convolutions followed by subpixel convolutions. These blocks are trained to detect and link key points. For architecture training, the COCO dataset was used, which can also be supplemented with synthetic data to adapt to a specific task. The method of the network retraining using only synthetic data with pretrained weights seems more promising.

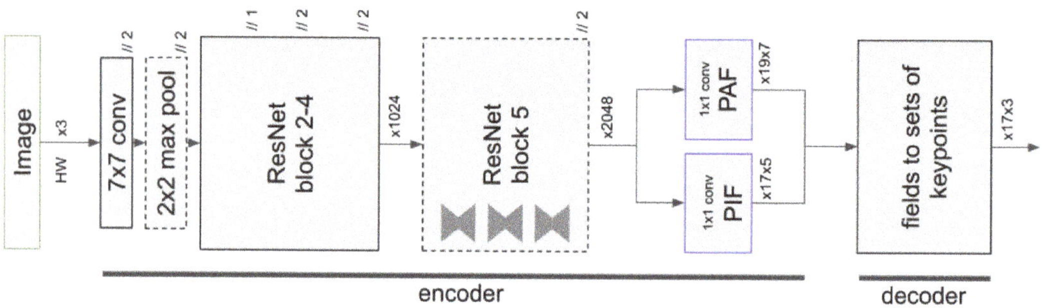

Figure 4. Keypoint detector architecture.

To obtain synthetic data, the Unity3D engine was used, as well as images obtained from the cameras of the observed scene. This approach allows you to adapt the algorithm to certain conditions.

Figure 5 shows the results of recognition on video surveillance cameras installed on the experimental site (a), as well as on synthetic data (b). As a result, the algorithm detected key points in the COCO format.

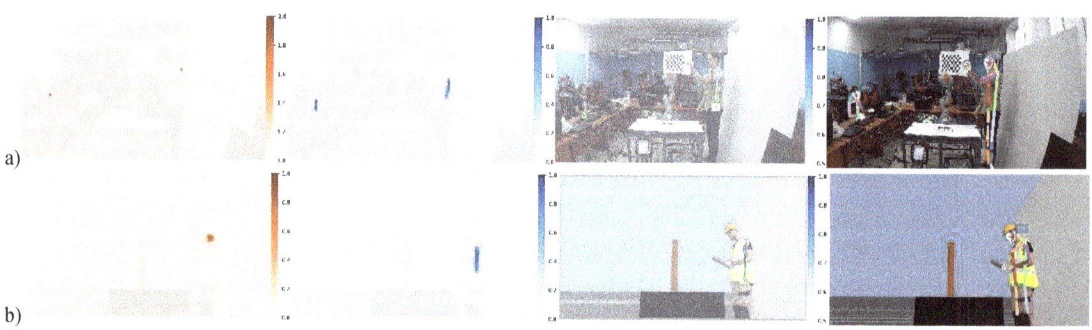

Figure 5. Keypoint detector: (**a**) real camera (**b**) synthetic data.

3.3. Mapping Keypoints in 3D Space

The result of the second stage was the keypoints of the operator detected on the images of each of the surveillance cameras. It was necessary to map 2D coordinates to the points in 3D space of the world coordinate system.

There are methods to calculate the 3D position of points for stereoscopic systems consisting of two cameras, the optical axes of which are parallel, and the straight line which passes through the optical centers is perpendicular to the optical axes. In this case, to obtain a point in three-dimensional coordinates, it is necessary to calculate the reference points using the following expression:

$$x_1^1 = \frac{f(X^w + \frac{b}{2})}{Z^2}, \quad x_2^2 = \frac{f(X^w - \frac{b}{2})}{Z^2}, \quad y_1^1 = y_2^2 = \frac{fY^w}{Z^w},$$

where X^w, Y^w, Z^w are world coordinates of the point, x_1, y_1 are the projection coordinates in the image plane of the first camera, while x_2, y_2 for the second camera. Then the point coordinates in three-dimensional space will be as follows:

$$X^{w^c} = b\frac{(x_1^1 + x_2^2)}{2(x_1^1 - x_2^2)}, \quad Y^{w^c} = b\frac{(y_1^1 + y_2^2)}{2(x_1^1 - x_2^2)}, \quad Z^{w^c} = \frac{fb}{x_1^1 - x_2^2},$$

where f is the distance from the optical center, b is the length of the straight line segment between the optical centres. In the case when the camera axes are not parallel and the direction of cameras optical center displacement is arbitrary, the calculation of the point coordinates for any camera presupposes that the following parameters should be calculated:

$$\mathbf{v}_1 = \frac{\mathbf{A}_1\mathbf{M}_1}{Z_1}, \quad \mathbf{v}_2 = \frac{\mathbf{A}_2\mathbf{M}_2}{Z_2},$$

$$\begin{bmatrix} Z_1^t \\ Z_2^t \end{bmatrix} = \begin{bmatrix} \mathbf{v}_1^T\mathbf{A}_1^{-T}\mathbf{A}_1^{-1}\mathbf{v}_1 & -\mathbf{v}_1^T\mathbf{A}_1^{-T}\mathbf{R}^T\mathbf{A}_2^{-1}\mathbf{v}_2 \\ -\mathbf{v}_1^T\mathbf{A}_1^{-T}\mathbf{R}^T\mathbf{A}_2^{-1}\mathbf{v}_2 & \mathbf{v}_1^T\mathbf{A}_1^{-T}\mathbf{A}_1^{-1}\mathbf{v}_1 \end{bmatrix}^{-1} \begin{bmatrix} \mathbf{v}_1^T\mathbf{A}_1^{-T}\mathbf{R}^T \\ \mathbf{v}_2^T\mathbf{A}_2^{-T} \end{bmatrix} \mathbf{t},$$

where $\mathbf{M}_1, \mathbf{M}_2$ characterize the coordinates of a certain point in three-dimensional space in the coordinate system in the system of the first and second cameras, $\mathbf{A}_1, \mathbf{A}_2$ are the matrix of internal parameters of the first and second cameras. $\mathbf{R}$ is an orthogonal matrix that describes the orientation of the coordinate system of the second camera relative to the first one, $\mathbf{t}$ is the translation vector that determines the position of the optical center of the second camera in the coordinate system of the first one. The obtained parameters can be used to get a vector of 3D point coordinates for any of the cameras: $\mathbf{M}_1^p = Z_1^t\mathbf{A}_1^{-1}\mathbf{v}_1$, $\mathbf{M}_2^p = Z_2^t\mathbf{A}_2^{-1}\mathbf{v}_2$.

4. Semi-Natural Experiment

The proposed approach was implemented in Python using the pytorch library. The following computer configuration was used for testing: Intel Core i5, 8 Gb RAM, Nvidia Geforce 1080 Ti. Figure 6 demonstrates the algorithm in action in the current industrial line of the robotics center.

It has been experimentally proven that this approach allows the pose to be restored when only a subset of keypoints are in the video scene, as well as when they partially overlap.

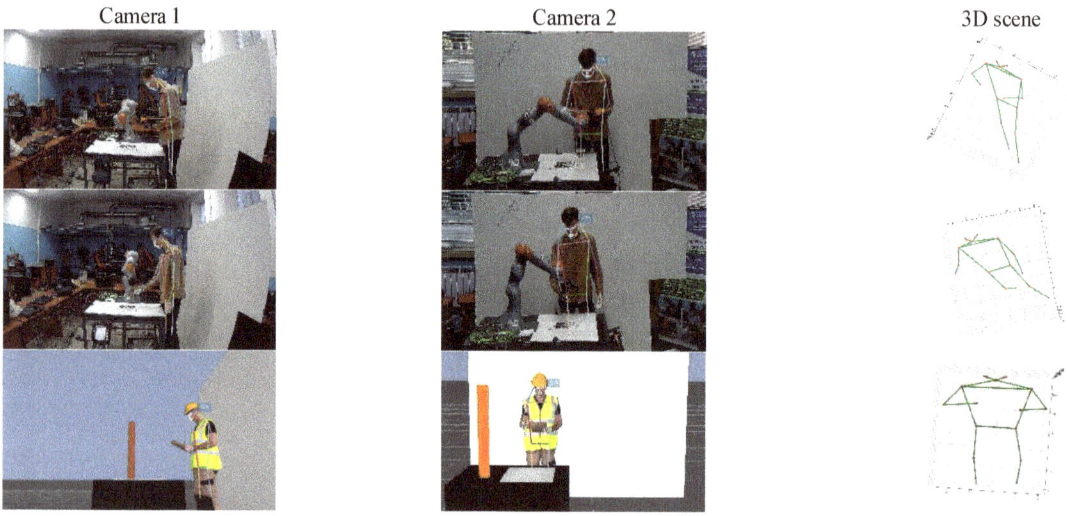

Figure 6. The proposed approach application.

5. Conclusions

This article solves the problem of determining the spatial position of the operator using keypoints with a multi-camera sensor system. The possibility of the proposed approach application to solve the problems of collaborative robotics has been demonstrated. The advantage of the algorithm is the use of a multi-chamber system that allows you to attach points to the local coordinate system of an industrial robotic complex. The obtained data will further be used in planning the trajectory of the robot to safely perform joint operations in the same workspace. As additional modifications, the possibility of installing a lidar to clarify the coordinates of the edges of the skeleton is being considered.

A research perspective is the development of an algorithm to predict the actions of the operator in the workspace and detect hazardous situations and possible intersections in the trajectories of a collaborative robot.

Author Contributions: Conceptualization, Y.I., S.Z. and M.G.; methodology, Y.I. and S.Z.; software, S.Z. and D.G.; validation, S.Z., M.G. and S.S.; writing—original draft preparation, Y.I. and S.Z.; writing, review and editing, Y.I. and D.G. All authors have read and agreed to the published version of the manuscript.

Funding: The work was supported by the Russian Science Foundation (project № 22-71-10093).

Institutional Review Board Statement: Not applicable.

Informed Consent Statement: Informed consent was obtained from all subjects involved in the study.

Data Availability Statement: Not applicable.

Conflicts of Interest: The authors declare no conflict of interest.

References

1. Ivanov, Y. S.; Kosichkov, A.O. Recognition and Control of the Athlete's Movements Using a Wearable Electronics System. In Proceedings of the International Multi-Conference on Industrial Engineering and Modern Technologies (FarEastCon), Vladivostok, Russia, 6–9 October 2020; pp. 1–5.
2. Amosov, O.S.; Amosova, S.G.; Ivanov, Y.S.; Zhiganov, S.V. Using the deep neural networks for normal and abnormal situation recognition in the automatic access monitoring and control system of vehicles. *Neural Comput. Appl.* **2020**, *33*, 3069–3083. [CrossRef]
3. Amosov, O.S.; Amosova, S.G.; Ivanov, Y.S.; Zhiganov, S.V. Using the Ensemble of Deep Neural Networks for Normal and Abnormal Situations Detection and Recognition in the Continuous Video Stream of the Security System. *Procedia Comput. Sci.* **2019**, *150*, 532–539. [CrossRef]

4. Sigal, L. *Human Pose Estimation*; Ikeuchi, K., Ed.; Computer Vision; Springer: Boston, MA, USA, 2014.
5. Gril, L.; Wedenig, P.; Torkar, C.; Kleb, U. A Tensor Based Regression Approach for Human Motion Prediction. 2022. Available online: https://arxiv.org/abs/2202.03179 (accessed on 15 June 2022).
6. Huck, T.P.; Ledermann, C.; Kröger, T. Testing Robot System Safety by Creating Hazardous Human Worker Behavior in Simulation. *IEEE Robot. Autom. Lett.* **2022**, *7*, 770–777. [CrossRef]
7. Makris, S.; Karagiannis, P.; Koukas, S.; Matthaiakis, A. Augmented reality system for operator support in human–robot collaborative assembly. *Cirp-Ann.-Manuf. Technol.* **2016**, *65*, 61–64. [CrossRef]
8. Park, J.S.; Park, C.; Manocha, D. I-planner: Intention-aware motion planning using learning-based human motion prediction. *Int. J. Robot. Res.* **2019**, *38*, 23–39. [CrossRef]
9. Zhang, J.; Liu, H.; Chang, Q.; Wang, L.; Gao, R.X.Recurrent neural network for motion trajectory prediction in human-robot collaborative assembly. *CIRP Ann.* **2020**, *69*, 9–12. [CrossRef]
10. Cheng, Y.; Sun, L.; Liu, C.; Tomizuka, M. Towards Efficient Human-Robot Collaboration With Robust Plan Recognition and Trajectory Prediction. *IEEE Robot. Autom. Lett.* **2020**, *5*, 2602–2609. [CrossRef]
11. Xu, Y.; Zhang, J.; Zhang, Q.; Tao, D. ViTPose: Simple Vision Transformer Baselines for Human Pose Estimation. 2022. Available online: https://arxiv.org/abs/2204.12484 (accessed on 6 July 2022).
12. Papers with Code. MPII Human Pose Benchmark (Pose Estimation). 2022. Available online: https://paperswithcode.com/sota/pose-estimation-on-mpii-human-pose (accessed on 15 July 2022).
13. Carraro, M.; Munaro, M.; Burke, J.; Menegatti, E. Real-time marker-less multi-person 3D pose estimation in RGB-Depth camera networks. In *Intelligent Autonomous Systems 15*; Springer: Berlin/Heidelberg, Germany, 2018.
14. GitHub, Inc. Open PTract v2. 2022. Available online: https://github.com/OpenPTrack/open_ptrack_v2 (accessed on 28 July 2022).
15. Elhayek, A.; de Aguiar, E.; Jain, A.; Thompson, J.I.; Pishchulin, L.; Andriluka, M.; Bregler, C.; Schiele, B.; Theobalt, C. MARCOnI—ConvNet-Based MARker-Less Motion Capture in Outdoor and Indoor Scenes. *IEEE Trans. Pattern Anal. Mach. Intell.* **2017**, *39*, 501–514. [CrossRef] [PubMed]
16. Lora, M.; Ghidoni, S.; Munaro, M.; Menegatti, E. A geometric approach to multiple viewpoint human body pose estimation. In Proceedings of the 2015 European Conference on Mobile Robots (ECMR), Lincoln, UK, 2–4 September 2015; pp. 1–6.
17. Takahashi, K.; Mikami, D.; Isogawa, M.; Kimata, H. Human Pose as Calibration Pattern: 3D Human Pose Estimation with Multiple Unsynchronized and Uncalibrated Cameras. In Proceedings of the 2018 IEEE/CVF Conference on Computer Vision and Pattern Recognition Workshops (CVPRW), Salt Lake City, UT, USA, 18–22 June 2018; pp. 1856–18567.
18. Papers With Code. Pose Estimation Subtasks. 2022. Available online: https://paperswithcode.com/area/computer-vision/pose-estimation (accessed on 3 August 2022).
19. COCO Dataset. COCO–Common Objects in Context. 2022. Available online: https://cocodataset.org/#home (accessed on 5 August 2022).
20. Ferrari, V.; Marín-Jiménez, M.J.; Zisserman, A. Progressive search space reduction for human pose estimation. In Proceedings of the 2008 IEEE Conference on Computer Vision and Pattern Recognition, Anchorage, Alaska, 23–28 June 2008; pp. 1–8.
21. Andriluka, M.; Pishchulin, L.; Gehler, P.V.; Schiele, B. 2D Human Pose Estimation: New Benchmark and State of the Art Analysis. In Proceedings of the 2014 IEEE Conference on Computer Vision and Pattern Recognition, Columbus, OH, USA, 23–28 June 2014; pp. 3686–3693.
22. Brown, D.C. Close-Range Camera Calibration. *Photogramm. Eng.* **1971**, *37*, 855–866.
23. Zauss, D.; Kreiss, S.; Alahi, A. Keypoint Communities. In Proceedings of the 2021 IEEE/CVF International Conference on Computer Vision (ICCV), Montreal, BC, Canada, 11–17 October 2021; pp. 11037–11046.
24. Kreiss, S.; Bertoni, L.; Alahi, A. OpenPifPaf: Composite Fields for Semantic Keypoint Detection and Spatio-Temporal Association. *IEEE Trans. Intell. Transp. Syst.* **2022**, *23*, 13498–13511. [CrossRef]
25. He, K.; Zhang, X.; Ren, S.; Sun, J. Deep Residual Learning for Image Recognition. In Proceedings of the 2016 IEEE Conference on Computer Vision and Pattern Recognition (CVPR), Las Vegas, NV, USA, 27–30 June 2016; pp. 770–778.
26. Ma, N.; Zhang, X.; Zheng, H.; Sun, J. ShuffleNet V2: Practical Guidelines for Efficient CNN Architecture Design. In Proceedings of the European Conference on Computer Vision (ECCV), Munich, Germany, 8–14 September 2018; pp. 116–131.

engineering proceedings

MDPI

Proceeding Paper

Deep Learning for Detecting Dangerous Objects in X-rays of Luggage †

Nikita Andriyanov

Data Analysis and Machine Learning Department, Financial University under the Government of the Russian Federation, Leningradsky pr-t 49, Moscow 125167, Russia; naandriyanov@fa.ru; Tel.: +7-927-818-8416

† Presented at the 15th International Conference "Intelligent Systems" (INTELS'22), Moscow, Russia, 14–16 December 2022.

Abstract: The investigation presented in this text is the study of object detection algorithms in the task of analyzing images of baggage and hand luggage. A modified version of the YOLOv5 convolutional neural network with additional rechecking based on the VGG-19 network is proposed. The modification is based on transfer learning from the available images. A comparison is made with other known algorithms. The article shows that the application of the proposed model made it possible to achieve the value of the mean average recall (mAR) at the level of 87% for dangerous objects of five classes.

Keywords: object detection; convolutional neural networks; aviation security; pattern recognition; computer vision

Citation: Andriyanov, N. Deep Learning for Detecting Dangerous Objects in X-rays of Luggage. *Eng. Proc.* **2023**, *33*, 20. https://doi.org/10.3390/engproc2023033020

Academic Editors: Askhat Diveev, Ivan Zelinka, Arutun Avetisyan and Alexander Ilin

Published: 13 June 2023

1. Introduction

At present, the issue of using various intelligent systems to ensure security in crowded places is quite acute [1–4]. Of course, optical cameras are an important source of information for preventing and detecting suspicious situations. At the same time, sufficient progress has been made in the processing of optical images, including the task of detecting objects [5,6].

Nevertheless, in the overwhelming majority of works, the marked problem of object detection is considered for optical images. In the case of developing a computer vision system for analyzing luggage images, such a system must be able to perform efficient processing of X-ray images. There is one important thing about such data. Unlike optical images, X-ray images have the property of "transparency", i.e., some objects can be seen against the background of others, and not overlapped by them. This complicates the task of analyzing such images. In the literature, there are works on the recognition of objects in X-ray images of luggage [7–9], but they consider that the image contains an object of any single class in the full image. Some solution to the problem of prohibited item detection is presented in [10]. At the same time, the value of the mean average precision (mAP) metric is about 64%. The main weakness of the work is the proposal of its own architecture of a convolutional neural network, which is insufficient for high-quality training.

In view of the foregoing, the task of detecting prohibited objects in X-ray images of baggage and hand luggage remains an urgent task. The solution to this problem will automate the routine activities of inspection operators and avoid errors caused by human factors. The solution to this problem will be described in the following sections so that the Section 2 will consider the data for which training and testing are carried out. Then, in the Section 3, special attention is paid to the proposed approach for detecting dangerous objects in X-ray images. Finally, the Section 4 is a description of the processing results for test images, accompanied by analysis and comparison with known algorithms. In conclusion, the main conclusions of the study are formulated and options for further improvement of the results obtained are proposed.

2. Description of the Baggage and Hand Luggage Images Set

Training data plays a critical role in deep learning and machine learning. For the purposes of the presented study, X-ray photographs of luggage were used, provided by the Ulyanovsk Civil Aviation Institute (UCAI) named after the Chief Marshal of Aviation Boris Pavlovich Bugaev (Ulyanovsk, Russia). In total, there were 1500 images in the database. Figure 1 shows some X-ray images to be processed by developed algorithms. It should be noted that the images of various prohibited items were taken for the purposes of the study and do not correspond to actual cases of attempted smuggling of prohibited items on board the aircraft.

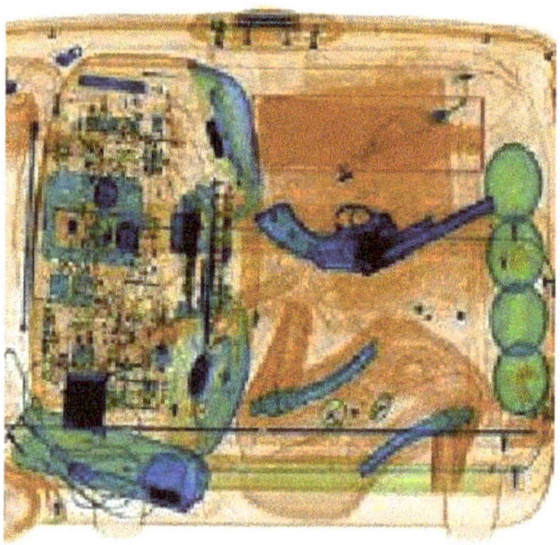

Figure 1. An example of an X-ray image of baggage with a dangerous object.

In Figure 1, it is easy to spot firearms. However, detailed thinking about the objects in Figure 1 provides the following conclusions. First of all, the definition of the image is quite small, and, secondly, the image itself has a specific nature when perceived by a person.

For further convenience and in order to train the verification network for pattern recognition, image preprocessing procedures were carried out, including the selection of only areas with dangerous objects. As part of this selection, five classes of hazardous objects were identified and 40 to 50 images were collected for each. Figure 2 shows examples of images for each selected class.

Figure 2. Examples of images for dangerous objects.

In Figure 2, from left to right, shockers, ammunition, grenades, firearms, and steel arms are presented.

Figure 3 shows the distribution of training images by class for the recognition task.

From Figure 3, it follows that a balanced dataset was prepared with a fairly small number of objects. As for processed images, the truth is that the good and bad pattern recognition in the form in which dangerous objects are presented in Figure 1 is possible only at the post-processing stages, where it is planned to use the recognizer. In addition, the objects are visually different.

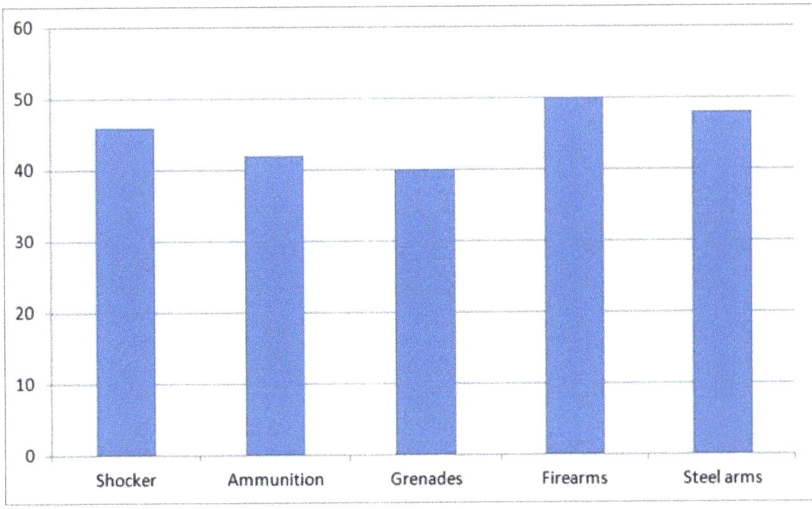

Figure 3. Distribution of objects of interest.

For images such as the one shown in Figure 1, manual labeling was performed using the CVAT tool [11] and RoboFlow [12]. Image markup example for the image from Figure 1 is shown in Figure 4. The accompanying files store the coordinates of the bounding box needed to train the neural networks.

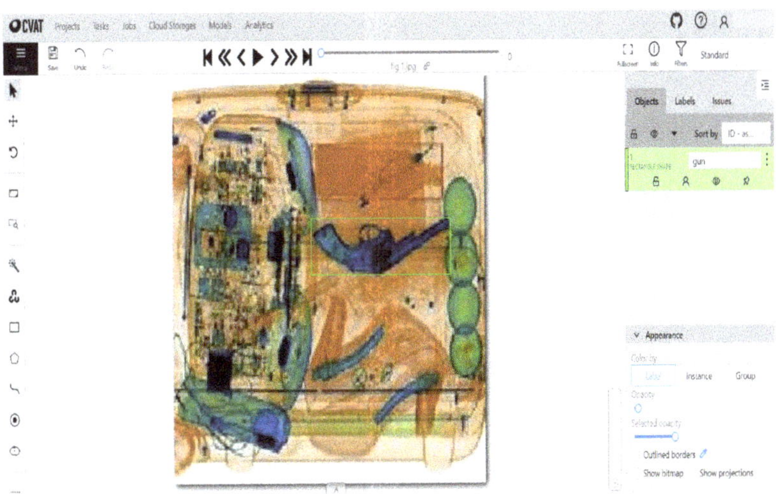

Figure 4. Markup example.

The number of images in the training set was 1200. The remaining 300 images were left for testing. Table 1 shows the distribution of objects in the images of the training and test sets.

Despite the fact that the analysis of the data in Table 1 shows that the set is not balanced enough, it was decided to train on such data.

Next, let us consider the following approaches. They are very useful and applied to detect dangerous objects in images.

Table 1. Number of dangerous objects of each class.

Class	Train, N = 1200 Samples	Test, N = 300 Samples
Shocker	122	34
Ammunition	109	28
Grenades	135	31
Firearms	180	46
Steel arms	166	37

3. Methods for Detecting Dangerous Objects in Images

As noted earlier, when describing the original images, the level of training was allocated for the detection task and the recognition task. This is due to the fact that more convenient images are formed when solving the recognition problem.

The convolutional neural network VGG-19 [13] was chosen as a model for recognition. The architecture of such a network is a convolutional network of 19 convolution blocks, pooling, trained under the ImageNet dataset. The transfer learning method was used with the following hyperparameters: batch size is 16, optimizer is ADAM, learning rate is 0.00001, activation function is softmax, loss function is cross-entropy.

The YOLOv5 architecture network [14] was used to train the solution to the detection problem. This network belongs to single-pass detectors, i.e., sets the class and bounding box for objects in one image processing pass. One more attractive thing about architecture is that the neural network is also convolutional. At the same time, the weights are adjusted in accordance with the COCO dataset, which includes 80 classes. The default hyperparameters are used.

Figure 5 shows the final scheme for processing new images.

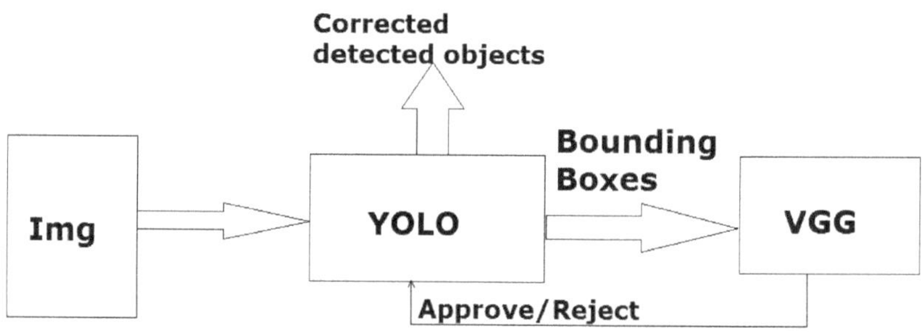

Figure 5. Baggage image processing scheme.

From Figure 5, it is clear that the transfer-trained YOLO model is used as the base detector. However, after the initial analysis, the images inside the bounding boxes are transferred to the transfer-trained VGG network. In this network, predictions are adjusted for one of the five classes according to the original dataset mentioned above.

Such an approach also has some drawbacks. For example, a special pre-trained model such as the VGG network in the diagram of Figure 5 is not capable of detecting non-hazardous items. However, this approach, taking into account the fact that the detection recall indicator is quite important in such a task, helps to detect more objects.

Next, let us consider the results obtained.

4. Results and Discussion

Since the problem of detecting dangerous objects has a higher priority than the error of missing the target, it is necessary to use the mAR metric as a comparison characteristic. The mean average recall is calculated as the average recall for each class in each image.

One of the properties of the developed approach is that training and inference took place on an NVIDIA GeForce GTX 1060 video card.

Figure 6 shows an example of image processing using the trained model.

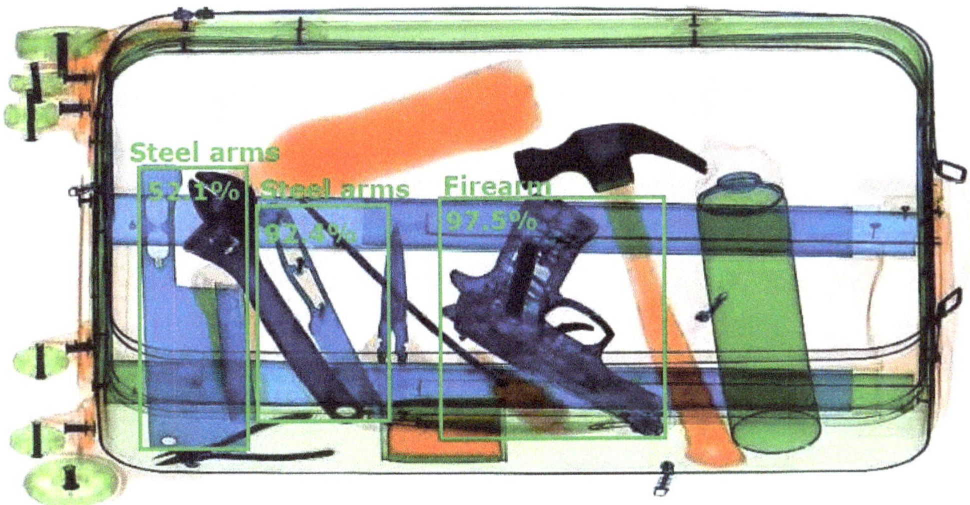

Figure 6. Processing results.

From Figure 6, it is clear that two steel arms and one firearm were found with class probabilities of 52.1%, 92.4%, and 97.5%, respectively.

Let us make a detailed comparison of research in terms of the proposed algorithm with other optical image-detecting methods transferred to X-ray images. The results for the mean average recall metric are presented in Table 2.

Table 2. Recall comparison by models.

Model	mAR
YOLOv5	0.843
SSD	0.672
DETR	0.697
Ours	0.871

Let us look at Table 2. It is interesting because the results show that the proposed model, because of using an additional level of verification based on the VGG network, allows us to increase the result compared to YOLO by 3%.

Finally, let us consider the completeness of each of their objects, as presented in Table 3.

Table 3. Recall comparison by models.

Class	Average Recall
Shocker	0.912
Ammunition	0.714
Grenades	0.839
Firearms	1
Steel arms	0.892

It can be seen from the presented results that firearms are detected in all cases. The worst of these things are those with the definition of ammunition.

5. Conclusions

This paper proposes a new algorithm for detecting dangerous objects on X-ray images of luggage, combining YOLO, and VGG. This approach provided an increase in the mean average recall of 3%. In the future, it is planned to use augmentation and tuning of model hyperparameters to obtain higher recall values.

Funding: This research was funded by the Russian Federation Presidental Scholarship, SP-3738.2022.5 and partly by the Russian Foundation for Basic Research, Project №19-29-09048.

Institutional Review Board Statement: Not applicable.

Informed Consent Statement: Not applicable.

Data Availability Statement: Not applicable.

Conflicts of Interest: The authors declare no conflict of interest.

References

1. Janssen, S.; van der Sommen, R.; Dilweg, A.; Sharpanskykh, A. Data-Driven Analysis of Airport Security Checkpoint Operations. *Aerospace* **2020**, *7*, 69–74. [CrossRef]
2. Kim, M.H.; Park, J.W.; Choi, Y.J. A Study on the Effects of Waiting Time for Airport Security Screening Service on Passengers' Emotional Responses and Airport Image. *Sustainability* **2020**, *12*, 10634. [CrossRef]
3. Asmer, L.; Popa, A.; Koch, T.; Deutschmann, A.; Hellman, M. Secure rail station—Research on the effect of security checks on passenger flow. *J. Rail Transp. Plan. Manag.* **2019**, *10*, 9–22. [CrossRef]
4. Janssen, S.; Berg, A.; Sharpanskykh, A. Agent-based vulnerability assessment at airport security checkpoints: A case study on security operator behavior. *Transp. Res. Interdiscip. Perspect.* **2020**, *5*, 1–14. [CrossRef]
5. Andriyanov, N.A.; Dementiev, V.E.; Tashlinskii, A.G. Detection of objects in the images: From likelihood relationships towards scalable and efficient neural networks. *Comput. Opt.* **2022**, 46, 139–159. [CrossRef]
6. Carion, N.; Massa, F.; Synnaeve, G.; Usunier, N.; Kirillov, A.; Zagoruyko, S. End-to-End Object Detection with Transformers. *arXiv* **2020**, arXiv:2005.12872. Available online: https://arxiv.org/abs/2005.12872 (accessed on 27 July 2022).
7. Andriyanov, N.A.; Volkov, A.K.; Volkov, A.K.; Gladkikh, A.A. Research of recognition accuracy of dangerous and safe X-ray baggage images using neural network transfer learning. *IOP Conf. Ser. Mater. Sci. Eng.* **2020**, *1061*, 012002. [CrossRef]
8. Hättenschwiler, N.; Michel, S.; Kuhn, M. A first exploratory study on the relevance of everyday object knowledge and training for increasing efficiency in airport security X-ray screening. *IEEE ICCSTAt* **2015**, *49*, 12–25.
9. Andriyanov, N.A.; Volkov, A.K.; Volkov, A.K.; Gladkikh, A.A.; Danilov, S.D. Automatic X-ray image analysis for aviation security within limited computing resources. *IOP Conf. Ser. Mater. Sci. Eng.* **2020**, *862*, 052009. [CrossRef]
10. Hassan, T.; Shafay, M.; Akçay, S.; Khan, S.; Bennamoun, M.; Damiani, E.; Werghi, N. Meta-Transfer Learning Driven Tensor-Shot Detector for the Autonomous Localization and Recognition of Concealed Baggage Threats. *Sensors* **2020**, *20*, 6450. [CrossRef] [PubMed]
11. Computer Vision Annotation Toolbox. Available online: https://cvat.ai/ (accessed on 28 July 2022).
12. Roboflow. Available online: https://app.roboflow.com/ (accessed on 28 July 2022).
13. Simonyan, K.; Zisserman, A. Very Deep Convolutional Networks for Large-Scale Image Recognition. *arXiv* **2015**, arXiv:1409.1556. Available online: https://arxiv.org/abs/1409.1556 (accessed on 29 July 2022).
14. YOLOv5. Available online: https://github.com/ultralytics/yolov5 (accessed on 29 July 2022).

engineering proceedings

MDPI

Proceeding Paper

An Overview of Object Detection and Tracking Algorithms †

Kehao Du * and Alexander Bobkov

Department of Informatics and Control Systems, Bauman Moscow State Technical University, 105005 Moscow, Russia; alexander.bobkov@bmstu.ru
* Correspondence: duhuhaizi@gmail.com; Tel.: +7-925-793-0588

† Presented at the 15th International Conference "Intelligent Systems" (INTELS'22), Moscow, Russia, 14–16 December 2022.

Abstract: With the development of information technology, the vision-based detection and tracking of moving objects is gradually penetrating into all aspects of people's lives, and its importance is becoming more prominent, attracting more and more scientists and research institutions at home and abroad to participate in research in this field. With in-depth research into vision-based object detection and tracking, various superior algorithms have appeared in recent years. In this article, we attempt to compare some of the classic algorithms in this area of detection and tracking that have appeared recently. This article examines and summarizes two areas: the detection and the tracking of moving objects. First, we divide object detection into one-stage algorithms and two-stage algorithms depending on whether a region proposal should be generated, and we accordingly outline some commonly used object detection algorithms. Second, we separate object tracking into the KCF and SORT algorithms according to the differences in the underlying algorithms.

Keywords: region proposal; R-CNN; Fast R-CNN; Faster R-CNN; YOLO; MOSSE; KCF; SORT

Citation: Du, K.; Bobkov, A. An Overview of Object Detection and Tracking Algorithms. *Eng. Proc.* **2023**, *33*, 22. https://doi.org/10.3390/engproc2023033022

Academic Editors: Askhat Diveev, Ivan Zelinka, Arutun Avetisyan and Alexander Ilin

Published: 13 June 2023

1. Object Detection Algorithms

Object detection algorithms can be divided into one-stage algorithms and two-stage algorithms. Two-stage object detection algorithms appeared earlier, so we start with them.

1.1. Two-Stage Object Detection Algorithms

The task of object detection in an image usually involves drawing bounding boxes and labels for each object, i.e., classification and localization for each object, not merely the main object.

If we use a sliding window for image classification and object localization, we would need to apply convolutional neural networks to many different objects in the image. Since convolutional neural networks will recognize every object in the image as either an object or the background, we would need to use convolutional neural networks in a large number of places and scales, but this is computationally intensive.

To solve this problem, neural network researchers propose using the concept of regions so that we can identify "blobs" of areas in an image that contain objects, which can be performed much faster. The first model is regions with CNN features (R-CNN) [1], and the principles of its algorithm are shown in Figure 1.

In R-CNN, the input image is first scanned for possible features using a selective search algorithm, resulting in roughly 2000 region proposals; then, we run a convolutional neural network on these region proposals; finally, the output of each convolutional neural network is fed into a support vector machine (SVM), which uses linear regression to narrow the feature's bounding box.

Essentially, we transform object detection into an image classification task. However, there are also several problems: the learning rate is low, it requires a large amount of disk space, and the inference speed is also low.

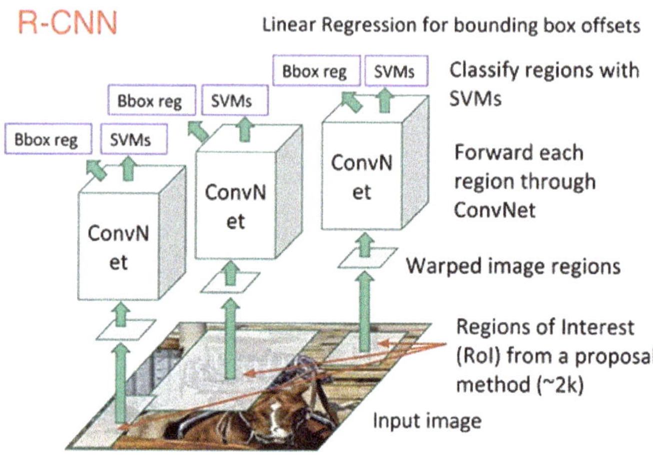

Figure 1. R-CNN architecture.

The first updated version of R-CNN is Fast R-CNN [2], which significantly improves the average precision through two improvements, which are shown in Figure 2.

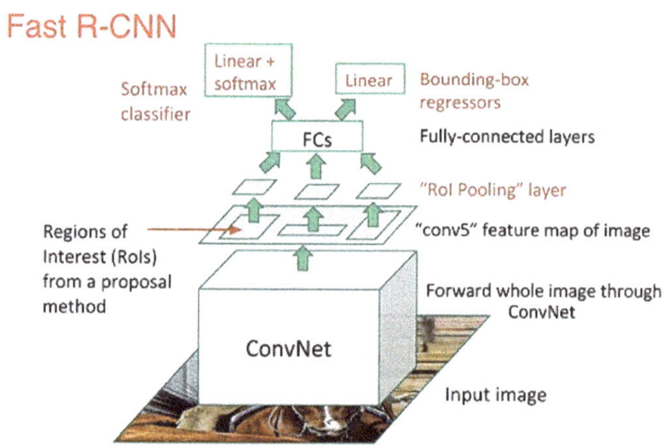

Figure 2. Fast R-CNN architecture.

Feature extraction is performed before creating a proposal of regions, so the convolutional neural network can only be run once for the entire image. We extend the neural network used for prediction by replacing the SVM with a SoftMax layer, instead of creating a new model.

Fast R-CNN is much faster than R-CNN because it only trains one CNN per image. However, the selective search algorithm still requires a long time to generate region proposals.

Faster R-CNN [3] is a typical case of object detection based on deep learning. The algorithm replaces the slow selective search algorithm with a fast neural network: by inserting a region proposal network (RPN), it predicts proposals based on features. The RPN decides "where" to look, and this reduces the amount of computation for the entire inference process. Once we have region proposals, we feed them directly into Fast R-CNN. Moreover, we add a pooling layer, several fully connected layers, a SoftMax classification layer, and a bounding box regressor, which are shown in Figure 3. In conclusion, Faster R-CNN is faster and more accurate.

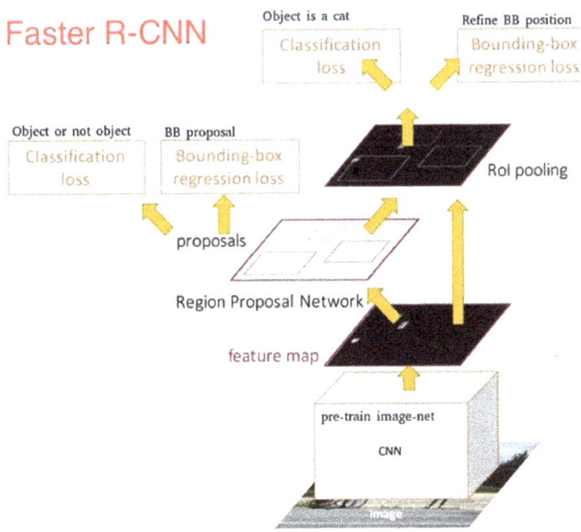

Figure 3. Faster R-CNN architecture.

1.2. One-Stage Object Detection Algorithms

Although Faster R-CNN has advanced greatly in terms of computational speed, it cannot meet the real-time detection requirements, so some researchers have proposed a regression-based method to directly regress the position and category of the object from the image. Two classic methods are YOLO [4] and SSD [5]. This article only discusses YOLO and its one-stage algorithm.

Different from the two-stage (two stages) detection algorithms represented by the R-CNN series, YOLO discards the region proposals and directly completes feature extraction, bounding box regression, and classification in the same non-branched convolutional network, which makes the network structure simpler and the detection speed almost 10 times higher than that of Faster R-CNN. This allows deep learning object detection algorithms to meet the needs of real-time detection tasks with improved processing power.

The specific steps of the algorithm are shown in Figure 4. We divide the input image into an S × S grid; each grid is responsible for determining what the objects in the grid are. We output the parameter information of the rectangles around the objects and the confidence score. The confidence score here refers to which features are contained in the grid and the predictive accuracy of each feature.

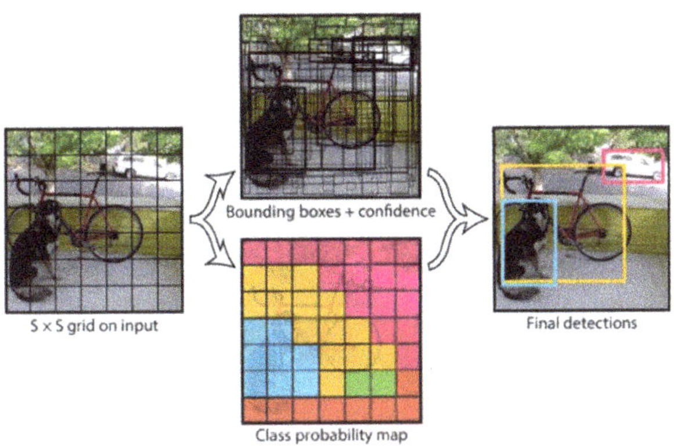

Figure 4. You Only Look Once (YOLO) model.

Since YOLO only looks at the image once, we do not use sliding windows. The feature map extracted after image convolution is divided into S × S blocks and then each block is classified with a superior classification model and each mesh is processed using non-maximum suppression to remove overlapping blocks; finally, we obtain our result. It can be seen that the whole process is very simple: there is no need for region proposals to find the object, and direct regression completes the determination of the position and category.

2. Object Tracking Algorithms

The next category concerns object tracking algorithms. It is generally considered that visual object tracking falls into two categories: generative model methods and discriminative model methods. Currently, the most popular is the discriminative method, also called tracking by detection. All the algorithms presented in this article are related to tracking by detection.

Apart from classical tracking algorithms, object tracking algorithms are divided into two categories: correlation-filter-based tracking (MOSSE) [6] and deep-learning-based tracking algorithms (SORT) [7].

The notion of correlation first appeared in the field of signal processing to measure the relationship between signals, including autocorrelation and cross-correlation. Researchers have introduced cross-correlation into the field of computer vision to analyze the relationships between factors; it is used to measure the degree of local matching of different images in a certain area. Correlation filtering is a method of searching for objects in an image frame based on the principle of cross-correlation.

Before the advent of correlation filtering and deep learning, the traditional visual tracking algorithm was slow in research and had poor tracking performance. MOSSE contributes to the development of the correlation filter tracking algorithm and also reflects the potential of correlation filter technology in the tracking domain for the first time. The frame rate of MOSSE is shown in Table 1.

Table 1. Compares the frame rates of MOSSE and other algorithm.

Algorithm	Frame Rate	CPU
Frag Track	realtime	Unknown
GBDL	realtime	3.4 Ghz Pent. 4
IVTL	7.5 fps	2.8 Ghz CPU
MILTrack	25 fps	Core 2 Quad
MOSSE Filters	669 fps	2.4 Ghz Core 2 Duo

Based on the principle of correlation filtering, aiming to measure the similarity of local pixels, MOSSE develops a filter that can realize the sum of the least squares of errors. After tracking the initialization of the first frame of a video sequence, MOSSE can generate a relatively stable filter to use a matching pattern to measure the similarity in subsequent video images. The tracking speed can reach 669 fps when limited; with the advantage of a high tracking speed, the method has good reliability. MOSSE uses the Fast Fourier Transform (FFT) for image and filter processing, converting the convolution of images and filters in the frequency domain into a mathematical product operation. After filtering for image detection, the point with the highest cross-correlation value, i.e., the center point of the object, will be found.

Considering that the correlation filter algorithm using raw pixels for feature extraction is too simple, the accuracy of the algorithm is not high. An improved CS correlation filter algorithm is proposed, also based on correlation filtering, which additionally provides higher accuracy while maintaining the speed advantage of correlation filtering. KCF [8] is an improvement of CSK [9], and CSK is also an improvement and addition based on MOSSE.

However, in terms of tracking, in order to reduce the computational complexity of the model, KCF proposes a cyclic matrix to reduce the computational complexity of the cyclic shifts, and the Fourier transform is used to switch the calculation process between the frequency domain and the spatial domain, so that matrix multiplication with higher computational complexity becomes computation point multiplication with less computational complexity. During the KCF tracking process, the detection model also learns to measure the similarity between the predicted object and the tracking object. For each object predicted by the tracking model, the detection model first measures the similarity between the predicted object and the tracking object, and then decides whether to update the tracking track.

Another algorithm, SORT, differs from the correlation filter algorithm in that it does not use any signs of the tracked object during tracking, but only uses the position and size of the detection frames to estimate the movement and match the data. It does not implement any re-detection algorithm, focusing instead on frame-to-frame comparisons.

SORT mainly consists of three parts: object detection, a Kalman filter, and the Hungarian algorithm. Object detection mainly uses algorithms such as YOLO and R-CNN. Kalman filtering can identify the "optimal" estimate using the value predicted by the mathematical model and the measured value of the observation to find the "optimum" estimate (optimal here refers to the smallest standard error).

The Hungarian algorithm is a data association algorithm; in essence, the tracking algorithm should solve the data association problem. Suppose that there are two sets S and T, and set S has m elements and set T has n elements. The Hungarian algorithm seeks to match elements in S with elements in T in pairs (although they may not be the same). Combined with tracking scenario S and T, we have frame t and frame t − 1. The task of the Hungarian algorithm is to match frame t with frame t − 1.

SORT, as reflected in its name, is simple and can correspond to real time. On the one hand, its principle is straightforward and easy to implement; on the other hand, it is extremely fast due to its simplicity, as shown in Figure 5. At the same time, thanks to the introduction of object detection technology, the tracking accuracy is also significantly improved, which means that it can achieve a good compromise between accuracy and speed.

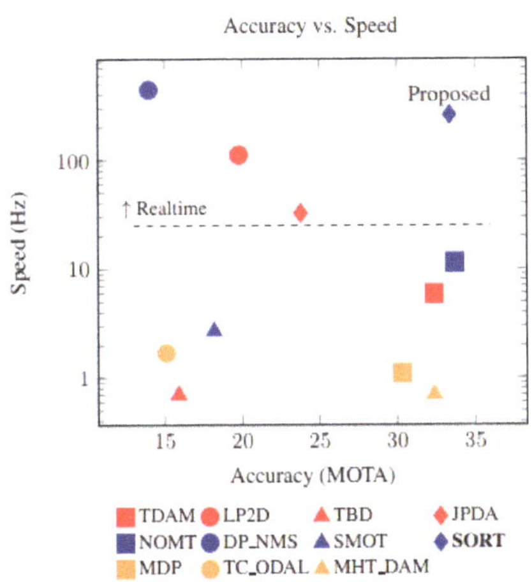

Figure 5. Benchmark performance of SORT.

3. Conclusions

In this article, modern algorithms for object detection and tracking are considered. Object detection is presented in terms of one-stage algorithms and two-stage algorithms. Object tracking via the SORT and KCF algorithms is introduced. Algorithms for object detection and tracking based on different concepts are compared, the unique and advantageous qualities of each concept are identified, and a promising direction for the development of future detection and tracking algorithms is explored. For example, self-driving cars always require improvements in processing speed and accuracy.

Author Contributions: Conceptualization, A.B.; methodology, A.B.; software, K.D.; validation, K.D.; writing—original draft preparation, K.D.; writing—review and editing, K.D. All authors have read and agreed to the published version of the manuscript.

Funding: This research received no external funding.

Institutional Review Board Statement: Not applicable.

Informed Consent Statement: Not applicable.

Data Availability Statement: Not applicable.

Conflicts of Interest: The authors declare no conflict of interest.

References

1. Girshick, R.; Donahue, J.; Darrell, T.; Malik, J. Rich feature hierarchies for accurate object detection and semantic segmentation. In Proceedings of the 2014 Computer Vision and Pattern Recognition (CVPR), Columbus, OH, USA, 23–28 June 2014; pp. 580–587.
2. Girshick, R. Fast R-CNN. In Proceedings of the 2015 International Conference on Computer Vision (ICCV), Santiago, Chile, 7–13 December 2015; pp. 1440–1448.
3. Ren, S.; He, K.; Girshick, R.; Sun, J. Faster R-CNN: Towards real-time object detection with region proposal networks. In Proceedings of the Advances in Neural Information Processing Systems, Montreal, QC, Canada, 7–12 December 2015; Volume 28.
4. Redmon, J.; Divvala, S.; Girshick, R.; Farhadi, A. You only look once: Unified, real-time object detection. In Proceedings of the 2016 Computer Vision and Pattern Recognition (CVPR), Las Vegas, NV, USA, 27–30 June 2016; pp. 779–788.
5. Liu, W.; Anguelov, D.; Erhan, D.; Szegedy, C.; Reed, S.; Fu, C.Y.; Berg, A.C. SSD: Single shot multibox detector. In Proceedings of the 2016 European Conference on Computer Vision (ECCV), Amsterdam, The Netherlands, 11–14 October 2016; pp. 21–37.
6. Bolme, D.S.; Beveridge, J.R.; Draper, B.A.; Lui, Y.M. Visual object tracking using adaptive correlation filters. In Proceedings of the 2010 Computer Vision and Pattern Recognition (CVPR), San Francisco, CA, USA, 13–18 June 2010; pp. 2544–2550.
7. Bewley, A.; Ge, Z.; Ott, L.; Ramos, F.; Upcroft, B. Simple online and realtime tracking. In Proceedings of the 2016 International Conference on Image Processing (ICIP), Phoenix, AZ, USA, 25–28 September 2016; pp. 3464–3468.
8. Henriques, J.F.; Caseiro, R.; Martins, P.; Batista, J. High-speed tracking with kernelized correlation filters. *IEEE Trans. Pattern Anal. Mach. Intell.* **2014**, *37*, 583–596. [CrossRef] [PubMed]
9. Henriques, J.F.; Caseiro, R.; Martins, P.; Batista, J. Exploiting the circulant structure of tracking-by-detection with kernels. In Proceedings of the 2012 European Conference on Computer Vision (ECCV), Florence, Italy, 7–13 October 2012; pp. 702–715.

engineering proceedings

MDPI

Proceeding Paper

Design and Development of Information and Computational System for Energy Facilities' Impact Assessment on Environment †

Vladimir R. Kuzmin *, Tatyana N. Vorozhtsova and Liudmila V. Massel

Melentiev Energy Systems Institute of the Siberian Branch of the RAS, Lermontova Str. 130, 664058 Irkutsk, Russia
* Correspondence: kuzmin_vr@isem.irk.ru; Tel.: +7-(3952)-500-646 (ext. 440)
† Presented at the 15th International Conference "Intelligent Systems" (INTELS'22), Moscow, Russia, 14–16 December 2022.

Abstract: In this article we consider authors' information and computational system for energy facilities' impact assessment on the environment. The necessity of such assessments and development of this system is substantiated. We developed this system as a Web application using the agent-service approach. To develop a database for the system, we utilized ontological engineering of energy and ecology. For assessments, we developed a set of information susbsystems that use approved regulatory methods. Our system can be used for assessment of the impact of both existing and planning energy facilities and also for planning measures to reduce the harmful impact of such facilities. We also performed a set of computational experiments in order to test the developed system. Experiments have shown the correctness of the methods used, and the results of one of them are presented in the article.

Keywords: energy facilities; information and computational system; impact assessment; environment

Citation: Kuzmin, V.R.; Vorozhtsova, T.N.; Massel, L.V. Design and Development of Information and Computational System for Energy Facilities' Impact Assessment on Environment. *Eng. Proc.* **2023**, *33*, 21. https://doi.org/10.3390/engproc2023033021

Academic Editors: Askhat Diveev, Ivan Zelinka, Arutun Avetisyan and Alexander Ilin

Published: 13 June 2023

1. Introduction

A necessity of impact assessment and requirement for a reduction in the impact of polluting (harmful) substances from industrial facilities (including energy facilities) on the environment are gaining more and more attention in the world. Currently, in the Russian Federation, the following have been approved and taken effect: "Energy strategy of the Russian Federation until 2035" [1], the national project "Ecology" [2], and "Strategy for socio-economic development of the Russian Federation with low greenhouse gas emissions until 2050" [3]. These documents envisage, among other things, a shift to environmentally friendly and resource-saving energy, the rational use of natural resources, and a reduction in dangerous polluting substances' emissions [4,5]. Various teams, both Russian [6,7] and foreign [8,9], perform research on assessing the impact of industrial facilities on the environment. To carry out impact assessments, the results of measurements of environmental elements (air, water, soil) are used, as well as monitoring and statistical information: state reports and reports of industrial facilities. If that information is not available, then the assessment of the impact of pollutants on the environment is carried out using officially approved regulatory methods, such as [10,11]. However, these methods are used separately and we did not find information about attempts to integrate them. Methods also require a significant amount of information about object of study, starting from the technical parameters of the facility (brand and model of the boiler, characteristics of the used fuel) and ending with information about the weather in the area where the facility is located (wind speed and direction, air temperature) and terrain data. Research on the impact assessment of energy facilities on the environment is interdisciplinary, as it requires the involvement of experts from various subject areas, such as energy, ecology, and economics (to assess the economic feasibility of measures to reduce the harmful effects of pollutants). When conducting research, it will also be advisable to be able not only to assess the current state of environmental pollution, but also to provide an opportunity to evaluate the effectiveness of measures to reduce the harmful effects of energy facilities and plan the placement of new facilities.

Based on the above, it is required to develop an information and computational system (ICS), which will allow for a comprehensive assessment of the impact of energy facilities on the environment and will include decision support tools to reduce the harmful effects of these facilities.

2. Methods and Tools Used for Development

To solve the problem mentioned in the introduction, we developed ICS WICS (Web-oriented Information and Computational System). ICS is based on the authors' methodical approach for impact assessment of energy facilities on the environment. A detailed description of this approach was given in [12]. ICS allows us to

- assess the amount of pollutants from energy facilities and dispersion of these pollutants in the air using approved regulatory methods [10,11,13,14];
- assess economical damage [15,16];
- work with the results of analysis of snow samples;
- visualize results.

The components responsible for work with the mentioned methods will be discussed in the next section.

While developing ICS, we decided to develop it as a multiagent system (or MAS). The concept of MAS is to increase system's performance (speed of processing and output results' quality) by distributing tasks [17]. In [18], a multiagent system is considered as a network of asynchronous objects that jointly solve problems that cannot be solved by a single agent. Thus, the system consists of decentralized autonomously operating elements (or agents). To develop ICS as an MAS, we used the authors' method based on the agent-service approach, which was given in a previous paper [19]. In this paper, we also considered information and the analytical system WIS, which is a conceptual prototype of the ICS described in this article. The main steps of the method are

- Give a description of the system, taking into account the characteristics of the problem. For this it is necessary to:
  - Determine the purpose of the ICS;
  - Define the set of tasks $\{T\}$ that the ICS must be able to solve;
  - Define the function set of the ICS $\{F\}$;
  - Create a list of agents $\{A\}$ of the ICS based on the $\{F\}$;
  - Develop set of basic components $\{C_B\}$.
- Develop agent scenarios:
  - Define agents' call order $\{P_A\}$;
  - Develop agent call scripts $\{S_A\}$;
  - Give a description of the developed scenarios using event models $\{E_S\}$.
- Develop an architecture of the ICS;
- Design the ICS;
- Implement the ICS.

In this article, we will not consider the second step because the description of their development requires an additional article. The purpose of the ICS, tasks, and functions were described in the introduction and the beginning of this section. The architecture and implementation of the ICS will be considered in the next section.

To store the information necessary for conducting research and research results, a database (DB) was developed, which is also used to store the knowledge base. The studies themselves require, as mentioned in the introduction, a significant amount of data and are interdisciplinary. Therefore, when formalizing knowledge in the considered subject areas, inaccuracies may occur, leading to the construction of incorrect models. Therefore, to formalize knowledge about the subject areas under consideration, ontological engineering of the considered subject areas was performed [20]. We also proposed a database design methodology for assessing environmental pollution by energy facilities based on ontologies. The main steps of the method are:

- Determine the energy facilities that will be assessed and establish their internal hierarchy based on the subsystem of ontologies of the energy facility;
- Create infological data models based on ontologies;
- Create appropriate tables in the database for each object in the hierarchy;
- Determine the characteristics of the object of assessment and create fields in the corresponding tables;
- Analyze the selected methods for calculating the quantitative indicators of emissions and calculating the dispersion of pollutants in the atmospheric air and create tables for these calculations that include the fields corresponding to the calculation formula;
- Determine the list of pollutants from the object of assessment based on the calculation method and the subsystem of ontologies for assessing emissions;
- Create tables of calculation results according to the methods and the list of pollutants;
- Based on the analysis of calculation methods, create tables for storing auxiliary data, for example, information about weather conditions and terrain data.

Next, we will consider the process of creating tables in a database using the constructed ontologies and the proposed methodology. Figure 1 shows an ontology describing the method from [10].

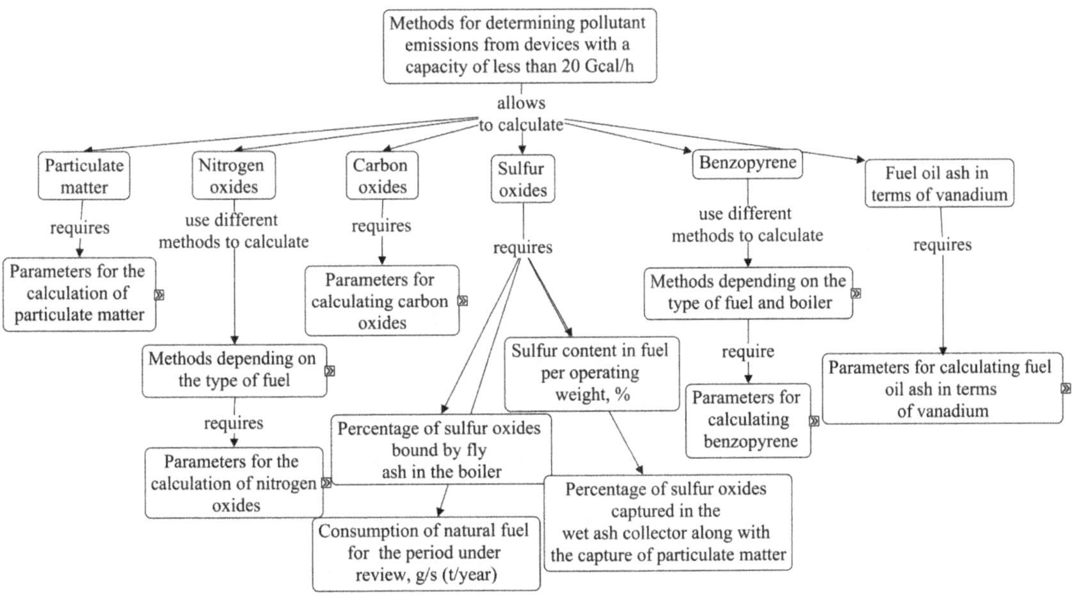

Figure 1. Ontology of the method for determining emissions of pollutants into the atmosphere during fuel combustion in boilers with a capacity of less than 30 tons of steam per hour or less than 20 Gcal per hour.

As can be seen from the ontology, in order to calculate the volume of emissions from an energy facility, it is required to know both the fuel parameters (for example, the lower heating value of the fuel) and the parameters of the boiler (fuel consumption, the fraction of particles captured in ash collectors). Based on the constructed ontologies, the corresponding tables in the database were developed. Figure 2 shows some of the tables.

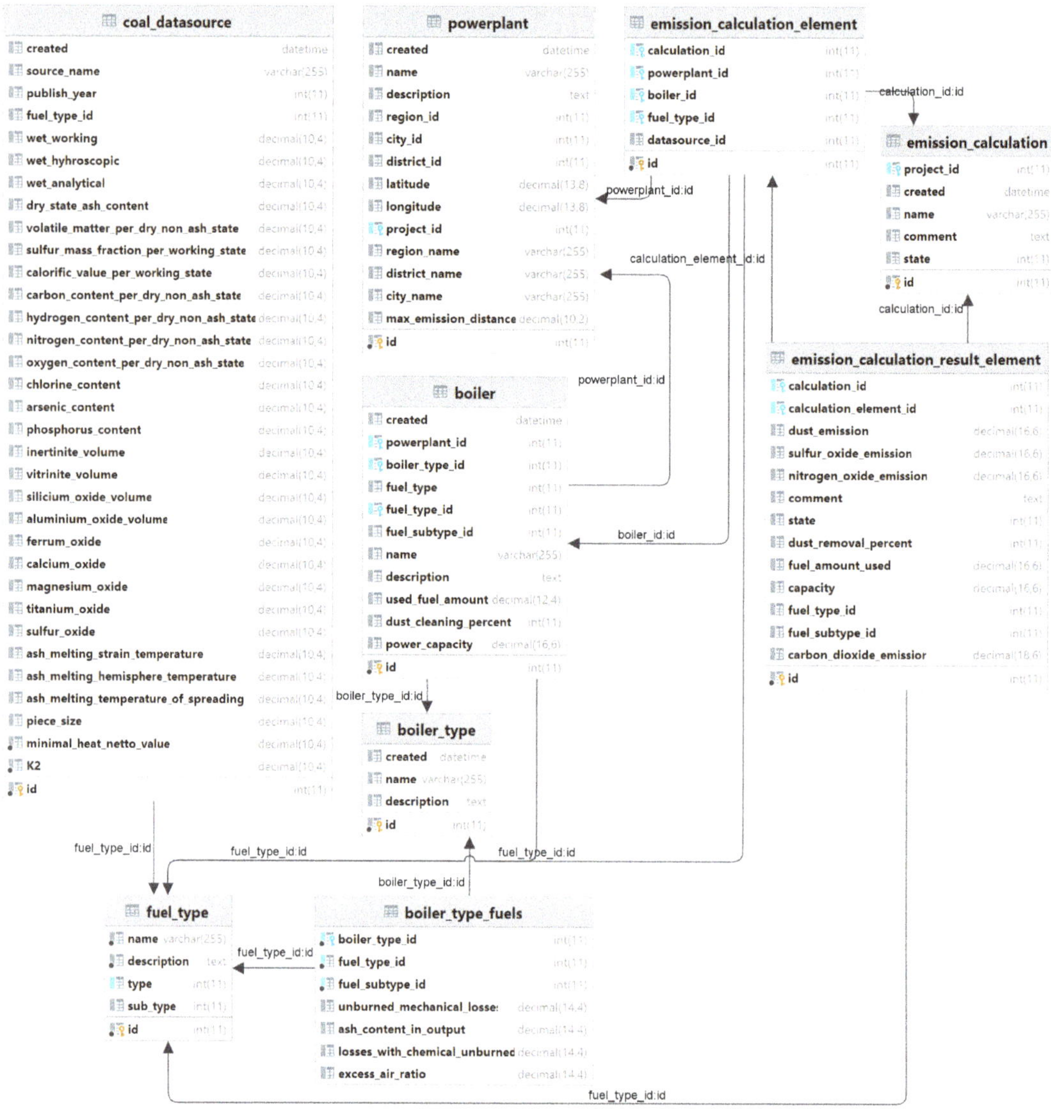

Figure 2. Tables in the database developed on the basis of ontologies.

Comments. The *powerplant* table stores basic information about the energy facility, such as name and location. The *boiler* table stores data on the boiler units installed at the energy facility—type of boiler, fuel used, volume of fuel burned, installed capacity. The *boiler_type_fuels* table contains information about the allowed types of fuel that can be burned in the boiler and technical parameters, for example, heat loss from mechanical incomplete combustion of the fuel (*unburned_mechanical_losses*). The *coal_datasource* table contains information about the parameters of the coal. The *emission_calculation* and *emission_calculation_element* tables store data on emission calculation and calculation element, respectively. They indicate which energy facilities will participate in the calculation. The *emission_calculation_result_element* table stores the results of the performed calculations. Next, we will consider in more detail the implementation of the proposed ICS: the architecture and main components, as well as the approbation of the ICS.

3. Implementation of ICS WICS

ICS WICS is implemented as a Web application, due to the following reasons:

- The ICS should provide the mutual work of several experts. The web application allows us to organize the storage of data and results in one place.
- Emission calculations and pollutant dispersion calculations, especially when dealing with a significant number of energy facilities, are complex, both due to the amount of data used and the complexity of the calculation methods. The implementation of the ICS as a Web application with an agent-service architecture allows us to speed up the process of calculations by distributing them among various agent-executors.
- The Web application helps us to reduce the requirements for experts' PCs, because in this case, the expert needs only a Web browser with Internet access to work with the ICS. It also reduces time needed for the development and testing of the ICS, since it is not required to compile for various operating systems and platforms.
- The process of supporting the ICS and adding new functionality to the system is simplified.

Figure 3 shows the client-server architecture of the ICS.

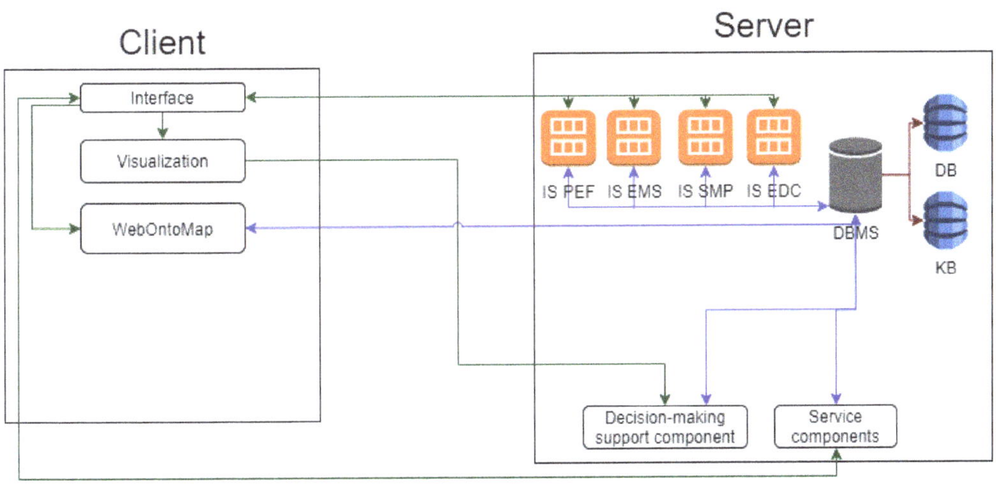

Figure 3. Architecture of ICS WICS.

The server side includes the following subsystems: calculation subsystem, DBMS subsystem, and subsystem with auxiliary components. The calculation subsystem consists of:

- IS PEF—implements calculations of quantitative indicators of pollutant emissions from an energy facility based on regulatory methods [10,13];
- IS EMS—implements pollutant dispersion calculations based on normative methods [7,10] using the results obtained from the IS PEF subsystem;
- IS SMP—allows the user to work with the results of the analysis of snow samples;
- IS EDC—provides work with calculations of economic damages based on methods [11,12] and uses the calculation results obtained from the IS PEF subsystem.

ICS WICS is a multiagent system, so it includes a main server that handles client requests and coordinates agents, responsible for calculations; auxiliary servers (four servers) that contain the mentioned calculation subsystems (currently, one server per subsystem) and a server with service components; database server.

The client part includes a user interface, tools for visualizing results (including geovisualization), as well as WebOntoMap, a component for working with ontologies stored in the knowledge base.

The developed ICS allows us to calculate pollutant emissions from energy facilities, assess the economic damage from emissions, calculate the dispersion of pollutants in the atmospheric air, and work with the results of analyzing snow samples for pollutant content. The system can be used to assess the current situation with environmental pollution, to assess the effectiveness of measures to reduce the harmful impact of energy facilities, for example, when planning the placement of new energy facilities.

4. Computational Experiment

To test the ICS WICS, computational experiments were carried out based on information about energy facilities (boiler houses) located in the Central Ecological Zone of the Baikal Natural Territory (CEZ BNT) [21]. Boiler houses have different installed capacity and equipment; coal is used as fuel (the results of calculations are given based on data for 2015). First, the quantitative indicators of pollutant emissions into the atmospheric air from the assessed facilities were calculated. In an aggregated form, a fragment of the results obtained is shown in Table 1.

Table 1. Fragment of the results of calculations of pollutant emissions into the atmospheric air.

Facility	Total Emission, tons/year	Particulate Matter, tons/year	SO, tons/year	NO_X, tons/year	Installed Capacity, Gcal/hour
Kudara	322.8	224.9	97.5	0.4	12.8
Tvorogovo, Shigaevo	77.5	53.9	23.4	0.1	4.4
Elantsy (central)	361.16	251.79	109.2	0.17	3.5

The geovisualization of the results was also performed in the form of a heat map using the the Yandex.Maps cartographic service. Figure 4 shows the visualization of particulate matter emissions.

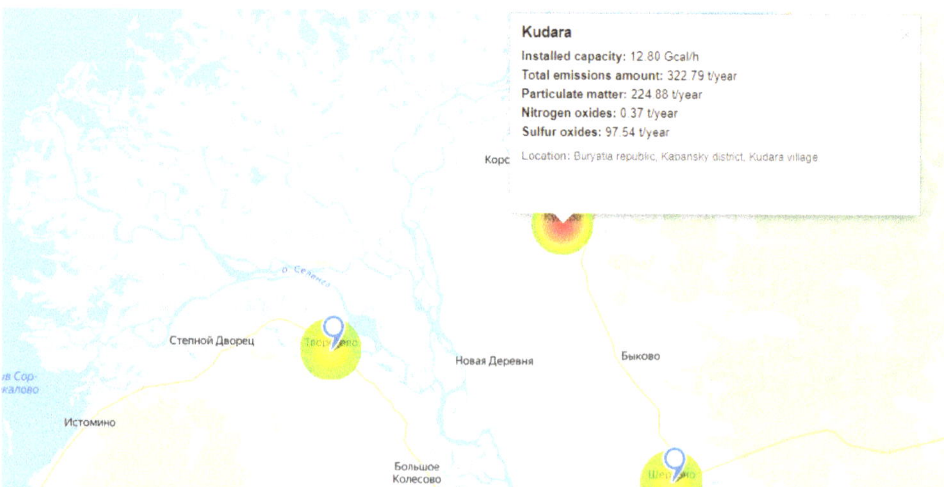

Figure 4. Geovisualization of the results of calculating the volume of pollutant emissions from energy facilities.

Blue marks indicate the assessed energy facilities. When a mark is clicked, a legend (detailed information about the facility) will be shown at the top left. The redder the heat map, the higher the volume of pollutant emissions relative to other facilities.

Based on the results of calculating the quantitative indicators of pollutant emissions, the calculation of the economic damage caused to the environment was carried out according to the methodology [11]. A fragment of the results is shown in Table 2.

Table 2. A fragment of the results of the calculation of economic damage.

Facility	Particulate Matter Damage, RUB	SO Damage, RUB	NO_X Damage, RUB	Total Damage, RUB
Kudara	817,600	354,000	1600	1,173,900
Tvorogovo, Shigaevo	209,000	90,000	600	301,000
Elantsy (central)	1,700,000	737,000	1400	2,439,500

The calculation of the dispersion of pollutants in the atmospheric air was also performed; Table 3 shows a fragment of the calculation results for particulate matter.

Table 3. Fragment of the results of the calculation of dispersion of particulate matter from energy facilities in the atmospheric air.

Facility	Min. One-Time Concentration, mg/m^3	Avg. One-Time Concentration, mg/m^3	Max. One-Time Concentration, mg/m^3	Min. Distance, km	Avg. Distance, km	Max. Distance, km
Kudara	65	4938	12,416	1.5	3.2	4.8
Tvorogovo, Shigaevo	529	2370	3039	0.4	0.5	15.3
Elantsy (central)	20	1957	10,487	1.6	4.1	5.0

The results of the dispersion calculation can also be geovisualized in the form of a heat map; an example is shown in Figure 5.

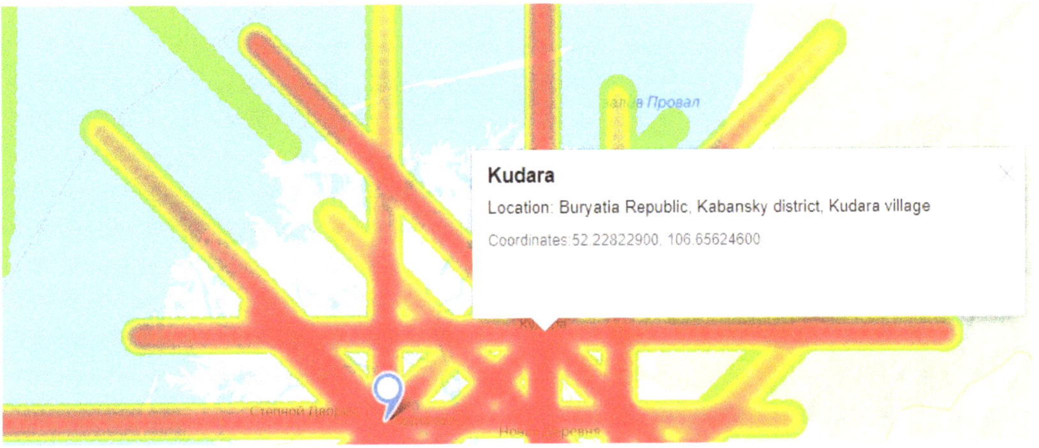

Figure 5. Geovisualization of the results of calculating the dispersion of pollutants (particle matter) for energy facilities located in the Kabansky district of the Republic of Buryatia.

The results of the analysis of snow samples for the content of pollutants were additionally uploaded to the system. Figure 6 shows the geovisualization of the results of the analysis of samples for the content of SO_4 in the form of a heat map after interpolation using Bayesian empirical kriging [22,23].

Figure 6. Interpolation of the results of the analysis of snow samples for SO_4 content.

Blue dots on the map indicate sampling sites, while red dots indicate energy facilities. The redder and brighter the heat map at sampling points, the higher the pollutant concentration.

The ICS also includes a subsystem for supporting the development of recommendations, which aggregates the results and visualizes them using infographics. An example of the subsystem operation is shown in Figure 7.

Figure 7. Decision-making support subsystem.

5. Conclusions

In this article, we substantiated the need for research on the impact assessment of energy facilities on the environment and described the ICS WICS developed for this purpose. The ICS allows us to assess the impact of both existing and planned energy facilities. The approaches and methods used in the development of the ICS are shown: the agent-service approach and ontological engineering. The ICS architecture is presented and its main blocks are described; examples of system interfaces are shown. The results of approbation confirming the correctness of the applied methods are presented.

In the future, as a part of the research on the sustainability of energy and socio-ecological systems, it is planned to integrate the ICS WICS with the INTEC-A software package (developed in a team represented by the authors [24]) to study the directions for the development of the fuel and energy complex, taking into account the requirements of energy security. This will make it possible to carry out comprehensive research on the directions of development of the fuel and energy complex of the Russian Federation, taking into account the impact of decisions made on the environment.

Author Contributions: Conceptualization, L.V.M.; investigation, V.R.K.; methodology, V.R.K. and T.N.V.; software, V.R.K.; validation, T.N.V. and L.V.M.; writing—original draft preparation, V.R.K. and T.N.V.; writing—review and editing, L.V.M. All authors have read and agreed to the published version of the manuscript.

Funding: The research was carried out under State Assignment Project (No FWEU-2021-0007 AAAA-A21-121012090007-7).

Institutional Review Board Statement: Not applicable.

Informed Consent Statement: Not applicable.

Data Availability Statement: Not applicable.

Conflicts of Interest: The authors declare no conflict of interest.

References

1. Energy Strategy of the Russian Federation for the Period up to 2035: Decree of the Government of the Russian Federation, 9 June 2020, No. 1523-r. Available online: http://static.government.ru/media/files/w4sigFOiDjGVDYT4IgsApssm6mZRb7wx.pdf (accessed on 15 July 2022). (In Russian)
2. Passport of the National Project "Ecology". Available online: http://static.government.ru/media/files/pgU5Ccz2iVew3Aoel5vDGSBjbDn4t7FI.pdf (accessed on 15 July 2022). (In Russian)
3. Strategy for Socio-Economic Development of the Russian Federation with Low Greenhouse Gas Emissions Until 2050: Decree of the Government of the Russian Federation, 29 October 2021, No. 3052-r. Available online: http://static.government.ru/media/files/ADKkCzp3fWO32e2yA0BhtIpyzWfHaiUa.pdf (accessed on 16 July 2022). (In Russian)
4. Egorchenkov, A.V.; Egorchenkov, D.A. Social aspects of environmental issues in the context of the national project "Ecology". *IOP Conf. Ser. Earth Environ. Sci.* **2020**, *579*. 012099. [CrossRef]
5. Alekseev, A.N.; Bogoviz, A.V.; Goncharenko, L.P.; Sybachin, S.A. A Critical Review of Russia's Energy Strategy in the Period until 2035. *Int. J. Energy Econ. Policy* **2019**, *9*, 95–102. [CrossRef]
6. Zorina, T.G.; Aleksandrovich, S.A.; Maysyuk, E.P.; Massel, A.G. The impact of energy on the geoecology of regions (Russian Federation and Republic of Belarus). *Inf. Math. Technol. Sci. Manag.* **2019**, 2, 151–161.
7. Khaustov, A.P.; Redina, M.M.; Nedostup, P.Y.; Silaev, A.V. Problems of Assessment and Management of Environmental Risks at Fuel and Energy Complex Enterprises. *Energy Secur. Energy Sav.* **2005**, *6*, 25–30. (In Russian)
8. Wyrwa, A.; Zyśk, J.; Mirowski, T. Assessment of Environmental Impacts of Energy Scenarios Using the πESA Platform. In *eScience on Distributed Computing Infrastructure*; Lecture Notes in Computer Science; Bubak, M., Kitowski, J., Wiatr, K., Eds.; Springer: Cham, Switzerland, 2014.
9. Schiavo, B.; Morton-Bermea, O.; Salgado-Martínez, E.; García-Martínez, R.; Hernández-Álvarez, E. Health risk assessment of gaseous elemental mercury (GEM) in Mexico City. *Environ Monit Assess.* **2022**, *194*, 19. [CrossRef] [PubMed]
10. *Methodology for Determining Emissions of Pollutants into the Atmosphere during Fuel Combustion in Boilers with a Capacity of Less Than 30 Tons of Steam per Hour or Less Than 20 Gcal per Hour;* State Committee for Environmental Protection of the Russian Federation (with the Participation of the Firm "Integral", St. Petersburg): Moscow, Russia, 1999; 53p. (In Russian)
11. On Approval of Methods for Calculating the Dispersion of Emissions of Harmful (Polluting) Substances in the Atmospheric Air: Decree of the Ministry of Natural Resources of Russia, 6 June 2017, No. 273. Available online: https://docs.cntd.ru/document/456074826 (accessed on 1 August 2022). (In Russian)

12. Kuzmin, V.R.; Massel, L.V. Methodical Approach for Impact Assessment of Energy Facilities on Environment. In *Information Systems and Design*; ICID 2021. Communications in Computer and Information Science 1539; Taratukhin, V., Matveev, M., Becker, J., Kupriyanov, Y., Eds.; Springer: Cham, Switzerland, 2022.
13. *Methodology for Determining Gross Emissions of Pollutants into the Atmosphere from Boiler Plants of Thermal Power Plants*; RD 34.02.305-98/VTI; PMB VTI: Moscow, Russia, 1998; 36p. (In Russian)
14. Berlyand, M.E. *Forecasting and Regulation of Atmospheric Pollution*; Gidrometeoizdat: Leningrad, Russia, 1985; 272p. (In Russian)
15. On Approval of the Methodology for Calculating the Amount of Damage Caused to Atmospheric Air as a Component of the Natural Environment: Decree of the Ministry of Natural Resources of Russia, 28 January 2021, No. 59. Available online: https://docs.cntd.ru/document/573536168 (accessed on 31 July 2022). (In Russian)
16. Temporary Standard Methodology for Determining the Economic Efficiency of the Implementation of Environmental Measures and Assessing the Economic Damage Caused to the National Economy by Environmental Pollution: Decree of the USSR State Planning Committee, the USSR State Construction Committee and the Presidium of the USSR Academy of Sciences, 21 October 1983 No. 254/284/134. Available online: http://www.consultant.ru/document/cons_doc_LAW_94300/ (accessed on 22 August 2022). (In Russian)
17. Alonso, E. From Articial Intelligence to Multi-Agent Systems: Some Historical and Computational Remarks. 2001. Available online: http://citeseerx.ist.psu.edu/viewdoc/summary?doi=10.1.1.26.3210 (accessed on 23 September 2014).
18. *Reducing the Risk of Cascading Accidents in Power Systems*; Voropai, N.I., Russian Academy of Sciences, Siberian Branch, ESI SB RAS, Eds.; SB RAS Publishing: Novosibirsk, Russia, 2011; 303p. (In Russian)
19. Kuzmin, V.R.; Zagorulko, Y.A. Usage of the Agent-Service Approach for the Development of Intelligent Decision Support Systems in the Energy Sector. *Vestn. NSU Ser. Inf. Technol.* **2020**, *18*, 5–18. (In Russian) [CrossRef]
20. Massel, L.V.; Ivanova, I.Y.; Vorontsova, T.N.; Maysyuk, E.P.; Izhbuldin, A.K.; Zorina, T.G.; Barseghyan, A.R. Ontological aspects of the study of the mutual influence of energy and geoecology. *Ontol. Des.* **2018**, *4*, 550–561. (In Russian) [CrossRef]
21. Kuzmin, V.R.; Zarodnyuk, M.S.; Massel, L.V. Impact assessment of emissions from energy facilities on the Baikal natural area. *iPolytech J.* **2022**, *26*, 70–80. (In Russian) [CrossRef]
22. Diggle, P.J.; Tawn, J.A.; Moyeed, R.A. *Model-Based Geostatistics*; Springer: New York, NY, USA, 2007; 230p.
23. Gribov, A.; Krivoruchko, K. Empirical Bayesian kriging implementation and usage. *Sci. Total. Environ.* **2020**, *722*, 137290. [CrossRef] [PubMed]
24. Massel, A.G.; Mamedov, T.G.; Pyatkova, N.I. Computational experiment technology in research of power industries when implementing threats to energy security. *Inf. Math. Technol. Sci. Manag.* **2021**, *3*, 62–73. (In Russian)

engineering proceedings

MDPI

Proceeding Paper

Comparative Analysis of Feature Extraction Methods for Intelligence Estimation Based on Resting State EEG Data †

Tatiana Avdeenko *, Anastasiia Timofeeva and Marina Murtazina

Applied Mathematics and Computer Science Department, Novosibirsk State Technical University, 20 Karla Marksa Ave., Novosibirsk 630073, Russia; a.timofeeva@corp.nstu.ru (A.T.); musla@inbox.ru (M.M.)
* Correspondence: tavdeenko@mail.ru
† Presented at the 15th International Conference "Intelligent Systems" (INTELS'22), Moscow, Russia, 14–16 December 2022.

Abstract: This paper presents a comparative study of relationship estimation between intelligence indicators and single-channel and multi-channel feature sets extracted from resting EEG data. In the first case, the power of four frequency bands (alpha, theta, beta, delta) calculated using the discrete Fourier transform (DFT) and the power spectral density (PSD) estimated through the Welch's method for each of the channels were extracted as features from the EEG signals. In the second case, Imaginary Coherence (iMOCH) measure values for a pair of channels in the frequency bands were extracted. Graph theoretical connectivity metrics were calculated for iMOCH. As part of the experimental part of the study, the data of the EEG records of 79 subjects at rest and the values of four IQ indicators (IQ2—ability to abstract; IQ3—verbal analogies and combinatorial abilities; IQ7—figure detecting, combinatorial abilities; IQ8—spatial imagination) of the structure of intelligence were analyzed by the Amthauer method. For relationship estimation, a principal component regression was used. The performance evaluation is based on the nested Monte-Carlo cross-validation. The single-channel feature set provides the smallest standard deviation of mean absolute error. For non-verbal intelligence, the results of the multi-channel approach are better. For verbal intelligence, on the contrary, the single-channel approach gives the best result.

Keywords: EEG; resting state; intelligence; feature extraction; frequency domain; connectivity; graph measures; principal component regression

Citation: Avdeenko, T.; Timofeeva, A.; Murtazina, M. Comparative Analysis of Feature Extraction Methods for Intelligence Estimation Based on Resting State EEG Data. *Eng. Proc.* **2023**, *33*, 25. https://doi.org/10.3390/engproc2023033025

Academic Editors: Askhat Diveev, Ivan Zelinka, Arutun Avetisyan and Alexander Ilin

Published: 14 June 2023

1. Introduction

Differences in human intelligence have been the focus of research for many years. Intelligence quotient (IQ) indicators are used as a measure of intelligence. An assessment of a subject's intelligence, obtained through a series of tests, shows the subject's intelligence relative to the intelligence of an average individual in a cohort. In order to obtain a qualitative assessment of IQ using a series of tests, it is necessary that the questionnaires be localized for different countries, taking into account their linguistic and cultural backgrounds. In recent years, new methods for assessing the level of intelligence, such as electroencephalography (EEG) [1–5], have become of particular interest to scientists around the world. Since the EEG method is able to capture individual differences in brain function, this method of studying the brain has come to be considered a powerful tool for studying the biological basis of intelligence. The first works in this area focused on studying the correlation between IQ test scores and brain rhythms. In [6], it was shown that Raven's Progressive Matrices positively correlated with the alpha frequency in the prefrontal and frontal areas, while the verbal subtests of the Amthauer Intelligence Structure Test positively correlated with mean and peak alpha frequency factors.

Numerous studies prove that IQ is related to the characteristics of EEG signals in the time and frequency domains [7,8]. In [9], an approach to predicting IQ is proposed involving the analysis of the ratio of the powers of subranges of brain waves and an artificial neural network. In [10], aspects of quantitative EEG prediction of intelligence are considered, which provide moderate to strong estimates of the size of the effect of cognitive functioning. In [11], the relationship between individual EEG characteristics at rest and the level of non-verbal intelligence is studied. In [12], a new non-invasive method for measuring human intelligence is proposed using a direct approach by classifying EEG data. In [13], an approach to predicting the IQ level from EEG data based on cross-relational analysis and the SVM classification model is presented. A promising method in the field of detecting relationships between IQ and EEG at rest is the graph-theoretic approach to brain network analysis [14]. The purpose of this study is to compare approaches to identifying the relationship between intelligence measures and EEG data at rest using single-channel and dual-channel feature sets. We considered four components of the intelligence test, two of which can be attributed to verbal intelligence (using words) and two to non-verbal intelligence (using geometric objects).

The paper is organized as follows. Section 1 substantiates the relevance of the topic. Section 2 presents the theoretical background. Section 3 contains a description of materials and methods. Section 4 describes the experimental results. Section 5 summarizes the work performed in this study.

2. Theoretical Background

2.1. EEG Signals

An EEG signal is composed of the integration of neuronal activity in various spatial and temporal scales. To take an EEG, a certain number of scalp electrodes are installed on the human scalp in accordance with a standardized scheme for applying electrodes. Figure 1 shows the layout of electrodes according to the international 10–20 system. The scalp electrodes' identifiers consist of letters indicating the region of the brain and numbers (odd numbers for the left hemisphere of the brain, even numbers for the right) or the letter "z" (for the midline). Electrodes A1 and A2 are connected to the left and right earlobes and are used as reference electrodes [15].

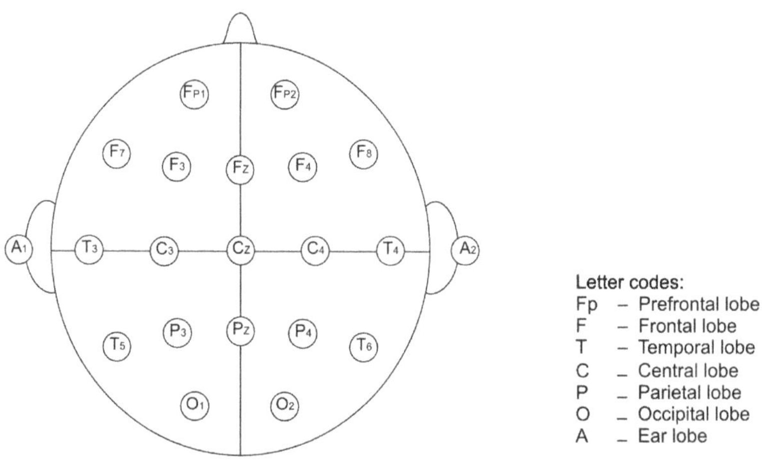

Figure 1. EEG electrode positions according to the International 10–20 System.

There are five major frequency bands. From low to high frequencies, they are called alpha (less than 4 Hz), theta (from 4 to less than 8 Hz), beta (from 8 to less than 13 Hz), delta (from 13 to less 30 Hz), and gamma (more than 30 Hz) [16].

2.2. EEG Signal Features

EEG signals are complex, so it is quite difficult to extract from them the information necessary to solve a particular problem. Currently, there are many features that can be extracted from EEG data. The set of features extracted from EEG data can be divided into several categories: measures that are calculated from the data of one channel and measures that take into account the data of two channels. The first category includes feature extraction methods based on transformation (for example, Fourier transform), time domain methods (for example, Hjort parameters), frequency domain (for example, band power), and complexity (for example, Petrosian fractal dimension) [17–19]. The second category includes measures such as cross-correlation and coherence [20].

The most demanded measure among researchers is the estimation of power in various frequency ranges. This feature extraction method requires the transformation of the EEG time series from the time domain to the frequency domain. To estimate the power spectral density (PSD), the Welch method is most often used, but the Discrete Fourier Transform (DFT) method can also be used. EEG coherence can provide information about the formation of the functional integration of brain regions. A measure of coherence can be used to determine whether two or more sensors or brain regions have the same oscillatory activity of neurons with each other [21].

2.3. Amthauer Intelligence Structure Test

One of the most popular tests for measuring general intelligence is the Amthauer intelligence structure test. In Russia, this test is used in adaptation [22]. The Amthauer intelligence structure test allows one to build an intelligence profile consisting of nine components:

- IQ1—completion of sentences, logical thinking, language skills;
- IQ2—word exception, ability to abstract;
- IQ3—verbal analogies, combinatorial abilities;
- IQ4—conceptualization, ability for abstract verbal thinking;
- IQ5—calculations, mathematical abilities;
- IQ6—number series completion, ability to operate with numbers and inductive thinking;
- IQ7—figure detecting, combinatorial abilities;
- IQ8—identification of cubes, spatial imagination;
- IQ9—remembering words, ability for short-term storage of information.

We used four subtests. The two selected subtests, IQ2 and IQ3, can be attributed to verbal intelligence, since they use words, while the other two subtests, IQ7 and IQ8, can be attributed to non-verbal intelligence, as they use geometric shapes. Non-verbal intelligence was studied in [14]. Thus, we want to confirm or not the previous results and also to compare the results for verbal and non-verbal intelligence.

3. Materials and Methods

3.1. Principal Component Regression

One of the approaches to regression analysis, used in conditions where a correlation is observed between input factors, is the construction of a principal component regression [23]. The idea is to use the principal component method to decompose the input variables into orthogonal factors.

Let the data matrix X contain n rows (observations) and m columns (input variables) normalized so that the mean is zero and the variance is equal to one. Then, the correlation matrix R can be represented as $R = X^T X$, where ^{T} means transpose. The correlation matrix R can be expressed by the matrix Λ of its eigenvectors (factor loadings) as follows:

$$R = \Lambda \Phi \Lambda^T \quad (1)$$

where the eigenvalues of R on the diagonal of the diagonal matrix Φ are the variances of the corresponding principal components.

Usually ($q << m$) principal components are used to represent correlation matrix R. Then, (1) can be rewritten as

$$\hat{R}_q = \Lambda_q \Phi_q \Lambda_q^T$$

where the index $_q$ means that instead of the m components, their smaller number q is taken. The principal components are ordered in descending order of their explanatory power to represent the correlation matrix R. Principal component scores F_q can be calculated as $F_q = X\Lambda_q$.

If we use the matrix of principal components F_q as the input matrix in the model, the problem of collinearity is weakened because of the orthogonality of columns of the matrix F_q. In order to go back from the q-dimensional space of principal components to the original m-dimensional space, it is necessary to multiply the resulting vector of estimates by the matrix Λ_q. Then, the final expression for the principal component regression estimates has the form

$$\hat{\beta} = \Lambda_q (F_q^T F_q)^{-1} F_q^T y \tag{2}$$

where y is a response vector of dimension $n \times 1$.

The covariance matrix for estimates (2) is expressed as follows:

$$var(\hat{\beta}) = \sigma^2 \Lambda_q (F_q^T F_q)^{-1} \Lambda_q^T$$

where σ^2 is the error variance.

The choice of q can be made on the basis of cross-validation. Hyperparameter selection should be separated from the model training procedure. Therefore, to evaluate the performance of the considered approaches, it is more correct to use nested cross-validation [24]. To do this, the sample is divided into three parts: training, validation, and testing. Here, they were divided in the ratio of 60%, 20%, and 20%, respectively. On the training sample, the principal component regressions are estimated for various values of the parameter q (from one to five). The validation sample is used to estimate the prediction error for various values of the parameter q and to choose the optimal value of the parameter q that provides the smallest average error. Furthermore, this optimal value is used in the procedure of estimating a principal component regression based on a sample, including training and validation together. Based on the obtained regression estimates, a prediction is built for the test sample. It is important that, here, the test sample is not involved in any way when choosing the parameter q. The procedure is based on the Monte-Carlo cross-validation [25]; that is, the division into three parts was carried out randomly.

3.2. Data Description and Model Structure

The data of EEG records of 79 subjects at rest were analyzed, as well as the values of their IQ2, IQ3, IQ7, and IQ8 components of the intelligence structure according to the Amthauer method. The sample included only female individuals aged 17 to 21. Individuals were grouped into age groups: 17 years old, 18 years old, 19 years old, and older. It turned out that the distribution of IQ2 component is not the same depending on age. Table 1shows the p-value of the two-sample Kolmogorov–Smirnov test for comparison IQ value in the age groups 18 years old, 19 years old, and older.

Table 1. The results of two-sample Kolmogorov–Smirnov test.

IQ Component	The p-Value
IQ2	0.0354
IQ3	0.5860
IQ7	0.4334
IQ8	0.6945

From the EEG data, the band power spectra calculated using DFT and PSD were extracted from various channels and frequency ranges (alpha, theta, beta, delta). There are 152 features in total. Since the records had different durations (minimum—2 min; maximum—6 min), the features were extracted for each minute. Observations for different minutes for one subject were considered as repeated. As a result, the sample size was 222.

To estimate synchronization between a pair of signals, Imaginary Coherence (iMOCH) was used, calculated using MNE Python software. The various values of the quantiles (from 10% to 80%) of strength of connections within the person were used as the threshold. Connection strengths lower than the threshold were removed. For four IQ indicators, correlations were calculated with each of the seven graph connectivity indicators (average and characteristic path length, transitivity, network modularity, diameter, eigenvector centrality, closeness centrality) in five frequency bands (alpha, theta, beta, delta, and gamma). Thus, a total of 35 extracted features were obtained.

During the analysis, it was revealed that the effect of the interaction of age and EEG features has a significant effect on IQ components. Therefore, the following model was taken as the basis:

$$y_i = \theta_0 + \sum_{j=1}^{J} \theta_j z_{ij} + \sum_{l=1}^{k} \alpha_l x_{il} + \sum_{l=1}^{k} \sum_{j=1}^{J} \gamma_{lj} x_{il} z_{ij} + \varepsilon_i \quad (3)$$

where y_i is the IQ value for the i-th observation, z_{ij} is the i-th value of the binary variable reflecting the subject's belonging to group j by age, x_{il} is the value of the l-th feature according to EEG data for the i-th observation, k is the number of features taken from the EEG data, ε_i is a random error, and $\theta_0, \ldots, \theta_J, \alpha_1, \ldots, \alpha_L, \gamma_{11}, \ldots, \gamma_{LJ}$ are the parameters to be estimated.

The model (3) is estimated using principal component regression described above.

4. Results

The results of the multi-channel approach turned out to depend on the way the brain network graph was constructed. Three alternatives were used: unweighted, weighted by the coherence values, and weighted by the reciprocal coherence values (weighted reversed). An unweighted graph has unit weights. Similar to how it was carried out in [26] to remove group bias [27], the iMOCH values were divided by the maximum value within each matrix.

Figure 2 shows that there are no significant advantages in using weighted graphs. A mean absolute error similar in magnitude can also be obtained using unit weights. Based on the results obtained, it is impossible to make recommendations regarding the choice of the threshold value. For IQ2 (Figure 2a), the minimum mean absolute error is reached at a small threshold value. At the same time, for IQ7 (Figure 2c), the error sharply increases with the threshold value 0.3. For IQ3 and IQ8 (Figure 2b,d), the minimum mean absolute error corresponds to the threshold value 0.7.

For comparison with the single-channel approach, the minimum error values that were achieved using the multi-channel approach are taken. Table 2 shows the comparison results. The standard errors of the mean are given in parentheses.

In all cases, the standard errors of the mean using the single-channel approach are almost two times smaller. This suggests that this approach allows for a more stable result. This is because when using the single-channel approach, one main component ($q = 1$) for PCR is always selected during nested cross-validation. Meanwhile, for the multi-channel approach, the number of components varies more, especially for IQ2.

Table 2. The mean absolute error using nested cross-validation.

IQ Component	Single-Channel	Multi-Channel
IQ2	5.663 (0.065)	5.902 (0.128)
IQ3	7.763 (0.078)	7.989 (0.153)
IQ7	6.860 (0.071)	6.588 (0.135)
IQ8	7.809 (0.089)	7.657 (0.165)

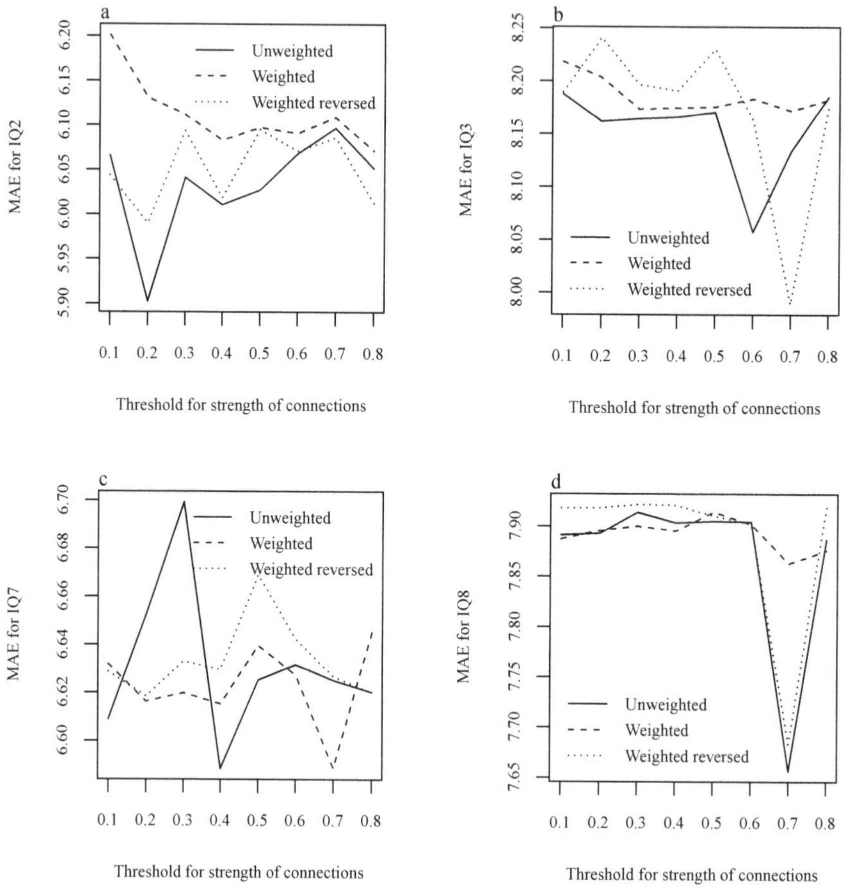

Figure 2. The mean absolute error for (**a**) IQ2; (**b**) IQ3; (**c**) IQ7; (**d**) IQ8 depending on threshold values.

For verbal intelligence (IQ2 and IQ3), the single-channel approach gives the best result. However, for non-verbal intelligence (IQ7 and IQ8), the results of the multi-channel approach are better. The complexity of the single-channel approach is that a very large number of features are extracted, which is a multiple of the number of channels. In the multi-channel approach, the number of features does not depend on the number of channels, since it takes into account the connections between them. However, there are difficulties with choosing a method for constructing a brain network graph and choosing an appropriate value of the threshold of connection strength.

Probably, a reasonable compromise would be to include both single-channel and multi-channel features as an initial subset of features in the principal component regression.

5. Conclusions

A single-channel feature set is easier to extract, but there are difficulties with the large dimensions of the feature space and with high correlations between features. To build a multi-channel feature set, it is necessary to determine the values of the weights of the graph edges, as well as to set the value for the coherence threshold. It turns out that the choice of these parameters affects the error of the IQ model built on the basis of feature set obtained using a multi-channel approach. Moreover, the empirical study results do not allow us to give recommendations on the optimal choice of the threshold. Nevertheless, it was found that it is possible to build an unweighted graph. As a result, the accuracy is not worse than in the case of a weighted graph. For verbal intelligence, the single-channel approach gives the best result. For non-verbal intelligence, on the contrary, the results of the multi-channel approach are better. This is consistent with the results of the article [14], where, for non-verbal intelligence, the graph-theoretic approach to the analysis of the brain network was used. A possible compromise is the combination of single-channel and multi-channel features in the overall principal component regression.

Author Contributions: Conceptualization, T.A.; methodology, T.A.; software, A.T. and M.M.; validation, A.T.; investigation, A.T.; writing—original draft preparation, A.T. and M.M.; writing—review and editing, T.A.; visualization, A.T. and M.M.; supervision, T.A.; project administration, T.A. All authors have read and agreed to the published version of the manuscript.

Funding: The research is supported by the Ministry of Science and Higher Education of the Russian Federation (project No. FSUN-2020-0009).

Institutional Review Board Statement: The study was conducted according to the guidelines of the Declaration of Helsinki and with the principles of the Code of Ethics of the Russian Psychological Society.

Informed Consent Statement: Informed consent was obtained from all subjects involved in the study.

Data Availability Statement: Not applicable.

Acknowledgments: The data for building the models were collected at the Department of Psychology and Pedagogy of the Novosibirsk State Technical University (NSTU) under the supervision of O.M. Razumnikova. The authors express their deep gratitude to the staff, students, and graduate students of the Department of Psychology and Pedagogy of the NSTU, who took part in the measurement of IQ, registration, and primary processing of the EEG.

Conflicts of Interest: The authors declare no conflict of interest.

References

1. Mustafal, M.; Taib, M.N.; Lias, S.; Murat, Z.H.; Sulaiman, N. AEEG spectrogram classification employing ANN for IQ application. In Proceedings of the 2013 the International Conference on Technological Advances in Electrical, Electronics and Computer Engineering (TAEECE), Konya, Turkey, 9–11 May 2013; pp. 199–203.
2. Mashiri, D. The relationship between projected IQ from QEEG and neurocognitive ability. *New Voices Psychol.* **2014**, *10*, 91–100. [CrossRef]
3. Anoor, M.M.; Anwar, K.K.H.; Ali, M.S.A.M.; Jahidin, A.H. Classification of students' IQ level using EEG-based intelligence classifier model. *J. Fundam. Appl. Sci.* **2017**, *9*, 684–694. [CrossRef]
4. Luo, S.; Chen, R.; Yang, Z.; Li, K. Intelligence level might be predicted by the characteristics of EEG signals at specific frequencies and brain regions. *J. Mech. Med. Biol.* **2021**, *21*, 2140047. [CrossRef]
5. Thatcher, R.W.; Palmero-Soler, E.; North, D.M.; Biver, C.J. Intelligence and eeg measures of information flow: Efficiency and homeostatic neuroplasticity. *Sci. Rep.* **2016**, *6*, 38890. [CrossRef]
6. Anokhin, A.; Vogel, F. EEG alpha rhythm frequency and intelligence in normal adults. *Intelligence* **1996**, *23*, 1–14. [CrossRef]
7. Firooz, S.; Setarehdan, S.K. IQ estimation by means of EEG-fNIRS recordings during a logical-mathematical intelligence test. *Comput. Biol. Med.* **2019**, *110*, 218–226. [CrossRef]
8. Avdeenko, T.V.; Timofeeva, A.Y.; Murtazina, M.S. Modified Correlation-Based Feature Selection for Intelligence Estimation Based on Resting State EEG Data. *Lect. Notes Comput. Sci.* **2022**, *13345*, 289–300.
9. Jahidin, A.H.; Taib, M.N.; Tahir, N.M.; Megat Ali, M.S.A.; Yassin, I.M.; Lias, S.; Isa, R.M.; Omar, W.R.W.; Fuad, N. Classification of intelligence quotient using EEG sub-band power ratio and ANN during mental task. In Proceedings of the 2013 IEEE Conference on Systems, Process & Control (ICSPC), Kuala Lumpur, Malaysia, 13–15 December 2013.
10. Thatcher, R.W.; North, D.; Biver, C. EEG and intelligence: Relations between EEG coherence, EEG phase delay and power. *Clin. Neurophysiol. Off. J. Int. Fed. Clin. Neurophysiol.* **2005**, *116*, 2129–2141. [CrossRef] [PubMed]
11. Stankova, E.P.; Myshkin, I.Y. Association between individual EEG characteristics and the level of intelligence. *Moscow Univ. Biol. Sci. Bull.* **2016**, *71*, 256–261. [CrossRef]
12. Kadam, S.T.; Dhaimodker, V.; Patil, M.M.; Reddy Edla, D.; Kuppili, V. EIQ: EEG based IQ test using wavelet packet transform and hierarchical extreme learning machine. *J. Neurosci. Methods* **2019**, *322*, 71–82. [CrossRef]
13. Ros Azamin, N.J.; Taib, M.N.; Jahidin, A.H.; Awang, D.S.; Megat Ali, M.S.A. IQ level prediction and cross-relational analysis with perceptual ability using EEG-based SVM classification model. *IAES Int. J. Artif. Intell.* **2019**, *8*, 436–443. [CrossRef]
14. Zakharov, I.; Tabueva, A.; Adamovich, T.; Kovas, Y.; Malykh, S. Alpha Band Resting-State EEG Connectivity Is Associated with Non-verbal Intelligence. *Front. Hum. Neurosci.* **2020**, *14*, 10. [CrossRef]
15. Sanei, S.; Chambers, J.A. *EEG Signal Processing*; John Wiley & Sons: Chichester, UK, 2007.

16. Nottage, J.F.; Horder, J. State-of-the-Art Analysis of High-Frequency (Gamma Range) Electroencephalography in Humans. *Neuropsychobiology* **2015**, *72*, 219–228. [CrossRef]
17. Mironovova, M.; Bíla, J. Fast fourier transform for feature extraction and neural network for classification of electrocardiogram signals. In Proceedings of the 2015 Fourth International Conference on Future Generation Communication Technology (FGCT), Luton, UK, 29–31 July 2015; pp. 1–6.
18. Jović, A.; Suć, L.; Bogunović, N. Feature extraction from electroencephalographic records using EEGFrame framework. In Proceedings of the 36th International Convention on Information and Communication Technology, Electronics and Microelectronics (MIPRO), Opatija, Croatia, 20–24 May 2013; pp. 965–970.
19. Bao, F.S.; Liu, X.; Zhang, C. PyEEG: An open source Python module for EEG/MEG feature extraction. *Comput. Intell. Neurosci.* **2011**, *2011*, 406391. [CrossRef]
20. Liew, S.H.; Choo, Y.H.; Low, Y.F.; Yusoh, Z.I.M.; Yap, T.B.; Muda, A.K. Comparing Features Extraction Methods for Person Authentication Using EEG Signals. In *Pattern Analysis, Intelligent Security and the Internet of Things. Advances in Intelligent Systems and Computing*; Abraham, A., Muda, A., Choo, Y.H., Eds.; Springer: Cham, Switzerland, 2015; Volume 355.
21. Bowyer, S.M. Coherence a measure of the brain networks: Past and present. *Neuropsychiatr. Electrophysiol.* **2016**, *2*, 1. [CrossRef]
22. Coemets, E.H.; Liimets, H.I. *Intellectual Tasks—Series 730. Russian Version of the Amthauer's Test Based on the Estonian Methodic*; Novosibirsk NSU Publisher: Novosibirsk, Russia, 1973; 24p.
23. Bair, E.; Hastie, T.; Paul, D.; Tibshirani, R. Prediction by supervised principal components. *J. Am. Stat. Assoc.* **2006**, *101*, 119–137. [CrossRef]
24. Krstajic, D.; Buturovic, L.J.; Leahy, D.E.; Thomas, S. Cross-validation pitfalls when selecting and assessing regression and classification models. *J. Cheminform.* **2014**, *6*, 10. [CrossRef] [PubMed]
25. Xu, Q.S.; Liang, Y.Z. Monte Carlo cross validation. *Chemom. Intell. Lab. Syst.* **2001**, *56*, 1–11. [CrossRef]
26. Mehraram, R.; Kaiser, M.; Cromarty, R.; Graziadio, S.; O'Brien, J.T.; Killen, A.; Taylor, J.-P.; Peraza, L.R. Weighted network measures reveal differences between dementia types: An EEG study. *Hum. Brain Mapp.* **2020**, *41*, 1573–1590. [CrossRef] [PubMed]
27. Onnela, J.P.; Saramäki, J.; Kertész, J.; Kaski, K. Intensity and coherence of motifs in weighted complex networks. *Phys. Rev. E* **2005**, *71*, 065103. [CrossRef] [PubMed]

engineering proceedings

MDPI

Proceeding Paper

Computer Simulation of Anti-Drone System †

Nikita Bykov [1] and Vadim Fedulov [2,*]

1 Russian University of Transport (MIIT), Obraztsova st., 9, Building 9, Moscow 127994, Russia; bykovnv@bk.ru
2 Bauman Moscow State Technical University (BMSTU), 2-nd Baumanskaya, 5, Moscow 105005, Russia
* Correspondence: vadimfedulov.bmstu@gmail.com; Tel.: +7-977-654-1722

† Presented at the 15th International Conference "Intelligent Systems" (INTELS'22), Moscow, Russia, 14–16 December 2022.

Abstract: In this article, we present the results of an anti–drone system simulation. The system is designed to counter mini unmanned aerial vehicles. A radar system with one or several antennas and an elimination system with one or more countermeasures are included in the system. The drones are destroyed by kinetic weapons. In the developed computer model, it is possible to simulate a raid of several drones against several countermeasures in an environment without obstacles. The computer model-specific feature is a discrete-event approach that provides higher calculating performance compared with the "soft time" method.

Keywords: anti-drone system; drones; UAV; simulation; modelling; radar; detection; elimination; countermeasures; discrete-event approach

1. Introduction

One of the most famous modern urban planning concepts is the concept of the Smart City. The main goal of the Smart City is effective city management and ensuring a high quality of life for citizens [1]. Urban specialists talk about the need to create a wide and advanced Internet of Things (IoT) infrastructure to implement this concept. It is expected that unmanned vehicles (cars, buses, trains, etc.) will be one of the most important parts of the IoT. For example, IoT will make it possible to improve city traffic management and to use unmanned public transport [2]. Moreover, small unmanned aerial vehicles (UAVs) may operate as a part of services for the delivery of various commodities [3]. Mini-UAVs are also widely used in modern military conflicts for aerial reconnaissance, guidance and fire adjustment of artillery [4].

However, there are a number of questions regarding the safety of using drones as part of the modern city system [5,6]. Potential security threats include drone hacking and cyber attacks on or using drones, which in turn can lead to attacks on critical urban infrastructure. Threats also exist in the field of combat application of mini-drones. One of the most effective ways to prevent illegal actions of drones is the physical destruction of drones. Such destruction methods in an urban environment should be carried out in a manner that is safe for people.

The relevance of our work can be attributed to the lack of studies on the effectiveness of complex anti-drone systems. There are few works on related topics. For example, the authors of the article [7] propose a method for determining the probability of hitting a drone using an assault rifle. They take the target geometry into account. A defence system based on the employment of cooperative interceptor drones is discussed in the study [8]. The interceptor drones operate in cooperation and can destroy the target through various countermeasures. In particular, in the article [9], the authors propose applying a group of interacting drones to physically neutralize the target using a usual net. The net blocks the target propellers. Nonetheless, there are no papers in which a complex system with countermeasure and detecting and localizing subsystems is analysed.

Citation: Bykov, N.; Fedulov, V. Computer Simulation of Anti-Drone System. *Eng. Proc.* **2023**, 33, 24. https://doi.org/10.3390/engproc2023033024

Academic Editors: Askhat Diveev, Ivan Zelinka, Arutun Avetisyan and Alexander Ilin

Published: 14 June 2023

We present the computer model of the mini unmanned aerial vehicle (UAV) countermeasures system (the anti–drone system) acting in 3D space. The model takes into account subsystems used to detect, localise, and eliminate targets. It allows us to solve a *direct task*. The direct task of a counteracting the drones' raid involves obtaining the counteraction results and accumulating statistics for given initial and boundary conditions.

2. Model Description

The anti-drone system (ADS) consists of

- The countermeasure subsystem with one or several countermeasures.
- The radar subsystem with one or more active antennas and one detector that are detect and localize targets.

The counteraction process is considered in empty 3D space. There are no obstacles, terrain folds or anything else that can hide observable objects in this space.

The computer model is developed using the Python 3 programming language. Its SimPy package [10] is applied for the discrete-event approach realization.

2.1. Simulation Process Description

The general scheme of the simulation process is shown in Figure 1.

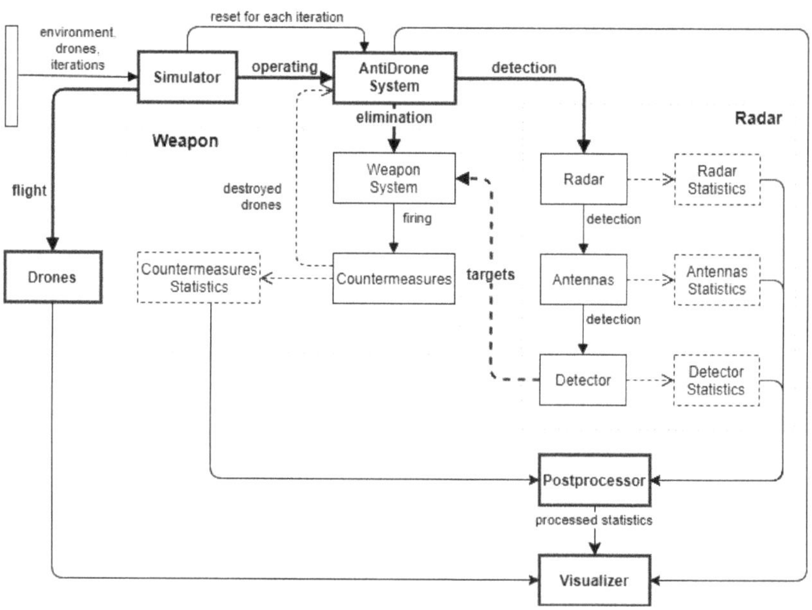

Figure 1. The general scheme of counteraction process.

Before modelling, instances of drones (`drones`), antennas (`Antennas`), detector (`Detector`), countermeasures (`Countermeasures`), weapon (`Weapon`) and radar (`Radar`) subsystems are initialized using the parameters given by the researcher. The weapon and radar subsystems instances form the anti–drone system model (`AntiDroneSystem`). This model contains the necessary fields and methods for starting and linking simulation processes. In addition, at this step, an instance of the environment (`environment`) is created to implement the discrete-event approach.

Initialized objects are passed to the simulator (`Simulator`) input. The simulator starts two parallel processes:

- The flight process (`flight`) of every drone.
- The ADS' operating process (`operating`).

The `AntiDroneSystem` instance starts two parallel processes, too:

- The countermeasures `elimination` processes.
- The radar detecting and localizing process (`detection`).

The `elimination` process begins the `firing` process of every counterweapon. The radar's `detection` process initializes the `detection` process of every antenna. The antennas, in turn, start the detector process `detection` when there is a desired signal in the detector's input. The detector tries to detect targets and, in case of successful detection, it puts the target into a shared list of targets (`targets`).

The dynamic list (`targets`) of detected targets shared between countermeasures system and radar is used by the targets distributor of the countermeasures system. The targets distributor activates the `firing` process of the selected countermeasure. When there are no targets in the `targets` list, the `firing` processes are inactive.

The antenna, detector, drones, weapon and other instances accumulate statistics. The simulation runs a given number of times (`iterations`). Before every run, states of all objects are reset and saved in files.

At the simulation end, accumulated statistics are passed in the postprocessor (`Postprocessor`) input. The postprocessor processes the data. The results are introduced to the visualization module (`Visualizer`).

2.2. The Drone Model

The model of an unmanned aerial vehicle is an entity that has geometric shape and flies along a trajectory. A trajectory is the third-ordered Bezier curve. The UAV's 3D geometry consists of parallelepiped that is the central body and four ellipsoids that are engines. The drone model is described in more detail in the article [11].

2.3. The Gun Model

The kinetic countermeasure is the same as in [11]. Additionally, the weapon model is described in more detail in those paper.

However, the weapon model presented in [11] has been improved; namely, it can interact with the radar subsystem. If the UAV is not detected by radars then the countermeasures have nothing to aim at. In cases in which the drone is detected, we make the assumption that the weapon subsystem has its own target tracking system; for example, an optoelectronic. Due to this system, the countermeasures are able to track the target and calculate the aiming point.

The criterion for destroying a target is at least one hit in its projection in the *picture plane*. The picture plane is perpendicular to the line of sight "weapon–target center of mass" and contains the UAV center of mass. The drone projection on the picture plane is calculated for each shot.

2.4. The Radar Model

The radar includes one or more antennas and one detector. The major function of the antenna is to determine the direction and distance to the target which flew into the radar area.

Coordinates from the antenna output are directed to the detector input. The purpose of the detector function is to establish the fact of detection and initiate the target track. It allows us to approximate the drone trajectory using the radar for the initial aiming of countermeasures. It requires N_{obs} detection in a row to initiate the target track. The target tracking is a necessary condition to input the target trajectory parameters into the weapon subsystem. If the target is not detected for N_{lost} times in a row then there is a mistrack. In this case, the algorithm of detection starts over [12,13].

The output voltage of the detector depends on many factors, including those of a random nature, as well as on the specific implementation of the radar data processing path [12,13]. Therefore, it was decided to simplify the detector model as follows.

The detector is characterized by probabilities of false alarm p_{FA} and correct detection p_{CD}. The random value of the output voltage U_D is generated according to the normal distribution law $U_D \sim \mathcal{N}(\mu_{U_D}, \sigma^2_{U_D})$ with the mean voltage value μ_{U_D} and the standard deviation σ_{U_D}. If $U_D > U_T$, where U_T is the threshold, then, taking into account the probability p_{CD}, a decision is made to detect the object. The target coordinates and velocity are measured. Otherwise, the target is not detected.

There is always a thermal noise in the detector input. This interfering signal negatively affects the accuracy of the detector. We describe this thermal noise as a white Gaussian noise. When there is no useful signal in the input, the output voltage value is distributed according to the normal law $U_{noise} \sim \mathcal{N}(0, \sigma^2_{U_{noise}})$ with the standard deviation $\sigma_{U_{noise}}$ [13].

The threshold voltage value U_T is determined according to the Neuman–Pirson criterion and based on a given level of false alarm probability p_{FA}. As a rule, p_{FA} is of order $10^{-2} \ldots 10^{-8}$ and p_{CD} is taken to be 0.9 [12]. The values of standard deviations σ_{U_D} and $\sigma_{U_{noise}}$ are derived experimentally or through computer simulation of the radar operating [13].

The antenna instance is initialized with distance range $[R_{min}; R_{max}]$, azimuth angle range $[\varphi_0; \varphi_1]$, elevation angle range $[\theta_0; \theta_1]$, rotation speed ω_a, etc. The R_{max} value depends on the radar cross-section (RCS) value of the drone. The antenna can operate in sector mode or a full overview (Figure 2). There are $R_{min} = 0$ and $\theta_0 = 0$ on the Figure 2.

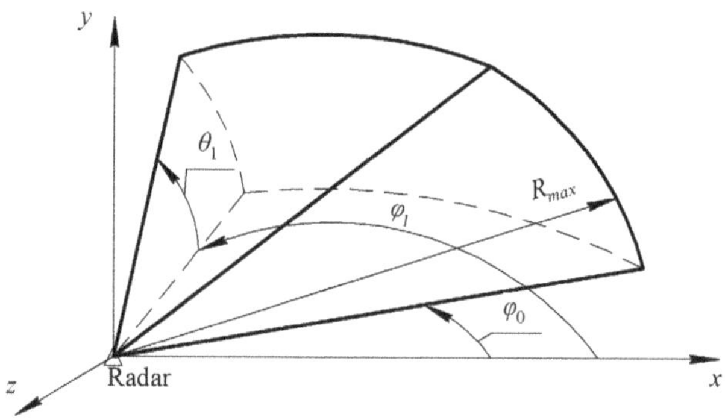

Figure 2. The antenna sector area.

3. Simulation Results

Flights of several (up to five) drones have been simulated. UAVs were eliminated by one or more countermeasures. In each run, the coordinates of the guns changed in the range ±500 m along the Ox and Oz axes; the coordinate along the Oy axis was constant and equalled 0. Each drone spawned at a random point in space with coordinates from the following ranges: $x_0 = -1000$ m, $y_0 \subset [250; 750]$ m and $z_0 \subset [-1000; 1000]$ m. For each drone, the end point of the trajectory was randomly generated with the next set of coordinates: $x_1 = 1000$ m, $y_1 \subset [250; 750]$ m and $z_1 \subset [-1000; 1000]$ m. The parameters of drones and countermeasures are presented in the Table 1. Table 2 contains the coordinates and dimensions of the drone components. All drones are the same. Parameters of the radar antenna are shown in the Table 3. A detection system with one antenna was modelled. When a target was hit, the target distributor assigned the closest tracked drones to the newly freed guns.

Table 1. The drone parameters.

Parameter Name	Distribution Law	Distribution Parameters
Initial speed	Normal	$\mathcal{N}(\mu = 21, \sigma = 3)$ m/s
Angle between initial velocity and *Ox* axis	Uniform	$[-30°; 30°]$
Initial drone angle of attack	Uniform	$[-10°; 10°]$
Final speed	Normal	$\mathcal{N}(\mu = 21, \sigma = 3)$ m/s
Angle between final velocity and *Ox* axis	Uniform	$[-30°; 30°]$
Final drone angle of attack	Uniform	$[-10°; 10°]$

Table 2. The drone geometry.

Airframe Part	Local Coordinates, mm	Axis Dimensions, mm
Central body (parallelepiped)	(0; 0; 0)	(400; 400; 250)
1st engine(ellipsoid)	(450; 300; 450)	(600; 50; 600)
2nd engine(ellipsoid)	(−450; 300; 450)	(600; 50; 600)
3rd engine(ellipsoid)	(450; 300; −450)	(600; 50; 600)
4th engine(ellipsoid)	(−450; 300; −450)	(600; 50; 600)

Table 3. The radar parameters.

Parameter Name	Unit Measure	Value
Position	m	(0; 0; 0)
Rotation speed	rad/s	$\pi/4$
Range	m	0 to 600
Azimuth angle range	—	360°
Elevation angle range	—	$[0; 90°]$
Detector mean output voltage	V	12
Detector output voltage standard deviation	V	3
Detector noise voltage standard deviation	V	1
Correct detection probability	—	0.9
False alarm probability	—	10^{-7}

The simulation results are shown in Figures 3–5. The number of iterations is 200,000.

The destroyed drones fraction shown in Figure 3 is the result of modelling the ADS without a radar subsystem. In this case, the one drone raid against an anti-drone system with one countermeasure was simulated. The countermeasure parameters are listed in Table 4. There are also accuracy variables in this table:

- σ_{aim} is a standard deviation of the aiming point at a distance of 1 m to the target;
- σ_b is a standard deviation of the mean aiming point of the burst at a distance of 1 m;
- σ_s is a standard deviation of each individual hit point at a distance of 1 m.

These parameters are used to generate random points of hits in a picture plane of a target, as described in [11]. Random points are distributed according to the normal law.

Table 4. The countermeasure parameters.

Parameter Name	Unit Measure	Value
Fire rate	shots/min	800
Burst length	—	10
Shots capacity	—	120
Aiming accuracy σ_{aim}	—	$(10^{-3}; 10^{-3})$
Burst shots accuracy σ_b	—	$(1.6 \times 10^{-3}; 1.2 \times 10^{-3})$
Individual shot accuracy σ_s	—	$(1.5 \times 10^{-3}; 1.5 \times 10^{-3})$

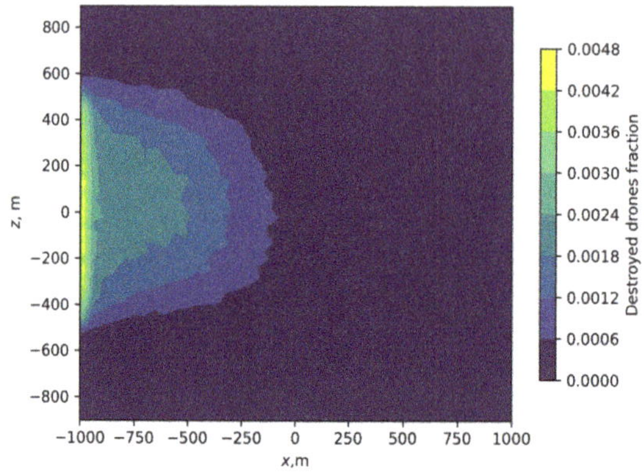

Figure 3. Spatial probability density of drones destroyed by one weapon without radar.

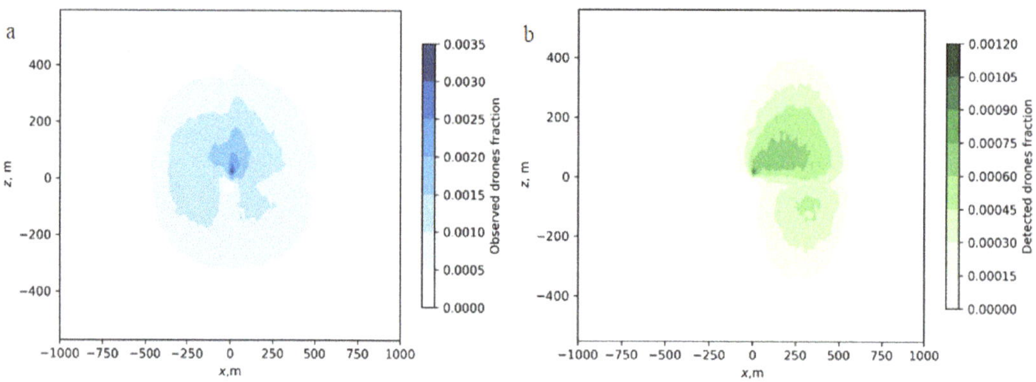

Figure 4. Spatial probability density of (**a**) observed and (**b**) detected drones.

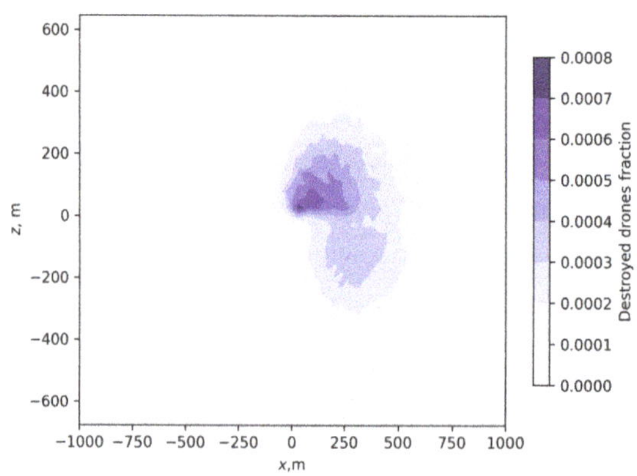

Figure 5. Spatial probability density of destroyed drones.

The contour graphs of observed and detected drones quantity are presented in Figure 4a,b. The *Oxz* plane corresponds to the top view. It can be seen from the figures that the graph in Figure 4b is shifted relative to Figure 4a in the direction of the drones' flight. This is due to the need to detect the target N_{obs} times in a row to capture its trajectory. In

Figure 5, the percentage of destroyed targets is shown. This shows that countermeasures are more effective against closer targets.

From Figures 3 and 5, it can be seen the inclusion of a radar in the counter-drone system significantly affects the position of target destruction zones in space.

The computer model also stores the simulation statistics; for example, the consumption of ammunition and the number of drones that flew to the end point [11], etc. In the future, this information can be used to calculate the efficiency criterion for anti–drone system. In addition, our computer model allows us to compare anti-drone systems that have radar subsystems with different accuracy.

4. Conclusions

The direct task of simulating the process of counteracting UAV raids can be solved employing the developed computer model. The discrete-event approach provides greater calculating performance and better scalability of the computer model.

The computer model of the counter-drone system will be improved. Obstacles and terrain folds will be added and intelligent agents of drone and detection and destruction subsystems control systems will be trained to more effectively control them. This will allow the application of machine learning or artificial neural networks and conduct research on an complex countermeasure system with intelligent agents. The intelligent agent of the countermeasure system will take into account the consequences of the UAV fall.

It will be possible to solve the inverse problem of modelling; namely, the need to create a more efficient anti-drone system structure.

Author Contributions: Supervision, N.B.; software, V.F.; writing—original draft preparation, V.F., writing—review and editing, N.B. All authors have read and agreed to the published version of the manuscript.

Funding: The work was supported by RFBR grant 19–29–06090 mk.

Data Availability Statement: Data sharing not applicable.

Conflicts of Interest: The authors declare no conflict of interest.

Nomenclature

$\mathcal{N}$	Normal distribution
μ	Mean value
σ	Standard deviation

References

1. Roldan, J.J.; Garcia-Aunon, P.; Pena-Tapia, E.; Barrientos, A. SwarmCity Project: Can an Aerial Swarm Monitor Traffic in a Smart City? In Proceedings of the 2019 IEEE International Conference on Pervasive Computing and Communications Workshops (PerCom Workshops), Kyoto, Japan, 11–15 March 2019; pp. 862–867.
2. Alsamhi, S.H.; Ma, O.; Ansari, M.S.; Almalki, F.A. Survey on Collaborative Smart Drones and Internet of Things for Improving Smartness of Smart Cities. *IEEE Access* **2019**, *7*, 128125–128152. [CrossRef]
3. Abualigah, L.; Diabat, A.; Sumari, P.; Gandomi, A.H. Applications, Deployments, and Integration of Internet of Drones (IoD): A Review. *IEEE Sens. J.* **2021**, *21*, 25532–25546. [CrossRef]
4. Calcara, A.; Gilli, A.; Gilli, M.; Marchetti, R.; Zaccagnini, I. Why Drones Have Not Revolutionized War: The Enduring Hider-Finder Competition in Air Warfare. *Int. Secur.* **2022**, *46*, 130–171. [CrossRef]
5. Laufs, J.; Borrion, H.; Bradford, B. Security and the smart city: A systematic review. *Sustain. Cities Soc.* **2020**, *55*, 102023. [CrossRef]
6. Ilgi, G.S.; Ever, Y.K. Critical analysis of security and privacy challenges for the Internet of drones: A survey. In *Drones in Smart-Cities*; Elsevier: Amsterdam, The Netherlands, 2020; pp. 204–214.
7. Racek, F.; Balaz, T.; Krejci, J.; Prochazka, S.; Macko, M. Tracking, aiming, and hitting the UAV with ordinary assault rifle. In Proceedings of the SPIE 10441, Counterterrorism, Crime Fighting, Forensics, and Surveillance Technologies, Warsaw, Poland, 11–14 September 2017; pp. 112–122.
8. Castrillo, V.U.; Manco, A.; Pascarella, D.; Gigante, G. A Review of Counter-UAS Technologies for Cooperative Defensive Teams of Drones. *Drones* **2022**, *6*, 65–101. [CrossRef]

9. Rothe, J.; Strohmeier, M.; Montenegro, S. A concept for catching drones with a net carried by cooperative UAVs. In Proceedings of the IEEE International Symposium on Safety, Security, and Rescue Robotics (SSRR), Würzburg, Germany, 2–4 September 2019; pp. 126–132.
10. SimPy. PyPI 2020. Available online: https://pypi.org/project/simpy/ (accessed on 30 August 2022).
11. Tovarnov, M.S.; Bykov, N.V.; Vlasova, N.S.; Fedulov, V.A.; Pozharsky, A.A. Computer simulation of the physical neutralization of drones in a Smart City. *J. Phys. Conf. Ser.* **2022**, *2308*, 012003. [CrossRef]
12. Bakulev, P.A. *Radar Systems. Textbook for High Schools*; Radiotechnics: Moscow, Russia, 2004.
13. Verba, V.S.; Gavrilov, K.Y.; Ilchuk, A.R.; Tatarsky, B.G.; Filatov, A.A. *Radiolocation for Everyone*; Technosphere: Moscow, Russia, 2020.

engineering proceedings

MDPI

Proceeding Paper

Integration of Mathematical and Cognitive Modelling in the Software Package "INTEC-A" †

Liudmila Massel, Aleksei Massel * and Timur Mamedov

Melentiev Energy Systems Institute Siberian Branch of the Russian Academy of Sciences (ESI SB RAS), Lermontov St., 130, Irkutsk 664033, Russia; lvmassel@gmail.com (L.M.); mamedowtymur@yandex.ru (T.M.)
* Correspondence: amassel@isem.irk.ru; Tel.: +7-395-250-0646; Fax: +7-395-242-6796

† Presented at the 15th International Conference "Intelligent Systems" (INTELS'22), Moscow, Russia, 14–16 December 2022.

Abstract: Studies of the directions of development of the energy sector (ES) are of a multivariate nature. With a combinatorial approach, in order to form possible options for the development of the energy sector, the number of options reaches several million, of which the researcher needs to select several for research. To ease the burden on the researcher, an IT environment was developed that supports a two-level technology for researching energy security problems, including the stages of qualitative and quantitative analysis using semantic modeling methods and numerical calculations. Now, the transition from semantic models to numerical calculations is carried out manually, therefore it is proposed to integrate semantic and mathematical modeling into the software packages (SP) "INTEC-A". This article discusses the integration of cognitive and mathematical modeling in the SP "INTEC-A".

Keywords: reengineering; energy sector; energy security; cognitive modeling; mathematical modeling

Citation: Massel, L.; Massel, A.; Mamedov, T. Integration of Mathematical and Cognitive Modelling in the Software Package "INTEC-A". *Eng. Proc.* **2023**, *33*, 26. https://doi.org/10.3390/engproc2023033026

Academic Editors: Askhat Diveev, Ivan Zelinka, Arutun Avetisyan and Alexander Ilin

Published: 14 June 2023

1. Introduction

Melentiev Energy Systems Institute (ESI SB RAS) actively conducts predictive studies of the energy sector (ES) of a country and its regions, taking into account the requirements of energy security (ESy) [1–4]. Energy experts built an economic and mathematical model of the energy sector for these studies. In a meaningful sense, the model is based on the traditional territorial-production model of the energy sector with blocks of electric power, heat, gas, and coal supplies, as well as oil refining (fuel oil supply). In the mathematical sense, the model is used for solving the general problem of linear programming. Several versions of software packages (SP) were developed to automate the computational experiment using this model. SP "INTEC-A" is a version obtained as a result of the direct reengineering of previous versions of the SP. The need for reengineering was due to the transition of previous versions of the SP into the category of legacy software. SP "INTEC-A" gives the user the ability to create models of the energy sector, set options for the development of the energy sector and their calculation, as well as interpreting the optimization results in the form of balance sheets. Balance tables display the presence or absence of a shortage of a certain fuel and energy resource for each region.

Such studies are multivariate in nature. To work with them, the Department of Artificial Intelligence Systems in the Energy Industry of the Institute of Energy Management and Energy of the Siberian Branch of the Russian Academy of Sciences proposed a two-level technology for studying ES problems. At the first level, using semantic modeling methods, the stage of qualitative analysis is performed. At the second level, a quantitative analysis is performed on the basis of calculations obtained using software systems [5–7].

The ESI SB RAS designed and developed an intelligent IT environment that supports a two-level research technology and includes semantic modeling tools and the SP "INTEC" [8]. The IT support environment for semantic modeling includes tools for ontological, cognitive, event, and Bayesian modeling. Currently, the transition from semantic models to numerical calculations is carried out manually; the authors proposed to automate this process within the framework of the SP "INTEC-A".

2. A Model for Predictive Studies of the Energy Sector, Taking into Account the Requirements of Energy Security

The model for optimizing the balances of fuel and energy resources with the allocation of territorial entities under conditions of possible disturbances in a meaningful sense is based on the traditional territorial production model of the energy sector with blocks of electric power, heat, gas, and coal supplies, as well as oil refining (fuel oil supply). In the mathematical sense, the model is used for solving the general problem of linear programming. When setting the problem, the constraints are written as a system of linear equations and inequalities. The objective function minimizes the amount of losses and damages from fuel and energy resources deficits among consumers. Figure 1 shows the meta-ontology of the model used.

Energy experts adopted a system of notation for variables and inequalities [9,10] to describe the models. The name of the variable (TRV) consists of seven characters (XXXYYZZ), where XXX is the code of the object (or group of objects) of extraction, production, transport, processing, and consumption of energy resources, YY is the area code, and ZZ is the code of the technology used at a particular object in a certain area. The name of the equation (IR) contains five characters (QQQYY), where QQQ is the code of the energy resource and YY is the code of the region. For example, variable D012101 stands for gas production (D01) in the northwest region (21) using technology 01. Inequality 00122 stands for natural gas surplus in the central region, where 001 is gas and 22 is the central region.

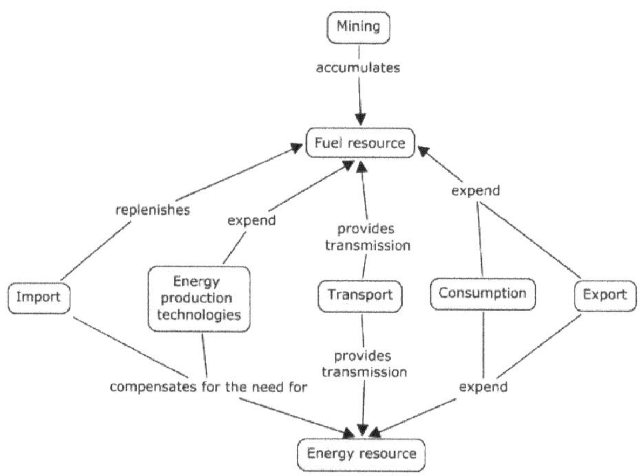

Figure 1. Metaontology of the energy sector model.

A computational experiment using this model includes the following steps: preparation of the base version of the model; preparation of options for the development of the scenario by adjusting the constraints of inequalities and variables; optimization with the help of the "solver" of each of the variants of the model; and presentation of the interpretation of the results of the balance sheets. To automate these studies, versions of the SP "INTEC" were developed at different times.

3. Versions of SP "INTEC"

SP "INTEC" is a complex of computer programs used for researching the directions of development of the energy sector taking into account the requirements of energy security. Versions of this SP have been developed and used since the 1980s on the BESM-6 and unified computer system [11,12]. SP "INTEC-A" is a new version developed on the basis of existing versions of "INTEC" and "INTEC-M' for personal computers [7,13]. A decision was made to reengineer them due to the transition of previously created SP versions into the category of legacy software. The main functionality of the new version of the SP "INTEC-A" includes: formation of energy sector models; optimization of the amount of costs and losses from fuel and energy shortages; and interpretation of optimization results in the form of balance tables. Figure 2 shows the architecture of the SP "INTEC-A".

The information model agent was designed for the convenient formation of an information model. The agent has the following functionality: formation of various patterns of the model; formation of technological dictionaries; formation of sets of variables, linear inequalities and objective functions; and multicriteria adjustments for variables and linear inequalities. The model pattern is understood as a system of notation for variables and inequalities/equations.

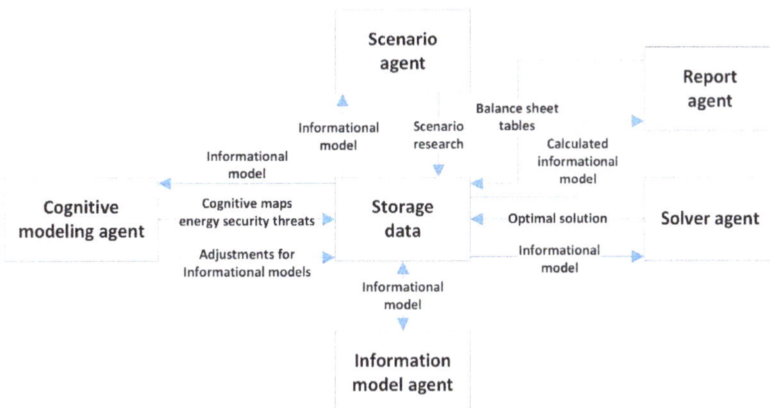

Figure 2. Architecture of SP "INTEC-A".

The scenario agent was designed to build a computational scenario, the elements of which are variants of the energy sector model. This agent acts as a coordinating agent. Its task is to set various situations and form separate scenarios, call other agents, and provide control over user actions. The agent supports multivariate studies.

The solver agent searches for the optimal solution for each version of the model using the double simplex method, designed to solve the general linear programming problem. The agent is implemented in Python using the PuLP library.

The report generation agent builds various tabular reports containing indicators of the results of a computational experiment that are of interest to the researcher, such as: balances of fuel and energy resources for each region, group of regions, and the country as a whole; interregional flows of various types of fuel; and assessment of the efficiency of energy resources and technological methods.

Integration of the cognitive modeling tool into the SP "INTEC-A" was proposed, to ease the load on the user in multivariate calculations. The following section shows the possibility of forming and correcting the computational scenario and displaying the results of its calculation using cognitive maps.

4. Development and Integration of a Cognitive Modeling Agent in the SP "INTEC-A"

Traditionally, a combinatorial approach to multivariate calculations has been used in studies of energy security problems, which assigned the expert with the task of choosing suitable solutions from a variety of solutions (up to several million options) [14]. In this regard, a two-level technology was proposed [8], integrating the stages of qualitative analysis (using semantic modeling tools) and quantitative analysis (using linear economic and mathematical models and traditional software systems, in this case, SP "INTEC-A"). It was assumed that using qualitative analysis significantly reduces the number of options that need to be calculated, which will help to significantly reduce the burden on the expert.

The intelligent IT environment integrates semantic and mathematical (INTEC-A) modeling tools and provides support for the proposed two-level research technology. At the first level, for semantic modeling, the use of event, Bayesian, ontological, and the cognitive models and tools supporting them, was proposed.

To support cognitive modeling, the CogMap tool was developed to create, view, edit, and analyze cognitive maps to determine:

- Concepts that affect the development of the energy sector or the energy system of the region.
- Strategic threats to ESy.
- Causal relationships between concepts and their weights.
- Preventive, operational, and liquidation measures that affect the scenarios for the development of the energy sector/power plant, directly for each threat [15].

At the second level, SP "INTEC-A" is used for quantitative analysis. Nowadays, the transition from cognitive maps to a SP is performed "manually". To automate the transition, it is required to develop a cognitive modeling support agent and implement it in the SP "INTEC-A". The development of a cognitive modeling support agent based on CogMap was proposed, taking into account modern requirements. The integration of the cognitive modeling agent into the SP "INTEC-A" will provide the following features:

- Alternative interface for working with the energy sector model.
- Interpretation of calculation results using cognitive models.
- Reducing the number of uncertainty factors due to expert assessments.
- Formation of a computational scenario using a cognitive model.
- Identification of implicit links, their presentation in an explicit form, and their formalization.

5. The Technology of the Computational Experiment after the Integration of the Cognitive Modeling Agent into the SP "INTEC-A"

The technology of the computational experiment after the integration of the cognitive modeling agent into the SP "INTEC-A" is shown in Figure 3 in the form of an algorithm. Let us consider the algorithm in more detail. An "empty" computational scenario is automatically created when starting SP "INTEC-A'. Next, the user either uploads an existing version of the model or creates it using the information modeling agent. The model variant is considered the base after adding the resulting model variant to an empty computational scenario. The corresponding concepts of the cognitive map are automatically generated for each element of the base version of the model. The user independently generates additional concepts of energy security threat factors or uses previously saved ones. The elements of the model are variables, inequalities, and elements of balance tables. Further, the user selects the concepts necessary for the study and initializes the values of the ES threat concepts, and then establishes links between the concepts to form a cognitive model. The generated basic version of the model is calculated using a solver agent. The values of the concepts of the cognitive model are initialized based on the optimization results. The user selects concepts, makes adjustments to their characteristics, and calculates the resulting version of the model using a solver agent to form a development option for the energy sector. The values of concepts and relationships are updated according to the results of calculations. A search is also automatically performed after calculating a version of the model for concepts that are not included in the cognitive model, but whose values have been changed as a result of the calculation. As a result, a set of cognitive models is obtained, each of which corresponds to a variant of the model used in the computational scenario.

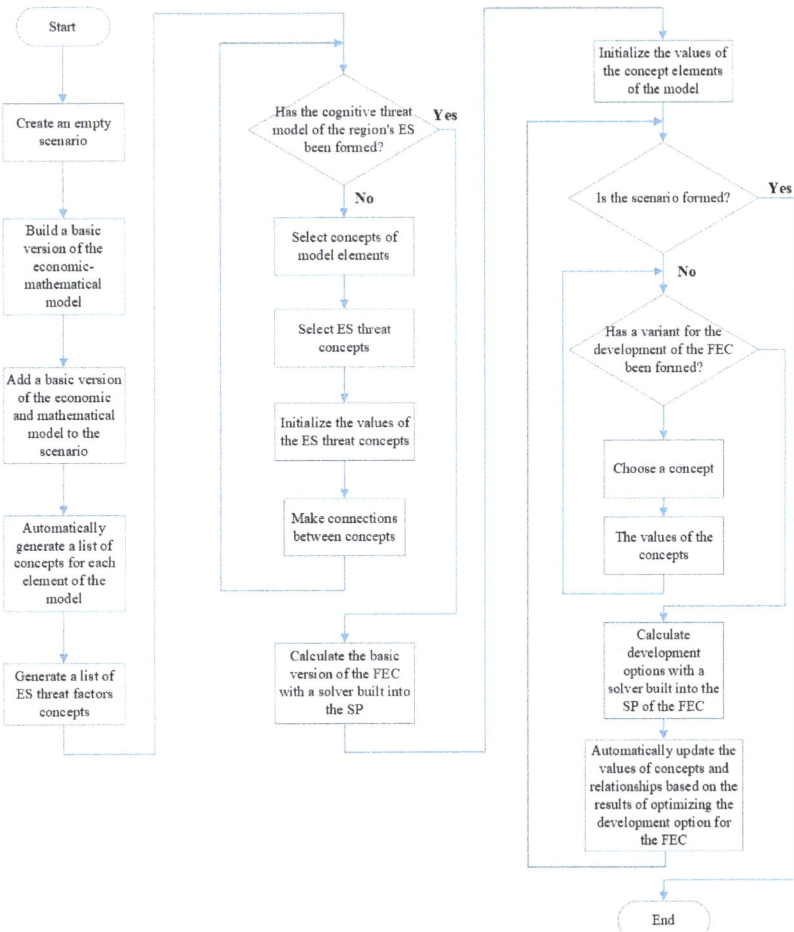

Figure 3. The technology of the computational experiment after the integration of the cognitive modelling agent into the SP "INTEC-A".

6. Development and Integration of a Cognitive Modeling Agent in the SP "INTEC-A"

Damage to the gas pipeline running from Western Siberia through the Urals to the European part of Russia is a very dangerous threat due to the dominance of natural gas in the production of electricity and heat. To demonstrate the possibilities of calculations, the situation of an accident at a gas pipeline section was considered.

Consider a 28% reduction in natural gas transport from the Urals Federal District (FD) to the Volga Federal District. Significant changes in the consumption of fuel resources should occur in the Central Federal District. Most cogeneration or combined heat and power (CHPPs) in the Central Federal District use gas as a fuel resource, most of the gas is produced in the Urals Federal District, and gas is transported from the Urals Federal District through the Volga Federal District to the Central Federal District.

The required concepts were identified in the model to study the possibilities of gas diversification for CHPPs, condensing power plants (CPPs), and boiler houses in the Central Federal District. The concept of the threat of failure of the gas pipeline threads, external to the model, was also added. The definition and initialization of the weight coefficients of links between concepts were carried out by an expert. Figure 4 shows a cognitive map (a graphic representation of a cognitive model) corresponding to this stage of the algorithm.

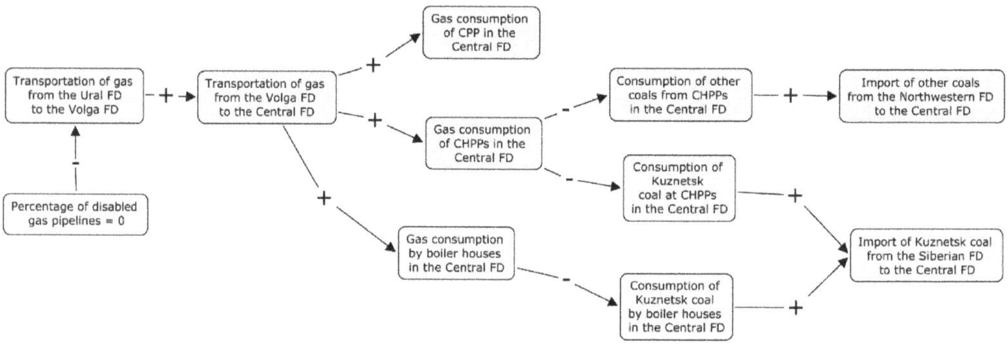

Figure 4. Cognitive map of the decline in natural gas transport from the Urals Federal District to the Volga Federal District.

The values of the model concepts were initialized in conventional units after using the solver to find the optimal solution for the base version of the model. At this stage, the cognitive map resembled that shown in Figure 5.

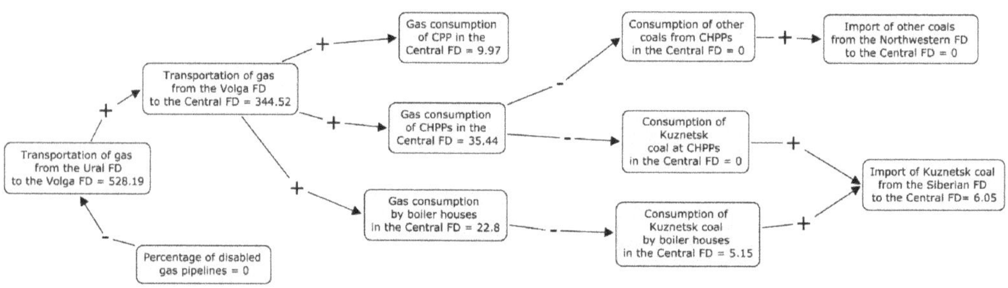

Figure 5. Map (Figure 4) indicating the values of the concepts of the model in arbitrary units.

It is necessary to identify a certain concept and make adjustments to its characteristics to form a model development option that characterizes an accident on a gas pipeline, using a cognitive model. In the corresponding concept, we indicated that the percentage of disabled gas pipeline threads has increased from zero to twenty-eight percent. As a result of the search for the optimal solution by the solver agent, we obtained the following changes in the values of the concepts presented in Figure 6.

As seen in Figure 6, according to the results of the computational experiment, if 28% of the gas pipeline from the Ural Federal District to the Volga Federal District fails, gas supplies to the Central Federal District can be compensated for by the supply of other coals from the Northwestern Federal District and the supply of Kuznetsk coal from the Siberian Federal District.

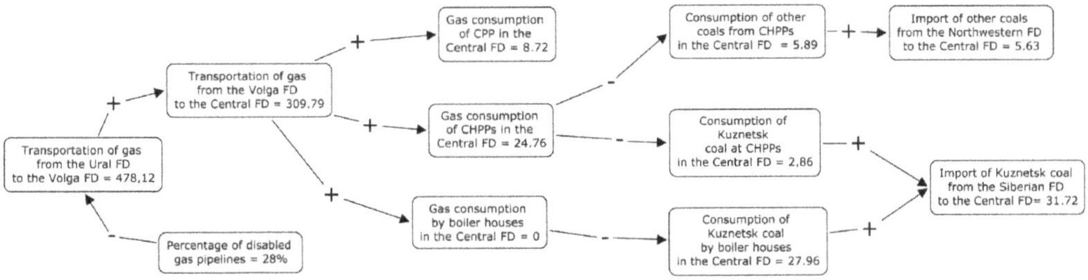

Figure 6. Map (Figure 5) with changes in the meanings of concepts.

7. Conclusions

This article considered the current state of predictive studies of the fuel and energy complex (FEC) of a country and its regions, taking into account the requirements of energy security (ES). The proposal was to move from a combinatorial approach to a computational experiment to a two-level research technology, where, at the first level, using semantic modeling methods, the stage of qualitative analysis was performed and, at the second level, quantitative analysis was carried out based on calculations obtained using software systems.

Based on the cognitive modeling tool CogMap, the development of a cognitive modeling agent and its integration into the SP "INTEC-A" were proposed. The possibilities of the developed agent were considered and the technology of the computational experiment after the integration of the cognitive modeling agent into the SP "INTEC-A" was proposed. An example of such an experiment was given for the scenario of reducing the transport of natural gas from the Ural Federal District to the Volga Federal District, using the construction of cognitive maps for this scenario.

Author Contributions: Conceptualization, L.M.; methodology L.M. and A.M.; software, T.M.; validation, A.M.; writing—original draft, A.M.; writing—review & editing, L.M. All authors have read and agreed to the published version of the manuscript.

Funding: The research was funded by Russian Science Foundation, according to the research project No. 22-21-00841.

Institutional Review Board Statement: Not applicable.

Informed Consent Statement: Not applicable.

Data Availability Statement: Not applicable.

Conflicts of Interest: The authors declare no conflict of interest.

References

1. Senderov, S.M.; Pyatkova, N.I.; Krupenev, D.S. Hierarchy of Models for the Study of National and Regional Energy Security. *Energy Secur. Energy Syst. Res.* **2019**, *2*, 32–42. (In Russian) [CrossRef]
2. Senderov, S.M.; Rabchuk, V.I. On the system of fundamental indicators for monitoring the fulfillment of the requirements of the energy security doctrine of the Russian Federation in terms of ensuring reliable fuel and energy supply to consumers of energy resources within the country. *E3S Web Conf.* **2019**, *139*, 84–93. (In Russian) [CrossRef]
3. Senderov, S.M.; Rabchuk, V.I.; Krupenev, D.S. *Reliability of Fuel and Energy Supply to Consumers from the Standpoint of Ensuring Energy Security*; Siberian Branch of the Russian Academy of Sciences: Siberia, Russia, 2022. (In Russian) [CrossRef]
4. Pyatkova, N.I.; Beresneva, N.M. Determination of critical elements of the fuel and energy complex from the standpoint of reliable energy supply. *Izv. Ross. Akad. Nauk. Energy* **2020**, *209*, 72–81. (In Russian) [CrossRef]
5. Edelev, A.V.; Pyatkova, N.I.; Chemezov, A.V. Software complex "Correction" for research on the long-term development of the fuel and energy complex of Vietnam. *Softw. Prod. Syst.* **2014**, *26*, 211–216. (In Russian)
6. Danilov, G.K.; Edelev, A.V. Development of a geoinformation system for the study of the survivability of energy systems. *Syst. Anal. Data Process.* **2022**, *1*, 41–58. (In Russian) [CrossRef]
7. Fartyshev D.A. Development of a multi-agent SP INTEC-M for research into the problem of energy security. *Softw. Prod. Syst.* **2020**, *22*, 126–129. (In Russian)

8. Massel A.G. Methodological approach to the organization of intellectual support for research into the problem of energy security. *Inf. Technol.* **2010**, *16*, 32–36. (In Russian)
9. Pyatkova, N.I.; Senderov, S.M. Methodological and model aspects of research into the functioning and development of electric power systems from the standpoint of energy security. *Electr. Transf. Distrib.* **2020**, 4, 50–53. (In Russian)
10. Pyatkova, N.I.; Beresneva, N.M. Model and tools for the study of critical energy infrastructures. In Proceedings of the Eleventh International Conference "Management of Large-Scale System Development" (MLSD'2018), Moscow, Russia, 1–3 October 2018; Volume 2, pp. 481–482. (In Russian)
11. Antonov, G.A.; Antonov, N.N.; Nemolyaev, S.I.; Fartyshev, N.N. Computing complex ENERGY. Features of implementation and development. In *Automation of Research in the Development of the Energy Complex*; SEI SB AS USSR: Irkutsk, Russia, 1986; pp. 5–21. (In Russian)
12. Antonov, G.A.; Antonov, N.N.; Belanova, G.A.; Nemolyaev, S.I. Software-computing complex for the analysis of development and functioning of EC taking into account survivability. In *Modern Problems of System Research in the Energy Sector*; SEI SB AS USSR: Irkutsk, Russia, 2019; pp. 5–14. (In Russian)
13. Massel, L.V.; Boldyrev, E.A. Modeling and development of modern software systems for energy research. *Comput. Technol.* **2020**, *7*, 59–70. (In Russian)
14. Edelev, A.V.; Zorkaltsev, V.I.; Van Binh, D.; Hoai Nam, N. Formation of options for the development of energy in Vietnam using combinatorial modeling methods. *Softw. Prod. Syst.* **2017**, *30*, 172–179. (In Russian)
15. Pyatkova, N.I.; Massel, L.V.; Massel, A.G. Methods of situational management in research of problems of energy security. *Izv. RAN Energy* **2016**, *4*, 156–163. (In Russian)

engineering proceedings

MDPI

Proceeding Paper

Application of Neural Networks to Power Analysis †

Alla Levina and Roman Bolozovskii *

Laboratory of Fundamentals of Intelligent Systems "LETI", Saint Petersburg Electrotechnical University, Professora Popova Str., 5, 197376 Saint-Petersburg, Russia
* Correspondence: bolozovskii@gmail.com

† Presented at the 15th International Conference "Intelligent Systems" (INTELS'22), Moscow, Russia, 14–16 December 2022.

Abstract: The purpose of this work is to research the possibility of a side-channel attack, more precisely power consumption attack (recovering the encryption key according to the board's power consumption schedule) on the AES-128 algorithm implemented in hardware. Basically, various methods can be used to make an attack, including SPA (Simple Power Consumption Attack) and DPA (Differential Power Consumption Attack). SPA methods involve a simple visual analysis of energy consumption graphs, while DPA involves the use of statistical methods to recover the encryption key. One way to make side-channel attacks more effective is to implement machine learning methods for the described purposes.

Keywords: cryptography; AES; side-channel attacks; power analysis attack; neural networks

Citation: Levina, A.; Bolozovskii, R. Application of Neural Networks to Power Analysis. *Eng. Proc.* **2023**, *33*, 27. https://doi.org/10.3390/engproc2023033027

Academic Editors: Askhat Diveev, Ivan Zelinka, Arutun Avetisyan and Alexander Ilin

Published: 15 June 2023

1. Introduction

The purpose of this research work is to study the possibility of implementing machine learning to side-channel attack, more precisely to speed up the process of analyses of the oscillogram of the power consumption (power consumption) of the encryption board (traces) [1].

Side-channel attacks (SCA) are a powerful type of cryptographic attack that takes advantage of physical leakages, such as electromagnetic radiation or changes in power consumption. Early SCAs involved timing attacks to break Rivest–Shamir–Adleman (RSA), Diffie–Hellman (DH), and Digital Signature Standard (DSS) encryption by measuring the execution time of different code branches and operations. Later, it was discovered that device power consumption leaks information about Data Encryption Standard (DES) and Advanced Encryption Standard (AES) by using both simple [2] and differential analysis [3–6]. Modern SCAs use many non-invasive ways to measure leakage, for example, electromagnetic signals [7,8] or acoustic cryptanalysis [9,10]. More works about different SCA on various cryptographic algorithms can be found in [11–14].

In this work, we will be focusing on the power consumption attack on algorithm AES-128. During the operation, bits of the key used by the encryption board will be restored from the original traces. As initial data, power consumption traces, and encryption keys are needed, they were used during the operation of the board. These data can be obtained using an oscillogram from a programmable logic integrated circuit, on which any encryption algorithm is above.

For the purposes of this work, a TinyAES dataset was selected, it was taken from the board on which the AES-128 algorithm is implemented. Connection data from plaintext, encryption key, ciphertext, and a set of 20,000 points representing a power consumption waveform. In the end, the implementation of research involves obtaining a ready-made algorithm. The extraction of the generated key from the original traces using statistical methods, as well as machine learning methods and neural networks.

The paper is organized as follows. Section 1 describes basic concepts for the work of neural networks and the method of its implementation, Section 2 demonstrates the results of the implementation of neural networks to power analysis attack on AES-128, and Section 4 summarizes the results of presented researches and plans for future studies.

2. Neural Networks

In this work, a machine learning method will be used as neural networks [15,16]. Neural networks are based on the principle of connectivism—a large number of relatively simple elements are connected in them, and learning comes down to building the optimal structure of connections and setting the connection parameters. To build an artificial neural network (ANN) we will use the same structure. Like a biological neural network, an artificial one consists of neurons interacting with each other, but it is a simplified model. So, for example, an artificial neuron that makes up an ANN has a much simpler structure, it has several inputs on which it receives various signals, transforms them, and transmits them to other neurons. In other words, an artificial neuron is such a function $R^T - > R$, which converts multiple inputs into one output.

As you can see in Figure 1, the neuron has n inputs x_i, each of which has a weight w_i, by which the signal passing through the connection is multiplied. After that, the weighted signals $x_i \cdot w_i$ are sent to the adder, which aggregates all the signals into a weighted sum. This sum is also called *net*. In this way:

$$net = \sum_{i=1}^{n} x_i \cdot w_i = x \cdot w^T. \tag{1}$$

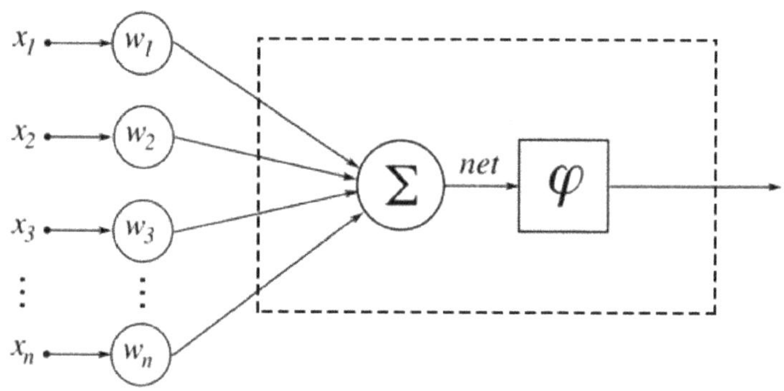

Figure 1. Scheme of an artificial neuron.

Just like that, it is rather pointless to transfer the weighted sum *net* to the output—the neuron must somehow process it and form an adequate output signal. For these purposes, an activation function is used, which converts the weighted sum into some number, which will be the output of the neuron. The activation function is denoted $\phi(net)$. Thus, the output of an artificial neuron is $\phi(net)$.

Neural network training is the search for such a set of weight coefficients, in which the input signal, after passing through the network, is converted into the output we need.

This definition of "training a neural network" is consistent with biological neural networks. Our brain consists of a huge number of interconnected neural networks, each of which individually consists of neurons of the same type (with the same activation function). Our brains learn by changing synapses, elements that increase or decrease the input signal.

If we train the network using only one input signal, then the network will simply "remember the correct answer", and as soon as we give a slightly modified signal, instead of the correct answer, we will receive nonsense. We expect the network to be able to generalize some features and solve the problem on various input data. It is for this purpose that training samples are created.

A training sample is a finite set of input signals (sometimes along with the correct output signals) on which the network is trained. After the network has been trained, that is, when the network produces correct results for all input signals from the training sample, it can be used in practice. However, before immediately using a neural network, the quality of its work is usually assessed on the so-called test set. A test sample is a finite set of input signals (sometimes together with the correct output signals), which are used to evaluate the quality of the network. Neural network training itself can be divided into two approaches, supervised learning, and unsupervised learning. In the first case, the weights are changed so that the network's answers are minimally different from the ready-made correct answers, and in the second case, the network independently classifies the input signals.

3. Results

In this research work, a model was developed based on the theory of neural networks. An implemented neural network model allows one to obtain information about the secret key that was used for encryption from the captured waveforms of the power consumed by the encryption device.

The initial oscillograms were loaded and normalized to values from −1 to 1, the graph of one of the traces is shown in Figure 2.

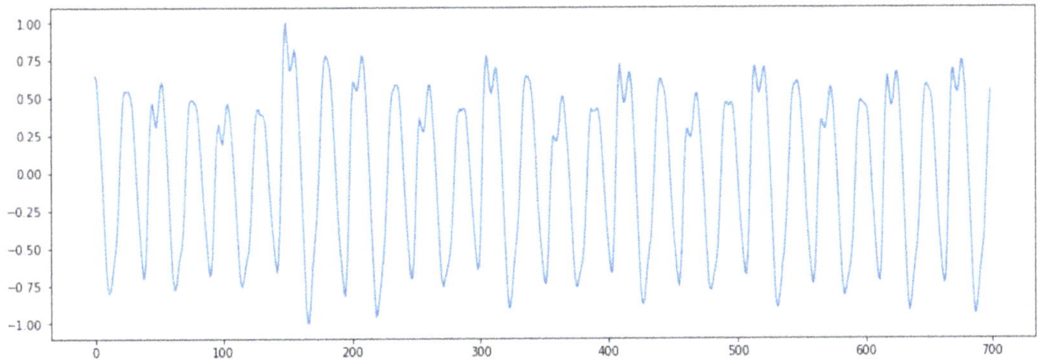

Figure 2. Example of an oscillogram of the power consumption of an encryption module.

Further, for each oscillogram, the bytes of interest were allocated-outputs from the Sbox for the third round. After that, these bytes were transferred to one main vector (vector with dimension 256, where 1 is at the position corresponding to the value of this byte and 0 in all other positions).

The resulting waveforms and one main vector were used to train the SCANet neural network. Graphs of changes in the loss function, and classification accuracy are shown in Figure 3.

To evaluate the final result, the prediction of the neural network, it is necessary to use the rank function, its graph is shown in Figure 4. This function indicates how reliable the values provided by the neural network are. The lower the rank, the easier it is to pick up the bytes that were used in the encryption. As can be seen from the graph, with an increase in the number of traces, the rank decreases, which means a decrease in the complexity of determining the initial key.

When using an untrained neural network to determine the initial key, the rank function does not fall with an increase in the number of traces, that is, the complexity of obtaining the initial key remains at the level of random predictions, which is shown in Figure 5.

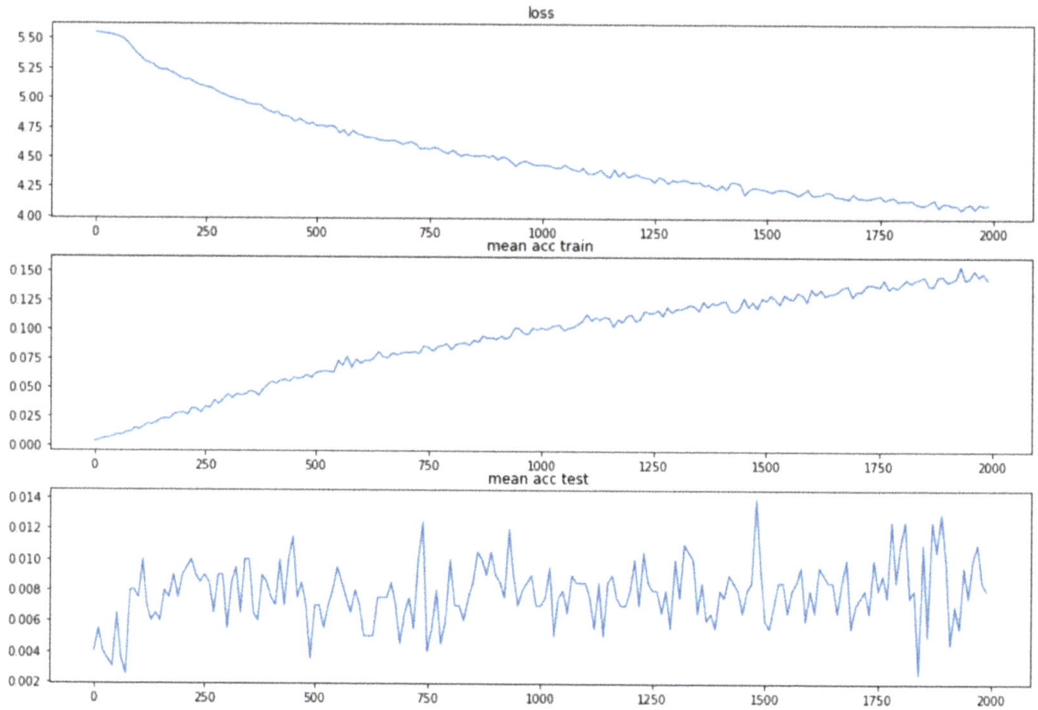

Figure 3. Loss function and classification accuracy plots.

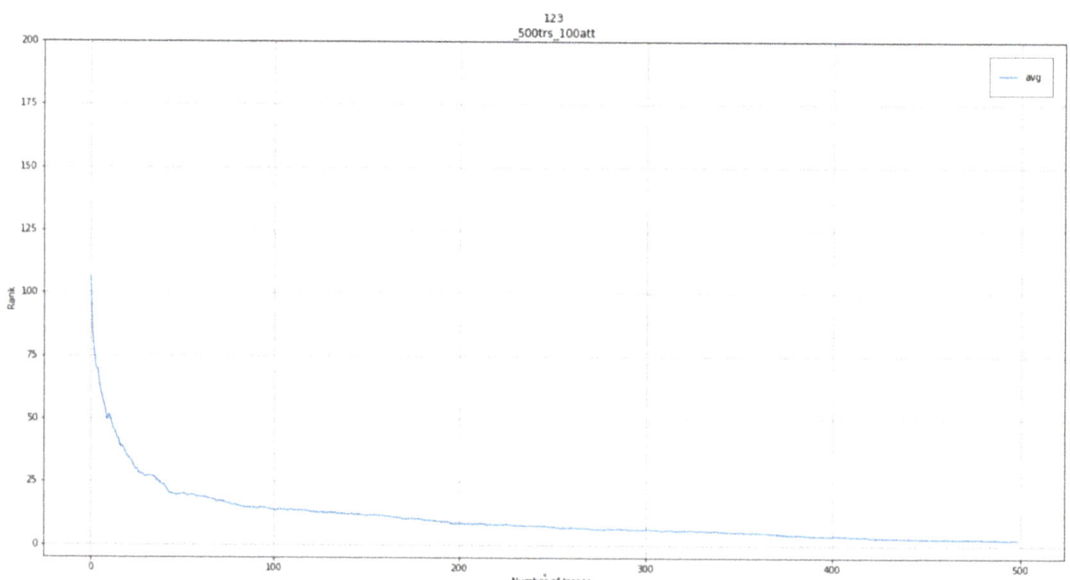

Figure 4. Rank function of the trained network.

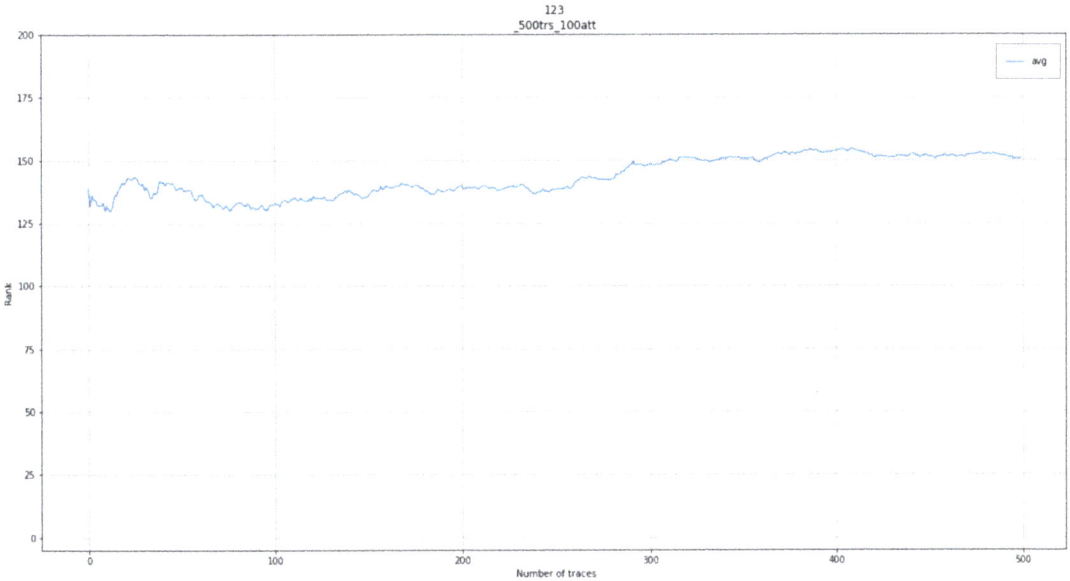

Figure 5. Rank function of an untrained network.

4. Conclusions

In the course of the research work, a neural network model was obtained, which allows, using the oscillogram of the power consumption of the encryption board, to restore the encryption key. The ASCAD dataset was used as the initial data, containing oscillograms, encryption keys, source texts, and points of interest—outputs from the Sbox function of the third round of the AES-128 cipher. The model is a convolutional neural network consisting of four convolutional and two fully connected layers, with Softmax activation function and CrossEntropyLoss as loss function. The quality of the resulting network was verified by the classical method for energy consumption analysis problems, namely, using the rank function.

Author Contributions: Conceptualization, A.L. and R.B.; methodology, R.B.; software, R.B.; validation, A.L. and R.B.; formal analysis, R.B.; investigation, R.B.; resources, R.B.; data curation, R.B.; writing—original draft preparation, A.L.; writing—review and editing, A.L. and R.B; visualization, A.L. and R.B; supervision, A.L.; project administration, A.L.; funding acquisition, A.L. All authors have read and agreed to the published version of the manuscript.

Funding: This work was supported by the Ministry of Science and Higher Education of the Russian Science Foundation (Project "Goszaadanie" No075-01024-21-02 from 29.09.2021, FSEE-2021-0015).

Informed Consent Statement: Not applicable.

Conflicts of Interest: The authors declare no conflict of interest.

References

1. Mahanta, H.J.; Azad, A.K.; Khan, A.K. Power analysis attack: A vulnerability to smart card security. In Proceedings of the 2015 International Conference on Signal Processing and Communication Engineering Systems, Guntur, India, 2–3 January 2015; pp. 506–510.
2. Bechtsoudis, A.; Sklavos, N. Side channel attacks cryptanalysis against block ciphers based on FPGA devices. In Proceedings of IEEE Computer Society Annual Symposium on VLSI, Kefalonia, Greece, 5–7 July 2010.
3. Alioto, M.; Giancane, L.; Scotti, G.; Trifiletti, A. Leakage power analysis attacks: A novel class of attacks to nanometer cryptographic circuits. *IEEE Trans. Circuits Syst. I Regul. Pap.* **2009**, 57, 355–367. [CrossRef]

4. Prouff, E.; Rivain, M.; Bevan, R. Statistical analysis of second order differential power analysis. *IEEE Trans. Comput.* **2009**, *58*, 799–811. [CrossRef]
5. Longo, J.; Mulder, E.D.; Page, D.; Tunstall, M. SoC it to EM: Electromagnetic side-channel attacks on a complex system-on-chip. In Proceedings of the International Workshop on Cryptographic Hardware and Embedded Systems, Saint-Malo, France, 13–16 September 2015; Springer: Berlin/Heidelberg, Germany, 2015; pp. 620–640.
6. Kocher, P.; Jaffe, J.; Jun, B. Differential power analysis. In Proceedings of the Annual International Cryptology Conference (CRYPTO), Santa Barbara, CA, USA, 15–19 August 1999.
7. National Security Agency. *NACSIM 5000 Tempest Fundamentals (Report)*; National Security Agency: Fort Meade, MD, USA, February 1982.
8. Genkin, D.; Pipman, I.; Tromer, E. Get your hands off my laptop: Physical side-channel key-extraction attacks on PCs. *J. Cryptogr. Eng.* **2015**, *5*, 95–112. [CrossRef]
9. Backes, M.; Dürmuth, M.; Gerling, S.; Pinkal, M.; Sporleder, C. Acoustic side-channel attacks on printers. In Proceedings of the 19th USENIX Conference on Security (USENIX Security'10), Washington, DC, USA, 11–13 August 2010.
10. Genkin, D.; Shamir, A.; Tromer, E. Acoustic cryptanalysis. *J. Cryptogr. Eng.* **2017**, *30*, 392–443. [CrossRef]
11. Levina, A.; Mostovoi, R.; Sleptsova, D.; Tcvetkov, L. Physical model of sensitive data leakage from PC-based cryptographic systems. *J. Cryptogr. Eng.* **2019**, *9*, 393–400 [CrossRef]
12. Ometov, A.; Orsino, A.; Andreev, S.; Levina, A.; Borisenko, P.; Mostovoy, R. Mobile social networking under side-channel attacks: Practical security challenges. *IEEE Access* **2017**, *5*, 2591–2601. [CrossRef]
13. Levina, A.; Varyukhin, V.; Kaplun, D.; Zamansky, A.; van der Linden, D. A Case Study Exploring Side-Channel Attacks On Pet Wearables. *IAENG Int. J. Comput. Sci.* **2021**, *48*, 878–883.
14. Levina, A.B.; Sleptsova, D. Correlation Side-Channel Attack on Mifare Classic Cards. In Proceedings of the ISPIT 2015 Information Security Conference, St. Petersburg, Russia, 5–6 November 2015; pp. 53–56.
15. Maghrebi, H. Deep Learning based Side-Channel Attack: A New Profiling Methodology based on Multi-Label Classification. *IACR Cryptol. Eprint Arch.* **2020**, *2020*, 436
16. Maghrebi, H.; Portigliatti, T.; Prouff, E. Breaking cryptographic implementations using deep learning techniques. In Proceedings of the International 46 Conference on Security, Privacy, and Applied Cryptography Engineering, Hyderabad, India, 14–18 December 2016; Springer: Berlin/Heidelberg, Germany, 2016; pp. 3–26.

engineering proceedings

MDPI

Proceeding Paper

Modified Evolutionary Test Data Generation Algorithm Based on Dynamic Change in Fitness Function Weights †

Tatiana Avdeenko ‡ and Konstantin Serdyukov *,‡

Applied Mathematics and Computer Science Department, Novosibirsk State Technical University, 20 Karla Marksa Ave., 630073 Novosibirsk, Russia; avdeenko@corp.nstu.ru

* Correspondence: zores@live.ru

† Presented at the 15th International Conference "Intelligent Systems" (INTELS'22), Moscow, Russia, 14–16 December 2022.

‡ These authors contributed equally to this work.

Abstract: In this paper, we investigate a modification of the method of data generation for multiple code paths within a single launch of the genetic algorithm. This method allows the consideration of the remoteness of paths initiated by different test cases by introducing an additional additive component into the fitness function. Previous studies have shown that the parameter defining the relationship between the different components of the fitness function has a rather strong effect on code coverage. To eliminate this effect, we propose the modification of the first component of the fitness function, which is responsible for path complexity. This modification is based on a dynamic change in code statement weights between generations to achieve greater population diversity. We propose several methods for implementing this modification, divided into two groups. In the first group, the statement weights change depending only on the fact of statement coverage in a generation, and the rate of change depends on the number of previous generations in which it was covered. In the second group, the rate of change depends on the proportion of statement coverage by the test sets in the previous generation. Each of the proposed methods is investigated to achieve complete coverage with different values of the parameter defining the ratio of the components of the fitness function. As a result, the best method is determined, which eliminates the need to determine this parameter for each testing code, thus achieving a greater universality for the algorithm.

Keywords: white box testing; test data generation; genetic algorithm; fitness function

Citation: Avdeenko, T.; Serdyukov, K. Modified Evolutionary Test Data Generation Algorithm Based on Dynamic Change in Fitness Function Weights. *Eng. Proc.* **2023**, *33*, 23. https://doi.org/10.3390/engproc2023033023

Academic Editors: Askhat Diveev, Ivan Zelinka, Arutun Avetisyan and Alexander Ilin

Published: 13 June 2023

1. Introduction

Testing is one of the most complex and time-consuming processes in software development, and is necessary to ensure high quality of the developed product [1]. Therefore, automating the testing process, or at least its subprocesses, is an important research task. One of the important subprocesses of the testing is the test data generation. An analysis of existing studies, methods and approaches in the field of the application of methods for automatic test data generation has shown [2–4] that in software development, a blind strategy of random data generation is mainly used. At the same time, an analysis of scientific studies has shown that there are approaches, the development and application of which can significantly improve the quality of generated tests cases, expressed in the degree of coverage of the code being tested [5,6].

Among such advanced approaches to test data generation, static methods of symbolic analysis of program code were historically the first [7,8]. The generation of test data as a result of such analysis was reduced to automatic generation and resolution in symbolic form of a system of equations and inequalities obtained by logical union and the intersection of all conditions of the software-under-test (SUT). The undoubted advantage of the static approach is that it obtains results in symbolic form, which makes it possible to analytically determine subareas of test-case values that guarantee the passage of calculations over the given parts of the code.

However, a significant limitation of the possibility to apply the static approach is the problem of computational complexity of symbolic computations even for tasks of relatively small dimensionality. Therefore, a dynamic approach based on the actual execution of the SUT with specially generated values of input variables and subsequent analysis of data flows is currently more realistic and efficient for practical use in software development companies.

The most promising methods for implementing a dynamic approach to test data generation are evolutionary optimization methods [9–11]. The evolutionary paradigm, which is the basis of the genetic algorithm (GA), uses a set of random test data generated at the initial stage, after which a sequential "evolution" of the data is performed to improve the coverage quality of the testing code. Therefore, the assumption arises that the GA can be adapted to implement the idea of the evolutionary improvement of test data in terms of maximizing the coverage of the testing code. However, the existing research focuses on solving local problems, such as finding a specific set of test data that covers some given statements. At the same time, to solve the practical task of comprehensive software testing, it is relevant to develop methods for generating test sets that provide the maximum coverage of the entire SUT, taking into account its multi-connected complex structure.

The paper is organized as follows. Section 1 gives an introduction into the problem. Section 2 describes the usage of the GA in terms of test data generation. In Section 3, we formulate the fitness function for achieving maximum coverage. In Section 4, we propose a modification of the algorithm with dynamic weight assignment. Section 5 provides the results. Section 6 is conclusion.

2. Theoretical Background

The genetic algorithm [12–14] works iteratively, performing consecutive steps at each iteration several until the completion conditions are reached. With each new iteration, GA creates a new generation of the population based on the previous (parent) one. The main GA cycle for test data generation includes the following stages, which, except for the first stage, are performed iteratively until a given coverage value or number of generations are reached:

1. Initialization. The initial population is formed randomly, taking into account the constraints on the values of input variables. The size of population m is chosen based on the size of the SUT (more specifically, the minimum number of different possible paths that the computation can take).
2. Fitness function calculation. Each chromosome in a population is evaluated by a fitness function.
3. Selection. The best 20% of chromosomes are selected unchanged for the next generation; the remaining 80% of chromosomes of the next generation will be obtained by crossover.
4. Crossover. Half of the chromosomes of the next generation are formed by randomly crossing 20% of the best chromosomes of the previous generation with each other. The remaining chromosomes will be obtained by randomly crossing all chromosomes of the previous generation with each other. Crossover occurs by choosing a random constant $\beta_i \in [0,1]$ for each $i = \overline{1,N}$ and subsequent crossing, where $i-th$ gene of the offspring is a linear combination of the corresponded parent genes:

$$var_i^{offspring} = \beta_i \times var_i^{mother} + (1-\beta_i) \times var_i^{father}, i = \overline{1,N}.$$

5. Mutation. With a given mutation probability (0.05) each gene can change its value at random within given constraints. The main goal of mutation is to achieve greater diversity.
6. Formation of the elite chromosome pool. In each generation, individuals of the population are selected into the elite chromosome pool. Only the chromosomes that provide additional code coverage compared to the previous coverage are included in the pool.

After all GA stages have been executed, it is determined whether the completion conditions are met, or whether the process proceeds to the next iteration. The iterability of GA is the factor that allows the obtaining of new solutions. Each new generation is formed based on the previous one, i.e., the test sets of the previous generation participate in the formation of new sets, thus providing the "evolution" of previously obtained solutions

3. Multi-Path Algorithm for Maximum Code Coverage

Input variables of the testing code are either the variables $var_i, j = \overline{1, N}$, which are part of the input statement, either input parameters of procedures and functions, initiating calculations along a certain code path. In this way, we can describe a vector of input variables as $(var_1, var_2, \ldots, var_N)$, and entire definition area as $D = D_1 \times D_2 \times \ldots \times D_N$, where D_i is definition area of the input variable var_i. When chromosome $x_i \in D$ is represented by a dimensional vector N $x_i = [var_1^i, var_2^i, \ldots, var_N^i]$.

The purpose of automatic test data generation is to find many test cases $\{x_1, x_2, \ldots, x_m\}$, which initiate passing through a given set of reachable paths, i.e., the paths that can be covered by the test sets. The main coverage criterion is the criterion of statement coverage [15]. We introduce notation $g(x_i)$ as a vector that is an indicator of the statement coverage initiated by a certain test set x_i:

$$g(x_i) = (g_1(x_i), g_2(x_i), \ldots, g_n(x_i)),$$

where n is the number of statements of the SUT, and

$$g_j(x_i) = \begin{cases} 1 & \text{if path initiated by the set } x_i \text{passes though the statement } j; \\ 0 & \text{otherwise.} \end{cases}$$

If we denote the vector of statement weights of the SUT as $(w_1, w_2, \ldots, w_n)$, then we can define the fitness function for a single chromosome x_i as follows

$$F(x_i) = \sum_{j=1}^{n} w_j g_j(x_i), \tag{1}$$

where w_j—weight of the statement j, g_j—value of the coverage indication, n—number of statements.

The greater the sum of the statement weights executed on the path initiated by the test case x_i, the greater the value of the fitness function $F(x_i)$. To ensure greater population diversity, a component is added to Formula (1) that allows the consideration of the remoteness of paths from each other. The remoteness of the paths is defined through the similarity operation. To calculate the j-th similarity coefficient $sim_j(x_{i_1}, x_{i_2})$ of two chromosomes x_{i_1} and x_{i_2}, check whether the j-th statement of SUT, whose coverage is marked by the indicator g_j, is at the intersection of both paths initiated by test cases x_{i_1} and x_{i_2}:

$$sim_j(x_{i_1}, x_{i_2}) = \overline{g_j(x_{i_1}) \oplus g_j(x_{i_2})}, j = \overline{1, n} \tag{2}$$

where the logical operations "negation" (NOT) and "exclusive OR" ($\oplus$, XOR) are used.

The more matching the covered statements at the intersection of two paths, the greater the similarity value between chromosomes. The following formula defines the similarity between two chromosomes as the weighted average of the similarity over all code statements:

$$sim(x_{i_1}, x_{i_2}) = \sum_{j=1}^{n} w_j \times sim_j(x_{i_1}, x_{i_2}) \tag{3}$$

The similarity value between chromosome x_i and the other chromosomes of the population is calculated as

$$f_{sim}(x_i) = \frac{1}{(m-1)} \sum_{s=1;s\neq i}^{m} sim(x_s, x_i), \tag{4}$$

where m is the number of chromosomes in the population.

Now, we can determine the average similarity value of paths in the entire population

$$\overline{f_{sim}} = \frac{1}{m} \sum_{i=1}^{m} f_{sim}(x_i). \tag{5}$$

and further formulate the additive component of the fitness function responsible for the diversity of paths in the population as the modulus of the difference between the average similarity of the population and the similarity of a particular chromosome

$$F_2(x_i) = \left| \overline{f_{sim}} - f_{sim}(x_i) \right|. \tag{6}$$

As a result, the resulting fitness function for chromosome x_i, taking into account the diversity of paths, is calculated by the formula

$$F(x_i) = F_1(x_i) + k \times F_2(x_i), \tag{7}$$

where $F_1(x_i)$ and $F_2(x_i)$ are defined by Formulas (1) and (6), respectively. Accordingly, the first component $F_1(x_i)$ determines the complexity of the path initiated by the chromosome x_i, and the second component $F_2(x_i)$ determines the remoteness of this path from all other paths in the population. The parameter k defines the relationship between the components.

Using Formula (7) as a fitness function leads to more diverse populations as a result of a single GA run. However, due to the use of a continuous version of the genetic algorithm in this research, the resulting diversity is not sufficient to fully cover the code within a single GA run.

The latter circumstance is related to the detected "swing effect" arising from the presence of indistinguishable chromosomes in the population. If indistinguishable chromosomes have a high value of fitness function in one generation, they will be selected for crossover. Then, their offspring will, with high probability, also be indistinguishable from their parents. The new generation, in this case, will consist of a greater number of indistinguishable chromosomes, and similarity in the population will depend more on them. For all these chromosomes, the value of the additive component F_2 of the fitness function will be reduced, and for chromosomes passing through other paths, it will be increased. Now, other chromosomes could be selected for crossing and will form a multitude of indistinguishable chromosomes for the next generation. Thus, the population will be cyclically first filled with indistinguishable chromosomes, which in the next generation will lead to a decrease in similarity value for them, and, accordingly, to a decrease of F_2, thus reducing the priority of indistinguishable chromosomes in the next iteration. A similar cycle will be repeated for different sets, and as a result, both path complexity (F_1) and similarity value (F_2) cease to play an important role in the formation of a new population, and different sets are constantly shuffled without investigation of the solution space.

To exclude "swing effect" it is proposed to use the indicator $ind(x_1, \ldots, x_i)$, which is determined by the number of chromosomes from the set $\{x_1, \ldots, x_(i-1)\}$ indistinguishable from the chromosome x_i:

$$\widetilde{F}(x_i) = F_1(x_i) + \frac{1}{1 + ind(x_1, \ldots, x_i)} \times F_2(x_i). \tag{8}$$

Indeed, the initial value $ind(x_1) = 0$, because the set in which the indistinguishable chromosomes are identified is empty at the first step. At each subsequent step, the value $ind(x_1, \ldots, x_i)$ can either increase by 1 if the next chromosome is indistinguishable from one of the previous ones, or keep the same value if the next chromosome is unique. This will allow chromosomes passing through different paths to be more evenly distributed throughout the population as a whole.

4. Modification of the Fitness Function Based on Dynamic Changes in Statements Weights

The studies carried out in the articles [16–18] showed a relatively strong influence of the value of k on the coverage of the SUT. At $k = 0$, the coverage was minimal, reaching its maximum value at $k = 10$, after which it began to decline. Obviously, choosing the right k can significantly affect the final results. The value of $k = 10$ obtained in the studies was optimal only within the tested programs; for others, this value may not be optimal. Therefore, to achieve greater universality of the algorithm, it would be preferable to reduce the influence of k.

For this purpose, we propose the modification of the F_1 component of the fitness function (8), so that a greater population diversity is achieved by it alone. The idea of modifying the F_1 component is inspired by other evolutionary methods. In studies [19–21] some of the evolutionary methods are used, in particular Particle Swarm Optimization (PSO), which is one of the Swarm Intelligence algorithms. However, the application of PSO in existing studies has been based more on comparing PSO and GA implementations than on the hybridization of approaches [22]. Other representatives of this family are Ant Colony Optimization (ACO), Artificial Bee Colony Algorithm (ABC), Cuckoo Search (CS), and many other algorithms based on the collective interaction of different particles or agents.

The ACO [23] is one of the methods that allows solving pathfinding problems on graphs. It is based on simulating the behavior of a colony of ants. The ants, passing along certain paths, leave a trail of pheromones behind them. The better the solution found, the more pheromones there will be on one or another path. In the next generation, ants already form their paths based on the number of pheromones—the more pheromones on a certain path, the more ants will be directed to that path and continue exploring it. In this way, the colony gradually explores the entire solution space, gradually reaching better and better paths.

It is not possible to directly apply the ACO to the test data generation problem, because the output to certain paths is initiated by different datasets, and the only way to change the path is to directly change the values of the test sets themselves. Nevertheless, the idea of using the "pheromones" model to prioritize pathfinding could have a positive effect in providing more diversity in the population. When applied to the problem of increasing the diversity of test sets, the idea of pheromones leads to the expediency of dynamically (from generation to generation) increasing or decreasing the weights of operators $w_j, j = \overline{1,n}$, depending on the number of chromosomes previously (in the previous generations) passed through these statements. The dynamic change in the weights of statements can be represented as

$$\widetilde{w}_j^{(q)} = Ph_j^{(q)} w_j, j = \overline{1,n}; q = \overline{1,Q} \tag{9}$$

where $\widetilde{w}_j^{(q)}$ is the weight assigned to the statement j in generation q, $Ph_j^{(q)}$ is weight multiplier of the statement j in generations q $(0 \le Ph_j^{(q)} \le 1)$, Q is number of generations (iterations of GA). Taking into account dependence (9) the dynamic variant of the F_1 fitness function component will have the form:

$$F_1^{(q)} = \sum_{j=1}^{n} \widetilde{w}_j^{(q)} g_j(x_i) = \sum_{j=1}^{n} Ph_j^{(q)} w_j g_j(x)i); q = \overline{1,Q}. \tag{10}$$

In expression (10) it is very important to determine the dependence of the multiplier $Ph_j^{(q)}$ $(0 \leq Ph_j^{(q)} \leq 1)$ on the arguments, so that the statements weights in the fitness function respond to operator coverage in the previous generations in time. The resulting diversity of the population of test datasets, and hence the degree to which they cover the SUT, depends on the choice of the variation method of $Ph_j^{(q)}$.

Two basic strategies were proposed for the initial behavior of the multiplier $Ph_j^{(q)}$ depending on the number of generations q—the direct and the reverse strategy. In the direct strategy, we assume $Ph_j^{(1)} = 0$ in the first generation and then this value increases (or remains the same) depending on the coverage (or non-coverage) of operator j in the previous generation. In the reverse strategy, on the contrary, in the first generation we assume $Ph_j^{(1)} = 1$ and then this value decreases (or remains the same), depending on the coverage (or non-coverage) of operator j.

In both strategies, the multiplier can reach the boundaries of the interval [0, 1]. Thus, in the direct strategy, the value of $Ph_j^{(q)}$ increases monotonically, but after reaching the limit value $Ph_j^{(q)} = 1$ (this value corresponds to the maximum priority of the operator j in the fitness function) it is necessary to begin its decrease to change the algorithm direction to other, still uncovered, statements. Then, after reaching the minimum possible value $Ph_j^{(q)} = 0$, corresponding to the non-inclusion of operator j in the fitness function, we start monotonic increasing again, and so on. In the reverse strategy, changes occur in opposite directions, first in the direction of decreasing, then in the direction of increasing, etc.

This fluctuating change in the multiplier $Ph_j^{(q)}$ between values 0 and 1 can occur with different rates, given by the parameter ΔPh, which affects the total number of fluctuations within the interval [0, 1] during the process of test data generations. Let $Trans_i^{(q)}$ be the number of complete passes from 0 to 1 or from 1 to 0 by the multiplier $Ph_j^{(q)}$ made to the current generation q. Then, the behavior of the multiplier for the direct strategy can be written as

$$Ph_j^{(q)} = x = \begin{cases} 0 & \text{if } q = 1, \\ Ph_j^{(q-1)} + \Delta Ph \times (-1)^{Trans_j^{(q)}} & \text{if } \widetilde{m}_j^{(q-1)} \neq 0, \\ Ph_j^{(q-1)} & \text{if } \widetilde{m}_j^{(q-1)} = 0, \end{cases} \tag{11}$$

and the behavior of the multiplier for the reverse strategy is in the form

$$Ph_j^{(q)} = x = \begin{cases} 1 & \text{if } q = 1, \\ Ph_j^{(q-1)} - \Delta Ph \times (-1)^{Trans_j^{(q)}} & \text{if } \widetilde{m}_j^{(q-1)} \neq 0, \\ Ph_j^{(q-1)} & \text{if } \widetilde{m}_j^{(q-1)} = 0, \end{cases} \tag{12}$$

where $\widetilde{m}_j^{(q-1)}$ is the number of chromosomes in a population consisting of m individuals that covered the operator j in a generation $(q-1)$.

The article comprises several methods for determining the rate parameter ΔPh. The $Half$ method assumes that one full pass of the multiplier (from 0 to 1 or from 1 to 0) with the rate ΔPh can be obtained by covering the operator j in half of the generations from the initially given number of generations Q $(Q/2)$, the *Quarter* method—in quarters of generations $(Q/4)$, *Tenth*—one tenth of all generations $(Q/10)$. The fewer generations (iterations) needed for one complete pass, the greater the rate parameter ΔPh, and the more often the multiplier will fluctuate between the limit values [0, 1]. Table 1 presents the main indicators used to implement the proposed methods for varying the multiplier $Ph_j^{(q)}$ using a constant rate of change ΔPh. Direct strategy methods are marked with a plus sign (+), and methods with a reverse strategy are marked with a minus sign (−).

Table 1. Methods for determining the rate parameter ΔPh of multiplier $Ph_j^{(q)}$.

Method	Formula for Calculating the Multiplier $Ph_j^{(q)}$	ΔPh
Half+	(11)	$2/Q$
Half−	(13)	$2/Q$
Quarter+	(11)	$4/Q$
Quarter−	(13)	$4/Q$
Tenth+	(11)	$10/Q$
Tenth−	(13)	$10/Q$

Another method of determining the multiplier $Ph_j^{(q)}$, which we called *Count*− (the method is based on the reverse strategy), involves changing it not by a constant value, but by a value depending on the coverage intensity of the operator j in the previous generation:

$$Ph_j^{(q)} = x = \begin{cases} 1 & \text{if } q = 1, \\ Ph_j^{(q-1)}(1 - \widetilde{m}_j^{(q-1)}/m & \text{if } \widetilde{m}_j^{(q-1)} \neq 0, \\ 1 & \text{if } \widetilde{m}_j^{(q-1)} = 0. \end{cases} \tag{13}$$

In contrast to the previously proposed method, in *Count*− the value of $Ph_j^{(q)}$ will decrease the stronger the more chromosomes in the previous generation were covered by operator j. There is no gradual increase in the multiplier in this case; instead, if the operator was not covered ($\widetilde{m}_j^{(q-1)} = 0$), then the maximum value $Ph_j^{(q)} = 1$ is set. Thus, often covered operators cease to play a significant role in the process of searching for test sets, and the algorithm will mostly try to generate sets for as yet uncovered paths.

5. Results

Let us compare the application of various methods for determining the multiplier $Ph_j^{(q)}$, using the methods proposed above for the test program SUT2 described in [24]. Figure 1 shows a comparison of the average coverage for different values of the parameter k of the components of the fitness function (8), in which F_1 is calculated either by Formula (1), i.e., without modification, or by Formula (10) when using modification by the methods *Half*+, *Quarter*+, *Tenth*+ and *Count*−. The average coverage is calculated based on 1500 runs. $Q = 50$ and $m = 25$ are chosen as the GA parameters, at which full coverage is relatively rarely achieved.

In Figure 1, red highlights the average coverage when using Formula (8) (static method), methods of monotonic change in Ph based on direct strategy are in shades of blue and black—*Count*− method. The methods for determining the parameter ΔPh based on the reverse strategy are not presented in the figure, but, in general, they have approximately similar values of average coverage.

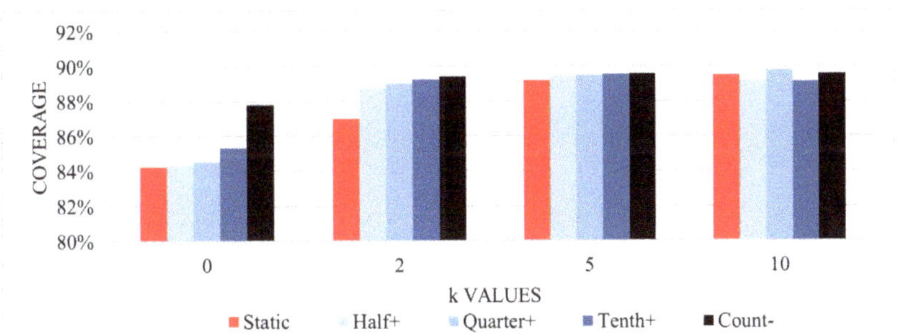

Figure 1. Comparison of different modifications ($Q = 50$, $m = 25$).

Each of the proposed methods for determining the multiplier $Ph_j^{(q)}$ showed a higher average coverage value than the static method (without modification) for each of the k values. Figure 1 shows that for the static method, the average coverage gradually increases with increasing k, while using modifications, the maximum average coverage is reached already at $k = 2$, and thereafter does not decrease. The best result among all proposed methods showed *Count*−, which is why exactly this method will be used in further research of this modification of the fitness function.

Analysis of the results presented in Figure 1 allows us to conclude that the modification allows the significant increase in the coverage even without using the previously determined optimal value of $k = 10$. At the same time, even at $k = 0$, i.e., without use of the additive component F_2 of the fitness function, a higher coverage is achieved than without modification. Comparison of average coverage without and with the best count modification method can be seen in Figure 2. It shows average coverage with algorithm parameters $Q = 50$, $m = 25$.

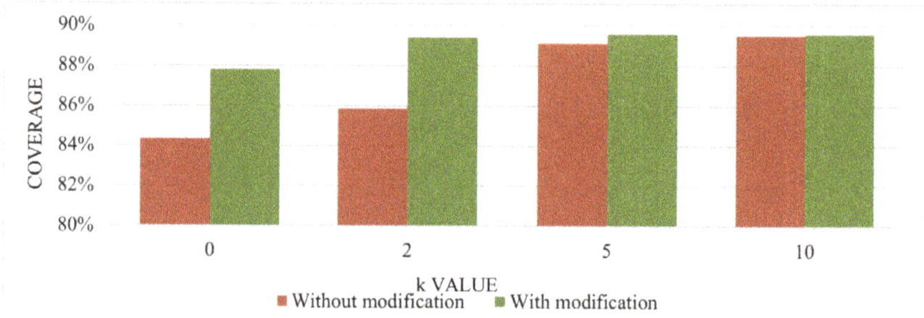

Figure 2. Comparison of coverage with and without modification ($Q = 50$, $m = 25$).

Thus, the use of the modification makes it possible to increase the average coverage, which is especially noticeable at $k = 0$. More importantly, the maximum coverage is achieved when using any non-zero value of k, i.e., the ratio parameter of the fitness function components k ceases to play a significant role in achieving the maximum coverage. As a result, the proposed modification based on the dynamic change in statement weights makes it possible to increase code coverage when generating test sets, as well as eliminate the need to determine the k value for each individual SUT.

6. Conclusions

The paper proposes a modification of the method for generating test data for multiple paths in one launch of GA. The initial problem, which consists of the necessity to determine the value of the ratio parameter of the fitness function components, is solved by dynamically changing the weights of statements between generations. The methods proposed in the paper eliminate the need to define the parameter for each individual program, and one of the methods, *Count*−, allows the achievement of greater coverage, even if the ratio parameter value is zero. Therefore, not only the original goal of implementing the modification is achieved, but also the diversity of generated test cases increases, so overall coverage has also improved.

Author Contributions: Conceptualization, T.A.; methodology, T.A.; software, K.S.; investigation, K.S.; validation, T.A.; writing—original draft preparation, T.A. and K.S.; writing—review and editing, T.A.; visualization, K.S.; supervision, T.A. All authors have read and agreed to the published version of the manuscript.

Funding: This research was funded by Ministry of Science and Higher Education of Russian Federation (project No. FSUN-2020-0009).

Institutional Review Board Statement: Not applicable.

Informed Consent Statement: Not applicable.

Data Availability Statement: Not applicable.

Conflicts of Interest: The authors declare no conflict of interest.

References

1. *ISO/IEC TR 19759*; Software Engineering—Guide to the Software Engineering Body of Knowledge (SWEBOK). ISO Copyright Office: Switzerland, Geneva, 2015.
2. Kumar, M.; Chaudhary, J. Reviewing Automatic Test Data Generation. *Int. J. Eng. Sci. Comput.* **2017**, *7*, 11432–11435.
3. Sui, J.; Gong, Y.; Jin, D.; Wang, Y. Statistical testing data generation for UAS. In Proceedings of the 3rd International Conference on Material Engineering and Advanced Manufacturing Technology, Shanghai, China, 26–28 April 2019; p. 7.
4. Xuan, J.; Jiang, H.; Ren, Z.; Hu, Y.; Luo, Z. A random walk based algorithm for structural test case generation. In Proceedings of the 2nd International Conference on Software Engineering and Data Mining, Chengdu, China, 23–25 June 2010; pp. 583–588.
5. Meudec, C. ATGen: Automatic Test Data Generation using Constraint Logic Programming and Symbolic Execution. *Softw. Test. Verif. Reliab.* **2001**, *11*, 81–96. [CrossRef]
6. Gerlich, R. Automatic Test Data Generation and Model Checking with CHR. *arXiv* **2014**, arXiv:1406.2122.
7. Clarke, L. A system to generate test data and symbolically execute programs. *IEEE Trans. Softw. Eng.* **1976**, *SE-2*, 215–222. [CrossRef]
8. Howden, W. Symbolic testing and the DISSECT symbolic evaluation system. *IEEE Trans. Softw. Eng.* **1977**, *SE-4*, 266–278. [CrossRef]
9. Girgis, M.R. Automatic Test Data Generation for Data Flow Testing Using a Genetic Algorithm. *J. Univers. Comput. Sci.* **2005**, *11*, 898–915.
10. Doungsa-ard, C.; Dahal, K.; Hossain, A.; Suwannasart, T. GA-based Automatic Test Data Generation for UML State Diagrams with Parallel Paths. In *Advanced Design and Manufacture to Gain a Competitive Edge: New Manufacturing Techniques and Their Role in Improving Enterprise Performance*; Springer: London, UK, 2008; pp. 147–156.
11. Sharma, A.; Patani, R.; Aggarwal, A. Software Testing Using Genetic Algorithms. *Int. J. Comput. Sci. Eng. Surv.* **2016**, *7*, 21–33. [CrossRef]
12. Holland, J.H. *Adaptation in Natural and Artificial Systems*; MIT Press: Cambridge, UK, 1975; p. 236
13. Mitchel, M. *An Introduction to Genetic Algorithms*; A Brad-Ford Book; The MIT Press: Cambridge, UK, 1999; p. 162
14. Simon, D. *Evolutionary Optimization Algorithms: Biologically-Inspired and Population-Based Approaches to Computer Intelligence*; John Wiley & Sons: Hoboken, NJ, USA, **2013**; p. 784
15. Spillner, A.; Linz, T.; Schaefer, H. Software Testing Foundations. In *A Study Guide for the Certified Tester Exam*; Rocky Nook: Kingston, MA, USA, 2014; p. 305
16. Avdeenko, T.V.; Serdyukov, K.E. Genetic algorithm fitness function formulation for test data generation with maximum statement coverage. *Lect. Notes Comput. Sci.* **2021**, *12689*, 379–389.
17. Avdeenko, T.V.; Serdyukov, K.E. Development and Research of the Test Data Generation Approach Modifications. In Proceedings of the 2021 International Conference on Information Technology and Nanotechnology (ITNT), Samara, Russia, 20–24 September 2021; pp. 1–6.
18. Avdeenko, T.V.; Serdyukov, K.E.; Tsydenov, Z.B. Formulation and research of new fitness function in the genetic algorithm for maximum code coverage. *Procedia Comput. Sci.* **2021**, *186*, 713–720. [CrossRef]
19. Singla, S.; Kumar, D.; Rai, M.; Singla, P. A hybrid PSO approach to automate test data generation for data flow coverage with dominance concepts. *J. Adv. Sci. Technol.* **2011**, *37*, 15–26.
20. Bueno, P.M.; Wong, W.E.; Jino, M. Automatic test data generation using particle systems. In Proceedings of the 2008 ACM Symposium on Applied Computing, New York, NY, USA, 16–20 March 2008; pp. 809–814.
21. Khan, S.A.; Nadeem, A. Automated Test Data Generation for Coupling Based Integration Testing of Object Oriented Programs Using Particle Swarm Optimization (PSO). In *Genetic and Evolutionary Computing, Proceedings of the Seventh International Conference on Genetic and Evolutionary Computing, ICGEC 2013, Prague, Czech Republic, 25–27 August 2013*; Springer International Publishing: Berlin/Heidelberg, Germany, 2014; Volume 238, pp. 115–124.
22. Dixit, S.; Tomar, P. Applying Computational Intelligence in Software Testing. *J. Artif. Intell. Res. Adv.* **2015**, *2*, 7–11.
23. Dorigo, M.; Birattari, M.; Stutzle, T. Ant colony optimization. *IEEE Comput. Intell. Mag.* **2015**, *1*, 28–39. [CrossRef]
24. Avdeenko, T.; Serdyukov, K. Automated Test Data Generation Based on a Genetic Algorithm with Maximum Code Coverage and Population Diversity. *Appl. Sci.* **2015**, *11*, 4673. [CrossRef]

engineering proceedings

MDPI

Proceeding Paper

Lightweight 2D Map Construction of Vehicle Environments Using a Semi-Supervised Depth Estimation Approach †

Alexey Kashevnik [1,*] and Ammar Ali [2]

1 SPC RAS, Saint Petersburg 199178, Russia
2 Information Technology and Programming Faculry, ITMO University, Saint Petersburg 197101, Russia
* Correspondence: alexey.kashevnik@iias.spb.su
† Presented at the 15th International Conference "Intelligent Systems" (INTELS'22), Moscow, Russia, 14–16 December 2022.

Abstract: This paper addresses the problem of constructing a real-time 2D map for driving scenes from a single monocular RGB image. We presented a method based on three neural networks (depth estimation, 3D object detection, and semantic segmentation). We proposed a depth estimation neural network architecture that is fast and accurate in comparison with the state-of-the-art models. We designed our model to work in real time on light devices (such as an NVIDIA Jetson Nano and smartphones). The model is based on an encoder–decoder architecture with complex loss functions, i.e., normal loss, VNL, gradient loss (dx, dy), and mean absolute error. Our results show competitive results in comparison with the state-of-the-art methods, as our method is 30 times faster and smaller.

Keywords: machine learning; deep learning; computer vision; monocular depth estimation; real-time depth estimation; driver assistant systems

1. Introduction

This paper tackles the problem of constructing a 2D map of the vehicle environment from a single monocular RGB image. This computer vision problem is essential for driver assistant systems because it could be used for dangerous situation detection, camera-based GPS, vehicle motion prediction, etc. We propose a method that covers 3D object detection, which provides information about vehicle, pedestrian, and bike locations and helps to estimate the vehicle orientation; semantic segmentation, which is one of the most famous computer vision problems that helps to identify the surrounding environment objects such as roads, trees, and buildings; and depth estimation, that will be used to understand the vehicle environment in the 3D space to build a 2D map with an approximate distance for each object identified by the 3D detector.

Depth distribution is subject to highly dynamic changes due to the object's location in the scene, for example, an image that includes multiple tiny objects on a table or an image of a park including people, a river, and sky. Differences between the object distance values make it a complex problem. To solve this problem, many researchers have proposed different approaches such as adaptive blocks or attention-based approaches. Transformers [1] have also had a huge impact on this particular problem because splitting the image into multiple chunks helps to isolate the jumps in the depth values. Unfortunately, the methods available at the moment have high complexities in terms of running time and memory.

The vehicle environment can be tracked simply and the camera has a fixed place. Thus, the depth gradient does not change a lot. However, it has an approximately static change (the positions of most objects, i.e., cars, people, buildings, and trees, change but within a lower range).

Instead of making the neural network architecture more complicated to adapt to the depth changes, we propose to use a simple encoder–decoder architecture with a hybrid loss function to obtain a light and accurate model.

Citation: Kashevnik, A.; Ali, A. Lightweight 2D Map Construction of Vehicle Environments Using a Semi-Supervised Depth Estimation Approach. *Eng. Proc.* **2023**, *33*, 28. https://doi.org/10.3390/engproc2023033028

Academic Editors: Askhat Diveev, Ivan Zelinka, Arutun Avetisyan and Alexander Ilin

Published: 15 June 2023

The main motivation of our work is that the currently existing methods are not suitable for real-time applications on mobile devices. Our general idea is to apply complex post-calculations (hybrid loss) to compute the loss from the output of a basic encoder–decoder architecture to obtain the best performance using light neural network models to reduce inference in real time.

The main contributions of the paper include the following: (1) a light neural network architecture with a hybrid loss function for training to obtain an accurate light model; (2) an open-source dataset that includes vehicle environment videos; and (3) comparison results between our approach and the state-of-the-art methods using a shared dataset.

2. Related Work

Depth estimation in images from a monocular image is a well-known task in computer vision. There are multiple approaches that have been considered using different training data. In paper [2], the authors proposed an approach to enhance self-learning for depth estimation using a reprojection loss. To solve the problem of gradient locality of bilinear samples and avoid local minima, they proposed a multi-scale image reconstruction and depth estimation. Thus, the loss is taken at each scale of the decoder. An auto masking loss helps to ignore the violated pixels by camera motion assumptions.

The authors of paper [3] introduced another approach for unsupervised learning. They used a Generative Adversarial Network (GAN) for the generation of synthetic data and combined these data with real images during the training process to determine the geometric information from a single image. To reduce the gap between synthetic data and real data they proposed an approach that maps the corresponding domain-specific information related to the primary task into shared information.

Paper [4] proposed further enhancements to self-learning techniques for depth estimation by introducing a self-attention layer, as well as a discrete disparity volume, based on the hypothesis that the robustness and sharpness of the depth estimation are allied with the possibility of predicting uncertain depth maps. These depth maps can be used by autonomous systems to improve decision making even in unsupervised depth estimation techniques.

The authors of paper [5] used video sequences and time recurrence in tandem with geometric constraints to estimate depth maps. They introduced a spatial reprojection layer to maintain the spatio-temporal consistency between the levels.

A tool for merging different depth estimation datasets for training even if the annotations are incompatible was introduced in paper [6], presenting a robust training objective against the changes in depth range and scale. They published their work under the name Midas.

The authors of paper [7] presented an approach for enhancing current research on the topic of monocular depth estimation. According to their observations, there is a trade-off between a consistent scene structure and high frequencies, allowing them to propose a simple depth merging network.

The proposed neural network architecture in paper [8] uses the transformer approach as a backbone, replacing the CNN from the encoder, although one version of the model (DPT-hybrid) still has a CNN encoder which is used with the transformer for feature extraction. However, the DPT-Large model uses only a transformer for feature extraction and obtained the best results with the validation data.

Paper [9] proposed a new loss function to enforce the geometric constraints. Virtual normal loss (VNL) allows aligning the spatial relationship between the prediction and the ground truth as point clouds. The prediction of the model results in accurate depth maps and point clouds.

Paper [10] presented a simple encoder/decoder architecture (encoder: EfficientNet b5 and decoder: UNet) for designing an adaptive block (AdaBins), using a transformer (mini Vit) to divide the depth ranges into bins then obtaining the depth map from the bins. This work is the current state-of-the-art method for the NYUv2 dataset.

We used the methods presented in the following papers [8,9] to annotate our dataset for the training of the neural network architecture presented in this paper. Furthermore, we compared the obtained results with the current state-of-the-art method [10] with the following parameters: accuracy, loss, running time, and the number of parameters used.

3. Method

To construct a 2D map from an RGB image, we propose the following main steps (see Figure 1). In the top of the figure, we show the full system architecture to build a 2D map, on the bottom we show our proposed approach for depth estimation: (1) design and train a neural network model for monocular depth estimation to obtain information about the depth of each object in the scene; (2) chose the best 3D object detection model to detect the dynamic objects and their orientation (vehicles, pedestrians, bikes, etc.); and (3) find and enhance a semantic segmentation model to obtain information about the static objects (trees, buildings, etc.).

In the scope of this paper, we mostly concentrated on the topic of depth estimation since it is the most important part to reduce running time and memory allocation for 2D map construction. We tried several neural network architectures and the best one is based on the EfficientNetb0 encoder and Nested UNet (UNet++) decoder. We developed the architecture and trained it on our dataset with a hybrid loss function that consists of multiple weighted loss functions (VNL loss, cosine similarity loss, gradient loss, and log MAE loss). We describe in detail the proposed neural network architecture as well as the hybrid loss function in Section 5.

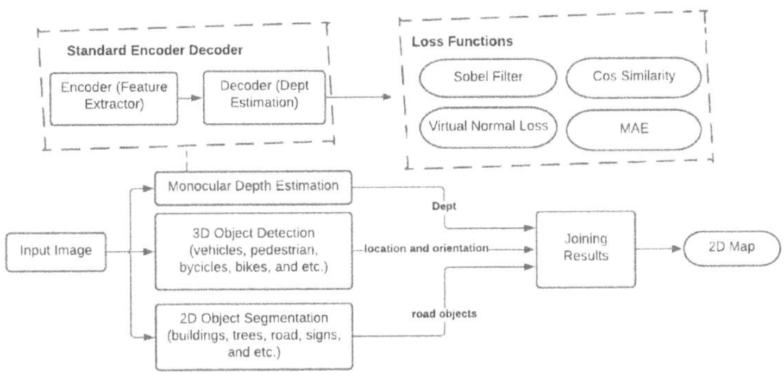

Figure 1. General scheme of the proposed method.

4. Dataset

We used a previously developed platform to collect the data for our dataset in a real environment. To record the data in the vehicle, we used the Drive Safely mobile system (http://mobiledrivesafely.com, accessed on 9 June 2023) developed for Android-based smartphones. The system is a driver assistant and monitoring system which is responsible for detecting dangerous situations in vehicle cabins and providing recommendations to the driver, as well as collecting all information on the cloud server [11]. For the presented paper, the most important information is the data from road cameras.

We collected the data from ten drivers that drove vehicles in St. Petersburg, Russia, for a few months. The collected dataset is publicly available at http://doi.org/10.5281/zenodo.8020598 (accessed on 9 June 2023). The image size from the road cameras is 480 × 640. For training, we scaled them down to 420 × 420 and applied center cropping to a 416 × 416 area to exclude border areas that contain noise and distortion. To annotate the data, we used pseudo-labeling and an ensemble of different open-source models that were trained on the KITTI dataset (see Figure 2). The ensemble is based on a weighted mean average using the following weights [0.4, 0.3, 0.2, 0.1] from top to bottom. For each image in the dataset, we used the following four models (LapDept, DTP Hybrid, BTS, and VNL) to obtain the predictions. After that, we took the weighted average between all four masks and saved the results as our ground truth.

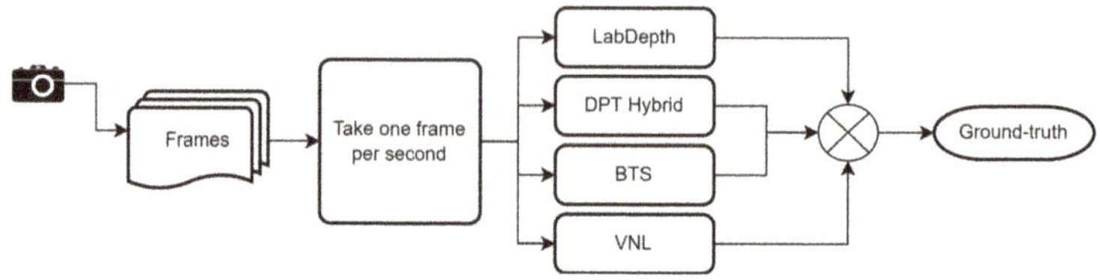

Figure 2. The proposed labeling process for the recorded dataset.

5. Monocular Depth Estimation Model Architecture

This section describes the proposed efficient depth estimation model with high performance for both evaluation metrics and loss computations, as well as running time, weight size, and parameters (applied to be run on modern smartphones and NVidia Jetson Nano devices). We propose a simple encoder–decoder architecture. The encoder (feature extractor) is MobileNetV3 for smartphones and EfficientNet-b0 [12] for the NVidia Jetson Nano. We chose the UNetPlusPlus architecture as a decoder to minimize the gap between encoder–decoder feature maps and to obtain a better gradient flow. Figure 3 shows the basic architectures of EfficientNet and nested UNet (UNetPlusPlus). We propose a similar architecture for smartphones but replaced EfficientNet-b0 with MobileNetv3. The top part of the figure presents the encoder (feature extractor) that is EfficientNet-b0. For our proposed model, we replaced it with MobileNetv3. The bottom part presents the UNetPlusPlus architecture that consists of similar multiple blocks with decreasing numbers of nodes. The black arrows indicate down-sampling, the red arrows indicate up-sampling, and the dotted arrows indicate skip connections. Each node represents a convolution layer [13].

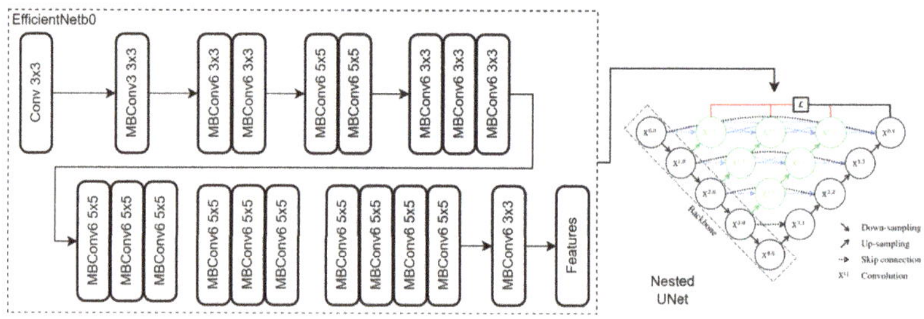

Figure 3. The proposed light-weight efficient depth estimation model [14].

5.1. UNetPlusPlus Architecture

We describe the UNetPlusPlus architecture in detail since it will have a significant effect on depth estimation. The main difference between UNet and nested UNet is that the latter has a loss estimated from four semantic levels (deep supervision named by authors). Nested UNet inherited the dense blocks from the DenseNet architecture as follows.

The output of the previous convolution layers is concatenated with the current convolution by upsampling the lower dense block. The multi-semantic levels will give Nested UNet a more adaptive property to handle the gradient changes in the depth mask that will help to make the architecture itself capable of performing depth estimations without adding more complex layers or heads.

5.2. Loss Functions

The loss function is an important key factor in obtaining a good performance of the neural network model. We suggest a combination of different loss functions to obtain the best performance using a light model: (1) Sobel filter, an image processing technique used for edge extractions by taking the derivatives along the x- and y-axes; (2) mean absolute error (MAE); (3) cos similarity loss, a calculation of the dot product of two vectors (predictions and ground truth); and (4) virtual normal loss (VNL), a method to construct the point cloud from the depth masks and use the camera parameters to project the 2D mask in the 3D space to assure high-order geometric supervision in the 3D space.

To calculate the approximated derivatives for horizontal and vertical changes, we propose a Sobel loss function that can express the changes as follows, where G_x, G_y are the approximated gradients in x, y directions and img is the input image. From the above, we calculate two losses. Multiplying by $\frac{1}{N}$ means taking the average value.

$$G_x = \begin{bmatrix} 1 \\ 2 \\ 1 \end{bmatrix} * \left(\begin{bmatrix} +1 & 0 & -1 \end{bmatrix} * \mathbf{img} \right) \tag{1}$$

$$G_y = \begin{bmatrix} +1 \\ 0 \\ -1 \end{bmatrix} * \left(\begin{bmatrix} 1 & 2 & 1 \end{bmatrix} * \mathbf{img} \right) \tag{2}$$

$$loss_{dx} = log(|(G_{x_{pred}} - G_{x_{gt}})| + 1) * \frac{1}{N} \tag{3}$$

$$loss_{dy} = log(|(G_{y_{pred}} - G_{y_{gt}})| + 1) * \frac{1}{N} \tag{4}$$

The log mean absolute error indicates taking the logarithm of the MAE between the predictions and ground truth (gt).

The cosine similarity loss is calculated by the following and we can calculate the loss as the distance.

$$similarity = \frac{pred.gt}{max(||pred||_2.||gt||_2, eps)}, loss_{cos} = |1 - similarity| * \frac{1}{N} \tag{5}$$

To calculate the virtual normal loss, we chose random multiple groups from the predicted depth mask and their correspondents from the ground truth. Each group consists of three points that are not co-linear (similar to making random triangulations for the depth map). Our goal is to minimize the difference between the centers of these triangles (see Figure 4). Let us assume that triangle ABC is from the predicted depth map and ADE is the corresponding triangle in the ground truth depth mask. F and G are the centers of ABC and ADE, respectively. We minimize the distance f.

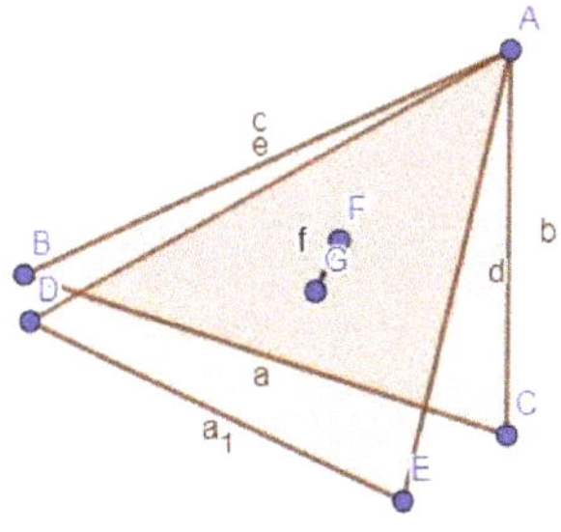

Figure 4. Virtual normal loss clarification.

Thus, we propose $loss_{vn} = \frac{1}{M} * \sum_{i=0}^{M} f_i$, where M is the number of the groups. We propose the hybrid loss to construct the loss function by the following: $loss = a1.loss_{vn} + a2.loss_{cos} + a3.loss_{log-mae} + a4.loss_{gradient-loss}$, where a1, a2, a3, and a4 are the coefficients for the weighted average loss. We calculate empirically that for the best results, the following values should be used (0.5, 4, 1, and 1).

Furthermore, we tried to add the chamfer distance to the losses list, which caused worse results by about 2%. This is because of the countering between the chamfer distance and the VN loss. The goal of the former is to minimize the distance between the centers of the triangles and the other is to move the triangles' corners. On the other hand, the pseudo-labeled ground truth has huge jumps between the depth levels, so it is not suitable to use the chamfer distance. Another idea was to add a discriminator to classify the predictions and the ground truth. The generator (our model) was updated using the proposed loss with the discriminator loss (semi-GAN approach). It did not impact the results, so we decided to remove it.

6. Experiments and Discussion

The main goal of the proposed experiments is to evaluate the proposed model on real data as well as to compare the results obtained with the AdaBins approach, which is the state-of-the-art approach at the moment for monocular depth estimation. We compared both on our recorded dataset as well as on the NYU dataset.

6.1. Running Time

In this subsection, we present the running time for different depth estimation methods as well as for our proposed method (see Table 1). This table shows the running time of different methods for monocular depth estimation on the GPU and CPU. Moreover, it shows the number of parameters for each model (the complexity of the model). The image's height and width are both equal to 416 in all experiments and the GPU used was a Tesla V100 16 Gb (accessible in Colab Pro). Furthermore, CPU experiments were performed on a Colab CPU. We use Colab for the following reasons: (1) reproducibility (anyone can run tests in the same environment and obtain the same results) and (2) ease of use for most developers.

Table 1. Running time of methods for monocular depth estimation.

Method	Parameters	CPU	GPU
AdaBins	78.257 M	15,000 ms	150 ms
DPT-Hybrid	123.146 M	4200 ms	52 ms
MiDas small	21.320 M	250 ms	17 ms
VNL	115.360 M	5500 ms	73 ms
LapDepth	73.131M	1700 ms	56 ms
Proposed (effb0)	6.569 M	850 ms	15 ms
Proposed (MNv3)	3.007 M	500 ms	10 ms

6.2. Evaluation on the NYU Dataset

We trained our model on over 50 K images from the unlabeled NYU dataset. Unfortunately, due to the difference between the chosen data for training between different methods, it is not a sufficient comparison but it can show an overall perspective of each method (see Table 2). The table shows the evaluation metrics for each method on the NYU dataset. For the RMSE (root mean square error) and ABS_REL (absolute relative error), lower is better, see [15]. log_10 is the log_10 difference between the target and prediction and (Delta1, Delta2, Delta3) presents the accuracy with different thresholds. Ours (large) is the model that uses EfficientNet-b0 as a feature extractor and Ours (small) is the model using MobileNetV3 as a feature extractor.

Table 2. Comparison of the different methods on the NYU dataset.

Method	Encoder	Decoder	RMSE	ABS_REL	log_10	Delta1	Delta2	Delta3
AdaBins	EfficientNet-b5	UNet + miniViT	0.364	0.103	0.044	0.903	0.984	0.997
DPT-Hybrid	ViT transformer	DPT	0.357	0.110	0.045	0.904	0.988	0.998
VNL	ResNext-101 (32 × 4d)	-	0.416	0.111	0.048	0.845	0.976	0.994
LapDepth	ResNext101	Laplacian	0.384	0.105	0.045	0.895	0.983	0.996
Ours (large)	EfficientNet-b0	UNet++	0.514	0.120	0.047	0.825	0.973	0.992

6.3. Evaluation of Recorded Dataset

We trained our model on 7000 images from the pseudo-labeled data for the presented dataset (see Section 4). The goal of this research is to achieve the best performance on the dataset not only in terms of accuracy but also in terms of running time and complexity. The results showed that the proposed model is more than 30 times faster and smaller and it also has a smaller ABS_REl than the state-of-the-art method (see Table 3), where RMSE is the root mean square error, ABS_REL is the absolute relative error (lower is better, see [15]), log_10 is the log_10 difference between target and prediction, and (Delta1, Delta2, Delta3) present the accuracy with different thresholds.

Table 3. Evaluation metrics for each method on our dataset.

Method	Encoder	Decoder	RMSE	ABS_REL	log_10	Delta1	Delta2	Delta3
AdaBins	EfficientNet-b5	UNet + miniViT	0.5136	0.1176	0.0477	0.8638	0.9658	0.9884
Ours (large)	EfficientNet-b0	UNet++	0.5960	0.1173	0.04885	0.8559	0.9597	0.9841
Ours (small)	MobileNetV3	UNet++	0.6088	0.1197	-	0.8494	0.9564	0.9821

6.4. Results

In this section, we present results examples and provide a visual comparison between our small and large models and with the existing methods that were trained on the KITTI dataset as well as on our data. In Figure 5, in the right image we present the depth map output calculated by our method. The first row presents the input images, the second illustrates the output of our small model (MobileNetv3 encoder), and the third is the output of our large model (EfficientNetB0 encoder). The left image presents a visual comparison with other methods.

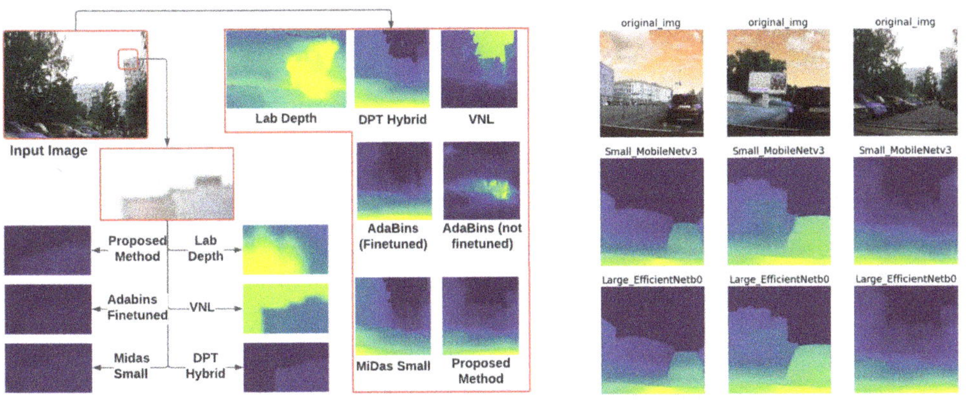

Figure 5. Left image: visual comparison between the methods; right image: depth map output calculated by our method.

7. Conclusions

We introduced a novel approach for real-time depth estimation that includes an efficient neural network architecture and a hybrid loss function. The approach shows competitive results in comparison to the state-of-the-art methods in terms of accuracy, and improvements in terms of running time and size. Based on the presented approach, we propose the method for 2D map construction of vehicle environments. In future work, we will add the Adabins block to our model and investigate its performance on our dataset using the proposed hybrid loss which obtained a less accurate model by 2% in Delta1. We will analyze the method for 3D construction of multiple images and build a model for semantic segmentation and depth estimation.

Author Contributions: Conceptualization, A.K.; methodology, A.K.; software, A.A.; validation, A.A.; writing—original draft preparation, A.A.; writing—review and editing, A.A. All authors have read and agreed to the published version of the manuscript.

Funding: This work was supported by the Russian Science Foundation (project # 18-71-10065). The dataset described in the paper (Section 4) was partially funded by Russian State Research FFZF-2022-0005.

Institutional Review Board Statement: Not applicable.

Informed Consent Statement: Not applicable.

Data Availability Statement: The dataset collected in scope of the paper preparation is available here: http://doi.org/10.5281/zenodo.8020598.

Conflicts of Interest: The authors declare no conflict of interest.

References

1. Subakan, C.; Ravanelli, M.; Cornell, S.; Bronzi, M.; Zhong, J. Attention is all you need in speech separation. In Proceedings of the 2021 IEEE International Conference on Acoustics, Speech and Signal Processing (ICASSP), Toronto, ON, Canada, 6–11 June 2021; pp. 21–25.
2. Godard, C.; Aodha, O.; Firman, M.; Brostow, G.J. Digging into self-supervised monocular depth estimation. In Proceedings of the IEEE/CVF International Conference on Computer Vision (ICCV), Seoul, Republic of Korea, 27 October–2 November 2019.
3. Koutilya, P.; Zhou, H.; Jacobs, D. Sharingan: Combining synthetic and real data for unsupervised geometry estimation. In Proceedings of the 2020 IEEE/CVF Conference on Computer Vision and Pattern Recognition (CVPR), Seattle, WA, USA, 13–19 June 2020; pp. 13971–13980.
4. Johnston, A.; Carneiro, G. Self-supervised monocular trained depth estimation using self-attention and discrete disparity volume. In Proceedings of the 2020 IEEE/CVF Conference on Computer Vision and Pattern Recognition (CVPR), Seattle, WA, USA, 13–19 June 2020; pp. 4755–4764.
5. Fonder, M.; Ernst, D.; Droogenbroeck, M. M4depth: A motion-based approach for monocular depth estimation on video sequences. *arXiv* **2021**, arXiv:2105.09847.
6. Ranftl, R.; Lasinger, K.; Hafner, D.; Schindler, K.; Koltun, V. Towards robust monocular depth estimation: Mixing datasets for zero-shot cross-dataset transfer. *IEEE Trans. Pattern Anal. Mach. Intell.* **2020**, *44*, 1623–1637. [CrossRef] [PubMed]
7. Miangoleh, S.M.H.; Dille, S.; Mai, L.; Paris, S.; Aksoy, Y. Boost- ing monocular depth estimation models to high-resolution via content-adaptive multi-resolution merging. In Proceedings of the 2021 IEEE/CVF Conference on Computer Vision and Pattern Recognition (CVPR), Nashville, TN, USA, 20–25 June 2021; pp. 9680–9689.
8. Ranftl, R.; Bochkovskiy, A.; Koltun, V. Vision transformers for dense prediction. *arXiv* **2021**, arXiv:2103.13413.
9. Yin, W.; Liu, Y.; Shen, C.; Yan, Y. Enforcing geometric constraints of virtual normal for depth prediction. In Proceedings of the 2019 IEEE/CVF International Conference on Computer Vision (ICCV), Seoul, Republic of Korea, 27 October–2 November 2019; pp. 5683–5692.
10. Bhat, S.; Alhashim, I.; Wonka, P. Adabins: Depth estimation using adaptive bins. In Proceedings of the 2021 IEEE/CVF Conference on Computer Vision and Pattern Recognition (CVPR), Nashville, TN, USA, 20–25 June 2021; pp. 4008–4017.
11. Kashevnik, A.; Lashkov, I.; Ponomarev, A.; Teslya, N.; Gurtov, A. Cloud-Based Driver Monitoring System Using a Smartphone. *IEEE Sens. J.* **2020**, *20*, 6701–6715. [CrossRef]
12. Tan, M.; Le, Q. Efficientnet: Rethinking model scaling for convolutional neural networks. In Proceedings of the 36th International Conference on Machine Learning, Volume 97 of Proceedings of Machine Learning Research, Long Beach, CA, USA, 9–15 June 2019; pp. 6105–6114; PMLR; Electronic Proceedings.
13. Howard, A.; Sandler, M.; Chen, B.; Wang, W.; Chen, L.; Tan, M.; Chu, G.; Vasudevan, V.; Zhu, Y.; Pang, R.; et al. Searching for mobilenetv3. In Proceedings of the 2019 IEEE/CVF International Conference on Computer Vision (ICCV), Seoul, Republic of Korea, 27 October–2 November 2019; pp. 1314–1324.
14. Available online: https://sh-tsang.medium.com/review-unet-a-nested-u-net-architecture-biomedical-image-segmentation-57be56859b20 (accessed on 9 June 2023).
15. Lyu, X.; Liang, L.; Wang, M.; Kong, X.; Liu, L.; Liu, Y.; Chen, X.; Yuan, Y. Hr-depth: High resolution self-supervised monocular depth estimation. In Proceedings of the AAAI Conference on Artificial Intelligence, virtually, 2–9 February 2021.

engineering proceedings

Proceeding Paper

Reinforcement Learning for Solving Control Problems in Robotics †

Askhat Diveev [1,*], Elena Sofronova [1], Sergey Konstantinov [2] and Viktoria Moiseenko [2]

1 Federal Research Center "Computer Science and Control" of the Russian Academy of Sciences, 44/2 Vavilova str., Moscow 119333, Russia; sofronova_ea@mail.ru
2 RUDN University, 6 Miklukho-Maklaya str., Moscow 117198, Russia; svkonstantinov@mail.ru (S.K.); vika_moiseenko1999@mail.ru (V.M.)
* Correspondence: aidiveev@mail.ru; Tel.: +7-905-711-4427
† Presented at the 15th International Conference "Intelligent Systems" (INTELS'22), Moscow, Russia, 14–16 December 2022.

Abstract: The use of reinforcement learning technology for the optimal control problem solution is considered. To solve the optimal control problem an evolutionary algorithm is used that finds control to ensure the movements of a control object along different trajectories with approximately the same values of the quality criterion. Additional conditions for passing the trajectory in the neighbourhood of given areas of the state space are included in the quality criterion. To build a stabilization system for the movement of an object along a given trajectory, machine learning control by symbolic regression is used. An example of solving the optimal control problem for a quadcopter is given.

Keywords: optimal control; evolutionary algorithm; reinforcement learning; symbolic regression

Citation: Diveev, A.; Sofronova, E.; Konstantinov, S.; Moiseenko, V. Reinforcement Learning for Solving Control Problems in Robotics. *Eng. Proc.* **2023**, 33, 29. https://doi.org/10.3390/engproc2023033029

Academic Editors: Ivan Zelinka, Arutun Avetisyan and Alexander Ilin

Published: 15 June 2023

1. Introduction

The optimal control problem with phase constraints often has a multi-modal functional. Therefore, with its numerical solution by direct approach, it is possible to obtain several control functions that ensure the movement of the object along different trajectories in the state space with approximately the same value of the control quality criterion which is close to the optimal.

A numerical solution to the optimal control problem leads to some difficulties. As a rule, in most optimal control problems, it is necessary to minimize not one but at least two criteria, reach the control goal or minimize the error of reaching the terminal state and still minimize the given quality criterion. Addition of weight coefficients into criteria does not significantly simplify the problem, since the problem of choosing weights arises.

Another search problem is defined as the loss of unimodality of the functional on the space of parameters of the approximating function. Even a piecewise linear approximation of the control function, when only one parameter needs to be found on each interval for each control component, does not guarantee the presence of a single minimum of the goal functional on the space of parameters.

The problem becomes more complicated in the presence of phase constraints that describe the areas of state space forbidden for the optimal trajectory. It is most likely that due to these reasons, and despite numerous attempts [1,2], a universal computational method for the optimal control problem has not been created.

Further studies have shown that if a strictly optimal solution is not needed and solutions close to the optimal are quite satisfactory, then evolutionary algorithms can be successfully applied to the optimal control problems [3].

Sometimes in practice the researcher knows how the object should move along the optimal trajectory, i.e., approximately knows the areas in the state space the optimal trajectory should pass through. If we introduce additional requirements in the form of passing through the given areas into the quality criterion, then the evolutionary algorithm should change the search area and look for a solution that satisfies the additional requirements. This approach is effective when using evolutionary algorithms. Due to the inheritance property, the improvement of the criterion value at each generation is performed on the basis of small evolutionary transformations of possible solutions to the previous generation. Therefore, if at some generation one of the possible solutions passes through the required areas specified by the researcher, then with a high probability the evolutionary algorithm will search for the optimal solution that preserves the obtained properties. A similar technique is used in machine learning with reinforcement [4,5], when the researcher awards the object by the change of the target functional value for the right actions. Currently, reinforcement learning is actively used in the practice of solving control problems [6]. The paper contains a formal description and practical application of reinforcement learning for solving the optimal control problem.

2. The Optimal Control Problem and Reinforcement Learning

Consider a formal statement of the optimal control problem.

The mathematical model of a control object is given in the Cauchy form of an ordinary differential equation system

$$\dot{\mathbf{x}} = \mathbf{f}(\mathbf{x}, \mathbf{u}), \tag{1}$$

where $\mathbf{x}$ is a state space vector, $\mathbf{u}$ is a control vector, $\mathbf{x} \in \mathbb{R}^n$, $\mathbf{u} \in \mathrm{U} \subseteq \mathbb{R}^m$, and U is a compact set.

The initial state is given by

$$\mathbf{x}(0) = \mathbf{x}^0 \in \mathbb{R}^n. \tag{2}$$

The terminal state is given by

$$\mathbf{x}(t_f) = \mathbf{x}^f \in \mathbb{R}^n, \tag{3}$$

where t_f is the time to reach the terminal state (3). Time t_f is not given, but it is limited to $t_f \leq t^+$, where t^+ is a given positive value.

The quality criterion is given by

$$J = \int_0^{t_f} f_0(\mathbf{x}, \mathbf{u})dt \rightarrow \min_{\mathbf{u} \in \mathrm{U}}. \tag{4}$$

Assume that the researcher knows the areas in the state space of where the optimal trajectory should be. Then, additional conditions are included in the quality criterion

$$J_1 = \int_0^{t_f} f_0(\mathbf{x}, \mathbf{u})dt + p \sum_{i=1}^{r} \psi_i(\mathbf{x}(t)) \rightarrow \min_{\mathbf{u} \in \mathrm{U}}, \tag{5}$$

where

$$\psi_i(\mathbf{x}) = \vartheta(\min_t\{\|\mathbf{z}^i - \mathbf{x}\|\} - \varepsilon_i)(\min_t\{\|\mathbf{z}^i - \mathbf{x}\|\} - \varepsilon_i), \tag{6}$$

p is a penalty coefficient, and $\vartheta(\alpha)$ is a Heaviside step function

$$\vartheta(\alpha) = \begin{cases} 1, \text{ if } \alpha > 0 \\ 0, \text{ otherwise} \end{cases}, \tag{7}$$

$\varepsilon_i, i = 1, \dots, r$ are given small positive values, and $\mathbf{z}^i, i = 1, \dots, r$ are the centres of known areas.

According to the introduced additional conditions, if an optimal trajectory does not pass near some given point $\mathbf{z}^i$, then value of criterion (5) will grow.

3. Computation Experiment

Consider the optimal control problem for the spatial movement of a quadcopter. The mathematical model of the control object is

$$\begin{aligned}\dot{x}_1 &= x_4,\\ \dot{x}_2 &= x_5,\\ \dot{x}_3 &= x_6,\\ \dot{x}_4 &= u_4(\sin(u_3)\cos(u_2)\cos(u_1)+\sin(u_1)\sin(u_2)),\\ \dot{x}_5 &= u_4\cos(u_3)\cos(u_1)-g_c,\\ \dot{x}_6 &= u_4(\cos(u_2)\sin(u_1)-\cos(u_1)\sin(u_2)\sin(u_3)),\end{aligned} \tag{8}$$

where $g_c = 9.80665$.

Control is constrained

$$u_i^- \le u_i \le u_i^+, \quad i = 1,\ldots,4, \tag{9}$$

where $u_1^- = -\pi/12$, $u_1^+ = \pi/12$, $u_2^- = -\pi$, $u_2^+ = \pi$, $u_3^- = -\pi/12$, $u_3^+ = \pi/12$, $u_4^- = 0$ and $u_4^+ = 12$.

The initial state is given by

$$\mathbf{x}^0 = [0\ 5\ 0\ 0\ 0\ 0]^T. \tag{10}$$

The terminal state is given by

$$\mathbf{x}^f = [10\ 5\ 10\ 0\ 0\ 0]^T. \tag{11}$$

The phase constraints are given by

$$\varphi_k(\mathbf{x}) = r_k - \sqrt{(x_1 - x_{k,1})^2 + (x_3 - x_{k,3})^2} \le 0, \tag{12}$$

where $k = 1, 2$, $r_1 = r_2 = 2$, $x_{1,1} = 2.5$, $x_{1,3} = 2.5$, $x_{2,1} = 7.5$, $x_{2,3} = 7.5$.

It is necessary to find a control function, taking into account the constraints in (9), that minimizes the following criterion

$$J_1 = t_f + p_1 \sum_{k=1}^{2} \int_0^{t_f} \vartheta(\varphi_k(\mathbf{x}))dt \to \min_{\mathbf{u}\in\mathrm{U}}, \tag{13}$$

where $t_f \le t^+ = 5.6$, $p_1 = 3$.

To solve the control problem numerically, let us use a piecewise linear approximation. The time axis is divided into equal intervals Δt, and the search for constant parameters is performed at the interval boundaries for each control component. Control is a piecewise linear function that consists of segments connecting points at the bounds of intervals. Given the control constraints, the desired control function is as follows

$$u_i = \begin{cases} u_i^+, \text{ if } \hat{u}_i \ge u_i^+ \\ u_i^-, \text{ if } \hat{u}_i \le u_i^- \\ \hat{u}_i, \text{ otherwise} \end{cases}, \ i = 1,\ldots,4, \tag{14}$$

where

$$\hat{u}_i = d_{i+(j-1)m} + (d_{i+jm} - d_{i+(j-1)m})\frac{t-(j-1)\Delta t}{\Delta t}, \ i = 1,\ldots,4, j = 1,\ldots,K, \tag{15}$$

and K is a number of time interval boundaries

$$K = \left\lfloor \frac{t^+}{\Delta t} \right\rfloor + 1 = \left\lfloor \frac{5.6}{0.4} \right\rfloor + 1 = 15. \quad (16)$$

When solving the problem by a direct approach, the condition of reaching the terminal state is included in the quality criterion

$$J_2 = t_f + p_1 \sum_{k=1}^{2} \int_0^{t_f} \vartheta(\varphi_k(\mathbf{x}))dt + p_2 \|\mathbf{x}^f - \mathbf{x}(t_f)\| \to \min_{\mathbf{u} \in \mathrm{U}}, \quad (17)$$

where $p_2 = 1$,

$$t_f = \begin{cases} t, \text{ if } t \le t^+ \text{and } \|\mathbf{x}^f - \mathbf{x}(t)\| \le \varepsilon = 0.01 \\ t^+, \text{ otherwise} \end{cases} . \quad (18)$$

To solve the problem, a hybrid evolutionary algorithm [7] is used.

Figures 1 and 2 show projections on the horizontal plane $\{x_1; x_3\}$ of the two found optimal trajectories. The big circles present the phase constraints in (12).

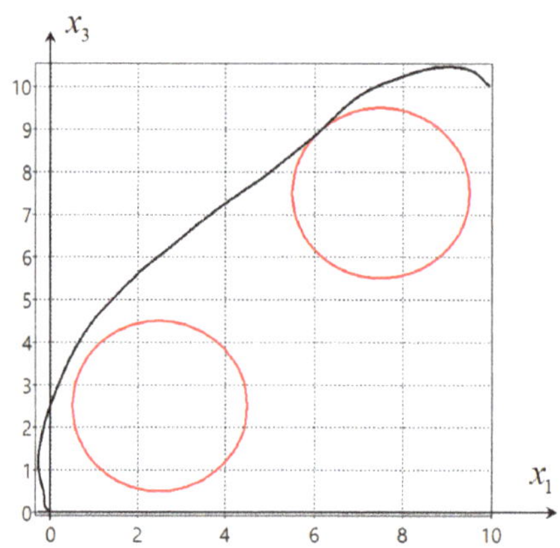

Figure 1. Projection of optimal trajectory 1 on the horizontal plane.

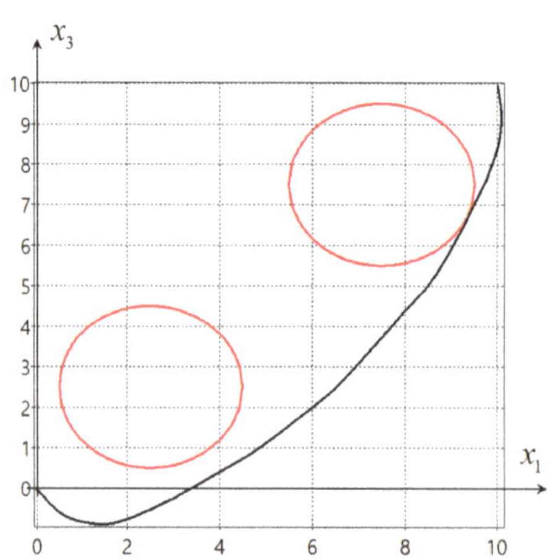

Figure 2. Projection of optimal trajectory 2 on the horizontal plane.

The criterion for the solutions found had the following values: for the solution in Figure 1 $J_2 = 5.6434$, for the solution in Figure 2 $J_2 = 5.6330$.

As can be seen from the experiment, the values of the criteria practically coincide, the solutions found ensure the movement of the object from the given initial state (10) to the given terminal state (11) without violation of the phase constraints. In the series of experiments, the hybrid evolutionary algorithm found solutions that bypass the phase constraints either from above, as in Figure 1, or from below, as in Figure 2.

Suppose that we need the control object to move between obstacles. For this purpose the desired areas on the horizontal plane are defined. It is known that in the presence of interfering phase constraints, the optimal trajectory should be close to the boundary of these constraints. For the given problem four desired areas are defined as

$$\begin{aligned} \mathbf{z}^1 &= [2.5\ 0.4]^T,\ \varepsilon_1 = 0.6, \\ \mathbf{z}^2 &= [4.5\ 2.5]^T,\ \varepsilon_2 = 0.6, \\ \mathbf{z}^3 &= [5.5\ 7.5]^T,\ \varepsilon_3 = 0.6, \\ \mathbf{z}^4 &= [7.5\ 9.6]^T,\ \varepsilon_4 = 0.6. \end{aligned} \tag{19}$$

The conditions for passing through the desired areas (19) are included in the quality criterion

$$J_3 = t_f + p_1 \sum_{k=1}^{2} \int_0^{t_f} \vartheta(\varphi_k(\mathbf{x}))dt + p_2\|\mathbf{x}^f - \mathbf{x}(t_f)\| + p_3 \sum_{i=1}^{4} \psi_i(\mathbf{x}) \to \min_{\mathbf{u}\in\mathrm{U}}, \tag{20}$$

where $p_3 = 3$.

The hybrid evolutionary algorithm found the following optimal solution $\mathbf{d} = [d_1 \dots d_{60}]^T$ $= [-11.1092, 5.9957, -0.0532, 7.0045, 17.5091, 18.1172, -1.6764, 18.7012, -16.0121, 10.5543, -19.9307, 12.1721, -6.0892, 0.5339, -0.8616, 19.2556, -13.7218, 15.1266, 0.3982, 14.2650, -1.1768, 2.9832, 4.3286, 15.1508, -8.9240, -19.6814, 4.5363, 15.9879, -0.0026, 1.1203, 13.2592, -6.6358, -6.2012, -0.5328, -0.0354, 4.2548, 11.6764, -4.3345, -6.7336, 19.8643, 0.3360, -8.9741, -2.6648, 12.5608, 19.6577, -19.9308, -1.6252, 19.3797, -1.1954, 2.2625, 5.9582, 16.0807, -0.8272, 2.3167, 0.9842, 14.2695, -6.3767, 2.3895, 0.3742, 16.2710]^T$.

Figure 3 shows the projection of the found optimal trajectory for the solution with quality criterion $J_3 = 5.5730$. Small dashed circles are the desired areas, while big circles are the constraints.

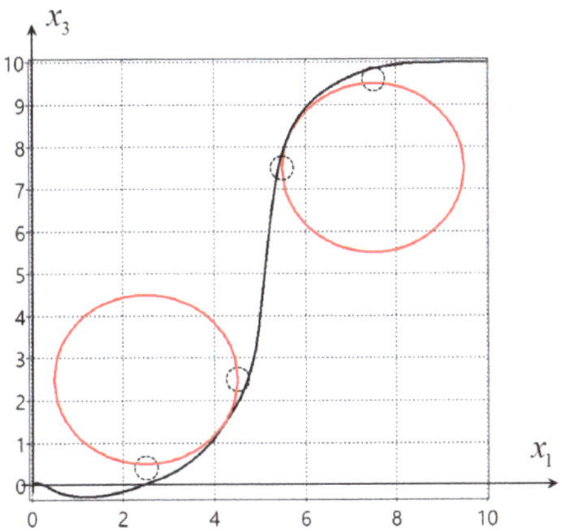

Figure 3. Projection on the horizontal plane of the optimal trajectory found by reinforcement learning.

To implement the obtained solution according to the extended statement of the optimal control problem, it is necessary to build a system to stabilize the movement of the object along the optimal trajectory [8]. For this purpose, machine learning control is used [9]. The control function structure search is carried out by symbolic regression [10].

The obtained solution is

$$u_i = \begin{cases} u_i^+, \text{if } \tilde{u}_i \geq u_i^+ \\ u_i^-, \text{if } \tilde{u}_i \leq u_i^- \\ \tilde{u}_i, \text{otherwise} \end{cases}, i = 1, \ldots, 4, \quad (21)$$

where

$$\tilde{u}_1 = \mu(G), \quad (22)$$

$$\tilde{u}_2 = (\tilde{u}_1 - \tilde{u}_1^3)\rho_{17}(A + \mu(G))\vartheta(F)\rho_{17}(x_4^* - x_4), \quad (23)$$

$$\tilde{u}_3 = \tilde{u}_2 + \tanh(\tilde{u}_1) + \rho_{19}(A + \mu(G)) + \rho_{17}(W), \quad (24)$$

$$\begin{aligned} \tilde{u}_4 \; = \; & \tilde{u}_3 + \ln|\tilde{u}_2| + \text{sgn}(A + \mu(G))\sqrt{|A + \mu(G)|} + \rho_{19}(A) + \arctan(C) + \\ & \text{sgn}(E) + \arctan(F) + \exp(q_2(x_2^* - x_2)) + \sqrt{q_1}, \end{aligned} \quad (25)$$

$$A = B\text{sgn}(D)\sqrt{|D|}\tanh(D)\exp(H), \; B = \exp(C) + \rho_{17}(F) + \cos(q_6(x_6^* - x_6)),$$

$$C = D + \tanh(E) + \rho_{18}(V), \; D = E + \sqrt[3]{F} + \sin(W),$$

$$E = F + G + \exp(H) - V, \; F = H + \text{sgn}(x_5^* - x_5) + (x_2^* - x_2)^3,$$

$$G = q_6(x_6^* - x_6) + q_3(x_3^* - x_3) + \text{sgn}(x_2^* - x_2)\sqrt{|x_2^* - x_2|},$$

$$H = \rho_{17}(q_6(x_6^* - x_6) + q_3(x_3^* - x_3)) + V^3 + W + q_6 q_5^2(x_5^* - x_5)^2 + (x_5^* - x_5)^2,$$

$$V = \sin(q_6(x_6^* - x_6)) + q_5(x_5^* - x_5) + q_2(x_2^* - x_2) + \cos(q_1) + \exp(x_5^* - x_5) + \vartheta(x_2^* - x_2),$$

$$W = q_4(x_4^* - x_4) + q_1(x_1^* - x_1) + \sin(q_6),$$

$$\mu(\alpha) = \begin{cases} \alpha, \text{if } |\alpha| \leq 1 \\ \text{sgn}(\alpha), \text{otherwise} \end{cases}, \; \rho_{17}(\alpha) = \text{sgn}(\alpha)\ln(|\alpha| + 1),$$

$$\rho_{18}(\alpha) = \text{sgn}(\alpha)(\exp(|\alpha|) - 1), \; \rho_{19}(\alpha) = \text{sgn}(\alpha)\exp(-|\alpha|),$$

where $q_1 = 13.02930$, $q_2 = 11.21509$, $q_3 = 15.91016$, $q_4 = 14.33447$, $q_5 = 14.67798$, $q_6 = 9.91431$ and $\mathbf{x}^* = [x_1^* \ldots x_6^*]^T$ is a state vector of the reference model.

Figure 4 shows the trajectories from eight initial states on the horizontal plane.

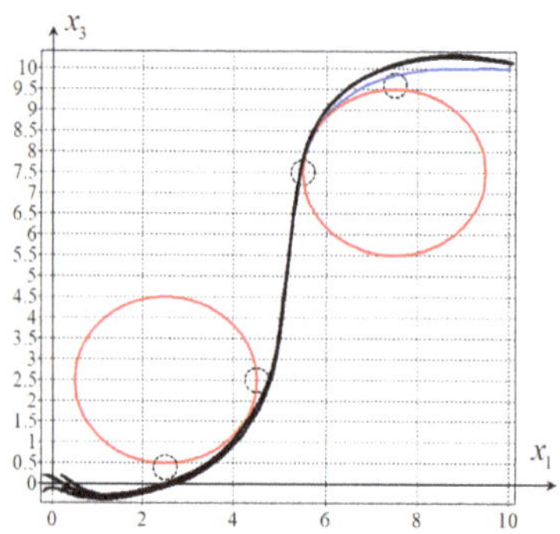

Figure 4. Projection on the horizontal plane of the optimal trajectories from eight initial states.

4. Results

The paper presents the use of machine learning technology with reinforcement to solve the optimal control problem with phase constraints using an evolutionary algorithm. To implement reinforcement learning, additional conditions defining the form of the optimal trajectory are introduced into the quality criterion. The optimal trajectory should pass through specified areas whose positions depend on the phase constraints. An example of solving the optimal control problem for a quadcopter by machine learning with reinforcement was given.

5. Discussion

The use of reinforcement learning technology to solve the optimal control problems of robotic devices is advisable, since in most cases the developer approximately knows the form of the optimal trajectory for the problem being solved.

Author Contributions: Conceptualization, A.D. and E.S.; methodology, A.D. and S.K.; software, A.D., S.K. and E.S.; validation, A.D. and V.M.; formal analysis, E.S.; investigation, S.K.; data curation, A.D., S.K. and E.S.; writing—original draft preparation, A.D, E.S. and V.M.; writing—review and editing, E.S.; visualization, V.M.; supervision, A.D. All authors have read and agreed to the published version of the manuscript.

Funding: The work was performed with partial support from the Russian Science Foundation, Project No 23-29-00339.

Institutional Review Board Statement: Not applicable.

Informed Consent Statement: Not applicable.

Data Availability Statement: The data that support the findings of this study are openly available in https://cloud.mail.ru/public/pKW4/fmrfqjkv3 (for optimal trajectory 1), https://cloud.mail.ru/public/mVZ5/Kr2yb9jJn (for optimal trajectory 2).

Conflicts of Interest: The authors declare no conflict of interest.

References

1. Rao, A.V. A Survey of Numerical Methods for Optimal Control. *Adv. Astronaut. Sci.* **2010**, *135*, 497–528.
2. Grachev, N.I.; Evtushenko, Y.G. A Library of Programs for Solving Optimal Control Problems. *USSR Comput. Math. Math. Phys.* **1980**, *2*, 99–119. [CrossRef]
3. Diveev, A.I.; Konstantinov, S.V. Study of the Practical Convergence of Evolutionary Algorithms for the Optimal Program Control of a Wheeled Robot. *J. Comput. Syst. Sci. Int.* **2018**, *4*, 561–580. [CrossRef]
4. Brown, B.; Zai, A. *Deep Reinforcement Learning in Action*; Manning, Publications Co.: Shelter Island, NY, USA, 2019; 475p.
5. Morales, M. *Grokking Deep Reinforcement Learning*; Manning Publications Co.: Shelter Island, NY, USA, 2020; 472p.
6. Duriez, T.; Brunton, S.; Noack, B.R. *Machine Learning Control–Taming Nonlinear Dynamics and Turbulence*; Springer: Cham, Switzerleand, 2017; 229p.
7. Diveev, A. Hybrid Evolutionary Algorithm for Optimal Control Problem. In *Lecture Notes in Networks and Systems*; Springer: Berlin/Heidelberg, Germany, 2023; Volume 543, pp. 726–738.
8. Diveev, A.; Sofronova, E. Synthesized Control for Optimal Control Problem of Motion Along the Program Trajectory. In Proceedings of the 2022 8th International Conference on Control, Decision and Information Technologies (CoDIT), Istanbul, Turkey, 17–20 May 2022; pp. 475–480.
9. Diveev, A.I.; Shmalko, E.Y. *Machine Learning Control by Symbolic Regression*; Springer: Cham, Switzerleand, 2021; 155p.
10. Koza, J.R.; Keane, M.A.; Streeter, M.J.; Mydlowec, W.; Yu, J.; Lanza, G. *Genetic Programming IV. Routine Human-Competitive Machine Intelligence*; Springer: Boston, MA, USA, 2003; 590p.

engineering proceedings

MDPI

Proceeding Paper

Designing a Digital Twin of a Wind Farm †

Liudmila Massel, Aleksei Massel *, Nikita Shchukin and Aleksey Tsybikov

Melentiev Energy Systems Institute Siberian Branch of the Russian Academy of Sciences (ESI SB RAS), Lermontov St., 130, Irkutsk 664033, Russia; lvmassel@gmail.com (L.M.); niksha14@mail.ru (N.S.); tsibikow@mail.ru (A.T.)

* Correspondence: amassel@isem.irk.ru

† Presented at the 15th International Conference "Intelligent Systems" (INTELS'22), Moscow, Russia, 14–16 December 2022.

Abstract: The article is devoted to the development of a prototype digital twin of a wind farm. An overview of existing works and solutions in the field of digital twins in wind energy is given. The approach to building a digital twin based on ontological engineering, which is widely used in the works of the authors, is considered in detail. An ontological approach is described, which the authors develop and use in the design and development of digital twins (the development is carried out on digital twins of wind farms and photovoltaic systems). The adapted stages of ontological engineering, examples of fragments of the ontological knowledge space in the field of wind energy and the ontology of tasks of the digital twin of a wind farm are given. The architecture of the digital twin prototype under development has been developed and proposed for consideration. The key parts in the structure of the developed digital twin are described. A mathematical model for determining the operation parameters of a wind farm is considered. Special attention is paid to the stages of implementation of the prototype of the digital twin of the wind farm.

Keywords: digital twin; wind farm; ontological engineering

Citation: Massel, L.; Massel, A.; Shchukin, N.; Tsybikov, A. Designing a Digital Twin of a Wind Farm. *Eng. Proc.* **2023**, *33*, 30. https://doi.org/10.3390/engproc2023033030

Academic Editors: Askhat Diveev, Ivan Zelinka, Arutun Avetisyan and Alexander Ilin

Published: 16 June 2023

1. Introduction

Currently, more and more questions are being raised about the need to build renewable energy facilities, in particular, wind farms. They are used to generate carbon-neutral energy. It should be noted that many European countries are already actively using wind power plants (WPP) [1]. Research [2] shows that in most cases the use of renewable energy sources is economically feasible.

Wind farms do not constantly produce electricity; because of this, it is desirable to consider electric energy storage devices. The tasks of power consumption management and activation of the role of consumers in this process are also relevant. It is advisable to use the technology of digital twins to consider all the proposed items together and their interactions with each other. A detailed description of the technology of digital twins is provided below.

The relevance of using renewable energy sources (RES) in the Russian Federation is due to the fact that a significant part of the territory of Russia is not covered by a centralized power supply. Areas of decentralized energy supply occupy about 60% of the area of the Russian Federation and are located mainly in the northern regions of the country [3]. There are many small isolated settlements in these areas. Their power supply is provided mainly on the basis of diesel power plants using expensive imported fuel.

At the same time, one main problem arising when using renewable energy sources is the volatility of electricity generation, in particular, solar and wind power plants are highly dependent on external factors, primarily on weather conditions. Accordingly, it is necessary to develop not only systems for managing RES, but also systems for predicting the behavior of these objects in changing conditions for the effective use of RES.

One of the technologies used in the digitalization of energy sector is the technology of digital twins (DT) [4]. A digital twin is a virtual prototype of a real object, with which one can conduct experiments and test hypotheses, predict the behavior of an object and solve the problem of managing its life cycle [5]. The digital twin saves equipment and system design costs as a whole, as well as reduces operating costs by improving object observability and the ability to simulate its life cycle, which contributes to timely diagnosis and troubleshooting. Consequently, the use of digital twins in the design of WPP has a positive impact at all stages of the life cycle of the object from the design stage to its operation.

The article considers the existing works in the field of building a DT WPP, the use of ontological engineering for designing a DT WPP, as well as the proposed information model and architecture of the DT WPP. Conclusion describes directions for further development of this work.

2. Analysis of Existing Solutions for the Development and Application of Digital Twins in Wind Energy

The problem of transition to high-tech and efficient production in recent decades has become particularly relevant in the conditions of general competition. It became possible to collect, store, transmit and analyze large amounts of data collected from real objects due to the rapid development of information technology. This revealed the need to revise the standard approaches to managing production processes in enterprises. These factors have prompted the development and adoption of strategic industrial development programs in a number of countries, such as Industry 4.0 Platform (Germany), Made in China 2025 (China), National Technology Initiative (Russia), etc. All proposed programs are aimed at increasing labor productivity, increasing the economic efficiency of production and introducing modern science-intensive technologies [6]. The implementation of these strategies is reflected in an increase in the level of automation of enterprises and in a wider digitalization of production processes. Such changes are due to the need for enterprises to quickly and accurately model the product and its production technology in order to increase profitability and save resources in a competitive environment.

A digital twin can help to solve this problem. DT combines both the virtual environment of the enterprise (data from sensors, mathematical and geometric models, etc.) and the physical environment (actuators, machines, circuits, etc.), and also describes the process of interaction between these environments and complements this with automation technologies. In 2018, Gartner, the world's leading research and consulting company, included digital twins in the list of 10 strategically important technologies. The company noted that "digital twins in the context of IoT projects are especially promising over the next three to five years" [7].

Currently, digital twin technology remains one of the most promising technologies that is widely researched and integrated into various areas of human activity [8–14]. The analyzed articles consider both the troubles in the development of digital twin for renewable energy facilities and the process of developing digital twin for energy facilities in general and for some equipment, for example, wind turbine blades.

The spread of the digital twins' concept began, in particular, with the automotive industry. Glaessgen and Stargel in one of their papers [15] explain the principle of using digital twins for vehicle certification and fleet management: "A digital twin is an integrated multiphysics, multiscale, probabilistic model of an assembled vehicle or system that uses the best available physical models, sensor data, and fleet history to simulate the condition of the original operating in real field conditions".

It can be noted that research and development in the field of digital twins in wind energy is aimed at the tasks of functioning at present [9–11,13,16–19], like our work. Relevance of the research is noted that in our time is widely conducted development of digital twins for objects wind energy [20,21].

Development and research in the field of digital twins of wind power plants are divided into two areas: the development of a digital twin of a separate part of a wind power plant, in particular, the development of a digital twin of the rotor blade of a wind power plant [11], and the development of a digital twin of a wind power plant as a whole [9,10].

The main difference of the presented work lies in the fact that it is proposed to use the information obtained at the stage of solving the problems of the functioning of objects to further solve the problems of developing wind turbines, integrating them into a common isolated power system based on digital twins, and subsequent integration in a single digital space. In addition to this work, the authors simultaneously solve the problems of designing a digital twin of a photovoltaic system [22].

3. Ontological Engineering in the Construction of DT WPPs

The use of ontologies in the structure of a digital database (DB) is given in [21]. Ontological engineering for the construction of the DT was also applied in the construction of the DT of the photovoltaic system [22]. The classical stages of ontological engineering regarding the development of DT WPP were used within the framework of this work.

Below are the results of these steps:

1. In the first stage the goal of building an ontology system was formulated. The goal is to extend the existing ontology of the fuel and energy complex to include the wind farm ontology system. In addition, at this stage it is necessary to formalize the knowledge of experts in the field of wind power for building a digital twin. Next, it is necessary to select the type of ontology and designate its scope. The developed ontology belongs to the type of light (heuristic) ontologies and will be used to build a digital shadow and a digital twin.
2. The second stage is aimed at highlighting and verbalizing the key knowledge of the subject area. After the above step, the objects related to the listed areas were selected: generation system; renewable energy sources; WPP; wind farm equipment.
3. The third step was to designate all the main levels of abstraction and the identification of a structured hierarchy. Based on the results from the previous step, the hierarchy was constructed as shown below:
    - Electric power systems;
        - Generation systems;
            - Renewable energy sources;
                - Wind power plants;
                - Wind farm equipment
4. The ontology system built on the basis of the previous steps has been modified by eliminating contradictions, duplications and synonymy. This step is necessary to improve the visual component of the constructed ontology by additional elaboration of concepts and relations between them.

Figures 1 and 2 present the wind farm ontology system built using the above algorithm.

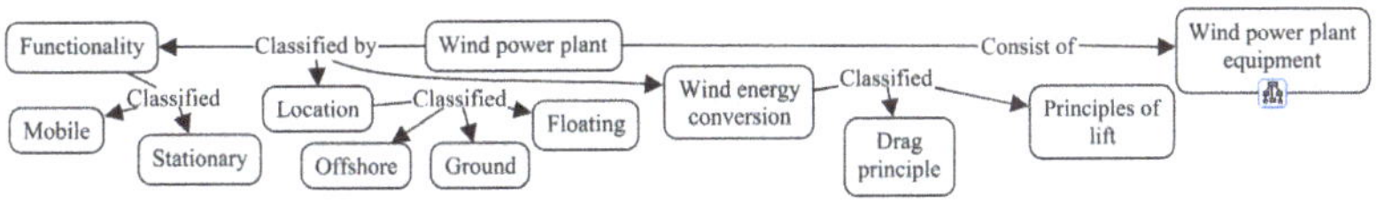

Figure 1. Ontology of wind farms.

Ontologies are used, on the one hand, for structuring the knowledge of the subject area, on the other hand, as the basis for developing a data model when designing DB. Figures 1 and 2 illustrate the main entities of the data model: wind farm equipment, functionality, location, wind power conversion.

The developing of the digital twin involves an ontology of the tasks of designing, functioning and developing a digital twin presented in Figure 3. It includes a description of the main tasks, methods and ways to solving these problems, and rather refers to the structuring of knowledge in the development of a digital database. Ontologies were built in the freely distributed tool CMapTools.

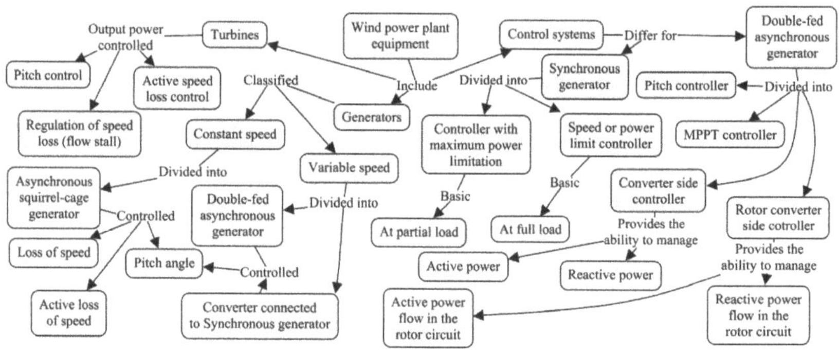

Figure 2. WPP equipment ontology.

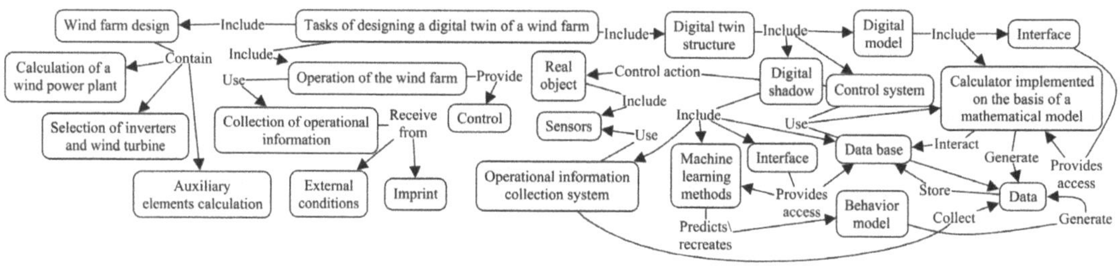

Figure 3. Ontology of tasks of the DT WPP.

4. Description of the Mathematical Model

In this paper, the approach proposed by E.V. Oganesyan, E.A. Bekirov, et al. [23] was used to build a mathematical model of DT. DT WPP based on mathematical model, description of which are given below for determining the parameters of the operation of a WPP. The calculation of the generated power of a WPP requires to use such parameters as: wind speed, wind energy utilization factor, rapidity wind turbine (rapidity), efficiency and swept surface area. The power characteristic of the wind turbine is calculated depending on the wind speed in Formula (1).

$$P_{WT}(V_i) = \rho C_P S_0 \eta \frac{V_i^3}{2} 10^{-3} \tag{1}$$

where ρ—air density (taken equal to 1.226 $\frac{\text{kg}}{\text{m}^3}$); C_P is the wind energy utilization factor; S_0 is the area of the swept surface of the wind turbine; η—total efficiency factor of WPP; V_i^3—wind speed.

Next, it is necessary to calculate the characteristic of the wind energy utilization factor C_P from the rapidity $Z(Z_{opt} < Z < Z_{max})$ (2).

$$C_P = C_{Pmax} - \frac{C_{Pmax}}{(Z_{max} - Z_{opt})^2}(Z - Z_{opt})^2, \tag{2}$$

where C_{Pmax} is the maximum wind power utilization factor; Z_{max}—maximum rapidity, selected based on the passport data of the blade; Z_{opt}—optimal rapidity, selected based on the passport data of the blade.

For $Z \leq Z_{opt}$:

$$C_P = C_{Pmax} - (\frac{Z}{Z_{opt}})^2(3 - 2\frac{Z}{Z_{opt}}), \tag{3}$$

The following expression is used to find the rapidity.

$$Z = \frac{\omega R}{V_0}, \tag{4}$$

where ω is the angular velocity; R is the radius of the wind wheel; V_0 is the speed of the incoming air flow. The effective output specific power ΔP per 1 m^2 swept surface of the wind turbine is calculated by the formula:

$$\Delta P = \Delta P_{WF} C_P \eta, \tag{5}$$

where $\Delta P_{WF} = \rho 10^{-3} \frac{V_{NOM}^3}{2}$—specific power of the wind flow at wind speed V_{NOM}; ρ—air density (winter—1.25, summer 0.72 $\frac{kg}{m^3}$); C_P is the average wind power utilization factor (assumed to be 0.31); $\eta = \eta_A \eta_{GE} \eta_{EG}$—total efficiency factor of WPP; η_A is the aerodynamic efficiency of the wind turbine (WT) (accepted within 0.91–0.916); η_{GE}—gear efficiency (accepted within 0.95–0.96); η_{EG}—the efficiency of the electric generator, depending on its power. After calculating the value of ΔP, the area of the swept surface of the WT is determined:

$$S_0 = \frac{P_{WTNOM}}{\Delta P}. \tag{6}$$

where P_{WTNOM} is the nameplate power of the wind turbine; ΔP—effective output specific power per 1 m^2 swept surface of the wind turbine.

Using data from the mathematical model allows the digital twin to predict the behavior of the object and solve the problem of managing its life cycle.

5. Design of DT WPP

It is necessary to use a large amount of data that needs to be organized and stored when designing and operating the digital twin of a wind farm. For example, it can be weather conditions, equipment characteristics, etc. The study uses the PostgreSQL relational database, which has the following advantages: functionality, convenience, and resource consumption.

The database design was based on ontologies (Figures 1–3) described in Section 3 "Ontological Engineering". The following entities were identified as a result of the design: weather characteristics; characteristics of the wind generator; characteristics of the network inverter; isolated system; network inverter; grid inverter calculation results; wind generator; wind generator calculation results; isolated weather system. Figure 4 shows relationships between entities.

Depending on the nature of the data flows between physical and digital objects, the authors distinguish three main stages in the construction of a DT: a digital model, a digital shadow, and a digital twin. A digital model is understood as an exhaustive description of a part or all of the physical object for which a DT is being developed. A digital model can be represented by both mathematical and other models, as well as the integration of several models. As a rule, a digital model cannot automatically exchange data with its real object. A digital shadow is a digital object to which there is a one-way data flow from the real object. In our case, the digital shadow is defined as a system of relationships and dependencies derived from a real object under normal operating conditions and contained in the Big Data that uses Industrial Internet technologies. In our case, the digital shadow is defined as a system of relationships and dependencies obtained from a real object under normal operating conditions and having Big Data, which use Industrial Internet technologies. DT construction is the development of a digital shadow, which is able to predict the behavior of a real object only in the conditions in which the data were collected, but does not allow the simulation of other situations [24]. The last stage of DT design involves organizing bi-directional automatic data communication between the physical object and the digital object, whereby the digital object can generate feedback to control the physical object. In this case, we can already talk about creating a digital duplicate of the real object.

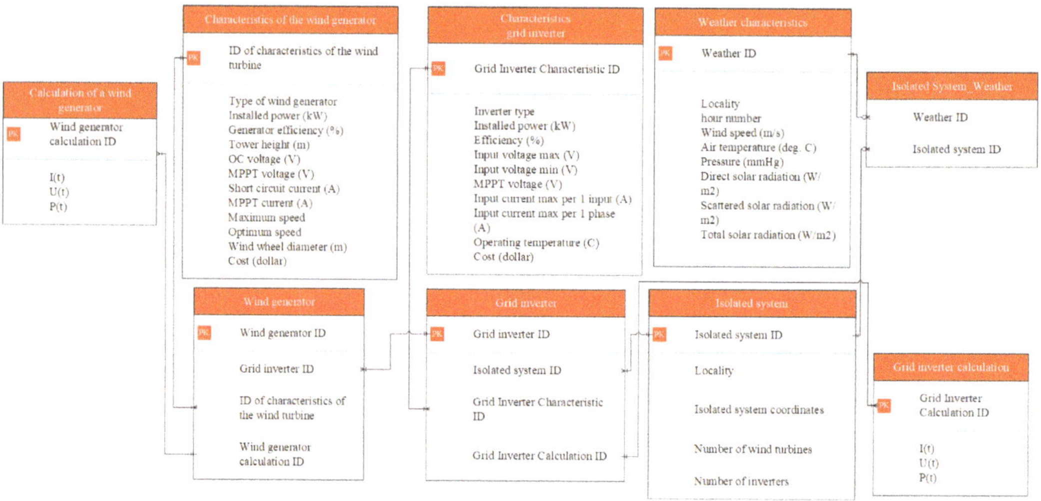

Figure 4. Database entities and relationships between them.

The digital shadow in the developed architecture of the DT includes a data model, methods for collecting information from a physical object and machine learning methods based on time series of weather characteristics and electricity consumption to predict the data needed by the digital twin for the operational management of a wind or solar power plant. The digital model, in turn, includes mathematical models of the behavior of a wind and solar power plant, on the example of which the development of DT is carried out. Research is aimed at developing a generalized methodology for creating a DT, which includes the following steps:

- Ontological engineering of the field in which the DT is developed.
- Construction of a data models based on the carried out ontological engineering.
- Study of the need to use machine learning methods in DT development.
- Implementation of the digital shadow of a physical object.
- Selection and/or development of mathematical and other models necessary to describe a physical object.
- Implementation of a digital model of a physical object.
- Determination of data emulation methods for testing the DT.
- Integration of a digital shadow and a digital model as parts of a DT, using data series obtained as a result of emulation.
- Development of a communication system with a physical object and testing a digital twin on real data.

Figure 5 shows the generalized architecture of the developed DT (solar and wind power plants). It was decided to implement calculations using mathematical models on the server side of the digital twin application given their complexity. It was also decided to store the database on the server side of the application.

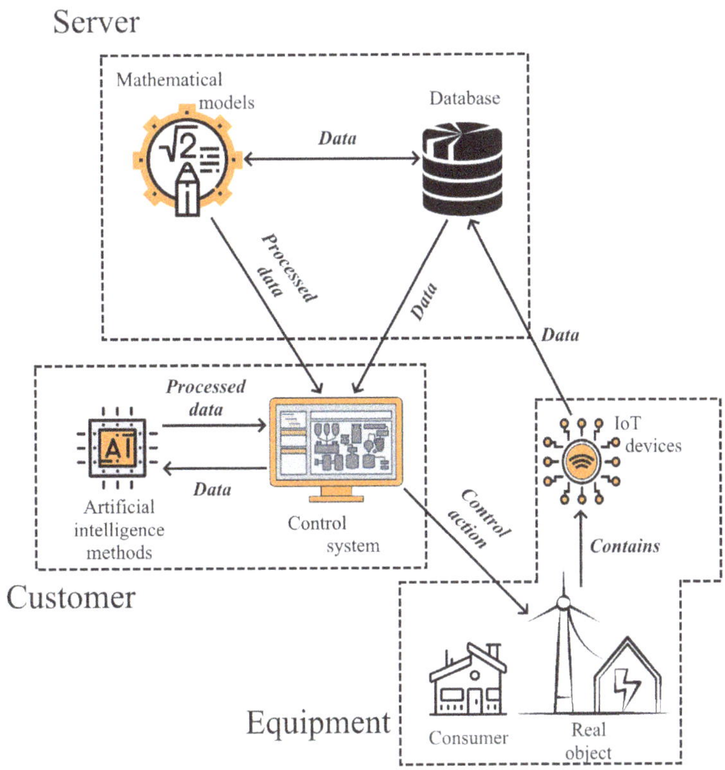

Figure 5. Generalized architecture of developed digital twins (solar and wind power plants).

6. Conclusions

The article considers the current state in the field of DT development, identifies the main shortcomings of these studies, primarily related to the difficulties of their integration with the DT of other systems. The ontological approach developed by the authors to the construction of the DT is considered and the results of its application in the development of the DT of a wind power plant are presented. It is worth noting that ontological engineering is a valuable step in the development of a DT. The structure of the database was developed on the basis of the ontological model and describes the main components and links between them. Integration and formal description of all components (models, databases, knowledge bases) is a key goal of ontologies in the development of a DT. The generation of data series made it possible to use machine learning methods in the implementation of the DT, which will be discussed in the following articles.

In the future, the digital twin will work extensively with the real object. The real object generates large amounts of data, which it needs to analyze. It follows that the digital twin must have the means and methods to perform this function. Big Data technology is used to solve this problem. Because of this, the data is processed within a reasonable timeframe.

As a result of the implementation of all these stages it will be possible to verify the DT prototype according to the data obtained from the real object, and, if necessary, its further adjustment.

Author Contributions: Conceptualization L.M. and A.M.; methodology L.M. and A.M.; software N.S. and A.T.; validation, A.M.; writing—original draft, A.M., N.S. and A.T.; writing—review & editing, L.M. All authors have read and agreed to the published version of the manuscript.

Funding: The research was carried out under State Assignment Project (no. FWEU-2021-0007) of the Fundamental Research Program of Russian Federation 2021–2030.

Institutional Review Board Statement: Not applicable.

Informed Consent Statement: Not applicable.

Data Availability Statement: Not applicable.

Conflicts of Interest: The authors declare no conflict of interest.

References

1. Ivan, K.; Guy, B.; Daniel, F.; Lizet, R. Wind Energy in Europe—2021 Statistics and the Outlook for 2022–2026. 2022. Available online: https://windeurope.org/intelligence-platform/product/wind-energy-in-europe-2021-statistics-and-the-outlook-for-2022-2026/#interactive-data (accessed on 25 March 2022).
2. IRENA. IRENA 2022 Renewable Capacity Highlights. 2022. Available online: https://www.irena.org/-/media/Files/IRENA/Agency/Publication/2022/Apr/IRENA_-RE_Capacity_Highlights_2022.pdf?la=en&hash=6122BF5666A36BECD5AAA2050B011ECE255B3BC7 (accessed on 25 March 2022).
3. Suslov, V. Development of power supply systems of Russian isolated territories using renewable energy sources. *Proc. Irkutsk. State Tech. Univ.* **2017**, *21*, 131–142. [CrossRef]
4. Collection of Legislation of the Russian Federation. Energy Strategy of the Russian Federation for the Period Up to 2035. 2020. Available online: http://static.government.ru/media/files/w4sigFOiDjGVDYT4IgsApssm6mZRb7wx.pdf (accessed on 25 March 2022).
5. Nikitina, E. Caught in the Net: How Digital Twins Work in the Electric Power Industry. 2022. Available online: https://pro.rbc.ru/news/5db1b59a9a79474bb142a3fe (accessed on 25 March 2022).
6. Magazine "ISUP". Digital Twin, Industry 4.0. Informatization and Control Systems in Industry. 2018. Available online: https://zen.yandex.ru/media/isup/cifrovoi-dvoinik-industriia-40-5b83b7155b279900a96c54e8 (accessed on 25 March 2022).
7. Garfinkel, J. Gartner Identifies the Top 10 Strategic Technology Trends for 2019 Gartner Tech. Rep. 2018. Available online: https://www.gartner.com/en/newsroom/press-releases/2018-10-15-gartner-identifies-the-top-10-strategic-technology-trends-for-2019 (accessed on 25 March 2022).
8. Ebrahimi, A. Challenges of developing a digital twin model of renewable energy generators. In Proceedings of the 2019 IEEE 28th International Symposium on Industrial Electronics (ISIE), Vancouver, BC, Canada, 12–14 June 2019; p. 1059.
9. Pimenta, F.; Pacheco, J.; Branco, M.; Teixeira, M.; Magalhães, F. Development of a digital twin of an onshore wind turbine using monitoring data. *J. Phys. Conf. Ser.* **2020**, *1618*, 022065. [CrossRef]
10. Jahanshahi, M.; Parvaresh, A.; Abrazeh, S.; Mohseni, S.-R.; Gheisarnejad, M.; Khooban, M.-H. Digital Twins-Assisted Design of Next-Generation Advanced Controllers for Power Systems and Electronics: Wind Turbine as a Case Study. *Inventions* **2020**, *5*, 19. [CrossRef]
11. Chetan, M.; Yao, S.; Griffith, T. Multi-fidelity digital twin structural model for a sub-scale downwind wind turbine rotor blade. *Wind Energy* **2021**, *24*, 1368–1387. [CrossRef]
12. Karl, M.; Valentin, C.; Paula, B.; Konstanze, K. A hierarchical supervisory wind power plant controller. *J. Phys. Conf. Ser.* **2021**, *2018*.
13. Solman, H.; Kirkegaard, K.; Smits, M.; Vliet, V.; Bush, S. Digital twinning as an act of governance in the wind energy sector. *Environ. Sci. Policy* **2022**, *127*, 272–279. [CrossRef]
14. Dembski, F.; Wössner, U.; Letzgus, M.; Ruddat, M.; Yamu, C. Urban Digital Twins for Smart Cities and Citizens: The Case Study of Herrenberg, Germany. *Sustainability* **2020**, *12*, 2307. [CrossRef]
15. Glaessgen, E.; Stargel, D. The Digital Twin Paradigm for Future NASA and U.S. Air Force Vehicles 53rd Structures. In Proceedings of the Structural Dynamics and Materials Conference, Honolulu, HI, USA, 23–26 April 2012.
16. GE Renewable Energy. Digital Solutions for Wind Farms. Available online: https://www.ge.com/renewableenergy/wind-energy/onshore-wind/digital-wind-farm (accessed on 25 March 2022).
17. Pargmann, H.; Euhausen, D.; Faber, R. Intelligent Big Data Processing for Wind Farm Monitoring and Analysis Based on Cloud-Technologies and Digital Twins. In Proceedings of the 2018 the 3rd IEEE International Conference on Cloud Computing and Big Data Analysis, Chengdu, China, 20–22 April 2018; pp. 233–237.
18. Olatunji, O.; Adedeji, A.; Madushele, N.; Jen, T.-C. Overview of Digital Twin Technology in Wind Turbine Fault Diagnosis and Condition Monitoring. In Proceedings of the 2021 IEEE 12th International Conference on Mechanical and Intelligent Manufacturing Technologies, Cape Town, South Africa, 13–15 May 2021; pp. 201–207.
19. Wagg, J.; Worden, K.; Barthorpe, J.; Gardner, P. Digital twins: State-of-the-art and future directions for modeling and simulation in engineering dynamics applications. *ASCE-ASME J. Risk Uncertain. Eng. Syst.* **2020**, *6*, 27. [CrossRef]
20. Andryushkevich, K. Approaches to the development and application of digital twins of energy systems. *Digit. Substation* **2019**, *12*, 247–255.
21. Kovalyov, P. Designing information support for digital twins of energy systems. *Syst. Means Inform.* **2020**, *30*, 66–81.

22. Ludmila, M.; Nikita, S.; Alexey, C. Digital twin development of a solar power plant. In Proceedings of the International Conference of Young Scientists "Energy Systems Research 2021", Irkutsk, Russia, 27–30 May 2021; Volume 289, p. 03002.
23. Oganesyan, V.; Bekirov, A.; Asanov, M. Mathematical model for determining the operating parameters of a wind turbine. *Constr. Ind. Saf.* **2016**, *55*, 82–86.
24. Borovkov, I. Digital twins: Definition, approaches and development methods. In *Digital Transformation of the Economy and Industry: Proceedings of the Scientific and Practical Conference;* Peter the Great St. Petersburg Polytechnical University: Saint Petersburg, Russia, 2019; pp. 234–245. (In Russian)

engineering proceedings

MDPI

Proceeding Paper

Decision Support in the Analysis of Cyber Situational Awareness of Energy Facilities †

Daria Gaskova * and Elena Galperova

Melentiev Energy Systems Institute Siberian Branch of the Russian Academy of Sciences, Lermontov St., 130, Irkutsk 664033, Russia; galper@isem.irk.ru

* Correspondence: gaskovada@isem.irk.ru; Tel.: +7-395-250-0646 (ext. 440); Fax: +7-395-242-6796

† Presented at the 15th International Conference "Intelligent Systems" (INTELS'22), Moscow, Russia, 14–16 December 2022.

Abstract: Cyber situational awareness is the result of both the analysis of cyber security and situational awareness studies and the line of research that uses artificial intelligence methods in the field of cybersecurity. It covers both methods of automatic detection of cyber threats in the network and methods of providing information to an analyst for further risk analysis and decision making to protect of the assets of the facility. Investigations of cyber situational awareness in the energy sector have become pertinent resulting from both the concept of the digital transformation of energy and the consideration of energy facilities and systems as cyber-physical systems. The problems of ensuring cybersecurity and raising awareness about the cyber environment of an energy facility are compounded by their high significance for the economies of countries. In this regard, such facilities are considered as critical infrastructure. The first part of this article discusses the basic concepts of cyber situational awareness, some knowledge representation models and some existing security metrics. The second part considers the use of frame, production and network models of knowledge representation in the analysis of the cyber situational awareness of energy facilities and the software components that implement them.

Keywords: semantic modeling methods; threat vectors; energy sector

Citation: Gaskova, D.; Galperova, E. Decision Support in the Analysis of Cyber Situational Awareness of Energy Facilities. *Eng. Proc.* **2023**, *33*, 31. https://doi.org/10.3390/engproc2023033031

Academic Editors: Askhat Diveev, Ivan Zelinka, Arutun Avetisyan and Alexander Ilin

Published: 16 June 2023

1. Introduction

Cyber situational awareness (CSA) studies are accompanied by planning strategies and the implementation of protection measures. These studies are not limited to searching for vulnerabilities, cyber threats in the cyber environment, tracking incidents or anomalies. Such studies include the activities of an analyst with a large amount of data, their analysis and the preparatory stages for decision making. This article describes methods and developed software tools to support decision making in the analysis of the CSA of energy facilities, although the classical model of cyber situational awareness presented by Mica Endsley [1] does not include the decision-making process.

1.1. Aspects of Cyber Situational Awareness

Cyber situational awareness includes awareness of any suspicious or atypical activities occurring in a cyber environment, where a cyber environment comprises any activities related to a computer network [2]. Cyber situational awareness is considered from both a technical and cognitive point of view.

The technical component is of great importance at the operational level of security incident management, which is characterized by some security elements. These include vulnerability management, patch management, event management, incident management, malware detection, asset management, configuration management, network management, license management and information management [3].

The cognitive component of cyber situational awareness concerns a person's ability to understand the technical implications and draw conclusions to make informed decisions [2]. Decision making plays an important role both on a tactical level and on a strategic level. The former is aimed at assessing the impact of the current state of the cyber environment on the normal functioning of the facility. The latter is related to identifying trends in malicious attacks and building effective protection plans.

Most of the existing security analysis tools are focused on detecting attacks [4]. Such tools are usually considered within the framework of the technical component of CSA. Safety assessments and decisions on the provision of safety measures refer to the prediction of what may happen in the future, which is closely related to the cognitive component of CSA. Observation of the current situation and its constituent factors is possible if there is an absence of security [5]. The assessment of the security of the cyber environment is accompanied by the choice and calculation of security metrics.

1.2. Security Metrics

Metrics are understood as "tools designed to facilitate decision making and improve performance and accountability through collection, analysis, and reporting of relevant performance-related data" [6]. Flater D. in [5,7] discusses the good and bad characteristics of security metrics. The main characteristics of bad metrics include the interpretation of safety as a quantity that can be measured; incorrect use of scales and the use of erroneous scales, the choice of which can mislead and lead to unreasonable or false conclusions; false precision of the assessment of the operability of products and technologies, in which the uncertainty of the measurement is ignored; combining disparate measures without substantiating how this gives a reliable measurement; and naive use of human input which ignores subjective factors.

Ref. [4] provides corporate network security metrics, where a scale of parameters and a method for calculating the indicator are defined for each metric. The remaining operability of a cyber asset after an attack or hack, the average length of attack paths and the percentage of compromised hosts at time t are some examples of such metrics. In [8], metrics are divided into calculating the metrics from the viewpoint of attacks, malefactors, network topologies and the cost characteristics of attacks and countermeasures, as well as the possibility of zero-day vulnerabilities. Paper [9] provides an extensive classification of model-based quantitative network security metrics and research papers that use them. The paper concludes that the development of metrics and models is aimed at supporting decision making, rather than forming an ideal conception of the level of network security.

The transition from awareness of the state of the cyber environment to decision making can be accomplished through the use of semantic modeling methods.

1.3. Semantic Modeling Methods and Some Their Challenges

Semantic modeling methods are used to analyze and model the behavior of systems, objects and situations that have a number of characteristics such as [10] (i) difficulties in formal description and accurate prediction of their behavior; (ii) absence or incompleteness of data; (iii) measurement of elements (factors) describing the behavior of systems, objects and situations cannot be performed with sufficient accuracy. Semantic models include ontologies, Bayesian belief networks and cognitive and event models [11]. Semantic models are subjective models, the construction of which is aimed at formalizing human (expert) knowledge in a certain problem area. Methods related to the formalization of human knowledge depend on such aspects as [12] (i) assumptions about the adequacy of knowledge accepted in theoretical models and (ii) the ability of analysts and experts to carry out adequate formalization in terms of the accepted theoretical model.

The problem of the subjectivity of expert assessments is also relevant in the development of security metrics. As mentioned in Section 1.2, one of the problems in building a good security metric is taking into account the human factor when evaluating security. In [13], three approaches to taking into account the subjective factor are described. These approaches actually relate to the construction of cognitive models; however, they can also be applied within the framework of constructing the causal topology of a Bayesian belief network.

1. The construction of collective casual maps, including the commonality of construction goals and the unity of decision-making strategies. Such an approach may be accompanied by the problem of mutual understanding of the parties involved in the modeling and the problem of differences in the key concepts of experts.
2. An approach for conflict situations, according to which the knowledge model includes several semantic models (cognitive maps) built by different experts with different goals and interests.
3. An approach to analyzing the validity of managerial decisions based on reflexive cognitive maps, aimed at assessing the adequacy of managerial decisions to find bottlenecks based on semantic models constructed by various experts with their own views but with a common goal.

The collective method of matching knowledge bases for various fields of science and education regarding the example of building ontologies is considered in [14]. Special attention is paid to the Cognitive Ergonomic Metric, the purpose of which is to evaluate two aspects: (i) correctness and depth of reflection of the subject area and (ii) the ergonomic aspect of the ontology representation from the point of view of the quality and speed of human perception.

The problem of the subjectivity of expert assessments arises both in the construction of semantic models and in the formation of probability distributions on the nodes of the Bayesian belief network. On the one hand, the subjectivity of a priori expert-derived probabilities can be partially offset by the use of publicly available statistical reports, and on the other hand, collective methods can lead to a consensus [15] in determining probabilistic characteristics. With the appearance of the necessary statistical data, it is possible to update the probability value to a plausible value for vulnerabilities and threats which are within a cyber environment. For instance, these include threats of cyber negligence and threats associated with the use of social engineering.

Models based on the Bayesian paradigm are widely used to solve predictive tasks of risk analysis. The purpose of this analysis is to provide decision support [16]. This paradigm fits into the context of the analysis of CSA in the application to cybersecurity.

The interpretation of large volumes of received data is a rather difficult task that requires a highly experienced analyst despite the intellectualization of security software in a computer network. The paper proposes to use frame, production and semantic models in the analysis of the CSA of an energy facility, which allows one to formalize expert knowledge and to replicate it in the future.

2. Analysis of Cyber Situational Awareness of Energy Facilities

The current research focuses on the cognitive component of CSA relying on existing systems for security incident detection, intrusion detection and anomaly detection in the network. A CSA analysis of an energy facility is proposed to be carried out in three stages:

1. Environment and situation analyses, including an analysis of detected vulnerabilities and threats.
2. Modeling of states and events in the cyber environment by constructing threat vectors in the cyber environment of an energy facility based on a Bayesian belief network.
3. Risk assessment in the cyber environment using the results of network reasoning.

Models of knowledge representation of experts from different fields are used for such analyses in stages 1 and 2. The stage of risk assessment in the cyber environment is usually accompanied by the assessment of the CSA level and the choice of security metrics. The key aspects of assessing the level of CSA are the definition of cyber threats and the choice of security metrics. Figure 1 shows the ontology of the stages of analysis of the CSA of an energy facility using frame, production and semantic models.

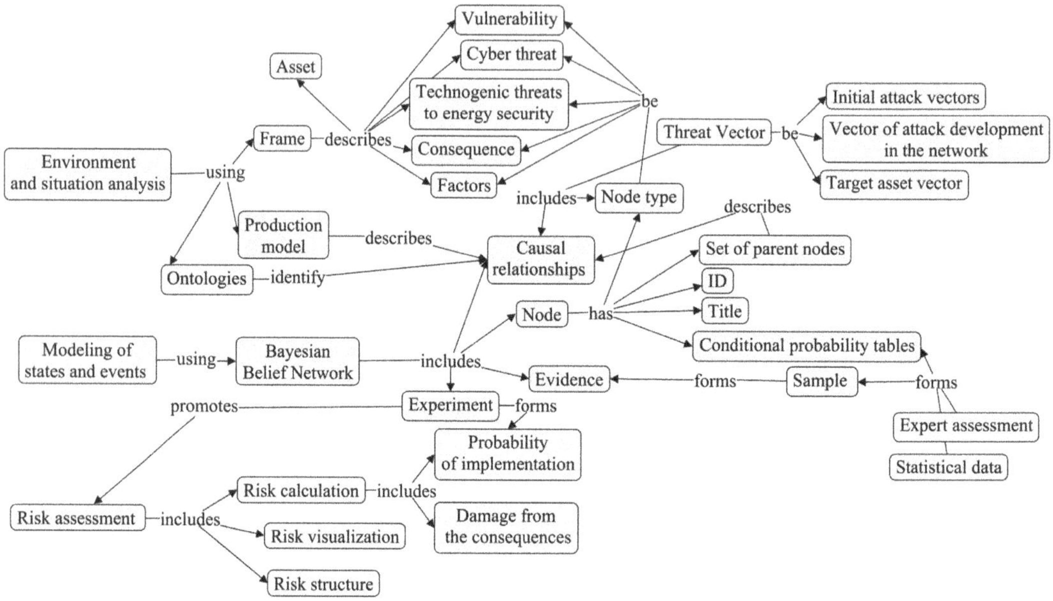

Figure 1. Stages of analysis of CSA of an energy facility using frame, production and semantic models.

2.1. Threat Vectors in the Cyber Environment of an Energy Facility

The importance of energy cyber-physical systems at the national, regional and local levels may motivate attackers to compromise such facilities [17]. The ways of compromising these systems are described by attack vectors. The International Organization for Standardization defines the attack vector as a "path or means by which an attacker can gain access to a computer or network server in order to deliver a malicious outcome" [18]. Attack vectors are often used to describe a targeted attack, which attackers carry out in several stages [19]. Attack vectors are often called threat vectors. There are many approaches to modeling cyber threats and the stages of attack vectors, for example, one of the most well known is the Cyber Kill Chain approach [20], including a life cycle model of cyber attacks.

The paper discusses threat vectors to describe the chains of causal relationships between the use of vulnerabilities of the cyber environment and cyber threats and technogenic threats to energy security and the consequences of the implementation of cyber threats. This characterizes a targeted attack, user actions with cyber negligent and the activities of hacktivists.

The semantic modeling and knowledge engineering methods aim to form the causal relationships in the framework of constructing scenarios of extreme situations in the energy sector caused by cyber threats.

2.2. The Main Stages of Building a Knowledge Representation Model for Threat Vectors

The construction of semantic models in general is reduced to the tasks of identifying the basic concepts, establishing relation between them and defining strategies for decision making in the developed model. It is quite difficult to develop a Bayesian belief network corresponding to the threat vector without prior preparation. In this regard, the construction of a threat vector model in the cyber environment of an energy facility is proposed to be carried out in several stages, presented in Table 1.

The knowledge model built at stages 1–3 is used in this work to achieve CSA goals from a cognitive aspect. Such a knowledge model has a number of limitations: (i) threat vectors are limited to the facility under consideration; (ii) threat vectors are considered from the point of view of disrupting business and technological processes that can cause an extreme situation to occur at the facility and/or deeper within the power system; and (iii) threat vectors are built at a sufficiently high level of abstraction and are designed to inform interested parties about possible security incidents from two aspects, i.e., building a threat model and conducting business games.

Table 1. The main stages of building a knowledge representation model for the threat vectors.

No	Stage Name	Description
1	Identification of the concepts of the subject area	Create frames for assets, vulnerabilities and threats and defining instances of such frames
2	Establishing relations between the concepts	Description of the causal relationships of elements of the threat vector using production rules. Construction of a graph model reflecting the topology of the Bayesian belief network
3	Defining strategies for decision making	Formation of network topology and probability distributions of network nodes, reasoning on the network by entry points and further by vectors. Security metric calculation

Stage 1. Identification of the concepts of the subject area. In this case, the understanding of the frame is used as a type of semantic network in which both declarative knowledge and structured procedural knowledge are used [21]. The hierarchical structure of the asset has the form: information asset, the software part and the hardware part. Additional fields include the device type, the segment of the computing network in which the asset is located, the list of incoming assets (for composite complex assets) and the list of related assets. A list of vulnerabilities and a list of threats inherited from the atomic components of the asset are also provided. The functions of adding a vulnerability to the list of vulnerabilities and adding related assets to the list of related assets are provided to perform rule-based reasoning. The vulnerability frame at a high level of abstraction includes the fields type of vulnerability, the asset in which it is contained, the level of danger and complexity of its use and a list of related vulnerabilities. The functions of adding threats to the list of cyber threats and adding vulnerabilities to the list of related vulnerabilities are provided. The cyber threat frame at a high level of abstraction includes the field type of threat, vulnerability that the threat can use, the level of danger and complexity of the attack and a list of related cyber threats. Each frame has a field for a verbal description. The functions of adding cyber threats and human-made threats to the list of related threats are designed. Environment and situation analyses are included in this stage.

Stage 2. Establishing relations between the concepts. Product rules of type 1 "Asset—Vulnerabilities" establish a relation between an asset and its vulnerabilities. Such rules allow one to output all the rules related to the fact of the asset type without introducing specific vulnerabilities. Type 2 "Vulnerability—Threats" rules associate the relations between asset vulnerabilities and cyber threats that can exploit these vulnerabilities. The type 3 "Threats—Threats" rules add the effect of the relationship between threats, including between a cyber-threat and a cyber-threat and between a cyber-threat and technological threat, on energy security. The generated rules store descriptions of the vector stage for the possibility of implementing the explanation subsystem in the expert system. This stage is in preparation for further topology formation of the Bayesian belief networks and it relates to environment and situation analyses.

Figure 2 shows the interface of the developed expert system that implements the frame and production models of knowledge representation.

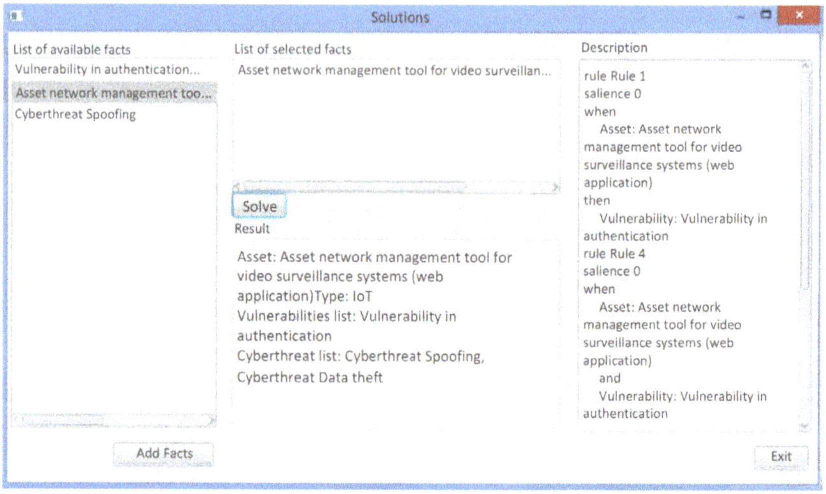

Figure 2. The interface of the developed expert system for constructing threat vectors of an energy facility.

Stage 3. Defining strategies for decision making. The stage of preparing information for defining strategies and choosing the appropriate metrics and measurement scales is a difficult task. The level of detail of the threat vector based on the Bayesian belief network apparatus depends on the expert's choice. The analysis of the change in the probability of the end nodes of the network corresponding to technogenic threats or consequences is supposed to be used as a part for managing risks in the cyber environment and justifying the choice of means of protection for the corresponding assets. The following aspects are being developed: (i) the construction of an information model of the threat vector in accordance with the topology of the Bayesian belief network based on the fields of frames containing descriptions of concepts, as well as a verbal description of the stages included in rules and (ii) an algorithm to evidence introduction into nodes, which are vector points of entry and further progress. This stage is performed when modeling states and events in the cyber environment and assessing the risks of the cyber environment.

Modeling scenarios of extreme situations in the energy sector caused by cyber threats based on Bayesian belief networks are presented in [22]. A joint probability distribution for the developed model of extreme situation scenarios in an energy facility caused by cyber threats is described in accordance with the chain rule for Bayesian networks:

$$P(X) = \prod_{i=1}^{n} P(X_i | pa(X_i)) \tag{1}$$

where $P(X)$ is the joint probability distribution for X, $pa(X_i)$ stands for the sets of parent nodes of X_i, n is a number of vertices of the graph and $X = X_1, \cdots, X_n$ denotes a set of discrete random variable such that:

$$X_i \in \{X^V \cup X^T \cup X^W \cup X^C\}, i = \overline{1,n} \tag{2}$$

where $X^V = \{X_h^V\}_{h=1}^H$ are discrete random variables corresponding to vulnerabilities, $X^T = \{X_k^T\}_{k=1}^K$ are discrete random variables corresponding to cyber threats, $X^W = \{X_m^W\}_{m=1}^M$ are discrete random variables corresponding to technogenic threats to energy security caused by cyber threats and $X^C = \{X_z^C\}_{z=1}^Z$ are discrete random variables corresponding to consequences.

The graph topology of the proposed model has the form:

$$pa(X_h^V) \in \{X_h^V\}_{h=1}^{H-1} \cup \{X_k^T\}_{k=1}^{K} \backslash \{X_h^V\}, \tag{3}$$
$$pa(X_k^T) \in \{X_h^V\}_{h=1}^{H} \cup \{X_k^T\}_{k=1}^{K-1} \backslash \{X_k^T\}, \tag{4}$$
$$pa(X_m^W) \in \{X_m^W\}_{m=1}^{M-1} \cup \{X_k^T\}_{k=1}^{K-1} \cup \{X_h^V\}_{h=1}^{H} \backslash \{X_m^W\}, \tag{5}$$
$$pa(X_z^C) \in \{X_m^W\}_{m=1}^{M}. \tag{6}$$

Moreover $\mathcal{P} = \{P(X_i|pa(X_i)) : X_i \in X\}$ is a conditional probability distribution for each variable from X, and if $pa(X_i) = \varnothing$, then $\mathcal{P}(X_i)$ are a priori probabilities of X_i.

The presented model is aimed at solving the predictive task of analysis and further assessment of the risk of cyber threats at energy facilities. In this paper, risk is considered as a probability distribution of damage [23]. This probability distribution corresponds to the a posteriori probability of a random variable corresponding to each of the simulated consequences, provided that a certain set of vulnerabilities and threats is implemented, i.e., $P(X_Z^C|X_1^V = v_1, \cdots, X_f^V = v_f, X_1^T = t_1, \cdots, X_l^T = t_l, X_1^W = w_1, \cdots, X_S^W = w_S)$, where $z = \overline{1, Z}, X^V, X^T, X^W$ are nodes with given evidence and v, t, w are observed values of the evidence variables with $v, t, w \in [0, 1]$. The ThreatNet component is implemented in the presented model.

2.3. Decision Making Based on a CSA Analysis of an Energy Facility

The scheme in Figure 3 shows the building of a knowledge representation model for threat vectors in the cyber environment of an energy facility and the stages of the CSA analysis of the energy facility.

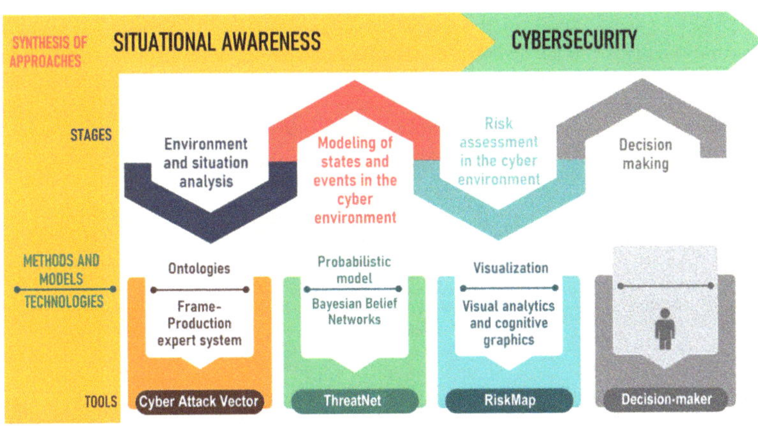

Figure 3. Scheme of analysis stages of the CSA of an energy facility.

The analysis of cyber situational awareness of an energy facility is a rather complex task combining at least three areas of knowledge: (i) cybersecurity of an energy facility; (ii) the aspect of energy security associated with the emergence of technogenic threats (accidents, explosions and fires) due to the possibility of influencing the functioning of energy units and installations from the cyber environment; and (iii) an assessment of the current situation and a risk analysis and decision making on securing the assets of the cyber environment with means of protection. In this regard, three groups of experts have been identified: (i) specialists in the field of cyber security; (ii) energy experts; and (iii) analysts.

Table 2 presents the conditions for assessing extreme situations in the energy sector caused by cyber threats.

Table 2. Conditions for assessing extreme situations in the energy sector caused by cyber threats.

Low Level of Danger	Middle Level of Danger	High Level of Danger
$\begin{cases} 0 \leq P(X_i^C)D_i < l_1 \\ 0 \leq P(X_i^C) \leq 1 \\ 0 \leq D_i \leq D^m \end{cases}$	$\begin{cases} l_1 \leq P(X_i^C)D_i < l_2 \\ 0 \leq P(X_i^C) \leq 1 \\ 0 \leq D_i \leq D^m \end{cases}$	$\begin{cases} l_2 \leq P(X_i^C)D_i \\ 0 \leq P(X_i^C) \leq 1 \\ 0 \leq D_i \leq D^m \end{cases}$

The semantic modeling of threat vectors is aimed at identifying critical assets and formalizing ways to compromise them. The constructed models contribute to a better understanding and choice of security methods such as secure data links based on VPN, an emulator of the software operating environment (honeypot), a demilitarized zone for services and tools, increasing the computer literacy of the organization's employees, etc. The task of selecting suitable security metrics is performed by the enterprise. The use of methods of system analysis and semantic modeling contributes to the selection of the necessary metrics for security assessment and decision making based on their results.

3. Conclusions

This article presents a study aimed at modeling possible illegitimate actions in the network and raising awareness about such behavior. Threat vectors to the cyber environment of an energy facility are built as a chain of possible interrelated events in the cyber environment under consideration in this regard. A Bayesian belief network allows the creation of such vectors and observation of the changes in the probability of possible events, turning them into security incidents.

Within the framework of a three-stage CSA analysis of an energy facility, it is planned to use topological metrics related to the number of vulnerabilities detected, the number of vulnerabilities involved in threat vectors, the complexity of their use of cyber threats, the number of vectors of attack development in the computer network and the amount of attacks on the target asset vectors. The joint use of safety metrics and approaches to assessing the adequacy and depth of models used in semantic modeling allow one to better improve the interaction of people involved in modeling and evaluate the quality of the resulting subjective model.

Author Contributions: Investigation, D.G.; methodology, D.G. and E.G.; software, D.G.; writing—original draft, D.G.; writing—review and editing, E.G. All authors have read and agreed to the published version of the manuscript.

Funding: The research was carried out under State Assignment Project (no. FWEU-2021-0003 and no. FWEU-2021-0007) of the Fundamental Research Program of Russian Federation 2021–2030 using the resources of the High-Temperature Circuit Multi-Access Research Center (Ministry of Science and Higher Education of the Russian Federation, project no 13.CKP.21.0038).

Institutional Review Board Statement: Not applicable.

Informed Consent Statement: Not applicable.

Data Availability Statement: Not applicable.

Conflicts of Interest: The authors declare no conflict of interest.

References

1. Edsley, M.R. Situation awareness global assessment technique (SAGAT). In Proceedings of the IEEE 1988 National Aerospace and Electronics Conference, Dayton, OH, USA, 23–27 May 1988; pp. 789–795.
2. Frank, U.; Brynielsson, J. Cyber Situational Awareness—A systematic review of literature. *Comput. Secur.* **2014**, *46*, 18–31. [CrossRef]
3. Bazrafkan, M.H.; Gharaee, H.; Enayati, A. National Cyber Situation Awareness Model. In Proceedings of the 2018 9th International Symposium on Telecommunications (IST), Tehran, Iran, 17–19 December 2018; pp. 216–220.
4. Cheng, Y.; Deng, J.; Li, J.; DeLoach, S.A.; Singhal, A.; Ou, X. Metrics of Security. In *Advances in Information Security*; Springer: Berlin/Heidelberg, Germany, 2014; Volume 62, pp. 263–295.

5. Flater, D. Bad Security Metrics Part 1: Problems. *IT Prof.* **2018**, *20*, 64–68. [CrossRef] [PubMed]
6. Metrics. NIST Glossary. Available online: https://csrc.nist.gov/glossary/term/metrics (accessed on 4 August 2022).
7. Flater, D. Bad Security Metrics Part 2: Solutions. *IT Prof.* **2018**, *20*, 76–79. [CrossRef] [PubMed]
8. Kotenko, I.; Polubelova, O.; Saenko, I.; Doynikova, E. The Ontology of Metrics for Security Evaluation and Decision Support in SIEM Systems. In Proceedings of the 2013 International Conference on Availability, Reliability and Security, Regensburg, Germany, 2–6 September 2013; pp. 638–645.
9. Ramos, A.; Lazar, M.; Filho, R.H.; Rodrigues, J. Model-Based Quantitative Network Security Metrics: A Survey. *IEEE Commun. Surv. Tutor.* **2017**, *19*, 2704–2734. [CrossRef]
10. Abramova, N.A. Human factors in the cognitive approach. *Manag. Large Syst.* **2007**, *16*, 5–25. (In Russian)
11. Massel, L.V.; Massel, A.G. Semantic technologies based on the integration of Ontological, Cognitive and Event modeling. In Proceedings of the III International Science and Technology Conference "OSTIS-2013", Minsk, Belarus, 21–23 February 2013; pp. 247–250. (In Russian)
12. Abramova, N.A. About some limitations of cognitive modeling methodology and ways to overcome them. In Proceedings of the 12th International Conference "Managing the Development of Large-Scale Systems" MLSD'2019, Moscow, Russia, 1–3 October 2019; pp. 335–337. (In Russian)
13. Abramova, N.A.; Kovriga, S.V.; Portsev, R.Y. On the development of a reflexive approach to the analysis of the substantiation of subjective managerial decisions through cognitive modeling. *Manag. Large Syst. Proc.* **2018**, *76*, 26–68. (In Russian)
14. Gavrilova, T.A.; Leshcheva, I.A. Building Collaborative Ontologies: A Human Factors Approach. In *Collaborative Knowledge in Scientific Research Networks*; IGI Publishing: Hershey, PA, USA, 2015; pp. 305–324.
15. Lindley, D.V. The Philosophy of Statistics. *J. R. Stat. Soc. Ser. D* **2000**, *49*, 293–337. [CrossRef]
16. Aven, T. Risk analysis and science. *Int. J. Reliab. Qual. Saf. Eng.* **2004**, *11*, 1–15. [CrossRef]
17. Zografopoulos, I.; Ospina, J.; Liu, X.; Konstantinou, C. Cyber-Physical Energy Systems Security: Threat Modeling, Risk Assessment, Resources, Metrics, and Case Studies. *IEEE Access* **2021**, *9*, 29775–29818. [CrossRef]
18. ISO. Online Browsing Platform. Available online: https://www.iso.org/obp/ui/#iso:std:iso-iec:27032:ed-1:v1:en:term:4.10 (accessed on 4 August 2022).
19. Encyclopedia of Kaspersky. Targeted Attacks. Available online: https://encyclopedia.kaspersky.ru/glossary (accessed on 4 August 2022). (In Russian)
20. Khan, M.S.; Siddiqui, S.; Ferens, K.A. Cognitive and Concurrent Cyber Kill Chain Model. In *Computer and Network Security Essentials*; Springer: Cham, Switzerland, 2018; pp. 585–602.
21. Ramirez, C.; Valdes, B. A General Knowledge Representation Model of Concepts. In *Advances in Knowledge Representation*; IntechOpen: London, UK, 2012; pp. 3–76.
22. Gaskova, D.A.; Massel, A.G. Modeling scenarios of extreme situations in the energy sector caused by cyber threats. In Proceedings of the International Conference of Young Scientists "Energy Systems Research" 2021, Irkutsk, Russia, 25–28 May 2021.
23. SRA Glossary. Society for Risk Analysis Glossary. Available online: https://www.sra.org/wp-content/uploads/2020/04/SRA-Glossary-FINAL.pdf (accessed on 4 August 2022).

engineering proceedings

MDPI

Proceeding Paper

Structural Analysis of Hydrodynamical Interaction of Full-Submerged Archimedes Screws of Rotary-Screw Propulsion Units of Snow and Swamp-Going Amphibious Vehicles with Water Area via Methods of Computer Simulation †

Svetlana Karaseva [1,*], Aleksey Papunin [1], Vladimir Belyakov [1], Vladimir Makarov [1] and Dmitry Malahov [2]

1 Department of Construction, Road and Carrying-and-Lifting Machines, Nizhny Novgorod State Technical University n.a. R.E. Alekseev, Minin Str., 24, Nizhny Novgorod 630092, Russia; lexa-lenia@rambler.ru (A.P.); belyakov@nntu.ru (V.B.); makvl2010@gmail.com (V.M.)

2 Department of Tow Trucks and Amphibious Vehicles, Moscow Automobile and Road Construction State Technical University, Leningradsky Prospect, 64, Moscow 125829, Russia; malahow_dm@mail.ru

* Correspondence: geliosveta@gmail.com; Tel.: +7-985-413-3988

† Presented at the 15th International Conference "Intelligent Systems" (INTELS'22), Moscow, Russia, 14–16 December 2022.

Abstract: The paper considers the problems of estimation and provision of efficiency of rotary-screw propulsion units (RSP) for snow and swamp-going amphibious vehicles. The performance curves of fully submerged Archimedes screws of RSP are presented with parameters typical for snow and swamp-going amphibious vehicles. These parameters were obtained from the results of computer simulation. For a wide range of performance modes, the impact of particular elements of Archimedes screws on the general efficiency of the propulsion unit is analysed. According to the results of analysis of data received, lines of further research will aim to increase the efficiency of RSP when moving afloat.

Keywords: rotary-screw propulsion units; amphibious vehicles; Archimedes screws

Citation: Karaseva, S.; Papunin, A.; Belyakov, V.; Makarov, V.; Malahov, D. Structural Analysis of Hydrodynamical Interaction of Full-Submerged Archimedes Screws of Rotary-Screw Propulsion Units of Snow and Swamp-Going Amphibious Vehicles with Water Area via Methods of Computer Simulation. *Eng. Proc.* **2023**, *33*, 32. https://doi.org/10.3390/engproc2023033032

Academic Editors: Askhat Diveev, Ivan Zelinka, Arutun Avetisyan and Alexander Ilin

Published: 19 June 2023

1. Introduction

Taking into account the ongoing reclamation of the Arctic, Siberia and the Far North, there is an increasing requirement for vehicles that can achieve high flotation over bearing surfaces such as ice, snow, water, brash ice and broken ice on water, and also over the low-load-bearing capacity soils, for example, over swamps, silt, liquid mud, etc. [1,2]. In such exploitation conditions, the snow and swamp-going amphibians with rotary-screw propulsion units prove themselves to be the best. When designing them, it is crucially important to define the optimal proportion of overwater and overland characteristics depending on the aim of the vehicle, the area of application and the planned region of operation to provide effective and rational use of it. The overland characteristics of RSP Archimedes screws are rather well researched. Meanwhile, there are very few data about the overwater characteristics and optimal geometrical parameters for water movement [3–5]. At the present time, research works on hydrodynamical characteristics of rotary-screw propulsion units are conducted in the department "Construction and road-building equipment" of Nizhny Novgorod State Technical University n.a. R.E. Alekseev and department "Truck tractors and amphibious vehicles" of Moscow Automobile and Road Construction State Technical University. The research on propulsive characteristics of RSP Archimedes screws with typical geometrical parameters for snow and swamp-going amphibious vehicles previously conducted by authors revealed the correspondences of thrust, torque and performance indicators for cruising and mooring modes [6–9]. These correspondences are shown in Figure 1.

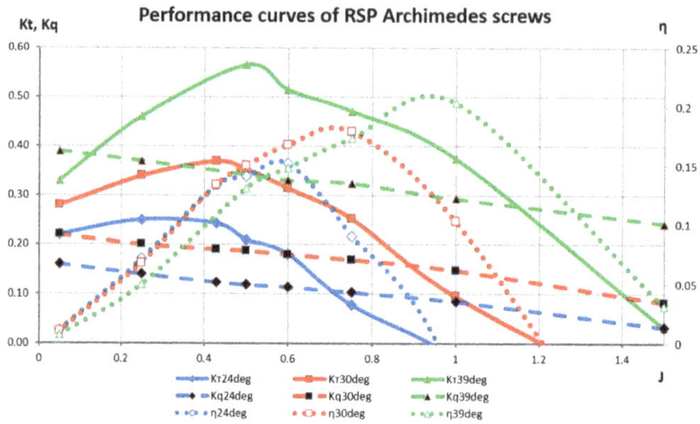

Figure 1. Performance curves.

Thrust coefficient

$$K_T = \frac{T}{\rho \cdot n^2 \cdot D^4}$$

Torque coefficient

$$K_Q = \frac{Q}{\rho \cdot n^2 \cdot D^5}$$

Propulsive coefficient

$$\eta = \frac{J}{2\pi} \cdot \frac{K_T}{K_Q}$$

Advance ratio

$$J = \frac{v_a}{n \cdot D}$$

Here, T is the propeller thrust (N), Q is the torque of the propeller shaft (N•m), n is the propeller rotational speed (rpm) and v_a is the speed of water flow (m/s). The analysis of visualization patterns of computer simulation results of Archimedes screws hydrodynamics proved the difficulty of the nature of their interaction with the water area and showed the essential non-uniformity of load distribution over particular elements. The results of quantitative analysis of pointed non-uniformity can be seen below. These allow us to estimate the contribution of Archimedes screw elements to useful thrust and structure of power loss to Archimedes screw rotation while moving afloat.

2. Representation of Geometrical Models and Statement of the Problem

Figure 2 shows the schemes of simulation objects. The research was conducted for three models of a three-start Archimedes screw with helix angles 24, 30 and 39 degrees. The other geometrical characteristics are identical. To determine the possibility of structural analysis of hydrodynamic interaction with the water area, the surfaces of Archimedes screw models were separated into elements. During simulation over each element, we registered the averaged values of thrust in the longitudinal direction and torsion torque, respectively, to the rotation axis. As a result, the Archimedes screw models consist of eight parts of helixes located on the basic cylinder of Archimedes screw, as follows: cylindrical helixes; eight parts of basic cylinder of the Archimedes screw; fore and aft cowls; three input and three output elements of helixes.

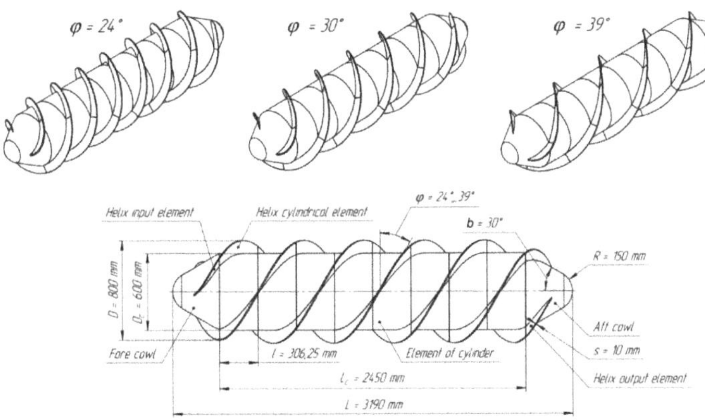

Figure 2. Schemes of researched models of Archimedes screw.

3. Results of Computer Simulation

These problems were solved with the aid of a computer simulation at the fixed value of flow speed v = 4 m/s and various values of rotation speeds of the Archimedes screw in the range n = 200...700 rpm. As a result, the essential dataset was received. Some visualization patterns of the results of computer simulation are shown in Figure 3.

Figure 4 presents the correspondences of thrust T_i and torque Q_i of basic elements thoroughly brought to the thrust of the entire Archimedes screw and torque, due to the advance ratio. Different elements can create either thrust or resistance. Because of this, the values of specific force factors shown in the diagrams in the first case can exceed one. In the second, the negative quantity is obtained.

Naturally, the main element generating the thrust is the cylindrical part of the helixes of Archimedes screws. In the entire range of helix angles, their thrust is close to the total thrust of the Archimedes screw. Meanwhile, with the increase in the advance ratio, the specific thrust and the torque of the cylindrical part of helixes simultaneously increase. The second element generating the thrust is the input helixes. On the ascending branch of efficiency of the Archimedes screw (see Figure 1), their stake in general thrust does not depend on an advanced ratio and varies in the range 0.5...0.8 (fewer values conform to the higher values of helix angles). Specific torque lies in the range 0.2...0.3, which reduces with the increase in the advance ratio. The main factor reducing the hydrodynamical efficiency of Archimedes screws is the resistance of the aft cowl. Its stake in general thrust increases with the reduction in the helix angle and the increase in the advance ratio. The critical excursion of specific resistance of the aft cowl is on the descending branch of efficiency of the Archimedes screw. The contribution of this element to power loss to Archimedes screw rotation is insignificant. The correspondences provided show that other structural elements slightly affect the general efficiency of fully submerged Archimedes screws in a frontal approach flow.

The results received for thrust and torque distribution in relation to the length of spiraling of Archimedes screws (Figure 5) are of particular interest. For clear presentation, the elements of helixes are numbered from fore to aft (the abscissa axis). The values "0" and "9" correspond to input and output helix elements.

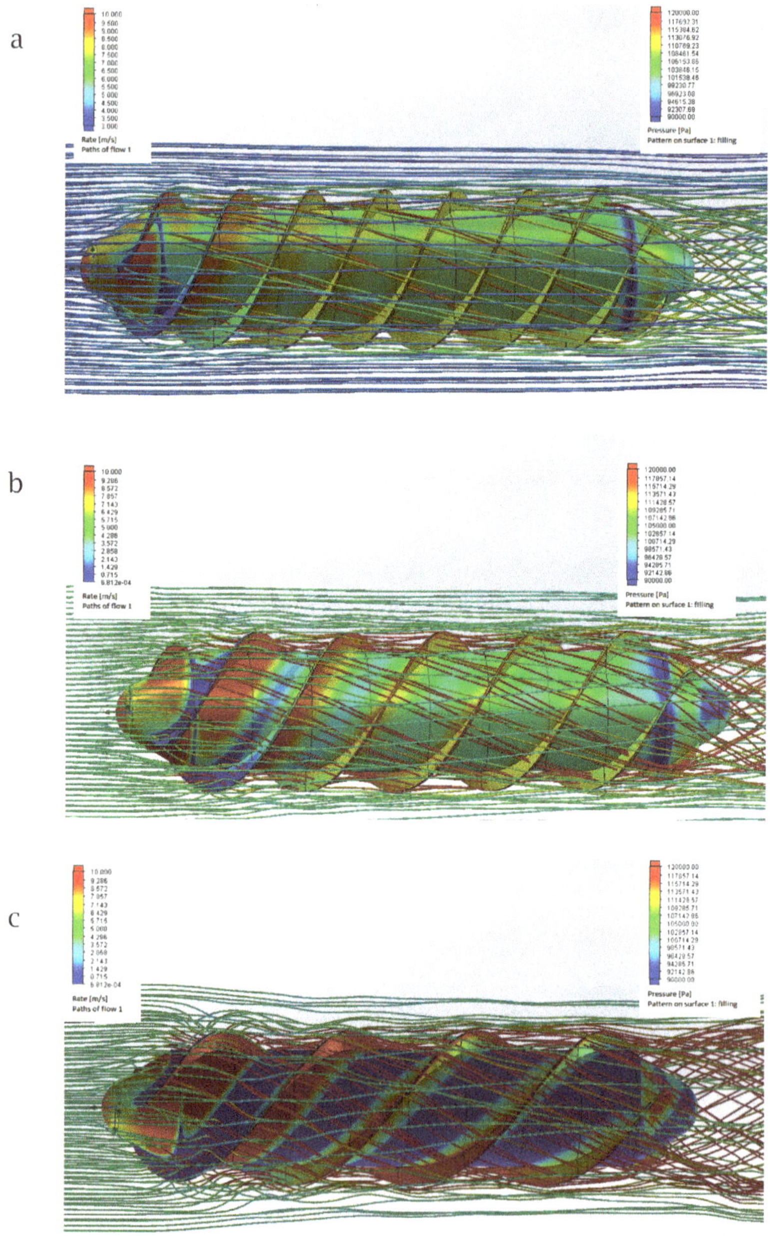

Figure 3. Visualization patterns of computer simulation of hydrodynamics of fully submerged Archimedes screw of RSP (v = 4 m/s, n = 500 rpm): (**a**) helix angle 24 deg; (**b**) helix angle 30 deg; (**c**) helix angle 39 deg.

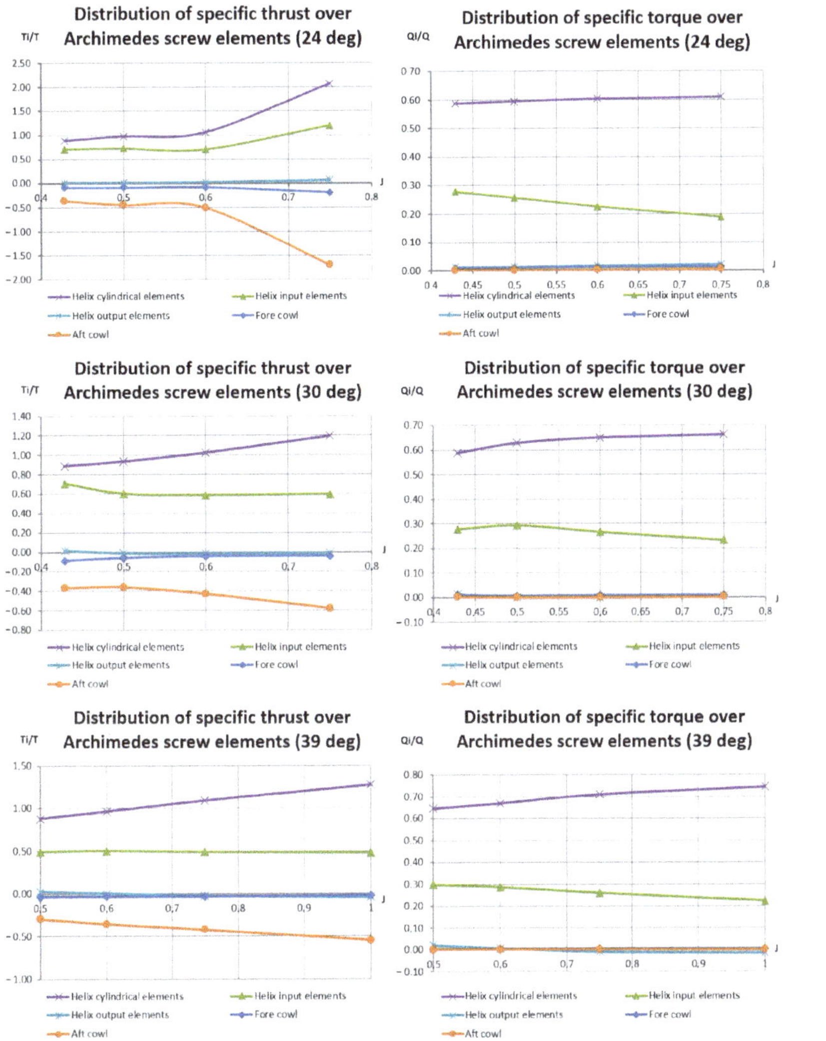

Figure 4. Impact of advance ratio on specific thrust and specific torque for frontally approaching flow to fully submerged Archimedes screw.

The correspondences given show that in the full range of helix angles, there is a conceptually identical pattern of distribution. In fact, the efficiency area is limited by the first four elements of helixes, including the input one. The contribution of the other parts to thrust and useful power is inessential, particularly at high rotation speed (for low values of advance ratio). The intriguing point is also the discovered effect of changeover of some helix parts (elements 5…8) to "turbine operation" appearing for high angles of spiraling (39 degrees).

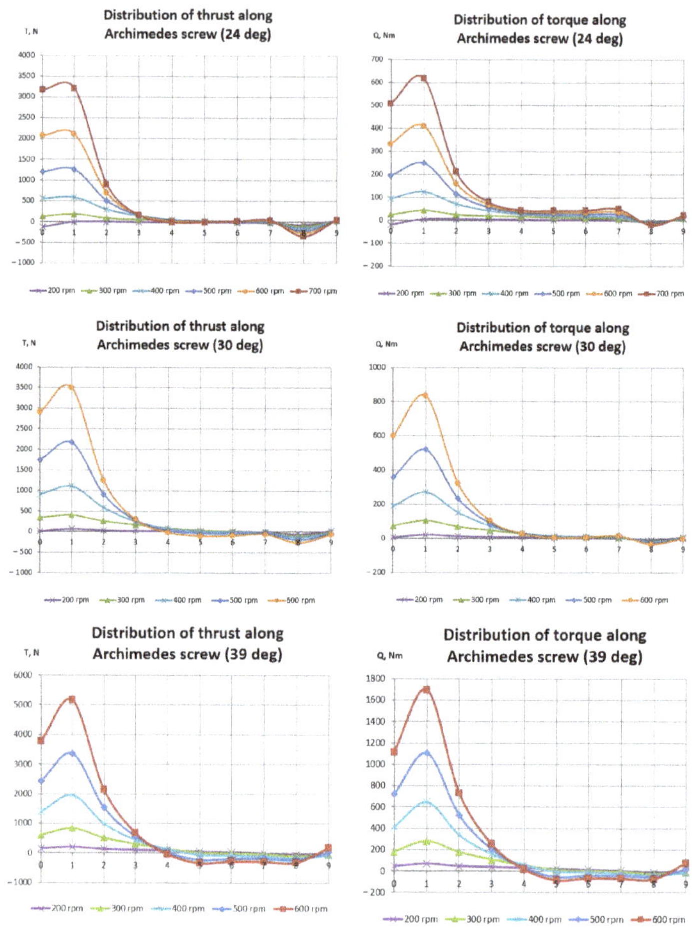

Figure 5. Thrust and torque distribution over the length of one helix of Archimedes screw.

4. Summary

According to the results of research on models of fully submerged Archimedes screws with the most typical geometrical characteristics for snow and swamp-going amphibious vehicles with rotary-screw propulsion units, the following main conclusions can be drawn:

- In the analyzed typical range of advance ratio J for frontally approaching water flow, the thrust is generated by input and cylindrical elements of Archimedes screw helixes. The fore cowl, output elements of helixes and cylindrical part (basic cylinder) of the Archimedes screw play a small part in thrust generation and input power costs. The main negative influence on the general thrust of the Archimedes screw has the resistance of the aft cowl.
- The input power and effective thrust are practically realized only on the input part of helixes and on the front part of cylindrical helixes with a length approximately equal to Archimedes screw diameter D.
- Archimedes screws in combination with their high hydrodynamical efficiency give evidence of the necessity of paying special attention to the construction development of these elements during the design process.

5. Conclusions

The observations about the contribution of Archimedes screw elements to its propulsive characteristics allows us to consciously form variants of design concepts targeted at increasing the hydrodynamical efficiency of RSP. As concepts, we can consider the mounting of optional equipment (deflectors, shrouds etc.) changing the behavior of flow around the particular parts of Archimedes screw and the shaping of cowls, input and output elements of helixes. While adding the optional equipment to Archimedes screw structure, it is necessary to design it in such a way that the application of equipment does not disimprove the overland characteristics of the rotary-screw propulsor. At the present time, the relevant research continues and the study of hydrodynamics of tandem Archimedes screws with counterrotation is being conducted.

Author Contributions: Conceptualization, V.B.; methodology, D.M.; formal analysis, D.M.; investigation, D.M.; writing—original draft preparation, S.K.; visualization, D.M.; supervision, V.M.; project administration, V.M.; funding acquisition, A.P. All authors have read and agreed to the published version of the manuscript.

Funding: The results of the given study have been obtained with the financial support of the grants of the President of the Russian Federation № MK-336.2022.4.

Data Availability Statement: The data presented in this study are available on request from the corresponding author. The data are not publicly available due to privacy policy of the Institute's scientific research.

Conflicts of Interest: The authors declare no conflict of interest.

References

1. Kristian, N. Requirements and Concepts for Arctic Evacuation: Department of Marine Technology. Master's Thesis, Norwegian University of Science and Technology, Kristiansand, Norway, 2011; 98p.
2. Lars, M.F.N. Modeling, Analysis and Joystick Control of the "AMV Oil Spill Fighter": Department of Marine Control Engineering. Master Thesis's, Norwegian University of Science and Technology, Kristiansand, Norway, 2011; 81p.
3. Kulyashov A.P.; Kolotilin V.E. *Ecological Safety of Propulsion Units of Transport and Technological Vehicles*; Mashinostroenie: Moscow, Russia, 1993.
4. Nikolaev A.F.; Kulyashov A.P. *Rotary-Screw Amphibian*; Gor'ky Polytechnic Institute n.a. A.A. Zhdanov, Volgo-Vyatskoe Book Publisher: Gorky, Russia, 1973.
5. Shapkin, V.A.; Vahidov, U.Sh.; Yerasov, I.A.; Koshurina, A.A.; Molev, Yu.I. *Introduction to Dynamic of Systems Snow-Covered Area—Rotary-Screw Vehicle: Monograph*; The Russian Presidental Academy of National Economy and Public Administration: Nizhny Novgorod, Russia, 2017.
6. Barnitsas, M.M.; Ray, D.; Kinley, P. *KT, KQ and Efficiency Curves for the Wageningen B-Series Propellers*; The University of Michigan: Ann Arbor, MI, USA, 1981.
7. Karaseva, S.; Makarov, V.; Malahov, D. Efficiency determinants of rotary-screw propulsion units of snow and swamp-going amphibious vehicles in afloat motion. In Proceedings of the 20th International Conference of the International Society for Terrain-Vehicle Systems, ISTVS 2021, Online, 27–29 September 2021.
8. Karaseva, S.; Belyakov, V.; Makarov, V.; Malahov, D. Analysis of Efficiency of Full-submerged Archimedes Screws of Rotary-screw Propulsion Units of Snow and Swamp-going Amphibious Vehicles. In Proceedings of the 8th International Conference on Vehicle Technology and Intelligent Transport Systems, Online, 27–29 April 2022; pp. 438–444. ISBN 978-989-758-573-9, ISSN 2184-495X.
9. Karaseva, S.A. The calculation of auger amphibian mover main characteristics. *Avtomob.' Doroga Infrastrukt.* **2014**, 2, 11.

engineering proceedings

Proceeding Paper

Algorithm for Determining the Singularity-Free and Interference-Free Workspace of a Robotic Platform for Fruit Harvesting †

Dmitry Malyshev, Larisa Rybak *, Elena Gaponenko and Artem Voloshkin

BSTU named after V.G. Shukhov, 46 Kostyukova Street, Belgorod 308012, Russia; malyshev.d.i@ya.ru (D.M.); gaponenkobel@gmail.com (E.G.); voloshkin.artem.a@gmail.com (A.V.)
* Correspondence: rlbgtu@gmail.com; Tel.: +7-4722-230530
† Presented at the 15th International Conference "Intelligent Systems" (INTELS'22), Moscow, Russia, 14–16 December 2022.

Abstract: This paper proposes a robotic system for fruit harvesting, which includes a mobile platform with a fruit basket, on which a parallel platform is installed, and in the center of the moving platform a telescopic link is installed for harvesting fruit from trees. Numerical algorithms are developed for determining the workspace of platforms with a parallel structure, represented as an ordered set of integers, and for transferring constraints from the platform orientation coordinate space to the end-effector coordinate space. The limits of the workspace are the permissible ranges of rod lengths and the condition that there are no singularities or link interference. The results of modeling are presented.

Keywords: workspace; parallel robot; fruit harvesting

Citation: Malyshev, D.; Rybak, L.; Gaponenko, E.; Voloshkin, A. Algorithm for Determining the Singularity-Free and Interference-Free Workspace of a Robotic Platform for Fruit Harvesting. *Eng. Proc.* **2023**, *33*, 33. https://doi.org/10.3390/engproc2023033033

Academic Editors: Askhat Diveev, Ivan Zelinka, Arutun Avetisyan and Alexander Ilin

Published: 19 June 2023

1. Introduction

The last few years have shown a sharp increase in demand for intelligent robots capable of performing complex tasks without any human intervention in almost all sectors of the national economy, including the agricultural industry. A lot of research is devoted to the development of autonomous robots for harvesting, identifying and removing weeds, spraying chemicals, etc. To date, several such robots have been developed, but most of them harvest fruits in ways that damage either the fruit or the tree, or both. In recent years, a number of results have been obtained in the field of robotic harvesting of fruits and vegetables. In [1], the optimization of robotic harvesting was performed by accurately determining the position of tomato fruits. In [2], an autonomous system is proposed for harvesting most types of crops with peduncles. A geometric approach was applied to obtain the cut point of the stem, based on the determination of the bounding box of the fruit using a neural network. The proposed architecture of the picking robot has two main modules: a module for detecting a point suitable for cutting on the stem and a gripping module that clamps the fruit and cuts the stem. In the article in [3], a prototype of a robot for picking apples was developed and tested. The robot was tested using a spin–pull apple picking pattern. The success rate for harvesting with the spin–pull scheme was 47.37% in the field and 78% in a simulated orchard, with a harvest cycle time of 4 s. The level of stem damage in the field garden was 11.11%. The developed picking prototype realized the task of picking apples, with a competitively short harvest cycle time. In [4], mathematical modeling of mechanical output link connections was carried out by analyzing the movement of the harvest, whereby the coordinates of the target fruits were parsed and analyzed in the context of a mathematical model for accurate location and harvesting. An autonomous system that harvest most types of agricultural fruit crops is discussed in [2]. The proposed grip is attached to a robot that uses the Robot Operation System. The installation was tested in laboratory conditions on various artificial plants. The method of apple harvesting using vacuum cups is discussed in [5]. The mechanical arm has four suction cups that deform over the surface of the apple. The parameters of the mechanical arm were determined, and laboratory tests were carried out. The article in [6] proposes a progressive analytical approach to the design and optimization of a citrus harvesting machine with a canopy. The approach was formulated using finite element methods (FEM) to find the optimal design parameters of the machine. The design parameters were determined in terms of the configuration (or stiffness) of the shaking rods and two operating parameters: the shaking frequency and the shaking amplitude. The formulated methodology consists of determining the properties of wood, the statistical modeling of tree branches, developing mechanical models, and performing optimization using FEM modeling. The proposed methodology uses quantitative estimation of objective functions, as well as Pareto search methods to find optimal constructions. In this study, three sets of device parameters were proposed to minimize tree damage and maximize fruit removal. These optimal parameters were proposed based on the configuration and distribution of the branches and fruits of a medium-sized tree.

Workspace determination is an important issue for robot design. Geometric and discrediting methods are used to determine the workspace. Geometric methods provide an accurate description, but they are applicable only to the simplest robots (some planar and the simplest spatial ones). The result obtained is an accurate analytical description of the workspace. Disadvantages: it is difficult to take into account all the constraints, and the result obtained is difficult to apply when planning the trajectory. A relatively simple definition of the workspace is possible for some robots, such works were carried out by Clavel [7] and Di Gregorio [8] for the Delta robot, Alizade [9] and Arun [10] for spherical robots, and Husty [11] for planar parallel robots. Discretization methods consist of determining a certain number of acceptable robot positions that form grid nodes. The calculation process is extremely time-consuming, but the results are easy to apply to trajectory planning. This method has been developed by many researchers, especially the work of Chablat [12]. In [13], we consider the non-uniform covering method for approximating the set of solutions to a system of nonlinear inequalities, as well as the application of this method to determine the workspace of some types of planar robots. In this paper, we solve the problem of determining the workspace of a robotic platform for fruit harvesting using the method of non-uniform covering and converting the covering set into a partially ordered set of integers [14].

2. Mathematical Model of RS

Consider the design of a fruit harvesting robotic system (RS) for harvesting fruit, the 3D model of which is shown in Figure 1a. The RS includes a mobile platform with a basket for picking fruit, on which a parallel mechanism is installed [15]. In the center of the moving platform, a telescopic link is installed for accessing fruits at high altitude.

The parallel mechanism consists of three drive RRPS-type kinematic chains and a central kinematic chain with a spherical joint rigidly connected to the fixed and movable platforms. Thus, the required rotation of the movable platform in all axes relative to the joint center C is provided by changing the lengths l_i of the A_iB_i rods. The fixed platform $A_1A_2A_3$ and the moving platform $B_1B_2B_3$ of the mechanism are regular triangles with radii R_1 and R_2, respectively.

The input coordinates are the lengths of the drive links l_1, l_2, l_3, the output coordinates are the coordinates of the point P of the end-effector x_p, y_p, z_p. The point P is located at a distance l_t from the center of the moving platform. Workspace determination has two stages. The first stage is determining the set of acceptable values of the angular coordinates of the moving platform relative to the joint C. The second stage is determinng the set of coordinates P of the end-effector for these values. The coordinates of P in the moving coordinate system X_2, Y_2, Z_2:

$$P^{(2)} = [\ 0 \quad 0 \quad (l_{c2} + l_t) \quad 1\ ]^T \tag{1}$$

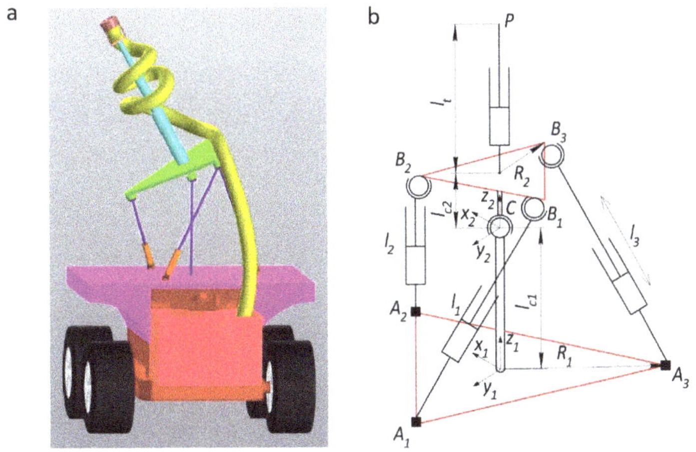

Figure 1. Conceptual design of the RS for fruit harvesting: (**a**) 3D model, (**b**) design scheme of the robotic platform.

The coordinates of the point P can be calculated in the fixed coordinate system $X_1Y_1Z_1$

$$P = M_{2\_1}P^{(2)} \tag{2}$$

where M is the transition matrix from the moving coordinate system $X_2Y_2Z_2$ to the stationary system $X_1Y_1Z_1$, which includes displacements along the X_1, Y_1, Z_1 axis and rotation using Euler angles α, β, γ, taking into account the orientation of the platform.

$$M_{2\_1} = \begin{bmatrix} C_\alpha C_\gamma - C_\beta S_\alpha S_\gamma & -C_\alpha S_\gamma - C_\beta C_\gamma S_\alpha & S_\alpha S_\beta & O \\ C_\gamma S_\alpha + C_\alpha C_\beta S_\gamma & C_\alpha C_\beta C_\gamma - S_\alpha S_\gamma & -C_\alpha S_\beta & O \\ S_\beta S_\gamma & C_\gamma S_\beta & C_\beta & l_{c_1} \\ 0 & 0 & 0 & 1 \end{bmatrix} \tag{3}$$

where $C_a = \cos\alpha, S_a = \sin\alpha, C_\beta = \cos\beta, S_\beta = \sin\beta, C_\gamma = \cos\gamma, S_\gamma = \sin\gamma$.

After the transformation, taking into account (1)–(3), we obtain

$$P = \begin{bmatrix} S_\alpha S_\beta (l_{c2} + l_t) \\ -C_\alpha S_\beta (l_{c2} + l_t) \\ l_{c1} + C_\beta (l_{c2} + l_t) \\ 1 \end{bmatrix} \tag{4}$$

Let us introduce restrictions on the geometrical parameters of the mechanism

$$l_{min} \le l_t \le l_{max} \tag{5}$$

where l_{min},l_{max} are determined by the design parameters of the mechanism; l_i is the length of the i-th rod

$$l_i = \sqrt{(x_{Bi} - x_{Ai})^2 + (y_{Bi} - y_{Ai})^2 + (z_{Bi} - z_{Ai})^2} \tag{6}$$

where x_{Ai}, y_{Ai}, z_{Ai}, and x_{Bi}, y_{Bi}, z_{Bi} are the coordinates of the centers of the joints A_i and B_i, respectively, in the fixed X_1, Y_1, Z_1 coordinate system. We can now define the coordinates of the joints B_i in the moving coordinate system $X_2Y_2Z_2$.

$$B_1^{(2)} = \begin{bmatrix} R_2 \\ 0 \\ l_{c2} \\ 1 \end{bmatrix}, B_2^{(2)} = \begin{bmatrix} 0.5R_2 \\ 0.5\sqrt{3}R_2 \\ l_{c2} \\ 1 \end{bmatrix}, B_3^{(2)} = \begin{bmatrix} -0.5R_2 \\ -0.5\sqrt{3}R_2 \\ l_{c2} \\ 1 \end{bmatrix} \quad (7)$$

Let us express the coordinates of the joints B_i in the X_1, Y_1, Z_1 coordinate system, taking into account (3) and (7)

$$\mathrm{B}_1 = \mathrm{M}_{2\_1}B_1^{(2)} = \begin{bmatrix} R_2(C_aC_\gamma - C_\beta S_a S_\gamma) + l_{c2}S_aS_\beta \\ R_2(C_\gamma S_a + C_aC_\beta S_\gamma) - l_{c2}C_aS_\beta \\ l_{c1} + R_2S_\beta S_\gamma + l_{c2}C_\beta \\ 1 \end{bmatrix}, \quad (8)$$

$$\mathrm{B}_2 = \mathrm{M}_{2\_1}B_2^{(2)} = \begin{bmatrix} 0.5R_2\left(C_aC_\gamma - C_\beta S_aS_\gamma - \sqrt{3}C_aS_\gamma - \sqrt{3}C_\beta C_\gamma S_a\right) + l_{c2}S_aS_\beta \\ 0.5R_2\left(C_\gamma S_a + C_aC_\beta S_\gamma + \sqrt{3}C_aC_\beta C_\gamma - \sqrt{3}S_aS_\gamma\right) - l_{c2}C_aS_\beta \\ l_{c1} + 0.5R_2\left(S_\beta S_\gamma + \sqrt{3}C_\gamma S_\beta\right) + l_{c2}C_\beta \\ 1 \end{bmatrix}, \quad (9)$$

$$\mathrm{B}_3 = \mathrm{M}_{2\_1}B_3^{(2)} = \begin{bmatrix} 0.5R_2\left(C_\beta S_aS_\gamma + \sqrt{3}C_aS_\gamma + \sqrt{3}C_\beta C_\gamma S_a - C_aC_\gamma\right) + l_{c2}S_aS_\beta \\ 0.5R_2\left(\sqrt{3}S_aS_\gamma - C_\gamma S_a - C_aC_\beta S_\gamma - \sqrt{3}C_aC_\beta C_\gamma\right) - l_{c2}C_aS_\beta \\ l_{c1} - 0.5R_2\left(S_\beta S_\gamma + \sqrt{3}C_\gamma S_\beta\right) + l_{c2}C_\beta \\ 1 \end{bmatrix}, \quad (10)$$

We define the coordinates of the joints A_t in the fixed coordinate system $X_1, Y_1\ Z_1$:

$$A_1 = \begin{bmatrix} R_2 \\ 0 \\ 0 \\ 1 \end{bmatrix}, A_2 = \begin{bmatrix} 0.5R_2 \\ 0.5\sqrt{3}R_2 \\ 0 \\ 1 \end{bmatrix}, A_3 = \begin{bmatrix} -0.5R_2 \\ -0.5\sqrt{3}R_2 \\ 0 \\ 1 \end{bmatrix}. \quad (11)$$

Thus, substituting (8)–(11) in (6), we obtain an analytical dependence of the form $l_i = f(\alpha, \beta, \gamma)$.

At the same time, we should take into account the presence of singularities in the mechanism, which significantly increase the dynamic loads on the links and cause the robot to lose control. We consider the method proposed in [16] to determine singularities, based on the analysis of the Jacobi matrix, whose determinant has the form

$$\det(J_A) = \begin{bmatrix} \frac{\partial l_1}{\partial \alpha} & \frac{\partial l_1}{\partial \beta} & \frac{\partial l_1}{\partial \gamma} \\ \frac{\partial l_2}{\partial \alpha} & \frac{\partial l_2}{\partial \beta} & \frac{\partial l_2}{\partial \gamma} \\ \frac{\partial l_3}{\partial \alpha} & \frac{\partial l_3}{\partial \beta} & \frac{\partial l_8}{\partial \gamma} \end{bmatrix}, \quad (12)$$

where l_i is defined taking into account the formulas (6), (8)–(11). Due to the cumbersome nature of the obtained formulas for each of the elements of the determinant, we give only the first one

$$\frac{\partial l_1}{\partial \alpha} = \frac{-(2s_1(s_2 - R_1) - 2s_1s_2)}{2\left(\left(l_{c1} + l_{c2}C_\beta + R_2S_\beta S_\gamma\right)^2 + (s_2 - R_1)^2 + s_1^2\right)^{0.5}} \quad (13)$$

where $s_1 = R_2(C_\gamma S_\alpha + C_\alpha C_\beta S_\gamma) - l_{c2} C_\alpha S_\beta$, $s_2 = R_2(C_\alpha C_\gamma - C_\beta S_\alpha S_\gamma) + l_{c2} S_\alpha S_\beta$

The condition for the presence of singularities has the form $\det(J_A) = 0$. It is necessary to ensure that the determinant of the Jacobi matrix is constant in sign to exclude singularities from the workspace. One of the following conditions must be added (5): $\det(J_A) < 0$ or $\det(J_A) > 0$, depending on the sign of the determinant.

We should also take into account the restriction associated with ensuring the orientation of the platform at which the telescopic link is directed up, since the orientation of the platform at which the telescopic link is directed down is not excluded by condition (5).

$$z_p > l_{c1} \tag{14}$$

The exclusion of link interference can be taken into account using the method described earlier by the authors in [17]. Thus, taking into account the link interference and other restrictions in accordance with (5), (12), and (14), the workspace in the coordinate space of the orientation of the moving platform of the parallel mechanism can be determined. The covering set of the workspace is obtained using the methods described in [13], and then transformed into a partially ordered set of integers [14].

3. Transferring Constraints to the Coordinate Space of the End-Effector

We can calculate the set of positions of the end-effector using formula (4). In this case, the telescopic link due to extension is an interval

$$l_{t,min} \le l_t \le l_{t,max} \tag{15}$$

Using the representation of the workspace as a partially ordered set of integers, we define set B of coordinates of the end-effector P in the space of integers. For this purpose, an algorithm is developed based on a modification of Bresenham's algorithm [18]. In [19], a modification of the algorithm was proposed for the three-dimensional case, but the coordinates of the beginning and end of the segments belong to the space of integers, which leads to a displacement of the segment and set B (Figure 2). Cells that intersect the orthosis are highlighted in red for coordinates represented as integers, yellow for coordinates represented as real numbers, and orange for both cases. As you can see from the figure, the application of integer coordinates does not allow us to accurately determine set B.

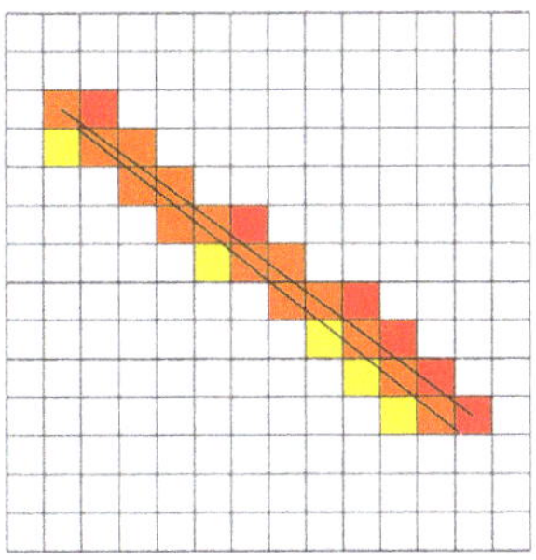

Figure 2. Offset of a segment of the trajectory depending on the initial data.

We modify Bresenham's algorithm, taking into account the use of initial data belonging to the three-dimensional space of real numbers (coordinates x_1, y_1, z_1 , y_1, z_1 of the end-effector at $l_{t,min}$ and x_2, y_2, z_2 at $l_{t,max}$). In this case, the coordinates must correspond to the covering set of the workspace, represented as a partially ordered set of integers, respectively. They must be obtained taking into account the accuracy of the approximation δ using formula (4)

$$x_1 = \frac{s_a S_\beta (l_{c2} + l_{\text{t,min}})}{\Delta}, y_1 = \frac{-c_a S_\beta (l_{c2} + l_{\text{t,min}})}{\Delta}, z_1 = \frac{l_{c1} + C_\beta (l_{c2} + l_{\text{t,min}})}{\Delta} \tag{16}$$

$$x_2 = \frac{s_a S_\beta (l_{c2} + l_{t,max})}{\Delta}, y_2 = \frac{-c_a S_\beta (l_{c2} + l_{t,max})}{\Delta}, z_2 = \frac{l_{c2} + C_\beta (l_{c2} + l_{t,max})}{\Delta} \quad (17)$$

The pseudocode of the modified Bresenham algorithm proposed by the authors is presented in [20].

4. Simulation Results

Let us perform a computational experiment. For this purpose, a software package was developed in the C++programming language. Parallel calculations were implemented using the OpenMP library. The visualization of link interference was performed using a developed Python script (Matplotlib and JSON libraries). The visualization of three-dimensional results was performed by exporting an ordered set of integers describing the work domain to the STL format. Simulation was performed for the following parameters: R_1 = 300 mm, R_2 = 200 mm, l_{min} = 360 mm, l_{max} = 600 l_{c1} = 500 mm, l_{c2} = 0 mm, $l_{t,min}$ = 500 mm, link diameter D_{link} = 20 mm, and minimum angle between links $\phi_{min} = 10°$. The workspace in the coordinate space α, β, γ of the moving platform orientation, without taking into account singularities, is shown in Figure 3a. Figure 3b shows the workspace in the coordinate space of the end-effector x_P, y_P, z_P.

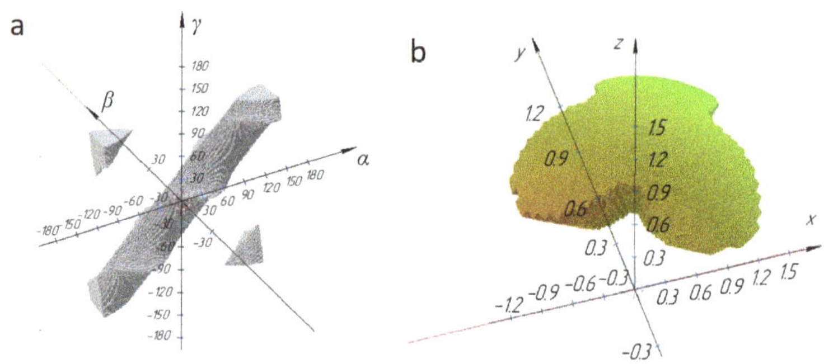

Figure 3. Platform workspace: (a) in coordinates (α, β, γ), (b) (x_p, y_p, z_P), mm*10^3.

The workspace was determined both with the positivity condition (Figures 4a and 5a) of the determinant and with the negativity condition (Figures 4b and 5b) to select the sign in the sign-constant condition of the determinant of the Jacobi matrix, in order to exclude singularities.

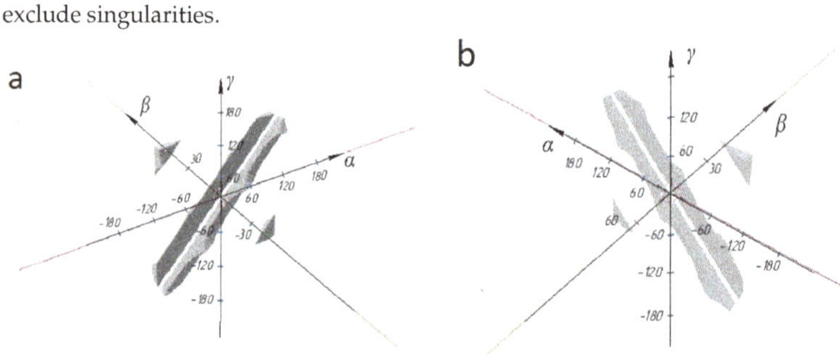

Figure 4. The workspace of the platform in coordinates (α, β, γ): (a) with a negative sign of the determinant of the Jacobi matrix, (b) with a positive sign.

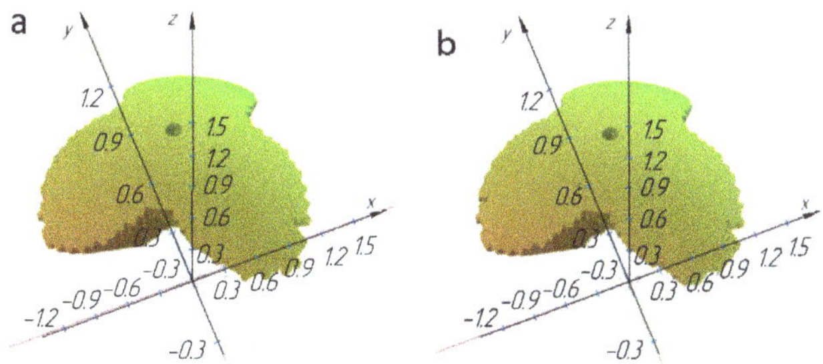

Figure 5. The workspace of the platform in coordinates (x_P, y_P, z_P), mm*10^3: (**a**) with a negative sign of the determinant of the Jacobi matrix, (**b**) with a positive sign.

Figures 4 and 5 show that the workspace in the coordinate space (α, β, γ) is divided into two parts, but the workspace in the coordinate space (x_P, y_P, z_P) practically coincides for different signs of the Jacobi matrix determinant. In both cases, the central zone in which the determinant of the Jacobi matrix is zero is excluded from the workspace. Thus, if the design of the RS gripper and fixing of the fruit harvesting tube will ensure fruit harvesting at any orientation of the end-effector, then both the condition of positivity and negativity of the determinant of the Jacobi matrix can be chosen. It is revealed that for the given initial data for modeling, there is no link interference. We will increase the ranges of changes in the length of the rods while maintaining the ratio l_{min}/l_{max} = 0.6 to ensure sufficient space inside the rod to accommodate ball–screw pairs. When the dimensions are increased to l_{min} = 420 mm, l_{max} = 700 mm, there is interference of the central kinematic chain on the moving platform (Figure 6).

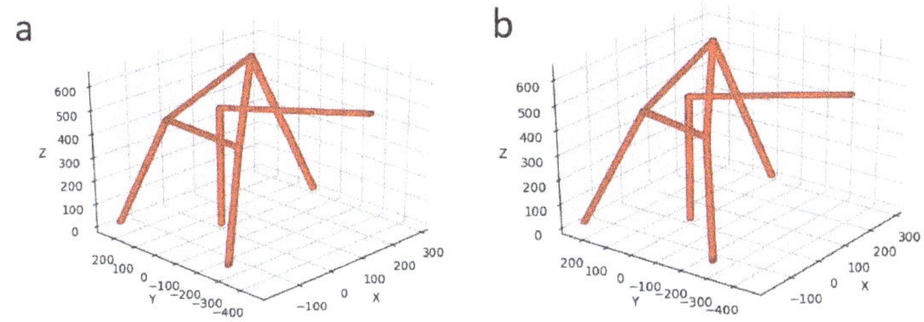

Figure 6. Examples of interference between a central chain and a moving platform.

Increasing the length of the rods allows us to increase the volume of the workspace, but it is important to exclude the interference zones of links from the workspace.

5. Conclusions

Effective numerical methods and algorithms for determining the workspace in the coordinate space of the orientation of the moving platform and the position of the end-effector, taking into account the singularities and interference of the links, are developed and tested for the proposed platform, which is part of the RS for fruit harvesting. It is revealed that if the design of the RS gripper and fixing of the fruit harvesting tube ensures fruit harvesting at any orientation of the end-effector, then both conditions of positivity and negativity of the determinant of the Jacobi matrix can be chosen to exclude singularities. It is shown that the presence of link interference depends on the value of the range of changes in the length of the rods. As part of future research, the task of optimizing the geometric parameters of the platform will be performed, taking into account the compactness of the design and providing the required workspace.

Author Contributions: Conceptualization, E.G., A.V. and D.M.; methodology, D.M.; software, D.M.; validation, L.R. and E.G.; formal analysis, D.M.; investigation, A.V. and D.M.; data curation, D.M. and E.G.; writing—original draft preparation, A.V. and D.M.; writing—review and editing, L.R. and E.G.; visualization, A.V.; supervision, L.R. and E.G.; project administration, E.G. All authors have read and agreed to the published version of the manuscript.

Funding: This research was funded by grant of the Russian Science Foundation No. 22-19-20153, https://rscf.ru/project/22-19-20153/ (accessed on 29 July 2022) and the Government of the Belgorod Region, Agreement No. 4.

Institutional Review Board Statement: Not applicable.

Informed Consent Statement: Not applicable.

Data Availability Statement: Not applicable.

Conflicts of Interest: The authors declare no conflicts of interest.

References

1. Rong, J.; Wang, P.; Wang, T.; Hu, L.; Yuan, T. Fruit pose recognition and directional orderly grasping strategies for tomato harvesting robots. *Comput. Electron. Agric.* **2022**, *202*, 107430. [CrossRef]
2. Zhang, T.; Huang, Z.; You, W.; Lin, J.; Tang, X.; Huang, H. An Autonomous Fruit and Vegetable Harvester with a Low-Cost Gripper Using a 3D Sensor. *Sensors* **2020**, *20*, 93. [CrossRef] [PubMed]
3. Hu, G.; Chen, C.; Chen, J.; Sun, L.; Sugirbay, A.; Chen, Y.; Jin, H.; Zhang, S.; Bu, L. Simplified 4-DOF manipulator for rapid robotic apple harvesting. *Comput. Electron. Agric.* **2022**, *199*, 107177. [CrossRef]
4. Lin, H.; Cai, K.; Chen, H.; Zeng, Z. Optimization design of fruit picking end-effector based on its grasping model. *INMATEH-Agric. Eng.* **2015**. *47*, 81–90.
5. Wang, M.; Yan, B.; Zhang, S.; Pan, F.; Zeng, P.; Shi, S.; Yang, F. Development of a Novel Biomimetic Mechanical Hand Based on Physical Characteristics of Apples. *Agriculture* **2022**, *12*, 1871. [CrossRef]
6. Gupta, S.; Ehsani, R.; Kim, N. Optimization of a Citrus Canopy Shaker Harvesting System: Mechanistic tree damage and Fruit Detachment Models. Trans. *ASABE* **2016**, *59*, 761–776.
7. Clavel, R. Conception d'un Robot Parallele Rapide a 4 Degres de Liberte. Ph.D. Thesis, EPFL, Lausanne, Switzerland, 1991.
8. Di Gregorio, R.; Zanforlin, R. Workspace analytic determination of two similar translational parallel manipulators. *Robotica* **2003**, *21*, 555–566. [CrossRef]
9. Alizade, R.; Tagiyev, N.; Duffy, J. A forward and reverse displacement analysis of an in-parallel spherical manipulator. *Mech. Mach. Theory* **1994**, *29*, 125–137. [CrossRef]
10. Arun, V. Determination of the workspace of the 3-dof double-octahedral variable-geometry-truss manipulator. In Proceedings of the 22nd Biennial Mechanisms Conference, Scottsdale, AZ, USA, 13–16 September 1992; American Society of Mechanical Engineers: New York, NY, USA, 1992; pp. 493–500.
11. Husty, M. On the workspace of planar three-legged platforms. *World Autom. Congr.* **1996**, *3*, 339–344.
12. Chablat, D.; Wenger, P. Moveability and collision analysis for fully parallel manipulators. In Proceedings of the 12th RoManSy, Paris, France, 6–9 July 1998; pp. 61–68.
13. Evtushenko, Y.; Posypkin, M.; Rybak, L.; Turkin, A. Approximating a solution set of nonlinear inequalities. *J. Glob. Optim.* **2018**, *71*, 129–145. [CrossRef]
14. Rybak, L.; Malyshev, D.; Gaponenko, E. Optimization Algorithm for Approximating the Solutions Set of Nonlinear Inequalities Systems in the Problem of Determining the Robot Workspace. In *Communications in Computer and Information Science*; Springer International Publishing: Cham, Switzerland, 2020; pp. 27–37.
15. Merlet J.-P. *Parallel Robots*, 2nd ed.; Springer: Berlin/Heidelberg, Germany , 2007.
16. Gosselin, C.; Angeles, J. Singularity analysis of closed-loop kinematic chains. *IEEE Trans. Robot. Autom.* **1990**, *6*, 281–290. [CrossRef]
17. Behera, L.; Rybak, L.; Malyshev, D.; Gaponenko, E. Determination of Workspaces and Intersections of Robot Links in a Multi-Robotic System for Trajectory Planning. *Appl. Sci.* **2021**, *11*, 4961. [CrossRef]
18. Rogers, D. *Procedural Elements for Computer Graphics*; Computer Science Series; McGraw-Hill, Inc.: New York, NY, USA, 1985.
19. Line3d-3D Bresenham's (a 3D Line Drawing Algorithm). Available online: ftp://ftp.isc.org/pub/usenet/comp.sources.unix/volume26/line3d (accessed on 29 July 2022).
20. Malyshev, D.; Cherkasov, V.; Rybak, L.; Diveev, A. Synthesis of trajectory planning algorithms using evolutionary optimization algorithms. *Commun. Comput. Inf. Sci.* **2022**, *1739*, 153–167.

engineering proceedings

MDPI

Proceeding Paper

Investigation of Internal Model for Unmanned Vehicle Control in Case of Its Aggressive Motion along a Spatial Trajectory †

Igor Prokopiev [1], Elena Sofronova [1,*] and Viktoria Moiseenko [2]

1 Federal Research Center "Computer Science and Control", Russian Academy of Sciences, 44 bld. 2, Vavilova Str., 119333 Moscow, Russia
2 Engineering Academy, RUDN University, 6 Miklukho-Maklaya St., 117198 Moscow, Russia
* Correspondence: sofronova_ea@mail.ru
† Presented at the 15th International Conference "Intelligent Systems" (INTELS'22), Moscow, Russia, 14–16 December 2022.

Abstract: The work is devoted to the study of methods that are used to control the movement of an object along a given trajectory. A control method involving an accurate internal model is proposed. This internal model was built on the basis of the object's mathematical model and real object, performed by artificial neural networks. For a limited period of time the model is able to determine the object state without surveillance system usage. The dynamic model of an unmanned vehicle was obtained by method developed at the Robotics Center of the FRC CSC RAS. This method acquires experimental data and performs model identification by means of a neural network. The trajectory is a set of spatial points generated by the developed real unmanned vehicle simulator. The control was carried out on the basis of PID-controller and model predictive control method. The comparison of control methods for a real and virtual unmanned vehicles was conducted in the simulator developed. The results of field experiments, during which control by internal model was applied, are presented.

Keywords: identification; unmanned vehicle; path tracking

Citation: Prokopiev, I.; Sofronova, E.; Moiseenko, V. Investigation of Internal Model for Unmanned Vehicle Control in Case of Its Aggressive Motion along a Spatial Trajectory. *Eng. Proc.* **2023**, *33*, 38. https://doi.org/10.3390/engproc2023033038

Academic Editors: Askhat Diveev, Ivan Zelinka, Arutun Avetisyan and Alexander Ilin

Published: 20 June 2023

1. Introduction

Unmanned vehicles control from the initial state to the final one should meet the desired quality requirements. At the same time, it is necessary to synthesize such control laws that ensure the achievement of the control goal in the wide class of the uncertainty of vehicle's dynamics. To synthesize the required control laws, the unmanned vehicles control system simulation should be performed. A mathematical model of a real system can be obtained with the help of well-known identification methods [1,2]. Herein the following system identification methods are used, ARMAX (autoregressive moving average models with exogenous inputs) and NARMAX (non-linear ARMAX). These methods allow to obtain linear and non-linear polynomial functions that model the relationship between the perception of the robot sensor and its motor response. These methods are extended to neural network structures [3] and are successfully applied to non-linear unmanned vehicles.

One of the approaches in the field of unmanned mobile systems, such as unmanned vehicles operating in automatic mode, is aimed at the development of non-linear control laws for real time trajectories tracking [4–6].

Unmanned vehicles' real-time trajectory control is usually divided into three sub-tasks, trajectory generation, position determination, and trajectory tracking.

This paper presents the results of model identification of a real unmanned vehicle. Kinematic equations of unmanned vehicle describe mass center position (x, y) and direction of movement (θ). These parameters depend on linear velocity (v) and rotation angle of front wheels (α). Unmanned vehicle and its kinematic movement control scheme are presented in Figure 1.

Artificial neural networks belong to modern model identification methods used in robotics. Combined with differential equations, ANNs provide satisfying results in real non-linear dynamics systems identification.

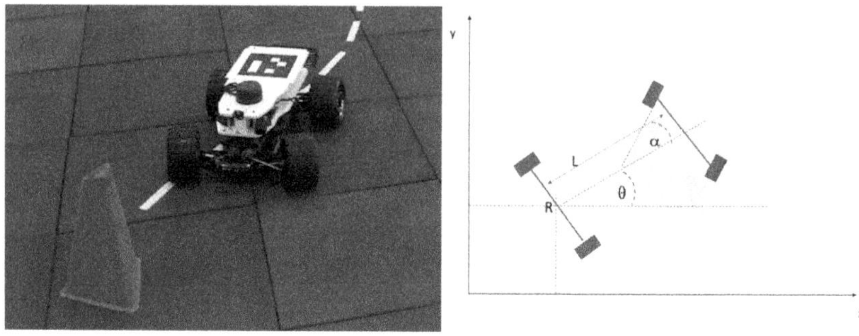

(**a**) (**b**)

Figure 1. Unmanned vehicle (**a**) and its kinematic movement control scheme (**b**).

The article presents a formal statement of the model identification problem. A universal strategy for its solution that uses ANNs is proposed. The obtained model, or so-called the internal one, is applied in a special simulator developed to automate path generation and the process of obtaining the state of the unmanned vehicle that is conducted in selected control methods in case of some parametrical changes.

2. Unmanned Vehicle Model Identification

The unmanned vehicle (UV) is built on the basis of an automobile chassis on a scale of 1:10, has a weight of 1.5 kg and can reach speed up to 10 m/s. The appearance of the UV is shown in Figure 1a. The front wheels of the UV implement steering, the rear ones monitor the state of the robot. Figure 1b shows the kinematic movement control scheme of the unmanned vehicle, where x, y refer to the coordinates; α is the angle of front wheels rotation; θ is the UV orientation relative to the x axis; L is the distance between the front and rear axes of the robot; and R is the base point located in the center of the robot's rear axis.

Experimental data on the vehicle movement was obtained according to a special technique developed in the Robotics Center of the FRC CSC RAS.

The vehicle model consists of hybrid finite-difference equations of the following form

$$\mathbf{x}(k) = \begin{bmatrix} x(k) \\ y(k) \\ \theta(k) \\ v(k) \end{bmatrix} = \begin{bmatrix} x(k-1) + \Delta t v(k-1)\cos(\theta(k-1)) \\ y(k-1) + \Delta t v(k-1)\sin(\theta(k-1)) \\ f^{\theta}(v(k-1), \theta(k-1), u(k), \omega(k), \Delta t) \\ f^{v}(v(k-1), \theta(k-1), u(k), \omega(k), \Delta t) \end{bmatrix}, k = 1, \ldots, K, \quad (1)$$

where $\mathbf{x}$(k) is a state vector,

$$\mathbf{x}(k) = \begin{bmatrix} x(k) & y(k) & \theta(k) & v(k) \end{bmatrix}^T = \begin{bmatrix} x(k) & y(k) & \mathbf{u}(k) \end{bmatrix}^T, \quad (2)$$

$v(k)$ is the velocity of the object, $\theta(k)$ is the UV orientation relative to the x axis, $\omega(k)$ is the component of control for orientation $\theta(k)$, $u(k)$ is the component of control for linear velocity $v(k)$, Δt is the time span, $f^{\theta}(v(k-1), \theta(k-1), u(k), \omega(k), \Delta t)$ and $f^{v}(v(k-1), \theta(k-1), u(k), \omega(k), \Delta t)$ are the outputs of a neural network, K is the number of control steps.

Object movement is determined by orientation $\theta(k)$ and velocity $v(k)$ being predicted by neural network. The inputs of the neural network include orientation and speed control signals $\omega(k)$ and $u(k)$ along with the previous state of orientation $\theta(k-1)$ and velocity $v(k-1)$. At the output, we obtain the current values of orientation $\theta(k)$ and velocity $v(k)$.

The neural network has the following structure

$$\mathbf{z}^{(l)} = \boldsymbol{\varphi}^{(l)}\left(\mathbf{W}^{(l)}\mathbf{z}^{(l-1)} + \mathbf{z}_0^{(l)}\right), l = 1, \ldots, L, \tag{3}$$

where $\mathbf{z}^{(l)}$ is the output vector of layer l, $\boldsymbol{\varphi}^{(l)}(\mathbf{s})$ is a vector activation function of layer l, $\mathbf{W}^{(l)}$ is a weight matrix of layer l, $\mathbf{W}^{(l)} = \left[w_{i,j}^{(l)}\right], i = 1, \ldots, n_l, j = 1, \ldots, n_{l-1}$, $\dim\mathbf{W}^{(l)} = n_l \times n_{l-1}$, $\mathbf{z}_0^{(l)}$ is a bias of l, $\mathbf{z}_0^{(l)} = \left[v(k-1) \quad \theta(k-1) \quad u(k) \quad \omega(k) \quad \Delta t\right]^T$, L is the number of layers, $\mathbf{z}^{(L)} = \left[\theta(k) \quad v(k)\right]^T, n_0 = 5, n_L = 2$.

Activation function $\boldsymbol{\varphi}^{(l)}(\mathbf{s})$ is applied to each component of the state vector

$$\boldsymbol{\varphi}^{(l)}(\mathbf{s}) = \left[\varphi_1^{(l)}(s_1) \quad \ldots \quad \varphi_{n_l}^{(l)}(s_{n_l})\right]^T.$$

The sought-for parameters of the neural network include the components of the weight matrix and bias of each layer. The total number of the parameters sought is $\sum_{l=1}^{L} n_l(n_{l-1}+1)$.

The quality criterion for neural network parameter optimization is the root-mean-square error of state prediction (2)

$$\sum_{k=1}^{K} \|\mathbf{z}^L(k) - \mathbf{u}(k)\|_2 \to \min. \tag{4}$$

For the experiments the following neural network was used: $L = 4, n_0 = 5, n_1 = n_2 = n_3 = 64, n_4 = 2$. All layers but the last used ReLu as activation function. To find the vector of optimal neural network parameters the stochastic gradient descent method ADAM was used [7].

To collect the required amount of data in the entire robot operating control range, VISLAM system was used. This system is based on RealSense t265 camera, that records data on control and state vectors at frequency of 20 Hz.

Figure 2 shows a trajectory obtained from a training sample (the light dotted line) and a trajectory obtained by internal model (1) with a neural network (the solid line).

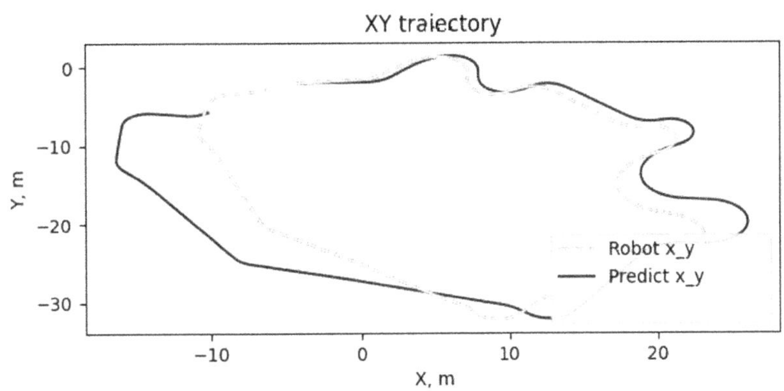

Figure 2. The trajectories of the vehicle movement on the plane $\{x, y\}$.

Figure 3 depicts the change of state vector components $v(t)$ and $\theta(t)$ over time. The light line shows the movement of a real vehicle, the solid one corresponds to the movement according to the model (1), and the dotted line corresponds to control components.

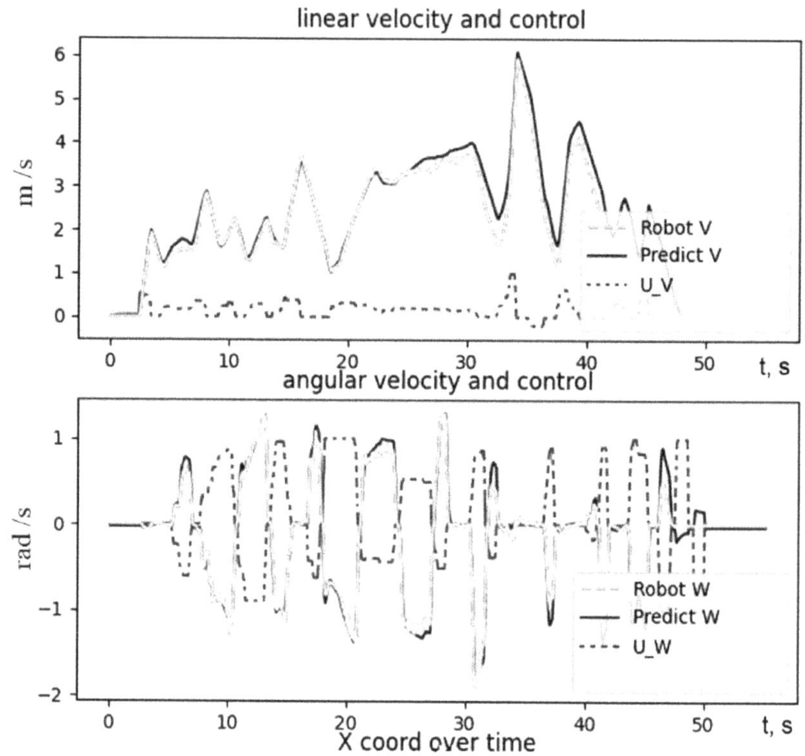

Figure 3. The change of state and control components over time.

3. Trajectory Tracking Scheme

To implement motion control along the trajectory, it is necessary to solve the control synthesis problem [8] and represent control as function that depends on the coordinates of the state space. In the majority of cases, a trajectory tracking error is calculated and a control signal, that works out this error, is set.

Depending on the task, the desired trajectory is set as a sequence of points on a plane or in three-dimensional space, and instead of calculating the distance to the trajectory, the distance to the points of the desired trajectory is determined. The block diagram of the unmanned vehicle control system, in which either PID-controller or model predictive control is used, is shown in Figure 4.

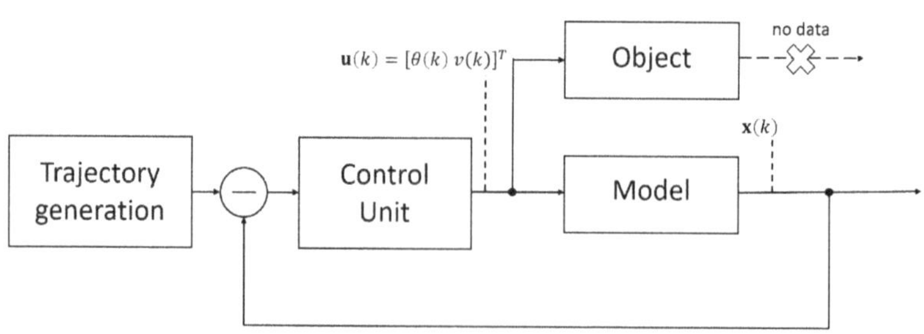

Figure 4. Block diagram of an internal model control system.

4. Controllers

4.1. Closed-Loop Control

In the steering control loop of the vehicle a P-controller was used. The control signal had the following form

$$\alpha = \arctan \frac{L\omega}{v}, \tag{5}$$

where L is the distance between the axes of the robot, v and ω is the linear and angular velocities of the robot,

$$\omega = \frac{vk(s)\cos(\theta_e)}{1 - k(s)e} - k_\theta \mid v \mid \theta_e - \left(k_e v \frac{\sin(\theta_e)}{\theta_e}\right)e, \tag{6}$$

where $k(s)$ is a path curvature, θ_e is an error in motion direction, e is a position error, and k_θ and k_e are controller coefficients.

Figure 5 shows the results of tracking the desired trajectory using P-controller and an internal model. The dashed line shows the desired trajectory, and the solid line shows the trajectory that has been tracked. The dots mark the constraints in the form of gates.

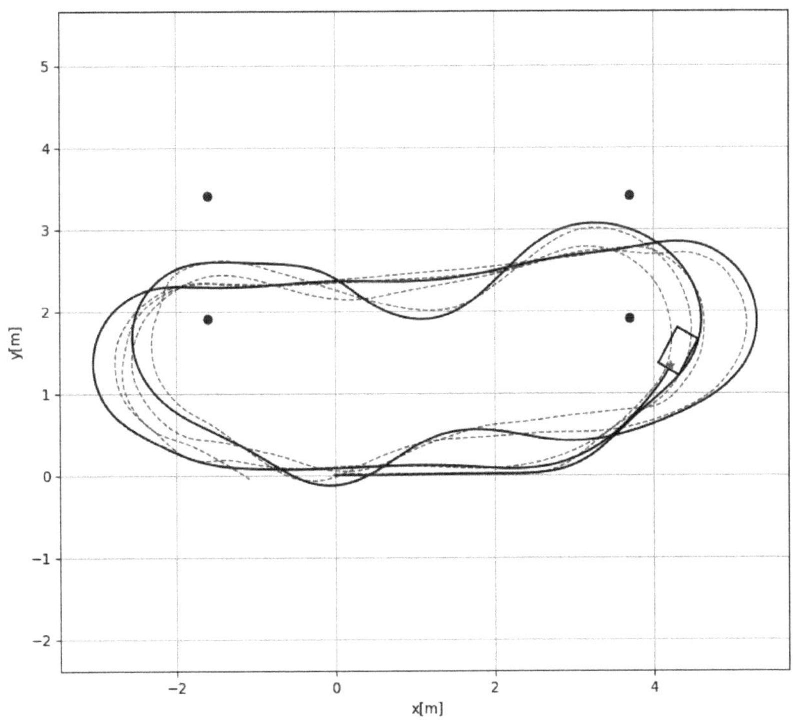

Figure 5. The movement along the trajectory with the help of internal model and P-controller.

4.2. Model Prediction Control

Predictive control is an optimization strategy that has a variety of advantages, such as accounting constraints imposed on system's state and control signals, and ability to work with non-linear objects. The optimal trajectory is the solution to the optimization problem. Predictive control includes object simulation under various control actions that ensure current problem solution, for example, bypassing the obstacles or passing through a gate. Among the resulting set of trajectories, one trajectory is selected according to some criterion. As a criterion, the squared error of achieving the goal and the squared control signal with some weighting factors are often used

$$J = \sum_{i=1}^{L} w_x (r_i - x_i)^2 + \sum_{i=1}^{L} w_u \Delta u_i^2, \tag{7}$$

where w_x and w_u are weights, r_i is a desired trajectory, x_i is a manipulated variable, u_i is a control signal, L is a control horizon.

The goal of predictive control lies in sequential search for optimal control strategy for a certain time interval that is called control horizon. Each time the first step of the strategy found is the only one applied. This step completed, the prediction horizon is being shifted and new optimization starts.

Online optimization taking place at every step has became possible due to the availability of high-performance computing facilities embedded on the vehicle's board.

For the control object model (1) and the prediction horizon value equal to 8, the desired and obtained trajectories are shown in Figure 6.

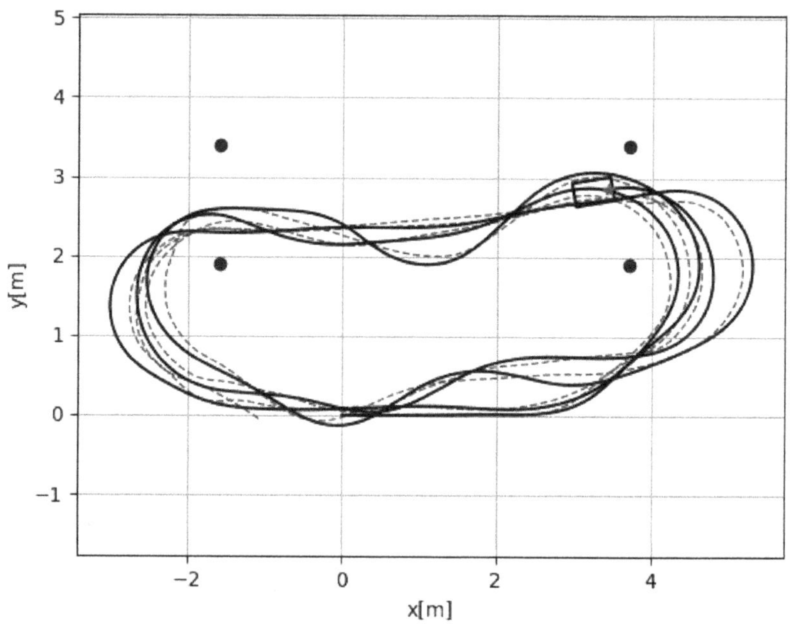

Figure 6. The movement along the trajectory with the help of internal model and model prediction control.

5. Results

To study the approaches proposed, an environment for unmanned vehicles simulation based on ROS1 and ROS2 operating systems, has been developed.

Experimental comparison of control methods that apply internal model for real-time control was conducted, with the problem of unmanned vehicle trajectory tracking served as experimental domain. The control was carried out on the basis of P-controller and model predictive control (MPC). The results of field experiments are shown in Table 1. The difference between the length of the desired trajectory S_1 and the length of the path S_2 covered under different velocity v was used as a comparison criterion. The desired trajectory length was 67.4 m.

Table 1. Results of field experiments.

Experiment Number	Speed v, m/s	Deviation When Using P-Controller S_1, m	Deviation When Using MPC S_2, m
1	1.41	4.4	3.5
2	1.12	2.1	1.9
3	0.88	1.6	1.7

According to the experiments the following conclusions may be performed:

1. Control based on P-controller is distinguished by the possibility of using relatively simple controllers;
2. With both control methods being used at high vehicle speeds, the deviation from the trajectory increases significantly;
3. High speeds can be achieved if controller parameters are fine-tuned or in case of applying an intelligent P-controller that adapts to the speeds and curvature of the trajectory;
4. High-performance computers are needed for MPC.

Author Contributions: Conceptualization, I.P. and E.S.; methodology, I.P.; software, I.P.; validation, I.P. and V.M.; formal analysis, E.S.; investigation, I.P.; data curation, I.P.; writing—original draft preparation, I.P., E.S. and V.M.; writing—review and editing, E.S.; visualization, V.M.; supervision, I.P. All authors have read and agreed to the published version of the manuscript.

Funding: This research received no external funding.

Institutional Review Board Statement: Not applicable.

Informed Consent Statement: Not applicable.

Data Availability Statement: The experimental data for both methods is available at https://cloud.mail.ru/public/raik/h6vjgsMkR (accessed on 19 June 2023).

Conflicts of Interest: The authors declare no conflict of interest.

References

1. Ljuing, L. *System Identification*; Prentice Hall PTR: Hoboken, NJ, USA, 1999.
2. Pupkov, K.A.; Egupov, N.D.; Trofimov, A.I. *Statisticheskie Metody Analiza, Sinteza i Identifikacii Nelinejnyh Sistem Avtomaticheskogo Upravleniya*; BMSTU: Moscow, Russia, 1998. (In Russian)
3. Liu, G.P. *Nonlinear Identification and Control: A Neural Network Approach*; Springer: Berlin/Heidelberg, Germany, 2012.
4. Paden, B.; Cáp, M.; Yong, S.Z.; Yershov, D.; Frazzoli, E. A Survey of Motion Planning and Control Techniques for Self-driving Urban Vehicles. *IEEE Trans. Intell. Veh.* **2016**, *1*, 33–55. [CrossRef]
5. Mehta, B.R.; Reddy, Y.J. Advanced process control systems. In *Industrial Process Automation Systems*; Mehta, B.R., Reddy, Y.J., Eds.; Butterworth-Heinemann: Oxford, UK, 2015; pp. 547–557. [CrossRef]
6. Huang, B.; Kadali, R. *Dynamic Modeling, Predictive Control and Performance Monitoring: A Data-Driven Subspace Approach*; Springer: Berlin/Heidelberg, Germany, 2008; 241p.
7. Kingma, D.P.; Ba, J. Adam: A Method for Stochastic Optimization. *arXiv* **2014**, arXiv:1412.6980v8.
8. Diveev, A.; Sofronova, E. The Synthesis of Optimal Control System by the Network Operator Method. In *IFAC Proceedings Volumes*; Elsevier: Amsterdam, The Netherlands, 2009; Volume 42, pp. 232–237.

engineering proceedings

MDPI

Proceeding Paper

Data Generation with Variational Autoencoders and Generative Adversarial Networks †

Daniil Devyatkin * and Ivan Trenev

V.A. Trapeznikov Institute of Control Sciences, Russian Academy of Sciences, 65 Profsoyuznaya Street, 117997 Moscow, Russia
* Correspondence: devyatkin.for.other@bk.ru; Tel.: +7-495-334-8910; Fax: +7-499-234-64-26
† Presented at the 15th International Conference "Intelligent Systems" (INTELS'22), Moscow, Russia, 14–16 December 2022.

Abstract: The paper considers the problem of modelling the distribution of data with noise in the input data. In this paper, we consider encoders and decoders, which solve the problem of modelling data distribution. The improvement of variational autoencoders (VAEs) is discussed. Practical implementation is performed using the Python programming language and the Keras framework. Generative adversarial networks (GANs) and VAEs with noisy data are demonstrated.

Keywords: machine learning; deep learning; autoencoders; generative adversarial network; MNIST

Citation: Devyatkin, D.; Trenev, I. Data Generation with Variational Autoencoders and Generative Adversarial Networks. *Eng. Proc.* **2023**, 33, 37. https://doi.org/10.3390/engproc2023033037

Academic Editors: Askhat Diveev, Ivan Zelinka, Arutun Avetisyan and Alexander Ilin

Published: 20 June 2023

1. Introduction to Variational Autoencoders

An autoencoder is a special architecture of a neural network, applying unsupervised learning using the backpropagation method. In other words, it is a neural network that has been trained to copy its input to output. The network consists of two parts: encoding functions $z = E(x, \theta_E)$ and decoding function $\hat{x} = D(z, \theta_D)$ [1]. Figure 1 demonstrates an example of an autoencoder with three hidden states. In the learning process, the autoencoder tries to learn the identical function $\hat{x} = D(E(x))$ by minimizing some loss function $L(\hat{x}, x)$. The solution to minimizing the factor can be consider as

$$[\theta_E, \theta_D] = arg\,min\, L(\hat{x}, x), \tag{1}$$

where θ_E and θ_D are the weights of the encoder and decoder, respectively.

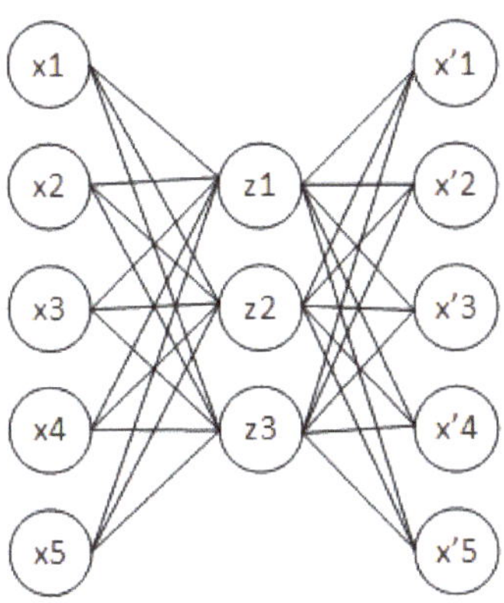

Figure 1. Autoencoder with three latent states.

There are many types of autoencoders: downscaling, where the dimension of z is less than of the input one; sparse, where a penalty is added to the loss function for large values of the latent representation z, etc. The main idea of autoencoder theory is the conjecture on the concentration of data in the neighbourhood of some low-dimensional manifold. Any autoencoder training procedure involves a compromise between two goals:

1. Training the latent representation z of the training sample x such it can be accurately reconstructed from z using the decoder. However, there is one critical factor: x must be chosen from the training dataset. This means that the autoencoders must not reconstruct $x$which are not possible with respect to the probability function.
2. Satisfaction of the constraints or regularizing penalty.

These two requirements force the latent representation to capture information concerning the structure of the distribution which generates the data. If we consider an image as an input, then it is obvious that not all pixels carry useful information, most of them are noise. For example, if you search for some object in the image, then the informative pixels are only those that form the desired object. Furthermore, if we consider the image of an object in a certain neighbourhood, then it can also be considered an object.

The main problem with these models is predicting a sample from the original sample based on its hidden representation; however, this is inconvenient since there is no knowledge of how much confidence was predicted. In other words, each sample has a hidden representation that can be depicted in some space, but we do not know if neighbourhood of this point is still an object similar to the predicted one. The solution to this problem is not to predict a sample, but some distribution of the latent variables.

As noted above, z is a vector of hidden variables that defines an object x from the sample [2]. Let $P(x)$ be a distribution function of the initial data, $P(z)$ denotes the latent factor distribution density, and $P(x \mid z)$ is the image probability distribution for given latent factors. Then the data generation process can be expressed as [3]:

$$P(x) = \int_z P(x \mid z) P(z)\, dz.$$

Let $P(x \mid z)$ be the sum of some generating function $f(z)$ and noise ϵ. We parametrize the generating function $f(z, \theta)$ by the vector θ in some space Θ, where $f : Z \times \Theta \to X$. Then the generation process takes the following structure:

$$P(x, \theta) = \int_z [f(z, \theta) + \epsilon] P(z)\, dz. \tag{2}$$

If we assume that the optimization is being carried out according to the L_2 metric, then the noise is normally distributed $\epsilon \sim \mathcal{N}(0, \sigma^2 E)$, and we obtain

$$P(x \mid z, \theta) = \mathcal{N}(x \mid f(z, \theta), \sigma^2 E),$$

where $f(z, \theta)$ is modelled by a neural network, E is the identity matrix, and σ^2 is a positive scalar.

Next, the parameters θ must be found to guarantee the maximum likelihood estimation in order to maximize the probability of occurrence for objects of the sample $P(x)$. In practice, for most z, the probability $P(x \mid z)$ tends to zero, and therefore, it contributes almost nothing to the estimate of $P(x)$. The key idea of the variational autoencoders (VAEs) is to try and find values of z which lead to x, enabling the calculation of the probability estimate $P(x)$ on x. To this, introduce a new function, $Q(z \mid x)$, must be introduced which can take values of x and construct a distribution for z that leads to x. The main hypothesis is the cardinality of a set with "good" z is much smaller than the cardinality of the set for all z. This distribution Q can be trained to assign high-probability values to those z that are highly likely to generate x. However, instead of maximizing (2), we need to maximize $\mathbb{E}_{z \sim Q}[\, P(x \mid z)\,]$.

We then write the Kullback–Leibler divergence between the $Q(z\,|\,x)$ and $P(x\,|\,z)$ distributions as follows

$$D_{\mathrm{KL}}[\,Q(z\,|\,x)\;\|\;P(z\,|\,x)\,] = \mathbb{E}_{z\sim Q}[\log Q(z\,|\,x) - \log P(z\,|\,x)],$$

where $P(z\,|\,x)$ is the real probability distribution of hidden factors for a given x. Next, by applying the Bayes formula to $P(z\,|\,x)$, we obtain

$$D_{\mathrm{KL}}[\,Q(z\,|\,x)\;\|\;P(z\,|\,x)\,] = \mathbb{E}_{z\sim Q}[\log Q(z\,|\,x) - \log P(x\,|\,z) - \log P(z)] + \log P(x). \quad (3)$$

In the expression in (3), $\log P(x)$ does not depend on z, so it leaves the mathematical expectation. Next, let us obtain one more Kullback–Leibler divergence

$$D_{\mathrm{KL}}[\,Q(z\,|\,x)\;\|\;P(z\,|\,x)\,] = D_{\mathrm{KL}}[\,Q(z\,|\,x)\;\|\;\log P(z)\,] - \mathbb{E}_{z\sim Q}[\log P(x\,|\,z)] + \log P(x). \quad (4)$$

The expression in (4) holds true for any $Q(z\,|\,x)$ and $P(z\,|\,x)$. By applying permutations to the terms of the equation, one can obtain the following expression

$$\log P(x) - D_{\mathrm{KL}}[\,Q(z\,|\,x)\;\|\;P(z\,|\,x)\,] = \mathbb{E}_{z\sim Q}[\log P(x\,|\,z)] - D_{\mathrm{KL}}[\,Q(z\,|\,x)\;\|\;\log P(z)\,]. \quad (5)$$

In Formula (5), the $Q(z\,|\,x)$ is the encoder and $P(z\,|\,x)$ is the decoder, and these distributions are modelled by a neural network; therefore,

$$Q(z\,|\,x) = Q(z\,|\,x,\theta_E), \;\; P(z\,|\,x) = P(z\,|\,x,\theta_D),$$

where θ_E and θ_D are the weights from Formula (1). The goal of training a VAE is to maximize $P(x)$. The right-hand side of (5) can be optimized by gradient methods. On the right, the first term denotes the quality of prediction x by the decoder from the values of the hidden variables z, and the second term is the Kullback–Leibner divergence between $P(z)$ and $Q(z\,|\,x)$. Let us predict Q as a normal distribution with the following parameters,

$$Q(z\,|\,x,\theta_E) = \mathcal{N}(\,\mu(x,\theta_E),\, \Sigma(x,\theta_E)),$$

where $\mu(x,\theta_E)$ is the mathematical expectation, and $\Sigma(x,\theta_E)$ is the covariance matrix. That is, the encoder for each x predicts two values: the mathematical expectation μ and the covariance matrix Σ. In other words, the encoder predicts some normal distribution. It is necessary for the distribution of $Q(z\,|\,x)$ to be similar to a normal distribution, that is, the Kullback–Leibner divergence tends to 0, and the quality of the data generated by the decoder is maximized. This means that two loss functions are used in the implementation of the model

$$D_{\mathrm{KL}}[\,Q(z\,|\,x,\theta_E)\;\|\;\mathcal{N}(0,\,E)\,],$$

$$||x - f(z)||,$$

where x is the true data sample, and $f(z) = D(E(x))$ is the predicted data by the VAE.

Here, random values $z \sim Q(z\,|x,\theta_E)$ are taken and passed to the decoder. It is impossible to propagate errors through random values directly, so the reparametrization trick is used. This method is based on the following formula:

$$\mathcal{N}(\mu(x\,,\theta_E),\, \Sigma(x\,,\theta_E)) = \mu(x\,,\theta_E) + \Sigma(x\,,\theta_E)\cdot\mathcal{N}(0,\,E),$$

where $\mu(x\,,\theta_E)$ is the expected value, $\Sigma(x\,,\theta_E)$ is the covariance matrix, and $\mathcal{N}(0,\,E)$ is a standard normal multivariate distribution (visualization in the Figure 2).

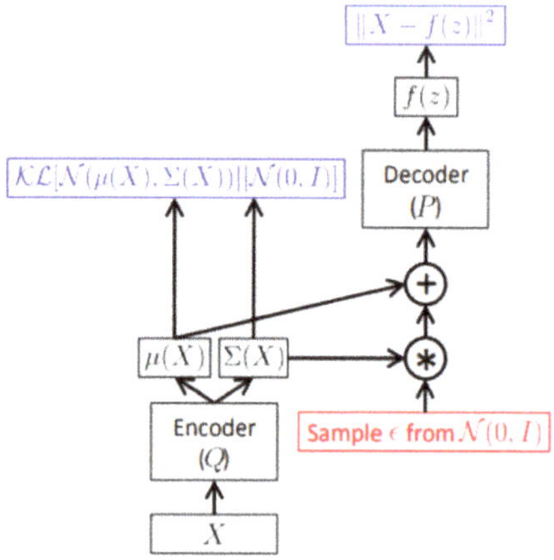

Figure 2. Block diagram of a variational encoder (VAE) architecture.

2. Generative Adversarial Networks

VAEs compare the original and generated objects based on the mean squares error and binary cross entropy if the labels are given as the input: this is a poor comparison. This disadvantage manifests to a lesser extent in another approach, generative adversarial networks (GANs). The overall goal of a GAN is to synthesize new data that has the same distribution as the training set.

There are two neural networks in the GAN model—generator and discriminator [4]. These models are trained in turn. After the model weights are initialized, the generator generates images, initially this is noise. After several iterations of training the generator, the discriminator starts working and the generator stops training. This network distinguishes differences between real images and images synthesized by the generator, and predicts whether the image is original or generated (0 if false, 1 if true). Networks have been trained to learn to solve their problems. Two networks play an adversarial game, where the generator learns to obtain its output and fool the discriminator. Discriminator detection becomes better at detecting synthesized images. The generator creates random numbers from a given distribution $P(z)$ and generates objects $X_p = G(z, \theta_g)$ from them, used as the input of the second network. The discriminator receives the objects from the training sample X_s and objects created by the generator X_p as the input; subsequently, it learns to predict the probability whether that particular object is real, giving the scalar $D(z, \theta_d)$. Let the generator be represented as a mapping $G(z, \theta_g)$, where G is a differentiable function, and θ_g are generator parameters. The discriminative model $D(z, \theta_d)$ is presented, the model predicts if the input is real from the training set or synthesized by the generator, where θ_d are the discriminator parameters [5].

Figure 3 shows the following: (Left) the training phase of the discriminator is shown: the gradient (red arrows) only flows from the loss function to the discriminator, where θ_d (green) is updated to reduce the loss function. The gradient from the right side of the loss function (object identification error) flows to the generator, only updating the θ_g generator weights (green) towards increasing the probability of the discriminator to make an error. During the training of the two models, it is necessary for the discriminator D to maximize the probability of correctly identifying objects using the training and generated samples, enabling the generator G to minimize $\log(1 - D(G(z)))$. In other words, it is necessary to reach the next criterion (see Figure 3)

$$\min_{G}\max_{D} V(D,G) = \underset{x\sim p_{data}}{E}[\log D(x)] + \underset{z\sim p_z}{E}[\log(1 - D(G(z)))].$$

Generally, the task of training the discriminator and generator is not to find the local or global minimum of a function, but to find an equilibrium point. In game theory, this point is called the Nash equilibrium point, where both players no longer benefit, although they follow the optimal strategy [6].

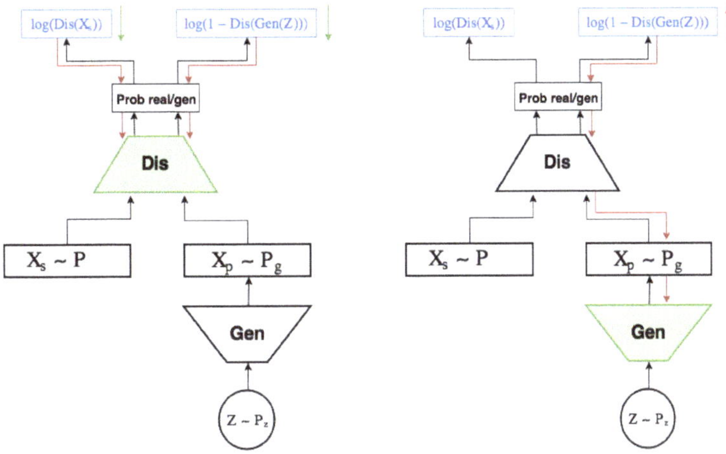

Figure 3. GAN training scheme.

3. Practical Implementation

The implementation of the described generative models is demonstrated using Python and the Keras framework [7–9]. In the first stage, data generation based on the VAE and GANs on the MNIST (Modified National Institute of Standards and Technology database) dataset is demonstrated [10]. MNIST is a well-known dataset for handwritten numbers.

The encoder architecture used two convolution layers and three fully connected layers. For the encoder, except the output, the activation function was ReLU and the output is sigmoid. The size of the hidden variables was 2. The decoder used a fully connected layer and three layers of transposed convolutions. Furthermore, the decoder output is sigmoid. Let us move on to the GAN architecture. The generator used a fully connected layer, a 2D convolution layer, and upsampling. The discriminator used a fully connected layer, 2D convolution, max pooling, and flattening [11].

Figure 4 shows an example of data generation. The image shows that the autoencoder performing as expected; however, the contours of the images are very blurry and some numbers are very similar (this is due to the similar hidden representation of these objects). Let us add noise to the data resulting in the following: neither the VAE nor the GAN are robust models. It is worth noting that the noise has a normal distribution with a mathematical expectation of 0 and a variance of 0.1. Figure 5 shows that in the presence of noise, data generation does not give the necessary results [12].

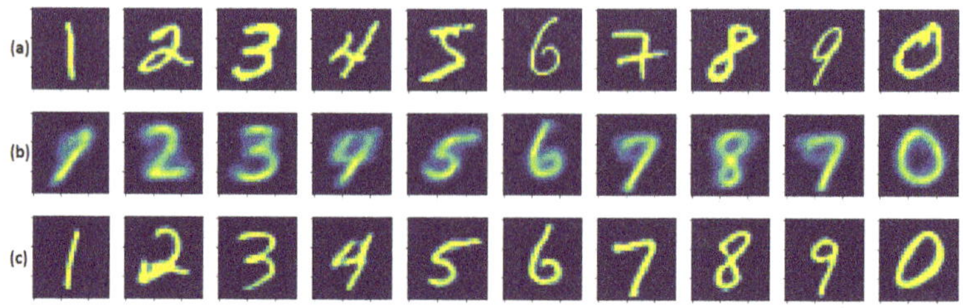

Figure 4. (**a**) Original MNIST dataset. (**b**) Synthesized images after autoencoder training. (**c**) Synthesized images by the GAN generator.

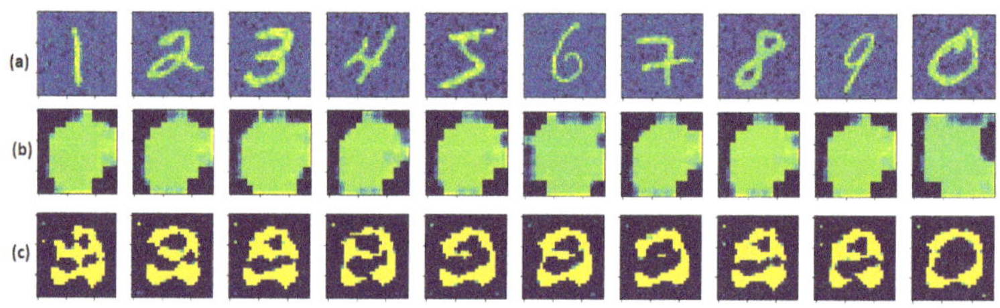

Figure 5. (**a**) Original MNIST dataset with noise. (**b**) Synthesized images after autoencoder training. (**c**) Synthesized images by the GAN generator.

4. Conclusions

This experiment shows that generative models demonstrate good results with pure data; however, if noise is added to the original sample, the results become unpredictable. To solve this problem, conditional generative models can be considered. In the case of GAN, one can consider their improvements: DCGAN and WGAN. DCGAN is a modification of the GAN algorithm based on convolutional neural networks (CNN). The task of finding a convenient representation of features on large volumes of unlabelled data is one of the most active areas of research, in particular, the representation of images and videos. One convenient way to find views can be this network. WGAN uses the Wasserstein loss metric inside the error function, allowing the discriminator to learn to identify repetitive outputs faster on which the generator stabilizes.

Author Contributions: Conceptualization, I.T.; methodology, I.T.; software, D.D.; validation, I.T.; visualization, D.D.; writing—original draft preparation, D.D.; writing—review and editing, D.D. All authors have read and agreed to the published version of the manuscript.

Funding: This research received no external funding.

Institutional Review Board Statement: Not applicable.

Informed Consent Statement: Not applicable.

Data Availability Statement: Not applicable.

Conflicts of Interest: The authors declare no conflict of interest.

References

1. Goodfellow, I.; Bengio, Y.; Courville, A. *Deep Learning*; The MIT Press: Cambridge, MA, USA, 2018; pp. 422–441.
2. Doersch, C. Tutorial on Variational Autoencoders. *arXiv* **2016**, arXiv:1606.05908
3. Ivchenko, G.I.; Medvedev, Y.I. *Introduction to Mathematical Statistics*; Publishing House LCI: Moscow, Russia, 2010; pp. 457–542.
4. Raschka, S. *Python Machine Learning*; Packt Publishing Ltd.: Birmingham, UK, 2020; pp. 513–551.
5. Goodfellow, I.J.; Pouget-Abadie, J.; Mirza, M.; Xu, B.; WardeFarley, D.; Ozair, S.; Courville, A.; Bengio, Y. Generative Adversarial Nets. *arXiv* **2014**, arXiv:1406.2661.

6. ITMO University. Generative Adversarial Nets (GAN). Available online: https://neerc.ifmo.ru/wiki/index.php?title=Generative_Adversarial_Nets_(GAN) (accessed on 12 July 2022).
7. Python Software Foundation. The Python Standard Library. 2020. Available online: https://docs.python.org/3/library/ (accessed on 12 July 2022).
8. Keras. Simple. Flexible. Powerful. Keras: The Python Deep Learning API. Available online: https://keras.io/ (accessed on 12 July 2022).
9. Chollet, F. *Deep Learning with Python*; Simon & Schuster: New York, NY, USA, 2018; pp. 269–314.
10. MNIST Handwritten Digit Database. The Mnist Database of Handwritten Digit. Available online: http://yann.lecun.com/exdb/mnist/ (accessed on 14 July 2022).
11. SImonyan, K.; Zisserman, A. Very Deep Convolutional Networks for Large-Scale Image Recognition. 2014. Available online: https://arxiv.org/abs/1409.1556 (accessed on 13 July 2022).
12. Mueller, A.; Guido, S. *Introduction to Machine Learning with Python*; O'Reilly Media Inc.: Sebastopol, CA, USA, 2016; pp. 180–221.

engineering proceedings

MDPI

Proceeding Paper

Anomaly Detection on Video by Detecting and Tracking Feature Points †

Ivan Fomin *, Yurii Rezets and Ekaterina Smirnova

Russian State Scientific Center for Robotics and Technical Cybernetics (RTC), Tikhoretsky Prospect, 21, St. Petersburg 194064, Russia
* Correspondence: i.fomin@rtc.ru; Tel.: +7-911-127-5876
† Presented at the 15th International Conference "Intelligent Systems" (INTELS'22), Moscow, Russia, 14–16 December 2022.

Abstract: There is a well-known problem of video sequence analysis when it is necessary to identify and localize areas of abnormal movement of objects. This is necessary to attract the attention of the operator in the process of work or when analyzing the archive of video recordings. One of the solutions is based on tracklet analysis using short segments of the object's trajectory that characterize its movement over a certain period of time, and then analysis of activity in various areas of the frame. Since the construction of the tracklet and trajectory is based on the optical flow, the quality and performance of the algorithm significantly depend on the choice and configuration of methods for detecting and tracking feature points. We have analyzed various combinations of these methods using the examples of test videos of normal and abnormal activity in a pedestrian zone. The necessity of a preliminary analysis of the methods used when setting up a video surveillance system to solve specific tasks is shown. Suitable combinations of methods are proposed.

Keywords: anomaly detection; tracklet analysis; feature points detection; feature points tracking; computer vision

Citation: Fomin, I.; Rezets, Y.; Smirnova, E. Anomaly Detection on Video by Detecting and Tracking Feature Points. *Eng. Proc.* **2023**, *33*, 34. https://doi.org/10.3390/engproc2023033034

Academic Editors: Askhat Diveev, Ivan Zelinka, Arutun Avetisyan and Alexander Ilin

Published: 19 June 2023

1. Introduction

Currently, the number of surveillance cameras are installed indoors and outdoors to monitor various aspects of industrial and public safety is actively increasing. Previously, viewing cameras and determining emergency situations were usually assigned to the operator, who had to activate an alarm or perform other active actions. Even when there were relatively few cameras, most often records of the incident were extracted from the archive later, if there was information available about where and when to look. Now, when there are so many cameras that the operator is not able to control them all, video surveillance systems of varying degrees of complexity are in great demand, from systems that are able to detect movement in the frame to those that can determine some complex emergency situations [1].

In this paper, the detection of anomalies in a video based on a sparse optical flow is considered. Here, the input data is a sequence of extracted video frames, and the output data is an area or a set of areas in the frame where abnormal movement occurs.

There are several approaches to anomalies detection, which are based on a grid pattern [2], global pattern [3], trajectories [4], etc. Each of them has advantages and disadvantages, but they are all relatively computationally complex. There are slightly simpler algorithms that perform just as well.

This paper considers the implementation of the method based on the analysis of local binary tracklets [5], which requires detection of feature points and construction of a trajectory for them using the optical flow method.

2. Related Works

2.1. Feature Points Detection Methods

To detect features, algorithms are designed to determine the points of a sharp change of an image value in more than one direction.

Harris corner detector [6] uses image change in a sliding window of a certain size when it is shifted relative to a given point in any direction.

GFTT (Good Features To Track) [7] works on the same principle, but it calculates the measure of features differently, defining it as the minimum of two eigenvalues in the matrix of pixels of a sliding window—if both are large enough, the window shift strongly affects both directions. FAST [8] performs pixel comparisons with those lying on a circle with a radius of 3 pixels, and this is a special point if 12 pixels in the circle are noticeably different from the central one. AGAST [9] uses an approach with a binary decision tree so that each point can be checked in literally two comparisons, significantly speeding up the procedure. MSER [10] creates several binary (black and white) images with different thresholds, using areas that change slightly when a threshold changes as features.

SIFT descriptor-based detector [11] uses Gaussian filter smoothing to simulate different scales of the image and more efficiently highlight feature points on both large and small objects. BRIEF [12] develops the idea of SIFT, significantly simplifying the calculation of feature points descriptors. ORB [13] is a newer alternative to the SIFT and BRIEF detector; it uses the TreeFAST detector and another descriptor to effectively describe feature points. BRISK [14] raises the quality of successful matching even higher at different scales using the AGAST detector and taking into account additional maxima on each of the octaves of Gaussians when comparing. Method KAZE [15] avoids the disadvantages of Gaussian filter and achieves high localization accuracy and distinctiveness. AKAZE [16] speeds up KAZE using a more computationally efficient FED (Fast Explicit Diffusion) platform. AKAZE uses rotation and scaling invariant M-LDB [16] as a descriptor.

2.2. Methods of a Local Optical Flow Construction

To track feature points, in particular, to build an optical flow between frames, several tracking methods have been developed. The basis of the most famous Lucas–Kanade algorithm [17] is the assumption that the value of pixels of one object varies slightly between frames, and using the least squares method, it finds a position that minimizes the discrepancy between the pixel values in neighboring frames.

The method of G. Farneback [18] applies the decomposition of the change in image intensities around a singular point in the Taylor series to the second term using a weight function to approximate the values of neighboring pixels. Based on an extensive analysis of the available data, a new modification of the Lucas–Kanade method—the RLOF [19] algorithm—was created. It uses a modified Hampel estimate with robust characteristics.

2.3. Anomaly Detection Algorithms

Let us also discuss the existing algorithms for abnormal behavior in a video sequence detection. Optical flow is directly widely used for behavior analysis [20,21], but is not able to analyze spatial relations. Particle flow [22] copes with the task a little better, but it is unable to fully analyze objects in space. Methods based on local spatio-temporal features, in the form of gradients [23] and motion histograms [24], can be applied to the analysis of poorly structured scenes (crowded scene for example). Methods based on tracklets (for example, [25,26]) are well suited to detecting short-term anomalies and analyzing dense crowds. In trajectory methods, tracking algorithms track the trajectory of an object, after which it is analyzed through clustering of string kernels [27], single-class SVM [28] or semantic scene learning [29]. Methods based on global patterns analyze the entire sequence using a dense optical flow [30]. There are Gaussian Mixture Model (GMM)-based methods [31], models of social interactions [3], or Principal Component Analysis (PCA) [32] and several others. Grid pattern-based methods split the frame into blocks (grid cells) and perform block-based analysis. There are methods based on the reconstruction of sparse textures [2], motion context descriptor [33] or spatio-temporal Laplace maps. In [34,35] hierarchical sparse coding is applied, and in [36] multilevel sparse coding is utilized, on top of which SVM is applied for classification.

3. The Tracklet Analysis Algorithm

In order to study the influence of feature points detection and tracking algorithms on anomaly detection algorithm quality, we selected the LBT [5] (Local Binary Tracklet analysis). The overall pipeline is shown in the Figure 1.

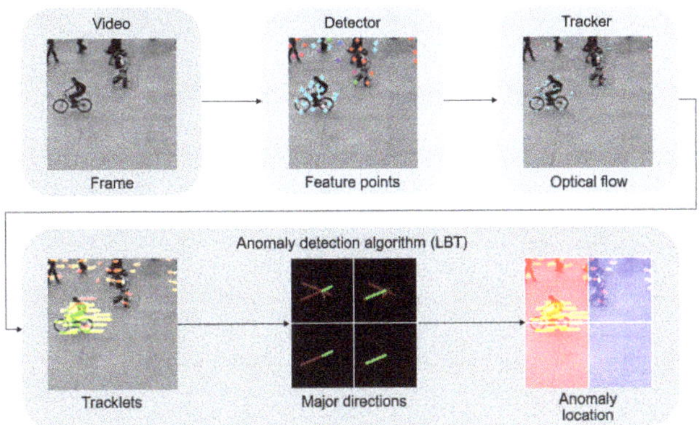

Figure 1. Data pipeline of the LBT algorithm.

The input is an ordered set of images extracted from the video sequence. Then, using one of the detectors described in Section 2.1, the detection of feature points is performed. Then, the tracking of points between frames is performed using one of the tracking algorithms (or trackers), e.g., Lucas–Kanade or RLOF, as described in Section 2.2. According to tracking results, the trajectory of a point is sequentially formed, and its last section with a length of a given number of points (parameter) is used as a tracklet for anomalies detection. Too short tracklets, as well as those where there is a sharp change in velocity or direction, are discarded as unreliable. We assume that the objects that give rise to the tracklets have some inertia. Then, if during the time period specified for the analysis, the point abruptly changed the direction or velocity, there is a tracking failure. Next, the anomaly detection is performed. To achieve this, the frame is evenly divided by a grid into a given number of areas horizontally and vertically (the number of areas is a parameter) and a certain number of characteristic directions are identified for each of the areas. If a new characteristic direction of movement appears in the area or the magnitude of the characteristic speed of movement changes significantly (the threshold is a parameter), then the algorithm designates this area of the frame as abnormal.

4. Experiments

4.1. Test Datasets

When selecting test datasets for the analysis of optical flow anomalies, datasets UMN [37] and UCSD [38] were found. For an experimental study of the algorithm, a video sequence from an UMN dataset was chosen, where a group of people run away, simulating a panic (UMN sequence in the tests). A sequence was also selected from the UCSD dataset, where vehicles are present in the pedestrian zone (the Ped sequence in the tests). Examples of the appearance of frames from both sequences are shown in the Figure 2.

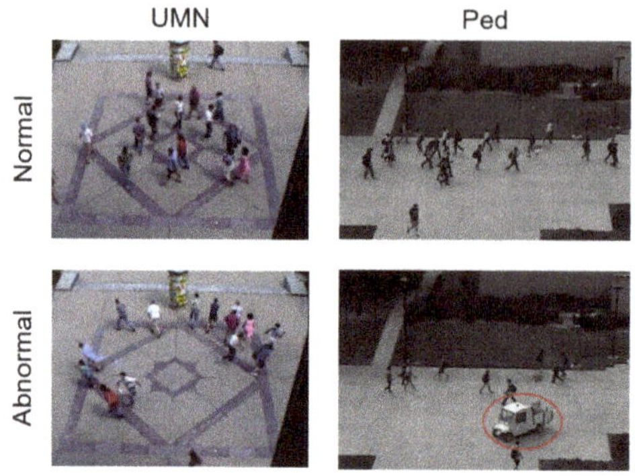

Figure 2. Examples of frames from selected video sequences.

4.2. Quality Metrics

To evaluate the influence of algorithms on the quality and performance, it is necessary to determine quality metrics. In the case of performance, the calculation was performed as follows. During the processing of a sequence of frames, the time required for processing was calculated and then divided by the frame number in the video sequence. Thus, the average duration for one frame is obtained, from which we determine the number of frames processed per second (FPS). The following approach was chosen to determine the quality. Once the task (to draw attention to a certain moment in the video, or to the certain camera at some point in time) was selected, the markup was implemented for each frame as a whole. A sequence of values was compiled for all frames, whether an anomaly was present or absent in the frames. The output of the algorithm is designed in the same format. Thus, it is possible to calculate the number of successfully detected abnormal frames (TP, true positive), erroneously detected abnormal frames (FP—false positive, type II errors) and erroneously rejected abnormal frames (FN—false negative, type I errors). Based on these data, it is possible to calculate the metrics of accuracy, completeness and F1-score (1) as a harmonic mean.

$$precision = \frac{TP}{TP + FP}, \; recall = \frac{TP}{TP + FN}, \; F1 = \frac{2TP}{2TP + FP + FN} \tag{1}$$

It is also possible to build a Precision–Recall Curve (PRC) and the area under this curve (Precision and Recall-Area Under Curve, PR-AUC). Methods for plotting this curve and calculating the area under it are presented in a variety of packages; for example, in this case, the version from the python library scikit-learn [39] was used.

For tracklets and feature points, some characterizing parameters can also be proposed. Despite the fact that for each feature point, only a small set of the last points of the trajectory is used for analysis, the total number of frames on which the point is successfully tracked is calculated for comparison of algorithms. Then, this indicator is averaged over all frames of the video and tracklets. In this way, the average lifetime of the tracklet is obtained, which shows the overall robustness of tracking using a given method. The number of detected and tracked feature points on each frame is also summed after the removal of those whose position does not change, provided that a significantly higher threshold (several thousand, for example) is detected. The value is averaged by the frames quantity in the sequence and the average number of detected points is obtained. It characterizes how well each of the algorithms detects feature points.

4.3. Configuring Algorithm Parameters

Before analyzing the dependence of performance and quality on the choice of algorithm, it was necessary to pre-configure the parameters of the algorithms used. Note that chosen videos and the algorithms themselves ensure the practical absence of interdependence, so parameters of the points detection, tracking and anomaly detection algorithms can be configured sequentially.

At the first stage, with fixed parameters of the algorithms for detecting and tracking feature points, the parameters of the algorithm for anomalies detection in the optical flow were configured. To evaluate the performance of the algorithm in descending order of priority, the following metrics were used: F1-score, PR-AUC, FPS, and the average lifetime of the tracklet (higher value equal to better result for all metrics) and the average number of detected points (with the same maximum for all methods). According to the research results, we found that the number of cells for analysis has the main influence. With a value of the parameter "8 cells vertically by 12 horizontally", the maximum of accuracy is obtained. Neither the maximum length of the tracklet nor the discretization of the direction histogram have a significant effect; these parameters were selected according to maximal accuracy and FPS. At the second stage, the parameters of each of algorithms for feature points detection and tracking were configured. In total, two algorithms for feature points tracking were considered: Lukas–Kanade (LK) and RLOF, described above in Section 2.2. Also eight classical algorithms for feature points detection were evaluated: GFTT, FAST, AGAST, SimpleBlob, SIFT, MSER, KAZE, and AKAZE, all described in Section 2.1. For each of the algorithms permissible limits of parameter changes were selected based on the information provided in the corresponding article. After that, the grid search for the best combinations of parameters for each algorithm was performed. Thus, sets of the best parameters for each of the algorithms were obtained to compare them with each other.

4.4. Comparison of Detection and Tracking Algorithms

Using configured methods for detecting and tracking feature points, as well as a customized anomaly detection algorithm, a study of the quality metrics described above was conducted. It was determined that choice of method has practically no effect on the maximal length of the trajectory; therefore, this parameter was discarded during the study of the results. Average number of detected points also does not directly depend on the method of feature points detection and does not correlate with quality. For any method, it can be changed depending on the requirements of the tracking algorithm or optical flow anomalies detection algorithm. For this reason, this parameter has also been omitted in this study. The summary tables of the results for the Lukas–Kanade tracking algorithm (Table 1) and RLOF tracking algorithm (Table 2) are shown below.

Table 1. Results for Lukas–Canade method.

				Ped				
	GFTT	**FAST**	**AGAST**	**Sblob**	**SIFT**	**MSER**	**KAZE**	**AKAZE**
F1	0.72	0.66	0.58	0.63	0.7	0.73	0.73	0.72
PR AUC	0.58	0.57	0.49	0.51	0.59	0.59	0.65	0.59
FPS	263	333	284	230	102	108	34	174
				UMN				
	GFTT	**FAST**	**AGAST**	**Sblob**	**SIFT**	**MSER**	**KAZE**	**AKAZE**
F1	0.91	0.88	0.82	0.8	0.87	0.81	0.89	0.81
PR AUC	0.94	0.88	0.89	0.88	0.9	0.9	0.92	0.84
FPS	252	398	291	247	102	178	36	176

Table 2. Results for RLOF method.

	Ped							
	GFTT	**FAST**	**AGAST**	**Sblob**	**SIFT**	**MSER**	**KAZE**	**AKAZE**
F1	0.7	0.53	0.41	0.63	0.69	0.7	0.71	0.71
PR AUC	0.61	0.37	0.3	0.51	0.54	0.57	0.62	0.6
FPS	259	322	274	210	99	108	34	165
	UMN							
	GFTT	**FAST**	**AGAST**	**Sblob**	**SIFT**	**MSER**	**KAZE**	**AKAZE**
F1	0.87	0.84	0.87	0.86	0.85	0.86	0.87	0.85
PR AUC	0.9	0.88	0.91	0.92	0.88	0.91	0.91	0.89
FPS	217	345	267	221	101	76	35	161

According to these tables, it is possible to conclude which method of detecting feature points is best suited for anomaly detection task. To do this, we present normalized graphs of the PR-AUC and FPS parameters for all combinations of the dataset and the tracking method (see Figure 3). The F1-score graph, with exclusion of insignificant details, coincides with the PR-AUC graph and is therefore omitted here.

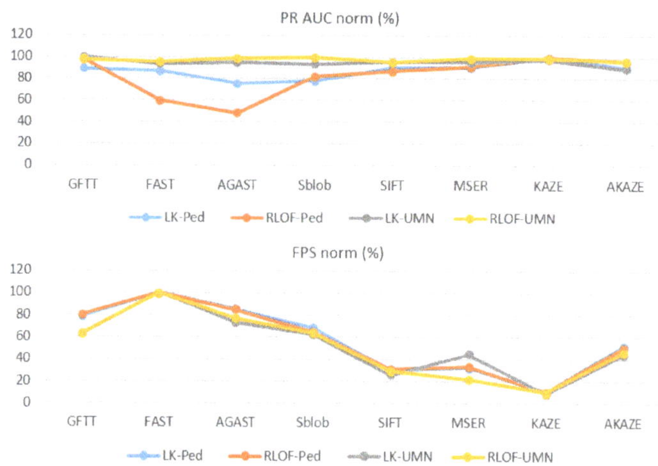

Figure 3. Normalized PR-AUC and FPS graphs.

From the graphs shown, demonstrating the ratio of quality metrics of different pairs of the tracking method and the dataset when using a particular detector, several conclusions can be drawn. The most significant difference between the methods is observed in the resulting FPS, but the dataset and chosen tracker does not have significant effect on relative values. Therefore, in general, when the algorithm must work faster, it is recommended to use one of the detectors from the left part of the figure—GFTT, FAST, AGAST or SimpleBlob. For the UMN dataset, due to the relative simplicity of the problem being solved, the choice of detector did not have a significant impact on quality. For the Ped dataset, the FAST and AGAST detectors showed slightly lower quality as simpler methods. We also noted that the quality of the algorithm when tracking points by the RLOF method depends on detected feature points and detector quality more than when using the Lukas–Kanade method. Comparison based on absolute values of quality metrics (Tables 1 and 2) shows the absence of significant differences in quality when using different trackers. For some items, a quality is higher when using RLOF. FPS differs insignificantly, by units and fractions of percent, so there is almost no difference between trackers for this indicator. The Lukas–Kanade method works faster, while RLOF is a little slower, but more accurate for a number of detectors. There is a possibility that the absence of significant differences in quality for detection methods is because of the selected dataset or quality metrics. The study of this set of methods on other tasks may become the goal of further research.

5. Conclusions

This paper considered several well-known algorithms for feature points detection and tracking and for anomaly detection in an optical flow constructed from a video sequence. After performing the analytical review, we chose algorithms: LBT for anomaly detection in optical flow; GFTT, FAST, AGAST, SimpleBlob, SIFT, MSER, KAZE, and AKAZE for detecting feature points; RLOF and Lucas–Kanade for feature points tracking. In the course of the study on the selected video sequences, significant differences in the performance of the algorithm were revealed depending on the selected feature point detector. There was no difference in quality, except for the FAST and AGAST detectors, which performed worse than the others on the Ped dataset with vehicles in the crowd. For use in such scenes, GFTT, SimpleBlob and AKAZE can be recommended as better methods in terms of FPS, precision and robustness. Tracking methods both have almost the same quality. The Lukas–Kanade tracking method has better performance and is more robust to poor quality of feature points; therefore, it can be recommended for use in scenes similar to considered dataset. In further research on this topic we plan to expand the variety of video sequences for analyzing methods, as well as to study existing neural network algorithms for feature points detection and tracking, and other methods for optical flow anomalies detection.

Author Contributions: Conceptualization, I.F. and Y.R.; methodology, Y.R.; software, Y.R.; validation, Y.R., I.F. and E.S.; formal analysis, Y.R. and I.F.; investigation, Y.R.; data curation, Y.R. and I.F.; writing—original draft preparation, I.F.; writing—review and editing, E.S.; visualization, I.F.; supervision, E.S. All authors have read and agreed to the published version of the manuscript.

Funding: The work was carried out as the part of the state task of the Russian Ministry of Education and Science for 2023 "Research of methods for creating self-learning video surveillance systems and video analytics based on the integration of technologies for spatiotemporal filtering of video stream and neural networks" (FNRG 2022 0015 1021060307687-9-1.2.1 №075-01595-23-00).

Institutional Review Board Statement: Not applicable.

Informed Consent Statement: Not applicable.

Data Availability Statement: Not applicable.

Conflicts of Interest: The authors declare no conflict of interest.

References

1. Rezaee, K.; Rezakhani, S.M.; Khosravi, M.R.; Moghimi, M.K. A survey on deep learning-based real-time crowd anomaly detection for secure distributed video surveillance. *Pers. Ubiquitous Comput.* **2021**, 1–17. [CrossRef]
2. Xu, J.; Denman, S.; Sridharan, S.; Fookes, C.; Rana, R. Dynamic texture reconstruction from sparse codes for unusual event detection in crowded scenes. In Proceedings of the 2011 Joint ACM Workshop on Modeling and Representing Events, Scottsdale, AZ, USA, 30 November 2011; pp. 25–30.
3. Mehran, R.; Oyama, A.; Shah, M. Abnormal crowd behavior detection using social force model. In Proceedings of the 2009 IEEE Conference on Computer Vision and Pattern Recognition, Miami, FL, USA, 20–25 June 2009; pp. 935–942.
4. Zhou, B.; Wang, X.; Tang, X. Understanding collective crowd behaviors: Learning a mixture model of dynamic pedestrian-agents. In Proceedings of the 2012 IEEE Conference on Computer Vision and Pattern Recognition, Providence, RI, USA, 16–21 June 2012; pp. 2871–2878.
5. Ravanbakhsh, M.; Mousavi, H.; Nabi, M.; Marcenaro, L.; Regazzoni, C. Fast but not deep: Efficient crowd abnormality detection with local binary tracklets. In Proceedings of the 2018 15th IEEE International Conference on Advanced Video and Signal Based Surveillance (AVSS), Auckland, New Zealand, 27–30 November 2018; pp. 1–6.
6. Harris, C.; Stephens, M. A combined corner and edge detector. In Proceedings of the Alvey Vision Conference, Manchester, UK, 31 August–2 September 1988; Volume 15, pp. 10–5244.

7. Shi, J. Good features to track. In Proceedings of the 1994 Proceedings of IEEE Conference on Computer Vision and Pattern Recognition, Seattle, WA, USA, 21–23 June 1994; pp. 593–600.
8. Rosten, E.; Drummond, T. Machine learning for high-speed corner detection. In Proceedings of the European Conference on Computer Vision, Graz, Austria, 7–13 May 2006; Springer: Berlin/Heidelberg, Germany, 2006; pp. 430–443.
9. Mair, E.; Hager, G.D.; Burschka, D.; Suppa, M.; Hirzinger, G. Adaptive and generic corner detection based on the accelerated segment test. In Proceedings of the European Conference on Computer Vision, Crete, Greece, 5–11 September 2010; Springer: Berlin/Heidelberg, Germany, 2010; pp. 183–196.
10. Matas, J.; Chum, O.; Urban, M.; Pajdla, T. Robust wide-baseline stereo from maximally stable extremal regions. *Image Vis. Comput.* **2004**, *22*, 761–767. [CrossRef]
11. Lowe, D.G. Distinctive image features from scale-invariant keypoints. *Int. J. Comput. Vis.* **2004**, *60*, 91–110. [CrossRef]
12. Calonder, M.; Lepetit, V.; Strecha, C.; Fua, P. Brief: Binary robust independent elementary features. In Proceedings of the European Conference on Computer Vision, Crete, Greece, 5–11 September 2010; Springer: Berlin/Heidelberg, Germany, 2010; pp. 778–792.
13. Rublee, E.; Rabaud, V.; Konolige, K.; Bradski, G. ORB: An efficient alternative to SIFT or SURF. In Proceedings of the 2011 International Conference on Computer Vision, Barcelona, Spain, 6–13 November 2011; pp. 2564–2571.
14. Leutenegger, S.; Chli, M.; Siegwart, R.Y. BRISK: Binary robust invariant scalable keypoints. In Proceedings of the 2011 International Conference on Computer Vision, Barcelona, Spain, 6–13 November 2011; pp. 2548–2555.
15. Alcantarilla, P.F.; Bartoli, A.; Davison, A.J. KAZE features. In Proceedings of the European Conference on Computer Vision, Florence, Italy, 7–13 October 2012; Springer: Berlin/Heidelberg, Germany, 2012; pp. 214–227.
16. Alcantarilla, P.F.; Solutions, T. Fast explicit diffusion for accelerated features in nonlinear scale spaces. *IEEE Trans. Patt. Anal. Mach. Intell.* **2011**, *34*, 1281–1298.
17. Lucas, B.D.; Kanade, T. An iterative image registration technique with an application to stereo vision. In Proceedings of the 7th International Joint Conference on Artificial Intelligence, Vienna, Austria, 23–29 July 2022.
18. Farnebäck, G. Two-frame motion estimation based on polynomial expansion. In Proceedings of the Scandinavian Conference on Image Analysis, Halmstad, Sweden, 29 June–2 July 2003; Springer: Berlin/Heidelberg, Germany, 2003; pp. 363–370.
19. Senst, T.; Eiselein, V.; Sikora, T. Robust local optical flow for feature tracking. *IEEE Trans. Circuits Syst. Video Technol.* **2012**, *22*, 1377–1387. [CrossRef]
20. Su, H.; Yang, H.; Zheng, S.; Fan, Y.; Wei, S. The large-scale crowd behavior perception based on spatio-temporal viscous fluid field. *IEEE Trans. Inf. Forensics Secur.* **2013**, *8*, 1575–1589. [CrossRef]
21. Benabbas, Y.; Ihaddadene, N.; Djeraba, C. Motion pattern extraction and event detection for automatic visual surveillance. *EURASIP J. Image Video Process.* **2010**, *2011*, 163682. [CrossRef]
22. Li, T.; Chang, H.; Wang, M.; Ni, B.; Hong, R.; Yan, S. Crowded scene analysis: A survey. *IEEE Trans. Circuits Syst. Video Technol.* **2014**, *25*, 367–386. [CrossRef]
23. Kratz, L.; Nishino, K. Tracking pedestrians using local spatio-temporal motion patterns in extremely crowded scenes. *IEEE Trans. Pattern Anal. Mach. Intell.* **2011**, *34*, 987–1002. [CrossRef] [PubMed]
24. Cong, Y.; Yuan, J.; Liu, J. Abnormal event detection in crowded scenes using sparse representation. *Pattern Recognit.* **2013**, *46*, 1851–1864. [CrossRef]
25. Lewandowski, M.; Simonnet, D.; Makris, D.; Velastin, S.A.; Orwell, J. Tracklet reidentification in crowded scenes using bag of spatio-temporal histograms of oriented gradients. In Proceedings of the Mexican Conference on Pattern Recognition, Querétaro, Mexico, 26–29 June 2013; pp. 94–103.
26. Zhou, B.; Wang, X.; Tang, X. Random field topic model for semantic region analysis in crowded scenes from tracklets. In Proceedings of the CVPR 2011, Colorado Springs, CO, USA, 20-25 June 2011; pp. 3441–3448.
27. Brun, L.; Saggese, A.; Vento, M. Dynamic scene understanding for behavior analysis based on string kernels. *IEEE Trans. Circuits Syst. Video Technol.* **2014**, *24*, 1669–1681. [CrossRef]
28. Piciarelli, C.; Micheloni, C.; Foresti, G.L. Trajectory-based anomalous event detection. *IEEE Trans. Circuits Syst. Video Technol.* **2008**, *18*, 1544–1554. [CrossRef]
29. Song, X.; Shao, X.; Zhang, Q.; Shibasaki, R.; Zhao, H.; Cui, J.; Zha, H. A fully online and unsupervised system for large and high-density area surveillance: Tracking, semantic scene learning and abnormality detection. *ACM Trans. Intell. Syst. Technol. (TIST)* **2013**, *4*, 1–21. [CrossRef]
30. Popoola, O.P.; Wang, K. Video-based abnormal human behavior recognition—A review. *IEEE Trans. Syst. Man Cybern. Part C (Appl. Rev.)* **2012**, *42*, 865–878. [CrossRef]
31. Yuan, Y.; Feng, Y.; Lu, X. Statistical hypothesis detector for abnormal event detection in crowded scenes. *IEEE Trans. Cybern.* **2016**, *47*, 3597–3608. [CrossRef] [PubMed]
32. Lee, Y.J.; Yeh, Y.R.; Wang, Y.C.F. Anomaly detection via online oversampling principal component analysis. *IEEE Trans. Knowl. Data Eng.* **2012**, *25*, 1460–1470. [CrossRef]
33. Cong, Y.; Yuan, J.; Tang, Y. Video anomaly search in crowded scenes via spatio-temporal motion context. *IEEE Trans. Inf. Forensics Secur.* **2013**, *8*, 1590–1599. [CrossRef]
34. Yu, K.; Lin, Y.; Lafferty, J. Learning image representations from the pixel level via hierarchical sparse coding. In Proceedings of the CVPR 2011, Colorado Springs, CO, USA, 20–25 June 2011; pp. 1713–1720.

35. Zhao, B.; Li , F.-F.; Xing, E.P. Online detection of unusual events in videos via dynamic sparse coding. In Proceedings of the CVPR 2011, Colorado Springs, CO, USA, 20–25 June 2011; pp. 3313–3320.
36. Xu, K.; Jiang, X.; Sun, T. Anomaly detection based on stacked sparse coding with intraframe classification strategy. *IEEE Trans. Multimed.* **2018**, *20*, 1062–1074. [CrossRef]
37. Ma, K.; Doescher, M.; Bodden, C. *Anomaly Detection in Crowded Scenes Using Dense Trajectories*; University of Wisconsin-Madison: Madison, WI, USA, 2015.
38. Mahadevan, V.; Li, W.; Bhalodia, V.; Vasconcelos, N. Anomaly detection in crowded scenes. In Proceedings of the 2010 IEEE Computer Society Conference on Computer Vision and Pattern Recognition, San Francisco, CA, USA, 13–18 June 2010; pp. 1975–1981.
39. Pedregosa, F.; Varoquaux, G.; Gramfort, A.; Michel, V.; Thirion, B.; Grisel, O.; Blondel, M.; Prettenhofer, P.; Weiss, R.; Dubourg, V.; et al. Scikit-learn: Machine learning in Python. *J. Mach. Learn. Res.* **2011**, *12*, 2825–2830.

engineering proceedings

Proceeding Paper

Feedback Linearization Control of Nonlinear System †

Ivan Sergeevich Trenev *,‡ and Daniil Dmitrievich Devyatkin ‡

V.A. Trapeznikov Institute of Control Sciences, Russian Academy of Sciences, 65 Profsoyuznaya Street, 117997 Moscow, Russia
* Correspondence: trenev.99@mail.ru
† Presented at the 15th International Conference "Intelligent Systems" (INTELS'22), Moscow, Russia, 14–16 December 2022.
‡ These authors contributed equally to this work.

Abstract: In this work, a neural network controller is developed for a wide class of nonlinear systems including dynamic systems in the Brunovsky canonical form and those with skew-symmetry properties and bounded nonlinearities. An example of the applicability of this controller to the control of the position of a magnet over an electromagnet is considered. The modeling has been provided through the Simulink environment.

Keywords: neural network; control; Matlab; Simulink

Citation: Trenev, I.S.; Devyatkin, D.D. Feedback Linearization Control of Nonlinear System. *Eng. Proc.* **2023**, *33*, 36. https://doi.org/10.3390/engproc2023033036

Academic Editors: Askhat Diveev, Ivan Zelinka, Arutun Avetisyan and Alexander Ilin

Published: 20 June 2023

1. Introduction

Since linearization is based on accurate knowledge of the nonlinearities of the system, the applicability of its approaches to the control of real systems is severely limited. To guarantee the existence of a closed system solution, the nonlinearities must satisfy certain conditions, and for linearization methods for feedback control, system controllability is an important factor [1]. In what follows, several adaptive schemes that allow for linear parametric uncertainties are presented, which will help loosen constraints on model matching. Due to the fact that neural networks are acceptable approximators of nonlinear functions, we can assume that the linearity conditions will not be imposed on the system under consideration [2].

The feedback linearization algorithm is centered around geometric methods. However, due to the fact that this algorithm is based on exact knowledge of the nonlinearities of the system, the applicability of these approaches to the control of real systems is limited. To loosen the limitations of the exact model-matching restrictions, several adaptive schemes have been introduced that allow for linear parametric uncertainties [3]. Due to the properties of the universal approximation, NNs are used to calculate nonlinearities, which means that system parameters are not required [4].

For the sake of simplicity, all components of the state vector are assumed to be measurable. If only some components of the state vector are measurable, which corresponds to the case of an output feedback control; then, an additional dynamic neural network is required to evaluate non-measurable states.

2. Brunovsky Canonical Form

To begin, some definitions are given that are necessary for further reasoning. Let $f, g : \mathbb{R}^n \to \mathbb{R}$ be the unknown smooth functions, $x = (x_1, \ x_2, \ \ldots, \ x_n)^T \in \mathbb{R}^n$ be the state vector, u be the control, and y be the output. The canonical Brunovsky form defines a special form of nonlinear dynamics of continuous time

$$\begin{cases} \dot{x}_1 = x_2, \\ \dot{x}_2 = x_3, \\ \dots\dots \\ \dot{x}_n = f(x) + g(x)u, \\ y = h(x). \end{cases} \tag{1}$$

As shown in Figure 1, each integrator stores information. Each of them requires an initial condition [5].

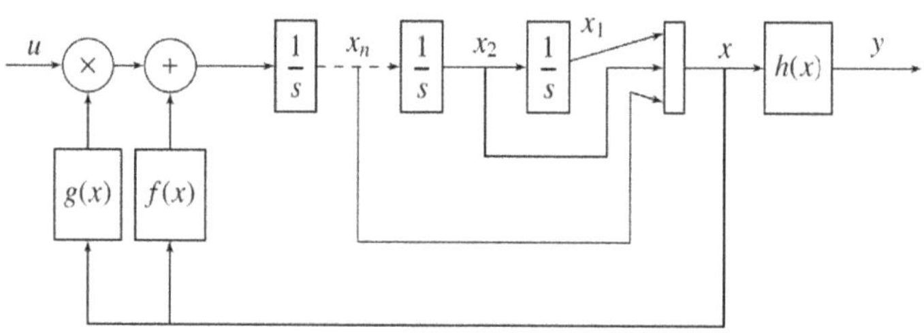

Figure 1. Continuous-time SISO (one input, one output) Brunovsky form.

Consider a SISO linearizable state feedback system with an unknown disturbance. This system has a representation in the state space in Brunovsky's canonical form (1). Then, the following assumption is true [6].

Proposition 1. *(Bounds on disturbance and function $g(x)$)*

1. *There exists a known upper bound b_d: $\|d(t)\|_2 \leqslant b_d, \forall t \in \mathbb{R}^+$;*
2. *Function $g(x)$ satisfies the condition: $|g(x)| \geqslant g_l > 0, \forall x \in \mathbb{R}^n$.*

3. Tracking Controller and Error Dynamics

To begin with, let us form a tracking goal: for a given output $y_d(t)$, it is necessary to find the control $u(t) : y(t) = x_1(t)$ (i.e., $h(x) = x_1$). This condition implies a desired trajectory with acceptable accuracy, with restrictions on the state and control vector components. Feedback linearization will be used to track the output here. Let us introduce some assumptions to design the tracking controller. The desired trajectory vector will look like this [7]:

$$x_d(t) = \left(y_d, \ \dot{y}_d, \ \dots, \ y_d^{(n-1)}\right)^T. \tag{2}$$

For the state vector approximation error $e = x - x_d$, define the filtered tracking error with its derivative as follows:

$$r = \Lambda^T e, \ \dot{r} = f(x) + g(x)u + d + Y_d, \tag{3}$$

where $Y_d = -x_d^{(n)} + \sum_{i=1}^{n-1} \lambda_i e_{i+1} = -x_d^{(n)} + \bar{\Lambda}^T e$ is a known signal and $\Lambda = (\lambda_1, \ \lambda_2, \ \dots, \ \lambda_{n-1} \ 1)^T$ is a well-chosen vector of coefficients, $\bar{\Lambda} = \left(\lambda_1, \ \lambda_2, \ \dots, \ \lambda_{n-1}\right)^T$.

For both cases, when $g(x)$ is known or unknown, define the control action in the following form:

$$\begin{aligned} u_{exact} &= \frac{1}{g(x)}(-f(x) - K_v r - Y_d), \\ u_c &= \frac{1}{\hat{g}(\hat{\Theta}_g, x)}\left(-\hat{f}(\hat{\Theta}_f, x) + v\right), \ \ v = -K_v r - Y_d. \end{aligned} \tag{4}$$

where the estimates $\hat{g}(\hat{\Theta}_g, x)$ and $\hat{f}(\hat{\Theta}_f, x)$ are to be designed using two neural networks with appropriate weights matrices Θ_g and Θ_f.

4. Neural Networks for Approximating Functions

Let neural networks be used to approximate unknown continuous functions. Then, the compact set $\Re^n$ has ideal target weights W_i and V_i ($i = f$ for the case of a known $g(x)$ and $i \in \{f, g\}$ for the case of an unknown $g(x)$):

$$i(x) = W_i^T \sigma(V_i^T x) + \varepsilon_i$$

where the estimation errors are bounded $\|\varepsilon_i\| < \varepsilon_{iN}$. Let the ideal weights be unknown and possibly not unique but satisfying the following assumption [8].

Proposition 2. *(Restriction on the neural network target weights) Ideal neural network weights for functions $f(x)$ and $g(x)$ are bounded on any compact subset of $\Re^n$:*

$$\|\Theta_i\| \leqslant \Theta_{im}, \text{ where } \Theta_i = \begin{pmatrix} V_i & 0 \\ 0 & W_i \end{pmatrix} \text{ are the weight matrices for } i, i \in \{f, g\}.$$

Then, using two neural networks, nonlinear functions $f(x)$ and $g(x)$, evaluations are obtained in the following form:

$$\hat{i}(x) = \hat{W}_i^T \sigma(\hat{V}_i^T x), \text{ where } \hat{\Theta}_i = \begin{pmatrix} \hat{V}_i & 0 \\ 0 & \hat{W}_i \end{pmatrix} \text{ are the actual value of the weight matrix.}$$

Using the expansion in a Taylor series, the functional errors of the estimate are as follows:

$$\tilde{i}(x) = i(x) - \hat{i}(x) = \tilde{W}_i^T (\hat{\sigma}_i - \hat{\sigma}_i' \hat{V}_i^T x) + \hat{W}_i^T \hat{\sigma}_i' \tilde{V}_i^T x + w_i,$$

where, for some C_j, it is true that $\|w_i(t)\| \leqslant C_0 + C_1 \|\tilde{\Theta}_i\|_F + C_2 \|r\| \cdot \|\tilde{\Theta}_i\|_F$. Estimates based on the boundedness of the standard activation functions together with their derivatives have the following form [7].

Lemma 1. *(Bounds on state vector and approximated function)*

a. *The inequality restricting the state space vector is valid for the computed constants d_0 and d_1: $\|x\| \leqslant d_0 + d_1 \|r\|$;*
b. *On any compact set, there are constants C_3 and C_4 that restrict the Euclidean norms of the approximated functions: $\|i(x)\| = \|W_i^T \sigma(V_i^T x) + \varepsilon_i\| \leqslant C_3 + C_4 \|r\|$.*

5. Controller Structure

Here, a tracking controller is designed for both cases, when $g(x)$ is known or unknown. The main goal of this section is to construct a neural network that approximates the unknown function $f(x)$ (and function $g(x)$) [9].

5.1. System with a Known Function $g(x)$

Figure 2 shows the resulting controller, described in Table 1.

Table 1. The resulting controller with known function $g(x)$.

Neural network controller:	$u = \frac{1}{g(x)}(-\hat{W}^T\sigma(\hat{V}^T x) - K_u r - Y_d + u_r).$
Robustifying Term:	$u_r = -K_z(\lVert\hat{\Theta}\rVert + \Theta_m).$
Neural network weight update law:	$\dot{\hat{W}} = M(\hat{\sigma} - \hat{\sigma}'\hat{V}^T x)r - \alpha\lvert r\rvert M\hat{W},$ $\dot{\hat{V}} = Nrx\hat{W}^T\hat{\sigma}' - \alpha\lvert r\rvert N\hat{V}.$
Parameters:	$\Lambda > 0$ — coefficient vector of the Hurwitz function; M, N — positive definite symmetric matrices; Θ_m — bound on the unknown target weight norms; $K_v, K_z > 0, \alpha > 0.$
Signals:	$e(t) = x(t) - x_d(t)$ — tracking error; $Y_d = -x_d^{(n)} + \bar{\Lambda}^T e$ — desired trajectory feedforward signal; $r(t) = \Lambda^T e(t)$ — filtered tracking error.

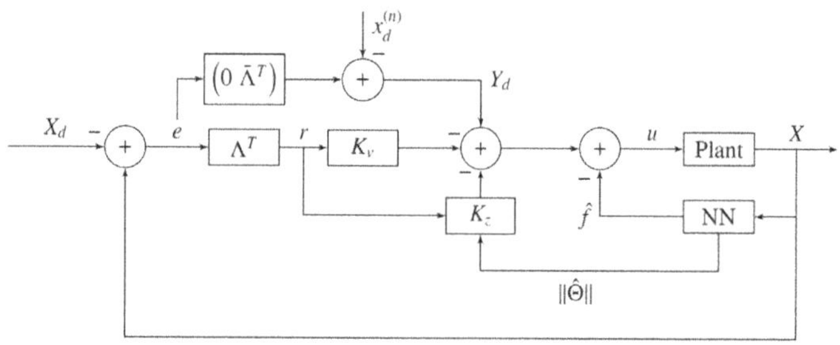

Figure 2. Neural network controller with known function $g(x)$.

5.2. System with an Unknown Function $g(x)$

Figure 3 shows the resulting controller, described in Table 2.

Table 2. The resulting controller with unknown function $g(x)$.

Neural network controller:	$u = \begin{cases} u_c + \frac{u_r - u_c}{2}\epsilon^{\gamma(\lvert u_c\rvert - s)}, \text{if } \hat{g} \geqslant g_l \text{ and } \lvert u_c\rvert \leqslant s, \\ u_r - \frac{u_r - u_c}{2}\epsilon^{-\gamma(\lvert u_c\rvert - s)}, \text{otherwise.} \end{cases}$ $u_c = \frac{1}{\hat{W}_g^T\sigma(\hat{v}_g^T x)}\left(-\hat{W}_f^T\sigma(\hat{V}_f^T x) + v\right).$
Robustifying Term:	$u_r = -\mu\frac{\lvert\hat{g}\rvert}{g_l}\lvert u_c\rvert\text{sign}(r),$ $v = -K_v r - Y_d,$ $K_v(t) = K_N + K_z((\lVert\hat{\Theta}_f(t)\rVert + \Theta_{fm}) + s(\lVert\hat{\Theta}_g(t)\rVert + \Theta_{gm})).$
Neural network weight update laws:	$\dot{\hat{W}}_f = M_f(\hat{\sigma}_f - \hat{\sigma}_f'\hat{V}^T + fx)r - \alpha\lvert r\rvert M_f\hat{W}_f,$ $\dot{\hat{V}}_f = N_f rx\hat{W}_f^T\hat{\sigma}_f' - \alpha\lvert r\rvert N_f\hat{V}_f,$ $\dot{\hat{W}}_g = \begin{cases} M_g\left((\hat{\sigma}_g - \hat{\sigma}_g'\hat{V}_g^T x)u_c r - \alpha\lvert r\rvert\lvert u_c\rvert\hat{W}_g\right), \text{if } \hat{g} \geqslant g_l \text{ and } \lvert u_c\rvert \leqslant s, \\ 0, \text{otherwise,} \end{cases}$ $\dot{\hat{V}}_g = \begin{cases} N_g u_c rx\hat{W}_g^T\hat{\sigma}_g' - \alpha\lvert r\rvert\lvert u_c\rvert N_g\hat{V}_g, \text{if } \hat{g} \geqslant g_l \text{ and } \lvert u_c\rvert \leqslant s, \\ 0, \text{otherwise.} \end{cases}$
Signals:	$e(t) = x(t) - x_d(t)$ — tracking error; $Y_d = -x_d^{(n)} + \bar{\Lambda}^T e$ — desired trajectory feedforward signal; $r(t) = \Lambda^T e(t)$ — filtered tracking error.
Parameters:	$K_N > 0,\ K_z > \max\left\{C_2, \frac{C_4}{s\gamma\epsilon\Theta_{gm}}\right\}, \alpha > 0,\ s > 0,\ \gamma < \ln\frac{2}{s},\ \mu \geqslant 2;$ Θ_m — bound on the unknown target weight norms; $\Lambda > 0$ — coefficient vector of the Hurwitz function; M_i, N_i — positive definite symmetric matrices.

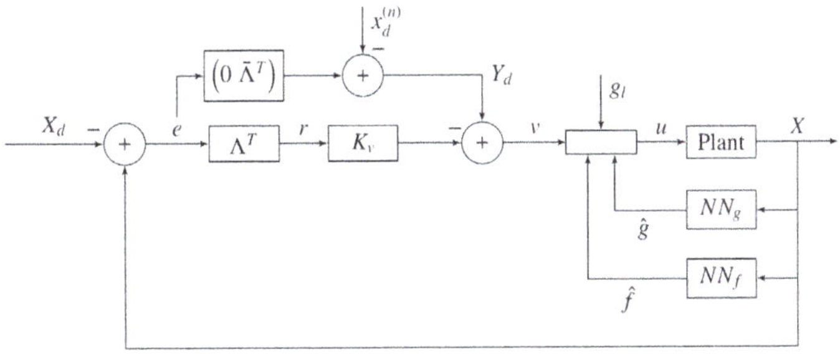

Figure 3. Neural network controller with unknown function $g(x)$.

6. Modelling

Consider the system shown in Figure 4 and a magnet suspended above an electromagnet. The magnet can only move in the vertical direction. The purpose of the control is to control the position of the magnet. The equation for motion in the described system is

$$\frac{d^2y(t)}{dt^2} = -g + \frac{\alpha}{M}\frac{i^2(t)}{y(t)} - \frac{\beta}{M}\frac{dy(t)}{dt}, \tag{5}$$

where g is the standard gravitational acceleration, $i(t)$ is the current flowing in the electromagnet, $y(t)$ is the distance above the electromagnet, and M is the mass of the magnet. The parameter β is the coefficient of viscous friction, and α is the field strength constant.

Figure 5 shows the results of modeling this system in the presence of two neural networks. (Red graph—desired trajectory; blue graph—approximation using neural networks.) The simulation was carried out in the Matlab/Simulink environment using additional neural network packages [10].

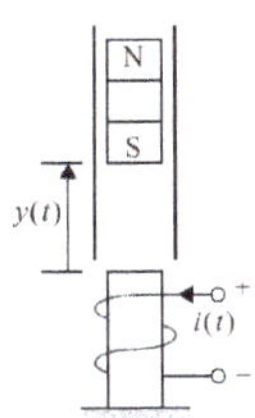

Figure 4. Electromagnet system.

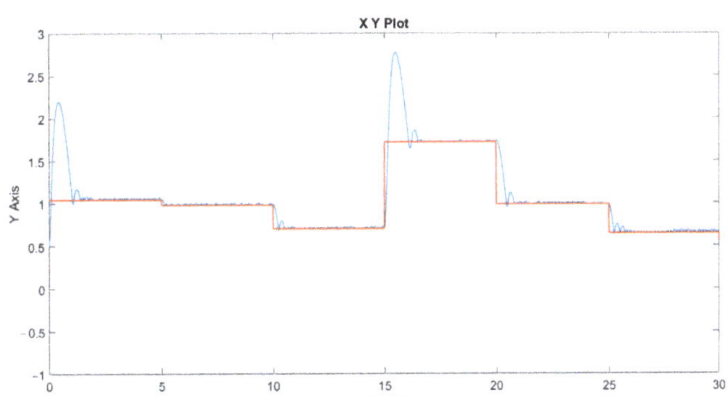

Figure 5. Simulation result.

7. Conclusions

In this paper, we showed how to design neural network controllers for systems in the Brunovsky form for the case of a known input influence function and the case of an unknown input influence function. For the case with an unknown input influence function, it was necessary to make certain efforts to ensure that control remained bounded. Based on the simulation results presented in the last section, we can conclude that neural networks are a universal approximator and can be used to simulate the behavior of a real object.

Author Contributions: These authors contributed equally to this work. All authors have read and agreed to the published version of the manuscript.

Funding: This research received no external funding.

Institutional Review Board Statement: Not applicable.

Informed Consent Statement: Not applicable.

Data Availability Statement: Data sharing not applicable.

Conflicts of Interest: The authors declare no conflict of interest.

References

1. Charlet, B.; Lévine, J.; Marino, R. On dynamic feedback linearization. *Syst. Control. Lett.* **1989**, *13*, 143–151. [CrossRef]
2. Ge, S.S.; Hang, C.C.; Lee, T.H.; Zhang, T. *Stable Adaptive Neural Network Control*; Springer Science and Business Media: Berlin/Heidelberg, Germany, 2013.
3. Campion, G.; Bastin, G. Indirect adaptive state feedback control of linearly parameterized nonlinear systems. *Int. J. Control Signal Proc.* **1990**, *4*, 345–358. [CrossRef]
4. Cui, R.; Yang, C.; Li, Y.; Sharma, S. Adaptive neural network control of AUVs with control input nonlinearities using reinforcement learning. *IEEE Trans. Syst. Man Cybern. Syst.* **2017**, *47*, 1019–1029. [CrossRef]
5. Yesildirek, A.; Lewis, F.L. Feedback linearization using neural networks. *Automatica* **1995**, *31*, 1659–1664. [CrossRef]
6. Hauser, J.; Sastry, S.; Kokotovic, P. Nonlinear control via approximate input-output linearization. *IEEE Trans. Automat. Control* **1989**, *37*, 392–398. [CrossRef]
7. Krener, A.J. Feedback linearization. In *Mathematical Control Theory*; Baillieul, J., Willems, J.C., Eds.; Springer: New York, NY, USA, 1999; pp. 66–98.
8. Lewis, F.W.; Jagannathan, S.; Yesildirak, A. *Neural Network Control of Robot Manipulators and Non-Linear Systems*; CRC Press: Boca Raton, FL, USA, 2020.
9. Ge, S.S. Robust adaptive NN feedback linearization control of nonlinear systems. *Int. J. Syst. Sci.* **1996**, *27*, 1327–1338. [CrossRef]
10. Demuth, H.; Beale, M.; Hagan, M. *Neural Network Toolbox. For Use with MATLAB*; The MathWorks Inc.: Natick, MA, USA, 1992.

engineering proceedings

MDPI

Proceeding Paper

Methods of Recognition Based on Wavelet Transform for Analysis of Characteristics of Spherical Quantum Dot †

Evgenia Kozhanova [1,2,*], Sergey Danilov [1,*] and Victor Belyaev [1,3,*]

1 Department of Nanotechnology and Microsystem Technology, Engineering Academy, RUDN University, Miklukho-Maklaya Str. 6, 117198 Moscow, Russia
2 Department of Information Security of Automated Systems, Institute of Electronic Engineering and Instrumentation, Yuri Gagarin State Technical University of Saratov, Politechnicheskaya Str. 77, 410054 Saratov, Russia
3 Department of Fundamental Physics and Nanotechnology, Faculty of Physics and Mathematics, State University of Education, Very Voloshinoi Str. 24, 141014 Mytischi, Russia
* Correspondence: kozhanovaer@sstu.ru (E.K.); danilov-si@rudn.ru (S.D.); vic_belyaev@mail.ru (V.B.)

† Presented at the 15th International Conference "Intelligent Systems" (INTELS'22), Moscow, Russia, 14–16 December 2022.

Abstract: The paper contains the results of a recognition technique based on the comparison of statistical and stochastic characteristics of the wavelet coefficients of energy density describing the emission energy of a nanocrystal with a quantum dot according to the Brus equation for traditional and perspective materials for quantum dots (CdSe, GaAs, CdTe, PbS) used in optoelectronic engineering and technology, in order to analyze their characteristics.

Keywords: recognition; wavelet transform; emission energy; spherical quantum dot; nanostructure; nanobject; optical properties.

Citation: Kozhanova, E.; Danilov, S.; Belyaev V. Methods of Recognition Based on Wavelet Transform for Analysis of Characteristics of Spherical Quantum Dot. *Eng. Proc.* **2023**, *33*, 35. https://doi.org/10.3390/engproc2023033035

Academic Editors: Askhat Diveev, Ivan Zelinka, Arutun Avetisyan and Alexander Ilin

Published: 20 June 2023

1. Introduction

The application of quantum structures in the design of optical devices and instruments offers an opportunity to significantly expand the range of parameters and purposes of traditional and perspective materials [1]. In this paper, we study the dependence of the total energy of a quantum dot (QD) and the emission energy of nanocrystals with QDs on its size (radius) (CdSe, GaAs, CdTe, PbS), representing nano-objects of spherical shape, bounded in three directions. The analysis is based on a complex approach integrating theoretical research methods, mathematical modeling and numerical methods, and wavelet transform methods.

Wavelet transform methods are a powerful tool to analyze and process signals and functions that are non-stationary in time or heterogeneous in space in order to identify local features. They are divided into continuous (for analyzing signals and time series) and discrete wavelet transforms (for analyzing signals and time series images) [1–3].

Three approaches [1,2], which are based on comparing wavelet coefficients (continuous wavelet transform (CWT)) and signal components (detailing and approximating coefficients in discrete wavelet transform (DWT), are used to recognize signals of a different nature:

- Statistical characteristics (mean, dispersion, standard deviation) are used to conduct a wavelet coefficient (CWT) and signal component (DWT) analysis.
- An energy spectrum analysis examines the energy spectrum and is used only for signal components (DWT).
- Stochastic characteristics (fractal dimensionality, Hurst index, correlation dimensionality and phase space dimensionality) evaluate the chaotic nature of wavelet coefficients (CWT) and signal components (DWT).

Continuous wavelet transform (CWT) is chosen as the analysis tool; the result is a wavelet spectrogram with statistical and stochastic characteristics used for recognition.

2. Theoretical Research and Analytical Part

Most modern, traditional and perspective materials used in optoelectronic engineering and technology are nanostructures with QDs, which determine the properties of materials, including their optical properties [4]. QDs are spherical-shaped nano-objects bounded in three directions. QDs are characterized by the effect of the dimensional quantization of energy states by changing the particle size. Decreasing the size of QDs leads to an increased degree of confinement, and the width of the band gap increases as a result of the formation of a bound electron-hole pair (exciton) with a higher energy.

The total energy of QD is the sum of the band gap energy between the occupied level and unoccupied energy level, confinement energies of the hole and the excited electron, and the bound energy of the exciton:

$$\Delta E(r)_{TOTAL} = Eg + (h^2\pi^2/2d^2) - (1/\varepsilon_r^2) \cdot (\mu/m_e) \cdot R_y \quad (1)$$

μ is the reduced mass of the exciton system; m_e is the effective mass of the electron; d is diameter of the confinement; ε_r is the size-dependent dielectric constant of the semiconductor; R_y is the Rydberg energy.

Consider the effect of the QD size on their optical properties by calculating the emission energy of a nanocrystal with QDs varying in radius from 2 to 10 nm, using the Brus equation:

$$\Delta E(r) = Eg + (h^2/8r^2) \cdot (1/m_e + 1/m_h) \quad (2)$$

where Eg is the band gap energy between occupied level and unoccupied energy level, h is the Planck constant ($h = 6.62607015 \cdot 10^{-34}$ J·Hz^{-1}), r is the radius of the spherical particle (2 to 10 nm), m_e is the effective mass of the electron, and m_h is the effective mass of the hole (Table 1).

Table 1. Parameters for calculating ΔE for different materials.

Nanocrystal Material	Effective Electron Mass, m_e	Effective Electron Mass, m_h	Effective Point Charge Mass, m^*	Transition Energy Eg, eV
PbS	$0.07m_0$	$0.09m_0$	$0.04m_0$	0.41
GaAs	$0.06m_0$	$0.51m_0$	$0.054m_0$	1.4
CdTe	$0.14m_0$	$0.35m_0$	$0.1m_0$	1.50
CdSe	$0.13m_0$	$0.45m_0$	$0.1m_0$	1.74

* $m_0 = 9.11 \times 10^{-31}$ (kg) is electron rest mass

The energy of the band gap (Eg) is calculated using the following equation:

$$Eq = (h^2n^2)/(8m^*r^2) \quad (3)$$

n is energy level (1, 2, 3...), h is the Planck constant ($h = 6.62607015 \times 10^{-34}$ [J·Hz^{-1}]), m^* is the effective point charge mass (Table 1), and r is the radius of the spherical particle.

Then, the Brus Equation (2) is as follows:

$$\Delta E(r) = (h^2n^2)/(8m_er^2) + (h^2/8r^2) \cdot (1/m_e + 1/m_h) \quad (4)$$

Table 2 presents the results of calculating the $\Delta E(r)$ value for CdSe, GaAs, CdTe, PbS, based on the data in Table 1 and Equation (4) [5].

Table 2. Results of calculations of ΔE for different materials (CdSe, CdTe, GaAs, PbS).

QD Radius,	CdSe		CdTe		GaAs		PbS	
nm	ΔE, eV	Wavelength, nm	ΔE, eV	Wavelength, nm	ΔE, eV	Wavelength, nm	ΔE, eV	Wavelength, nm
2.0	2.67	465	2.42	507	3.11	394	2.74	448
2.5	2.34	532	2.09	587	2.49	491	1.90	645
3.0	2.15	577	1.91	642	2.16	567	1.44	848
3.5	2.04	608	1.80	681	1.96	626	1.17	1050
4.0	1.97	630	1.73	708	1.83	671	0.99	1240
4.5	1.92	646	1.68	729	1.74	705	0.87	1410
5.0	1.89	658	1.65	744	1.67	732	0.78	1570
5.5	1.86	667	1.62	755	1.63	753	0.72	1710
6.0	1.84	674	1.60	765	1.59	770	0.67	1830
6.5	1.83	680	1.59	772	1.56	784	0.63	1940
7.0	1.82	684	1.57	778	1.54	796	0.60	2040
7.5	1.81	688	1.57	783	1.52	805	0.58	2130
8.0	1.80	691	1.56	786	1.51	813	0.56	2210
8.5	1.79	694	1.55	790	1.49	820	0.54	2270
9.0	1.79	696	1.55	793	1.48	825	0.52	2330
9.5	1.78	698	1.54	795	1.48	830	0.51	239
10.0	1.78	699	1.54	797	1.47	834	0.50	2430

Visible spectrum light includes light from violet to red, and a wavelength from 380 nm to 750 nm, which corresponds to the energy range of the photon from 2.75 eV to 1.98 eV. Table 2 shows that the wavelength of the emitted photon depends on the QD radius. Therefore, a smaller point radius is closer to the blue part of the spectrum, while a larger point radius is closer to the red part of the spectrum. Consequently, changing and controlling the size of QDs can help to obtain a certain wavelength of emissions, which determines the color of light.

Below are graphs of the emission energy of a nanocrystal $\Delta E(r)$ as a function of the radius of a spherical QD (Figure 1a) and graphs of the wavelength dependence from the radius of a spherical QD (Figure 1b).

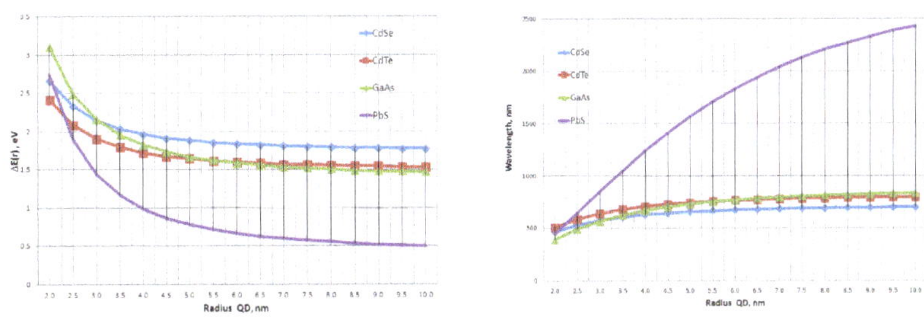

(**a**) (**b**)

Figure 1. Graph of the dependence of the emission energy of a nanocrystal $\Delta E(r)$ on the radius of a QD (**a**) and a graph of the dependence of wavelength on the radius of a QD (**b**) for various materials (CdSe, CdTe, GaAs, PbS)

The analysis of the dependencies (Figure 1a,b) shows that with increasing radius QD the emission energy of the nanocrystal decreases and the wavelength increases. PbS material containing lead (Pb) differs from others (CdSe, CdTe, GaAs) in the type of dependence (Figure 1a,b—graphs have a high rate of change; Figure 1b—graphs have a different tangent sign of the tangent angle) and values (Figure 1a—a large spread of the values of the characteristic $\Delta E(r)$ in the range from 2 to 7; Figure 1b—a large spread of the wavelength value over the entire interval).

The materials considered in this paper are of interest in terms of optical properties describing the emission energy of the nanocrystal and the wavelength, which depend on the size (radius) of the QD, so we calculate the energy density of the signal given the Formula [3]:

$$P_w f(a,b) = |Wf(a,b)|^2 \tag{5}$$

where the continuous wavelet transform (CWT) is given by [3,4]:

$$Wf(a,b) = |a|^{-1/2} \int_{-\infty}^{\infty} f(t)\psi\left(\frac{t-b}{a}\right)dt \tag{6}$$

where $\psi(t)$ is a real wavelet-forming function, often referred to simply as a wavelet; a is the scaling parameter (the inverse of the frequency); b is the shift parameter (analog of time).

For effective data analysis with continuous wavelet transformation, it is necessary to carefully select a family of wavelets, according to the specifics of the research, in order to obtain the best conversion result. This is one of the reasons for the significant variety of wavelet functions, many of which are focused on the wavelet transformation of certain classes of signals and other analysis and synthesis tasks [1,2,6].

Below is the signal energy density based on a continuous wavelet transform (CWT) with a maximum scale of 64 and a step of 1 using a Gaussian wavelet of the 1st order (Table 3).

Table 3. Energy density of the dependences of the emission energy of a nanocrystal $\Delta E(r)$ (Figure 1a) based on a continuous wavelet transform and its visualizations-scalograms (n = 64; dn = 1).

Material QD	Wavelet Function			
	Haar Wavelet	**Gaussian Wavelet of the 1st Order**	**Wavelet "Mexican Hat"**	**Daubeshi Wavelet of the 4th Order**
PbS	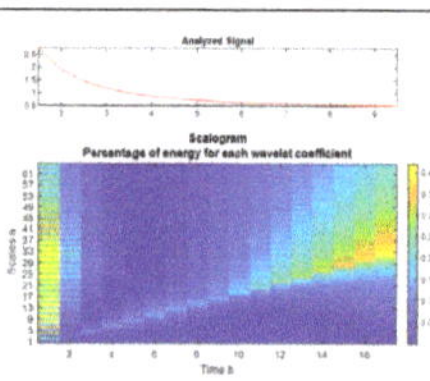	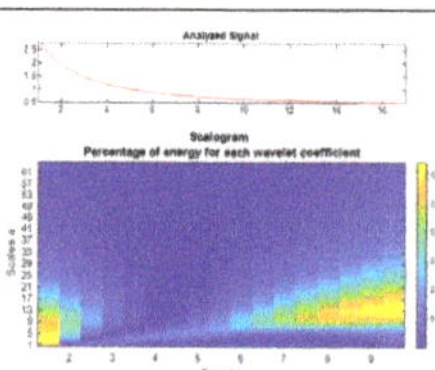	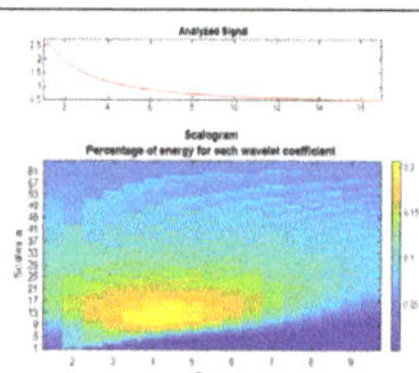	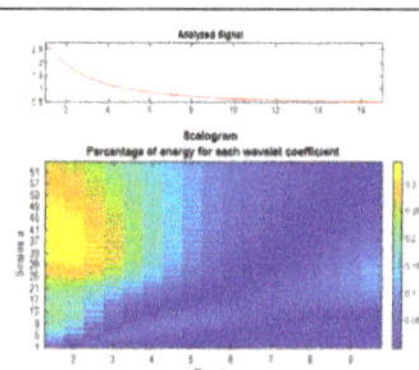
GaAs	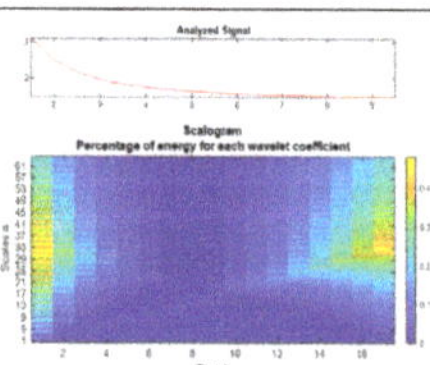	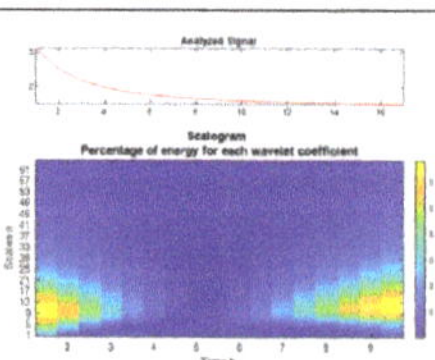	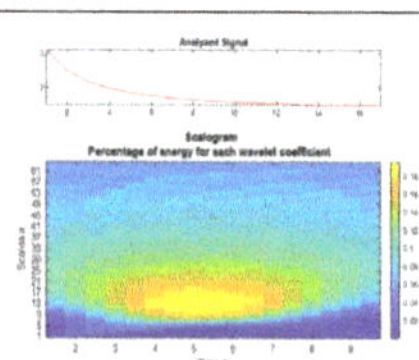	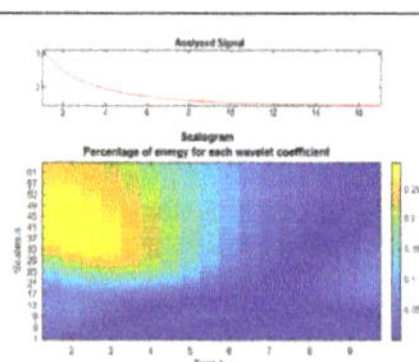
CdTe	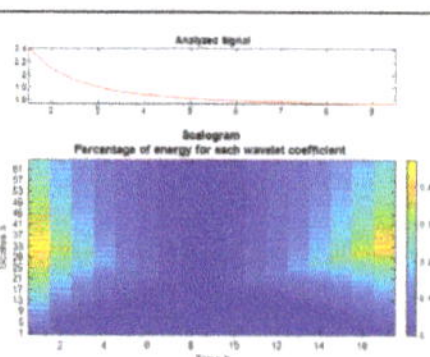	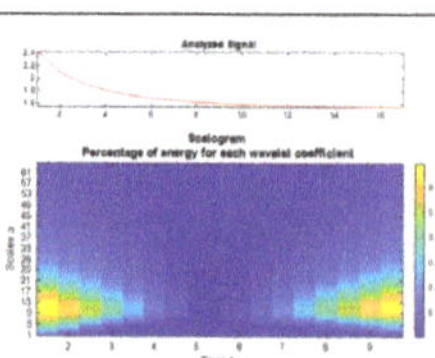	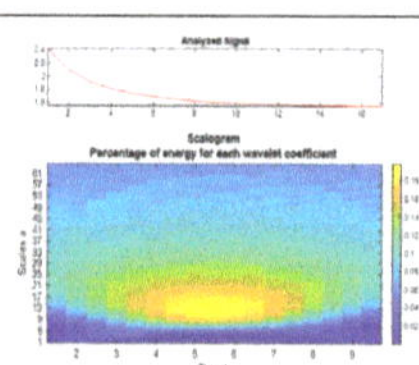	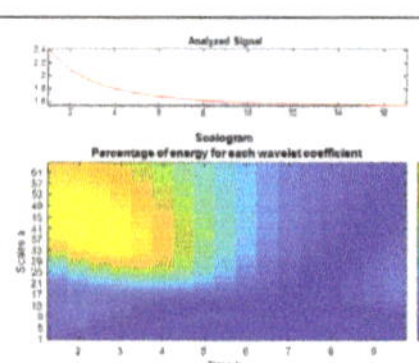
CdSe	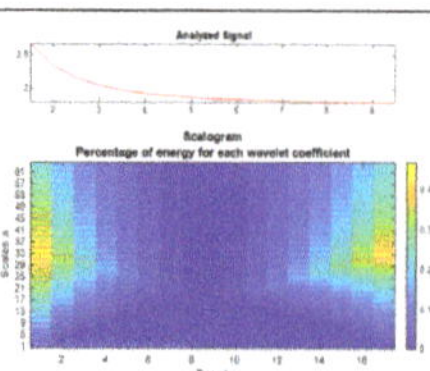	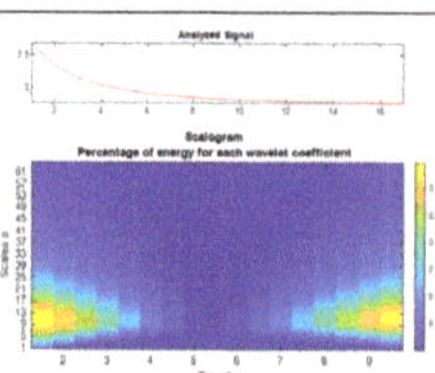	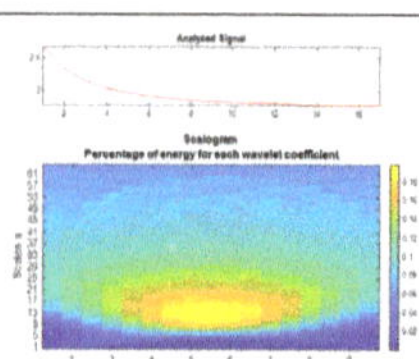	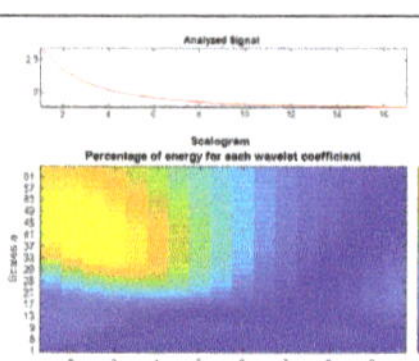

For analysis and recognition, we first use statistical characteristics; however, before that, we consider the cross sections of the resulting surface, and the energy density of each material at different scale values to justify the feasibility of application (Figure 2).

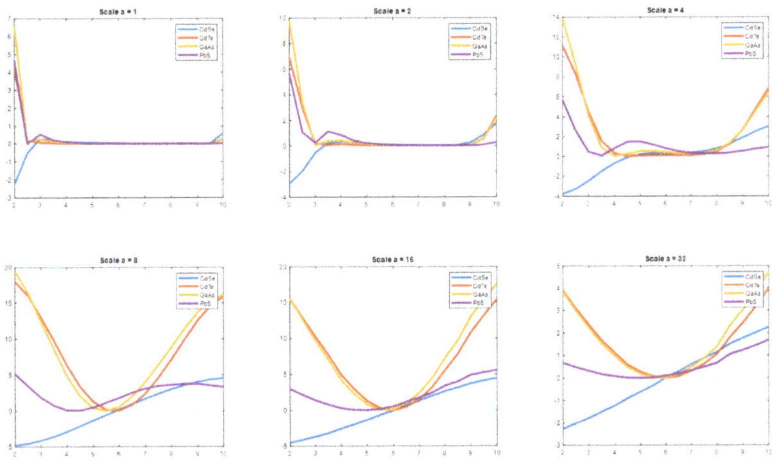

Figure 2. Graphs of cross sections of signal energy density at scales 1, 2, 4, 8, 16, 32 (Gaussian wavelet of the 1st order): horizontal axis—radius QD (nm), vertical axis—values of wavelet coefficients of energy density.

For each section, we calculate statistical indicators and present the results in the form of graphs and radar charts grouped by statistical characteristics and cross-section scales (1, 2, 4, 8, 16, 32) (Figures 3 and 4).

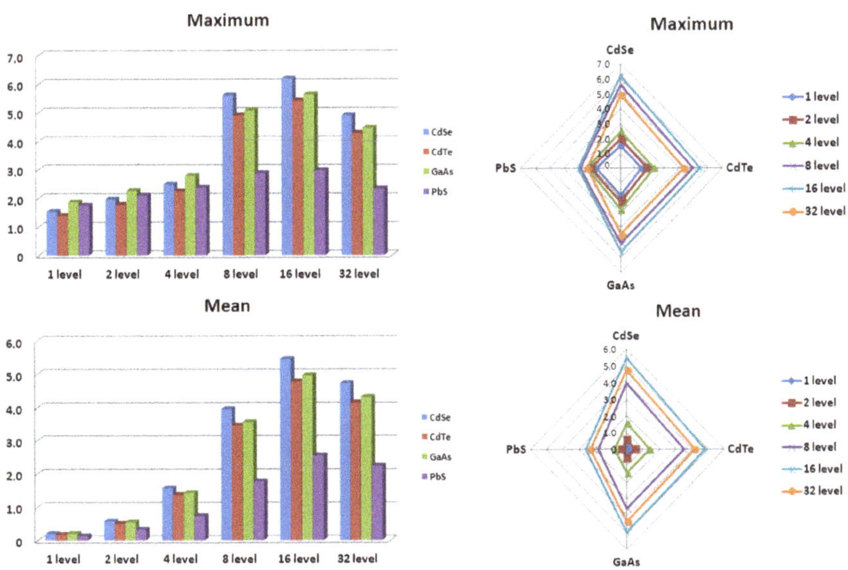

Figure 3. Graphs and radar charts of statistical characteristics (maximum and mean) of cross sections (Figure 2) of signal energy density at scales 1, 2, 4, 8, 16, 32.

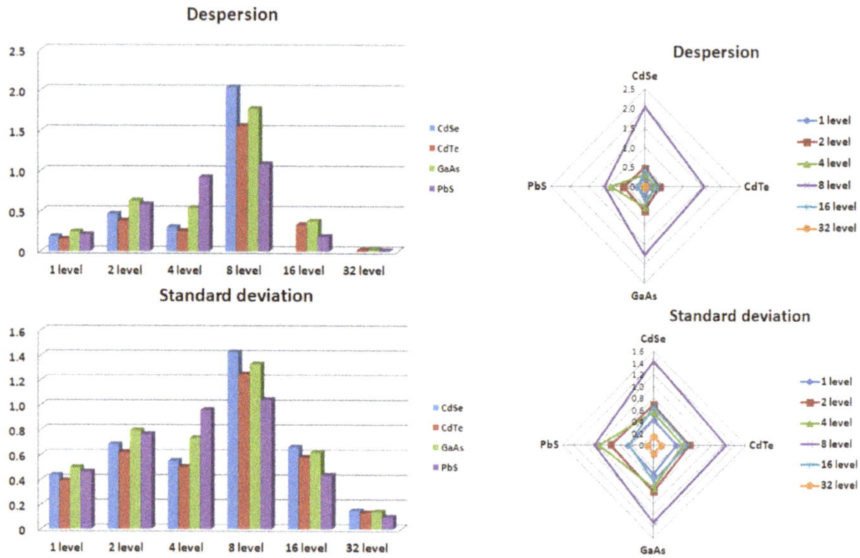

Figure 4. Graphs and radar charts of statistical characteristics (Dispersion and Standard deviation) of cross sections (Figure 2) of signal energy density at scales 1, 2, 4, 8, 16, 32.

The analysis of the graphs in Figures 3 and 4 shows that the mean will be the most informative and has the highest diagnostic ability if a scale equal to one or two is excluded from the analysis.

The obtained characteristics strongly depend on the wavelet function, which is used for continuous wavelet transformation. Figure 5 shows graphs of the cross sections of the signal energy density at scales 1, 2, 4, 8, 16, 32 and the same dependencies, except that the "Mexican Hat" wavelet is taken as a wavelet function.

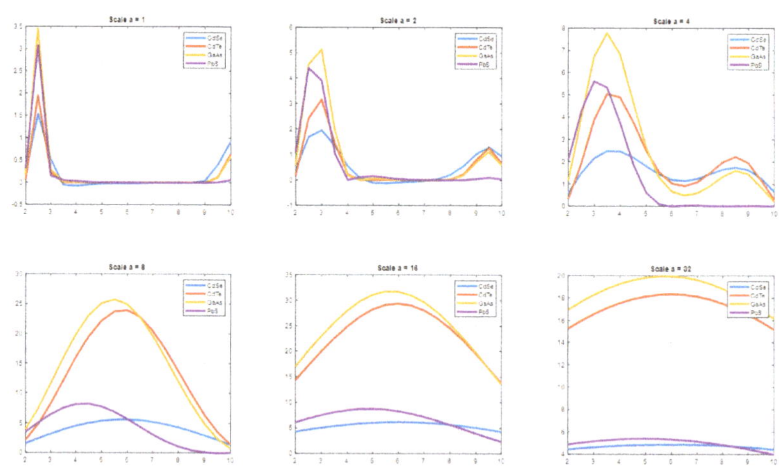

Figure 5. Graphs of cross sections of signal energy density at scales 1, 2, 4, 8, 16, 32 ("Mexican Hat" wavelet): horizontal axis—the radius of the quantum dot (nm), vertical axis—the values of the wavelet coefficients of energy density.

In the next step, for analysis and recognition, we will use stochastic characteristics, in the form of entropy values, for the energy density obtained using various wavelet functions (Table 4).

Table 4. The value of the energy density entropy for various wavelet functions and calculation algorithms (setting—Shanon, LogEnerge).

Wavelet	Setting	CdSe	CdTe	GaAs	PbS
Gaussian of the 1st order	Shanon	−89.8469	−56.4606	−75.7172	−17.6034
	LogEnerge	22.4197	17.8532	18.7774	−7.3786
Gaussian of the 2nd order	Shanon	−89.8469	−56.4606	−75.7172	−17.6034
	LogEnerge	22.4197	17.8532	18.7774	−7.3786
Mexican Hat	Shanon	−89.8469	−56.4606	−75.7172	-17.6034
	LogEnerge	22.4197	17.8532	18.7774	−7.3786
Daubechies of the 4th order	Shanon	−89.8469	−56.4606	−75.7172	−17.6034
	LogEnerge	22.4197	17.8532	18.7774	−7.3786

The analysis of the values in Table 4 shows that the entropy value does not depend on the wavelet function, which is used in the continuous wavelet transform. The obtained values have a diagnostic ability.

3. Conclusions

This paper presents the results of a recognition technique based on a comparison of statistical and stochastic characteristics of the wavelet coefficient of energy density describing the emission energy of a nanocrystal with a quantum dot based on the Brus equation for traditional and perspective materials for quantum dots (CdSe, GaAs, CdTe, PbS), used in optoelectronic engineering and technology [7]. An analysis of techniques has shown that the statistical characteristics strongly depend on the wavelet function, which is used to implement a continuous wavelet transform, while the stochastic characteristic, in the form of entropy, has good diagnostic ability and does not depend on the type of wavelet function. Further research is planned for the application of wavelet transforms to extended spectrum materials on spherical QDs used in optoelectronics, as well as research and comparisons of the total energy of quantum dots (1) using wavelet analysis for materials such as CdSe and CdTe.

Author Contributions: Conceptualization, V.B.; methodology, E.K.; software, E.K.; validation, S.D.; writing—original draft preparation, S.D.; writing—review & editing, V.B. All authors have read and agreed to the published version of the manuscript.

Funding: The paper is prepared under partial support of Russian Science Foundation (RSF) (project No. 22-29-00466, 14 January 2022).

Institutional Review Board Statement: Not applicable.

Informed Consent Statement: Not applicable.

Data Availability Statement: Not applicable.

Conflicts of Interest: The authors declare no conflict of interest.

References

1. Smolentsev, N.K. *Fundamentals of Wavelet theory. Wavelets in MATLAB*; DMK Press: Moscow, Russia, 2008.
2. Zakharov, A.A.; Kozhanova, E.R.; Tkachenko, I.M. *Comparative Characteristics of Classical Fourier Transform and Continuous Wavelet Transform Applications: Tutorial*; Saratov State Technical University: Saratov, Russia, 2012.
3. Lazorenko, O.V.; Lazorenko, S.V.; Chernogor, L.F. Wavelet analysis of model signals with features. 1. Continuous wavelet transform. *Radiophys. Radio Astron.*, **2007**, *12*, 182–204.
4. Brkić, S. Optical properties of quantum dots. *Eur. Int. J. Sci. Technol.* **2016**, *5*, 98–107.
5. Belyaev, V.V.; Kozhanova, E.R.; Tkachenko, I.M.; Marusin, A.V. Mathematical modeling of optical amplifier processes with modification of the active area. In Proceedings of the 2022 International Conference Laser Optics (ICLO), St.Petersburg, Russia, 20–24 June 2022; p. 1. [CrossRef]

6. Tkachenko, I.M.; Kozhanova, E.R.; Belyaev, V.V.; Yazbeck, H. Comparative analysis of application of wavelet analysis for the recognition of element composition nanostructures. *J. Phys. Conf. Ser.* **2019**, *1309*, 012020. [CrossRef]
7. Belyaev V.; Yazbek H. Application of quantum dot technology for electro-optical and optoelectronic devices. *Electron. Sci. Technol. Bus.* **2020**, *9*, 131–139.

engineering proceedings

MDPI

Proceeding Paper

Anisotropy-Based Estimation for Sensor Network with Non-Centered Disturbance †

Alexander V. Yurchenkov * and Arkadiy Yu. Kustov

V.A. Trapeznikov Institute of Control Sciences of Russian Academy of Sciences, 65 Profsoyuznaya Street, 117997 Moscow, Russia; arkadiykustov@yandex.ru

* Correspondence: alexander.yurchenkov@gmail.com; Tel.: +7-495-198-17-20; Fax: +7-495-334-93-40

† Presented at the 15th International Conference "Intelligent Systems" (INTELS'22), Moscow, Russia, 14–16 December 2022.

Abstract: This paper concerns the anisotropy-based estimation design for sensor networks with coloured external disturbance. The boundedness criterion of anisotropic norm for estimation problems in network systems relies on the analysis of multiplicative noise systems in the framework of anisotropy-based theory. The solution of the considered problem is reduced to a convex optimization problem.

Keywords: anisotropy-based theory; multiplicative noise system; estimation; convex optimization

Citation: Yurchenkov, A.V.; Kustov, A.Y. Anisotropy-Based Estimation for Sensor Network with Non-Centered Disturbance. *Eng. Proc.* **2023**, *33*, 39. https://doi.org/10.3390/engproc2023033039

Academic Editors: Askhat Diveev, Ivan Zelinka, Arutun Avetisyan and Alexander Ilin

Published: 20 June 2023

1. Introduction

The estimation problem has remained a fundamental one in control and filtering theory since the 1960s, when R. Kalman proposed an optimal filtering approach based on prediction and correction concepts [1]. The other well known $\mathcal{H}_2$ and $\mathcal{H}_\infty$ methods use different assumptions on both system and input properties in order to design controllers and/or filters in the presence of measurement noise, exogenous disturbances, system uncertainties, etc. [2,3]. Each of these theories has its own benefits as well as disadvantages. For this reason, many attempts to unify and generalize $\mathcal{H}_2$ and $\mathcal{H}_\infty$ theories have been made over the last 30 or more years [4–6]. Initially solved in Riccati equation terms, the $\mathcal{H}_2$, $\mathcal{H}_\infty$ and mixed $\mathcal{H}_2/\mathcal{H}_\infty$ problems were then solved via linear matrix inequalities (LMI) techniques [7–9]. This brings the study into the new era of approaches and methods applicable to practice.

In 1994, I. Vladimirov suggested an approach to generalize (in a stochastic and robust sense) $\mathcal{H}_2$ and $\mathcal{H}_\infty$ filtering and control theories. This approach takes into account stochastic disturbance uncertainty described by means of an information criterion called anisotropy [10,11]. Introducing a certain performance gain, this theory succeeded in generalizing the mentioned $\mathcal{H}_2$ and $\mathcal{H}_\infty$ theories. The problem of analysis for time-varying systems was studied in [12], and the filtering problem for this type of system was considered in [13], where an optimal solution of the filtering problem was obtained in terms of the Riccati equation solution. Widely used in control theory, the LMI approach was adapted in anisotropy-based theory later in [14]. In [15], a general case of anisotropic norm boundedness sufficient conditions for the estimation problem were derived using forward-time difference LMIs. An anisotropy-based analysis for non-zero disturbances with constraint on the first and second moments of input disturbance was considered in [16], and a design was presented in [17].

Despite the fact that many problems have successfully been solved in anisotropy-based theory, only deterministic linear systems have been the subject of study. The corresponding analysis for stochastic systems was studied in [18]. It helped to take into account multiplicative noise systems. This kind of system describes financial and population models, mechanical and hybrid systems, sensor networks, etc. The boundedness criterion for the anisotropic norm of a multiplicative noise system is suggested in [19]. After that, the anisotropy-based analysis yields a solution of the estimation problem for time-varying sensor networks. The case of network systems was studied in [20], and the case of dropouts with corrections was considered in [21]. Both papers contained a convex optimization approach to calculate the minimal value of xanisotropic norm upper bound for an input-to-estimation error system. A multiplicative noise system analysis for non-centered disturbance was described in [22]. A natural continuation of sensor network research is to extend anisotropy-based estimation for sensor networks with non-centered disturbance, as far as possible.

In this paper, the problem of anisotropy-based estimation for a sensor network system with non-centered disturbances is considered. The external disturbances are assumed to be sequences of random vectors with bounded anisotropy and additional constrain on two stochastic moments are given.

2. Background

The basic concepts of anisotropy-based theory are the anisotropy of a random vector and the mean anisotropy of a random vector sequence. If a time-varying system with finite time horizon is considered, then the anisotropy of an extended vector of input is used in the analysis. Likewise, the mean anisotropy of a random sequence is used when time-invariant systems are under research.

The anisotropy of a real-valued random vector W with a finite second moment is defined in [13] as $\mathbf{A}(W) = \inf_{\lambda>0} \mathbf{D}(f||p_{m,\lambda}) = \frac{m}{2}\ln\left(\frac{2\pi e}{m}\mathbf{E}|W|^2\right) - h(W)$, where $\mathbf{D}$ is the Kullback–Leibler information divergence, $\mathbf{E}|\cdot|$ is the expectation, f denotes the probability density function (pdf) of W w.r.t. Lebesque measure in $\mathbb{R}^m$, $h(W) = -\int_{\mathbb{R}^m} f(w)\ln f(w)dw$ is the differential entropy of W, and λ is considered to bea positive parameter which defines the following pdf of etalon vectors: $p_{m,\lambda}(x) = (2\pi\lambda)^{-m/2}\exp\left(-\frac{|x|^2}{2\lambda}\right)$. The function $p_{m,\lambda}(x)$ corresponds to isotropic Gaussian distribution with zero mean and scalar covariance matrix λI_m, where I_m denotes an m-dimensional identity matrix. The anisotropy of a random vector has a simple meaning: it shows the measure of the difference between the pdf of a certain vector and a set of vectors that have isotropic Gaussian pdfs parameterized by λ. It can be easily calculated that, if random vector has zero mean Gaussian distribution with covariance matrix Σ, the precise formula for its anisotropy (as given in [12]) is of the following form:

$$\mathbf{A}(W) = -\frac{1}{2}\ln\det\left(\frac{m\Sigma}{\mathrm{tr}(\Sigma)}\right).$$

Anisotropy-based analysis for time-varying systems with non-centered disturbance was studied in [16,23]. As derived in [24], the anisotropy of an m-dimensional random vector with expectation μ and covariance matrix Σ is equal to the following formula:

$$\mathbf{A}(W) = -\frac{1}{2}\ln\det\left(\frac{m\Sigma}{\mathrm{tr}(\Sigma) + |\mu|^2}\right).$$

The performance criterion in anisotropy-based theory is caused by using the information functional for input disturbance. The anisotropic norm of the linear discrete time-varying system is induced by another norm. Let us consider the matrix $F \in \mathbb{L}^{p\times m}$ of a linear operator mapping for input vectors W into output vectors Z: for random input $W \in \mathbb{L}_2^m$, the output $Z \in \mathbb{L}_2^p$ is defined as $Z = FW$. In this paper, the components of vector W and matrix F are considered to be mutually independent. The root mean square (RMS) gain of the matrix F is defined as $Q(F,W) = \sqrt{\frac{\mathbf{E}|FW|^2}{\mathbf{E}|W|^2}} = \sqrt{\frac{\mathrm{tr}(\Lambda\Sigma)}{\mathrm{tr}(\Sigma)}}$, where $\Lambda = \mathbf{E}|F^{\mathrm{T}}F|$, $\Sigma = \mathbf{E}|WW^{\mathrm{T}}|$. The supremum of RMS gain coincides with the maximum singular value of the matrix Λ, and it will further be referred to as the $\mathcal{H}_\infty$ norm of F:

$\sup_{W \in \mathbb{L}_2^m} Q(F, W) = \sqrt{\max_{i=\overline{1,m}} \lambda_i(\mathbf{E}[F^{\mathrm{T}}F])} = \|F\|_\infty$. If the input vector W has zero mean Gaussian pdf with scalar covariance matrix, the RMS gain is equal to the stochastic analogue of the Frobenius norm of matrix F: $Q(F, W) = \frac{\|F\|_2}{\sqrt{m}} = \sqrt{\frac{\mathrm{tr}\Lambda}{m}}$.

To denote the set of random vectors with anisotropy by a, we use the following notation: $\mathbf{W}_a = \{W \in \mathbb{L}_2^m : \quad \mathbf{A}(W) \leqslant a\}$. With that, the anisotropic norm of F is defined by the following formula:

$$|||F|||_a = \sup_{W \in \mathbf{W}_a} Q(F, W). \tag{1}$$

Now, let us briefly consider the case of non-centered disturbances, see [16,23] for more details. If a random vector W is a sum of the zero-mean vector $\widetilde{W}$ and the constant vector $\mu = \mathbf{E}[W]$, the RMS gain of the system (operator) with such an input is expressed by the following equation: $Q(F, W) = \sqrt{\frac{\mathbf{E}[|F\widetilde{W}|^2] + \mathbf{E}[|F\mu|^2]}{\mathbf{E}[|\widetilde{W}|^2] + |\mu|^2}}$. Consequently, the anisotropic norm of F in this case is defined as follows:

$$|||F|||_a = \sup_{W \in \mathbf{W}_a} \sqrt{\frac{\mathrm{tr}(\Lambda\Sigma) + \mu^{\mathrm{T}}\Lambda\mu}{\mathrm{tr}\Sigma + \mu^{\mathrm{T}}\mu}}. \tag{2}$$

Since both anisotropy and anisotropic norm are invariant under the scaling of input W, one can consider only the case when the two following constraints are given: $|\mu| \geqslant \tau \in [0, 1)$ and $\mathrm{tr}\Sigma \leqslant \sigma$. In [16,23], it was shown that the considered constrains for the first and second moments can be equivalently replaced by those that satisfy $\sigma + \tau^2 = 1$. This condition allows (2) to be modified in the following form:

$$|||F|||_{a,\tau} = \sup_{\Sigma = \Sigma^{\mathrm{T}} \succ 0} \left\{ (\mathrm{tr}(\Lambda\Sigma) + \sup_{e_0 \neq 0} \mathbf{E}[|Fe_0|^2]\tau^2)^{1/2} - \frac{1}{2}\ln\det(m\Sigma) \leqslant a,\ \mathrm{tr}(\Sigma) = 1 - \tau^2 \right\}, \tag{3}$$

where e_0 corresponds to unit vector $e_0 = \mu/|\mu|$. In (3), the second term inside the supremum over Σ takes the maximum value when unit vector e_0 corresponds to the eigenvector of the maximum eigenvalue of Λ, hence giving $\sup_{e_0 \neq 0} \mathbf{E}[|Fe_0|^2] = \|F\|_\infty^2$. The first term corresponds to the problem of local maximum finding of functional $\mathrm{tr}(\Lambda\Sigma)$ subject to one equality constraint $\mathrm{tr}\Sigma = 1 - \tau^2$ and one inequality constraint $-\frac{1}{2}\ln\det(m\Sigma) \leqslant a$. Note that $\tau \in [0; 1)$ is assumed to be given. This problem can be solved by means of the Lagrange method, and all details can be found in [16,23]. Based on [22,23], the problem of anisotropy-based estimation is studied in this paper for multiplicative noise systems.

3. Problem Statement

In this section, a model of a sensor network with random failures is considered. The description of the communication scheme for the sensors is associated with an adjacency matrix of the corresponding communication graph.

Let us consider the following linear discrete time-varying (LDTV) system:

$$\begin{aligned} x_{k+1} &= A_k x_k + B_k w_k, \\ z_k &= M_k x_k + N_k w_k, \\ y_{j,k} &= \lambda_{j,k} C_{j,k} x_k + D_{j,k} w_k, \end{aligned} \tag{4}$$

with zero initial condition $x_0 = 0$ and finite-time horizon N_h, i.e., $k = 0, 1, \ldots, N_h$. The state is denoted by $x_k \in \mathbb{R}^{n_x}$, the input disturbance $w_k \in \mathbb{R}^{m_w}$ belongs to sequence of random vectors with bounded anisotropy level a of extended input vector $W_{0:N_h} = (w_0^{\mathrm{T}}, \ldots, w_n^{\mathrm{T}})^{\mathrm{T}}$, and the non-zero expectation $|\mathbf{E}W_{0:N_h}| > \tau$ and the constrained covariance $\mathrm{tr}(\mathrm{cov}(W_{0:N_h})) < 1 - \tau^2$ are given. For each jth sensor, a measurement $y_{j,k} \in \mathbb{R}^{p_y}$ is provided. The random variable $\lambda_{j,k}$ has Bernoulli distribution with given probabilities $P(\lambda_{j,k} = 1) = p_j$ and

$P(\lambda_{j,k} = 0) = 1 - p_j = q_j$. The output to be estimated is denoted by $z_k \in \mathbb{R}^{p_z}$. All matrices $A_k, B_k, M_k, N_k, C_{j,k}, D_{j,k}$, $j = \overline{1,n}$ are known for all time instants and have appropriate dimensions. The communication scheme is defined by means of an adjacency matrix $\boldsymbol{a}$ with conditions $\boldsymbol{a}_{ji} \geqslant 0$, $\sum_{i=1}^{n} \boldsymbol{a}_{ji} = 1$, $\boldsymbol{a}_{jj} = \max_i \boldsymbol{a}_{ji}$; see [21] for more details.

To implement the results from anisotropy-based analysis to LDTV systems, we will identify the norm of such a system F with the norm of the corresponding transfer matrix $F_{0:N_h}$ associated with the mapping $W_{0:N_h} \mapsto F_{0:N_h} W_{0:N_h} = Z_{0:N_h}$ between fragments $W_{0:N_h}$ and $Z_{0:N_h}$ of the input and output sequences, respectively, where for the sake of example $W_{s:t} = (w_s^{\mathrm{T}}, w_{s+1}^{\mathrm{T}}, \ldots, w_t^{\mathrm{T}})^{\mathrm{T}}$ for any $s \leqslant t$. So, the anisotropic norm $|||F|||_a$ of the LDTV system F operating over a finite-time horizon $k = 0, 1, \ldots, N_h$ should be understood as an anisotropic norm $|||F_{0:N_h}|||_a$ of its transfer matrix $F_{0:N_h}$.

The problem studied in this paper is finding the linear estimation $\widehat{z}_k$ of output z_k for system (4) using measurements $y_{j,s}$, $s \leqslant k$, such that anisotropic norm of input-to-estimation error system is bounded by a chosen threshold $\gamma > 0$.

4. Main Result

To obtain sufficient conditions of anisotropic norm boundedness for the systems with multiplicative noises and noncentered disturbances, let us firstly discuss some results of [16,21–23].

In [16,23], the anisotropy-based analysis problem was studied for non-centered disturbance with constraints on the first and second moments of the input. The central result is provided by the following statement based on the idea that for computation of the (a, τ)-anisotropic norm for the case of noncentered disturbance one needs to shift the value of anisotropy and then calculate the anisotropic norm of the corresponding system with centered disturbance, making additional corrections in the resulting value.

Lemma 1 ([16]). *Let us consider a linear discrete time-varying system F of the form* (4) *but without measurements $y_{j,k}$, $j = \overline{1,n}$, and assume the parameters $\tau \in [0;1)$, $a \geqslant -\frac{l}{2}\ln(1-\tau^2)$, $l = m_w(N_h+1)$ to be given. The (a,τ)-anisotropic norm* (3) *of the system F is bounded by the given $\gamma > 0$ if there exist $\gamma_1 > 0$, $\gamma_2 > 0$, such that $|||F|||_b \leqslant \gamma_1$, $\|F\|_\infty \leqslant \gamma_2$, $\gamma_1^2(1-\tau^2) + \gamma_2^2\tau^2 \leqslant \gamma^2$, where $b = a + \frac{l}{2}\ln(1-\tau^2)$.*

Insofar as the input-to-estimation error system for the filtering problem is reduced to a multiplicative noise system form, the conditions of anisotropic norm boundedness are associated with an anisotropy-based analysis given in [22]. Let us consider the multiplicative noise system $\widetilde{F}$ of the form

$$
\begin{aligned}
x_{k+1} &= (A_{0,k} + \sum_{i=1}^{r} \xi_{i,k} A_{i,k}) x_k + B_k w_k, \\
z_k &= M_k x_k + N_k w_k,
\end{aligned}
\tag{5}
$$

with zero initial condition $x_0 = 0$. The matrices $A_{i,k}$, $i = \overline{0,r}$, are given, and the dimensions of matrices are the same as (4). Random variables $\xi_{i,k}$, $i = \overline{1,r}$ have the same distribution with zero mean, and covariances σ_i^2, and considered to be independent of each other as well of the input w_k.

Theorem 1. *Let us consider an LDTV system $\widetilde{F}$ with multiplicative noises* (5)*, and $\tau \in [0;1)$ and $a \geqslant -\frac{l}{2}\ln(1-\tau^2)$, $\gamma > 0$ are given. The (a,τ)-anisotropic norm of $\widetilde{F}$ is bounded by a given threshold γ if there exist some $\gamma_1 > 0$, $\gamma_2 > 0$ and the parameter $q \in [0; ||\widetilde{F}||_\infty^{-2})$, such that the following inequalities hold true:*

$$\begin{aligned} &R_k \succ A_{0,k}^{\mathrm{T}} R_{k+1} A_{0,k} + \sum_{i=1}^{r} \sigma_i^2 A_{i,k}^{\mathrm{T}} R_{k+1} A_{i,k} + q M_k^{\mathrm{T}} M_k + L_k^{\mathrm{T}} S_k^{-1} L_k, \\ &S_k = (I_{m_w} - B_k^{\mathrm{T}} R_{k+1} B_k - q N_k^{\mathrm{T}} N_k)^{-1}, \\ &L_k = S_k (B_k^{\mathrm{T}} R_{k+1} A_{0,k} + q N_k^{\mathrm{T}} M_k). \end{aligned} \tag{6}$$

$$\begin{aligned} &\widetilde{R}_k \succ A_{0,k}^{\mathrm{T}} \widetilde{R}_{k+1} A_{0,k} + \sum_{i=1}^{r} \sigma_i^2 A_{i,k}^{\mathrm{T}} \widetilde{R}_{k+1} A_{i,k} + q M_k^{\mathrm{T}} M_k + \widetilde{L}_k^{\mathrm{T}} \widetilde{S}_k^{-1} \widetilde{L}_k, \\ &\widetilde{S}_k = (I_{m_w} - B_k^{\mathrm{T}} \widetilde{R}_{k+1} B_k - q N_k^{\mathrm{T}} N_k)^{-1}, \\ &\widetilde{L}_k = \widetilde{S}_k (B_k^{\mathrm{T}} \widetilde{R}_{k+1} A_{0,k} + q N_k^{\mathrm{T}} M_k), \end{aligned} \tag{7}$$

$$\sum_{k=0}^{N} \ln \det S_k^{-1} \geqslant m \ln(1-\tau^2) + 2b, \tag{8}$$

and $\gamma_1^2(1-\tau^2) + \gamma_2^2\tau^2 \leqslant \gamma^2$. *Here,* $b = a + \frac{l}{2}\ln(1-\tau^2)$, $R_{N_h+1} = 0$, $\widetilde{R}_{N+1} = 0$, *and matrices* $R_k, \widetilde{R}_k, S_k$ *are all positively defined for* $k = 0, \dots, N_h$.

Proof. This theorem immediately follows from the results of [16,22,23]. □

Theorem 2. *The anisotropic norm of the system* $\widetilde{F}$ (5) *with non-centered disturbance* w_k *with parameter* $\tau \in [0;1)$ *and a bounded anisotropy level of extended vector* $\mathbf{A}(W_{0:N_h}) \leqslant a$ *is bounded by a given* γ, *i.e.,*

$$|||\widetilde{F}|||_a \leqslant \gamma,$$

if the following inequalities

$$\begin{aligned} &\begin{bmatrix} R_k - M_k^{\mathrm{T}} M_k & * & * & * & \dots & * \\ N_k^{\mathrm{T}} M_k & \eta^2 I_{m_w} - N_k^{\mathrm{T}} N_k & * & * & \dots & * \\ R_{k+1} A_{0,k} & -R_{k+1} B_k & R_{k+1} & * & \dots & * \\ \sigma_1 R_{k+1} A_{1,k} & 0 & 0 & R_{k+1} & \dots & * \\ \vdots & \vdots & \vdots & \vdots & \ddots & \vdots \\ \sigma_n R_{k+1} A_{n,k} & 0 & 0 & 0 & \dots & R_{k+1} \end{bmatrix} \succ 0, \\ &\begin{bmatrix} \widetilde{R}_k - M_k^{\mathrm{T}} M_k & * & * & * & \dots & * \\ N_k^{\mathrm{T}} M_k & \gamma_2^2 I_{m_w} - N_k^{\mathrm{T}} N_k & * & * & \dots & * \\ \widetilde{R}_{k+1} A_{0,k} & -\widetilde{R}_{k+1} B_k & \widetilde{R}_{k+1} & * & \dots & * \\ \sigma_1 \widetilde{R}_{k+1} A_{1,k} & 0 & 0 & \widetilde{R}_{k+1} & \dots & * \\ \vdots & \vdots & \vdots & \vdots & \ddots & \vdots \\ \sigma_n \widetilde{R}_{k+1} A_{n,k} & 0 & 0 & 0 & \dots & \widetilde{R}_{k+1} \end{bmatrix} \succ 0, \end{aligned} \tag{9}$$

$$\begin{bmatrix} \eta^2 I_{m_w} - \Psi_k - N_k^{\mathrm{T}} N_k & * \\ R_{k+1} B_k & R_{k+1} \end{bmatrix} \succ 0. \tag{10}$$

have positive definite solutions $R_k, \widetilde{R}_k, \Psi_k$, $k = \overline{0, N_h - 1}$, *with boundary constraints*

$$\begin{aligned} &\begin{bmatrix} R_{N_h} - M_{N_h}^{\mathrm{T}} M_{N_h} & * \\ N_{N_h}^{\mathrm{T}} M_{N_h} & \eta^2 I_{m_w} - N_{N_h}^{\mathrm{T}} N_{N_h} \end{bmatrix} \succ 0, \\ &\begin{bmatrix} \widetilde{R}_{N_h} - M_{N_h}^{\mathrm{T}} M_{N_h} & * \\ N_{N_h}^{\mathrm{T}} M_{N_h} & \gamma_2^2 I_{m_w} - N_{N_h}^{\mathrm{T}} N_{N_h} \end{bmatrix} \succ 0, \end{aligned} \tag{11}$$

$$\eta^2 I_{m_w} - \Psi_{N_h} - N_{N_h}^{\mathrm{T}} N_{N_h} \succ 0, \tag{12}$$

and special type inequalities

$$\sum_{k=0}^{N_h} \ln\det\Psi_k \geqslant 2b + l\ln(\eta^2 - \gamma_1^2), \tag{13}$$

$$\gamma_1^2(1-\tau^2) + \gamma_2^2\tau^2 \leqslant \gamma^2 \tag{14}$$

hold true, $b = a + \frac{l}{2}\ln(1-\tau^2)$.

Proof. Inequalities (9)–(12) are derived by using Schur's complement formula. Inequality (13) follows from substituting the anisotropic norm computation for the non-centered input problem for the corresponding one with a centered input and a modified anisotropy level of the extended input vector, while (14) is related to the boundedness criteria of $|||\widetilde{F}|||_{a,\tau}$, $|||\widetilde{F}|||_b$, $\|\widetilde{F}\|_\infty$ for a given system $\widetilde{F}$. □

For implementing the results of Theorem 2, some transformation of system (4) is necessary. Let us introduce a set of virtual objects as follows:

$$\begin{aligned} x_{j,k+1} &= A_k x_{j,k} + B_k w_k, \quad x_{j,0} = 0, \\ z_{j,k} &= M_k x_{j,k} + N_k w_k, \\ y_{j,k} &= \lambda_{j,k} C_{j,k} x_{j,k} + D_{j,k} w_k, \end{aligned} \tag{15}$$

where $x_{j,k} \in \mathbb{R}^{n_x}$ denotes the virtual state, $z_{j,k} \in \mathbb{R}^{p_z}$ is the output to be estimated, $y_{j,k} \in \mathbb{R}^{p_y}$ denotes measurement of the jth sensor. In this representation, each sensor has its own virtual dynamics, and the original sensor network is virtually separated to independent objects.

Let us choose an estimation model for every virtual system (15) in the following form:

$$\begin{aligned} \widehat{x}_{j,k+1} &= \sum_{i=1}^{n} \boldsymbol{a}_{ji}(A_k\widehat{x}_{i,k} + H_{ji,k}(y_{i,k} - \widehat{y}_{i,k})), \quad \widehat{x}_{j,0} = 0, \\ \widehat{z}_{j,k} &= \sum_{i=1}^{n} \boldsymbol{a}_{ji} M_k \widehat{x}_{i,k}, \end{aligned} \tag{16}$$

where $\widehat{y}_{i,k} = p_i C_{i,k}\widehat{x}_{i,k}$. Matrices H_{ji} are considered to be found. The input-to-estimation error system, which is using estimation model (16), has to have a bounded anisotropic norm.

The state estimation error and output estimation error are depicted by the following notations: $\widetilde{x}_{j,k} = x_{j,k} - \widehat{x}_{j,k}$, $\widetilde{z}_{j,k} = z_{j,k} - \widehat{z}_{j,k}$, $j = \overline{1,n}$. The dynamics of each virtual object error are as follows:

$$\begin{aligned} \widetilde{x}_{j,k+1} &= \sum_{i=1}^{n} \boldsymbol{a}_{ji}\left(\mathcal{A}_{ji,k}x_{i,k} + \widetilde{\mathcal{A}}_{ji,k}\widetilde{x}_{i,k} + \mathcal{B}_{ji,k}w_k\right), \\ \widetilde{z}_{j,k} &= M_k x_{j,k} - \sum_{i=1}^{n} \boldsymbol{a}_{ji} M_k \widetilde{x}_{i,k} + N_k w_k, \end{aligned} \tag{17}$$

where $\mathcal{A}_{ji,k} = A_k\delta_{ji} - \boldsymbol{a}_{ji}\left(A_k + H_{ji,k}\left(\lambda_{i,k}C_{i,k} - p_i C^p_{i,k}\right)\right)$, $\widetilde{\mathcal{A}}_{ji,k} = \boldsymbol{a}_{ji}\left(A_k - p_i H_{ji,k}C_{i,k}\right)$, $\mathcal{B}_{ji,k} = B_k\delta_{ji} - \boldsymbol{a}_{ji}H_{ji,k}D_{i,k}$. Hereafter, the Kronecker symbol δ_{ji} is used. Let us assemble the total vectors, $x_{j,k}$, $j = \overline{1,n}$, to the extended state vector $\overline{x}_k^{\mathrm{T}} = \left(x_{1,k}^{\mathrm{T}}, x_{2,k}^{\mathrm{T}}, \ldots, x_{n,k}^{\mathrm{T}}\right)^{\mathrm{T}}$, and similarly the extended vectors of state error and output error are denoted as follows: $\widetilde{x}_k^{\mathrm{T}} = \left(\widetilde{x}_{1,k}^{\mathrm{T}}, \widetilde{x}_{2,k}^{\mathrm{T}}, \ldots, \widetilde{x}_{n,k}^{\mathrm{T}}\right)^{\mathrm{T}}$, $\widetilde{z}_k^{\mathrm{T}} = \left(\widetilde{z}_{1,k}^{\mathrm{T}}, \widetilde{z}_{2,k}^{\mathrm{T}}, \ldots, \widetilde{z}_{n,k}^{\mathrm{T}}\right)^{\mathrm{T}}$. Hence, the dynamics of the extended state error and the output error are described by the following system:

$$\begin{aligned} \widetilde{x}_{k+1} &= \left(\overline{A}_k - \overline{W}_k - \overline{H}_k\left(\overline{C}_k^{\lambda} - \overline{C}_k^{p}\right)\right)\overline{x}_k + \left(\overline{W}_k - \overline{H}_k\overline{C}_k^{p}\right)\widetilde{x}_k + (\overline{B}_k - \overline{H}_k\overline{D}_k)w_k, \\ \widetilde{z}_k &= (\overline{M}_k - \overline{V}_k)\overline{x}_k + \overline{V}_k\widetilde{x}_k + \overline{N}_k w_k, \end{aligned}$$

where matrices are as follows:

$$
\begin{gathered}
\overline{A}_k = I_n \otimes A_k,\ \overline{B}_k = [1,\dots,1]^{\mathrm{T}} \otimes B_k,\ \overline{M}_k = I_n \otimes M_k,\ \overline{N}_k = [1,\dots,1]^{\mathrm{T}} \otimes N_k \\
\overline{C}_k^{\lambda} = \underset{j=\overline{1,n}}{\operatorname{diag}}\left(\lambda_{j,k} C_{j,k}\right),\ \overline{C}_k^{p} = \underset{j=\overline{1,n}}{\operatorname{diag}}\left(p_j C_{j,k}^{p}\right),\ \overline{W}_k = \boldsymbol{a} \otimes A_k, \\
\overline{H}_k = \underset{j,i=\overline{1,n}}{\operatorname{block}}\left(\boldsymbol{a}_{ji} H_{ji,k}\right),\ \overline{V}_k = \boldsymbol{a} \otimes M_k,\ \overline{D}_k = \left[D_{1,k}^{\mathrm{T}},\dots,D_{n,k}^{\mathrm{T}}\right]^{\mathrm{T}}.
\end{gathered}
$$

Notation $\otimes$ is assigned to the Kronecker product, and the block diagonal and block matrix are denoted by $\operatorname{diag}(\cdot)$ and $\operatorname{block}(\cdot)$, respectively.

To derive the dynamics of the input-to-estimation error system, let us define the extended state vector as $\zeta_k = [\overline{x}_k^{\mathrm{T}}, \widetilde{x}_k^{\mathrm{T}}]^{\mathrm{T}}$. Then, the dynamics of the input-to-estimation error system can be described by the following expression:

$$
\begin{aligned}
\zeta_{k+1} &= (\mathcal{A}_{0,k} + \sum_{i=1}^{n} \xi_{j,k} \mathcal{A}_{j,k})\zeta_k + \mathcal{B}_k w_k, \\
\widetilde{z}_k &= \mathcal{M}_k \zeta_k + \mathcal{N}_k w_k,
\end{aligned} \tag{18}
$$

where $\xi_{j,k} = \lambda_{j,k} - p_j$ for $j = \overline{1,n}$, $\mathcal{M}_k = [\overline{M}_k - \overline{V}_k \quad \overline{V}_k]$, $\mathcal{N}_k = \overline{N}_k$, $G_{j,k} = \underset{i=\overline{1,n}}{\operatorname{diag}} \delta_{ji} C_{j,k}$, and

$$
\mathcal{A}_{0,k} = \begin{bmatrix} \overline{A}_k & 0 \\ \overline{A}_k - \overline{W}_k & \overline{W}_k - \overline{H}_k \overline{C}_k^{p} \end{bmatrix},\ \mathcal{A}_{j,k} = \begin{bmatrix} 0 & 0 \\ -\overline{H}_k G_{j,k} & 0 \end{bmatrix},\ \mathcal{B}_k = \begin{bmatrix} \overline{B}_k \\ \overline{B}_k - \overline{H}_k \overline{D}_k \end{bmatrix}.
$$

As one can see, (18) presents the multiplicative noise system dynamics, so the statement of Theorem 2 can be applied to this object. However, inequalities (9) and (10) contain non-linear terms, therefore corresponding variables change should be used to provide an effective computational algorithm. Let us define the following set of matrices: $U_k = \begin{bmatrix} Y_k & 0 \\ 0 & \overline{H}_k \end{bmatrix}$, $X_k = R_{k+1} U_k$, $\widetilde{X}_k = \widetilde{R}_{k+1} U_k$, where $k = 0,\dots,N_h$ and Y_k are considered to be real-valued matrices of corresponding dimensions. So, after the variables change, inequalities (9) and (10) become of the form

$$
\begin{gathered}
\begin{bmatrix}
R_k - \mathcal{M}_k^{\mathrm{T}} \mathcal{M}_k & * & * & \dots & * \\
\mathcal{N}_k^{\mathrm{T}} \mathcal{M}_k & \eta^2 I_{m_w} - \mathcal{N}_k^{\mathrm{T}} \mathcal{N}_k & * & \dots & * \\
R_{k+1} \mathcal{A}_{00,k} + X_k \mathcal{A}_{01,k} & -R_{k+1} \mathcal{B}_{00,k} - X_k \mathcal{B}_{01,k} & R_{k+1} & \dots & * \\
\sigma_1 X_k \mathcal{A}_{11,k} & 0 & 0 & \dots & * \\
\vdots & \vdots & \vdots & \ddots & \vdots \\
\sigma_n X_k \mathcal{A}_{n1,k} & 0 & 0 & \dots & R_{k+1}
\end{bmatrix} \succ 0, \\
\begin{bmatrix}
\widetilde{R}_k - \mathcal{M}_k^{\mathrm{T}} \mathcal{M}_k & * & * & \dots & * \\
\mathcal{N}_k^{\mathrm{T}} \mathcal{M}_k & \gamma_2^2 I_{m_w} - \mathcal{N}_k^{\mathrm{T}} \mathcal{N}_k & * & \dots & * \\
\widetilde{R}_{k+1} \mathcal{A}_{00,k} + \widetilde{X}_k \mathcal{A}_{01,k} & -\widetilde{R}_{k+1} \mathcal{B}_{00,k} - \widetilde{X}_k \mathcal{B}_{01,k} & \widetilde{R}_{k+1} & \dots & * \\
\sigma_1 \widetilde{X}_k \mathcal{A}_{11,k} & 0 & 0 & \dots & * \\
\vdots & \vdots & \vdots & \ddots & \vdots \\
\sigma_n \widetilde{X}_k \mathcal{A}_{n1,k} & 0 & 0 & \dots & \widetilde{R}_{k+1}
\end{bmatrix} \succ 0,
\end{gathered} \tag{19}
$$

$$
\begin{bmatrix}
\eta^2 I_{m_w} - \Psi_k - \mathcal{N}_k^{\mathrm{T}} \mathcal{N}_k & * \\
R_{k+1} \mathcal{B}_{00,k} + X_k \mathcal{B}_{01,k} & R_{k+1}
\end{bmatrix} \succ 0. \tag{20}
$$

Boundary conditions (11) and (12) do not change.

If the obtained system of inequalities (11)–(14), (19), and (20) has a solution, then the estimator (16) designed for system (4) satisfies the condition of suboptimality in the sense of the anisotropic norm. The optimization problem can be stated as follows: $\gamma^2 \longrightarrow \min$ over (11)–(14), (19) and (20) w.r.t. $R_k, \widetilde{R}_k, X_k, \widetilde{X}_k, \eta^2, \Psi_k, \gamma_1^2, \gamma_2^2, \tau$.

5. Conclusions

In this paper, an estimation problem of a sensor network system is solved using an anisotropy-based approach. External disturbances belong to sequences of random vectors with bounded anisotropy levels of the extended input vector. By using virtual objects for every sensor, the augmented system is introduced and the input-to-error system is derived. The system is described by multiplicative noise state space description, since anisotropy-based analysis can be applied. An anisotropic norm boundedness criterion for the input-to-error system is implemented, and the non-linear system to be optimized is obtained. Appropriate variables change is used, and the optimization problem becomes linear. The parameters of the time-varying anisotropy-based estimator are calculated after proper inverse change when a suboptimal problem is numerically solved. The non-zero mean of disturbance determines the dimension rising of the convex optimization problem, since the modification of a previously developed anisotropy-based algorithm was achieved.

Author Contributions: Conceptualization, A.V.Y.; formal analysis A.Y.K., methodology, A.Y.K.; writing—original draft preparation, A.V.Y., writing—review and editing, A.Y.K. All authors have read and agreed to the published version of the manuscript.

Funding: This research received no external funding.

Institutional Review Board Statement: Not applicable.

Informed Consent Statement: Not applicable.

Data Availability Statement: Not applicable.

Conflicts of Interest: The authors declare no conflicts of interest.

References

1. Kalman, R. A New Approach to Linear Filtering and Prediction Problems. *J. Basic Eng.* **1960**, *82*, 35–45. [CrossRef]
2. Hassibi, B.; Sayed, A.; Kailath, T. *Indefinite Quadratic Estimation and Control: A Unified Approach to $\mathcal{H}_2$ and $\mathcal{H}_\infty$ Theories*; SIAM: Philadelphia, PA, USA, 1999.
3. Simon, D. *Optimal State Estimation: Kalman, $\mathcal{H}_\infty$, and Nonlinear Approaches*; John Wiley and Sons: Hoboken, NJ, USA, 2006.
4. Haddad, W.M.; Bernstein, D.S.; Mustafa, D. Mixed-norm $\mathcal{H}_2/\mathcal{H}_\infty$ regulation and estimation: The discrete-time case. *Syst. Control Lett.* **1991**, *16*, 235–247. [CrossRef]
5. Zhou, K.; Glover, K.; Bodenheimer, B.; Doyle, J. Mixed $\mathcal{H}_2$ and $\mathcal{H}_\infty$ performance objectives I: Robust performance analysis. *IEEE Trans. Autom. Control* **1994**, *38*, 1564–1574. [CrossRef]
6. Khargonekar, P.P.; Rotea, M.A.; Baeyens, E. Mixed $\mathcal{H}_2/\mathcal{H}_\infty$ filtering. *Int. J. Robust Nonlinear Control* **1996**, *6*, 313–330. [CrossRef]
7. Scherer, C.W. Mixed $\mathcal{H}_2/\mathcal{H}_\infty$ Control. In *Trends in Control*; Isidori, A., Ed.; Springer: Berlin/Heidelberg, Germany, 1995; pp. 173–216.
8. Balandin, D.V.; Kogan, M.M. LMI-based $\mathcal{H}_\infty$-optimal Control with Transients. *Int. J. Control* **2010**, *83*, 1664–1673. [CrossRef]
9. Palhares, R.M.; Peres, P.L.D. LMI approach to the mixed $\mathcal{H}_2/\mathcal{H}_\infty$ filtering design for discrete-time uncertain systems. *IEEE Trans. Aerosp. Electron. Syst.* **2001**, *37*, 292–296. [CrossRef]
10. Semyonov, A.V.; Vladimirov, I.G.; Kurdyukov, A.P. Stochastic Approach to $\mathcal{H}_\infty$-optimization. In Proceedings of the 33th Conference on Decision and Control, Lake Buena Vista, FL, USA, 12–14 December 1994; pp. 2249–2250.
11. Vladimirov, I.G.; Kurdyukov, A.P.; Semyonov, A.V. Anisotropy of Signals and Entropy of Linear Stationary Systems. *Dokl. Math.* **1995**, *51*, 388–390.
12. Vladimirov, I.G.; Diamond, P.; Kloeden, P. Anisotropy-based Robust Performance Analysis of Finite Horizon Linear Discrete Time Varying Systems. *Autom. Remote Control* **2006**, *67*, 1265–1282. [CrossRef]
13. Vladimirov, I.G.; Diamond, P.; Kloeden, P. *Anisotropy-Based Robust Performance Analysis of Linear Discrete Time Varying Systems*; CADSMAP Research Report 01-01; The University of Queensland: Brisbane, Australia, 2001.
14. Timin, V.N.; Tchaikovsky, M.M.; Kurdyukov, A.P. A Solution to Anisotropic Suboptimal Filtering Problem by Convex Optimization. *Dokl. Math.* **2012**, *85*, 443–445. [CrossRef]
15. Timin, V.N.; Kurdyukov, A.P. Suboptimal Anisotropic Filtering in a Finite Horizon. *Autom. Remote Control* **2016**, *77*, 1–20. [CrossRef]
16. Kustov, A.Y.; Timin, V.N. Suboptimal Anisotropy-based Control for Linear Discrete Time Varying Systems with Noncentered Disturbances. *IFAC-PapersOnline* **2017**, *50*, 6296–6301. [CrossRef]
17. Yurchenkov, A.V. On the Control Design for Linear Time-Invariant Systems with Moments Constraints of Disturbances in Anisotropy-based Theory. *IFAC-PapersOnLine* **2018**, *51*, 160–165. [CrossRef]

18. Kustov, A.Y. State-Space Formulas for Anisotropic Norm of Linear Discrete Time Varying Stochastic System. In Proceedings of the 15th International Conference on Electrical Engineering, Computing Science and Automatic Contro, Mexico City, Mexico, 5–7 September 2018; pp. 1–6.
19. Belov, I.R.; Yurchenkov, A.V.; Kustov, A.Y. Anisotropy-Based Bounded Real Lemma for Multiplicative Noise Systems: The Finite Horizon Case. In Proceedings of the 27th Mediterranean Conference on Control and Automation, Akko, Israel, 1–4 July 2019; pp. 148–152.
20. Kustov, A.; Yurchenkov, A. Finite-horizon Anisotropy-based Estimation with Packet Dropouts. *IFAC-PapersOnLine* **2020**, *53*, 4516–4520. [CrossRef]
21. Kustov, A.Y.; Yurchenkov, A.V. Finite-horizon Anisotropic Estimator Design in Sensor Networks. In Proceedings of the 59th IEEE Conference on Decision and Control, Jeju, Republic of Korea, 14–18 December 2020; pp. 4330–4335.
22. Yurchenkov, A.V. Lemma on Boundedness of Anisotropic Norm for Systems with Multiplicative Noises under a Noncentered Disturbance. *Autom. Remote Control* **2021**, *82*, 51–62. [CrossRef]
23. Kustov, A.Y.; Timin, V.N. Anisotropy-based Analysis for Finite Horizon Time-varying Systems with Non-centered Disturbances. *Autom. Remote Control* **2017**, *78*, 974–988. [CrossRef]
24. Kustov, A.Y. Anisotropy-based Analysis and Synthesis Problems for Input Disturbances with Non Zero Mean. In Proceedings of the 15th International Carpathian Control Conference, Velke Karlovice, Czech Republic, 28–30 May 2014; pp. 291–295.

engineering proceedings

MDPI

Proceeding Paper

Evaluation of Computer Technologies for Calculation of Exact Approximations of Statistics Probability Distributions †

Andrey Melnikov [1], Ilya Levin [2], Aleksey Dordopulo [2,*] and Lyubov Slasten [2]

1 "Computational Solutions" JSC, Varshavskoe Highway 125/17, Moscow 119991, Russia; niimvs@yandex.ru
2 Supercomputers and Neurocomputers Research Center, Italyansky Lane, 106, Taganrog 347900, Russia; levin@superevm.ru (I.L.); lmslasten@yandex.ru (L.S.)
* Correspondence: dordopulo@superevm.ru; Tel.: +7-8634-612111
† Presented at the 15th International Conference "Intelligent Systems" (INTELS'22), Moscow, Russia, 14–16 December 2022.

Abstract: This paper is devoted to the evaluation of the hardware resources of computer systems for solving a computationally expensive problem such as the calculation of the probability distributions of statistics by the second multiplicity method based on Δ-exact approximations of samples with a size of 320–1280 characters and an alphabet power of 128–256 characters. The accuracy is $\Delta = 10^{-5}$ and the total solution time should not exceed 30 days or 2.592×10^6 seconds for 24/7 computing. Owing to the use of the properties of the second multiplicity method, the computational complexity of the calculations can be brought within the range of 9.68×10^{22}–1.60×10^{52} operations with the number of tested vectors at 6.50×10^{23}–1.39×10^{50}. The solution of this problem for the specified parameters of samples during the given time requires hardware resources which cannot be provided by modern computer technology such as processors, graphics accelerators and programmable logic integrated circuits. Therefore, in this paper, we analyze the possibilities of promising quantum and photon technologies for solving the problem with the given parameters. The main advantage of quantum computer systems is the high speed of calculations for all possible parameter values. However, quantum acceleration will not be achieved to calculate the probability distributions of statistics due to the need to check all the obtained solutions. Here, the number of obtained solutions corresponds to the dimension of the problem. In addition, due to the current development level of quantum hardware components, it is impossible to create and use 120 qubit quantum computers for the solution of the considered problem. Photon computers can provide high computation speeds at low power consumptions and require the smallest number of nodes to solve the considered problem. However, unsolved problems with the physical implementation of efficient memory elements and the lack of available hardware components make the use of photon computer technologies impossible for calculating the probability distributions of statistics in the near future (5–7 years). Therefore, it is most reasonable to use hybrid computer systems containing nodes of different architectures. To solve the problem on various hardware platforms (general purpose processors, GPUs and FPGAs) and configurations of hybrid computer systems, we suggest to use the architecture-independent high-level programming language SET@L. The language combines the representation of calculations as sets and collections (based on the alternative set theory of P. Vopenka), the absolutely parallel form of the problem represented as an information graph and the paradigm of aspect-oriented programming.

Keywords: probability; statistic; exact distribution; exact approximation; algorithmic complexity; quantum calculations; photonic technologies; architecture-independent programming; Set@l language

Citation: Melnikov, A.; Levin, I.; Dordopulo, A.; Slasten, L. Evaluation of Computer Technologies for Calculation of Exact Approximations of Statistics Probability Distributions. *Eng. Proc.* **2023**, *33*, 40. https://doi.org/10.3390/engproc2023033040

Academic Editors: Askhat Diveev, Ivan Zelinka, Arutun Avetisyan and Alexander Ilin

Published: 21 June 2023

1. Introduction

In general, criteria that minimize the number of false decisions concerning the truth of tested hypotheses are used to solve application problems that require statistical processing of texts as meaningful character sequences. The criteria based on the exact distributions of reference statistics [1] have the greatest relative efficiency, but the calculation of exact distributions is a computationally laborious task [1,2], depending on the power of the alphabet N and the sample size n (the length of the text sequence).

We can reduce the computational complexity of the problem if we use exact approximations (limit distributions) instead of exact distributions which minimally reduce the efficiency of the used criteria [1,2]. As exact approximations of distributions of reference statistics, we use Δ-exact distributions [2] which differ from the initial ones by an arbitrarily small Δ. To calculate exact distributions, there are methods such as the calculation method of the exact distributions statistics of the Kolmogorov–Smirnov type [3,4], the well-known classical Monte Carlo method [5], etc. The most preferable is the second multiplicity method [2], which provides calculations of the exact approximations of distributions for the maximum values of the samples' parameters with the same resource. The second multiplicity method, based on solving systems of linear equations, has polynomial complexity, but its computational complexity for real application problems is still quite large [2], so calculation of exact approximations in a reasonable time using modern computational means is difficult. An evaluation of the required computing resources and the problem solution time by means of modern processors, graphics accelerators and FPGAs for the distribution parameters required in practice is given in [2]. In this paper, we consider an evaluation of the required resources and possible problem solution times for calculating exact approximations of probability distributions of statistics by means of promising computer technologies: quantum and photon computer systems.

2. Setting of the Problem

We consider probability distributions of statistics for an alphabet $A_N = \{a_1, \ldots, a_N\}$ with a power $\mid A_N \mid = N$ and its sample n. To calculate the exact approximations of statistics probability distributions $P_\Delta\{S_{N,n} \geq c\}$, which differ from the exact distributions $P_T\{S_{N,n} \geq c\}$ by a specified arbitrary small value Δ

$$\mid P_T\{S_{N,n} \geq c\} - P_\Delta\{S_{N,n} \geq c\} \mid \leq \Delta, \quad (1)$$

we use the second multiplicity method (SMM) based on the solution of a system of linear equations

$$\begin{cases} \mu_0^{(v)} + \mu_1^{(v)} + \ldots + \mu_n^{(v)} = N, \\ 1 \cdot \mu_1^{(v)} + 2 \cdot \mu_2^{(v)} + \ldots + n \cdot \mu_n^{(v)} = n \end{cases} \quad (2)$$

where $\mu_i^{(v)}$ is the number of characters that occurred j times in the sample v of the alphabet A_N and n is the size of the sample v.

The SMM [2] is based on sequential enumeration of the vectors $\mu^{(v)} = \{\mu_0^{(v)}, \mu_1^{(v)}, \ldots, \mu_n^{(v)}\}$ and determination of whether each $\mu^{(v)}$ is a solution of the system of linear Equations (2) or not. A detailed theoretical review of the use and implementation of the SMM is given in [2]. The SMM algorithm's complexity is defined by

$$C_{MVK}(P_T\{S_{N,n} \geq c\}) = L_{\mu(N,n,r)} \times 5 \cdot r + K_{\mu(N,n,r)} \cdot (5 \cdot (r+1) + 2(N+r) + 3) + 2 \cdot K_{\mu(N,n,r)} \cdot \log_2 K_{\mu(N,n,r)} + 2 \cdot K_{\mu(N,n,r)}, \quad (3)$$

where $L_{\mu(N,n,r)}$ is the reduced number of tested vectors of possible solution candidates of (2) with the limitations $\{r \leq n \mid \forall i = \overline{r+1,n},\ \mu_i^{(v)} = 0\}$ and $K_{\mu(N,n,r)}$ is the number of non-negative integer solutions of (2) with the limitations r. According to [2], the main complexity of (3) depends on the first term of the polynomial

$$L_{\mu(N,n,r)} = (N+1)^{min([n/N,r])+1} \cdot \frac{(min([n,N],r))!}{(n+min([n,N],r))!} \cdot \frac{(n+r)!}{r!} \quad (4)$$

It follows from (3) and (4) that the algorithmic complexity of calculating the exact approximations $C_{MVK}(P_\Delta\{S_{N,n} \geq c\})$ is a polynomial which depends on both the parameters of the sample N and n, and on the limitation parameter r, which also functionally depends on the sample's parameters and the accuracy Δ, i.e., $r = m(N, n, \Delta)$.

The most laborious part of the SMM is the procedure of calculating and testing solution candidate vectors. For practical problems, the power of the alphabet N belongs in the range from 128 to 256, and the sample sizes are in the range from 320 to 1280, when the required total solution time does not exceed 30 days. Therefore, to evaluate the algorithmic complexity [2], we specified the following samples as (the alphabet power, the sample size): (256, 1280), (128, 640), (128, 320) and (192, 320) with the accuracy $\Delta = 10^{-5}$. The algorithmic complexity and the required performance of the task with these sample parameters are given in Table 1.

According to Table 1, the computational complexity is in the range from 9.68×10^{22} to 1.60×10^{52} operations, the average complexity is about 4.55×10^{25} operations, and it is necessary to test from 6.50×10^{23} to 1.39×10^{50} vectors and to obtain from 4.67×10^{12} to 5.60×10^{25} solutions.

The number of tested variables in (2) does not exceed $(r+1)$, i.e., it does not exceed 24 for the parameters given in Table 1. Accordingly, $\mu^{(v)} = \{\mu_0^{(v)}, \mu_1^{(v)}, \ldots, \mu_n^{(23)}\}$, and all other variables are equal to zero: $\{\mu_j^{(v)} = 0 | j = 24, \ldots, n\}$. The values of $\mu_i^{(v)}$ are integers and belong to the range $\{0 \leq \mu_j^{(v)} \leq 23 | i = \overline{0, 23}\}$. One $\mu_j^{(v)}$ is no less than 5 bits ($2^4 < 23 < 2^5$), and the whole vector $\mu^{(v)}$, which contains 24 $\mu_i^{(v)}$, is no less than $5 \times 24 = 120$ bit.

The important property of the method is its parallelization by data because any μ^i and μ^j can be tested independently and concurrently when $i \neq j$.

Table 1. Characteristics of the calculation method of exact approximations for various parameters of samples.

№	Parameters of the Sample N/n	Limitation Parameter r	Reduced Number of Tests $L_{\mu(N,n,r)}$	Number of Solutions $K_{\mu(N,n,r)}$	Total Complexity $C_{MVK}(P_T\{S_{N,n} \geq c\})$	Required Performance for Calculation for 30 days (op/s)
1	256/1280	23	1.39×10^{50}	5.60×10^{25}	1.60×10^{52}	6.15×10^{45}
2	128/640	22	2.67×10^{27}	1.76×10^{20}	2.94×10^{29}	1.13×10^{23}
3	192/320	14	6.50×10^{23}	4.67×10^{12}	4.55×10^{25}	1.75×10^{19}
4	128/320	16	1.21×10^{21}	2.23×10^{13}	9.68×10^{22}	3.72×10^{16}

3. Use of General Purpose Processors to Calculate Exact Approximations of Distributions

The $\times 86$ processors with the traditional von Neumann architecture and 32-bit and 64-bit data processing compose the main and most widespread general purpose computing architecture for the design of cluster multiprocessor high-performance systems (MHPS) [6].

For calculating exact approximations, the performance P_{CPU} of a CPU-based MHPS with unlimited scalability is

$$P_{CPU} = N_{CPU} \cdot K_{CPU} \cdot H_{1\_CPU}, \tag{5}$$

where N_{CPU} is the number of processors; K_{CPU} is the number of provided parallel threads; and H_{1_CPU} is the frequency of one processor.

To estimate the number of processors needed to solve the problem with time constraints, let us consider a hypothetical CPU that supports eight parallel threads at 3000 MHz, i.e., $K_{CPU} = 8$ and $H_{1\_CPU} = 3 \times 10^9$. Then, to achieve the minimal performance (according to Table 1) of $P_{CPU} \geq 1.75 \times 10^{19}$ op/s, it is necessary to use

$$N_{CPU} \geq \frac{1.75 \times 10^{19}}{K_{CPU} \cdot H_{1\_CPU}} = \frac{1.75 \times 10^{19}}{8 \times 3.0 \times 10^9} = 7.29 \times 10^8,$$

i.e., about 729 million hypothetical processors. The obtained number exceeds the number of cores (not the number of processors!) of the most high-performance supercomputers in the world (such as Fugaku [6,7] with 7 million cores and Sunway TaihuLight [6,7] with 10 million cores, included in TOP500 [6]) by 1.5–2 decimal orders.

To calculate exact approximations for all values of the samples' parameters $\{N = \overline{2,256},\ n = \overline{1,5N}\}$, the number of required nodes is

$$N_{CPU} \geq \frac{6.15 \times 10^{45}}{K_{CPU} \cdot H_{1\_CPU}} = \frac{6.15 \times 10^{45}}{8 \times 3.0 \times 10^{9}} = 5.56 \times 10^{35},$$

which cannot be achieved if we take into account the modern level of processor architectures and technologies in cluster MHPS designs. The performance of modern CPUs is insufficient for the solution of the considered problem. Furthermore, calculations of exact approximations cannot be provided if a computer system is based only on modern CPUs because hundreds of million CPUs are needed to solve the problem even in its minimal form.

4. Use of Graphics Accelerator Technology to Calculate Exact Approximations of Distributions

The development of GPU (Graphic Processing Unit) technology [7], originally designed to calculate 3D graphics in real time, has led to their application in high-performance computing. Modern graphics accelerators, containing thousands of special purpose cores, provide a high degree of parallelism and can perform tasks in a multithread parallel mode. For example, the standard GEFORCE RTX-3090 game graphics card contains 10,496 NVIDIA CUDA cores operating at 1.70 GHz (1.7×10^{9} op/s) [8]. We can roughly define the performance of a GPU-based computer system P_{GPU} as the following product

$$P_{GPU} = N_{GPU} \cdot K_{CP} \cdot H_{CP}, \tag{6}$$

where N_{GPU} is the number of graphics accelerators GPU in the system; K_{CP} is the number of cores of each accelerator; and H_{CP} is the clock frequency.

To provide $P_{GPU} \geq 1.75 \times 10^{19}$, the system based on a GEFORCE RTX-3090 must contain no less than

$$N_{RTX3090} \geq \frac{1.75 \times 10^{19}}{K_{RTX} \times H_{RTX}} = \frac{1.75 \times 10^{19}}{1.05 \times 10^{4} \times 1.7 \times 10^{9}} = 9.80 \times 10^{5}$$

nodes, and the system based on an NVIDIA Quadro K6000 [9] with the performance

$$K_{K6000} \cdot H_{K6000} = 16.3 \times 10^{12} \text{ op/s}$$

of one videocard, must contain no less than

$$N_{K6000} \geq \frac{1.75 \times 10^{19}}{K_{K6000} \cdot H_{K6000}} = \frac{1.75 \times 10^{19}}{1.63 \times 10^{13}} = 1.07 \times 10^{6}$$

nodes. Despite the difference in the properties of the considered graphics accelerators, the number required for calculation of exact approximations is estimated at about one million, which is impossible with the current level of development of graphics accelerator architectures and design technologies of GPU-based computer systems. Thus, due to the insufficient performance of modern graphics accelerators, it is impossible to use them as the only basis to calculate exact approximations.

5. Use of Parallel Pipeline FPGA Technologies to Calculate Exact Approximations of Distributions

The FPGA (Field Programmable Gate Array) technology combines the capabilities of parallel and pipeline computing. In contrast to computer systems with a fixed architecture designed on a CPU and GPU basis, it allows reconfiguration [10] of the architecture of an FPGA-based computer system to the architecture of the problem to be solved. Taking into account information dependencies in the structure of the application [10], it is possible to provide high real performance for labor-intensive problems in various fields of science and technology [11]. For example, the computational block (CB) of the latest Seguin system based on the UltraScale+ FPGAs (3U height, 96 XCVU9P-1FLGC2104E chips designed using 16 nm technology) achieves the performance of 240 Tflops (2.4×10^{14} op/s) [12].

For the problem of calculating exact approximations, the performance of an FPGA-based reconfigurable multiprocessor computer system P_{FPGA} can be roughly estimated as the product of the number of computational blocks (CB) N_{CB_FPGA} in the system and the performance of one CB P_{CB_FPGA}.

$$N_{CB\_FPGA} \geq \frac{1.75 \times 10^{19}}{P_{CB\_FPGA}} = \frac{1.75 \times 10^{19}}{2.4 \times 10^{14}} = 7.29 \times 10^{4}. \quad (7)$$

To calculate exact approximations for all values of the considered parameters of the samples $\{N = \overline{2,256}, n = \overline{1,5N}\}$, it is necessary to use

$$N_{CB\_FPGA} \geq \frac{6.15 \times 10^{45}}{P_{CB\_FPGA}} = \frac{6.15 \times 10^{45}}{2.4 \times 10^{14}} = 2.56 \times 10^{31}$$

computational blocks. Parallel pipeline FPGA technologies with the real performance of up to 10^{14} operations per second for one computational block have the greatest potential for implementation of the second multiplicity method for the largest values of the sample parameters. However, a computer system, which is based only on FPGA technologies and implements the considered problem, cannot be easily designed because it requires a very large number of computational blocks.

According to the analysis of the capabilities of computer technologies to calculate exact approximations of distributions, none of the existing computer technologies can provide the required computing resources. To solve this computationally expensive problem during the specified time, it is necessary to analyze the capabilities of promising computer technologies such as quantum and photon computers.

6. Use of Quantum Computer Technologies to Calculate Exact Approximations of Distributions

Quantum computer technologies were proposed in the 1980s by a number of famous scientists, such as Richard F. Feynman [13], Paul Benioff [14], K.A. Valiev and A.A. Kokin [15] and Yu. I. Manin [16]. The main idea of quantum computing is the following: a quantum system of q concurrently operating qubits has 2^q linearly independent states. According to the principle of quantum superposition, the state space of such a quantum register is a 2^qD Hilbert space [17]. One operation on a group of q qubits is calculated immediately for all its possible values, unlike a group of classical bits, when only one current value exists. This ensures the maximum possible parallelization of data calculations and an increase in performance to 2^q, which is called "quantum acceleration". Any object with two quantum states, such as polarization states of photons, electronic states of isolated atoms or ions, spin states of atomic nuclei, etc., can be a physical system implementing qubits. The structure of the quantum computer proposed by Russian scientists K.A. Valiev and A.A. Kokin [15] is shown in Figure 1. According to the principle formulated by R. Feynman [13], for any algorithm, it is possible to obtain an implementation on a quantum system, which will be no worse than its implementation on the classical von Neumann system. At the same time, it is shown [16] that "quantum acceleration" is not possible for any algorithm and for an arbitrary algorithm, the possibility of quantum acceleration is not guaranteed. A feature of quantum calculations is the probabilistic nature of the result of calculations, i.e., the result is true only with some probability.

Quantum computers were designed at Harvard University (51 qubit system) [18] and at the Polytechnic School of the University of Paris-Saclay (70 qubit system) [19]. The American company IonQ [20] announced the first commercial quantum computer based on ion traps, which contains 160 qubits, but for quantum operations only 79 qubits are used, and only 11 qubits are used for implementations of arbitrary quantum algorithms. In 2020, IBM [21] introduced the most powerful 64 qubit quantum computer. In Russia, the design of a general purpose quantum computer of 100 qubits is expected by 2024 [19]. At present, a quantum computer capable of operating with 120 bit data does not exist in the world.

Let us evaluate the possibility of calculating exact approximations of distributions on a quantum computer. The structure of the vector testing procedure of the considered problem [2] corresponds to the structure of the quantum computer (Figure 1), which allows us to expect a significant effect when solving the problem on quantum computers. According to the minimum evaluation, 120 bits are required to place the test vector candidate $\mu^{(v)}$; therefore, the quantum computer must contain 120 qubits operating concurrently. Let us suppose that a quantum computer operating with 120 qubits (let us call it QC-120) exists, and the problem of valid results is solved. If exact approximations are calculated on QC-120, we obtain all $\{\mu^{(1)}, \mu^{(2)}, \ldots, \mu^{(2^{120})}\}$ for the 120-bit test vector $\mu^{(v)} = \{\mu_0^{(v)}, \mu_1^{(v)}, \ldots, \mu_{23}^{(v)}\}$, i.e., ($2^{120} \cong 1.3292 \times 10^{36}$) possible solutions. Then, it is necessary to select $K_\mu(N, n, r)$ real solutions. For $N = 256$ and $n = 1280$, we obtain $K_\mu(256, 1280, 23) = 5.60075 \times 10^{25}$ from Table 1.

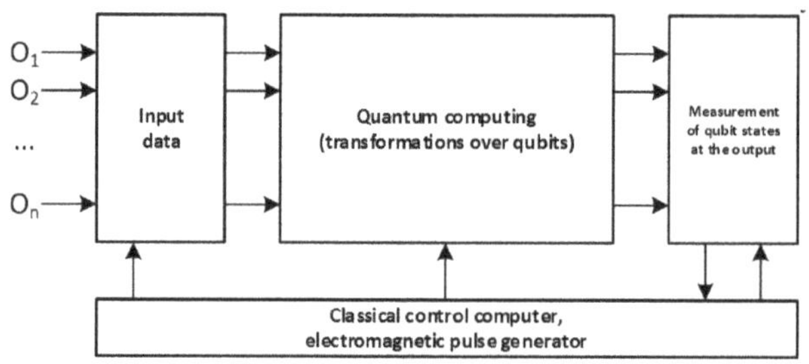

Figure 1. The structure of the quantum computer.

Thus, if we calculate the values of i-th possible solution, we obtain the SLE state $\underbrace{\{\mu^{(i)}, \mu^{(i)}, \ldots, \mu^{(i)}\}}_{2^{120}}$. To calculate the $(i+1)$-th possible solution, we obtain the SLE state $\{\mu^{(1)}, \mu^{(2)}, \ldots, \mu^{(2^{120})}\}$ and read the $(i+1)$-th possible solution. To test all possible solutions and to obtain $K_\mu(N, n, r)$ real solutions, we need 2^{120} accesses to QC-120, which drastically reduces the effect of concurrent calculations of 2^{120} possible solutions.

The need to check all the obtained solutions, the number of which corresponds to the dimension of the problem, does not allow achieving quantum acceleration and is a significant and fundamental limitation of the use of promising quantum computer technologies for calculating accurate approximations of distributions. The lack of a technical and technological base not only in the Russian Federation, but also in the world, is an additional, technological limitation that does not allow creation of a quantum computer system operating 120 qubits required to solve the problem of calculating exact approximations of statistical probability distributions. Therefore, the prospect of using quantum computer systems at the existing level of development of science and technology is quite modest for the considered problem, despite the potentially high performance. Perhaps the given evaluations for the considered problem can be revised with the development of quantum computing.

7. Use of Photon Computers to Calculate Exact Approximations of Distributions

Another relevant area for the development of promising computer facilities based on new physical principles is the design of computer systems that use the effects of interactions of coherent light waves generated by laser radiation and its carriers, i.e., photons [22,23].

The structure of the photon computer developed at the All-Russian Research Institute of Experimental Physics (Sarov) [22] is shown in Figure 2. The photon computer (Figure 2) consists of four large units: a unit for converting a task into a program for the photon computer (UCT), a laser radiation source (LRS), an input–output unit (IOU) and a photon processor (PP). In turn, the photon processor contains four processor elements which comprise arithmetic logic devices (ALU), control devices (CD) and switches (SW). The units of the photon computer are interconnected by electronic and optical channels, and the components of the processor elements are connected only by optical channels.

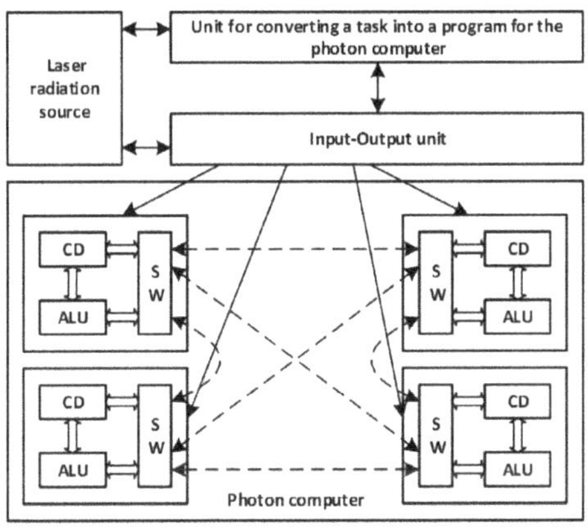

Figure 2. The structure of the photon computer.

The UCT transfers the photon program to the LRS which generates laser radiation and supplies it to the IOU, where it is divided into light rays according to the number of digits simultaneously supplied to the PE. The IOU generates a photon program and transmits it to a photon processor where calculations are performed by the processor elements. Light beams interact within the photon processor, and optical delay lines [23] perform synchronization. Within the photon processor, the PEs can be connected by optical channels to a multiprocessor environment of any topology [23]. A low power consumption and a high performance are the important advantages of photon computer systems.

It has been shown that the performance of a photon computing node can reach up to 10^{18} op/s per 100 W of power consumption [24].

To achieve the required performance of the photon computer P_{PHOTON}, it is necessary to use N_{PH_ND} nodes with the performance $P_{PH\_ND} = 10^{18}$ op/s each:

$$P_{PHOTON} = N_{PH\_ND} \times P_{PH\_ND}. \tag{8}$$

The performance level $P_{PHOTON} \leq 1.75 \times 10^{19}$ for calculating exact approximations (the parameters of the sample №3 in Table 1) is provided by

$$N_{PH\_ND} \geq 1.75 \times 10^{19} / P_{PH\_ND} = 1.75 \times 10^{19} / 1.0 \times 10^{18} \cong 17.5$$

nodes, i.e., no less than 18 photon computing nodes. To achieve the performance for calculating exact approximations of the whole spectrum of the parameters of the considered samples $\{N = \overline{2,256},\ n = \overline{1,5N}\}$, it is necessary to use no less than

$$N_{PH\_ND} \geq 6.15 \times 10^{45} / P_{PH\_ND} = 6.15 \times 10^{45} / 1.0 \times 10^{18} = 6.15 \times 10^{27}$$

computing nodes. This is much less than in all technologies that were reviewed earlier.

Unlike quantum computer systems, there is no available information about existing prototypes of digital photon computer systems. According to most works [22–24], scientific teams simulate the functioning of individual nodes and then evaluate the possible parameters of the calculator. There are works of academician V. A. Soifer [25–27] and some other scientists in the field of analog photonics. There, each computing node is created for a task with certain parameters, and the technology of developing and creating an analog photon computing node for an arbitrary task requires a whole cycle of research and development work. Therefore, there are no active prototypes of photon computer systems, technological bases and/or commercially available components today. Due to the current development level of science and technology, in the next 5–7 years it is impossible to consider the use of photon computer systems to solve the problem of calculating exact approximations of statistical probability distributions, although potentially such computer systems will have very high performances and require the smallest number of low-power computing nodes. Perhaps, owing to the development of photon computer technologies, it will be possible to revise and improve the presented evaluations for the considered task in the next few years.

8. Architecture-Independent Representation of the Exact Approximation Calculation Problem for Hybrid Computer Systems

According to the reviewed computer technologies and their current level of development, a possible solution is to design hybrid or heterogeneous computer systems containing the computing nodes of modern architectures (such as general purpose processors, graphics accelerators and FPGAs) capable of solving the problem with the given parameters in a reasonable time. Since the calculation of distributions is carried out by one and the same method for all parameters of samples, the possibility of programming different computer architectures in a single loop, which is provided in the architecture-independent programming paradigm, is especially important for such a system [28]. Taking into account the possible use of promising architectures of quantum computers and/or photon computing nodes, we consider an architecture-independent representation of the problem of calculating exact approximations for hybrid computer systems as especially significant.

Such capabilities are provided by the architecture-independent aspect-oriented language Set@l, which allows the developer to focus their attention on parallelizing methods when solving a problem, and not on their technical implementation in the selected computer architecture. The Set@l language reduces the problem of software transfer between different configurations and architectures of the HCS to the creation of aspects (descriptions) of technical means, which describe the key points of parallelization of the method on these technical means. At the same time, the source program implementing the data processing method remains unchanged [28].

The language Set@l is based on the paradigm of aspect-oriented programming (AOP) [29], according to which the algorithm of an application problem and the peculiarities of its implementation are described in the form of separate program modules. The Set@l program represents the information graph of the computational problem in the form of sets, whose partitioning and typing specify different parallelization options and other aspects of the implementation of the algorithm. Unlike other programming languages based on the set theory, for example, the languages SETL, SETL2 and SETLX, the Set@l language uses typing of sets according to different criteria and allows operations with fuzzy collections in accordance with the alternative set theory [30].

The Set@l language provides separate descriptions of the algorithm and the peculiarities of its implementation on a computer system by the use of the AOP paradigm. According to this paradigm, the cross-cutting concern of the program, which causes the negative effects of code tangling and scattering, is presented in the form of separate program modules (aspects). The source code, implementing the main functionality of the program, contains the user's markup, which determines its interaction with aspects during translation or execution. Analyzing the markup and source code, the preprocessor translator forms a new executable cross-cutting concern program. As a rule, the use of AOP technology simplifies the development and further support of software and increases the adaptability of programs to various modifications.

To implicitly describe various methods of parallelizing algorithms, the language of architecture-independent programming Set@l provides classification of collections by the type of parallelism of their elements during processing. When the parallelism nature of the set's elements is clearly defined, the types "par" (parallel-independent processing), "seq" (sequential processing), "pipe" (pipeline processing) and "conc" (parallel-dependent processing by iterations) are used. However, in some aspects of the program, the type of parallelism of a number of sets cannot be determined uniquely, since the architecture of the computer system on which the algorithm will be implemented is unknown. In this case, a special type of "imp" (implicit) is used, and the typing of collections is refined in other aspects of the program using special syntactic structures. If the aspects of the Set@l program do not change the algorithm in the process of its adaptation to the computer system's architecture, then the solution to the problem can be presented according to the classical Cantor–Bolzano set theory. However, in some cases, it is reasonable to modify the algorithm in accordance with the peculiarities of its implementation on the computer system with a certain architecture. In such cases, some collections are fuzzy and are not sets, so they cannot be specified as objects of the classical set theory. Using the Set@l language, we can describe different implementations of the same algorithm in a single-aspect-oriented program. To do this, classification of collections by the definiteness of their elements is performed according to the alternative set theory of P. Vopenka. The type "set" (set) corresponds to the classical clearly distinguished collection of elements, the type "semi-set" (sm) corresponds to a fuzzy collection and the type "class" (cls) corresponds to a set of objects, the type and partition of which cannot be uniquely determined at a given level of abstraction. An example of a computational problem that requires the Set@l fuzzy collections for its description is the Jacobi SLAE algorithm [31]. Unlike other approaches to parallel programming of high-performance computer systems, the Set@l programming language specifies not only boundary cases of the algorithm implementation, but also a family of intermediate options. They cannot be distinguished from the point of view of procedural programming but provide continuity of the description of the calculation model. Owing to the use of the Set@l language, it is possible to synthesize many variants of the problem solution and to switch between its elements depending on the architecture and configuration of the computer system. The program of fast Fourier transform illustrates this possibility. According to the available memory of the computer system, complex coefficients W are calculated in advance and loaded from memory or calculated with the help of basic and auxiliary components [32].

Thus, the algorithm of the problem in the Set@l language is presented as an architecture-independent source code by means of aspect-oriented programming, set theoretical code representation and relational calculus. The peculiarities of the algorithm's implementation are represented as the separate aspects that define the division into subsets and typing of the main collections of the program. The program can be quickly ported between computer systems with different architectures and adapted to any changes in hardware resources due to the use of aspects of the processing method, architecture and configuration. Owing to the suggested approach to parallel programming of high-performance computer systems in the Set@l language, there are new prospects for the development of architecture-independent and resource-independent software.

The language Set@l has been successfully used to solve algorithmically complex and resource-intensive problems, such as solving SLAE by the Gaussian [28] and Jacobi [31] methods and implementing the fast Fourier transform algorithm [32] and others with the same structure.

Therefore, the use of the Set@l language for the calculating exact approximations of distributions on a hybrid computer system will significantly decrease the programming complexity of computing nodes of various architectures such as general purpose processors, graphics accelerators, FPGAs, quantum computers and/or photon computing nodes.

9. Conclusions

In this paper, we have analyzed the use of promising computer technologies to solve the computationally expensive problem of calculating Δ-exact approximations of statistical probability distributions by the second multiplicity method based on solving the SLAE of the second multiplicity. The method has polynomial complexity and allows parallelization into fragments by data. We have evaluated the possibilities of promising computer technologies based on quantum and photon computers for calculating distributions by the second multiplicity method. An analysis of the capabilities of quantum and photon computer technologies shows that they have great potential for solving this problem. However, at present, these technologies cannot provide a solution to the problem of calculating exact approximations of sample distributions with an alphabet power less than 256, a size less than 1280 characters and an accuracy of $\Delta = 10^{-5}$.

We have performed a theoretical analysis of quantum computer systems and their applicability for the problem, which showed that quantum acceleration is impossible. Thus, the current level of development of quantum computers is insufficient, and there are also no algorithms and criteria for choosing a suitable solution from a huge number of obtained solutions to the problem. In addition, the level of development of photon computers does not allow creating a computer with the required number of computing nodes.

Author Contributions: Conceptualization, A.M., I.L. and A.D.; methodology, A.M. and I.L.; validation, A.M., A.D. and L.S.; formal analysis, I.L.; investigation, A.M.; resources, A.M.; writing—original draft preparation, A.D. and L.S.; writing—review and editing, A.D. and L.S.; supervision, A.M.; project administration, I.L.; funding acquisition, I.L. All authors have read and agreed to the published version of the manuscript.

Funding: This research was funded by the Russian Foundation for Basic Research, project number 20-07-00545.

Institutional Review Board Statement: Not applicable.

Informed Consent Statement: Not applicable.

Data Availability Statement: Not applicable.

Conflicts of Interest: The authors declare no conflict of interest.

References

1. Melnikov, A. Complexity of calculating exact probability distributions of symmetric additive separable statistics and application of limit distributions. *Proc. TUSUR* **2017**, *20*, 126–130. [CrossRef]
2. Melnikov, A.; Levin, I.; Dordopulo, A.; Pisarenko, I. Analysis of advanced computer technologies for calculation of exact approximations of statistics probability distributions. *Izvestia SFedU* **2021**, 7, 6–19.
3. Timonin, V.; Chernomordik, O. Method of calculating the exact distribution of statistics of the Kolmogorov–Smirnov type under Lehman alternatives. *Probab. Theory Its Appl.* **1985**, *30*, 572–573.
4. Tyannikova, N.; Timonin, V. Method of calculation of exact distributions of Kolmogorov–Smirnov type statistics in case of violation of homogeneity and independence of analyzed samples. *Sci. Educ.* **2014**, 11, 227–237.
5. Cramer, G. *Mathematical Methods of Statistics*; Princeton University Press: Princeton, NJ, USA, 1946.
6. TOP500. 2022. Available online: https://www.top500.org (accessed on 21 September 2022).
7. Antonov, A.; Afanasiev, I.; Voevodin, V. High-performance computer platforms: Current status and trends. *Numer. Methods Program.* **2021**, *22*, 135–177.
8. GeForce RTX 3090 NVIDIA. 2022. Available online: https://3dnews.ru/1021405/obzor-videokarti-nvidia-geforce-rtx-3090 (accessed on 21 September 2022).
9. NVIDIA Quadro RTX 6000. 2022. Available online: https://3dnews.ru/1027931/nvidia-obyavila-o-dostupnosti-sverhmoshchnoy-videokartirtx-a6000-s-48-gbayt-gddr6-i-tsenoy-5500 (accessed on 21 September 2022).
10. Levin, I.; Fedorov, A.; Doronchenko, Y.; Raskladkin, M.; Guzik, V.; Kalyaev, I. *Reconfigurable Computer Systems*; SFedU Publishing: Taganrog, Russia, 2016.

11. Levin, I.I.; Dordopulo, A.I.; Fedorov, A.M.; Kalyaev, I.A. Reconfigurable computer systems: From the first FPGAs towards liquid cooling systems. *Supercomput. Front. Innov.* **2016**, *3*, 22–40.
12. Levin, I.; Fedorov, A.; Doronchenko, Y.; Raskladkin, M. Promising high-performance reconfigurable computers with immersion cooling. In *Supercomputer Technologies*; SFedU Publishing: Taganrog, Russia, 2020; pp. 29–34.
13. Feynman, R. Simulating physics with computers. *Int. J. Theor. Phys.* **1982**, *21*, 467–488. [CrossRef]
14. Benioff, P. Quantum mechanical hamiltonian models of turing machines. *J. Stat. Phys.* **1982**, *29*, 515–546. [CrossRef]
15. Valiev, K.; Kokin, A. *Quantum Computers: Hopes and Reality*; RHD: Izhevsk, Russia, 2004.
16. Manin, Y.I. *Computable and Non-Computable*; Soviet Radio: Moscow, Russia, 1980.
17. Moren, K. *Hilbert Space Methods*; Mir: Moscow, Russia, 1965.
18. Physicists from Russia and the United States Have Created the First 51-Qubit Quantum Computer. 2022. Available online: https://ria.ru/20170714/1498476410.html? (accessed on 2 May 2022).
19. How Russia Will Spend 15 Billion to Create a Quantum Computer. 2022. Available online: https://www.cnews.ru/articles/2020-01-29_kak_rossiya_potratit_15_mlrd_na_sozdanie?ysclid=liy7s0gbbk340669962 (accessed on 2 May 2023).
20. IonQ Announced the Creation of the World's Most Powerful Quantum Computer. 2022. Available online: https://3dnews.ru/1060908/kompaniya-ion-q-soobshchila-o-sozdanii-samogo-moshchnogo-v-mire-kvantovogo-kompyutera? (accessed on 2 May 2022).
21. IBM Demonstrated a Quantum Computer with "Quantum Volume" of 64. 2022. Available online: https://habr.com/ru/company/raiffeisenbank/news/t/516062/? (accessed on 2 May 2022).
22. Stepanenko, S. Photon computer. Design principles. Parameters evaluation. *Proc. RAS* **2017**, *476*, 389–394.
23. Stepanenko, S. Photon computer: Structure and algorithms, parameters evaluation. *Photonics* **2017**, *7*, 72–83. [CrossRef]
24. Stepanenko, S. Interference logic elements. *Proc. RAS Math. Inform. Control Process.* **2020**, *493*, 64–69.
25. Soifer, V.A.; Kharitonov, S.I.; Khonina, S.N.; Strelkov, Y.S.; Porfirev, A.P. Spiral caustics of vortex beams. *Photonics* **2021**, *8*, 24. [CrossRef]
26. Kashapov, A.I.; Doskolovich, L.L.; Bezus, E.A.; Bykov, D.A.; Soifer, V.A. Spatial differentiation of optical beams using a resonant metal-dielectric-metal structure. *J. Opt.* **2021**, *23*, 023501. [CrossRef]
27. Golovastikov, N.V.; Doskolovich, L.L.; Bezus, E.A.; Bykov, D.A.; Soifer, V.A. An Optical Differentiator Based on a Three-Layer Structure with a W-Shaped Refractive Index Profile. *J. Exp. Theor. Phys.* **2018**, *127*, 202–209. [CrossRef]
28. Levin, I.; Dordopulo, A.; Pisarenko, I.; Melnikov, A. Approach to architecture-independent programming of computer systems in aspect-oriented SET@L language. *Izvestiya SFedU* **2018**, *3*, 46–58.
29. Safonov, V. *Using Aspect-Oriented Programming for Trustworthy Software Development*; John Wiley & Sons.: New York, NY, USA, 2008.
30. Vopenka, P. *The Alternative Set Theory: A New Look at Infinity*; Institute of Mathematics Publishing: Novosibirsk, Russia, 2004.
31. Levin, I.; Dordopulo, A.; Pisarenko, I.; Melnikov, A. Description of Jacobi algorithm for solution of linear equation system in architecture-independent SET@L programmning language. *Izvestiya SFedU* **2018**, *5*, 34–48.
32. Levin, I.; Dordopulo, A.; Pisarenko, I.; Melnikov, A. Architecture-independent program of Fast Fourier transform in SET@L programming language. *Her. RSREU* **2019**, *68*, 28–36. [CrossRef]

engineering proceedings

MDPI

Proceeding Paper

Automating the Study of a Linearized Model of Diabetes Mellitus and Tuning a PID Controller for This Model †

Denis Andrikov and Sinan Kurbanov *

The Patrice Lumumba Peoples' Friendship University of Russia, 6 Miklukho-Maklaya St., Moscow 117198, Russia; andrikov-da@rudn.ru
* Correspondence: ya.sinan@yandex.ru; Tel.: +7-963-756-06-97
† Presented at the 15th International Conference "Intelligent Systems" (INTELS'22), Moscow, Russia, 14–16 December 2022.

Abstract: This paper proposes a simulation linear model created in the MATLAB environment, which provides a process for regulating blood sugar levels. The controller is built for the need for any type of diabetes to control and normalize the blood sugar content of the patient in order to eliminate the differences in the quality of life of a diabetic patient and a healthy person. The linearization of the nonlinear model was performed, and the adequacy of the linearized model was verified and confirmed using the MATLAB simulation. The choice of the PID controller and the CHR method for its adjustment was justified and MATLAB tools were used to show the implementation of these methods. The model of the patient with the controller has been built; the algorithm for the automatic adjustment of the PID controller parameters has been developed and realized. The directions for continuation of the work on this problem regarding regulation in the system under study are proposed.

Keywords: diabetes mellitus; insulin pump; automatic control system; insulin–glucose interaction models; PID controller; linearization; parameter setting

Citation: Andrikov, D.; Kurbanov, S. Automating the Study of a Linearized Model of Diabetes Mellitus and Tuning a PID Controller for This Model. *Eng. Proc.* **2023**, *33*, 42. https://doi.org/10.3390/engproc2023033042

Academic Editors: Askhat Diveev, Ivan Zelinka, Arutun Avetisyan and Alexander Ilin

Published: 27 June 2023

1. Introduction

Diabetes mellitus (DM) is a group of metabolic diseases characterized by chronic hyperglycemia, which is the result of impaired insulin secretion, the action of insulin, or both. Chronic hyperglycemia leads to damage and dysfunction of various organs, including the eyes, kidneys, nerves, heart, and blood vessels. In any type of diabetes, the control and normalization of blood sugar becomes one of the main tasks of the patient and his attending physician. When the sugar level is closer to the normal limits, the risk of complications is lower, and there are fewer symptoms of diabetes as well as fewer differences between the quality of life of a patient with diabetes and a healthy person. The current state of medicine and science in general allows the treatment of diabetes mellitus on an outpatient basis, mainly through the use of an individual control system, that is, an insulin pump. The algorithms of such a system can be implemented using a controller.

According to research published in the tenth edition of the Diabetes Atlas of the International Diabetes Federation (IDF), 537 million people aged from 20 to 79 years have diabetes, and this number is steadily increasing every year [1], including in the Russian Federation [2]. This is why the topic of developing new and improving existing automatic control systems for regulating blood sugar levels in patients with diabetes mellitus is becoming increasingly relevant. The aim of the work is the practical implementation of a simulation model in MATLAB for regulating blood sugar levels using an actuator, namely an insulin pump, in order to be able to use this technique when such compact medical devices are put into operation. To achieve this goal, the following tasks were set and solved:

1. Study and analysis of existing mathematical models of diabetes mellitus in terms of applying the methods of the theory of linear automatic control systems to them [3–15].

2. Analysis of the selected model for the possibility of its linearization [15].
3. Obtaining the transfer functions of the sugar level by food intake and insulin injection, building a mathematical model of the control object (human) in the MATLAB/Simulink environment.
4. Testing of transient graphs of a patient and a healthy person in the original and linearized models during the transient time.
5. Evaluation of the suitability of a linearized model for the development of a PID controller for an insulin pump control system.
6. Study of various engineering methods for the synthesis of controllers and the choice of the most adequate method for the synthesis of the PID controller by the CHR method, taking into account the features of the linearized system.
7. Synthesis of the PID controller by the CHR method.
8. Creation of a control system model with the PID controller in the MATLAB/Simulink environment.
9. Development of an algorithm for determining controller parameters from the transient curve in a model of a patient, with the possibility of automatically adjusting these parameters to obtain a given blood sugar level.
10. Performing simulations with various parameters of the linearized mathematical model to check the operability of this model.

2. Materials and Methods

Rationale for Choosing a Mathematical Model

After the possibility of an external control for blood glucose levels in patients with type I diabetes mellitus by monitoring blood glucose levels and injecting insulin was experimentally proven, it became necessary to develop a safe and effective algorithm for controlling the devices that solve the problem of diabetes for patients. When developing a mathematical model of the interaction between insulin and glucose in the human body, pharmacokinetic and physiological models were considered. The main difference between pharmacokinetic models and physiological ones is that they contain fewer parameters, so they are easier to use for interpreting experimental data. Physiological models are characterized, as a rule, by a large number of parameters. Therefore, Sorensen's model [8] has 21 states and 22 metabolic functions that describe the dynamics of glucose, insulin, and glucagon. This model is widely used in glucose monitoring and has been criticized for inaccurately representing the observed changes in glucose [9]. Another physiological model, from Kobelli [10], is nonlinear and consists of glucose, insulin, and glucogenic subsystems; it has nine states, 23 metabolic functions, and 46 parameters. Therefore, only some pharmacokinetic models were considered as prototypes for modeling: the Bergman model [11], the Hovorka model [12], and the simplest basic differential mathematical model. The simplest basic differential mathematical model is of the following form:

$$\begin{cases} \frac{dx}{dt} = -a_1 xy + a_2(x_0 - x) + a_3 P(t) \\ \frac{dy}{dt} = b_1(x - x_0)H(x - x_0) - b_2 y + b_3 u(t) \end{cases} \quad (1)$$

where $x(t)$ is the blood sugar level, $y(t)$ is the level of insulin in the blood, and x_0 is the fasting blood sugar level. The constants a_1, a_2, and a_3, as well as b_1, b_2, and b_3, are positive and are the sensitivities of the sugar and insulin gradients, respectively. The function $H(x)$ is a single step, $P(t)$ describes changes in sugar levels from food intake, and $u(t)$ describes changes in insulin levels. The undoubted advantages of this model are its simplicity and compactness, consisting in a minimum number of equations and parameters. The adequacy of this model is proved in practice [14]. The model has two input variables, $P(t)$ and $u(t)$, and two state variables, $x(t)$ and $y(t)$. The state variables are the main variables of the model. They are quantities that uniquely determine the current state of the control object. These variables can be changed (i.e., controlled) in clinical practice. As follows from the equations in (1), the mathematical model is nonlinear. The equations in (1) include the product of the state variables; in both equations, the coefficients on the state variable x depend on this variable, and this dependence is also nonlinear because one of the co-multipliers is a step function. The deviations of the state variables from the initial value ("starvation") are commensurate with the value itself [14,15] and, therefore, cannot be considered small. In this connection, the decomposition of nonlinear functions into a Taylor series for the purpose of linearization of the object will be ineffective because, to achieve acceptable accuracy, the account of a large number of row members will be required. The attempt to linearize differential equations at steady-state constant values for "hunger" x_0, y_0 via a Taylor series expansion is impossible because of the infinite derivatives of the step functions $H(x - x_0)$ at point x_0. As stated in [16], for this kind of linearization, the derivatives must have a single and finite value; otherwise, the equation is nonlinearizable. The application of I.A. Vyshnegradsky's hypothesis for small deviations of all variables from steady-state values seems to be an acceptable method of linearization [16]. The steady-state value is logically considered to be the function of the state variables for a healthy person. The model was linearized by entering a new variable—the deviation of the sugar level from zero, corresponding to hunger:

$$z = x - x_0$$

and the following matrices of the linear system in the state space were obtained

$$\mathrm{A} = \begin{pmatrix} -a_2 H(-z) & -a_1 x_0 \\ b_1 H(-z) & b_2 \end{pmatrix}; \mathrm{B} = \begin{pmatrix} a_3 & 0 \\ 0 & b_3 \end{pmatrix}; \mathrm{C} = \begin{pmatrix} 1 & 0 \end{pmatrix}; \mathrm{D} = \begin{pmatrix} x_0 \end{pmatrix}.$$

The transfer function with a second-order characteristic polynomial was derived from the representation in the state space

$$Z(s) = -\frac{a_1 x_0 b_3}{s^2 + b_2 s + a_1 x_0 b_1} U(s) + a_3 \frac{s + b_2}{s^2 + b_2 s + a_1 x_0 b_1} P(s)$$

The simulation showed a practical coincidence of the transition functions of the linearized and the original system within two to three hours from the moment of eating (Figure 1), so, the linear model was accepted for further research provided that the maximum of the transition function of the system with the controller is reached at the specified time.

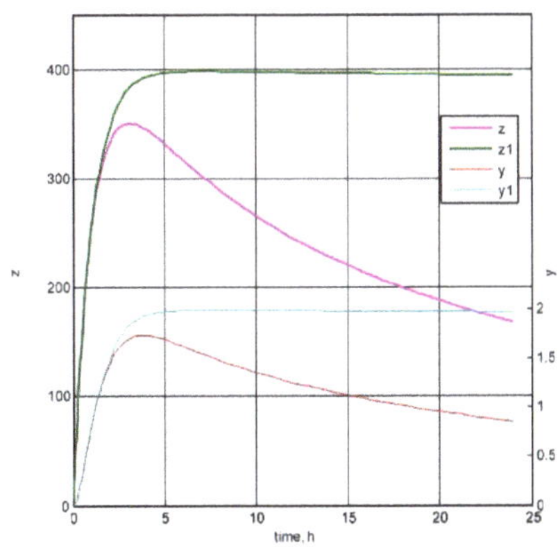

Figure 1. Comparison of transients in the initial (z,y) and linearized (z_1,y_1) models. The control object is a patient with diabetes mellitus (DM) [17].

When choosing the controller configuration method, the main preference was given to engineering methods that allow for calculation of the controller parameters based on the characteristics of the transient process graph. To date, the most widespread use is made of PID controllers. Practical experience shows that the use of PID controllers in complex systems (devices) provides a sufficiently small control error and the transition process required by the operating conditions of most control objects. Therefore, the PID regulation will be the priority in the control task under study. Currently, there are quite a lot of different methods for tuning the parameters of PID controllers; still, the most common is the Ziegler–Nichols tuning method [18], proposed by its authors in 1942. This method is simple in application but gives not very good results. As a rule, after setting the regulator parameters, one has to make manual adjustments in order to improve the quality of regulation. Nevertheless, this method is still often used in practice, although many more accurate methods have become available. Ziegler and Nichols proposed two methods for tuning PID regulators [19]. One of them is based on the response parameters of the object to a single jump, and the other is based on the frequency characteristics of the control object. To calculate the PID controller parameters according to the first Ziegler–Nichols method, only two parameters are used: the negative coordinate "a" of the point of intersection of the tangent to the transient curve with the maximum slope, and the coordinate "L" of the point of intersection of the same tangent and the time axis. The formulas for calculating PID controller coefficients are summarized in Table 1.

Table 1. Formulas for calculating regulator coefficients by the Ziegler–Nichols method.

Controller Type	Step Response K	T_i	T_d	Frequency Parameters K	T_i	T_d
P	$1/a$			$0.5/K_{180}$		
PI	$0.9/a$	$3L/K$		$0.4/K_{180}$	$0.8T_{180}/K$	
PID	$1.2/a$	$0.9L/K$	$0.5L/K$	$0.6/K_{180}$	$0.5T_{180}/K$	$0.125T_{180}/K$

The Ziegler–Nichols method does not take into account in any way the requirements for the stability margin of the system, which is its significant disadvantage. In addition, the Ziegler–Nichols method gives parameters that are far from optimal. The second Ziegler–Nichols method (the frequency method), as input data for calculations, uses the frequency ω_{180}, at which the phase shift in the open circuit reaches 180°, and the modulus of the loop gain at this frequency is K_{180}. Knowing the parameter ω_{180}, first, find the period of natural oscillations of the system, and then, from Table 1, determine the parameters of the regulator. The accuracy of the controller tuning and the disadvantages of both Ziegler–Nichols methods are the same. The CHR method, which uses a similar method for calculating the parameters of the PID regulator, allows us to bypass the drawbacks of the Ziegler–Nichols method. The authors of this method, Chien, Hrones, and Reswick (CHR) [20], in contrast to Ziegler and Nichols, used the criterion for the maximum slew rate in the absence of overshoot or with no more than 20% overshoot. This criterion gives a larger stability margin than in the Ziegler–Nichols method. The CHR method gives two different systems of regulator parameters. One is obtained by observing the response to the setpoint change (Table 2), and the other is obtained by observing the response to external disturbances (Table 3). Which parameter system to choose depends on what is more important for a particular controller: the quality of control when changing the setpoint or the attenuation of external disturbances. If both are important, it is necessary to use regulators with two degrees of freedom [21].

Table 2. Formulas for calculating regulator coefficients by CHR method, by response to set point change.

Controller Type	No Overshoot K	T_i	T_d	20% Overshoot K	T_i	T_d
P	$0.3/a$			$0.7/a$		
PI	$0.35/a$	$1.2L/K$		$0.6/a$	$1.0L/K$	
PID	$0.6/a$	$1.0L/K$	$0.5L/K$	$0.95/a$	$1.4L/K$	$0.47L/K$

Table 3. Formulas for calculating regulator coefficients by CHR method, by response to external perturbations.

Controller Type	No Overshoot K	T_i	T_d	20% Overshoot K	T_i	T_d
P	$0.3/a$			$0.7/a$		
PI	$0.6/a$	$4L/K$		$0.7/a$	$2.3L/K$	
PID	$0.95/a$	$2.4L/K$	$0.42L/K$	$1.2/a$	$2.0L/K$	$0.42L/K$

The CHR method uses an approximation of the object by a first-order delayed model. CHR uses the same initial parameters a and L as the Ziegler–Nichols method. Thus, each system requires a different approach to adjust the controller parameters. A comparative analysis of the methods was carried out, and the CHR method was adopted as providing a better quality of the transient process and a greater margin of stability of the system with a controller compared to the Ziegler–Nichols method. A method was developed for determining the parameters of the PID controller based on a single input variable—the angle of inclination of the tangent to the transient graph. The method was based on the approximation of the transition process graph by a polynomial, followed by analytical differentiation of this polynomial, in order to accurately determine the angle of the tangent. The following algorithm of PID controller tuning was developed:

1. Obtaining a tabular function (graph) of the transient with zero controller settings, with individual parameters a_i and b_i of the model.
2. Determination of the boundary value a_{max} of the regulator.
3. Calculation of the regulator's coefficients.
4. Calculation of the transient process with the regulator.
5. Comparison of the transient process of step 4 of the algorithm with the transient process of a "healthy person".
6. If the difference exceeds the admissible one, decrease by "a" and pass to step 3.
7. Otherwise, the tuning is completed.

Step 2 of the algorithm—the task of tangent construction, or the task of numerical differentiation of a tabularly defined function—requires detailed consideration. The simplest solution to this problem is based on determining the derivative of the function:

$$y' = \lim_{\Delta x \to 0} \frac{\Delta y}{\Delta x}$$

and uses various finite difference formulas [22]; however, the finite difference method is associated with known difficulties in estimating the error of the result. As stated in [22], the main components of the error of numerical differentiation are the approximation error (also called the truncation error) and rounding errors in computer calculations. The approximation error is determined by the magnitude of the residual term—the difference between the approximated and actual values of the derivative. It is noted in [22] that the analysis of the residual term is nontrivial and that the approximation error tends to decrease as the step Δx decreases. In contrast to the approximation error, the rounding error increases as the step Δx decreases. Therefore, the total error of numerical differentiation can decrease as the step decreases only up to some limiting value, after which a further decrease in the step will, at best, not improve the accuracy of the results. In addition, finite difference formulas usually use a constant step, and the resulting simulation function may be a variable step. Optimal accuracy can be achieved by regularizing the numerical differentiation procedure. The simplest way to regularize is to choose a step Δx such that the inequality is as follows:

$$|f(x+h) - f(x)| > \varepsilon, \tag{2}$$

where $\varepsilon > 0$ is some small number. When calculating the derivative, this eliminates the subtraction of closely related numbers, which usually leads to an increase in errors. This is all the more dangerous when subsequently dividing the increment of a function by a small number Δx. This method cannot be applied to our problem for the reason mentioned above for the variable step Δx in the original table function.

Another method for regularization is to smooth the tabulated values of the function by selecting some smooth approximating function, for example, a polynomial. Further differentiation of the polynomial using the derivatives table does not present computational difficulties and does not introduce additional errors. This method was used to solve the problem.

The polynomial calculated by the built-in polyfit function will be used as the interpolation polynomial. This choice among many variants [22]—Lagrange, Hermite, and other polynomials—is dictated by the simplest implementation in the program. The transient graph is a fairly smooth curve. By plotting the polynomial trend on this graph, the degree of the polynomial that showed the best coincidence with the graph of the original function was determined. It was found that the polynomial should be of degree 10. This does not exclude a lower degree polynomial; the change in degrees was implemented in the program. The following algorithm was developed to form a set of graph points to build the approximating polynomial:

1. The first point of the set, obviously (0; 0), was marked as "extreme"; the current set length is $k = 1$.
2. Since the maximum value of the transient function was about 400, we took the value $\varepsilon = 0.5$, which is small enough in comparison with the maximum and simultaneously much larger than the accuracy of calculations in MATLAB (it is equal to the system constant $eps = 2.2204 \cdot 10^{-16}$); in this way, the gross rounding errors were leveled.
3. The "outermost" point of the set was assigned to the current point.
4. We proceed to the next point of the transient table function and check if condition (2) holds.
5. If condition (2) is unfulfilled, then we proceed to item 4.
6. If condition (2) is satisfied, the point was added to the set—it becomes a "last" point, and the length of the set increased by $k = k + 1$.
7. The condition $k = (n + 1)$ was checked—for the uniqueness of the approximating polynomial the set must contain at least $(n + 1)$ points, where n is the degree of the polynomial.
8. If the condition $k = (n + 1)$ is unfulfilled, pass to item 3.
9. Otherwise, the set is formed—end.

After forming a vector of approximating polynomial coefficients by this set, this vector was transformed into a vector of derivative coefficients according to the following scheme:

$$\begin{pmatrix} polynom \\ a_n \\ a_{n-1} \\ \cdots \\ a_1 \\ a_0 \end{pmatrix} \rightarrow \begin{pmatrix} derivative \\ na_n \\ (n-1)a_{n-1} \\ \cdots \\ 1a_1 \\ 0 \end{pmatrix} \rightarrow \begin{pmatrix} derivative \\ na_n \\ (n-1)a_{n-1} \\ \cdots \\ 1a_1 \end{pmatrix}$$

and the value of this derivative $y'(0)$ at the point of food intake point was calculated. Since the food intake point has the coordinates (0;0), the equations of the tangent are as follows:

$$y(x) = y'(0) - x$$

and

$$a = y'(0) - L$$

The parameter L has a clear biophysical meaning and is the average time elapsed from the beginning of a meal to the feeling of fullness, which is the signal of an increase in blood sugar level because of the digestion of nutrients. The study of this parameter and the scientific substantiation of its value is a topic for a separate study that is beyond the scope of this work, so we took an a priori value of 15 min or

$$L = 0.25h.$$

Each control system requires its own approach for setting the controller parameters. In this case, the setup process is hampered by some factors, namely:

1. Since the PID controller is an external device to the control object, it cannot be implemented in the control object;
2. The food intake and insulin production (or injection) subsystems in the diabetic patient are practically unrelated to each other;
3. The characteristics of each diabetic patient's model are individual and can vary quite widely.

Considering these factors, the MATLAB/Simulink software package was selected to study the linearized insulin injection control system and test the PID controller tuning algorithm. The development of the model in Simulink is characterized by relative simplicity and clarity, and the management of this model from the Matlab script (m-file) allows for automation of the process of setting model parameters and processing simulation results. In the process of research, it was necessary to take into account a significant variation in the transfer function coefficients of the model. Therefore, the model of diabetes mellitus itself was selected in the mathematical modeling program. Its visual implementation was carried out by means of Simulink. The model is shown in Figure 2. In accordance with the type of transfer function, the model has two channels—one for nutrition and another for insulin injection. The transfer function of each channel is recorded with parametrization through the components of the coefficient vectors $a(i)$ and $b(i)$. The PID controller as a block from the Simulink library is included in the feedback loop of the channel for insulin injection. Physically, the PID controller is an insulin pump. The nutrition function according to [4] has the form

$$P(t) = \begin{cases} 0, & t < t_0 \\ Qe^{-K(t-t_0)}, & t \geq t_0 \end{cases}$$

where Q is the amount of food, K is a parameter that characterizes the type of sugar that comes with food, and t_0 is the meal time. This function is implemented in the MATLAB function block. The signals from the channel outputs are summed according to the expression for the transfer function $Z(s)$. The transfer to MATLAB of the following simulation results is provided—the model time is represented as t, and the insulin level is represented as y.

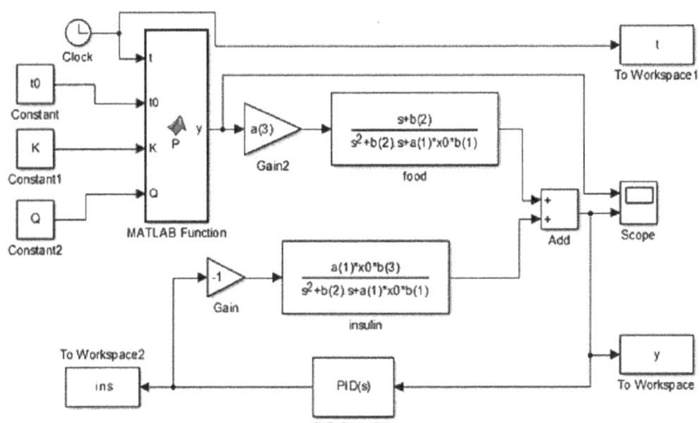

Figure 2. Development of a control object model in MATLAB/Simulink [17].

The interaction algorithm between the model and the MATLAB script that calls it includes the following steps:

1. Initialization of model parameters—assignment of the specific values for $a(i)$ and $b(i)$;
2. Modeling;
3. Processing of simulation results (vectors [t, y] returned by the model).

Then, steps 1–3 are repeated according to the purpose of the study; i.e., these items are executed the required number of times or until another termination condition is met [17].

3. Results & Discussions

The criterion for completing the adjustment was the specified excess of the maximum insulin level over the level of a healthy person, while the level of a healthy person was calculated by a direct numerical solution of the original nonlinear differential in Equation (1). A number of computational experiments were carried out on the model, as a result of which the limits of changing the parameters of the PID controller were determined:

- Gain $K = 12\ldots24$;
- Time constant of the integrator $T_i = 0.01\ldots0.05$ h;
- Derivation time constant $T_d = 0.004\ldots0.008$ h.

A transient graph with the controller settings $K = 18$, $T_i = 0.0333$ h, and $T_d = 0.0058$ h, for a patient with an insulin pump (IP), is shown in Figure 3. For comparison, the same graph shows the transition process in the "healthy" model.

As a result of the work completed, the prospects of applying the considered approach to the development of controllers for insulin pumps can be considered confirmed. This is evidenced by the results of modeling and comparing the parameters of the transition process (see Figure 3). The advantage of using a linear model is the simplification of the controller tuning procedure, and linearization allows you to automate this procedure. Thus, it is possible to achieve the dynamics of the blood sugar level of the "patient" to those characteristic of the "healthy".

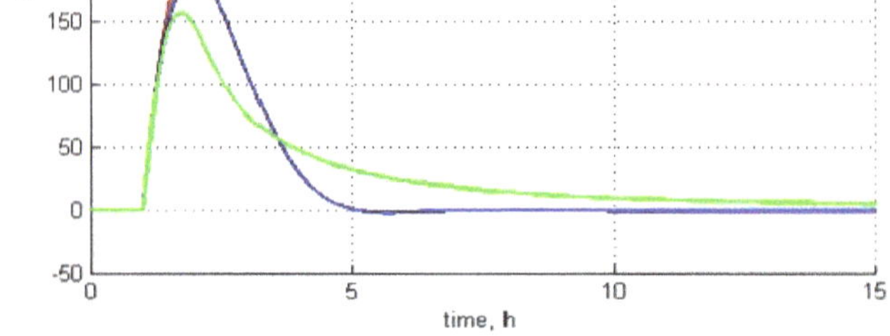

Figure 3. Demonstration of transient modeling results [17].

4. Conclusions

To sum up, we can say that the goal of regulation set in the work has been achieved—a simulation model has been created in MATLAB as a result of the following stages:

1. Based on the study and analysis of existing mathematical models of diabetes mellitus, in terms of the application (to them) of methods of the theory of linear automatic

control systems, a non-linear model was chosen that is consistent with the known physiological facts while having a minimal set of equations and parameters.
2. The selected model, with some assumptions, has been modified and linearized.
3. A mathematical model of the control object (human) was built in the MATLAB/Simulink environment, and, based on the simulation results, a conclusion about the suitability of the linearized model for the development of the PID controller of the insulin pump control system was formed.
4. In the MATLAB/Simulink environment, a model of a control system with a PID controller was built; an algorithm, implemented as an m-file, was created to determine the parameters of the controller; and the adequacy of the algorithm was experimentally verified.
5. The impact of changing the model parameters during the control process was evaluated due to the possible significant variation in these parameters when moving from one patient (digital model) to another.
6. The insignificant influence of sufficiently large deviations of the model parameters on the controller parameters was experimentally established, and a conclusion was formed about the stability of the control object with a PID controller to changes in the parameters of the control object model.

These results suggest the following directions for further research:

1. Improving regulation by introducing a relay element into the control loop that turns off the insulin pump when a certain relatively low sugar level is reached in order to prevent hypoglycemia [5];
2. Study of the system with a regulator with multiple periodic meals to study the effect of breaks between meals;
3. Development of a more convenient interface for the controller parameter setting program.

Author Contributions: Conceptualization, D.A.; Formal analysis, S.K.; Investigation, S.K.; Methodology, D.A.; Project administration, D.A.; Software, S.K.; Validation, D.A.; Visualization, S.K.; Writing—original draft, S.K.; Writing—review and editing, D.A. All authors have read and agreed to the published version of the manuscript.

Funding: This research received no external funding.

Institutional Review Board Statement: The study was conducted according to the guidelines of the Declaration of Helsinki, and approved by the Institutional Review Board (or Ethics Committee) of The Patrice Lumumba Peoples' Friendship University of Russia (protocol code 200, 2006).

Informed Consent Statement: Written informed consent has been obtained from the patient(s) to publish this paper.

Data Availability Statement: The study did not report any data.

Conflicts of Interest: The authors declare no conflict of interest.

References

1. Dedov, I.I.; Shestakova, M.V.; Aleksandrov, A.A.; Galstyan, G.R.; Grigoryan, O.R.; Esayan, R.M.; Kalashnikov, V.Y.; Kuraeva, T.L.; Lipatov, D.V.; Mayorov, A.Y.; et al. *Algorithms of Specialized Medical Care for Patients with Diabetes Mellitus: Clinical Guidelines*, 10th ed.; Diabetes Mellitus: Moscow, Russia, 2021; 14p.
2. Shapoval, I.N.; Nikitina, S.Y.; Ageeva, L.I. (Eds.) *Healthcare in Russia. 2019: Statistical Collection*; Rosstat: Moscow, Russia, 2019; 170p.
3. Pfeiffer, E.F.; Thum, C.; Clemens, A.H. The artificial beta cell-a continuous control of blood sugar by external regulation of insulin infusion (glucose-controlled insulin infusion system). *Horm. Metab. Res.* **1974**, *6*, 339–342. [CrossRef] [PubMed]
4. Clemens, A.H.; Chang, P.H.; Myers, R.W. The development of Biostator, a glucose controlled insulin infusion system (GCIIS). *Horm. Metab. Res.* **1977**, *7*, 23–33.
5. Steil, G.M.; Rebrin, K.; Darwin, C.; Hariri, F.; Saad, M.F. Feasibility of automating insulin delivery for the treatment of type 1 diabetes. *Diabetes* **2006**, *55*, 3344–3350. [CrossRef] [PubMed]

6. Davoust, A.; Gavigan, P.; Ruiz-Martin, C.; Trabes, G.; Esf, iari, B.; Wainer, G.; James, J. An architecture for integrating BDI agents with a simulation environment. In *Engineering Multi-Agent Systems*; Springer International Publishing: Cham, Switzerland, 2020; pp. 67–84.
7. Mikhalsky, A.I.; Novoseltseva, J.A. Modeling of a continuous glycemic curve. In Proceedings of the 12th International Conference "Managing the Development of Large-scale Systems", Moscow, Russia, 1–3 October 2019; IPU RAS: Moscow, Russia, 2019.
8. Sorensen, J.T. A Physiologic Model of Glucose Metabolism in Man and Its Use to Design and Assess Improve Insulin Therapies for Diabetes. Ph.D. Thesis, Department of Chemical Engineering, MIT, Cambridge, MA, USA, 1985.
9. Steil, G.M.; Clark, B.; Kanderian, S.; Rebrin, K. Modeling insulin action for development of a closed-loop artificial pancreas. *Diabetes Technol. Ther.* **2005**, *7*, 94–108. [CrossRef] [PubMed]
10. Cobelli, C.; Federspil, G.; Pacini, G.; Salvan, A.; Scandellari, C. An integrated mathematical model of the dynamics of blood glucose and its hormonal control. *Math. Biosci.* **1982**, *58*, 27–60. [CrossRef]
11. Bergman, R.N.; Ider, Y.Z.; Bowden, C.R.; Cobelli, C. Quantitative estimation of insulin sensitivity. *Am. J. Physiol. Endocrinol. Metab.* **1979**, *236*, E667. [CrossRef] [PubMed]
12. Hovorka, R.; Canonico, V.; Chassin, L.J.; Haueter, U.; Massi-Benedetti, M.; Federici, M.O.; Pieber, T.R.; Schaller, H.C.; Schaupp, L.; Vering, T.; et al. Nonlinear model predictive control of glucose concentration in subjects with type 1 diabetes. *Physiol. Meas.* **2004**, *25*, 905–920. [CrossRef] [PubMed]
13. Doodnath, A.; Kong, A.; Sastry, M.K.S. Optimal Linear Control of Blood Glucose. In Proceedings of the WRI World Congress on Computer Science and Information Engineering 2009, Los Angeles, CA, USA, 31 March–2 April 2009; Volume 5.
14. Davis, M.J. Differential model of diabetes mellitus. In *Mathematical Modelling*; Andrews, J.G., McLone, R.R., Eds.; Butterworths: London, UK, 1976; Volume 260, pp. 128–139.
15. Kozeko, L.G.; Tsirulyova, V.M.; Shapovalova, I.A. The Study and Numerical Implementation of the Mathematical Model of the Task of Correcting Blood Sugar Levels. Study and Implementation of Numerical Mathematical Model the Task of Adjusting the Level of Sugar in the Blood. *TVSU Ser. Chem.* **2018**, *1*, 169–178.
16. Kuropatkin, P.V. *Theory of Automatic Control*; A Textbook for Electrotechnics Specialties of Universities; Vysshaya Shkola: Moscow, Russia, 1973; 527p.
17. Andrikov, D.A.; Kurbanov, S.V. Development of a Linear Model of Diabetes Mellitus with a PID Controller and a Method for Automatic Adjustment of the PID Controller. In *Intelligent Sustainable Systems*; Lecture Notes in Networks and Systems; Nagar, A.K., Jat, D.S., Marín-Raventos, G., Mishra, D.K., Eds.; Springer: Singapore, 2022; Volume 333.
18. Denisenko, V.V. PID controllers: Questions of realization the modification no 1. *Mod. Autom. Technol.* **2008**, *4*, 86–99.
19. Ziegler, J.G.; Nichols, N.B. Optimum settings for automatic controllers. *Trans. ASME* **1942**, *64*, 759–768. [CrossRef]
20. Chien, K.L.; Hrones, J.A.; Reswick, J.B. On automatic control of generalized passive systems. *Trans. ASME* **1952**, *74*, 175–185. [CrossRef]
21. Denisenko, V.V. PID controllers: Principles of construction and modification. *Mod. Autom. Technol.* **2006**, *4*, 66–74. (In Russian)
22. Turchak, L.I. *The Fundamentals of Numerical Methods*; Nauka: Moscow, Russia, 1987; 320p.

engineering proceedings

MDPI

Proceeding Paper

Extracting and Processing of Russian Unstructured Clinical Texts for a Medical Decision Support System †

Irina Bolodurina [1,2], Alexander Shukhman [1], Leonid Legashev [1,*], Lyubov Grishina [1] and Arthur Zhigalov [1]

1 Research Institute of Digital Intelligent Technologies, Orenburg State University, Prosp. Pobedy 13, 460018 Orenburg, Russia
2 Department of Public Health and Healthcare, Orenburg State Medical University, Sovetskaya Street 6, 460000 Orenburg, Russia
* Correspondence: silentgir@gmail.com
† Presented at the 15th International Conference "Intelligent Systems" (INTELS'22), Moscow, Russia, 14–16 December 2022.

Abstract: The rapid growth in the volume of medical data is pushing the development and implementation of artificial intelligence (AI) tools. One of the directions of the application of AI in the field of healthcare is the use of natural language processing methods to build medical decision support systems based on electronic medical record (EMC) data. As a result of this study, a module for the extraction and pretreatment of patients' EMC was developed. In addition, an approach was implemented to extract features from the unstructured textual information of patient admission protocols, with the formation of an appropriate vector representation of data. Predictive models for the diagnosis of groups of diseases based on the logistic regression model and BERT were developed. The highest efficiency in the experiments was shown by the logistic regression model, with a F1-score of 0.81 and Matthews correlation coefficient of 0.75. The obtained results have been posted for public access based on the django framework and can be used for preliminary assessment of patient health status, as well as integrated into existing medical decision support systems.

Keywords: electronic health records; medical decision support system; natural language processing; BERT; logistic regression

Citation: Bolodurina, I.; Shukhman, A.; Legashev, L.; Grishina, L.; Zhigalov, A. Extracting and Processing of Russian Unstructured Clinical Texts for a Medical Decision Support System. *Eng. Proc.* **2023**, *33*, 41. https://doi.org/10.3390/engproc2023033041

Academic Editors: Askhat Diveev, Ivan Zelinka, Arutun Avetisyan and Alexander Ilin

Published: 26 June 2023

1. Introduction

Cardiovascular diseases (CVD) continue to be the most urgent health problem in most countries of the world, including in the Russian Federation. In 2020, cardiovascular diseases became the most common cause of death (47%) and claimed the lives of more than 900 thousand Russians. In this regard, it is necessary to develop new approaches to reducing this indicator.

From an economic point of view, the direct costs of public health for the treatment and diagnosis of CVD amount to about RUB 220 billion. This indicator is 8-times higher than the cost of screening and prevention, with which 40% of CVD cases can be prevented with a proper assessment of the risks of development [1,2].

The large growth of medical data is pushing the development of AI tools, for implementation, processing, and analysis. One of the directions of the application of AI in the field of healthcare is the use of NLP methods to build systems to support medical decision-making based on electronic medical records. One of the tasks of a medical decision-making system is the task of determining a diagnosis according to the ICD and based on patient complaints. Thus, the task of multiclass classification based on the text documents of the EMC arises.

2. Related Work

Currently, natural language processing (NLP) methods allow analyzing unstructured information and building highly efficient AI models [3]. In this regard, scientists around the world are engaged in the development and application of NPL methods in the field of digital medicine.

Thus, the study in [4] presented an approach to the processing and analysis of electronic medical records (EHR) of patients based on NLP and deep learning methods for prediction in healthcare. The presented methodology can be used to evaluate various health indicators and in subsequent decision-making. In [5], this approach was also highlighted as the main tool for developing end-to-end applications using multimodal data (images, quantitative analysis data, etc.).

The authors of the study in [6] proposed a deep learning model for predicting heart failure according to EHR data in the UK. However, the resulting model in testing demonstrated an AUROC equal to 0.6965, which generally does not correspond to predictive models of high accuracy.

An effective convolutional neural network (CNN) model for estimating the costs and duration of hospital stays was presented in [7]. The peculiarity of this model is its ability to extract potential knowledge from clinical data with low-frequency medical events.

In [8], an algorithm for predicting a diagnosis based on a deep neural network and by analyzing the data of the EMC of a department of pediatrics was proposed. In their study, the authors used an unstructured and unbalanced data set to build a model using bidirectional recurrent neural networks. The accuracy of the predictive model was 80.9 according to the precision metric.

The study in [9] presented an approach to deep learning for identifying risk factors for cardiovascular diseases based on EHR analysis. The experimental results showed that the proposed models for the binary classification of the presence of CVD using several individual factors (smoking, diabetes, genetic predisposition, etc.) had a high accuracy of prediction (from 0.81 to 0.96).

Thus, at the moment, research in the field of the diagnosis of diseases and risk factors is often based on NLP and deep learning methods. In addition, the results of evaluating the quality of AI models for solving similar problems showed high predictive power. In this regard, within the framework of this work, a study was conducted on the effectiveness of various models for classifying groups of diagnoses of diseases based on the textual information of patient complaints from EHRs.

3. Problem Statement

At the moment, a medical information system (MIS) is a comprehensive software product, the main purpose of which is to automate the main processes related to the work of medical institutions of general and narrow specialization.

The problem of developing an intelligent decision support system (DSS) for the operational interaction of the patient and the doctor at the reception is as follows: the MIS databases with information about the protocols of visits, the results of additional examinations, etc., are stored in a distributed manner, and additional tools for extraction, transformation, and structuring are needed for the implementation of AI models.

In this regard, within the framework of this study, a methodology of interaction with the regional MIS is presented from the stage of information extraction to the implementation of forecasting results, which is schematically presented in Figure 1.

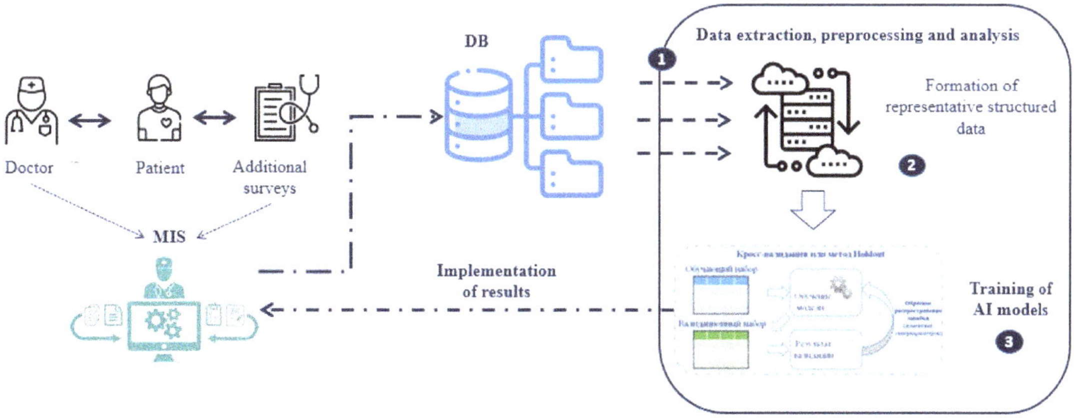

Figure 1. Methodology of the development and implementation of models in MIS.

Based on the presented scheme, there are three main modules for the development of intelligent DSS:

1. A module for extracting and preprocessing large depersonalized data of the electronic medical records of patients;
2. A module for extracting signs from the unstructured textual information of patient visits to a medical organization and for forming a vector representation of data;
3. A module for diagnosing disease groups, to support medical decision-making.

The data storage structure describes for each patient a "case diagnosis" and a "treatment step", which are related to the actual "services rendered". For patients with a history of CVD, within the framework of this study, all protocols of visits to a medical organization, including diagnoses of other groups of diseases, were uploaded. Thus, the purpose of this study is to extract and process unstructured clinical texts in Russian, to build a prognosis of a group of diseases and integrate the results into a medical decision support system.

4. Development of a Module for Extracting and Processing Electronic Health Records

The Medical Information and Analytical Center (MIAC) of Orenburg provided an opportunity to connect to the regional MIS to download depersonalized data from their server. After a preliminary analysis of the data, more than 1 million records of various protocols of medical visits were found, with institutions of patients with CVD (diaries of patient appointments, conclusions of additional examinations (Electrocardiography, blood test, etc.).

For the provided xlm documents, it is necessary to automatically recognize the most informative blocks suitable for building AI models. A characteristic problem of this stage is the availability of documents of various structures - due to the possibility of correction by the doctor of the template of visit protocols, individual concepts for filling in information about additional medical examinations. laboratories, etc. Thus, it is necessary to develop a unified approach to the processing of heterogeneous documents and their informative blocks of textual information.

Within the framework of this, study, methods of parallel reading and processing of the data stream, models and methods of deep learning, as well as NLP text information processing methods were used to build DSS in the diagnosis of groups of diseases.

4.1. Extracting MIS Data

When visiting a patient of a medical organization, the doctor fills out a diary, which consists of data from the objective study of the patient, anamnesis of life, and the complaints of the patient. The anamnesis of life contains information about heredity, bad habits, etc. The generated xml documents with patient examination data do not have a single structure and are modified directly by the doctor.

In this regard, for the development of predictive AI models, the following modules of interaction with the regional MIS of the city of Orenburg were implemented: the XMLParseModule module loads impersonal protocols via the MIAC API in xml format; and for processing heterogeneous templates, a DictParseModule module for automatic conversion of xml documents was developed.

The DictParseModule module for extracting information from heterogeneous xml protocols is based on an approach to recursive node search, with sequential analysis of the contents (Figure 2). A distinctive feature of the proposed approach is the creation of a service record tree in the MOD, which allows analyzing the relationship of certain factors within the document.

Thus, the presented modules convert the xml documents of patient visits into a json file that contains information about complaints at the reception, test results, lifestyle information, etc. into the "key" format:value". As a result of the work of this xml parser module for September–December 2021, 364,020 protocols were uploaded in xml format for patients diagnosed with CVD. The volume of xml files ranged from 3 KB to 1008 KB. The dataset was preprocessed; missing values and records in which the length of the patient's complaint line was less than 100 characters were removed. The final distribution of patient complaint protocols at the reception by disease group is shown in Figure 3.

It should be noted that in addition to cardiovascular diseases, the control group of patients also reported "Acute respiratory infections of the upper respiratory tract" (J0), as well as "New diagnoses of unclear etiology" (U0). These diagnoses were considered in the general order and included in the predictive model.

4.2. Text Information Preprocessing

For AI models, the EMC data obtained after processing xml documents in the form of textual information had to be represented at the input by a feature vector. Let us consider several approaches to feature extraction.

At the first stage, we perform numerical encoding of the target variable—the names of six groups of diseases according to the ICD—and also determine the dictionary of stop words from the Russian-language corpus of the nltk library and the minimum and maximum length of n-grams from 1 to 4.

The first approach to extracting features is as follows: First, operations are performed to convert tokens to lowercase, and remove punctuation, stop words, accents, etc. Next, a collection of unstructured text documents with patient complaints is converted into a matrix of the number of tokens using the CountVectorizer method (bag-of-words model). The resulting vector text embeddings are divided into training and test samples and are used subsequently to train a logistic regression model with cross-validation support.

In addition, within the framework of this study, an approach using Russian-language models of BERT transformers on unstructured medical texts was considered, which consisted in forming a vector representation of tokens of medical texts. In this case, the maximum size of the dictionary num_words = 15,000 and the maximum message length max_len = 200 in tokens were set, and then the sentences of the original dataset were aligned to the same length (padding='post'). Tokenization of the training sample was performed using the EnRuDR-BERT model [10], pretrained on a collection of consumer reviews about taking medications. To solve the problem of classifying a group of diseases, the output layer is represented by six outputs in accordance with the ICD codes. Thus, as a result of training, an attention mask is created for each sample: those tokens that need to be taken into account when training and calculating gradients are filled with units, those tokens that should be skipped are filled with zeros.

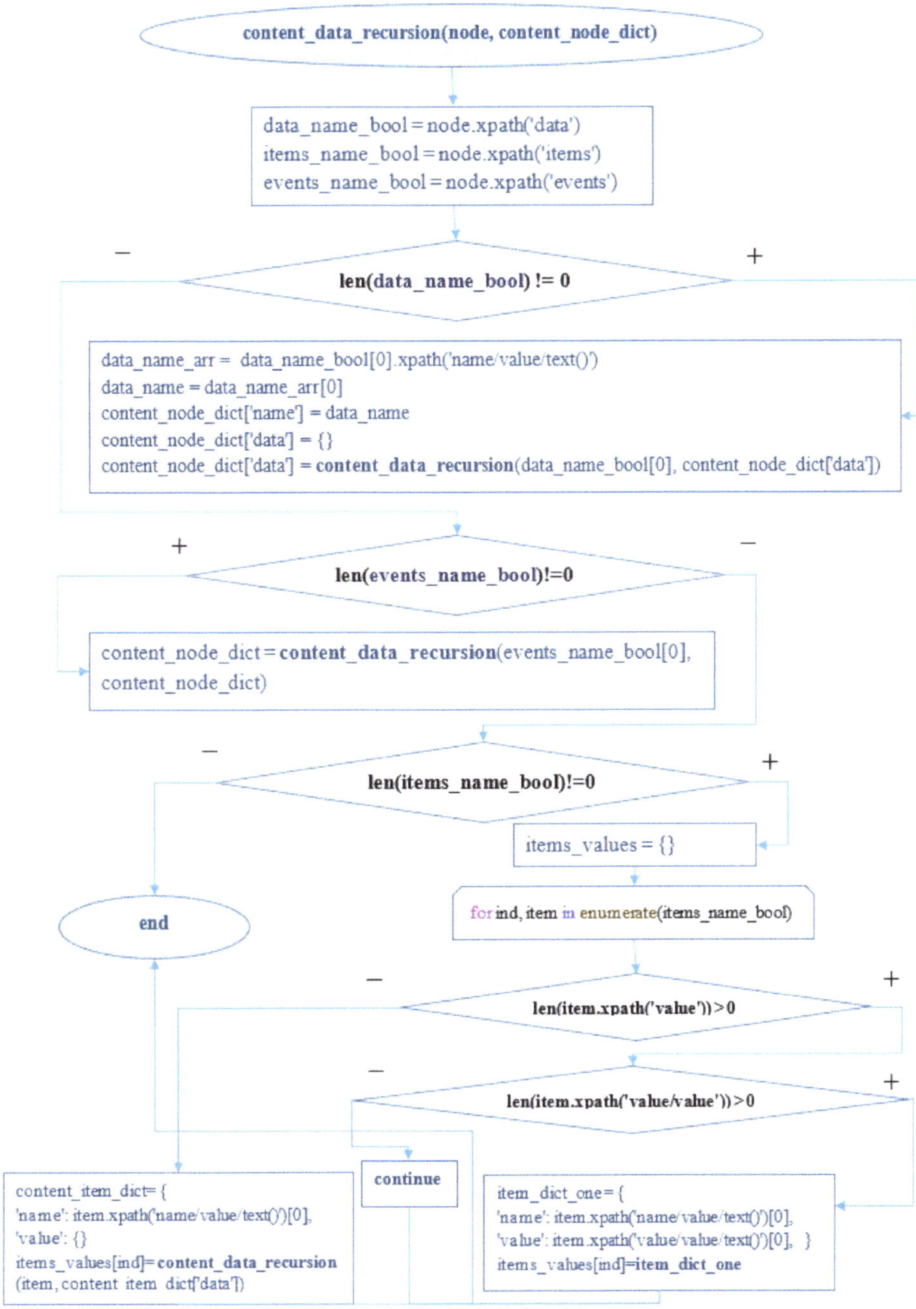

Figure 2. DictParseModule for automatic xml document conversion.

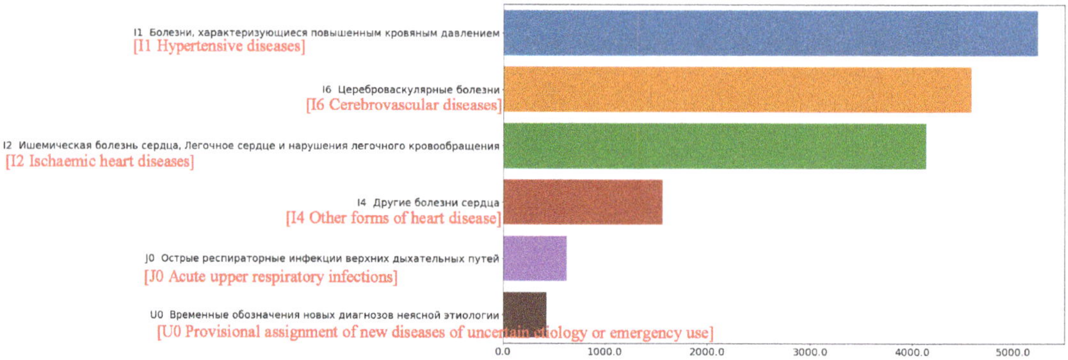

Figure 3. Distribution of patient complaint protocols by disease group.

5. The Training of AI Models

The next stage of constructing a prognosis of a patient's disease group is the training of AI models. Schematically, the process of learning a logistic regression model for classifying groups of diseases is shown in Figure 4. Note that for each sample object, when using this approach, the probability of belonging to one of the six groups of diseases according to the ICD is calculated.

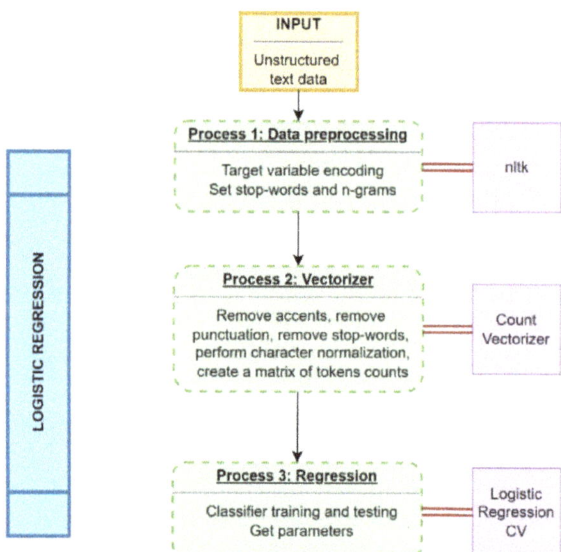

Figure 4. General scheme of the logistic regression approach.

The process of training a BAT-based model is shown in Figure 5. Embeddings are formed using the input layer of the neural network based on a list of dictionary numbers of text tokens. The model was trained and tested. The number of epochs was selected experimentally (epoch = 2). As a result, the error on the training and test dataset had the following values: train_loss: 0.5425, val_loss: 0.5644. The softmax function of the torch library was used to obtain the predicted probability of a sample belonging to one of the six groups of diseases according to the ICD.

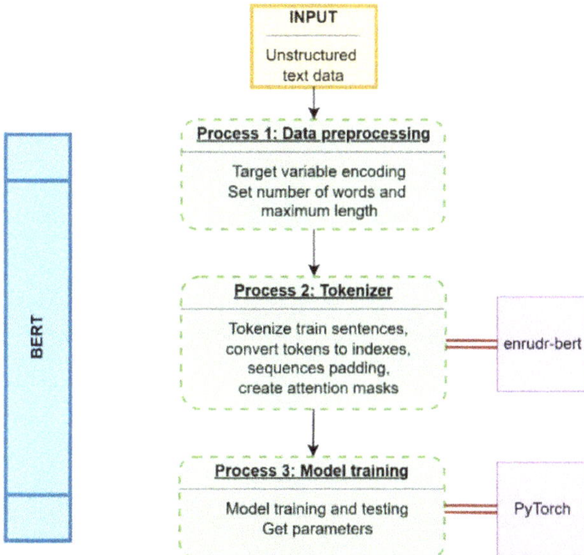

Figure 5. General scheme of the BERT-based approach.

The results of comparing the proposed approaches to predicting groups of diseases using precision, F1-score, and Matthews correlation coefficient (MCC) metrics are presented in Table 1.

The logistic regression approach showed the best results for all metrics. At the same time, the BERT-based approach functioned a little worse, which may indicate the need to retrain the model on specialized medical texts.

Table 1. An example of a table.

Algorithm	Precision	F1-Score	MCC
Logistic Regression	0.8187	0.8161	0.7551
BERT	0.8095	0.8088	0.7450

6. Implementation of a Medical Decision Support System Prototype

A demo version of the logistic regression model is available for general use at http://osudeepai.com/services/disease-ml (accessed on 4 June 2022), implemented using the django framework. The characteristics of the software provided by the provider are as follows: Intel(R) Xeon(R) Gold 6240R processor, 2.40 GHz CPU and 128 GB RAM.

Examples of the probability distribution of AI model classes for the complaints of patients with coronary heart disease are shown in Figure 6. According to the therapist's comments, the spread of probability classes occurred due to the fact that the listed complaints in some cases may relate to several groups of diseases, and in practice the doctor makes the final decision based on personal experience and, possibly, the results of additional examinations.

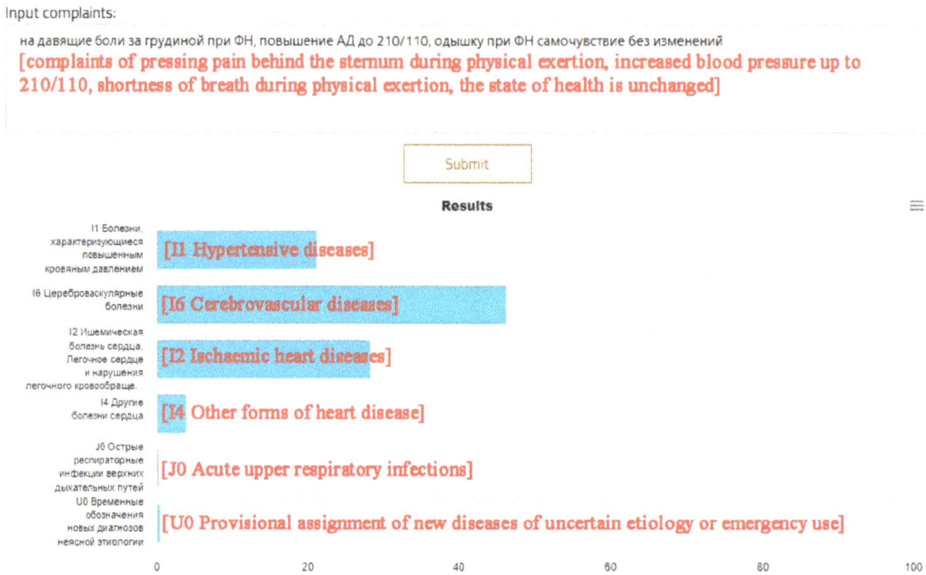

Figure 6. An example of the demo of the DSS based on logistic regression.

7. Conclusions

As a result of this study, a predictive model for the diagnosis of disease groups based on a logistic regression model was developed, which in the experiments showed a value of 0.81 for F1-score and 0.75 for MCC. To train the model, depersonalized regional MIS data obtained by extracting and preprocessing the patients' EMC were used. In addition, an approach to extracting features from the unstructured textual information of patient admission protocols and the formation of an appropriate vector representation of data was additionally implemented. The presented model of disease group prediction can be used for preliminary assessment of a patient's health status and also integrated into existing medical decision support systems. In the future, it is planned to implement a separate AI model for the DSS, which will check the data entered by the user for its relevance in relation to the service used. In addition, one of the areas of further research includes expanding the data set with examples of protocols with other groups of diseases, to scale the results obtained.

Author Contributions: Supervision, I.B.; conceptualization, I.B.; methodology, A.S.; software, A.Z.; investigation, L.L.; validation, A.Z.; writing—original draft preparation, L.G., writing—review and editing, L.G. All authors have read and agreed to the published version of the manuscript.

Funding: The research was carried out within the framework of the Priority 2030 program (Agreement No. 075-15-2021-1171/2 dated 11 May 2022), as well as scholarships of the President of the Russian Federation to young scientists and graduate students (No. SP-919.2022.5).

Institutional Review Board Statement: Not applicable.

Informed Consent Statement: Not applicable.

Data Availability Statement: Data sharing not applicable due to privacy restrictions.

Conflicts of Interest: The authors declare no conflict of interest.

References

1. Federal State Statistics Service. Health Care. 2021. Available online: https://rosstat.gov.ru/folder/13721 (accessed on 3 June 2022).
2. British Heart Foundation. Heart and Circulatory Disease Statistics. 2021. Available online: https://www.bhf.org.uk/what-we-do/our-research/heart-statistics/heart-statistics-publications/cardiovascular-disease-statistics-2020 (accessed on 3 June 2022).
3. Basques, R. What Is Natural Language Processing? 2020. Available online: https://towardsdatascience.com/what-is-natural-language-processing-86a7123a076b (accessed on 3 June 2022).
4. Jain, K.; Prajapati, V. NLP/Deep Learning Techniques in Healthcare for Decision Making. *Prim Health Care* **2021**, *11*, 1–4.
5. Zhou, B.; Yang, G.; Shi, Z.; Ma, S. Natural Language Processing for Smart Healthcare. 2021. Available online: http://arxiv.org/abs/2110.15803 (accessed on 4 June 2022).
6. Denaxas, S.; Stenetorp, P.; Riedel, S.; Pikoula, M.; Dobson, R.; Hemingway, H. Application of Clinical Concept Embeddings for Heart Failure Prediction in UK EHR Data. 2018. Available online: http://arxiv.org/abs/1811.11005 (accessed on 4 June 2022).
7. Feng, Y.; Min, X.; Chen, N.; Chen, H.; Xie, X.; Wang, H.; Chen, T. Patient outcome prediction via convolutional neural networks based on multi-granularity medical concept embedding. In Proceedings of the IEEE International Conference on Bioinformatics and Biomedicine (BIBM), Kansas City, MO, USA, 13–16 November 2017; pp. 770–777.
8. Shi, J.; Fan, X.; Wu, J.; Chen, J.; Chen, W. DeepDiagnosis: DNN-Based Diagnosis Prediction from Pediatric Big Healthcare Data. In Proceedings of the Sixth International Conference on Advanced Cloud and Big Data (CBD), Lanzhou, China, 12–15 August 2018; pp. 287–292.
9. Chokwijitkul, T.; Nguyen, A.N.; Hassanzadeh, H.; Perez, S. Identifying Risk Factors for Heart Disease in Electronic Medical Records: A Deep Learning Approach. In Proceedings of the BioNLP 2018 Workshop 2018, Melbourne, Australia, 19 July 2018; pp. 1–10.
10. Tutubalina, E.; Alimova, I.; Miftahutdinov, Z.; Sakhovskiy, A.; Malykh, V.; Nikolenko, S. The Russian Drug Reaction Corpus and Neural Models for Drug Reactions and Effectiveness Detection in User Reviews. *Bioinformatics* **2020**, *37*, 243–249. [CrossRef] [PubMed]

engineering proceedings

MDPI

Proceeding Paper

Control of Unmanned Vehicles in Smart Cities Using a Multi-Modal Brain–Computer Interface †

Daniyar Wolf , Mark Mamchenko * and Elena Jharko

V.A. Trapeznikov Institute of Control Sciences of Russian Academy of Sciences, Profsoyuznaya Street 65, Moscow 117342, Russia; runsolar@mail.ru (D.W.); zharko@ipu.ru (E.J.)

* Correspondence: markmamcha@gmail.com; Tel.: +7-495-198-1720 (ext. 1850)

† Presented at the 15th International Conference "Intelligent Systems" (INTELS'22), Moscow, Russia, 14–16 December 2022.

Abstract: The article presents an overview of several studies in the field of Brain–Computer Interfaces (BCIs), the requirements for the architecture of such promising devices, as well as multi-modal BCI for drone control in a smart-city environment. Distinctive features of the proposed solution are the simplicity of the architecture (the use of only one smartphone for both receiving and processing bio-signals from the headset and transmitting commands to the drone), an open-source software solution for signal processing, generating, and sending commands to the unmanned aerial vehicle (UAV), as well as multimodality of the BCI (the use of both electroencephalographic (EEG) and electrooculographic (EOG) signals of the operator). For bio-signal acquisition, we used the NeuroSky Mindwave Mobile 2 headset, which is connected to an Android-based smartphone via Bluetooth. The developed Android application (Tello NeuroSky) processes signals from the headset and generates and transmits commands to the DJI Tello UAV via Wi-Fi. The decrease (depression) and increase of α- and β-rhythms of the brain, as well as EOG signals that occur during blinking were the triggers for UAV commands. The developed software allows the manual setting of the minimum, maximum and threshold values for the processed bio-signals. The following commands for the UAV were implemented: take-off, landing, forward movement, and backwards movement. Two threads of the smartphone's central processing unit (CPU) were utilized when processing signals in the software to increase the performance: for signal processing (1-D Daubechies 2 (db2) wavelet transform) and updating data on the diagrams, and for generating and transmitting commands to the drone.

Keywords: Brain–Computer interface (BCI); Brain–Machine Interface (BMI); unmanned vehicle; unmanned aerial vehicle (UAV); smart city

Citation: Wolf, D.; Mamchenko, M.; Jharko, E. Control of Unmanned Vehicles in Smart Cities Using a Multi-Modal Brain–Computer Interface. *Eng. Proc.* **2023**, *33*, 43. https://doi.org/10.3390/engproc2023033043

Academic Editors: Askhat Diveev, Ivan Zelinka, Arutun Avetisyan and Alexander Ilin

Published: 28 June 2023

1. Introduction

The current concept of a smart city assumes the increase of self-sufficiency of unmanned vehicles (UVs), as well as the amount of data generated and transmitted. The decisions on changing the trajectory and modes of movement will be made either by the UVs themselves or by traffic control centers. The operators in these centers will monitor and control current road/air situations. In addition, in the case of emergencies, the operators will be able to take control of one or more UVs to prevent traffic/air incidents. The rapid development in BCI-based control of robotics, UAVs, and other objects (including in smart environments [1–4]) supposes the implementation and use of such a way of control of UVs in traffic control centers, including as a backup option. UV control using BCI has the potential to reduce the time of transmitting the commands, as well as to ensure simultaneous control of multiple vehicles by one operator. Thus, the problem of developing methods, techniques, algorithms, and software for the control of UVs in smart cities using BCI is relevant. It should be noted that this paper focuses exclusively on the control of aerial objects—UAVs—but the given solution can be adapted for other similar objects, including ground and surface UVs.

It is customary to allocate two main data-processing and transmission nodes in the loop of UV/UAV control using the operator's bio-signals:

- BCI, designed to acquire, convert, and process these bio-signals, classify and detect features;
- Computer–Machine Interface (CMI), designed to convert the output of BCI to drone/robot/machine-compatible control commands and transmit them to the control object.

Although the general solution for operating a robot/UAV using human bio-signals would be better called the Brain–Machine Interface, the literature also refers to it as the BCI (as a combination of the Brain–Computer Interface itself and the CMI). Hereinafter, the term BCI will be used [5].

The work is structured as follows. Section 2 presents a description of existing BCI-based UAV control solutions (related work), the types of bio-signals used, and their analysis. Section 3 provides the requirements for promising BCIs for UV control in a smart-city environment, as well as the aim of the research. Section 4 presents the description of the architecture of the proposed BCI, its hardware, and software. Section 5 presents a discussion on the current state of BCI-based solutions for UAV control, including in smart cities.

2. Related Work

One of the main steps in the development of BCI is the choice of the bio-signal to process, identify changes, and match them with a certain command for the control object. In particular, such signals may include [1–7]:

1. EOG, from blinking and eye movement;
2. Electromyographic (EMG) data resulting from tension/relaxation of facial muscles;
3. EEG, e.g., the increase of α-rhythms and decrease of *beta*-rhythms due to mental relaxation (meditation) or decrease of α-rhythms and increase of *beta*-rhythms when concentrating.

The latter type of signals includes:

- Steady-state visual-evoked potentials (SSVEPs) generated by the brain in response to visual stimulation of a certain frequency (flashes, brightness changes, etc.);
- Motor imagery (MI)—signals that occur during the imagination of performing motor movements;
- Visual imagery (VI)—signals that occur during the visual representation of objects in the absence of appropriate real visual stimuli;
- Speech imagery (SI)—signals that occur during the mental pronouncing of letters or words;
- Mental commands to the control object—signals that occur when imagining the desired state or action of the object (movement, rotation, etc.).

In addition, it is possible to select and use several types of signals; in this case, such BCIs are referred to as multi-modal [6,7]. Below is a description of several existing BCI-based solutions for UAV control, corresponding approaches, methods, algorithms, hardware, and software, as well as the results achieved.

2.1. EOG and EMG

In paper [8], a Parrot Mambo Fly UAV was controlled based on the changes in EOG and EMG signals of the operator. These signals were transmitted to a personal computer (PC) for processing, including signal filtering, the use of a convolutional neural network, and various classifiers (random forest (RF), nearest neighbors, and convolutional) for feature detection. The drone was controlled as follows: the move axis was selected by raising the eyebrows, and the blinking of the left and right eyes corresponded to forward and backward flight along the selected axis. The advantages of the proposed approach are high recognition accuracy (higher than 80%) and low time response (42.21 ms). However, the description of the process of actual application of the proposed BCI architecture for the control of the Parrot Mambo Fly UAV was insufficient, and the overall drone control process is relatively complex.

The authors of [9] presented a method of UAV control using widely available and relatively inexpensive equipment. In particular, the researchers used the TGAM module sensor made by NeuroSky Shennian Technology with two electrodes. Data transfer from BCI to Arduino Pro Mini with NodeMcu for further processing was carried out via Bluetooth (HC-05 module). The UAV (DJI Tello) was connected and controlled via Wi-Fi. EMG and EOG signals were used to form commands. The EMG triggers are eyebrow pick, single, double, and triple blink, corresponding to four commands for the UAV: forward, backward, left, and right. EMG signals were used to prevent losing control of the drone. Several other commands are stated to be implemented, including take-off, landing, move up, and move down. The authors note high control signal delay, low recognition accuracy, and inflexible switching between actions for the proposed solution as the disadvantages of the proposed solution. In addition, it is generally not clear how the drone was operated, i.e., what signals (except EMG) were eventually used to form other commands for the UAV.

2.2. *Mental Concentration and Relaxation*

The authors of [10] proposed an approach to control a virtual drone using Emotiv Insight EEG headset by changing the states of mental concentration and relaxation of the operator. A virtual scene was implemented in the Unity Game Engine environment. The virtual UAV was controlled using proprietary Emotiv Emokey software, which emulates keystrokes. Forward acceleration on the predefined path (track) was realized when the operator enters the active mental state. In a neutral state (mental relaxation), the drone slowed down until it stopped. As a disadvantage, there was the implementation of only two commands and a fixed route for the drone.

2.3. *SSVEPs*

The authors in [11] presented the method of UAV control and the corresponding BCI based on the Unicorn software and hardware, and the g.tec. Hybrid Black EEG headset. The EEG signal was transmitted via Bluetooth to the Unicorn Speller PC. UAV control commands were formed using SSVEPs: The operator focused on certain flashing characters in the Unicorn Speller program. Data from the Unicorn Speller were then passed to another PC via UDP, converted to UAV-compatible commands using Python API, and then transmitted via Wi-Fi to the Parrot Bebop 2 drone. The researchers managed to introduce 12 SSVEP-based commands (take-off, right, left, up, down, move forward, backward, take picture, start video stream, pause, land, and emergency stop). A disadvantage is the bulky hardware architecture of the solution, as well as delays in the transmission of control signals.

Another approach to UAV control using an SSVEP-based BCI is given in article [12]. Emotiv Epoch headset and Easycap devices were used to collect EEG data. Visual stimuli at four frequencies (5.3, 7, 9.4 and 13.5 Hz) were used to form UAV control commands. The 8th-order Butterworth bandpass filter and Fast Fourier Transform algorithm were used for data filtering and SSVEP feature detection, respectively. The use of four SSSVEPs allowed the implementation of the appropriate number of commands for UAV control—take-off, land, move forward, and right turn. The researchers managed to reach a 92.5% value of feature-detection accuracy. However, the architecture of the proposed solution and the stack of tools used is quite complicated: the user is supposed to use the EEG headset, Easycap device, and one PC to form SSVEPs and another one to process the EEG signals.

The authors in [13] presented a BCI for drone control capable of operating in VR and AR environments using head-mounted displays. A DSI VR300 device was used to acquire EEG signals. A virtual scene was developed in Unity, and OpenViBE software was used to communicate with the EEG headset. Eight control commands (turn right, turn left, move up, move down, move left, move right, move forward, and move backward) were formed using SSVEPs (interface buttons flashing with different frequencies). It is possible to control a virtual drone and a real DJI Tello UAV in VR and AR modes, respectively. However, the disadvantage of the proposed solution is similar to previous SSVEP-based BCIs [11,12]—bulky architecture and a large stack of hardware used.

2.4. Motor Imagery

The authors in [5] presented a BCI based on motor imagery (μ brain-wave response) for drone control purposes. The OpenBCI hardware platform was used to acquire the EEG signals. Common Spatial Patterns and Linear Discriminant Analysis methods were used to process EEG data. The authors proposed an algorithm for forming commands for the drone based on the classifier. To verify the proposed approach and algorithm, a virtual simulator with AR Drone 2.0 UAV (including its dynamic model) was used. However, only two commands were implemented—turn left and turn right—by changing the yaw angle, while the UAV in the simulator was moving with a constant velocity.

The authors in [14] proposed the use of LabVIEW and MATLAB software and the Undecimated Wavelet Transform algorithm for noise reduction and resolution analysis. A method for extracting the features of EEG signals using Independent Component Analysis and the coefficient of determination was developed. A hybrid neural network was developed to classify sensorimotor rhythms (MI) in EEG signals. The obtained maximum classification rate result was 95.67%. The LabVIEW environment was used to control a virtual drone.

2.5. Mental Commands

The authors in [15] presented an approach to UAV control based on the mental commands (mind concentration and relaxation) of the operator. The authors presented a mathematical model for processing EEG data in a MATLAB framework. The Emotiv Insight BCI headset was used to acquire the EEG signal, and a PC with the EmotivBCI application was used to process signals and form commands for the DJI Tello UAV via the appropriate API. Three commands were mentioned in the paper—move left, move up, and move down—but only two latter commands were described. A similar approach is given by the authors in [16], its fundamental difference being the purposeful use of consumer-grade EEG headsets for BCI. In particular, Emotiv Insight and Muse (Interaxon) devices were used to acquire EEG signals. The paper also proposes the use of a machine-learning model for classification and feature extraction, considering two classifiers: RF and Support Vector Machine (SVM). The use of the Muse EEG headset and the SVM classifier resulted in 70% feature extraction accuracy. However, the movement of the Crazyflie 2.0 drone in the practical experiment was realized only using two commands (move forward and move backward), while there is no description of other commands (choosing the direction of movement, turns, altitude changes, etc.).

The authors in [17] presented a similar approach to UAV control. Emotiv Insight headset was used for EEG signal retrieval. The object of control was the Parrot Rolling Spider quadcopter. The MATLAB EEGLAB toolbox and BCI2000 software were used to process EEG signals. Feature extraction and training were carried out using the Emotiv Xavier Control Panel software. Five commands for the UAV are available: take-off, move up, move down, change of pitch angle with right turn, and change of pitch angle with left turn. The main disadvantage is the lack of practical testing results of the proposed solution, as well as the use of proprietary systems for feature extraction and training.

2.6. Multi-Modal BCIs

The authors in [18] proposed a BCI that combines the use of MI and SSVEPs. Geodesic Sensor Net (Electrical Geodesics Inc.) was used to acquire an EEG signal. The processed data were transmitted with a fixed time interval to the DJI Matrice 100 quadcopter via Wi-Fi. As a result of the processing of detected features, the following commands for the UAV were implemented: left-forward, right-forward, move up, and move down. Eye-blinking was used to switch between the two flight modes. The main shortcoming of the proposed approach was the relative complexity of the architecture (the use of an EEG headset, one PC for signal processing, another PC, and a monitor for generating and showing visual stimuli).

The authors in [19] proposed an approach to the combined use of MI, VI, and SI techniques for EEG-based control of the drone swarm. A headset with 64 electrodes and BrainVision Recorder (BrainProducts GmbH) software was used to retrieve the EEG signal. The left and right hands were assigned to choose the left or right direction of the swarm of drones, respectively, and arms and legs were assigned to high- and low-altitude flights, respectively (MI usage). Smarm maneuvering, merging, and splitting were carried out via VI, and flight control was conducted using SI. Machine-learning algorithms were used to classify patterns in EEG signals.

The authors in [20] presented an approach to UAV control using BCI based on the recognition of the mental commands of the operator, and EMG signals resulting from the changes in the facial expression of the operator. An Emotiv Insight headset was used to acquire the EEG signal, which was then transferred via Bluetooth to a PC for further processing. Preliminary control signals were transmitted to the Raspberry Pi board, which converted them into drone-compatible commands and sent them to the Parrot Mambo MiniDrone UAV via Wi-Fi. Five commands were implemented: move backward, move forward, turn left, turn right, and land. The authors claim 88% precision recognition of mental commands of the operator. However, the influence of facial motor skills on the EEG signals in the practical experiment was not fully disclosed. Moreover, the architecture of the proposed solution was relatively complex.

The authors in [21] proposed EEG-based drone control algorithms based on monitoring the operator's blinking and mental concentration. A NeuroSky device was used for EEG signal retrieval, and the data acquired were then transferred to a PC via Bluetooth for further processing. The classification system was based on a neural network, the SVM classifier, the Linear Regression Method, and the dynamic concentration threshold value. Based on the number of operator blinks within five seconds, a 4-bit sequence corresponding to a specific command for the drone was generated. Thus, it is possible to form up to 16 commands, and the following nine commands were implemented in the paper: take-off, land, move up, move down, move forward, move backward, move to the right, move to the left, and stop. However, it should be noted that the article does not fully disclose the description of the problem of generating false commands for the UAVs when generating several control signals within less than five seconds.

2.7. Analysis of the Proposed Solutions

In general, based on the studies considered, it is possible to identify the following main shortcomings of the existing BCI solutions for UAV control:

1. The complexity of the control process and switching between actions, insufficient number of implemented commands, as well as a priori simplification of the UAV control (e.g., moving in a straight line, with a fixed velocity, etc.).
2. The complexity of the architecture of the solution (the use of one or more PCs, Arduino/Raspberry Pi boards, as well as various software), which leads to high control signal delay.
3. Proprietary and open-source software solutions for signal retrieval and processing are deployed mainly on PCs.
4. The need to use additional hardware for bio-signal processing (for example, SSVEP-based interfaces may require an additional monitor and a PC to display visual stimuli).

Based on the identified shortcomings, it is possible to form requirements for a promising BCI-based solution for drone control, as well as to state the problem of the research.

3. Requirements for a Promising BCI and Problem Statement

Based on the review of the considered studies that propose different approaches, methods, and algorithms for BCI-based drone control, as well as the shortcomings identified, it is advisable to highlight the following four main requirements for the promising solutions:

1. The implementation of at least 10 control commands (take-off, land, move forward, move backward, move left, move right, move up, move down, turn left, and turn right).
2. Ease of control and switching between actions, for example, using different bio-signals to generate different commands within multi-modal BCI.
3. Simplicity of BCI architecture, and the use of the minimum necessary hardware and software stack for drone control. It is preferable to use open-source software solutions.
4. Taking into account the unique features of the brain activity of each operator, including adaptive adjustments via the use of machine-learning methods when processing bio-signals.

Problem Statement

In this paper, we aim to develop the architecture of the BCI and its corresponding software that meet the third and second (partially) items of the above requirements. The aim of the work is to simplify the hardware components of the BCI, as well as to develop an open-source solution for processing bio-signals and generating commands for a drone. The multimodality of the developed BCI involves the use of both EEG and EOG signals.

4. BCI Architecture, Hardware and Software

This paper uses the proposed concept of non-invasive BCI, as well as methods of EEG signal retrieval and processing [22,23]. To acquire EEG and EOG signals, we used the NeuroSky Mindwave Mobile 2 headset, which was connected to an Android-based smartphone via Bluetooth. The developed Android application—Tello NeuroSky (v. 1.0)—processes the signals received from the headset, generates drone-compatible commands, and transmits them to the DJI Tello UAV via Wi-Fi. The architecture of the proposed BCI and the equipment used are shown in Figures 1 and 2, respectively.

The developed software allows the use of the increase and decrease of α- and β-rhythms and EOG signals from blinking as triggers to generated commands for UAV. The application implements manual adjustment of minimum, maximum, and threshold values for α- and β-rhythms, as well as for EOG signals. The following commands have been implemented: two blinks in a row—take-off; three blinks in a row—land; increase of β-rhythms (above the threshold and α-rhythms)—move forward by 20 cm; increase of α-rhythms (above the threshold and β-rhythms)—move backward by 20 cm. Due to the high computational load on the smartphone CPU, the software uses two separate threads: one for signal processing (1-D Daubechies 2 (db2) wavelet transform) and visualizing data on charts, and another one for generating and transmitting commands to the drone. Figure 3 shows the graphical user interface (GUI) of the developed software.

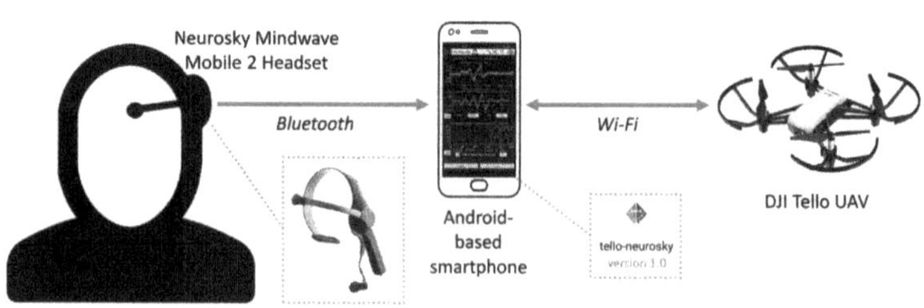

Figure 1. The architecture of the proposed BCI-based UAV control solution.

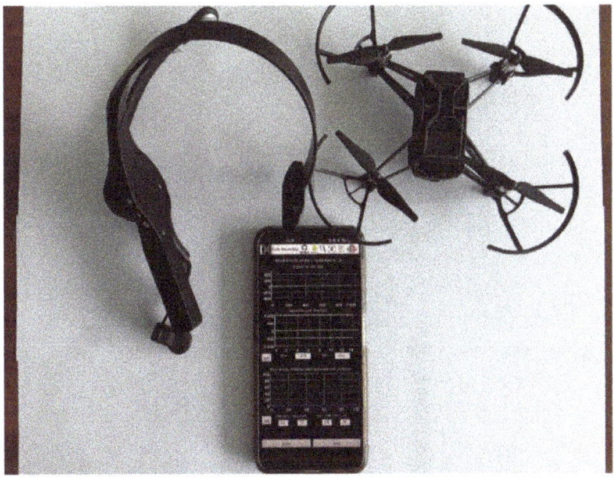

Figure 2. BCI equipment and hardware.

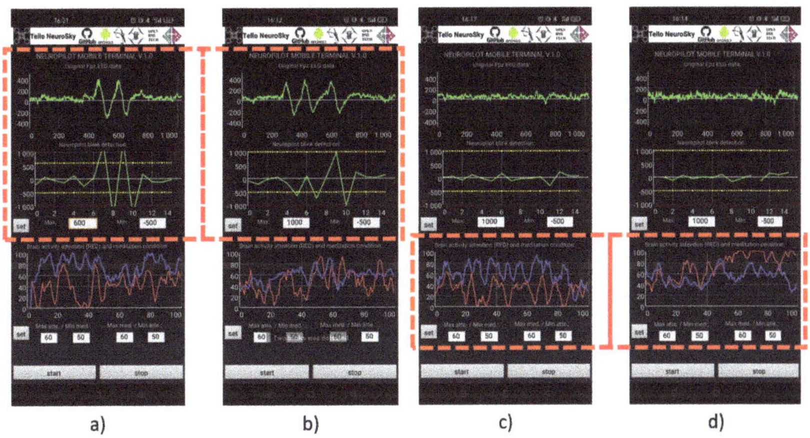

Figure 3. Software GUI (**a**) detection of two blinks in a row; (**b**) detection of three blinks in a row; (**c**) detection of α-rhythm amplification and β-rhythm depression; (**d**) detection of β-rhythm amplification and α-rhythm depression.

5. Discussion

Within the formed list of requirements for a promising BCI-based solution for UAV control (including in smart cities), we managed to implement the third as well as part of the second requirement. The simplicity of the architecture of the proposed solution lies in the use of only one smartphone for retrieving and processing bio-signals, and for transmitting control commands to the drone; the software solution is open-source and free to the public. The developed BCI is multi-modal: the tracking of changes in α- and β-rhythms and eye-blinking are used to form different commands for the UAV. We plan to introduce additional commands for the UAV: changes in μ-rhythms of the brain can be used to implement left/right movement. In addition, machine-learning methods and a neural network will be introduced to better detect blinking and take into account the individual characteristics of the bio-signals of each operator. This paper focuses exclusively on the control of aerial objects—UAVs—though the proposed solution can be adapted for other objects, including ground and surface UVs. The software code for the presented solution is available at [24].

6. Conclusions

The article presents an overview of the current state of research in BCI for UAV control and requirements for similar promising solutions, and proposes multi-modal BCI to control drones in a smart-city environment. The developed software allows the use of the depression and amplification of α- and β-rhythms of the brain and EOG signals that occur during blinking as commands to control the flight of aerial objects. Four commands are implemented: take-off, land, move forward, and move backward.

The distinctive features of the proposed solution are the simplicity of the architecture. Only one Android-based smartphone is used for both receiving and processing signals from the headset and generating and transmitting commands to the drone. The above functions are implemented in the developed open-source publicly available software. The proposed BCI is multi-modal since both EEG and EOG signals are being processed. It is planned to introduce additional commands for a UAV (move left/right) and implement a neural network and machine-learning methods to take into account the individual characteristics of bio-signals of each operator. The presented solution can be adapted to control other mobile objects in smart cities, including ground and surface UVs.

Author Contributions: Data curation, D.W.; formal analysis, D.W. and M.M.; funding acquisition, E.J.; investigation, D.W.; methodology, E.J.; project administration, E.J.; resources, D.W.; software, D.W.; supervision, E.J.; validation, M.M.; visualization, M.M.; writing—original draft, M.M.; writing—review and editing, M.M. All authors have read and agreed to the published version of the manuscript.

Funding: The reported study was partially funded by RFBR, project number 19-29-06044.

Institutional Review Board Statement: Not applicable.

Informed Consent Statement: Informed consent was obtained from all subjects involved in the study.

Data Availability Statement: Data available in a publicly accessible repository that does not issue DOIs. URL: https:/github.com/Runsolar/tello-neurosky.

Conflicts of Interest: The authors declare no conflict of interest.

References

1. Latif, M.Y.; Naeem, L.; Hafeez, T.; Raheel, A.; Saeed, S.M.U.; Awais, M.; Alnowami, M.; Anwar, S.M. Brain computer interface based robotic arm control. In Proceedings of the 2017 International Smart Cities Conference (ISC2), Wuxi, China, 14–17 September 2017; pp. 1–5.
2. Al-Turabi, H.; Al-Junaid, H. Brain computer interface for wheelchair control in smart environment. In Proceedings of the Smart Cities Symposium 2018, Zallaq, Bahrain, 22–23 April 2018; pp. 1–6.
3. Li, Y.; Zhang, F.; Yang, Y. Smart House Control System Controlled by Brainwave. In Proceedings of the 2019 International Conference on Intelligent Transportation, Big Data & Smart City (ICITBS), Changsha, China, 12–13 January 2019; pp. 536–539.
4. Thum, G.E.; Gaffar, A. The future of brain-computer interaction: How future cars will interact with their passengers. In Proceedings of the 2017 IEEE SmartWorld, Ubiquitous Intelligence & Computing, Advanced & Trusted Computed, Scalable Computing & Communications, Cloud & Big Data Computing, Internet of People and Smart City Innovation (SmartWorld/SCALCOM/UIC/ATC/CBDCom/IOP/SCI), San Francisco, CA, USA, 4–8 August 2017; pp. 1–5.
5. Duarte, R.M. Low Cost Brain Computer Interface System for AR.Drone Control. Master's Thesis, Universidade Federal de Santa Catarina Centro Tecnológico Programa de Pós-Graduação em Engenharia de Automação e Sistemas, Florianópolis, Brazil, 2017.
6. Hekmatmanesh, A.; Nardelli, P.H.J.; Handroos, H. Review of the State-of-the-Art of Brain-Controlled Vehicles. *IEEE Access* **2021**, *9*, 110173–110193. [CrossRef]
7. Vărbu, K.; Muhammad, N.; Muhammad, Y. Past, Present, and Future of EEG-Based BCI Applications. *Sensors* **2022**, *22*, 3331. [CrossRef] [PubMed]
8. Villegas, I.D.; Camargo, J.R.; Perdomo, C.C.A. Recognition and Characteristics EEG Signals for Flight Control of a Drone. *IFAC-PapersOnLine* **2021**, *54*, 50–55. [CrossRef]
9. Sun, S.; Ma, J. Brain Wave Control Drone. In Proceedings of the 2019 International Conference on Artificial Intelligence and Advanced Manufacturing (AIAM), Dublin, Ireland, 16–18 October 2019; pp. 300–304.
10. Tezza, D.; Garcia, S.; Hossain, T.; Andujar, M. Brain eRacing: An Exploratory Study on Virtual Brain-Controlled Drones. In *Virtual, Augmented and Mixed Reality. Applications and Case Studies, Proceedings of the 11th International Conference, VAMR 2019, Held as Part of the 21st HCI International Conference, HCII 2019, Orlando, FL, USA, 26–31 July 2019*; Springer International Publishing: Cham, Switzerland, 2019; Volume 11575, pp. 150–162.

11. Al-Nuaimi, F.A.; Al-Nuaimi, R.J.; Al-Dhaheri, S.S.; Ouhbi, S.; Belkacem, A.N. Mind Drone Chasing Using EEG-Based Brain Computer Interface. In Proceedings of the 2020 16th International Conference on Intelligent Environments (IE), Madrid, Spain, 20–23 July 2020; pp. 74–79.
12. Chiuzbaian, A.; Jakobsen, J.; Puthusserypady, S. Mind Controlled Drone: An Innovative Multiclass SSVEP based Brain Computer Interface. In Proceedings of the 2019 7th International Winter Conference on Brain-Computer Interface (BCI), Gangwon, Republic of Korea, 18–20 February 2019; pp. 1–5.
13. Kim, S.; Lee, S.; Kang, H.; Kim, S.; Ahn, M. P300 Brain—Computer Interface-Based Drone Control in Virtual and Augmented Reality. *Sensors* **2021**, *21*, 5765. [CrossRef]
14. Dumitrescu, C.; Costea, I.-M.; Semenescu, A. Using Brain-Computer Interface to Control a Virtual Drone Using Non-Invasive Motor Imagery and Machine Learning. *Appl. Sci.* **2021**, *11*, 11876. [CrossRef]
15. Reddy, M.H.H.N. Brain Computer Interface Drone. In *Brain-Computer Interface*; IntechOpen: Rijeka, Croatia, 2021; pp. 1–19. [CrossRef]
16. Peining, P.; Tan, G.; Aung, A.; Phyo Wai, A. Evaluation of Consumer-Grade EEG Headsets for BCI Drone Control. In Proceedings of the IRC Conference on Science, Engineering, and Technology, Singapore, 10 August 2017; pp. 1–6.
17. Rosca, S.; Leba, M.; Ionica, A.; Gamulescu, O. Quadcopter control using a BCI. *IOP Conf. Ser.* **2018**, *294*, 0120485. [CrossRef]
18. Duan, X.; Xie, S.; Xie, X.; Meng, Y.; Xu, Z. Quadcopter Flight Control Using a Non-invasive Multi-Modal Brain Computer Interface. *Front. Neurorobot.* **2019**, *13*, 23. [CrossRef] [PubMed]
19. Lee, D.-H.; Ahn, H.-J.; Jeong, J.-H.; Lee, S.-W. Design of an EEG-based Drone Swarm Control System using Endogenous BCI Paradigms. In Proceedings of the 2021 9th International Winter Conference on Brain-Computer Interface (BCI), Gangwon, Republic of Korea, 22–24 February 2021; pp. 1–5.
20. Marin, I.; Al-Battbootti, M.J.H.; Goga, N. Drone Control based on Mental Commands and Facial Expressions. In Proceedings of the 2020 12th International Conference on Electronics, Computers and Artificial Intelligence (ECAI), Bucharest, Romania, 25–27 June 2020; pp. 1–4.
21. Abdulwahhab, A.H. Improved Algorithms for EEG-Based BCI Application. Master's Thesis, Istanbul Gelisim University, Institute of Graduate Studies, Istanbul, Turkey, 2021.
22. Turovskiy, Y.; Volf, D.; Iskhakova, A.; Iskhakov, A.Y. Neuro-Computer Interface Control of Cyber-Physical Systems. In *High-Performance Computing Systems and Technologies in Scientific Research, Automation of Control and Production, Proceedings of the 11th International Conference, HPCST 2021, Barnaul, Russia, 21–22 May 2021*; Springer International Publishing: Cham, Switzerland, 2022; Volume 1526, pp. 338–353.
23. Kharchenko, S.; Meshcheryakov, R.; Turovsky, Y.; Volf, D. Implementation of Robot—Human Control Bio-Interface When Highlighting Visual-Evoked Potentials Based on Multivariate Synchronization Index. In Proceedings of the 15th International Conference on Electromechanics and Robotics "Zavalishin's Readings", Ufa, Russia, 15–18 April 2020; Smart Innovation, Systems and Technologies; Springer: Singapore, 2021; Volume 187, pp. 225–236.
24. GitHub Repository. Tello-Neurosky. 2022. Available online: https://github.com/Runsolar/tello-neurosky (accessed on 1 September 2022).

engineering proceedings

MDPI

Proceeding Paper

An Intelligent Gait Data Processing Algorithm Based on Mobile Phone Accelerometers †

Nikolay Dorofeev [1,]* and Anastasya Grecheneva [1,2]

1 Department of Management and Control in Technical Systems, Vladimir State University, Orlovskaya st. 23, 602264 Murom, Russia; grechenevaav@yandex.ru
2 Department of Applied Informatics, Moscow Timiryazev Agricultural Academy, Timiryazevskaya st. 49, 127434 Moskow, Russia
* Correspondence: dorofeevnv@yandex.ru; Tel.: +7-49234-77-236

† Presented at the 15th International Conference "Intelligent Systems" (INTELS'22), Moscow, Russia, 14–16 December 2022.

Abstract: This paper describes an algorithm for extracting human gait movements in data obtained from accelerometer sensors of a mobile phone, provided that the mobile phone is used in the usual mode. The algorithm also performs a classification of the selected movements based on a feed-forward neural network. The developed algorithm selects the best areas in the accelerometer data, which reflect individual steps, according to the optimality criterion. For the selected area, the optimality criterion is the maximum value of the correlation coefficient with all other data segments. The selected plots are used as templates. Changing the parameters of patterns over time is necessary to assess changes in the individual rate of the functioning of the musculoskeletal system. Due to the correction of tolerance limits at the segmentation stage, the algorithm adapts to the change in gait speed.

Keywords: gait; accelerometer; mobile phone; wearable device; automation; algorithm

Citation: Dorofeev, N.; Grecheneva, A. An Intelligent Gait Data Processing Algorithm Based on Mobile Phone Accelerometers. *Eng. Proc.* **2023**, *33*, 44. https://doi.org/10.3390/engproc2023033044

Academic Editors: Askhat Diveev, Ivan Zelinka, Arutun Avetisyan and Alexander Ilin

Published: 3 July 2023

1. Introduction

Biometric information has long been an important component in control and security systems. Increasingly more attention is being paid to biometric data in medicine, statistics, and economics. Modern information technology tools provide the ability to quickly collect and process biometric data, including in a covert mode. Human gait parameters are an example of biometric data that are analyzed in a covert manner. From the point of view of biometrics, gait parameters have attracted researchers for a long time; many approaches and algorithms have been developed for assessing and analyzing gait parameters, but the relevance of this area is not decreasing. This is due to the sharp growth in mobile, personal, and miniature information and technical means in control and authentication; the development of a personalized approach to assessing human health; and innovations in the gaming world [1–3]. Current trends in the processing and analysis of human gait data are aimed at identifying changes in motor patterns and the factors that cause these changes [1,4–6]. Technical means that are used to record gait parameters can be divided into the following categories: stationary systems and systems using wearable sensors or sensors sewn into clothing, video recording systems, and systems based on wearable mobile devices [7–12]. The last category includes fitness bracelets, smart watches, smartphones, and—devices that include an acceleration or spatial orientation sensor. This category is the most interesting from the position of quasi-continuous measurements of gait parameters in the background due to prolonged contact with a person during the day. However, the apparent simplicity of measurements is complicated by the technical features of the wearable devices and the peculiarities of their use. Unlike stationary systems or systems based on wearable sensors, the measurement of gait parameters using wearable devices is possible only at one point, which can move relative to the human body during the measurement process (for example, when moving a smartphone). In addition to the random movement of the smartphone, there are factors that affect measurements (various technical and design features of the device) and human gait in general (for example, physiological ones) [1,13]. Thus, the estimation of human movement (gait) parameters based on measurements at one point in real conditions leads to the need to solve an incorrect inverse problem, where, according to noisy data from one point, it is necessary to roughly estimate the values of parameters in a multi-parameter model of human gait, which evaluates the features of the functioning of the support motor apparatus and the reasons that change its work. All this indicates the need to develop methodological and algorithmic support for the processing and analysis of data on the parameters of a person's approach obtained using wearable devices. Increasing the reliability of the obtained estimates of the values of biometric gait parameters is important for the development of mobile information and analytical systems. The aim of this work is to increase the reliability of the estimated biometric gait parameters according to an acceleration sensor built into the smartphone by developing a neural network algorithm with preliminary correlation processing.

2. Data Processing

The assessment of gait parameters using wearable devices, in particular a smartphone, is carried out in several stages, which in general can be divided into the following: primary processing of the initial data of the acceleration sensor; the step of determining segment repeating areas that may correspond to walking cycles in the analyzed data; evaluation of the parameters of the selected data sections; and comparison of the obtained results with an individual model that was obtained during previous measurements or based on the base model. Taken together, the steps described are computationally expensive, so some of them can be performed on remote servers. At the stage of primary processing of the initial data, noise and high-frequency noise are eliminated using a set of filters, which can be window functions or wavelet filters [14]. Similarly, negative effects are eliminated, for example, accidental movement of the smartphone, the appearance of bias errors, etc., while smoothing the useful signal and reducing the individual features of measurement and gait, so more complex data processing and analysis are required to eliminate interference and noise [1]. At this stage, the user's activity level is also detected to separate the gait phases from random movements and amplitude normalization is performed. Depending on the implementation and processing algorithms, some stages are combined, for example, the user activity assessment performed at the primary processing stage is combined with the stage of identifying repeating sections (the so-called data segmentation), or is performed after segmentation, which makes it possible to pre-separate the selected segments by movement type. Separation by types of movement can be carried out on the basis of threshold criteria (selection of action or inaction) on the basis of informative features using classifiers (including neural networks), as well as on the basis of the autocorrelation function. The selection in these areas corresponding to one of the types of cyclic movements (for example, a single or double step when walking) is performed by different algorithms, which are divided into two groups. The first group includes algorithms that are based on the cyclicity of movements and the search for extremums in the data, crossing zero, which correspond to the boundaries of the elements of the cycle (single or double step). This group also includes algorithms that are based on the analysis of the phase of movements, dynamic changes, and correlation analyses. The search for similar areas is carried out with time normalization, for example, using the DWT time warp method. The second group includes algorithms that allow estimating the boundaries of the segments on the "fly" within the time window, which negatively affects the accuracy of marking data sections when changing the temporal parameters of the gait and its individual phases. Selected data sections are analyzed for the presence of informative features and subjected to statistical evaluation of informative time and frequency parameters. Based on the results obtained, an individual model of a person's gait is built, which is used in the procedures for authentication and evaluation of the parameters of the functioning of the musculoskeletal system. The input segments under study are compared with model templates using correlation and other statistical apparatus or using neural network technologies [15]. It is necessary to evaluate and take into account the influence of external and internal factors on the measurement results when analyzing the received data due to the peculiarities of measuring gait parameters using a smartphone. According to the works of S. Sprager and M.B. Juric [1], the measurement results are influenced by factors that can be classified as informative (useful) and factors that negatively affect the accuracy of the results and which must be compensated for or taken into account in data processing and analysis. Beneficial factors relate to human physiology—these are changes that occur in the human body during its life. These include uncontrolled changes (physiological and psychological) and controlled physiological and psychological changes [16–18]. Negative factors include the type of clothing and footwear, the parameters of the base on which the person walks, the uniformity of distribution of the additional load on the person or its absence, etc. Negative factors distort the measurement results, for example, when the phone makes random movements in the pocket of loose clothing or a bag. At the same time, some factors may also be useful (informative), corresponding to a certain style of gait in certain conditions. For example, the type of footwear is a negative factor and changes the natural gait of a person (transition from boots to shoes with long heels), but under these conditions, the gait still remains individual. For example, Figure 1 shows a graph of the change in the value of the correlation coefficient when comparing the waveform of the accelerometer of a smartphone (placed in the front pocket of a pair of trousers) when walking in trousers with different degrees of tightness. All measurement conditions, except for the different types of pants, were the same and the results were averaged. According to the results of the analysis of the graph in Figure 1, it follows that the type of pants when walking affects the shape of the accelerometer signal, distorting the signal by up to 30%. This can be misinterpreted and a decision may be made on the distortion of gait parameters, while the gait remains unchanged. Figure 2 also shows a graph of the change in the value of the correlation coefficient when comparing the accelerometer signals of a smartphone (placed in the front pocket of a pair of trousers) when walking in shoes of various types. Changing shoes distorts the shape of the accelerometer signal by up to 12%. Figures 1 and 2 show special cases from the data of one person. The degree of distortion of the accelerometer signal shape in other subjects under the same conditions is different, but the general trend in changes corresponds to the results presented. All this means that, depending on the tasks and scope, it is necessary to correctly classify factors and take this into account when processing and analyzing data [12]. Features of measuring gait parameters using a smartphone must be taken into account in the models and algorithms used, or classified in a template database. The data presented in Figures 1 and 2 were obtained under idealized conditions, when the number of interfering factors was minimized or their influence was the same.

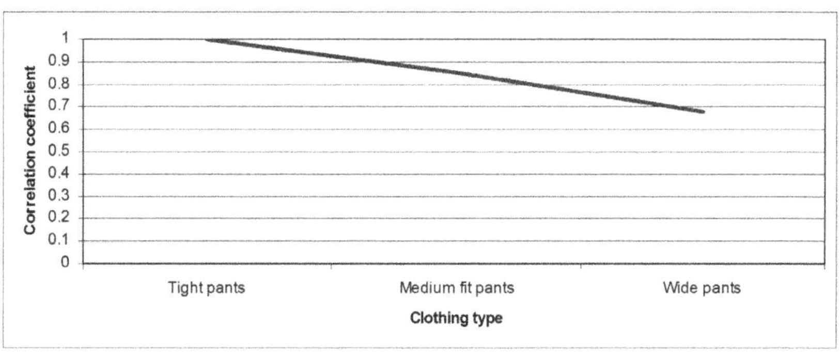

Figure 1. Distortion of the accelerometer waveform with different levels of trouser tightness.

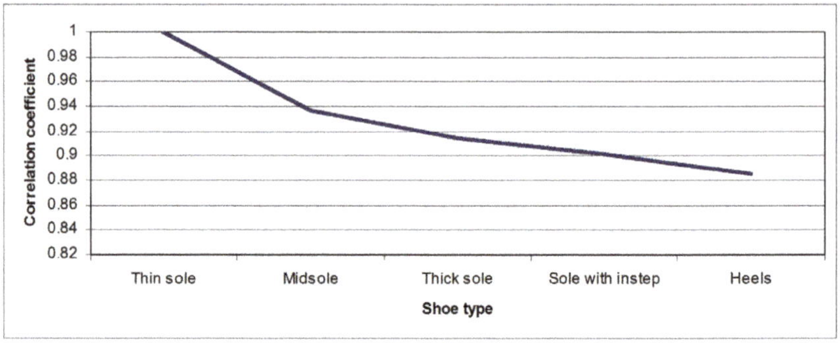

Figure 2. Distortion of the accelerometer waveform for different types of shoes.

In everyday life, such situations do not occur, so different factors can appear at random times and have a random degree of impact, i.e., in processing and analysis, it is difficult or impossible to accurately assess the components that affect the measurement results. From the position of automating the processing of measurement results, the main task is to select data sections corresponding to gait or other analyzed movements. Figure 3 shows a simple example of a time series of smartphone accelerometer data. The central section corresponds to the walking of a person, and in the extreme sections there are random, interfering movements. When segmenting this time series (for example, by extrema), interfering movements often fall into the category of informative signals, which correspond in duration to informative sections (in Figure 4 the boundaries of informative sections are marked in red).

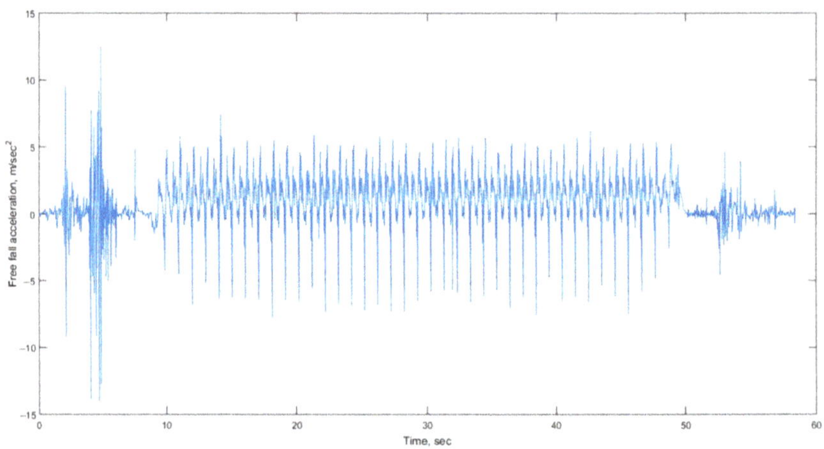

Figure 3. Time series example.

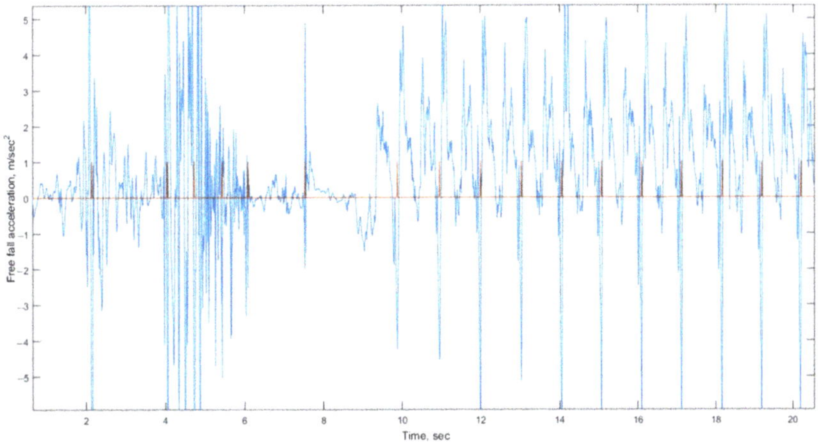

Figure 4. An example of non-informative areas getting into the selection results.

3. Processing Algorithm

In addition to this, when processing a time series with a window function during segmentation, situations arise when one movement corresponds to two time windows with its subsequent division into two movements. An example can be seen in Figure 5 (flat peak of the red mark), where the central section corresponding to one movement falls into two time windows due to the edge of the time window hitting the extremum. To reduce the problems described above, the authors propose an algorithm (Figure 6) for processing smartphone accelerometer measurements. For the high-quality operation of the algorithm, it is necessary to accumulate measurements from the smartphone accelerometer over time T. In authentication systems and personalized medicine, in terms of gait parameters, even a few minutes is not critical, since time series and statistical indicators of gait parameters are analyzed. As specified earlier, in such systems, more attention is paid to the reliability of the estimates obtained, which means a decrease in the number of negative factors in the processing results. The developed algorithm selects areas with the best quality for analysis and accumulation in the template database. The quality criterion is the maximum value of the correlation coefficient. The data section, which in the time domain is most similar to others, carries the least distortion and reflects the constancy in gait over the analyzed period. The quality of the selected areas simplifies the correction of the individual model and improves the accuracy of clustering and classification. In the first steps, the algorithm performs data preprocessing within the analyzed time T; the standard deviation STD is found after subtracting the constant component, the period is determined, and the local time window $Temp$ is formed based on the amplitude spectrum. The local window $Temp$, the doubled period Tsr of the harmonic Fsr, which has the maximum amplitude in the spectrum, is taken:

$$Temp =]2 \times Tsr \times Fd[, \tag{1}$$

where Fd is the measurement frequency.

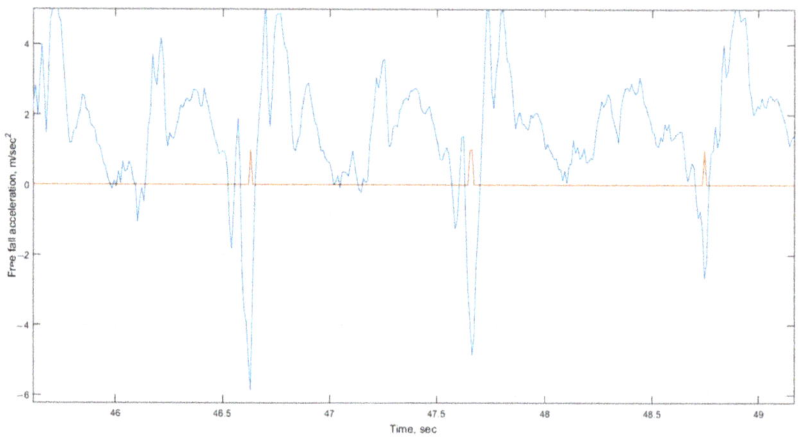

Figure 5. Movement separation example.

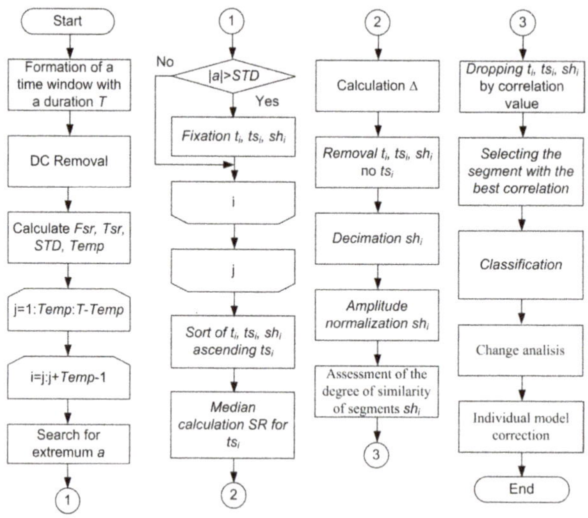

Figure 6. Processing algorithm.

Such a size of the local time window is guaranteed to capture the entire area corresponding to the movement. By moving the local $Temp$ window along the analyzed sequence T with a step equal to $Temp$, the search for extremums (minimums or maxima) within the local window is performed. The found extremum and its time position ti are saved if its value is greater than the standard deviation STD, which makes it possible to discard noise components, including non-informative movements. Two adjacent selected extrema form the boundaries of the sh_i segment with a duration of ts_i. This approach to selecting segments does not distort the data, which is typical for filtering algorithms based on moving averages. The resulting durations ts_i of the selected sections are arranged in ascending order with sorting of the corresponding values of t_i, sh_i, ts_i. According to the found median of the sorted set of durations ts_i, and taking into account the possible deviation in the duration of the segments by Δ, segments are rejected based on the expression $SR - \Delta < ts_i < SR + \Delta$:

$$\pm\Delta = SR \pm \left]\frac{SR \times Int}{100}\right[, \tag{2}$$

where *Int* is the allowable percentage deviation in the duration of the found segments from the obtained median in percent. This percentage adjusts to the person's gait and at the initial stage can take a value of 10%.

The remaining sections are compressed by the decimation procedure to the size $\min(ts_i)$ of the shortest section sh_i. Then, by calculating the value of the correlation coefficient, a search is made for the area most similar to the others, and the areas that have a correlation coefficient of less than a given threshold are removed (in this work, the value of the correlation coefficient equal to 0.9 was taken as a threshold). As a result, the sections of the time series and their parameters (the value of the correlation coefficient, the mean and standard deviation) which remain are fed to the neural network classifier. In this work, the neural network classifier was built on the basis of a feed-forward network. At the output of the classifier, the number of the movement is obtained, which corresponds to a particular section, as well as the degree of correspondence of the movement to a particular person. In the case of the expected outcome, the individual model is adjusted and the trend in individual gait measures is determined.

4. Results and Conclusions

The application of the developed algorithm increased the reliability of the obtained estimates and the quality of intermediate data at the stage of primary processing (Figure 7). At the stage of primary processing, data sections were selected that relate only to useful data (analyzed movements during walking) and no duplication of sections was found. As a result of the correlation processing of the selected sections, heavily noisy sections were removed that corresponded to random movements not related to the analyzed ones, as well as sections that had distortions, reflecting stumbling while walking, stopping and changing trajectory, the type of movement, etc. When constructing the neural classifier, feed-forward networks were used; their training was carried out by the gradient descent method with cross-entropy as an optimization criterion. The activation function for the hidden layers was the sigmoid and for the output layer it was the normalized exponential function. Testing a neural network classifier with at least 100 hidden layers on a test set showed a correct decision probability of 0.95. Reducing the number of hidden words to 10 led to a decrease in the probability of correct decision making to 0.9 (Figure 8). The classification of the subject's own movements was 100%.

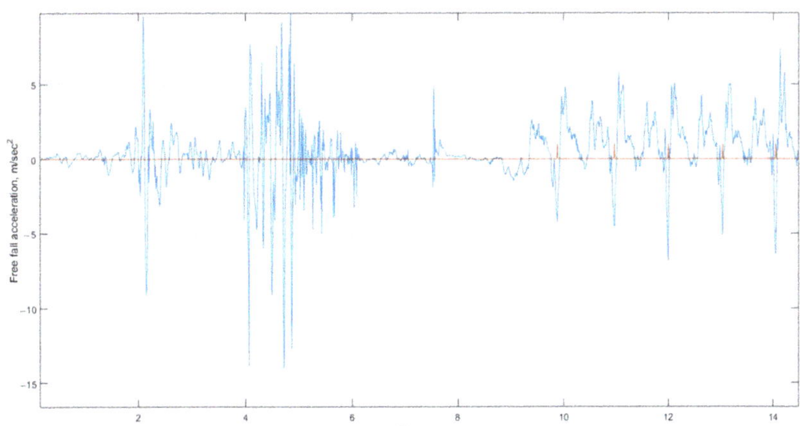

Figure 7. The result of primary processing of the time series.

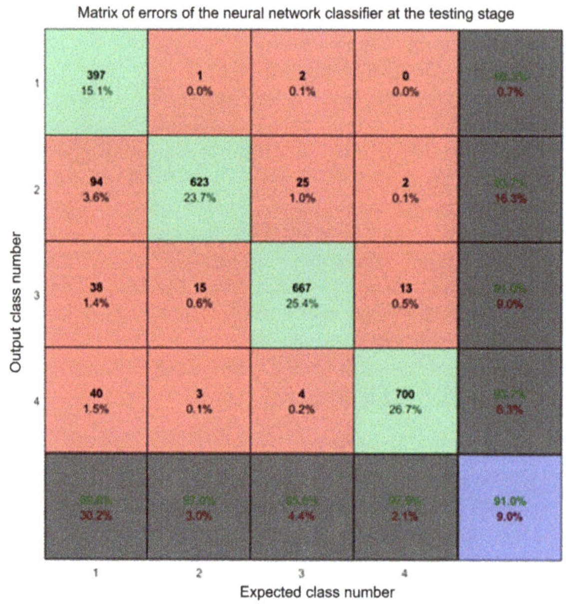

Figure 8. Classification results into four groups.

The application of the developed algorithm in practice made it possible to automate the process of selecting the best informative areas and forming a database of human movement patterns. Automatic adjustment of the allowable limit of the size of the data section allows one to track the change in gait speed and adjust the size of the local time window when selecting sections. The neural network classifier determines the type of movement and it belonging to a particular person. The developed algorithm is not inferior to similar solutions in terms of the quality of selection and discrimination of movements in the accelerometer data of a mobile phone.

Author Contributions: Conceptualization, N.D. and A.G.; methodology, N.D.; software, N.D.; validation, N.D. and A.G.; formal analysis, A.G.; investigation, N.D.; resources, N.D.; data curation, A.G.; writing—original draft preparation, A.G.; writing—review and editing, N.D.; visualization, N.D.; supervision, A.G.; project administration, A.G.; funding acquisition, A.G. All authors have read and agreed to the published version of the manuscript.

Funding: This research was supported by the grant of the President of the Russian Federation No. MK-1558.2021.1.6.

Institutional Review Board Statement: Not applicable.

Informed Consent Statement: Not applicable.

Data Availability Statement: Not applicable.

Conflicts of Interest: The authors declare no conflict of interest.

References

1. Sprager, S.; Juric, M.B. Inertial Sensor-Based Gait Recognition: A Review. *Sensors* **2015**, *15*, 22089–22127. [CrossRef] [PubMed]
2. Connor, P.; Ross, A. Biometric recognition by gait: A survey of modalities and features. *Comput. Vis. Image Underst.* **2018**, *167*, 1–27. [CrossRef]
3. Guelta, B.; Tlemsani, R.; Chouraqui, S.; Benouis, M. An Improved Behavioral Biometric System based on Gait and ECG signals. *Int. J. Intell. Eng. Syst.* **2019**, *12*, 147–156. [CrossRef]
4. Ren, Y.; Chuah, M.C. User Verification Leveraging Gait Recognition for Smartphone Enabled Mobile Healthcare Systems. *IEEE Trans. Mobile Comput.* **2015**, *14*, 1961–1974. [CrossRef]

5. Khabir, K.M.; Siraj, M.S.; Ahmed, M.; Ahmed, M.U. Prediction of gender and age from inertial sensor-based gait dataset. In Proceedings of the Joint 2019 8th International Conference on Informatics, Electronics & Vision (ICIEV) & 3rd International Conference on Imaging, Vision & Pattern Recognition (IVPR, Spokane, WA, USA, 30 May–2 June 2019; pp. 371–376.
6. Shema-Shiratzaky, S.; Beer, Y.; Mor, A.; Elbaz, A. Smartphone-based inertial sensors technology—Validation of a new application to measure spatiotemporal gait metrics. *Gati Posture* **2022**, *93*, 102–106. [CrossRef] [PubMed]
7. Garufov, D.; Bours P. User authentication based on foot motion. *SIViP* **2011**, *5*, 457–467.
8. Reyes, O.C.; Vera-Rodriguez, R.; Scully, P.J.; Ozanyan K.B. Analysis of Spatio-Temporal Representations for Robust Footstep Recognition with Deep Residual Neural Networks. *IEEE Trans. Pattern Anal. Mach. Intell.* **2018**, *4*, 1–13.
9. Sokolova, A.I.; Konushin, A.S. Methods of human identification by gait in video. *Proc. ISP RAS* **2019**, *1*, 1–19.
10. Helvas, A.V.; Belyaikina, N.G.; Gilya-Zetinov, A.A.; Chernikova, D.D.; Shabunin, V.M.; Yapryntsev E.O. Gesture recognition using a neural network and the application of this approach to create a new generation of gaming gadgets. *Proc. Mosc. Inst. Phys. Technol.* **2017**, *2*, 1–7.
11. Teh, P.S.; Zhang, N.; Tan, S.-Y.; Shi, Q.; Khoh, W.H.; Nawaz, R. Strengthen user authentication on mobile devices by using user's touch dynamics pattern. *J. Ambient Intell. Humaniz. Comput.* **2020**, *11*, 4019–4039. [CrossRef]
12. Zhang, X.; Yao, L.; Huang, C.; Gu, T.; Yang, Z.; Liu Y. DeepKey: A Multimodal Biometric Authentication System via Deep Decoding Gaits and Brainwaves. *ACM Trans. Intell. Syst. Technol.* **2020**, *11*, 3393619. [CrossRef]
13. Grecheneva, A.V.; Dorofeev, N.V.; Goryachev M.S. Estimation of human biomechanics during registration with a wearable device. *J. Phys. Conf. Ser.* **2021**, *2096*, 012117. [CrossRef]
14. Lu, H.; Huang, J.; Saha, T.; Nachman L. Unobtrusive gait verification for mobile phones. In Proceedings of the 2014 ACM International Symposium on Wearable Computers, Seattle, WA, USA, 13–17 September 2014; pp. 91–98.
15. Oguz, A.; Ertugrul O.F. Human identification based on accelerometer sensors obtained by mobile phone data. *Biomed. Signal Process. Control* **2022**, *77*, 103847. [CrossRef]
16. Tandon, R.; Javid, P.; Giulio I.D. Mobile phone use is detrimental for gait stability in young adults. *Gati Posture* **2021**, *88*, 37–41. [CrossRef] [PubMed]
17. Pierce, A.; Ignasiak, N.K.; Eiteman-Pang, W.K.; Rakovsky C. Mobile phone sensors can discern medication-related gait quality changes in Parkinson's patients in the home environment. *Comput. Methods Programs Biomed. Update* **2021**, *1*, 100028. [CrossRef]
18. Lunardini, F.; Malavolti, M.; Pedrocchi, A.L.G.; Borghese, N.A.; Ferrante S. A mobile app to transparently distinguish single-from dual-task walking for the ecological monitoring of age-related changes in daily-life gait. *Gait Posture* **2021**, *86*, 27–32. [CrossRef] [PubMed]

engineering proceedings

MDPI

Proceeding Paper

From Matryoshka and Fukuruma to Hierarchies of Criteria and Ranking Methods in Multi-Criteria Problems of Analysis and Decision Making †

Nicolay Klevanskiy [1,*], Victor Glazkov [2], Yermek Saparov [2] and Vladimir Mavzovin [3]

1 Department of Economic Cybernetics, Saratov State University of Genetics, Biotechnology and Engineering Named after N.I. Vavilov, 1, Teatralnaya Square, Saratov 410012, Russia
2 Department of Applied Information Technologie, Saratov State Technical University Named after Yu.A. Gagarin, 77, Politechnicheskaya, Saratov 410054, Russia
3 Department of Mathematics, National Research Moscow State Construction University, 26, Yaroslav Highway, Moscow 129337, Russia
* Correspondence: nklevansky@yandex.ru
† Presented at the 15th International Conference "Intelligent Systems" (INTELS'22), Moscow, Russia, 14–16 December 2022.

Abstract: Compound toys simulate the hierarchy of single inclusion. We propose classification of criteria and ranking methods for multi-criteria problems of analysis and decision making. The definitions of subjective and objective criteria are introduced. The application of two methods of cognitive visualization to identify the structure of ranking algorithms is represented: the method of separation of the ranking scheme and that of the color shading of individual elements of the scheme. A multiple-inclusion hierarchy was identified for the criteria and ranking methods. Visualization of the identified hierarchies of multiple inclusions was performed using circular representations similar to Euler–Venn diagrams. The use of objective criteria and their application in ranking algorithms was shown on the example of solving the problems of forming schedules and calendar schedules.

Keywords: criterion; ranking algorithm; inclusion hierarchy; multi-criteria ranking; multi-vector ranking; hypervector ranking; multi-criteria analysis and decision-making problems; timetable problem; priority rule; cognitive visualization

Citation: Klevanskiy, N.; Glazkov, V.; Saparov, Y.; Mavzovin, V. From Matryoshka and Fukuruma to Hierarchies of Criteria and Ranking Methods in Multi-Criteria Problems of Analysis and Decision Making. *Eng. Proc.* **2023**, *33*, 45. https://doi.org/10.3390/engproc2023033045

Academic Editors: Askhat Diveev, Ivan Zelinka, Arutun Avetisyan and Alexander Ilin

Published: 4 July 2023

1. Introduction

There are several versions of the origin of the Russian matryoshka (Figure 1a). According to the most plausible version, the idea—matryoshka—was borrowed from the compound Japanese toy Fukuruma (Figure 1b). The ancient elder Fukuruma is a revered Japanese pantheon deity of the seven gods responsible for luck, happiness and prosperity. The good sage's appearance is quite specific: an armless and legless body, shaped like an oval body-head, slightly stretched up and decorated with a beard and eyes. The eight copies inside are eight human bodies. This is allegorical, since the ancient sages believed that each person has several "shells". In the context of this study, both the Russian matryoshka and the Japanese Fukuruma model a single-inclusion hierarchy, in which each object contains only one object of the lower level (Figure 1c). The smallest object is monolithic.

To solve practical problems, one needs to study multiple-inclusion hierarchies in which any object of a higher level can contain an arbitrary number of objects of different lower levels. That is, the multiple-inclusion hierarchy is an asymmetric analogy of the generalization hierarchy, for example, with multiple inheritances in the object-oriented programming language class hierarchy. Both toys model a person for whom decision making is one of the main forms of activity. The compound construction of toys symbolizes the possibility of decision making at each level of the hierarchy, which allows us to consider both toys as models of system analysis related to decision theory [1]. Multiple-criteria decision making (MCDM) and multiple-criteria decision analysis (MCDA) solve problems [2] of selection, comparison, classification using multi-criteria ranking methods. When studying problems in well-structured systems of various types, selection criteria are usually formed based on certain estimates (indicators, characteristics) [3] with vector and multi-vector components. A criterion is some function of a decision made. This function allows for the quantification of the feasibility of a solution. The specific value of a criterion characterizes the level of the goal achievement, as well as the effectiveness of the methods and means used in this process. If the goal generally indicates the direction of action, the criterion complements the concept of the goal and indicates an effective way to achieve it. Criteria in decision support systems can be divided into subjective, set by the decision maker (DM), and objective, depending only on the characteristics and properties of the system under study. It should be noted that MCDM and MCDA software solutions are focused on DM's subjective criterial; therefore, they are implemented interactively. The possibility of embedding ranking algorithms in the developed software can be provided by using objective criteria. This fact requires a special analysis of ranking criteria and algorithms.

Figure 1. (**a**) matryoshka with the largest number of figures in Russia; (**b**) Japan's Fukuruma; (**c**) cross section of toys.

Criteria can be classified [4] as follows:

- Vector criterion—an ordered set of scalar components (estimates);
- Multi-vector criterion—an ordered set of vector criteria;
- Hypervector criterion—an ordered set of multi-vector criteria.

The proposed names of the introduced criteria are arbitrary and can be replaced with more preferred names. The criteria classification already contains an inclusion hierarchy. We also introduce the following designations:

- Scalar—S;
- Vector criterion—$VC^l = \{S_i\}, i = 1,2,\ldots,l$;
- Multi-vector criterion—$MVC^k = \left\{VC_i^l\right\}, i = 1,2,\ldots,k$;
- Hypervector criterion—$HVC^n = \left\{MVC_i^k\right\}, i = 1,2,\ldots,n$.

To demonstrate the relationships of the criteria, the concept of Euler–Venn diagrams was used. Circular representations of criteria (Figure 2) contain criteria symbols and specified quantities of their components. Figure 2 illustrates a multiple-inclusion hierarchy in which an object of each level contains set numbers of objects of lower levels. The hierarchy of criteria inclusion can be expanded by sequentially adding new levels to form criteria with a more complex structure. For example, you can enter a supervector criterion that contains a predetermined number of hypervector criteria, etc. Structuring objects of lower levels will allow us more fully to take into account the features of the subject area of the system under study.

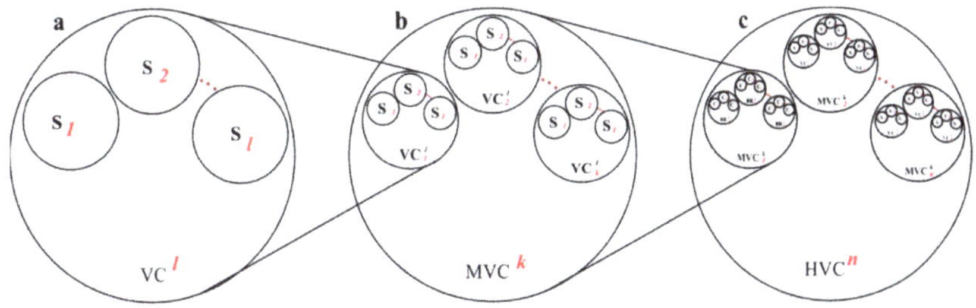

Figure 2. Hierarchy of multiple inclusions of criteria: (**a**) vector criterion; (**b**) multi-vector criterion; (**c**) hypervector criterion. The quantities and numbers of the criteria components are highlighted in red.

The study of criteria and algorithms for their ranking is associated with constantly developing multi-criteria decision-making methods [5,6] and artificial intelligence.

The purpose of this study is to analyze the structures of ranking algorithms for multi-criteria decision-making methods in systems of various types, including complex object management systems [1].

2. Methodology

The proposed approach to the analysis of criteria and structures of ranking algorithms in multi-criteria decision-making problems is based on cognitive processes. Cognitive theory covers different types of mental activities, such as observing different phenomena in the environment, pattern recognition, and problem presentation analytical reasoning. Beliefs, images, and roles can be products of cognitive processes, which are forms of heuristic judgment. One method of finding heuristic judgments is the visualization of non-trivial graphic representations [7,8]. The analysis of these representations is based on figurative thinking mechanisms. Visualizing information means transforming character data into geometric shapes that help a person form a mental image of these data.

Computer visualization is a direct route to the acquisition of information, and the application of visualization within multi-criteria methods greatly facilitates their use [3]. In this study, two approaches are considered to facilitate the perception of the features of ranking algorithms and their structure:

- Separating image ranked criteria and their ranks;
- Colored presentation of ranked criteria and their ranks, similar to the practice of using color plating graphs [9,10].

Examples of using criteria and ranking methods are taken from efforts to solve schedule problems of various types. Any schedule task is the task of determining the sequence of the specified actions in systems with limited resources. The formation of objective criteria in schedule tasks is based on the concepts of workload and uniformity [11]. Workload criteria represent the demand of applications or collections of applications for system resources. The uniformity criteria determine the consumption of system resources in the schedule interval. Both concepts are used in two resource-oriented scheduling strategies of different types [11]—constructive and optimizing. The first strategy is focused on obtaining an initial schedule. The second strategy optimizes the initial schedule and can be used for multi-pass solutions. Each strategy contains two priority rules. The first set of rules defines the busiest demand or set of demands or the most uneven activity or set of schedule activities. The second set of priority rules ensures uniform resource consumption when selecting timeslots for inclusion in the initial schedule in the first strategy or for permutation in the second strategy. Each priority rule uses the ranking algorithm it needs.

3. The Analysis of Ranking Methods

The results of the work of the investigated algorithms are the ranks of the criteria. The forward (ascending) and reverse (descending) ranking types will differ.

Here, we add the following symbols:

- $MCR_k^l()$—multi-criterion ranking function k vector criteria VC^l;
- $MVR_n^k()$—multi-vector ranking function of n multi-vector criteria MVC^k;
- $HVR_m^n()$—hypervector ranking function m hypervector criteria HVC^n;
- $SVR_o^m()$—supervector ranking function o supervector criteria SVC^m;
- R^{VC}—ordered set of ranks of vector criteria;
- R^{MVC}—ordered set of ranks of multi-vector criteria;
- R^{HVC}—ordered set of hypervector criteria.

3.1. Multicriteria Ranking

When forming the schedule of the examination session [12], each demand for the examination at the university is characterized by a vector criterion containing assessments of the workload of a group of students, a teacher and an audience. The selection of the busiest application to be included in the initial exam schedule is determined by multi-criteria ranking, i.e., ranking vector criteria containing ordered sets of scalar components.

The multi-criteria ranking is as follows.

Let there be a set of vector criteria VC^l (Figure 3). To take into account the relative importance [13] of scalar components, it is necessary to enter a set of importance coefficients $A = \{a_i\}$, $i = 1, 2, \ldots, l$, for which the condition $\sum_{i=1}^{l} a_i = 1$ is met. On the basis of the analysis of results of paired comparison of the scalar component criterion values $\left(VC^l\right)_k$ and $\left(VC^l\right)_i$, an estimated matrix $\|C_{ki}\|$ is formed. Matrix elements C_{ki} allow for the unambiguous determination of the relationship between k-th and i-th vector criteria. The values of the elements C_{ki} are selected so as to further separate ineffective ones, i.e., those not included in the Pareto set solutions, from the effective ones in the estimated matrix.

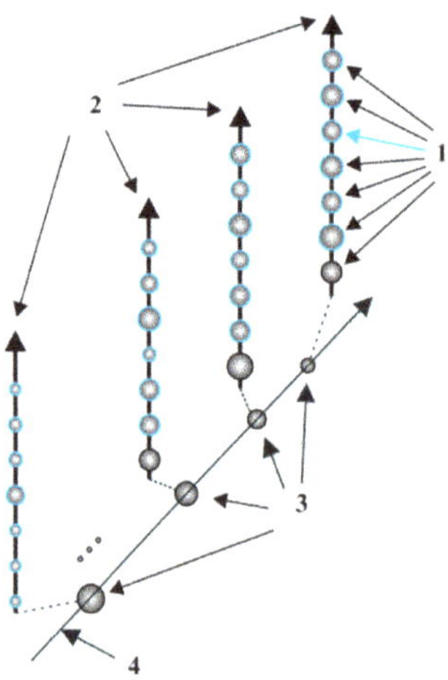

Figure 3. Multi-criteria ranking scheme MCR_k^l: 1—l scalar components of the vector criterion VC^l; 2—k ranked vector criteria VC^l; 3—k vector criteria ranks R^{VC}; 4—vector criteria rank vector.

For the organization of ranking of effective solutions, characteristic numbers are determined: H_i—the number of elements in the i-th column of the estimated matrix, the value of which is more than one; M_i—the number of elements in the i-th column of the same matrix, whose value is less than one; and C_{kimax}—the maximum value of the element in the i-th column of the matrix $\|C_{ki}\|$.

Mathematically, the meaning of the characteristic numbers is as follows:

- H_i shows how many criteria from the set in question exceed the i-th criterion;
- M_i shows how many criteria the i-th criterion dominates;
- C_{kimax} determines how many times the k-th criterion dominates the i-th criterion.

The cyclic processing of the estimated matrix involves redefining the characteristic numbers for all columns of the estimated matrix and then determining the best criterion, for example, in the j-th column with the entry of this criterion in the Pareto tuple. After the estimated matrix processing is completed, the j-th column and the j-th row of the estimated matrix are deleted. The criteria of the Pareto tuple are "rigidly" ranked, with only the direct dominance of one criterion by another being possible. Taking this into account, the author called this method of multi-criteria ranking "rigid" ranking [4].

3.2. Multi-Vector Ranking

The formation of the transport schedule [14] is related to the concept of the route, that is, the sequence of loading/unloading points and the paths between these points. The route load criterion contains two vectors with load estimates of loading/unloading points and hauls, respectively. That is, the route load criterion is a multi-vector criterion. The determination of the busiest route is achieved by multi-vector ranking.

Multi-vector ranking MVR_n^k is the ranking of multi-vector criteria MVC^k, representing ordered sets of vector criteria VC^l, with each vector criterion being an ordered set of scalar components. The analysis of the multi-vector ranking scheme (Figure 4a) allows for the division of its structure into two parts, the first of which (Figure 4b) contains all vector criteria, and with the second part (Figure 4c) there containing rank vectors R^{VC} and ranks R^{MVC}. Figure 4c shows that ranks R^{MVC} are the result of multi-criterion ranking of rank vectors R^{VC}. Partitioning the multi-vector ranking scheme, on the one hand, is a method of cognitive analysis, and, on the other hand, leads to the following definition. Multi-vector ranking of n multi-vector criteria with k vector criteria (Figure 4a) is performed as follows: initially k multi-criteria rankings $MCR_k^l\left(VC_i^l\right)$, $i = 1, 2\ldots, k$ are produced in respect to corresponding groups of vector criteria (Figure 4b) with formation of ranks R^{VC} (Figure 4c); then, multi-criterion ranking of n rank vectors $MCR_n^k\left(R^{VC}\right)$ (Figure 4c) generates the ranks of multi-vector criteria R^{MVC}K^MVC (Figure 4c).

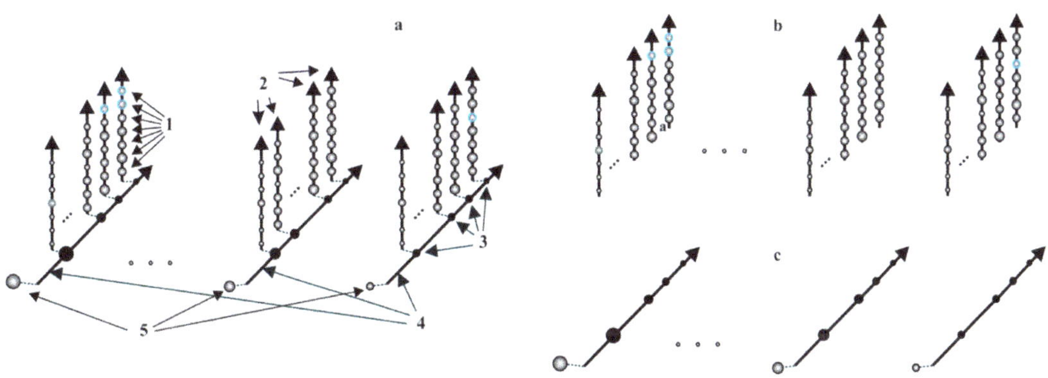

Figure 4. (**a**) multi-vector ranking scheme MVR_n^l 1—l scalar components of the vector criterion VC^l; 2—ranked vector criteria $VC_1^l, VC_2^l, \ldots, VC_k^l$ 〖BK〗 _1^l, 〖BK〗 _2^l, … , 〖BK〗 _k^l; 3—k vector criteria ranks R^{VC}; 4—n vector criteria rank vectors; 5—n ranks of multi-vector criteria; (**b**) vector criteria; (**c**) ranks of vector and multi-vector criteria.

Consideration of the relative importance of criteria in multi-vector ranking is carried out as follows. For each group of vector criteria $\left(VC_j^l\right)$, $j = 1, 2\ldots, k$ multiple coefficients of scalar components are introduced: $A_i^c = \left\{a_{i,j}^c\right\}, i = 1, 2,\ldots,l;\ j = 1, 2,\ldots,k,$ and $\bigvee j \sum_{i=1}^{l} a_{i,j}^c = 1$, as well as many coefficients of importance of vector criteria $A_j^{VC} = \left\{a_j^{VC}\right\}$, $j = 1, 2,\ldots, k$, under a condition $\sum_{j=1}^{k} a_j^{BK} = 1$.

3.3. *Hypervector Ranking*

The most complex workload criteria are the project criteria in multi-project planning [15]. Estimates of the workload demands for each type of resource form their vector loading criteria and form multi-vector loading criteria for the paths of the project graph. The set of graph path criteria generates a hypervector criterion for project load. The choice of the busiest project is determined by hypervector ranking.

Hypervector ranking HVR_m^n is a ranking of m hypervector criteria HVC^n, representing n ordered sets of multi-vector criteria MVC^k, containing ordered sets of k vector criteria VC^l, and each vector criterion includes an ordered set of l scalar components (Figure 5). In this case, the red color plating of the ranks of the multi-vector criteria is used to identify a method for determining the ranks of hypervector criteria. This allowed us to draw the following conclusions: the ranks of multi-vector criteria (Figure 5(5)) form vectors (Figure 5(6)), the multi-criterion ranking of which determines the ranks of hypervector criteria (Figure 5(7)).

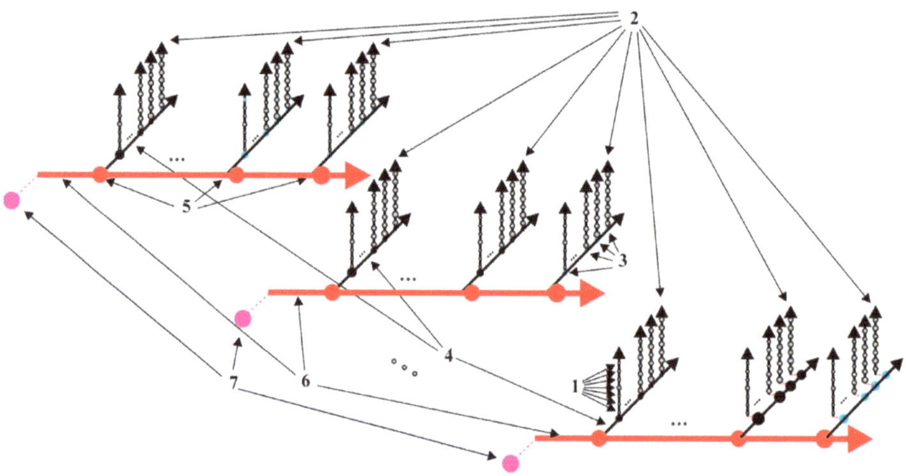

Figure 5. Hypervector ranking scheme HVR_m^n: 1—l scalar components of vector criterion VC^l; 2—ranked vector criteria VC_1^l, $VC_2^l,\ldots,$ VC_k^l; 3—k ranks of vector criteria R^{VC}; 4—n vector criteria rank vectors; 5—n ranks of multi-vector criteria R^{MVC}; 6—multi-vector criteria rank vectors; 7—m ranks of hypervector criteria R^{HVC}.

Thus, hypervector ranking HVR_m^n includes n multi-vector rankings, defined in Section 3.2, and multi-criterion ranking of multi-vector criteria rank vectors $MCR_n^k\left(R^{VC}\right)$ with the formation of hypervector criteria ranks. Consideration of the relative importance of criteria in hypervector ranking is as follows. In each group of multi-vector criteria (MBK_j^k, $j = 1, 2,\ldots, n$), the actions presented in Section 3.2 are performed. A plurality of importance coefficients of the multi-vector criteria $A_j^{MVC} = \left\{a_j^{MVC}\right\}$, $j = 1, 2,\ldots, n$, under a condition $\sum_{j=1}^{n} a_j^{MVC} = 1$.

4. Some Considerations about Ranking Criteria and Methods

The results obtained in Sections 3.2 and 3.3 provide the basis for a geometric interpretation of the hierarchy of ranking methods (Figure 6). The hierarchy diagram of multiple inclusions of ranking methods, which is also a criteria diagram (Figure 2), contains circular representations. The diagram includes the forecast of the supervector ranking method (Figure 6c).

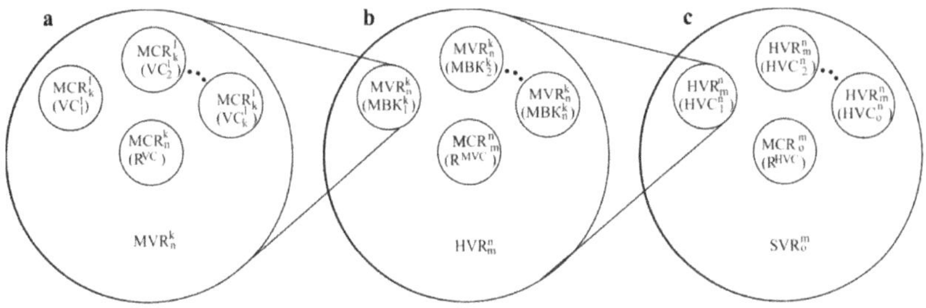

Figure 6. Hierarchy of multiple inclusions of ranking methods (**a**) multi-vector ranking; (**b**) hypervector ranking; (**c**) supervector ranking.

5. Conclusions

As a result of the analysis of ranking criteria and methods, the following judgments are made:

- Classification of criteria and ranking methods for multi-criteria problems of analysis and decision making is proposed.
- A multiple-inclusion hierarchy is identified for the criteria and ranking methods.

Author Contributions: Conceptualization, N.K. and V.G.; methodology, N.K.; software, Y.S.; validation, V.M. and Y.S.; formal analysis, N.K.; investigation, V.G.; resources, V.M.; data curation, Y.S.; writing—original draft preparation, N.K.; writing—review and editing, N.K.; visualization, Y.S.; supervision, V.G.; project administration, N.K. All authors have read and agreed to the published version of the manuscript.

Funding: This research received no external funding.

Institutional Review Board Statement: Not applicable.

Informed Consent Statement: Not applicable.

Data Availability Statement: Not applicable.

Conflicts of Interest: The authors declare no conflict of interest.

References

1. Mesarovic, M.D.; Macko, D.; Tanahara, Y. *Theory of Hierarchical, Multilevel, Systems*; Elsevier Science: Amsterdam, The Netherlands, 1970.
2. Hodgett, R.E. Comparison of Multi-Criteria Decision-Making Methods for Equipment Selection. *Int. J. Adv. Manuf. Technol.* **2016**, *85*, 1145–1157. [CrossRef]
3. Lotov, A.V.; Pospelova, I.T. *Multi-Criteria Problems of Decision Making*; Max Press: Moscow, Russia, 2008. (In Russian)
4. Safronov, V.V. *Basics of System Analysis: Multi-Vector Techniques Optimization and Ranking*; Scientific Book: Saratov, Russia, 2009; 329p. (In Russian)
5. Saaty, T. *Analytic Hierarchy Process*; McGraw-Hill: New York, NY, USA, 1980.
6. Roy, B. The outranking approach and the foundations of electre methods. *Theory Decis.* **1991**, *31*, 49–73. [CrossRef]
7. Zenkin, A.A. *Cognitive Computer Graphics*; Nauka: Moscow, Russia, 1991; 192p. (In Russian)
8. Klevanskiy, N.N.; Antipov, M.A.; Krasnikov, A.A. Cognitive Aspects of Timetable Visualization: Support Decision Making. *Procedia Comput. Sci.* **2017**, *103*, 94–99. [CrossRef]

9. Juhos, I.; T'oth, A.; van Hemert, J. Binary merge model representation of the graph colouring problem. In *Evolutionary Computation in Combinatorial Optimization*; Raidl, G., Gottlieb, J., Eds.; Springer: Berlin/Heidelberg, Germany, 2004; Volume 3004 of LNCS, pp. 124–134.
10. Juhos, I.; T'oth, A.; van Hemert, J. Heuristic colour assignment strategies for merge models in graph colouring. In *Evolutionary Computation in Combinatorial Optimization*; Springer: Berlin/Heidelberg, Germany, 2005; Volume 3448 of LNCS, pp. 132–143.
11. Klevanskiy, N.N. Timetable generation paradigm. *J. Adv. Res. Tech. Sci.* **2017**, *6*, 70–75. (In Russian)
12. Klevanskiy, N.N.; Mavzovin, V.S. High school exam timetabling problem. *Mod. High Technol.* **2018**, *5*, 97. (In Russian)
13. Podinovski, V.V. The quantitative importance of criteria for NCDA. *J. Multi-Criteria Decis. Anal.* **2002**, *11*, 1–15. [CrossRef]
14. Klevanskiy, N.N.; Antipov, M.A. New approaches to development of transport schedules. *World Transp. Transp.* **2016**, *14*, 18–27. (In Russian) [CrossRef]
15. Klevanskiy, N.N.; Tkachev, S.I.; Voloshchouk, L.A. Multi-Project Scheduling: Multicriteria Time-Cost Trade-Off Problem. *Procedia Comput. Sci.* **2019**, *150*, 237–243. [CrossRef]

engineering proceedings

MDPI

Proceeding Paper

A Review of Engineering Techniques for EEG Processing †

Veronica Tolmanova and Denis Andrikov *

Academy of Engineering, RUDN University, Mikluho-Maklaya Str. 6, Moscow 117198, Russia; 1042210065@rudn.ru
* Correspondence: andrikov-da@rudn.ru

† Presented at the 15th International Conference "Intelligent Systems" (INTELS'22), Moscow, Russia, 14–16 December 2022.

Abstract: This article provides not only a description of an electroencephalogram (EEG), the principle of its operation, conditions of use, methods of decoding but also studies aimed at improving this procedure and facilitating the work of highly professional employees studying EEG results. A study of published articles related to electroencephalograms was carried out in order to trace the entire path of development and use of EEGs.

Keywords: electroencephalogram; EEG; brain; waves; sleep depth; epilepsy

Today, the word "electroencephalogram" is not something unknown. Many people, whether for medical or research purposes, have come across it. An electroencephalogram (EEG) is a physiological assessment method that is used to study the functioning of the nervous system by recording the electrical activity of the brain, especially the cerebral cortex. Thus, with the help of an EEG, it is possible to detect the normal pattern of our brain functioning and the activation of the brain or its individual parts when faced with external or internal stimulation. Measurements can be taken while awake or asleep, depending on the purpose of the test. Through the electrodes, the measuring system captures the radiation of brain waves and their rhythm, shape, duration, and frequency of radiation.

The captured waves can be alpha, beta, theta, and delta. Each of them will cause the EEG to draw one or another pattern of wave frequencies.

- Alpha waves appear in moments of relaxation or before tasks that do not require concentration or effort.
- Beta waves usually reflect the results of intense mental effort, usually occurring while awake or during REM sleep.
- Theta waves are a type of brainwaves that are commonly observed during periods of relaxation, drowsiness, and deep meditation, and they play a significant role in memory consolidation and creative thinking.

Finally, delta waves are associated with deep sleep, as they have traditionally been associated with rest and the repair of nerve tissue.

With an encephalogram, you can measure both the general pattern of brain functioning and the differences between some areas and others by analyzing the differences in voltage between different areas. With regard to its use in medicine, it can be used to determine the normal functioning of the brain, the control of the state of consciousness during surgery, or the presence of changes in the nature of the radiation of waves. In this aspect, this type of technique tends to be used for suspected disorders such as epilepsy (seizures are provoked voluntarily to fix what is happening and how), dementia, encephalopathies, typical outbreaks of some mental disorders, and even differentiation between coma and brain death (while the former has brain activity, the latter will show a flat electroencephalogram). It is also widely used to analyze sleep problems and disorders. Throughout the history of the existence of the electroencephalogram, various studies have been carried out not only

Citation: Tolmanova, V.; Andrikov, D. A Review of Engineering Techniques for EEG Processing. *Eng. Proc.* **2023**, 33, 46. https://doi.org/10.3390/engproc2023033046

Academic Editors: Askhat Diveev, Ivan Zelinka, Arutun Avetisyan and Alexander Ilin

Published: 10 July 2023

to detect diseases but also to improve the technology itself. Let us take a look at some research examples.

A group of scientists, which included: Bob Kemp, Aeilko H. Zwinderman, Bert Tuk, Hilbert A C Kamphuisen, and Josefien J L Oberye, published an article in *IEEE Transactions on Biomedical Engineering* in October 2000 entitled "Analysis of a sleep-dependent neuronal feedback loop: The slow-wave microcontinuity of the EEG". In this article, the authors talked about the experiment they conducted, the conditions for its conduct, the participants, the equipment, and the results obtained. As a result of the study, they obtained an estimate of the feedback gain and applied it to quantify the depth of sleep [1].

In addition, in their article, the authors referred to the article by Allan Rechtschaffen and Antony Kales "A Manual of Standardized Terminology, Techniques and Scoring System for Sleep Stages of Human Subjects", published in 1968 in the journal *Psychiatry and Clinical Neurosciences*. This guide to standard terminology and scoring reflects the consensus of a group of researchers, each with years of experience in evaluating polygraph sleep recordings. The development of this guide required extensive discussion of all emerging issues. This guide should be seen as a working tool and not a teaching aid, based on which the study was carried out. It involved 22 people (both men and women) of different ages; each of them took temazepam (benzodiazepine—psychoactive substances that are part of a wide group of central nervous system depressants) once, and they were observed for two nights. It was found that in men, the power of slow waves is two times lower than in women. However, no influence of gender on slow-wavelength microcontinuity was found. All this leads to the conclusion that slow-wave microcontinuity based on the physiological model of sleep reflects the depth of sleep more accurately than the power of slow waves [2].

A new paper came out in 2020 related to the two previous research topics. It was published by Mostafa Neo Mohsenvand, Mohammad Rasool Izadi, and Pattie Maes in *Proceedings of Machine Learning Research*. The article is called "Contrastive Representation Learning for Electroencephalogram Classification". Here, the authors present a model for learning the representation of electroencephalogram signals using contrast learning. Reflecting on the complex task of EEG interpretation and labeling facing an experienced specialist, the authors increase the number of samples per record by recombining the channels of the records. Thus, the amount of labeled data required for tasks such as emotion recognition, anomaly classification, and sleep stage assessment is significantly reduced. Moreover, the model showed improved performance compared to supervised and self-observed models [3].

In 1937, "Cerebral states during sleep, as studied by human brain potentials" was published in *The Journal of Experimental Psychology* by Alfred L Loomis, E Newton Harvey, and Garret A Hobart. To study potential sleep patterns, a new type of push–pull amplification system was used, specifically designed to accurately record large slow potentials. It has been discovered that during sleep, there is a continuous fluctuation of states in a person, which can be associated with stimulus recognition or occur without external influences. During sleep there was a constant shifting of states up and down, sometimes associated with recognizable stimuli, sometimes without any external stimuli, but probably as a result of internal stimuli [4].

Based on the results (Figure 1) of the above article, Hubert Banville, Omar Chehab, Aapo Hyvarinen, Denis-Alexander Engemann, and Alexandre Gramfort published a study in 2020 entitled "Uncovering the structure of clinical EEG signals with self-supervised learning", which aimed to investigate self-supervised learning. It is a method for detecting structure in unlabeled data to study the presentation of EEG signals. The authors worked with two of the most significant problems—determining the stage of sleep based on electroencephalograms and detecting pathologies. The experiment was conducted with several public datasets containing thousands of records. As a result, the authors concluded that this algorithm may be the beginning of a wider use of deep learning models on EEG data [5].

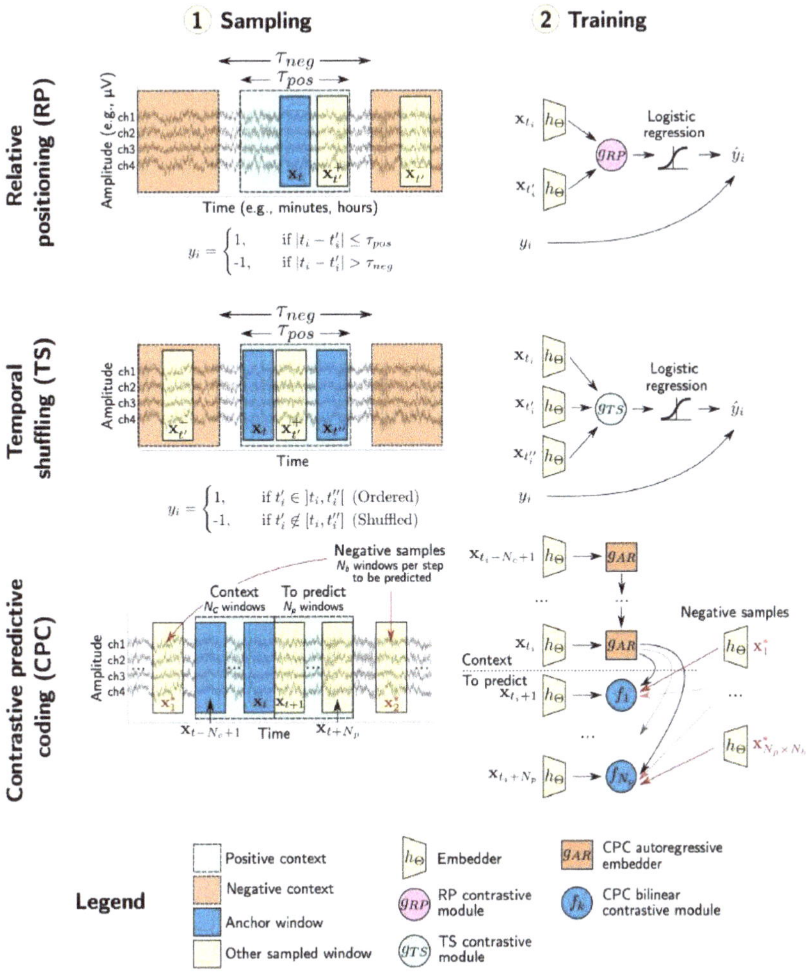

Figure 1. Visual explanation of the three tasks used in this study. The first column illustrates the sampling process by which examples are obtained in each of the tasks. The second column describes the learning process in which sample examples are used to train the hΘ extractor end-to-end.

In 2017, Albert Vilamala, Kristoffer H Madsen, and Lars K Hansen published an article entitled "Deep convolutional neural networks for interpretable analysis of EEG sleep stage scoring", in which they drew attention to the importance of sleep studies for diagnosing sleep disorders such as insomnia, narcolepsy, and sleep apnea. Despite advances in medicine, sleep assessment from raw polysomnography signals remains a tedious visual task for highly trained professionals, and various attempts have been made recently to conduct a number of studies to create an automatic assessment based on machine learning methods. The authors turn to multi-cone spectral analysis to generate visually interpretable images of sleep patterns from EEG signals as input to a deep convolutional network trained to solve visual recognition problems. In this article, emphasis was placed not only on the medical component but also on the mathematical one [6]. Thus, the authors refer to L. Fejér's article "Über trigonometrische Polynome", published in *The Journal für die reine und angewandte Mathematik* (*Crelles Journal*) in 1916, which describes the supertrigonometric polynomial [7]. Additionally mentioned is the article by G. Szego "Beiträge zur Theorie der Toeplitzschen Formen", which was published in *Mathematische Zeitschrift* back in 1920. It was in this article that the Hellinger–Toeplitz theorem applied to the Hilbert space was described [8]. Another article they linked to was D.J. Thomson's "Spectrum estimation and harmonic analysis", published in *Proceedings of the IEEE* in 1982. This article presented a method based on a "local" eigendecomposition for estimating the spectrum in terms of solving an integral equation [9].

In the book "*Automatic Diagnostics and Processing of EEG*", written by A. Khramov, A. Koronovskii, V. Makarov, A. Pavlov, and E. Sitnikova, one of the modern methods of time series analysis is considered—continuous wavelet analysis. In short, it outlines the general information and concepts of wavelet transform, the mathematical apparatus, the method for the numerical implementation of wavelet transform, the wavelet analysis of random processes, and the methods for applying wavelet transform to the analysis of nonlinear systems of various nature. In addition, aspects related to the study of spatially distributed systems, and, accordingly, structures that arise both in time and space, using wavelet analysis, are touched upon. In the world literature, a large number of different methods for recognizing peak–wave discharges are mentioned. The authors of this book propose to divide them all into three classes:

- The first one uses a non-linear dynamics approach and quantifies various characteristics of EEG signals, such as Lyapunov exponents or entropy;
- The second one uses artificial neural networks and learning algorithms for data classification;
- The third formalizes the definition of SWD (peak–wave discharges) and compares the statistical features of different typical epochs in the EEG.

In addition, the authors note a number of shortcomings of the above methods, namely:

- They are not applicable for real-time signal processing. For example, methods based on the calculation of Lyapunov exponents or entropy.
- Low accuracy. For example, methods for estimating the statistical properties of different epochs.
- Unstable operation, requiring individual adjustment of EEG parameters. For example, methods based on artificial neural networks.

Consider wavelet transform. Its use is a natural way to detect an increase in power in the gamma range. The main essence of the method is to estimate the instantaneous energy of the wavelet spectrum in the gamma band and compare it with the threshold: if the energy exceeds a certain critical value—the threshold—then the expert system concludes that SWD is present.

Figure 2 shows a typical energy distribution of the E(f) wavelet spectrum for SWD, sleep spindle, and background EEG. The main part of the SWD fluctuations is about 8 Hz, which corresponds to the strongest peak of the wavelet spectrum. However, at low frequencies ($f < 15$ Hz) this peak overlaps with the spindle wave spectral peak, which can confuse SWD with spindle events. However, the SWD epochs have more power at higher frequencies $f > 15$–20 Hz, i.e., in the gamma frequency range, due to the presence of the second and third harmonics of the fundamental SWD frequency (the harmonics are marked with arrows in Figure 2). Thus, each individual peak, except for the peak–wave complex, is displayed in the wavelet spectrum as a local burst of wavelet power [10].

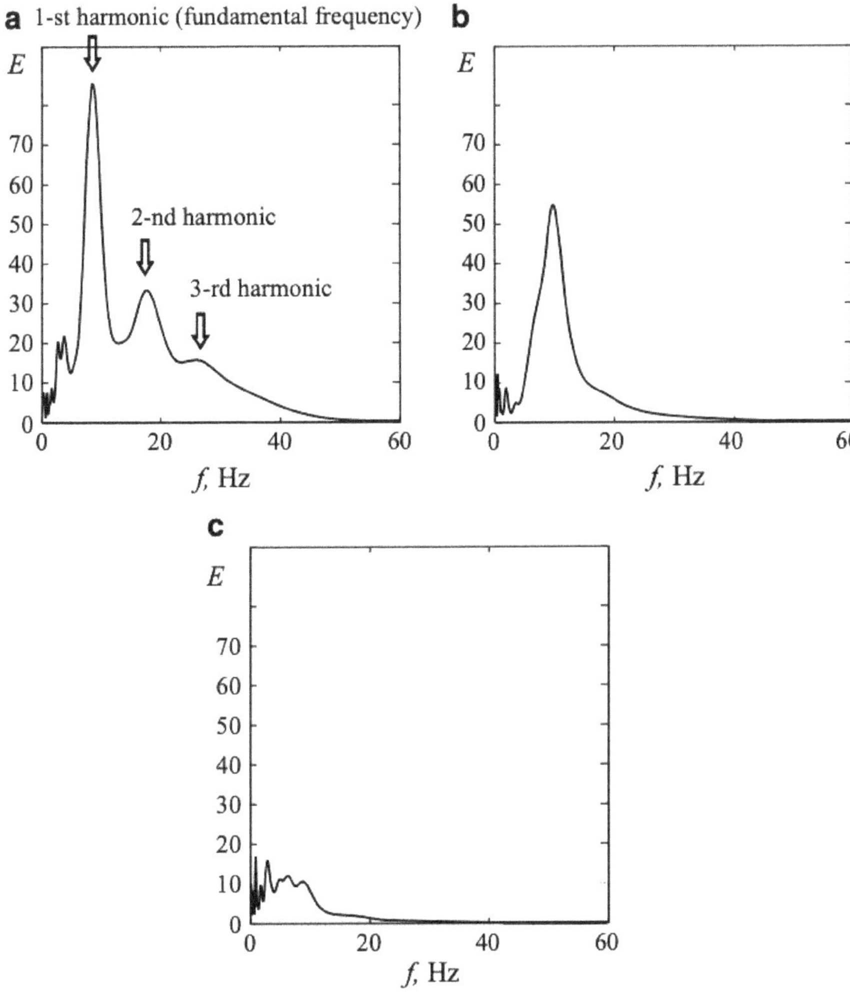

Figure 2. Typical frequency distributions of the wavelet spectrum energy E(f) characterizing various oscillatory patterns in the EEG of WAG/Rij rats. (**a**) Spike wave discharge. Arrows mark the fundamental frequency of SWD oscillations and its harmonics. (**b**) Sleep spindles. (**c**) Baseline EEG. The wavelet spectra were calculated using the Morlet wavelet.

The article "Review on Mathematical Modelling of Electroencephalography (EEG)" discusses the mathematical aspects of EEG modeling. When modeling an electroencephalogram, the greatest interest is not only the human head but also the various tissues inside it. As a rule, two types of head models can be found in the literature—a classical spherical model and a more realistic one. The second type of model allows you to more accurately describe the geometry of the structure of the head since it can be obtained by combining magnetic resonance imaging (MRI) and computed tomography (CT) [11].

The article "A review of parametric modelling techniques for EEG analysis" discusses autoregressive modeling. Thus, the AR modeling technique can be defined in the private domain as a spectral matching problem or in the time domain as a linear prediction problem. The authors of the article choose the latter approach since an important advantage of the model based on linear prediction is the relative simplicity of parameter estimation using linear signal processing procedures [12].

This article describes various methods for more accurate EEG removal. In addition to describing algorithms that help in solving the above problem, the authors talk about platforms for analyzing biosignals, one of which is EEG. Let us look at some of them. BioSig is an open-source software library for processing biomedical signals, including, for example, the analysis of biosignals such as an electroencephalogram (EEG), electrocorticogram (ECoG), electrocardiogram (ECG), electroculogram (EOG), electromyogram (EMG), respiration, and so on. Another platform, MATLAB—with powerful filters for importing and exporting data, algorithms for extracting functions, classification methods and powerful software for viewing and evaluating—allows you to work with EEG data [13].

There are many approaches for EEG processing; we have looked at many of them and for the engineering approach within the framework of medical cybernetics, we plan to develop a simple software for EEG processing that does not require special knowledge in IT from a doctor.

Summing up, I would like to draw attention to the fact that in this article, the articles that were analyzed covered a period of a little more than a hundred years. Each of the above articles is a reflection of the development of science of its time. Thus, we not only followed the first publications on EEGs but also saw the development and application of electroencephalograms, ways to decipher them, and the possibility of automatically reading their results, which will contribute to more accurate diagnoses in the future and avoid medical errors (since the risk of human error will be minimal). All this in general will help relieve medical personnel from unnecessary work on deciphering EEG results and direct all efforts to treating patients.

Author Contributions: Conceptualization, D.A.; methodology, D.A.; validation, D.A.; investigation, V.T.; writing—original draft preparation, V.T.; writing—review and editing, D.A.; visualization, V.T. All authors have read and agreed to the published version of the manuscript.

Funding: This research received no external funding.

Institutional Review Board Statement: The data was taken from open sources such as ResearchGate.

Informed Consent Statement: Not applicable.

Data Availability Statement: Not applicable.

Conflicts of Interest: The authors declare no conflict of interest.

References

1. Kemp, B.; Zwinderman, A.H.; Tuk, B.; Kamphuisen, H.A.C.; Oberye, J.J.L. Analysis of a sleep-dependent neuronal feedback loop: The slow-wave microcontinuity of the eeg. *IEEE Trans. Biomed. Eng.* **2000**, *47*, 1185–1194. [CrossRef] [PubMed]
2. Rechtschaffen, A.Q.; Kales, A.A. *A Manual of Standardized Terminology Techniques and Scoring System for Sleep Stages of Human Subjects*; U.S. National Institute of Neurological Diseases and Blindness, Neurological Information Network: Bethesda, MD, USA, 1968.
3. Mohsenvand, M.N.; Izadi, M.R.; Maes, P. Constrastive Representation Learning for Elecroencephalogram Classification. *Proc. Mach. Learn. Res.* **2020**, *136*, 238–253.
4. Loomis, A.L.; Harvey, E.N.; Hobart, G.A. Cerebral states during sleep, as studied by human brain potentials. *Eur. J. Exp. Psychol.* **1937**, *21*, 127. [CrossRef]
5. Banville, H.; Chehab, O.; Hyvarinen, A.; Engemann, D.-A.; Gramfort, A. Uncovering the structure of clinical eeg signals with self-supervised learning. *arXiv* **2020**, arXiv:2007.16104. [CrossRef] [PubMed]
6. Vilamala, A.; Madsen, K.H.; Hansen, L.K. Deep convolutional neural networks for interpretable analysis of eeg sleep stage scoring. In Proceedings of the IEEE International Workshop on Machine Learning for Signal Processing, Tokyo, Japan, 25–28 September 2017.
7. Fejer, L. Uber trigonometrische Polynome. *J. Reine Angew. Math.* **1916**. [CrossRef]
8. Szego, O. Beitrage zur Theorie der Toeplitzschen Formen. *Math. Z.* **1920**, *6*, 167–202. [CrossRef]
9. Thomson, D.J. Spectrum Estimation and Harmonic Analysis. *Proc. IEEE* **1982**, *70*, 1055–1096. [CrossRef]
10. Hramov, A.E.; Koronovskii, A.A.; Makarov, V.A.; Pavlov, A.N.; Sitnikova, E. Automatic Diagnostics and Processing of EEG. *SSSYN* **2015**. [CrossRef]

11. Darbas, M.; Lohrengel, S. Review on Mathematical Modelling of Electroencephalography (EEG). *Jahresber. Dtsch. Math.-Ver.* **2018**, *121*, 3–39. [CrossRef]
12. Pardey, J.; Roberts, S.; Tarassenko, L. A review of parametric modelling techniques for EEG analysis. *Med. Eng. Phys.* **1996**, *18*, 2–11. [CrossRef] [PubMed]
13. Balam, V.P.; Plawiak, P.; Prakash, A.J. A Brief Review on EEG Signal Pre-processing Techniques for Real-Time Brain-Computer Interface Applications. *TechRxiv* **2021**. [CrossRef]

engineering proceedings

MDPI

Proceeding Paper

Exponential Particle Swarm Optimization Algorithm for Complexly Structured Images Segmentation †

Samer El-Khatib [1], Yuri Skobtsov [2,*] and Sergey Rodzin [1]

1 ICTIS Department, Southern Federal University, 105/42 B. Sadovaya, Rostov-on-Don 344006, Russia; samer_elkhatib@mail.ru (S.E.-K.); srodzin@yandex.ru (S.R.)
2 Institute of Information Technologies and Programming, Saint Petersburg State University of Aerospace Instrumentation, 67 B. Morskaya Str., Saint-Petersburg 190000, Russia
* Correspondence: ya_skobtsov@list.ru; Tel.: +7-931-580-5059
† Presented at the 15th International Conference "Intelligent Systems" (INTELS'22), Moscow, Russia, 14–16 December 2022.

Abstract: Image segmentation is the process of dividing an image into homogeneous regions according to certain features and is widely used in image processing. Complexly structured images usually contain complex and essential objects. These images are non-linear structural images and they contain a large number of elements with required specifications. The main goal of the proposed EPSO (Exponential Particle Swarm Optimization) algorithm is to prevent local solutions and find the exact global optimal solutions for the task of segmenting medical images. The execution time is compared with well-known segmentation algorithms. The EPSO method is superior to the segmentation methods studied, including the graph algorithm. Comparisons were made with existing segmentation algorithms (Grow cut, Random Walker, DPSO, K-means PSO, and hybrid-K-means ant colony optimization algorithm) in tabular form.

Keywords: complexly structured image segmentation; swarm intelligence; particle swarm optimization algorithm

Citation: El-Khatib, S.; Skobtsov, Y.; Rodzin, S. EPSO Algorithm for Complexly Structured Images Segmentation. *Eng. Proc.* **2023**, *33*, 47. https://doi.org/10.3390/engproc2023033047

Academic Editors: Askhat Diveev, Ivan Zelinka, Arutun Avetisyan and Alexander Ilin

Published: 13 July 2023

1. Introduction

The development of image recognition methods is one of the difficult and involved tasks in AI. The development of image recognition methods is well studied in theory; however, there is no general method to solve it and the practical solution seems to be very difficult [1,2].

Complexly structured images often contain difficult and essential objects. These images are images with a difficult structure and they contain a large number of elements with required specifications. Images have a complex structure that contains many different semantics. These images contain the following attributes:

- Contains a lot of dissimilar objects;
- Objects on images are different;
- Each object has different properties that must be considered;
- Morphologically complex structures;
- Object and background usually have almost the same color distribution.

Examples of such images include geographic or topographical maps, remote sensing images of the earth, etc. Examples of such images are also medical images, including magnetic resonance imaging called MRI images or computed tomography images.

Recent research findings have demonstrated the potential for using techniques inspired by nature, such as ant, particle swarm (PSO), and bee colony optimizations. The application of biologically inspired techniques, including PSO, is still under-studied and more research is needed. In the article, it is proposed to modify the PSO algorithm to a Modified Exponential Particle Swarm Optimization algorithm (EPSO).

2. Particle Swarm Optimization

The PSO approach employs a collection of particles, each of which has a unique local solution [3,4]. According to its own habits and those of its neighbors, the particle's behavior varies every time it enters the search zone. Each particle keeps track of its own position coordinates with the best objective function and best neighbor, from which the best overall solution is derived.

Each particle stores the best fitness value and coordinates. This fitness value can be expressed as y_i and named as a cognitive component. Similar to this, let us indicate the best global optimum obtained by all particles as $\hat{y}(t)$ and call it the social component.

Each i-th particle has properties such as velocity $v_i(t)$ and position $x_i(t)$ a time t. Particle location changes based on

$$x_i(t+1) = x_i(t) + v_i(t+1), \tag{1}$$

where $x_i(0) \sim U(x_{min}, x_{max})$.

$$v_{ij}(t+1) = v_{ij}(t) + c_1 r_{1j}(t)[y_{ij}(t) - x_{ij}(t)] + c_2 r_{2j}(t)[\hat{y}_j(t) - x_{ij}(t)]. \tag{2}$$

The best position (gbest) at a time (t + 1) can be obtained as follows

$$y_i(t+1) = \begin{cases} y_i(t) & if\ f(x_i(t+1)) \geq f(y_i(t)) \\ x_i(t+1) & if\ f(x_i(t+1)) < f(y_i(t)), \end{cases} \tag{3}$$

where $f : R^{n_\infty} \to R$ is target function, that says if current position is optimal. $\hat{y}_j(t)$ (pbest) at a time t can be calculated as follows

$$\hat{y}(t) \in \{y_0(t) \dots y_{n_s}(t)\} | f(\hat{y}(t)) = min\{f(y_0(t)) \dots f(y_{n_s}(t))\} \tag{4}$$

where n_s is the sum of all swarm particles.

3. EPSO Algorithm for Complexly Structured Images Segmentation

To obtain better segmentation results, a combined method has been proposed which utilizes all the advantages of K-means and PSO algorithms.

The EPSO algorithm is basically similar to the mixed ACO K-means algorithm [5].

Each swarm particle x_i represents N areas (clusters) such as $x_i = (m_{i1}, \dots, m_{ij}, \dots, m_{iN})$ where m_{ij} is center for area j for swarm particle i. Target function can be represented as follows

$$f(x_i, Z_i) = \omega_1 \bar{d}_{max}(Z_i, x_i) + \omega_2(z_{max} - d_{min}(x_{min})), \tag{5}$$

where $z_{max} = 2^s - 1$ for image with s-bits; Z is representative table for connectivity between pixel and center of the area for particle i.

The table indicates if point z_p is in area c_{ij} for swarm particle i. Constant values ω_1 and ω_2 are user-defined, $\bar{d}_{max}$—max average Euclidian distance from swarm particles to linked areas. It can be measured as follows:

$$\bar{d}_{max}(Z_i, x_i) = \max_{j=1,\dots,N}\{\sum_{\forall z_p \in c_{ij}} d(Z_p m_{ij}/|c_{ij}|)\}, \tag{6}$$

$$d_{min}(x_i) = \min_{j_1, j_2, j_1 \neq j_2}\{d(m_{ij1}, m_{ij2})\} \tag{7}$$

Formula (7) contains minimal Euclidian distance among each pair of the area centers.

In the next task, the swarm is used to achieve good clustering using the passed parameters. It was achieved through self-study. Each particle in the PSO algorithm represents a pixel. The pixel intensity is used as an input parameter for the PSO algorithm.

The Algorithm 1 includes the following steps:

Algorithm 1 Exponential PSO segmentation algorithm

1. Presented swarm m. Set the number of swarm particles, personal and global acceleration rates c_1 and c_2, max allowed number of iterations N_{max}, parameters for target function *f(5)*.
2. *For i = 1, . . . , m*(for each particle)
   2.1 Initialize starting position of the particle using vector x_i.
   2.2 Starting position of the particle is currently known best position $y_i = x_i$.
   2.3 If $f(y_i) < f(\hat{y})$, then update best swarm's value replacing $\hat{y}$ to y_i.
   2.4 Randomly initialize velocities of the particles v_i.
3. Current number of iterations $N = 1$.
4. *For i = 1, . . . , m* (for each particle)
5. *For j = 1 . . . , n* (fitness function parameters)
   5.1 Update particle's velocity v_{ij} and position according to $x_{ij} = x_{ij} + v_{ij}$.
6. If $f(x_i) < f(y_i)$, then replace best local solution for particle $y_i = x_i$, otherwise return to Section 4.
7. If $f(x_i) < f(\hat{y})$, update best global swarm's solution $\hat{y} = x_i$, otherwise return to Section 4.
8. Inrease Number of iterations on 1: N = N + 1.
9. If $N \leqslant N_{max}$, then return to Section 4, otherwise $\hat{y}$ contains best found solution.
10. Initialize K centers of the clusters using best particles positions.
11. Calculate pixel's belonging to cluster (according to distance to the center).
12. Using (5) recalculate clusters centers. If they are not equal to previous, then repeat Section 11.
13. Save best individual solution for each particle (pbest (3)).
14. Save best common solution for m particles (gbest (4)).
15. Update clusters centers.
16. If centers have changed, then return to Section 12.

Thanks to the particle optimization algorithm, all particles tend to fly directly to the best location found by the best particle. This approach enables the rapid discovery of potential solutions. Using this mechanism, particles often accumulate in a local minimum instead of a global minimum, resulting in suboptimal solutions. To avoid this effect, El-Desouky [6] suggested making ω linear, for example

$$\omega = (\omega - \omega_1)\frac{(n_{max} - n)}{n_{max}} + \omega_{min}, \tag{8}$$

where n_max is maximum number of iterations, n is the number of the current iteration. Recommended values are $\omega_{max} = 0.9$; $\omega_{min} = 0.4$. ω can be decreased down to ω_{min} over 1500 iterations. In this article, we propose to change ω exponentially. In the presented algorithm, we propose to change ω in the following way:

$$\omega = (\omega - \omega_1)e^{\frac{(n_{max} - n)}{n_{max}}} + \omega_{min} \tag{9}$$

4. Testing EPSO Segmentation Method

To evaluate the efficiency of the algorithm, numerical experiments were provided. Six segmentation methods were considered. Among them are FC-Means [7], Darwinian PSO [8], PSO modification—K-means PSO [8], Grow cut [9], Modified HACO-K-means algorithm—K-means ACO [10,11], and Random Walker [12].

Table 1 represents the running time for three images from well-known image dataset [11] (Figures 1–3).

Table 1. Time taken for each image tested using different algorithms.

Image	K-Means-PSO	EPSO	ACO-K-Means	FC-Means	Grow Cut	Random Walker	Darwinian PSO
1	7.48	7.34	12.14	9.49	14.78	5.01	11.85
2	0.19	0.18	0.93	0.87	1.35	2.2	16.34
3	17.5	17.5	24.04	12.14	45.30	14.2	15.95

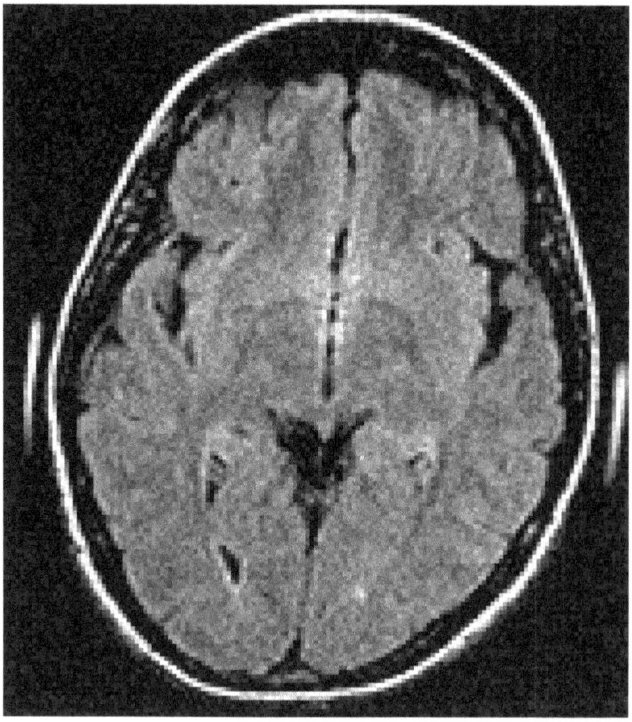

Figure 1. MRI image of the Brain. Params—421 × 392, 8 bits per pixel.

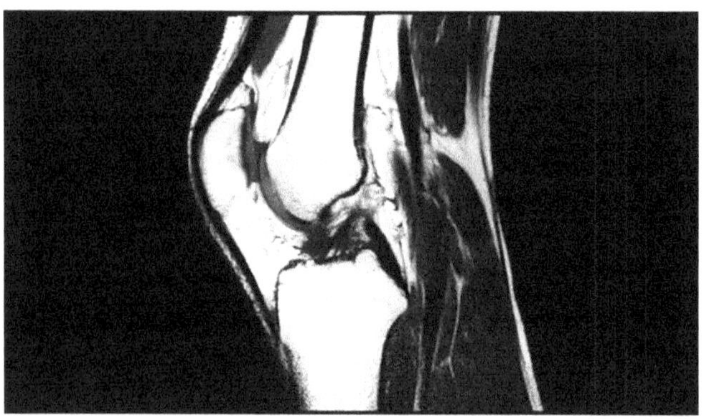

Figure 2. Noised brain image. Parameters: 151 × 166, 8 bits per pixel.

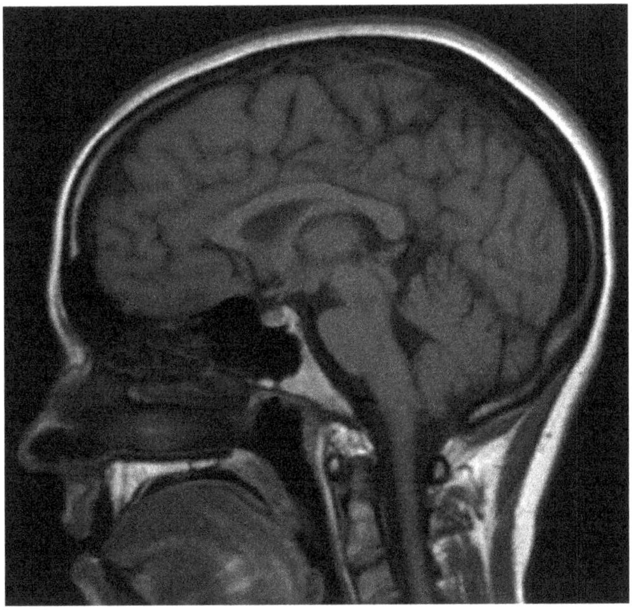

Figure 3. Sagittal MRI image of the knee. Parameters 802 × 450, 8 bits.

From Table 1 it can be observed that the proposed method outperforms all existing modifications of ACO and PSO and graph-based methods, except for Random Walker (in this case, the execution time affects the segmentation quality—EPSO- segment quality exceeded Random Walker by 15%).

5. Conclusions

In the presented paper, the modified EPSO algorithm for segmenting complexly structured images was introduced. A comparison of the results of the algorithm with other methods has been presented. In all cases, the algorithm produces a better final sharding time than the studied techniques (with the exception of the Random Walker algorithm, which has over 15% lower segmentation quality). All test results were obtained using the Ossirix MRI image dataset and own software products. The obtained results show that the transform EPSO algorithm can be used in digital image processing for images with complex structures.

Author Contributions: Co-authors: S.E.-K., Y.S., S.R. Description of the contribution of co-authors: Conceptualization, Y.S.; software, S.E.-K.; methodology, S.R.; validation, Y.S., S.R.; writing—original draft, Y.S.; investigation, S.E.-K.;writing—review and editing, S.E.-K.; formal analysis, S.E.-K.; methodology, S.R. All authors have read and agreed to the published version of the manuscript.

Funding: Presented research was supported by Russian Foundation for Basic Research according to the research project № 19-07-00570 "Bio-inspired models of problem-oriented systems and methods of their application for clustering, classification, filtering and optimization problems, including big data".

Informed Consent Statement: Not applicable.

Data Availability Statement: Not applicable.

Conflicts of Interest: The authors declare no conflict of interest.

References

1. Gonzalez, R.C.; Woods, R.E. *Digital Image Processing*, 3rd ed.; Prentice-Hall: Hoboken, NJ, USA, 2008.
2. Kennedy, J.; Eberhart, R.C. Particle swarm intelligence. In Proceedings of the IEEE International Joint Conference on Neural Networks, Perth, WA, Australia, 27 November–1 December 1995; pp. 1942–1948.
3. El-Khatib, S.; Rodzin, S.; Skobtcov, Y. Investigation of Optimal Heuristical Parameters for Mixed ACO-k-means Segmentation Algorithm for MRI Images. In *Proceedings of the III International Scientific Conference on Information Technologies in Science, Management, Social Sphere and Medicine (ITSMSSM 2016)*; Part of Series Advances in Computer Science Research; Atlantis Press: Amsterdam, The Netherlands, 2016; Volume 51, pp. 216–221; ISBN 978-94-6252-196-4. [CrossRef]
4. Saatchi, S.; Hung, C.C. Swarm intelligence and image segmentation swarm intelligence. *ARS J.* **2007**. [CrossRef]
5. Das, S.; Abraham, A.; Konar, A. Automatic kernel clustering with a Multi-Elitist Particle Swarm Optimization Algorithm. In *Pattern Recognition Letters*; Elsevier B.V.: Amsterdam, The Netherlands, 2008; Volume 29, pp. 688–699.
6. Ossirix Image Dataset. Available online: http://www.osirix-viewer.com/ (accessed on 10 May 2023).
7. Skobtsov, Y.A. *Speransky DV Evolutionary Computation: Hand Book.-Moscow*; The National Open University "INTUIT": Moscow, Russia, 2015; 331p. (In Russian)
8. El-Khatib, S.; Skobtsov, Y.; Rodzin, S.; Potryasaev, S. Theoretical and Experimental Evaluation of PSO-K-Means Algorithm for MRI Images Segmentation Using Drift Theorem. In *Artificial Intelligence Methods in Intelligent Algorithms. CSOC 2019. Advances in Intelligent Systems and Computing*; Silhavy, R., Ed.; Springer: Cham, Switzerland, 2019; Volume 985.
9. El-Khatib, S.; Skobtsov, Y.; Rodzin, S. Improved Particle Swarm Medical Image Segmentation Algorithm for Decision Making. In *Intelligent Distributed Computing XIII. IDC 2019. Studies in Computational Intelligence*; Kotenko, I., Badica, C., Desnitsky, V., El Baz, D., Ivanovic, M., Eds.; Springer: Cham, Switzerland, 2020; Volume 868.
10. Bozhenyuk, A.; El-Khatib, S.; Kacprzyk, Ja Knyazeva, M.; Rodzin, S.I. Hybrid Ant Fuzzy Algorithm for MRI Images Segmentation. In *Lecture Notes in Computer Science*; Including Subseries Lecture Notes in Artificial Intelligence and Lecture Notes in Bioinformatics; Springer: Cham, Switzerland, 2019; Volume 11509, pp. 127–137. [CrossRef]
11. Rother, C.; Kolmogorov, V.; Blake, A. GrabCut: Interactiveforeground extraction using iterated graph cuts. *ACM Trans. Graph.* **2004**, *23*, 309–314.
12. El-Khatib, S.A.; Skobtsov, Y.A.; Rodzin, S.I. Theoretical and Experimental Evaluation of Hybrid ACO-k-means Image Segmentation Algorithm for MRI Images Using Drift-analysis. *Procedia Comput. Sci.* **2019**, *150*, 324. [CrossRef]

engineering proceedings

Proceeding Paper

Methodological Approach to Controlling the Degree of Intentions about Novel Knowledge for the Digital Economy †

Alexander A. Zatsarinnyy and Alexander P. Shabanov *

Federal Research Center "Computer Science and Control" of the Russian Academy of Sciences, 44 Vavilova St. corpus 2, Moscow 119333, Russia; alex250451@mail.ru
* Correspondence: apshabanov2017@gmail.com; Tel.: +7-905-508-03-23

† Presented at the 15th International Conference "Intelligent Systems" (INTELS'22), Moscow, Russia, 14–16 December 2022.

Abstract: The methodological approach to controlling the level of intentions about new knowledge in the field of artificial intelligence, which are expressed by organizational systems in the scientific industry and in the real sector of the economy, is investigated. Control is carried out in a digital platform based on semantic databases, which are structured under control concepts, under data entity accounting strategies, under the functions of the implementation management process and under the requirements for the timeliness of data transmission. The novelty lies in the ranking of studies depending on the degree of coincidence of intentions. The practical significance is manifested in the priority provision of new knowledge, which is most significant in the digital economy.

Keywords: data conversion; information transfer; control loop; digital platform; innovative solutions

Citation: Zatsarinnyy, A.A.; Shabanov, A.P. Methodological Approach to Controlling the Degree of Intentions about Novel Knowledge for the Digital Economy. *Eng. Proc.* **2023**, *33*, 48. https://doi.org/10.3390/engproc2023033048

Academic Editors: Askhat Diveev, Ivan Zelinka, Arutun Avetisyan and Alexander Ilin

Published: 17 July 2023

1. Introduction

This study is a continuation of the works devoted to the problem of increasing the effectiveness of the scientific industry by increasing the number of scientific papers that are in demand in transposition structures—industry, regional, cross-border and other clusters, corporate and social systems, innovative and other associations of organizational systems, both conducting research and implementing it. Thus, the scientific significance of the research conducted in [1] lies in the possibility of creating a unified information management environment for real-time interaction of autonomous management systems of organizational systems that are consolidated on the basis of a digital platform.

As a result of the research presented in [2], innovative solutions have been developed to ensure the possibility of coordinated centralized management in a single information and control environment of a transposition structure.

The creation of a digital platform model placed in a transpositional structure and having the ability to reproduce an automated data routing process based on a classifier of standards allows interactions to be intensified between producers of scientific research, innovations and inventions, and their consumers. The description of this digital platform model is given in [3].

At the same time, the main problematic issue, which has a production and technological nature, still remains the need for expanded reproduction of technological solutions of artificial intelligence. Such solutions make it possible to simulate human cognitive functions and obtain results when performing specific tasks that are comparable, at least, with the results of human intellectual activity.

As part of the solution to this problem, the task is to develop a novel electronic model for the Intent Level Control process, implemented on the basis of a digital platform.

The solution of this problem was carried out using a methodological approach to managing the level of intentions about new knowledge in the field of artificial intelligence.

These intentions are expressed by organizational systems in the scientific industry and in the real sector of the economy, and are presented in the form of the following sets of data:

A_i—a lot of intentions that are expressed by establishments of the scientific branch;

B_j—a lot of intentions that are expressed by enterprises of the real industry;

C—multiple levels of intentions that are ranked for any $c_k \subset A_i \wedge c_k \subset B_j$.

The set of data C contains the initial data that are used during calculations to determine the levels of coincidence of intentions between the interacting parties in the scientific industry and in the real sector of the economy.

The entities that are involved in the Intent Level Control process are shown in Figure 1.

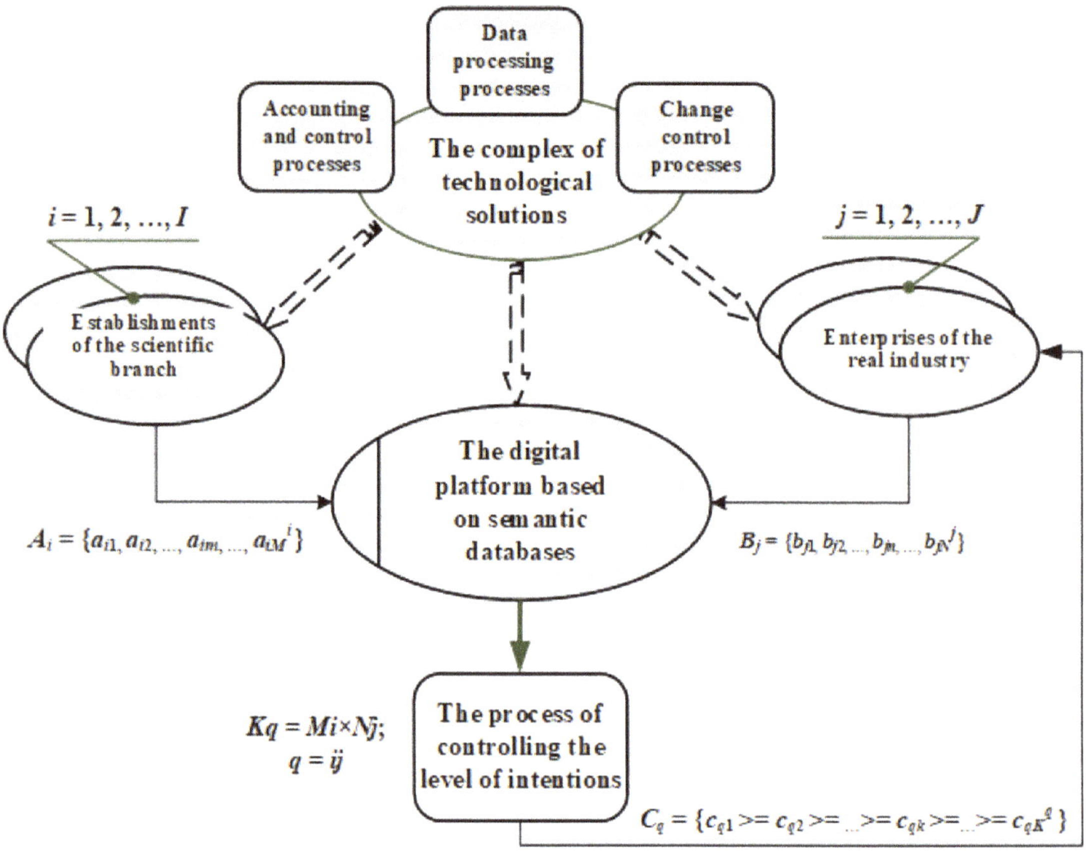

Figure 1. Entity diagram for the Intent Level Control process. A_i—a lot of intentions that are expressed by establishments of the scientific branch; B_j—a lot of intentions that are expressed by enterprises of the real industry; C—multiple levels of intentions that are ranked for any $c_k \subset A_i \wedge c_k \subset B_j$.

2. The Model of the Process of Control over the Degree of Coincidence of the Intentions of the Producer of New Knowledge and Their Potential Consumer

Intent management is carried out in a digital platform based on the use of tools for semantic databases. Semantic databases are structured to solve the tasks that are assigned to the management concept, to the strategies of accounting and control of data entities, to the functions of the change management process and to the requirements for the timeliness of data transmission.

Figure 2 shows the scheme of the Intent Level Control process model, which at the same time represents the scheme of the program for determining the degree of coincidence of intentions.

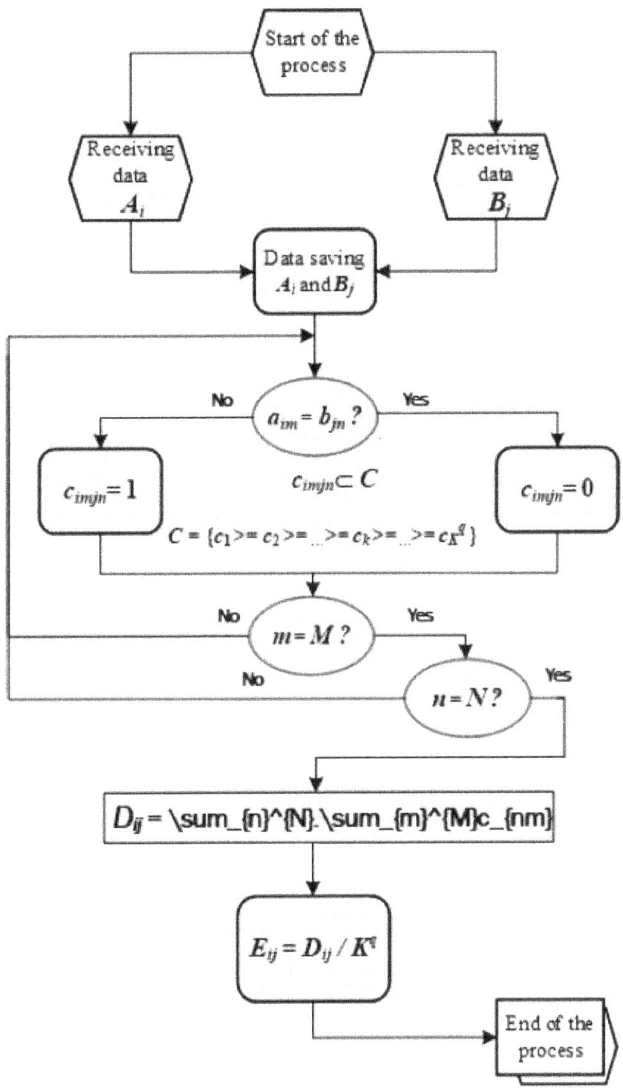

Figure 2. Scheme of the Intent Level Control process model.

The scientific novelty of the Intent Level Control process model (Figure 2) consists in creating a software and algorithmic reserve to improve the quality of new knowledge provided to the real sector of industrial artificial intelligence through priority selection for the most popular research results.

Thus, actions to introduce those innovative solutions that are most in demand in the real sector of the economy by manufacturers of microelectronic components and devices are undoubtedly capable of having a significant impact on the development of mass industrial production and the creation of competitive high-tech electronic products, which are the basis of components for artificial intelligence complexes.

The development presented in the Intent Level Control process model in Figure 2 used the following methodological approaches:

- Comprehension of the results of published studies that relate to (a) information interaction between organizational systems that carry out their activities as part of transposition structures [4–19], (b) semantic databases in digital platforms [20–25], which are carriers of electronic (digital) models of processes, objects, templates for machine learning and the relationships between them [26–34];
- Cognitive analysis of known research results that belong to the class of software-defined networks [35–48], including models that are presented in works [1–3] and can be correlated with this class;
- Development of new computer algorithms and mathematical-software approaches that are aimed at implementing the Intent Level Control process model using well-known artificial intelligence technologies [49–56].

3. On the Choice of a Strategy for the Implementation of the Accounting and Control System

Based on the analysis of the practical goal of the task being solved—the introduction of the Intent Level Control process into the IT infrastructure of a digital platform as part of a transposition structure—it can be argued that, in order to achieve it, strategies must be correctly evaluated, both by the producer of new knowledge and their consumers, in relation to the implementation of the Intent Level Control process model in real production.

The management of the activities of information and telecommunication technologies in contemporary enterprises, with which the Intent Level Control process that is presented in this article is undoubtedly correlated, is also characterized by process technology. Process technology is carried out according to regulated procedures and with the assistance of automatized systems for collecting, accounting and structuring data on the parameters of control objects and the external environment.

The process technology in combination with the above systems allows the following:

- To exercise control over the objects of control and the external environment based on the analysis of retrospective and factual information;
- To make and execute decisions in a timely manner that are in demand in a specific operational situation.

At the same time, automatized accounting and control systems can be used, which are located in autonomous organizational systems that consolidate their actions as part of transposition structures. These automatized accounting and control systems were created in accordance with the requirements of organizational systems based on ready-made software products developed, in turn, for the implementation of well-known models for the management of information technology services (ITSM) [26].

The centralized nature of the digital platform, within which the Intent Level Control process should be reproduced, provides an opportunity to technologically justify an approach to building a digital platform, including a new component—an automatized accounting and control subsystem. In this variant, this subsystem at the operational stage should carry out the following:

- Informational interaction with automatized accounting and control systems that are located in autonomous organizational systems;
- Functional compatibility with the subsystem that implements the Intent Level Control process.

When choosing a centralized option, there are different approaches to the process automation strategy for user support services and operational services that provide information and telecommunications services based on a digital platform. Consider the following strategies for building a subsystem that implements the Intent Level Control process in interaction with the accounting and control subsystem:

- The strategy of a manufacturer of the product (services);
- The strategy of a consumer of the product (services);
- The strategy of limited resources.

The first two strategies are extreme options. The third strategy occupies an intermediate value between them.

Table 1 shows the features of strategies and the main conditions under which one or another strategy can be chosen for a digital platform as an organizational and technical complex.

Table 1. Selection criteria for choosing a digital platform development strategy.

Strategy Name	Selection Criteria	Main Features
The strategy of a manufacturer of the product (services)	The readiness of the digital platform to formalize processes and develop procedures based on electronic models that underlie the selected implementation of the process.	The processes and procedures for which the models and the selected set of programs (product) were developed should be introduced into the digital platform. The goal is to introduce the product and formalize the activity "for this product".
The strategy of a consumer of the product (services)	The presence of formalized management processes and regulated procedures in the digital platform. Willingness to allocate the necessary financial resources to automate existing procedures.	The processes and procedures existing in the digital platform are accepted as specified. The goal is to automate new processes, including accounting and control processes. To do this, the most acceptable products are selected in combination with the development of new, missing programs.
The strategy of limited resources	The presence of informal processes in the digital platform. The readiness of the enterprise to correct some processes and to make changes in the regulations of procedures. Lack of funds for the development of new programs.	The processes existing in the digital platform are adjusted taking into account the principal feasibility with the help of selected well-known products. The compositions of the active elements, their interrelationships and the built-in functions of the product are selected and adjusted to the adjusted processes of the digital platform.

The analysis of the strategies given in the table on the array of information contained in published sources that relate to the topic of digital platforms allows us to draw the following conclusions.

1. Solutions for building a subsystem that implements the Intent Level Control process in accordance with the strategy of the product manufacturer are limited by the nomenclature of known products:
    - The products are provided with the necessary instructions and manuals;
    - Manufacturers, as a rule, have a network of partners to promote products on the market and to introduce them at enterprises;
    - The models that were the basis for the development of these products are sufficiently fully covered in publicly available sources.
2. Solutions for building a subsystem that implements the Intent Level Control process in accordance with the strategy of the consumer of the product are limited by the nomenclature of well-known design programs:

- Due to their individuality, their own models should be created for the digital platform in the transposition structure;
- Therefore, innovative solutions developed based on this strategy can be considered exclusive.

3. Solutions for the construction of a subsystem implementing the Intent Level Control process in accordance with the strategy of limited resources are practically not covered in the literature and, at the same time, may be most in demand at the present time, when many enterprises have realized the need to introduce modern methods of managing the provision of information and telecommunications services.

In this regard, a typical solution has been developed to build an electronic model for a new process, which should be adapted to the needs of specific organizational and technical structures, in our case, to the needs for the implementation of the Intent Level Control process. This typical solution has the following features:

- When developing the solution, an approach to process automation based on the use of boundary conditions for the stability of organizational and technical structures, and an approach to the formalization of the accounting and control process focused on the use in the class of organizational and technical structures whose activities are aimed at providing repetitive services were used.
- A description of the services that are the subjects of consumer contracts with service providers is compiled. Each service is described by the composition and necessary information about the consumers of the service, the means by which the service is provided, the composition of the work required to perform the provision of the service, the control parameters of the service (including information about the service provider and the service that provides the service) and the types of requirements that initiate the service, as well as other necessary information services.
- The solution is based on a three-level design architecture—activity processes, objects and procedures for implementing processes, and forms (templates) for working with information. This architecture ensures the adaptability of the Intent Level Control process and other processes to the features of the organizational and technical structures on which it is implemented, the automation of new processes and the regulation of procedures for their implementation (Figure 3).

Procedures implementing the above processes include the following main types of actions:

- Initiation of the user's request (request);
- Registration of the request;
- Classification of the problem;
- Registration of a work order;
- Recording information about the work performed;
- Recording information about the resolved problem;
- Review;
- Closing the request.

To implement each typical process, the procedures include all of the above types of actions, as well as actions to notify the participants of the process and transfer control over the request to them. At the same time, individual procedures differ from each other in the types of their production focus, and are provided with their own information and reference materials.

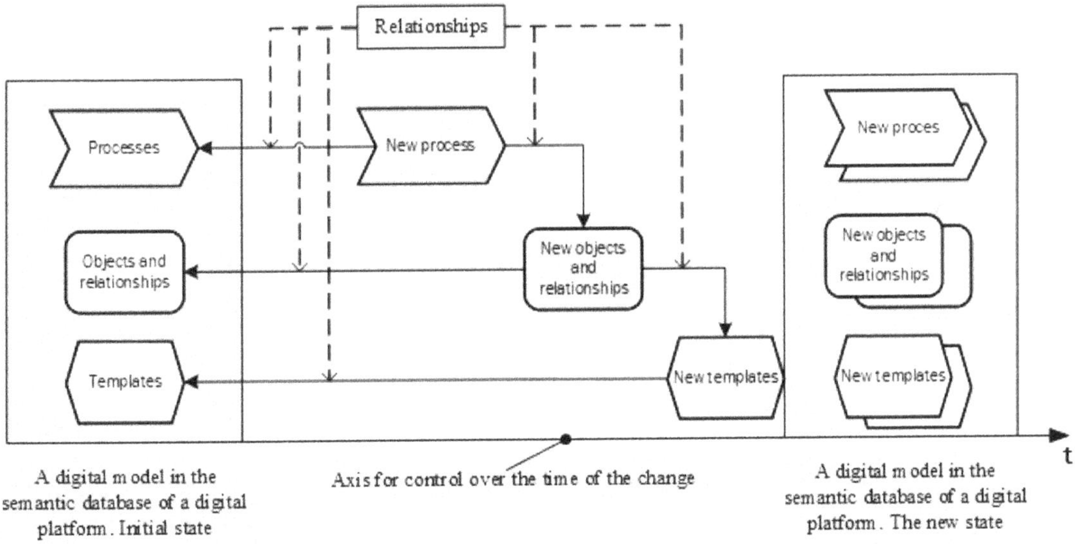

Figure 3. The general scheme of implementing a digital model for the Intent Level Control process in the database.

The main structural elements for users are operational accounting and control cards, and the necessary reference information is entered into accounting and control cards from information cards in accordance with the established procedure. Changes are recorded in the accounting and control cards, and work with them is included in the work of the relevant services. Information cards in their entirety are directories of services, facilities, employees, services and other entities. Information cards are filled in as needed.

Based on the model solution discussed above, an automated activity support system was developed and operated in the Information Technology Department of RAO UES of Russia [33]. The effectiveness of this solution is to reduce service time by automating accounting and control functions by an average of 50% in the customer support service and by 10% in operational services. At the same time, specific values are determined for each service and the type of requirements that caused it, and depend on the ratio of the time spent by performers on performing accounting and control functions and performing production work.

Thus, the use of a typical solution for the introduction of the Intent Level Control process digital platform into the IT infrastructure, functionally compatible with the accounting and control subsystem, and adapted to the local conditions of the digital platform application, is able to provide the following:

- Reducing the time required to transfer information (new knowledge) to consumers, in fact, by several times, due to the systematization of known information in reference books and convenient access to them. At the same time, almost all such requests are served in the support service automatically.
- Reducing the downtime of equipment and the time of inactivity of software in the digital platform due to the accumulation and use of retrospective information about previous work; a special effect is observed when new employees come to operational services.
- The degree of control of services reaches a new level—the quality of service can be quantified by statistics collected based on a comparison of real-time service and normalized in contracts.

- Consumers and service providers (of new knowledge) have a tool for obtaining reliable information about the real time and material resources spent on performing work on services.
- A high degree of automation of accounting and control functions creates prerequisites for reducing the cost of services for the transfer of new knowledge from their source to the consumer.

The main tool for introducing the Intent Level Control process into the enterprise infrastructure is the change management process.

4. Functional-Role Methodological Approach to Managing the Implementation of Intent Level Control Process

Along with a methodical approach to implementing a digital model for the Intent Level Control process in the database of the digital platform (Figure 3), undoubtedly, a methodological approach to the implementation of, in fact, the Intent Level Control process in the IT infrastructure of the digital platform is also necessary.

Thus, the methodological approach to managing the implementation of the Intent Level Control process is composite and includes the following methodological approaches.

1. Methodical management approach implementing a digital model for the Intent Level Control process in the database of a digital platform in a transpositional structure.
2. Methodological approach to the management of implementing the Intent Level Control process into the IT infrastructure of a digital platform in a transpositional structure.

The main provisions of the first methodological approach are set out in the previous section of the article.

The second methodological approach is based on the concept of a document model in terms of semantic modeling [26]. Within the framework of this concept, a special place is occupied by the change management process, which is compared with the management process of implementing the Intent Level Control process into the IT infrastructure of a digital platform in a transpositional structure. At the same time, in accordance with the research topic discussed in this article, a change is understood as a set of formalized procedures for implementing the Intent Level Control process into the IT infrastructure of a digital platform in a transpositional structure.

The use of formalized approval and control procedures in the implementation management process ensures that the negative impact on the quality of other services provided to users of the digital platform during the changes related to the implementation of the Intent Level Control process is minimized. The following roles are installed:

- Initiator of changes;
- Change coordinator;
- Advisory council on changes;
- Process manager;
- Council for urgent changes;
- Unified dispatch service.

Roles in the implementation management process are performed by robot programs or employees of the organizational system that is the operator of the digital platform. It depends on the stage of the project and on the state of operability of individual components of the digital platform. Organizational and analytical tasks are performed by employees.

Initiator of changes is a role as a result of which a new change is initiated, while the following functions are implemented: formation of a change request; registration of a change request; providing additional information regarding the requested change; visualization of the results of the change and closing the request.

Change coordinator is a role that results in the following functions: preliminary evaluation of a change request during registration; classification (risk and scale assessment) of changes; preliminary approval, decision-making, decision-making on standardization, analysis of the changes made—for changes in the category "requires approval"; implementation of the escalation of the change request—to change the category "requires approval"; evaluation of the results of the change—for the category "requires approval"; analyzes identified unauthorized changes in the category "requires approval".

Advisory council on changes is a role that unites service center employees. The advisory council on changes, if necessary, provides consultations within the framework of the analysis of the identified unauthorized changes in the category "requires approval".

Process Manager is a role performed by an assigned employee. Functions: monitoring compliance with the requirements of the regulations; providing, if necessary, advice and recommendations to the participants of the process; formation of an advisory council on changes.

Council for urgent changes performs the following main functions: analysis of submitted change requests; clarification of categorization and classification of changes; assessment of the consequences of changes and refusal to carry them out; assessment of the impact of changes; approval of changes.

Unified dispatch service performs the following main functions: interaction with the initiator of the request (receiving the request, notification); registration of change requests; sending change requests to the change coordinator to classify changes.

The general scheme of the implementation management process of the Intent Level Control process includes the following subprocesses.

1. Registration of change requests: the purpose of this subprocess is to ensure efficient processing of information about necessary changes in the digital platform IT infrastructure.
2. Classification of changes: the purpose of this subprocess is to perform the correct classification of the upcoming change for effective planning of the necessary resources and deadlines, determining the category of change, impact and risk.
3. Preliminary approval of changes: the purpose of this subprocess is to assess the necessity and expediency of implementing the change from the point of view of different levels of competence and authority.
4. Making a decision on making a change: the purpose of this subprocess is to assess the necessity and expediency of implementing the change, taking into account the developed preliminary implementation plan.
5. Making a decision on standardization of the change: the purpose of this subprocess is to decide whether to classify a certain change as "standard", authorizing such changes in the future without the need for approval. Changes for which a positive decision on standardization has not been made are non-standard. Such changes require approval.
6. Evaluation of the results of the change and closing the request: the purpose of this subprocess is to conduct an analysis, assess the quality of the execution of the change request and assess the compliance of the implementation of the change with the norms and requirements of the implementation management process.

A common functional and role-based scheme of actions for the implementation of the Intent Level Control process into a digital platform is shown in Figure 4.

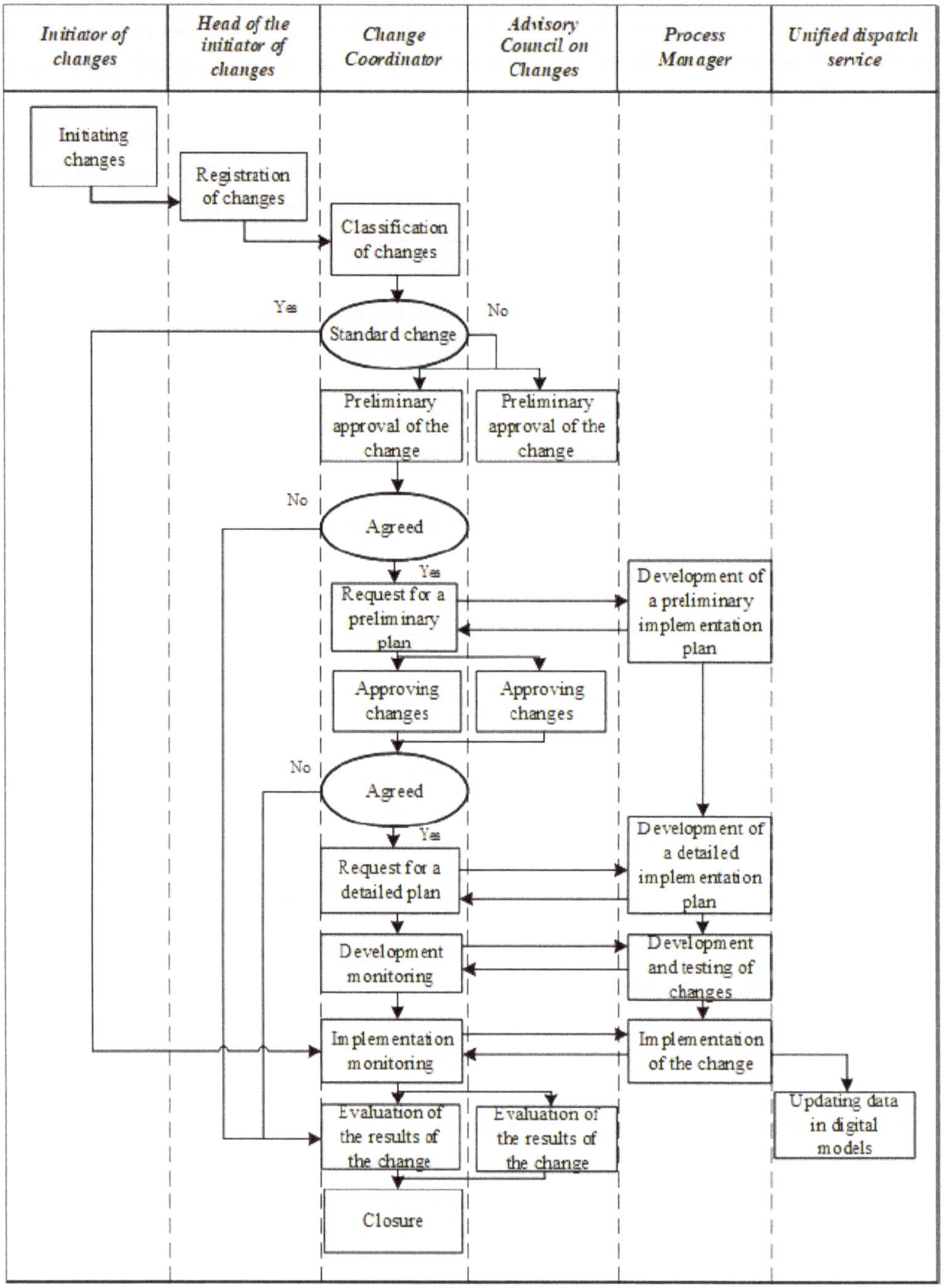

Figure 4. A common functional and role-based scheme of actions for the implementation of the Intent Level Control process.

5. Conclusions

As part of the work to solve the problem of expanded reproduction of innovative technological solutions in the field of artificial intelligence, the scientific task of developing a new process model for centralized reproduction in a digital platform of information transfer processes about new knowledge necessary for the reproduction of high-tech products for the digital economy has been set and solved.

The task was solved in the course of scientific research to create a technological reserve in the form of models of software and hardware complexes that ensure the implementation of end-to-end technologies, to ensure the availability of modern design and production tools, to repeatedly expand the use of production, scientific and engineering resources, and to develop and unify means to ensure the interoperability of organizational systems in the transposition structures of the digital economy.

A new Intent Level Control process model has been developed, which is a representative of a class of software-defined network models. The novelty of the model is determined by the availability of functionality to determine the level of coincidence of intentions available to the organizational system—the producer of new knowledge, and the organizational system—the potential consumer of new knowledge.

Methodological approaches to the implementation of the Intent Level Control process in the IT infrastructure of the digital platform have been developed and the practical feasibility of ensuring the functional compatibility of this process with the accounting and control subsystem has been substantiated, which can reduce the time of information delivery and the inactivity of the process, and create prerequisites for reducing the cost of services for the transfer of new knowledge from their source to the consumer.

Thanks to the use of the Intent Level Control process model, the effectiveness of the scientific industry and the productive sector of the economy can be increased by increasing the number of studies that are most in demand for manufacturers of high-tech products in the field of artificial intelligence.

Author Contributions: These authors contributed equally to this work. All authors have read and agreed to the published version of the manuscript.

Funding: This research received no external funding.

Institutional Review Board Statement: Not applicable.

Informed Consent Statement: Not applicable.

Data Availability Statement: Not applicable.

Conflicts of Interest: The authors declare no conflict of interest.

References

1. Zatsarinnyy, A.A.; Shabanov, A.P. Innovative Approach to Updating the Digital Platform Ecosystem. In Proceedings of the 2020 22nd International Conference on Data Analytics and Management in Data Intensive Domains, DAMDID/RCDL, Voronezh, Russia, 13–16 October 2020; CEUR Workshop Proceedings; Volume 2790, pp. 156–170.
2. Zatsarinnyy, A.A.; Shabanov, A.P. Method of centralized reproduction of information transmission pro-cesses in the digital platform control loop. *Procedia Comput. Sci.* **2021**, *186*, 63–69. [CrossRef]
3. Zatsarinnyy, A.A.; Shabanov, A.P. Disciplinary Routing as a Factor of Activization of Scientific Mobility. In *Information Systems and Design*; Taratukhin, V., Matveev, M., Becker, J., Kupriyanov, Y., Eds.; Communications in Computer and Information Science; Springer: Cham, Switzerland, 2021. [CrossRef]
4. Andreeva, T.; Astanina, L. Analysis of the industrial structure of clusters included into the register of the Ministry of Industry and Trade of Russia. *Trends Manag.* **2018**, *4*, 111–127. [CrossRef]
5. Rezanov, V.K.; Rezanov, K.V.; Zvereva, E.V.; Tai, Y. Cross-border umbrella structures as the basis of a cross-border international cluster. *Power Adm. East Russ.* **2018**, *85*, 90–99. [CrossRef]
6. Tashenova, L.V.; Babkin, A.B. Typology and structure of industrial clusters. *Manag. Russ. Abroad* **2019**, *1*, 4–14.
7. Balash, O.S.; Manukyan, M.M.; Yuklasova, A.V. Development of innovative activity of in-tegrated industrial structures (on the basis of aerospace cluster of the sa-mara region). *Vestn. Samara Univ. Econ. Manag.* **2018**, *9*, 7–11. [CrossRef]
8. Rezanov, V.K.; Zhang, J. The umbrella structure as a basis for sustainable devel-opment of international cooperation and cross-border oil and gas cluster. *Far East Probl. Dev. Arch.-Tural Constr. Complex* **2019**, *1*, 334–338.

9. Mrochkovsky, N.S. Developing models of organization management on the basis of their integration in digital clusters. *Vestn. Plekhanov Russ. Univ. Econ.* **2020**, *1*, 159–163. [CrossRef]
10. Suboch, F. Prospects for development and features of the associative concept in the construction of the latest transpositional structures, including clusters. *Agric. Econ.* **2020**, *298*, 20–40.
11. Matyukin, S.V. Models of Transformation of the Holding Companies Structures into Clusters: Scenarios under the Context of the Russian Economy. *Vlast'* **2020**, *28*, 143–151. [CrossRef]
12. Serikova, N.V. Clusters as a way to implement network communications of business and other structures in the context of digital transformation. *Econ. Entrep.* **2020**, *125*, 739–744. [CrossRef]
13. Novikova, I.V.; Sanko, G.G.; Timofeeva, Y.A. Cluster as a network structure and factor of economic growth of national economy. *Tr. BGTU* **2018**, *214*, 22–27.
14. Kudryashov, V.S. The structure of the formation and functioning of territorial cluster at the regional level. *Uchenii Zap. Tambov. Otd. RoSMU* **2018**, *9*, 56–68.
15. Khalilov, N.R. The structure of the unified information platform of interaction in machine building cluster. *Probl. Sovrem. Ekon.* **2018**, *66*, 207–210.
16. Bezpalov, V.V.; Skripnik, O.B.; Lochan, S.A.; Petrosyan, D.S. Regional clusters: Concept, structure and tendencies of development. *Audit. Finans. Anal.* **2018**, *3*, 153–161.
17. Napolskich, D.L. Analysis of the dynamics of changes in the organizational structure of innovation clusters in the Russian Federation. *Innov. Technol. Upr. Prava* **2019**, 24, 18–26.
18. Romanova, A.T.; Popova, M.V. Cluster as a large-scale project to increase the competitiveness of produc-tion and economic structures. *Vestn. Mosk. Gumanit.-Ekon. Instituta* **2018**, *1*, 80–86.
19. Avetisyan, A.A.; Drobyshevskiy, M.D.; Turdakov, D.Y. Methods for Information Spread Analysis. *Tr. ISP RAN/Proc. ISP RAS* **2018**, *30*, 199–220. [CrossRef]
20. Kuznetsova, S.; Markova, V. The Problems of Formation a Business Ecosystem based on a Digital Platform: Using the Example of 1C Company Platform. *Innovations* **2018**, *232*, 55–60.
21. Raunio, M.; Nordling, N.; Kautonen, M.; Rasanen, P. Open Innovation Platforms as a Knowledge Triangle Policy Tool – Evidence from Finland. *Foresight STI Gov.* **2018**, 12, 62–76. [CrossRef]
22. Ognivtsev, S. The concept of the digital platform of the agro-industrial complex. *Int. Agric. J.* **2018**, *362*, 16–22.
23. Grigoriev, M.; Maksimtsev, I.; Uvarov, S. Digital platforms as a resource for improving the competitiveness of supply chains. *News St. Petersburg State Univ. Econ.* **2018**, *110*, 7–11.
24. Makimtsev, I. Digital platforms and digital finance: Problems and development prospects. *News St. Petersburg State Univ. Econ.* **2018**, *109*, 7–9.
25. Korovin, Y.; Tkachenko, M. Software and Hardware Platform for Building a Digital Field System. *Oil Econ.* **2017**, *1*, 84–87.
26. Mantsivoda, A.; Ponomaryov, D. A formalization of document models with semantic modeling. *Bull. Irkutsk State Univ. Ser. Math.* **2019**, *27*, 36–54. [CrossRef]
27. The All-Russian Classifier of Standards. Available online: https://elib.kuzstu.ru/method2/mks.pdf (accessed on 30 July 2022).
28. Yin, L.; Gao, Q.; Zhao, L.; Zhang, B.; Wang, T.; Li, S.; Liu, H. A review of machine learning for new gen-eration SMART dispatch in power systems. *Eng. Appl. Artif. Intell.* **2020**, *88*, 103372. [CrossRef]
29. Mahdavinejadab, M.; Rezvan, M.; Adibi, P.; Sheathe, A.; Barekatain, M.; Barnaghi, P. Machine learning for internet of things data analysis: A survey. *Digit. Commun. Netw.* **2018**, *4*, 161–175. [CrossRef]
30. Jia, Z.; Wang, J.; Chen, X.; Sheng, Y.; Zheng, K. An SDN-based measurement scheme to build delay data-base for time-sensitive network scheduling. *Int. J. Innov. Comput. Inf. Control* **2019**, *15*, 1271–1286. [CrossRef]
31. Dos Santos, P.; Travassos, G. Scientific Knowledge Engineering: A conceptual delineation and overview of the state of the art. *Knowl. Eng. Rev.* **2016**, 31, 167–199. [CrossRef]
32. Yu, H.; Zhang, B. A Hybrid MADM Algorithm Based on Attribute Weight and Utility Value for Hetero-geneous Network Selection. *J. Netw. Syst. Manag.* **2019**, *27*, 756–783. [CrossRef]
33. Shabanov, A.P. Approach to Formalizing Coaches Controlling the Process in Resolving the Problem of Minimizing Resources Organizational Structure. In *Modern Complex Control Systems, Proceedings of the VII International Conference, Voronezh, Russia, 2005*; Voronezh State University of Architecture and Civil Engineering: Voronezh, Russia, 2005; pp. 186–190.
34. Bolotova, L.S.; Karasev, A.A.; Starykh, V.A. Formalization of the expert knowledge for incident's management of information system based on the ontological approach. *Inf. Technol.* **2014**, *6*, 3–10.
35. Rafiq, A.; Mehmood, A.; Khan, T.A.; Abbas, K.; Afaq, M.; Song, W.C. Intent-based end-to-end network service orchestration system for multi-platforms. *Sustainability* **2020**, *12*, 2782. [CrossRef]
36. Mansour, D.; Tschudin, C.; Osman, H. Load Balancing in the Presence of Services in Named-Data Net-working. *J. Netw. Syst. Manag.* **2020**, *28*, 298–339. [CrossRef]
37. Peter, K.; Marhold, K. Technological competence leveraging projects via intermediaries: Viable means to outbound open innovation and mediated capability building? *Int. J. Proj. Manag.* **2021**, *39*, 196–208. [CrossRef]
38. Füller, J.; Hutter, K.; Kröger, N. Crowdsourcing as a service – from pilot projects to sustainable innovation routines. *Int. J. Proj. Manag.* **2021**, *39*, 183–195. [CrossRef]

39. Mishra, D.; Kumar, S.; Sharma, R.R.K.; Dubey, R. Outsourcing decision: Do strategy and structure really matter? *J. Organ. Chang. Manag.* **2018**, *31*, 26–46. [CrossRef]
40. Burt, R.S.; Merluzzi, J. Embedded Brokerage: Hubs Versus Locals. In *Contemporary Perspectives on Organizational Social Networks (Research in the Sociology of Organizations)*; Emerald Group Publishing Limited: Bingley, UK, 2014; Volume 40, pp. 161–177. [CrossRef]
41. Liu, J.; Yan, J. Filling structural holes? Guanxi-based facilitation of knowledge sharing within a destination network. *J. Organ. Chang. Manag.* **2021**, *35*, 264–279. [CrossRef]
42. Stock, G.N.; Tsai, J.C.A.; Jiang, J.J.; Klein, G. Coping with uncertainty: Knowledge sharing in new product development projects. *Int. J. Proj. Manag.* **2021**, *39*, 59–70. [CrossRef]
43. Liu, L.; Zhao, M.; Fu, L.; Cao, J. Unraveling local relationship patterns in project networks: A network motif approach. *Int. J. Proj. Manag.* **2021**, *39*, 437–448. [CrossRef]
44. Odziemkowska, K.; Henisz, W.J. Webs of influence: Secondary stakeholder actions and cross-national corporate social performance. *Organ. Sci.* **2021**, *32*, 233–255. [CrossRef]
45. Roth, S.; Valentinov, V.; Kaivo-Oja, J.; Dana, L.P. Multifunctional organization models: A systems—The-oretical framework for new venture discovery and creation. *J. Organ. Change Manag.* **2018**, *31*, 1383–1400. [CrossRef]
46. Ståhle, M.; Ahola, T.; Martinsuo, M. Cross-functional integration for managing customer information flows in a project-based firm. *Int. J. Proj. Manag.* **2019**, *37*, 145–160. [CrossRef]
47. Ruiz-Martin, C.; Poza, D.J. Project configuration by means of network theory. *Int. J. Proj. Manag.* **2015**, *33*, 1755–1767. [CrossRef]
48. Xiao, S.; Lu, Z.; Xu, L. Global sensitivity analysis based on random variables with interval parameters by metamodel-based optimization. *Int. J. Syst. Sci. Oper. Logist.* **2018**, *5*, 268–281. [CrossRef]
49. Potapov, S. Research of process of information transfer on virtual routes in the radio network of the com-munication system with mobile objects. *Radio Commun. Theory Equip.* **2019**, *3*, 11–23.
50. Danilova, S.; Mikhail, I.; Rakitskiy, S.; Rakitskiy, D. Variant of robotic system with protected control channel. *I-Methods* **2019**, *11*, 1–10.
51. Arendt, C.; Nötzel, J.; Boche, H. Reliable Communication under the Influence of a State-Constrained Jammer: An Information-Theoretic Perspective on Receive Diversity. *Probl. Inf. Transm.* **2019**, *55*, 3–27. [CrossRef]
52. Kozlov, S.V.; Kubankov, A.N.; Shabanov, A.P. Innovations in control systems of actions of robotic objects in the field of emergency response. In Proceedings of the 2019 Wave Electronics and its Application in Information and Telecommunication Systems (WECONF), St. Petersburg, Russia, 3–7 June 2019; IEEE: New York, NY, USA, 2019; p. 18994090. [CrossRef]
53. Zatsarinnyy, A.A.; Shabanov, A.P. Methods of Computer Simulation Based on Shared Digital Platform. In Proceedings of the V International Conference on Information Technologies and High-Performance Computing (ITHPC-2019), Khabarovsk, Russia, 16–19 September 2019; CEUR Workshop Proceedings; Volume 2426, pp. 17–23.
54. Zatsarinnyy, A.A.; Shabanov, A.P. Model of a Prospective Digital Platform to Consolidate the Resources of Economic Activity in the Digital Economy. *Procedia Comput. Sci.* **2019**, *150*, 552–557. [CrossRef]
55. Kozlov, S.V.; Kubankov, A.N.; Shabanov, A.P. On the transformation of research data transmission pro-cesses in the digital platform. In *Wave Electronics and Its Application in Information and Telecommunication Systems (WECONF 2021)—Conference Proceedings, Saint-Petersburg, Russia, 31 May–4 June 2021*; IEEE: New York, NY, USA, 2021; p. 9470757. [CrossRef]
56. Kozlov, S.V.; Kubankov, A.N.; Shabanov, A.P. On the role of the semantic knowledge model in ensuring the stability of reproduction of data transmission processes. In *Wave Electronics and Its Application in Information and Telecommunication Systems (WECONF 2020), Saint-Petersburg, Russia, 1–5 June 2020*; IEEE: New York, NY, USA, 2020; p. 9131521. [CrossRef]

engineering proceedings

MDPI

Proceeding Paper

A Neuro-Fuzzy System for Power Supply Control †

Sergey Morozov ‡,* and Mikhail Kupriyanov ‡

Department of Computer Science and Engineering, Saint Petersburg Electrotechnical University, "LETI", ul. Professora Popova 5, 197376 St. Petersburg, Russia; mikhail.kupriyanov@gmail.com

* Correspondence: frostsergei01@gmail.com; Tel.: +7-921-779-2577

† Presented at the 15th International Conference "Intelligent Systems" (INTELS'22), Moscow, Russia, 14–16 December 2022.

‡ These authors contributed equally to this work.

Abstract: Automatic control of power supply sources is considered. Different types of batteries behave differently inside a device and require specific optimal conditions to perform efficiently. Using the wrong operating mode of a battery can lead to malfunctions within it, which may affect the device's performance. Adaptive methods for optimizing power modes are presented. The adaptivity of the method is achieved by using a baseline, which is based on artificial intelligence concepts. These concepts are neural networks and fuzzy logic, which are used to create systems capable of learning and being interpretable. This method is able to execute control over a batteries state and take actions which are required to save a battery and prolong its lifespan. The adaptivity of the system is demonstrated by the ability to take information about the environment into account.

Keywords: neural networks; fuzzy logic; neuro-fuzzy systems; power optimization; battery management systems

Citation: Morozov, S.; Kupriyanov, M. A Neuro-Fuzzy System for Power Supply Control. *Eng. Proc.* **2023**, *33*, 49. https://doi.org/10.3390/engproc2023033049

Academic Editors: Askhat Diveev, Ivan Zelinka, Arutun Avetisyan and Alexander Ilin

Published: 17 July 2023

1. Introduction

A high-quality power supply is essential for electrical devices to work properly. All devices can use a default voltage source, while some devices also use batteries in order to be mobile. Different types of batteries can be used as a power source for devices in different areas. A bad power supply can lead to malfunctions in a device or permanently break it. Using the wrong kind of batteries can lead to devices not working properly. In addition, the quality of batteries tends to worsen over time due to different reasons.

Different methods are used within devices to save battery quality over time. Some of them involve using concepts which are based on artificial intelligence (AI). In order to optimize the power consumption of a device, it is required to monitor a battery's condition and operate it in the most efficient way. Two of the AI-based concepts are neural networks and fuzzy logic, which are used to make systems trainable and interpretable. A combination of these concepts provides a neuro-fuzzy system—a hybrid AI system—which is designed to create adaptive and interpretable systems.

A neuro-fuzzy approach for battery control is presented. An intelligent system can read and process information, which is required to save batteries and prolong their lifespan. AI-based systems can detect malfunctions and specific conditions and act accordingly, for example, by informing the user about any malfunctions or executing specific actions which are required.

2. Battery Types

Devices in different areas have different requirements for the power supply. Some devices require powerful batteries in order to provide power for high-consuming devices. Other devices require batteries which can keep their charge for long periods of time while not being used. In some cases, the battery must be able to work at temperatures below zero.

Sometimes, a rechargeable battery is the most important aspect. Some devices require that the battery provides a fixed voltage, not depending on its charge.

The most important two parameters, which tend to worsen with time, are the battery capacity and inner resistance. Both of these are the main reasons behind battery degradation. Lowering the capacity leads to keeping less charge, while increasing the inner resistance provides less current, which means less power.

2.1. Lithium-Ion Batteries

Lithium-ion (Li-ion) batteries are the most well-known type of rechargeable battery, and are used for complex devices. They are used in most types of devices: smartphones, different vehicles, power banks and others. This type of battery has many subtypes, the most well known of which are lithium polymer batteries. These batteries are common because they have small sizes, a low self-discharge and a long lifespan. However, they have many downsides, like a poor performance at low temperatures [1]. The typical cell voltage is 3.6 V, but 12 V is also common.

Li-ion batteries are rechargeable, which means that they can be reused after a charging procedure. This procedure has to be performed carefully in order to save the battery. The specifics of the power element must be taken into account. For example, it is recommended to keep the charge of a Li-ion battery in the 40–80% range in order to prevent it from losing capacity [2]. They also must never be completely discharged, or they will not work again. Other types of batteries can have other specifics.

Lithium-polymer (Li-Pol) batteries are a variety of Li-ion batteries. They are very similar in terms of their chemical processes. However, Li-Pol batteries are more versatile (they can be molded into different shapes), lightweight and portable, while having lower energy levels and being more expensive then Li-ion batteries. Therefore, they are used in different areas; Li-ion batteries are used for more heavy usages, while Li-Pol batteries are preferred for mobility purposes.

2.2. Lithium Thionyl Chloride Batteries

Lithium thionyl chloride (Li-$SOCl_2$) batteries are commonly used in industrial devices, like systems of commercial accounting of energetic resources. The main reason for this is their high capacity, extremely low self-discharge and ability to work at low temperatures. They are unrechargeable, and therefore are used once. The most common nominal voltage for this type of battery is 3.6 V. This is the operating voltage for many low-power microcontrollers [3].

These batteries keep their charge very well due to the passivation effect, which is exhibited as a rise in the inner resistance. A passivated battery may not able to provide enough current to power up the device, while having enough charge to do it. This happens if the battery is not used for a long time. Depassivation is required to return a battery to a working state. This process provides a specific load, which is used to return the inner resistance to the normal value, while not losing too much charge. This can be achieved by connecting a positive battery contact to the ground through a transistor through a resistor. A transistor is required to turn the depassivating process on and off, while a resistor is used to control the current in the circuit. Mounting a resistor must be performed with special caution, since it will explode if a short circuit occurs. The same approach can be used in order to prevent passivation. The battery will lose some charge during the process, but it will work without problems [4].

Another feature of Li-SOCl2 batteries is providing a fixed voltage which does not depend on the charge of the battery. It is important for devices which use reference voltages. The downside of this feature is that it is difficult to monitor the real charge of the battery. Batteries' voltages can drop when a small amount of charge is left, which can be used to notify the user about battery replacement.

3. Control of the Batteries

Approaches which are dedicated to controlling batteries require reading a batteries state and interacting with it. Systems which do this are known as battery management systems (BMSs) [5]. Hardware tools are required for this. When it is necessary to check the output voltage of a battery, an ADC can be used. Special microchips, like MCP37811/2 [6], can be used as a specialized tool to control the charging process of Li-ion and Li-Pol batteries. Other tools may be required for other goals like external temperature monitoring or preventing the passivation process.

Hardware tools are required to manipulate a battery, while software is required to execute the control. The creation of a complex battery control system is required as an efficient monitoring and control tool. Artificial intelligence is used to create such systems. Using the concepts of artificial intelligence leads to creating adaptive systems. Systems for battery control should be interpretable in order to easily read the state of each battery. Since different types of batteries require specific operating conditions for better performances, machine learning can also be used. Fuzzy logic is used to create interpretable systems, while neural networks provide machine learning. A combination of these concepts provides a neuro-fuzzy system [7–9].

Control procedures are executed by monitoring the environment and modifying the system parameters. They can be used to stop the battery from operating in the wrong mode and to optimize the power consumption. For example, if a device periodically performs some tasks, and the results of some of these tasks do not change much, it is useful to reduce the frequency of executing these tasks to save power. In some cases, different features of devices can be manually turned on and off in order to save power. This can be useful to reduce the power consumption of high-power features like network connections and LCD displays. This is the purpose of sleep mode.

A neuro-fuzzy system can be used to optimize a device's power consumption. It can process information about several important elements:

- Power consumption at each moment.
- Processed tasks and power consumption of each of them.
- Information about usage (elapsed time and number of recharges).
- State of the battery (charge and health).
- Environmental factors (temperature and humidity).

Systems for battery control must be able to monitor these elements and adapt to them. Having information about processed tasks and the power consumption at the moment is essential in order to prioritize them. Tasks can be power consuming, and if these tasks are low priority, then it is required to decrease their power consumption. Processing information about the current battery state and environment is important to monitor operating conditions [10].

A neuro-fuzzy baseline can also be used for the modelling of a battery. It is essential to monitor the chemical processes in a power source to determine the optimal mode for it. Models can be used as an efficient development tool for testing a control system, since a good model is able to reproduce the required behavior of a battery.

4. Structure of Neuro-Fuzzy Battery Control Systems

Since different types of batteries require different care and operating conditions, it is impossible to create a control system for all power supply elements. However, it is possible to create a common structure for a system, which has to be trained to work with specific battery types. It can monitor every required parameter and adjust the system to them.

Hardware and software must be designed to create a control system. Battery circuits must be created in order to interact with the battery. These circuits must read the information from the environment and about the battery. The state of the environment is read by specific sensors which interact with a CPU by using a defined interface. A structure which demonstrates how sensors can be used to read data about the environment is shown in Figure 1. It shows how the device should utilize circuits with sensors and how to connect an ADC to interact with a battery. Power is connected both as a supply and as an analog signal, which can be read either by an external ADC or by microcontrollers. A fuse is used as a protection tool. A power management system is implemented within a CPU.

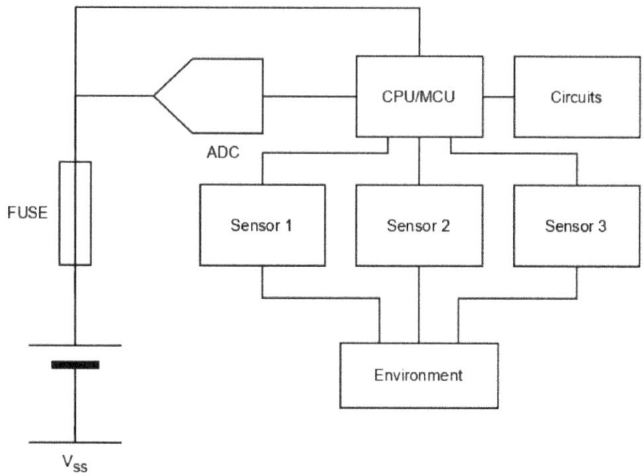

Figure 1. Hardware structure of battery optimization system.

Specialized microchips exist to control the power. They can be used to monitor malfunctions within a battery and switch between power modes. An example circuit for battery switching is shown in Figure 2. It uses a switch, which can generate fault signals, a linear regulator and a charge management controller. This circuit can be used to switch between an external USB source and a rechargeable Li-Pol battery. A power management controller is used to charge the battery and is not required if a non-rechargeable battery is used. A linear regulator is used as a tool to obtain the required voltage levels. If the protection of circuits is required, isolated DC/DC converters can be used, since they use galvanic isolation. A switch is used as a tool to select a power source.

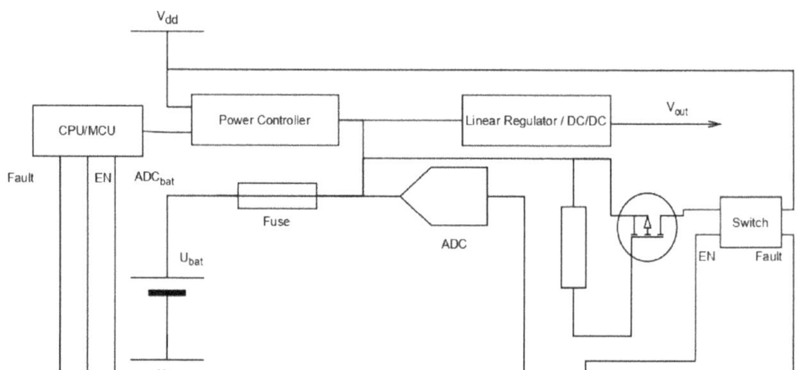

Figure 2. Battery management circuit.

Software is required to carry out the calculation processes of the neuro-fuzzy system. An adaptive neuro-fuzzy inference system (ANFIS) [11] is used. The ANFIS obtains values of the required parameters and transforms them into fuzzy values (obtains values of defined membership functions), which are processed by a neural network, which provides a fuzzy output value. The output value can be defuzzified if required. The structure of a neuro-fuzzy system for processing multiple parameters (for example, charge, temperature and external humidity) is shown in Figure 3. The neuro-fuzzy system can be implemented in a single CPU or in multiple microchips. This intellectual system is able to process multiple parameters and provide the required management accordingly. It can notify users about malfunctions, place microchips in an idle mode or reduce the amount of calculations in order to optimize the power consumption. A battery controller can also be used in this circuit. Some of the CPUs periphery can be used (for example, an ADC). Every power source can be connected to an ADC in order to detect malfunctions or change the device's operating mode. Connections to the ground and battery protection components (fuses) are omitted. The control of a single battery is demonstrated. The interaction of cells must be considered if a multiple number of them are used. Multiple subsystems for single battery control can be used, but it is important to determine how one battery can affect another.

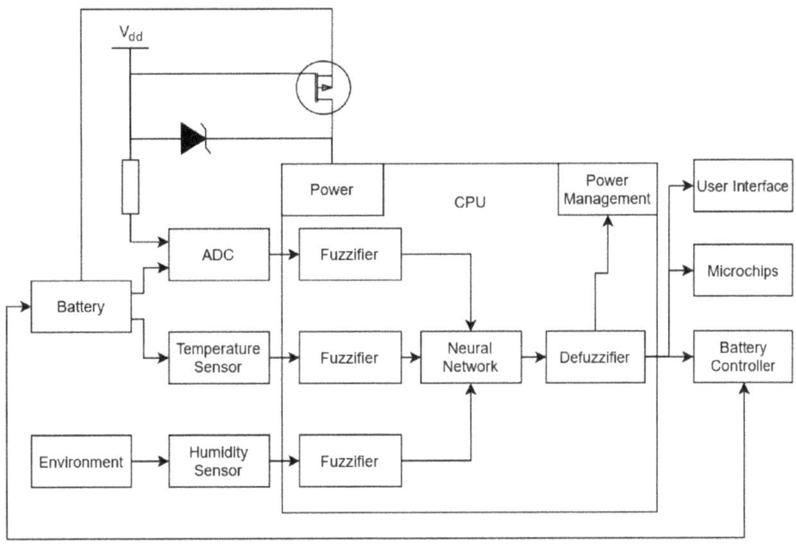

Figure 3. Circuit structure for a neuro-fuzzy system.

Fuzzy sets for the control system can be defined manually. Every fuzzy set must represent the conditions which the battery works in. Fitting and unfitting conditions must be described. A step into developing self-organization can be taken by developing an algorithm for automating the generation of fuzzy sets. A neural network for the system must be trained with parameters from the environment. Training is important for taking every parameter into account, while also looking at how their interaction can affect the system. If many parameters are considered, then the neural network can be an essential component. The demonstrated system is shown to produce a single result, but it can produce multiple decisions depending on the cell number and the monitored parameters. It can be changed by modifying the neural network.

The system is easily adjustable to process more parameters or change them. Using ANFIS-based control systems allows for a simple interpretation of a model for any type of battery. If some common aspects are found in the battery models, then this knowledge can be used in other systems for optimizing purposes. This can be reasonable if multiple types of batteries are used in a system.

5. Conclusions

The presented neuro-fuzzy approach allows creating adaptive subsystems to control batteries. The system is able to process information about the current environment and power consumption. A neuro-fuzzy baseline can be implemented within a device as a power control module. The intelligent system is able to monitor a battery and take actions in order to prolong the lifespan of a battery. If taking actions is impossible, the system can give recommendations about what to do with a specific battery. Each type of battery has to be taken care of individually, since each of them have different specifics and features.

Advanced approaches for creating intelligent BMSs are to be inspected in future works. Multiple batteries can be used in a system, and it is essential for them not to introduce errors into the work of another. Thus, the protection of batteries from each other must be considered.

Author Contributions: Conceptualization, S.M. and M.K.; methodology, S.M.; validation, M.K.; formal analysis, S.M.; investigation, S.M.; resources, S.M.; data curation, M.K.; writing—original draft preparation, S.M.; writing—review and editing, M.K.; visualization, S.M.; supervision, M.K.; project administration, M.K. All authors have read and agreed to the published version of the manuscript.

Funding: This research received no external funding.

Institutional Review Board Statement: Not applicable.

Informed Consent Statement: Not applicable.

Data Availability Statement: Data sharing not applicable.

Conflicts of Interest: The authors declare no conflict of interest.

References

1. Hamidi, S.A.; Manla, E.; Nasiri, A. Li-ion batteries and Li-ion ultracapacitors: Characteristics, modeling and grid applications. In Proceedings of the 2015 IEEE Energy Conversion Congress and Exposition (ECCE), Montreal, QC, Canada, 20–24 September 2015; pp. 4973–4979.
2. Li, H. Practical Evaluation of Li-Ion Batteries. *Joule* **2019**, *3*, 911–914. [CrossRef]
3. Astafev, E.A. Electrochemical noise measurement of a Li/$SOCl_2$ primary battery. *J. Solid State Electrochem.* **2018**, *22*, 3569–3577. [CrossRef]
4. Morozov, S.M. Neuro-fuzzy Approach for Batteries Depassivation. In Proceedings of the 2022 Conference of Russian Young Researchers in Electrical and Electronic Engineering (ElConRus), Saint Petersburg, Russia, 25–28 January 2022; pp. 1244–1247.
5. Patel, R.; Talmale, S. Battery Management System—Hardware Design. *Int. Res. J. Eng. Technol. (IRJET)* **2022**, *9*, 512–515.
6. Microchip Technology Inc. Simple, Miniature Single-Cell, Fully Integrated Li-Ion/Li-Polymer Charge Management Controllers. Datasheet for MCP73811/2. Available online: https://ww1.microchip.com/downloads/en/DeviceDoc/22036b.pdf (accessed on 15 June 2022).
7. Gorrostieta, E.; Pedraza, C. Neuro fuzzy modelling of control systems. In Proceedings of the 6th IEEE International Conference on Electronics, Communications and Computers (CONIELECOMP 2006), Puebla, Mexico, 1 March–27 February 2006.
8. Panickerm, D.; Kamthem, M.A.; Kapoor, D. A Study on Battery Management Systems. In Proceedings of the National Conference on Innovative Engineering Technologies & Management (IETM-2021), Pune, India, 11–12 November 2021.
9. Ashok, J.; Thirumoorthy, P. Neuro Fuzzy Controller for Automatic Battery Charge-Discharge Management system of a Wireless Sensor Node. *Aust. J. Basic Appl. Sci.* **2014**, *8*, 636–641.
10. Liu, W.; Placke, T.; Chau, K.T. Overview of batteries and battery management for electric vehicles. *Energy Rep.* **2022**, *8*, 4058–4084.
11. Shen, W.X.; Chan, C.C.; Lo, E.W.C.; Chau, K.T. Adaptive neuro-fuzzy modeling of battery residual capacity for electric vehicles. *IEEE Trans. Ind. Electron.* **2002**, *49*, 677–684. [CrossRef]

engineering proceedings

MDPI

Proceeding Paper

Functional Converter for Intelligent Sensor and Its Layout Design †

Olga Bureneva [1,*], Sergey Mironov [1], Nikolay Safyannikov [1] and Zhanna Sukhinets [2]

1 Department of Computer Science and Engineering, Saint Petersburg Electrotechnical University LETI, St. Petersburg 197022, Russia; semironov@etu.ru (S.M.); nmsafyannikov@etu.ru (N.S.)
2 Department of Telecommunication Systems, Ufa State Aviation Technical University, Ufa, The Republic of Bashkortostan 450008, Russia; sukhinets@mail.ru
* Correspondence: oibureneva@etu.ru; Tel.: +7-812-234-6077

† Presented at the 15th International Conference "Intelligent Systems" (INTELS'22), Moscow, Russia, 14–16 December 2022.

Abstract: Recently, the number of cyber-physical systems and systems with embedded sensors has been increasing. In order to minimize the amount of information transmitted, a new paradigm is being developed that involves moving data processing as close as possible to the point of its generation. At the same time, there is a need to create devices that will provide computations near the sensors. The article considers an approach to the primary processing of a quasi-digital sensor signal. The authors show a variant of the organization of calculations in pulse form, based on the method of small increments. The method is focused on the usage of simple logical elements with the realization of a transfer function in the base of increment/decrement operations. The solution proposed by the authors is effective in terms of simplicity of hardware implementation and can be used to connect sensors with pulse output to digital processing systems. The considered circuit solutions can be used for the implementation in programmable logic device of different levels of complexity, as well as for the manufacture of the transducer in the form of a custom circuit.

Keywords: intelligent sensor; PWM signal converter; sensor transducers; layout design; negative feedback

Citation: Bureneva, O.; Mironov, S.; Safyannikov, N.; Sukhinets, Z. Functional Converter for Intelligent Sensor and Its Layout Design. *Eng. Proc.* **2023**, *33*, 50. https://doi.org/10.3390/engproc2023033050

Academic Editors: Askhat Diveev, Ivan Zelinka, Arutun Avetisyan and Alexander Ilin

Published: 17 July 2023

1. Introduction

Recently, the number of cyber-physical systems and systems with embedded sensors is increasing. This leads to an increased load on the means of communication between such systems. To minimize the amount of information transferred, a new paradigm of edge computing is being developed. It implies moving data processing as close as possible to the point where it is received [1,2]. In the context of this paradigm, there is a need to develop special devices that provide computations near sensors. These should be compact, energy-efficient and fault-tolerant modules, which can be manufactured as separate chips or integrated with sensing elements in a single chip [3,4]. Hardware implementation of simple functional signal conversions is available in analog, digital, pulse, and mixed formats. Obviously, performing calculations in the formats of signals obtained at the outputs of sensitive elements will allow us to avoid additional operations related to format conversions. Sensors often generate a frequency output signal [5,6]. Also, analog-to-frequency converters can be installed in chips of sensors with an analog output [7,8]. The frequency output signal can be represented as a stream of pulses [9] or pulse-width modulated (PWM) signals [10]. In a pulse stream, the carrier of information is the number of pulses per unit time. In a PWM signal, the information is associated with the relative duration of the pulses. In both cases, the amplitude of the signals is the same and can be considered as logic level "1". The considered pulse forms of signal representation, also called bit-stream, allow the application of digital logic elements for primary data processing. In this case, programmable logic-integrated circuits of various levels of complexity can be used as the element base for the implementation of near-sensor computing modules [11,12]. In some cases, it is economically justified to manufacture the sensor signal primary converters in the form of application-specific integrated circuits (ASICs) [13]. If silicon technology allows, the sensing element, analog-to-frequency converter and modules for primary computation can be integrated in a single silicon chip. In both the FPGA basis and the ASIC implementation, various functional transformations can be performed. This can be realized by integrating various signal processing circuits for data reception, primary converters, digital signal processing circuits, memory, and nodes for wired or wireless communication. As a result of this integration, there will be a sensor-on-chip system, such as [14]. Based on the above facts, we can conclude that the development of digital hardware modules that process pulse signals and provide a connection between the pulse output of the sensor and the input of the digital information processing system is factual. In this paper, we focus on a new approach to the organization of calculations in a digital PWM signal converter. The traditional approach to signal processing implies conversion of analog or pulse signal into digital form with subsequent calculations using traditional units (adders, multipliers, dividers, etc.) or using a microcontroller, as shown in [15]. In the proposed transducer, all operations are performed in pulse form on the basis of simple logic elements, binary counters and registers, and the multiplication and division operation of pulse signals is implemented in the basis of operations "+1" and "-1". The proposed solution is effective in terms of simplicity of hardware implementation and can be used to connect sensors with pulse output to digital processing systems.

2. Design of the Converter

Transformations of pulse signals from sensors can be performed in different ways. Let us consider the implementation of the converter, for example, for a temperature sensor, which forms a signal in PWM form. This signal is formed by a semiconductor sensing element and is digitized with a first-order sigma–delta modulator. The output signal is a periodic signal represented by time intervals T_1 (high level) and T_2 (low level). The conversion function of the temperature sensor is as follows:

$$T(^\circ\mathrm{C}) = a_1 + a_2\frac{T_1}{T_2}. \tag{1}$$

where a_1 and a_2 are constants determined by the sensor model.

2.1. Implementation Based on Traditional Calculations

The sensor output signal is quasi-digital and can be processed by digital elements. Sensor manufacturers recommend two approaches to temperature calculation based on the parameters of the output signal [15]. Firstly, it can be a software implementation oriented to the usage of the microcontrollers. Most microcontrollers allow direct connection of sensitive elements, forming a quasi-digital output signal, to the input ports. Using a microcontroller allows you to implement simple, inexpensive solutions, but errors in the result may occur. The source of such errors is software-dependent effects. The analysis and options for compensating such effects are given in [16]. Secondly, the transformations can be performed by hardware using typical computing units. The schematic recommended by the manufacturer of a temperature sensor with output characteristic (1) is given in [15]. The circuit measures the duration of the output single signals from the sensor with a resolution of ±1 µs. For this purpose, a cascade connection of 8-bit counters is used. The counters start with the rising edge of the sensor signal and count clock pulses from an external 1 MHz oscillator. The counting results are stored in the output registers. The falling edge of the signal from the sensor forms the reset signal for the counters. After that, they start counting the next output phase—the duration of the zero value of the sensor output signal. The calculated value at the beginning of the new period is also stored in the registers. Then, the obtained values T_1 and T_2 are used to calculate the temperature by hardware or to transmit it to the software computing core.

2.2. Implementation Based on the Method of Small Increments

Consider the pulse stream implementation of expression (1). A schematic of the n-digit device realizing this operation is shown in Figure 1a. A reference pulse stream is sent to the input F and the generator G forms n reference pulse streams based on the F. Generation of reference pulse streams is performed as follows (Figure 1b). Each pulse of the F stream increases the value of the CT counter by one. At the same time, one of the outputs of this counter switches from the state "0" to "1". The only exception is the counter change from the state "all ones" to the state "all zeros", when none of the outputs switch from "0" to "1". The output signals of the counter are used as clock signals for the triggers T. In these triggers, a one is written when a rising edge occurs at the counter output. Obviously, for every F-flow pulse a one is written to no more than one trigger. After the falling edge of the F-flow pulse all triggers are reset. Thus, the sum of pulses in all generated streams is equal to 2^n–1 during the device operation period.

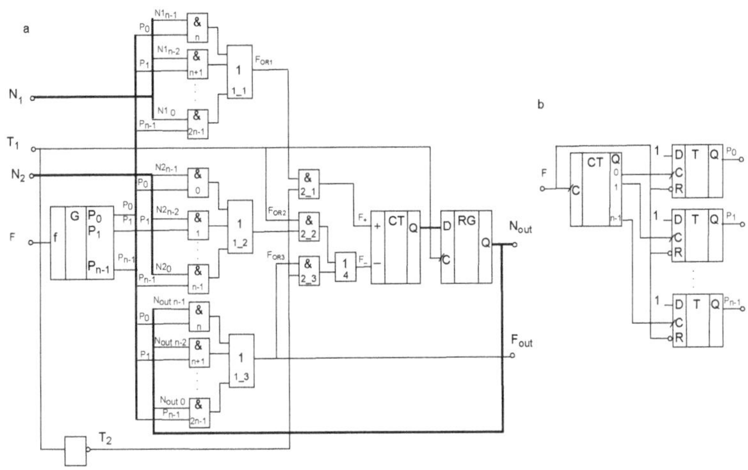

Figure 1. (**a**) General circuit diagram of the converter; (**b**) circuit diagram of the reference pulse signal generator G.

The average frequencies of P_i streams over the period of device operation are determined as follows:

$$P_i = \frac{F}{2^n} 2^{n-i-1}$$

where $i \in [0 : n-1]$.

Logic elements $\&_0 - \&_{n-1}$, connected to the outputs of generator G, implement the function of multiplication of streams P_i by bits of binary code N. The resulting streams are summed by the OR logic element. The logic expression for the output of this element is as follows:

$$F_{OR1} = L_0 \vee L_1 \vee L_2 \cdots L_{n-1}.$$

The function L in the general case for each i-th input of element OR1 is formed on the corresponding elements of $\&_i$:

$$L_i = N_{1_{n-i-1}} P_i 2^{-n},$$

where $N_{1_{n-i-1}}$ is the logical state of the n-i-1 digit of the code N_1; P_i is the pulse stream from the generator G output.

The pulse streams at the outputs of elements $\&_0 - \&_{n-1}$ have the following frequencies:

$$L_0 = N1_{n-1}P_0, \quad L_1 = N1_{n-2}P_1, \quad L_2 = N1_{n-3}P_2, \ldots, L_{n-1} = N1_0 P_{n-1}.$$

The total value of the frequency F_{OR1} at the output of the OR1 element is determined as follows:

$$F_{OR1} = N1_{n-1}P_0 + N1_{n-2}P_1 + N1_{n-3}P_2 + \cdots + N1_0 P_{n-1} =$$
$$F_2^{-n}\left(N1_{n-1}2^{n-1} + N1_{n-2}2^{n-2} + N1_{n-3}2^{n-3} + \cdots + N1_0 2^0\right).$$

In this expression, the sum of the products of the current code bit $N1_i$ by the weight of the n-bit code bit is the binary code N_1. Thus, the average frequency of the pulse stream at the output of the OR1 logic element can be written as follows:

$$F_{OR1} = \frac{F}{2^n}N_1$$

Similarly, the value of frequencies F_{OR2} and F_{OR3} at the outputs of elements OR2 and OR3, respectively, is determined as:

$$F_{OR2} = \frac{F}{2^n}N_2 \quad \text{and} \quad F_{OR3} = \frac{F}{2^n}N_{out}$$

Consider the operation of the converter shown in Figure 1a. Let the initial moment of time zero code be stored in the counter and register, i.e., $N_{out} = 0$. At digital inputs N_1 and N_2, the binary codes corresponding to constants a_1 and a_2 of Equation (1) are set. The reference pulse stream is supplied to input F; the PWM signal from the sensor output is received at input T_1. Elements 1_1 and 1_2, under the influence of codes N_1 and N_2, form the streams F_{OR1} and F_{OR2}. During the first period, as long as the input T_1 holds signal value "1", the stream F_{OR2} passes through the element $\&2_2$ and 1_4 to the counter input marked as "$-$". When the value at input T_1 becomes zero, the stream F_{OR2} is blocked. At this time, the stream F_{OR1} starts flowing through element $\&2_1$ to the counter input marked "+". The stream F_{OR3} is not generated because of the zero code N_{out}. When the signal "1" appears on the input of T_1 again, the counted number of pulses from the counter CT is rewritten into the register RG. While $T_1 = 1$ at the counter input "$-$" the stream F_{OR2} passes, when $T_1 = 0$ at the counter input "+" the stream F_{OR1} passes, and at the counter input "$-$" the F_{OR3} stream is formed on the basis of the zero code $N_{out} = 0$. Negative feedback is used in the device, and as a result, the dynamic equilibrium mode is reached. In this mode, the number of pulses coming to the adding N_+ and subtracting N_- inputs of the counter CT during the period T is the same, i.e.,

$$N_- = N_+ \quad \text{or} \quad \bar{F}_- = \bar{F}_+ \tag{2}$$

where $\bar{F}_+$ and $\bar{F}_-$ are the average values of pulse flux frequencies at the counter inputs "$-$' and "+", respectively.

The "+" counter input receives pulses during the device's operating period T. The number of pulses is determined by the following expression:

$$F_+ = F_{OR1}T_2 = \frac{F}{2^n}N_1T_2. \tag{3}$$

The number of pulses that come to the input of the counter "$-$" during the period of operation of the device T is calculated as follows:

$$F_- = T_1F_{OR2} + T_2F_{OR3} = \frac{F}{2^n}(T_1N_2 + T_2N_{out}) \tag{4}$$

In the dynamic equilibrium mode, the characteristic (2) with regard to (3) and (4) will have the following form:

$$\frac{F}{2^n}N_1T_2 = \frac{F}{2^n}(T_1N_2 + T_2N_{out})$$

From the last expression, we get the equation:

$$N_{out} = N_1 - N_2 \frac{T_1}{T_2}$$

That is, the characteristic of the device corresponds to the function of the sensor.

3. Converter Implementation

To study the functioning of the proposed circuit, we developed a description of the device in VerilogHDL.

3.1. Simulation of the Converter

The input signals for the program are the reference frequency signal F and the input signal T consisting of two parts: T_1 and T_2. For the correct operation of the model, an additional reset signal is used to ensure that the counters and registers are set to a zero state. However, given that the device has a steady state and always tends to it, it can start from any random state, so the reset signal is optional. The input signals for the program are the reference frequency signal F and the input signal T, consisting of two parts: T_1 and T_2. For the correct operation of the model, an additional reset signal is used to ensure that the counters and registers are set to a zero state. However, given that the device has a steady state and always tends to it, it can start from any random state, so the reset signal is optional. The constants in the implemented expression are set at the stage of synthesis; in the program, they are described as parameters. When creating a universal device, additional digital inputs can be provided for these constants. Figure 2 shows the operation of the device configured to process the signal from the AD TMP03 temperature sensor.

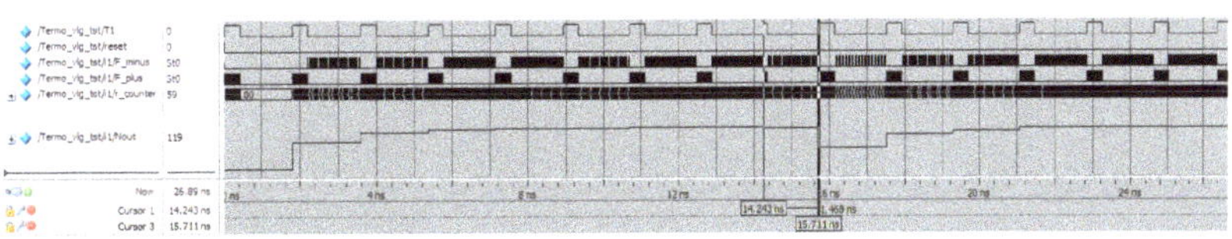

Figure 2. Result of the converter simulation.

The diagram shows the transient process, during which the transducer reaches the equilibrium mode. Starting from the fifth period of the input signal T_1, the transducer keeps the output code unchanged. This code is the result of the execution of transformations. At the moment of 14.243 ns, an interference occurs, which leads to a shortening of the signal duration T_1. After three periods, the converter returns to the equilibrium mode, compensating for the failure by triggering the feedback.

3.2. Custom Circuit Design

Despite the fact that the layout design of elements such as registers, counters, or logic primitives is quite well developed, research on topology optimization continues. This research is related to the search for original solutions in order to optimize the chip area, power and speed [17,18]. When minimizing the circuit layout design, we are looking for solutions that minimize the main component of pulse stream information processing devices—the binary counter. Solutions based on the principle of "first-zero search" are proposed. The circuit diagram of the counter cell based on the "first zero search" principle is shown in Figure 3. Its current–discharge inversion control circuit is built on pass-through keys (as in Manchester adder transfer circuits). The cells are combined into four-digit sections with an inverter at the input and at the output. At the input of each i-th bit from the previous one comes the INV_IN_i signal to invert the value of the current bit. If the inversion control signal is changed, the pass-through key in the circuit is opened. This cuts off the low-order bits and generates a new INV_OUT_i control signal value (cancelling the inversion of the current bit) in the cell. This value is sent to the next (high) counter digit.

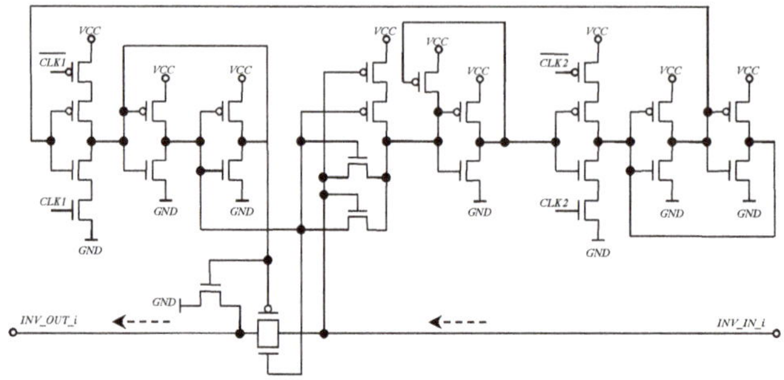

Figure 3. Electrical diagram of the single-digit counter cell "+1".

The considered circuit solution provides not only high speed and low hardware costs, but also a linear dependence of the delay time and on-chip area on the bit resolution of the device. Moreover, if necessary, the proposed circuit solution can significantly increase the speed of circuits with a relatively small increase in the area on the chip. Figure 4 shows an example of topological implementation of the converter considered above (Figure 1) for 4-bit data. Matrixing of the presented topology along the ordinate axis with an appropriate matrixing coefficient makes it possible to obtain a device of the required digit capacity (a multiple of four). The topology was developed using the original computer-aided design tools [19] in a technologically invariant concept, which allows the design to be adjusted to the design and technological requirements of the chosen manufacturing enterprise.

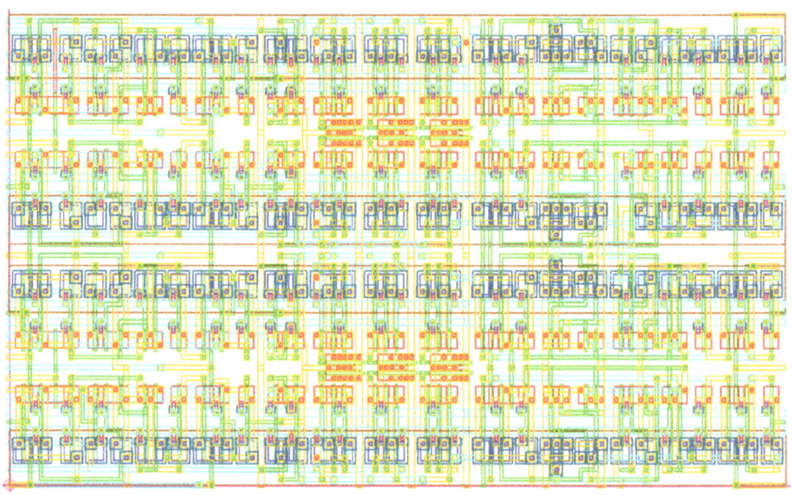

Figure 4. Layout design of the 4-digit calculator section.

From the presented material, we can see that the implementation of such calculators in the form of independent fragments of integrated circuits is compact and simple. In addition, it has greater speed than devices based on universal computing units.

4. Conclusions

A large number of sensors provide information in pulse form. This is due to the simplicity of obtaining a pulse signal at the output of the sensing element, as well as the following advantages of the pulse form:

- Energy efficiency of the presence of pauses between pulses, which reduces the average power consumption;
- High reliability of information transmission by a stream of pulses, since the loss of one pulse when transmitting a number of pulses is equivalent to the loss of one low-order bit of the binary code, while the loss of a pulse when transmitting a binary code, even in sequential format, is equivalent to the loss of a value with the weight 2^k, where k is a bit number in the binary code.

In this paper, we proposed a new method of functional processing of pulse streams based on elementary logical and arithmetic operations (increment/decrement) and implemented using simple digital elements: logic primitives, counters and registers without using an adder, multiplier, divider or other complex nodes. The considered device provides division of PWM signals directly in pulse form without additional conversion into digital binary codes. This device can be implemented based on PLD and used as an interface for sensors that generate PWM output signal and require multiplying and dividing operations. Small corrections will allow the proposed structure to be used to process signals of a large number of existing quasi-digital sensors with frequency and pulse-width modulated outputs, which are available at the moment.

Author Contributions: Conceptualization, O.B. and N.S.; methodology, O.B.; software, S.M.; validation, S.M., Z.S. and N.S.; formal analysis, O.B.; investigation, S.M. and Z.S.; resources, N.S.; writing—original draft preparation, O.B.; writing—review and editing, Z.S.; visualization, S.M.; supervision, N.S.; project administration, O.B.; funding acquisition, O.B. All authors have read and agreed to the published version of the manuscript

Funding: This work was supported by the Ministry of Science and Higher Education of the Russian Federation by the Agreement No. 075-15-2022-291 dated 15 April 2022 on the provision of a grant in the form of subsidies from the federal budget for the implementation of state support for the establishment and development of the world-class scientific center «Pavlov center Integrative physiology for medicine, high-tech healthcare, and stress-resilience technologies».

Institutional Review Board Statement: Not applicable.

Informed Consent Statement: Not applicable.

Data Availability Statement: Not applicable.

Conflicts of Interest: The authors declare no conflict of interest.

References

1. Jain, R.; Tata, S. Cloud to Edge: Distributed Deployment of Process-Aware IoT Applications. In Proceedings of the 2017 IEEE International Conference on Edge Computing (EDGE), Honolulu, HI, USA, 25–30 June 2017; pp. 182–189.
2. Singh, S. Optimize cloud computations using edge computing. In Proceedings of the 017 International Conference on Big Data, IoT and Data Science (BID), Pune, India, 20–22 December 2017; pp. 49–53.
3. Feichi, Z.; Yang, C. Near-sensor and in-sensor computing. *Nat. Electron.* **2020**, *3*, 664–671.
4. Yang, C. In-sensor computing for machine vision. *Nature* **2020**, *579*, 32–33.
5. Ueno, K.; Asai, T.; Amemiya, Y. Temperature-to-frequency converter consisting of subthreshold mosfet circuits for smart temperature-sensor LSIs. In Proceedings of the TRANSDUCERS 2009-2009 International Solid-State Sensors, Actuators and Microsystems Conference, Denver, CO, USA, 21–25 June 2009; pp. 2433–2436.
6. Beshliu, V.S.; Kantser, V.G.; Beldiman, L.N.; Beshliu, V.V.; Coban, R.A. Integral gauge pressure sensor with frequency output signal. In Proceedings of the CAS '99 Proceedings. 1999 International Semiconductor Conference (Cat. No.99TH8389), Sinaia, Romania, 5–9 October 1999; pp. 491–494.
7. Darwish, H.; Reig, C.; Cubells-Beltrán, M.D.; Leger, G.; De Marcellis, A. CMOS Capacitance-to-Time Converter-Based Interface for Differential Capacitive Sensors. In Proceedings of the 2020 Global Congress on Electrical Engineering (GC-ElecEng), Valencia, Spain, 4–6 September 2020; pp. 61–64.
8. Kokolanski, Z.; Gavrovski, C.; Dimcev, V.; Makraduli, M. Simple Interface for Resistive Sensors Based on Pulse Width Modulation. *Trans. Instrum. Meas.* **2013**, *62*, 2983–2992. [CrossRef]
9. Arbet, D.; Kováč, M.; Stopjaková, V.; Potočný, M. Voltage-to-Frequency Converter for Ultra-Low-Voltage Applications. In Proceedings of the 2019 42nd International Convention on Information and Communication Technology, Electronics and Microelectronics (MIPRO), Opatija, Croatia, 20–24 May 2019; pp. 53–58.

10. Areekath, L.; George, B.; Reverter, F. An Auto-Balancing Capacitance-to-Pulse-Width Converter for Capacitive Sensors. *IEEE Sens. J.* **2021**, *21*, 765–775. [CrossRef]
11. Safyannikov, N.; Bureneva, O. Bit-Stream Functional Converters for Decentralized Sensor Systems. In Proceedings of the 2020 9th Mediterranean Conference on Embedded Computing (MECO), Budva, Montenegro, 8–11 June 2020; pp. 1–4.
12. Santos, E.J.P.; Silva, L.B.M. FPGA-based smart sensor implementation with precise frequency to digital converter for flow measurement. In Proceedings of the 2010 VI Southern Programmable Logic Conference (SPL), Ipojuca, Brazil, 24–26 March 2010; pp. 21–26.
13. Tirupathi, R.; Kar, S.K. On-Chip Implementable Autocalibration of Sensor Offset for Differential Capacitive Sensor Interfaces. *IEEE Trans. Instrum. Meas.* **2021**, *70*, 1–9. [CrossRef]
14. Yoshii, Y.; Nakajo, A.; Abe, H.; Nimomiya, K.; Miyashita, H.; Sakurai, N.; Kosuge, M.; Hao, S. 1 chip integrated software calibrated CMOS pressure sensor with MCU, A/D convertor, D/A convertor, digital communication port, signal conditioning circuit and temperature sensor. In Proceedings of the International Solid State Sensors and Actuators Conference (Transducers '97), Chicago, IL, USA, 19 June 1997; pp. 1485–1488.
15. Serial Digital Output Thermometers. *Analog Device*. Available online: https://www.analog.com/media/en/technical-documentation/data-sheets/ (accessed on 10 October 2021).
16. Reverter, F.; Jordana, J.; Pallas-Areny, R. Program-dependent uncertainty in period-to-code converters based on counters embedded in microcontrollers. In Proceedings of the 20th IEEE Instrumentation Technology Conference (Cat. No.03CH37412), Vail, CO, USA, 20–22 May 2003; pp. 977–980.
17. Thakur, A.; Mehra, R. Power and speed efficient ripple counter design using 45 nm technology. In Proceedings of the 2016 IEEE 1st International Conference on Power Electronics, Intelligent Control and Energy Systems (ICPEICES), Delhi, India, 4–6 July 2016; pp. 1–4.
18. Putra, H.D.; Pratiwi, S.E.; Jamilah, W. The synthesis of counter circuit layout design based on CMOS technology 0.35 micro-m. In Proceedings of the 2017 Second International Conference on Informatics and Computing (ICIC), Jayapura, Indonesia, 1–3 November 2017; pp. 1–5.
19. Mironov, S.E.; Zibarev, K.M. Management of Layout Matching of Objects of Complex Microelectronic Systems with Uncertainty of Design Rules. In Proceedings of the 2019 III International Conference on Control in Technical Systems (CTS), St. Petersburg, Russia, 30 October–1 November 2019; pp. 69–73.

engineering proceedings

MDPI

Proceeding Paper

Application of a Deterministic Optical Network Model for the Implementation of an Expert System Knowledge Base for Information Transmission Failure Management †

Vladislav Kosyanchuk [1], Nikolay Selvesyuk [1], Evgeniy Zybin [1], Valeriy Novikov [1,*], Valentin Olenev [2], Andrey Solovyov [3] and Mikhail Semyonov [3]

1 State Research Institute of Aircraft Systems (GosNIIAS), 7 Victorenko St., 125167 Moscow, Russia; vvk@gosniias.ru (V.K.); nis@gosniias.ru (N.S.); eyzybin@2100.gosniias.ru (E.Z.)
2 Department of Aerospace Computer and Program Systems, Saint-Petersburg State University of Aerospace Instrumentation (SUAI), 67 Bolshaya Morskaya St., 197000 Saint-Petersburg, Russia; valentin.olenev@guap.ru
3 Department of Digital Technologies, Voronezh State University (VSU), 1 Universitetskaya pl., 394018 Voronezh, Russia; darkzite@yandex.ru (A.S.); mkl150@mail.ru (M.S.)
* Correspondence: vmnovikov@2100.gosniias.ru; Tel.: +7-499-157-70-47

† Presented at the 15th International Conference "Intelligent Systems" (INTELS'22), Moscow, Russia, 14–16 December 2022.

Abstract: This article presents an approach for the construction of an expert system, which helps to make the decisions when parrying failures during information exchange in real-time systems based on a deterministic optical network. The authors analyze the requirements for deterministic data exchange environments and consider the trends for the development of software and hardware for deterministic networks. The main goal of this article is to propose a concept for the implementation of a deterministic control unit based on an all-optical network with spectral division multiplexing. Two main steps for the implementation of such a system are described. The first is the implementation of the DOS protocol stack and its model, which was used in several protocol development stages. This model is a part of a proposed concept for an automation design and development system for information exchanges. The second presented step is the development of an expert decision support system, based on the principles of the supervisory approach. This article presents a neurocontroller, which provides hardware support for the implementation of individual modes of operation of an expert system.

Keywords: distributed information computer network architecture; deterministic optical network; network model; dynamic reconfiguration; design automation system; expert system; decision support system; neurocontroller; WDM network; system on a chip

Citation: Kosyanchuk, V.; Selvesyuk, N.; Zybin, E.; Novikov, V.; Olenev, V.; Solovyov, A.; Semyonov, M. Application of a Deterministic Optical Network Model for the Implementation of an Expert System Knowledge Base for Information Transmission Failure Management. *Eng. Proc.* **2023**, *33*, 51. https://doi.org/10.3390/engproc2023033051

Academic Editors: Askhat Diveev, Ivan Zelinka, Arutun Avetisyan and Alexander Ilin

Published: 19 July 2023

1. Introduction

The main goal of this article is to consider the development and operation of a deterministic control unit (DCU) for an arbitrary technological process in real time (RT). Modern onboard equipment and networks increase in their complexity from year to year, with the number of tasks undertook by such systems growing significantly. Therefore, the industry needs to ensure a high level of reliability during the operation of such systems. For this purpose, aircraft onboard networks' developers need effective mechanisms for ensuring deterministic data delivery. DCU is a deterministic real-time technological complex, which controls a technological process (TP). Its operation depends on external influencing factors, which move DCU from an internal state to a uniquely defined (deterministic) new state in a specific predetermined time interval. DCU is based on a distributed information computing environment (RIVS) [1] and built of the computing modules (CM) of the same type. A network should connect these models. There are several elements, which are the typical

RIVS modules: N computing modules, K control loops for technological systems, L data concentrators. All of these modules have access to a local information system (LIS). Most TP problems are solved using RIVS, but the time for solving the task should not exceed a certain maximum task duration time. The time for each task from the control loop is determined from the permitted operating time of a separate system control loop and the entire DCU. The key DCU functioning process is the exchange of the information between its subsystems. Transmitted or received data determine the correct operation of functional software applications (FSP) and the whole TP.

2. Local Information System Requirements

Several DCU LIS features distinguish it from other real-time and information networks:

- A limited number of CMs (50–100);
- The total amount of FSPs, which, for functionality, does not exceed 256 programs;
- All information flows operate in strong correlation with specific communication protocols;
- The network provides the transmission of control signals in the hard RT mode [2–4].

Such control signals are time markers used for the synchronization of the system, and control and notification signals [5–10].

Typical LISs face problems of data transmission delays, which makes the operation of the system unpredictable and leads to packet collisions. Of course, it is possible to significantly increase the data speed or use specialized type of quality of service (QoS). This QoS is scheduling, which ensures data transmission in accordance with a schedule specified for the whole network during the configuration phase. However, unfortunately, during the operation of the DCU, unexpected delays and collisions occur. This produces an effect on the work of the scheduler, so existing networks cannot provide complete determinism, which is very important for onboard networks. Regarding this, the main sources of collisions should be excluded from the network architecture to ensure complete determinism. Most such delays are caused by switches and routing mechanisms.

The construction of LIS DCU should be based on the following principles: the "point-to-point" organization of the data exchange between the FSP of different CMs; using broadcast mode for sending packets as the main mode; using of distributed memory for data exchange between FSPs. Such LIS principles could be implemented in deterministic real-time optical networks (DOS RV). Such networks transmit the packets without switching and use distributed memory principles for data exchange between end systems. All-optical networks are a type of networks where the switching, multiplexing, and data transmission are organized by purely optical technologies, rather than by electronic technologies. DOS RVs are built as an information transmission system with two types of topologies: a star and a ring. They contain N VMs with the FSPs function and a spectral SDA (SSDA). Figure 1 shows a block diagram of the DOS RV.

Some key principles should be taken into account to build an effective DOS RV [11,12]. The optical technology of wavelength division multiplexing (WDM) should be used as a transport architecture. WDM consists in the transmission of many different data flows over the single optical fiber, but in separate optical λ-channels. Each channel has its own optical frequency called the λ-channel. Information exchange is based on the principle of distributed shared memory (DSM). Memory access is performed via the simple "read" and "write" commands. Each FSP has a particular memory area, which is used to exchange data with external subscribers of the fiber optic medium. It is performed via the λ-channel specially allocated for this FSP, and this frequency is fixed for the data flow in terms of the whole network.

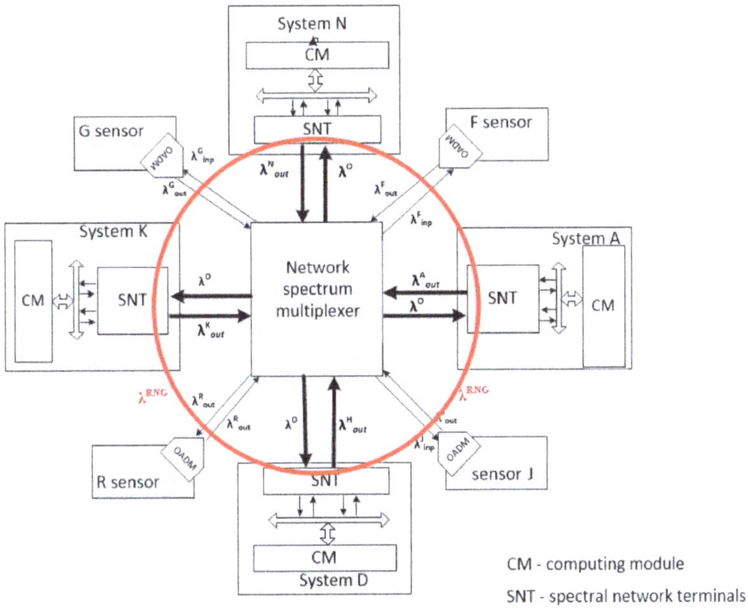

Figure 1. DOS RV structure.

There are a few important steps to organize the work of DOS RV:

- Form a list of RIVS systems, which participate in the data exchange of the real-time information transmission system—N;
- Form a private list of FSPs operating in each of the VM of RIVS modules. These modules could be logically split into two groups. The first group includes the total number of FSPs transmitting information L_{out}. The second group includes the total number of FSPs receiving information l_{inp}. Both groups form a general list of FSPs transmitting information throughout the RIVS—L_{out};
- Set the number of channels that are required for all information exchange between FSPs. $L_{out} \leq \lambda^0$ must be true, where λ^0 is the total number of spectrally multiplexed λ-channels in DOS RV;
- For each system, determine the total number of output λ–channels. For example, for system A, the group of channels is defined as $\lambda^A_{out} = U\{\lambda_1 \ldots \lambda_{nA}\}$. These groups have their own set of wavelengths for a given SSDA;
- In the same way, determine the total number of input l_{inp} in the SSDA of each RIVS component.

3. The DOS Model in the Automation Design and Development System for Information Exchanges of Real-Time Systems

To design communication protocols that implement information exchange in the DOS RV networks, modern methods should be used. Such a methods should cover all the system development stages (see Figure 2).

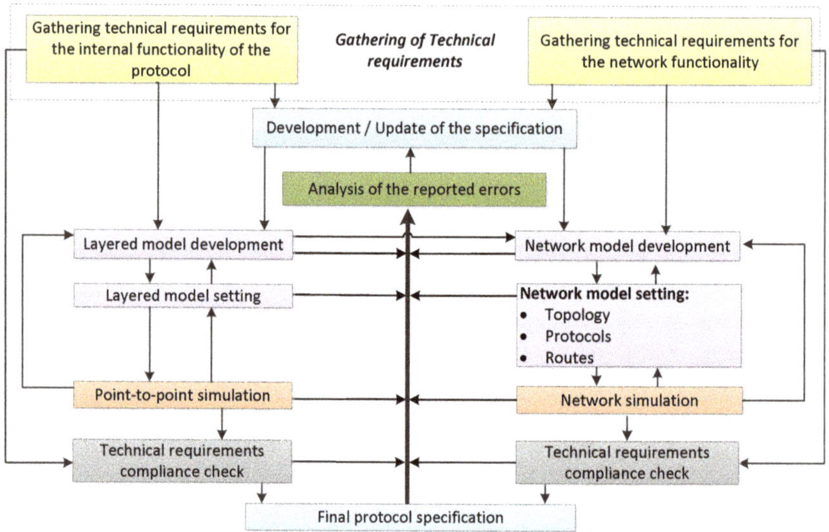

Figure 2. Methodology for development of communication protocols for real-time systems.

The design process begins with the gathering of technical requirements for the future protocol and their analysis. Technical requirements are divided into requirements for the internal mechanisms of the protocol and requirements for the operation of devices based on this protocol. Regarding this, the first version of the specification of the future protocol was developed. A simulation was used to test and verify the developed protocol.

The modeling process creates layered and network models. The network model is created after at least one successful version of the layered model has been created. Each of these types of models solves its own problems and checks the operation of different mechanisms of the protocol. After creating the models, it is necessary to configure them, and then carry out detailed modeling. If an error is found at any of the parallel stages, it is necessary to create a bug report, send it to the expert group for review, and stop work in the parallel branch until the specification is checked or clarified. After analyzing the error, the specification is refined and remodeled. When both simulation teams report that no more errors have been found, the simulation results are compared to the original technical requirements. If the requirements are not met, the specification is sent back for revision, otherwise the document is finalized and can be sent to further stages of the project, for example, hardware implementation.

The gathering of technical requirements is carried out by an expert group, which includes representative customers, developers, researchers who have the necessary knowledge and information on the further use of the protocol. The completeness of the specifications is an important issue, as the requirements will be the basis for deciding whether there are errors in the protocol.

Next, the specification of the designed communication protocol is written. Such a document usually includes a general description of the protocol stack (or a single protocol), data transfer formats, protocol mechanisms and algorithms in accordance with the layered architecture, interlayer interfaces, and examples of the application of the above mechanisms or the protocol as a whole. Once this task is completed and the initial draft of the specification is ready, the document is submitted for modeling. Modeling is carried out in order to check the underlying algorithms and can be carried out already in the process of writing the specification. To do this, an intermediate document is already given for modeling; it also passes through various checks, but is returned back to the revision stage if such a specification does not meet the technical requirements.

Further modeling should be carried in two parallel branches: layered protocol modeling (light gray blocks in Figure 2) and the system of devices modeling (dark gray blocks in Figure 2). When modeling a layered protocol stack, the model layers fully correspond to the protocol structure, and the interfaces between these layers correspond to service access points. Thus, to carry out the simulation, it is sufficient to fully describe the operation of the protocol in relation to one device. To obtain a result, it is necessary to organize data transfer between two devices only, since, for the task of analyzing the protocol, network interaction and the mechanisms of operation as part of a larger system do not matter. When writing a device interaction model, the network operation and the exchange of various information packets are modeled. With this approach, the interaction of the constituent elements of the processes inside the device is not considered. In order to effectively use the communication protocol modeling, both of these approaches should be used. This is the only way to detect most of the errors in the specification.

The development of a layered model can be carried out with software and formal models can be created using well-known mathematical tools. For example, the Petri nets theory, the automata theory, queuing systems, etc. can be used. Programming languages based on formal methods could be used, for example, SDL, GPSS, etc. Setting up such a model consists of setting protocol parameters, and other properties that should be described in the specification.

For modeling a system of devices, programming languages are chosen that can build large systems with the use of multiple modules, ensuring the parallel operation of these modules. One of the most efficient languages for these purposes is SystemC. It covers a wide range of protocol development areas and can be used for architecture design, hardware and software simulation, system behavior description, and functional verification. Finally, SystemC supports hardware modeling and design. Using the SystemC network model, you can set various configuration parameters, such as the data rate, number of nodes and switches, time delays, routing tables, number of ports in switches, etc. Also, on network models, it is possible to automate the process of designing a network topology, the tracking of routes, and setting various network properties.

During the tuning of the model, some errors in the model itself can be detected and, as a result, a request is sent to correct the model. During the model development phase and the modeling phase, errors in the designed specification can be found. In this case, a detailed description of the error is generated. The team developing another model in a parallel methodology branch is notified that the modeling process should be temporarily suspended. Next, an expert evaluates what changes are needed in the specification, and then corrects it. When changes are made to the models, the specification is resubmitted for modeling.

Once the simulation process is completed successfully and no more specification errors are found, the simulation results are compared to the original specifications. In case of discrepancy, the specification is sent for correction. Otherwise, the project team creates the final version of the specification and submits it for implementation [13].

Since modeling is such an important part of the communication protocol development process, it is necessary to have an automated system that allows one to set up a software model and run it for execution. Such CAD systems are becoming an integral part of the technology, since they can significantly speed up the development and reduce the cost of the process of creating new technological solutions. Let us consider what the process of creating such a CAD system for DOS RV networks could look like.

Modern onboard networks consist of a large number of elements with different functionality. All these elements are interconnected through the infrastructure of the onboard network. Such networks are easy to operate as long as they are small enough, and as long as they have predictable behavior and are used for a simple set of information flows without strict requirements and restrictions. However, as the network grows larger, information flows become denser, requirements and constraints become more stringent, and design becomes a challenge. In such cases, there is a need to create a software packages for computer-aided design and network modeling.

For DOS RV networks, there is a need to create an automated design and simulation system that will cover the full cycle of development and simulation of the onboard network, starting with the design of the network topology and ending with obtaining network simulation results and statistical data (see Figure 3).

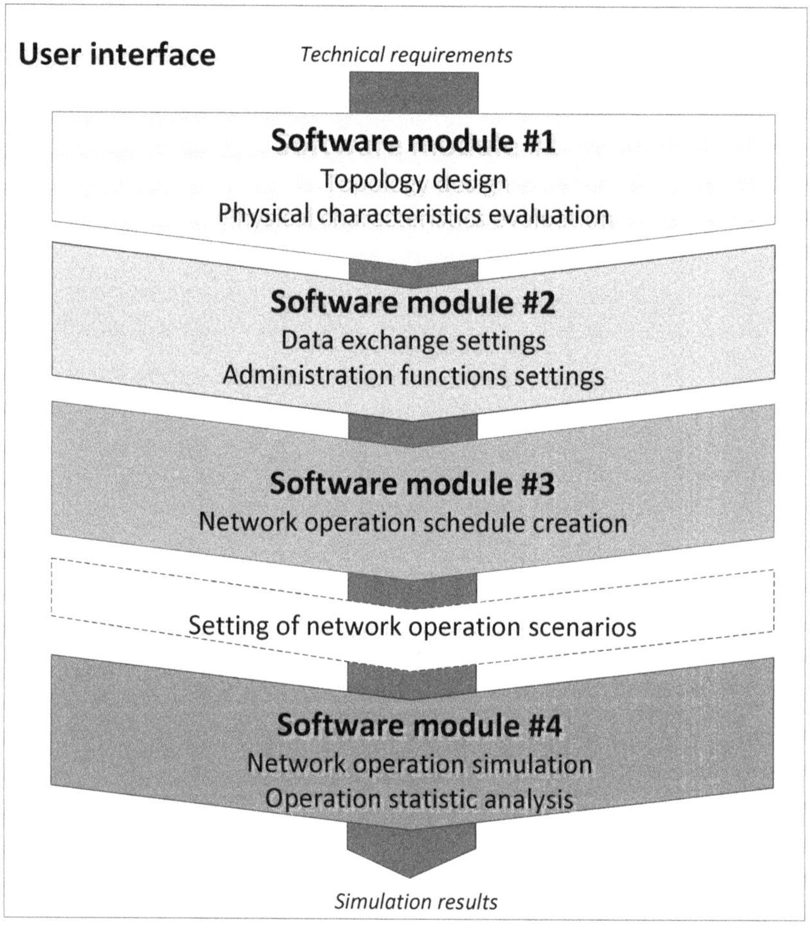

Figure 3. Architecture of CAD for DOS RV.

The designed CAD DOS RV will use a graphical interface that allows for an optical onboard network to be assembled from a variety of network nodes and a spectral multiplexer (in the future, several multiplexers) corresponding to the physical implementations that are available for building networks. A configuration interface will be provided for setting parameters, configuring protocols, and test scenarios.

Software modules (SM) will perform functions corresponding to the various stages of the onboard network design. SM-1 is responsible for designing the network topology and the evaluation of the physical characteristics, such as the mass of cables and devices, power consumption, and the evaluation of the required number of data exchange channels in the network. SM-2 is responsible for complete network configuration, including functions for configuring administration protocols. SM-3 is used to adjust the "Scheduling" quality of service implemented in the DOS RV protocols. At SM-3, the design work is completed, and at SM-4, the entire network is launched for simulation. The simulation proceeds in accordance with the specified work scenarios, which imply turning off devices, introducing errors into the network, changing settings, etc. Upon completion of the simulation, the user has a complete set of diagnostic information on the operation of the DOS RV and graphical information on statistics.

To implement the simulation functions, it is necessary to create an effective simulator of the DOS RV operation. This model can be used to solve the problem of modeling protocol stack specifications, testing mechanisms, and testing prototypes of future real networks on the model. The developed model allows one to work with the simulation environment in a graphical user interface, perform simulations of various scenarios, and generate detailed diagnostic information. As the first step in the implementation of the DOS RV protocols, a point-to-point network model was created. It consists of two nodes and a data transmission channel. Using the developed model, a detailed study of the mechanisms of the transport layer, the link layer, and the administration layer was carried out and the efficiency of the developed mechanisms was shown. To do this, we tested the transmission of various packets with different qualities of service, interrupts, time-codes, commands for working with distributed memory, and requests for the administration level [14].

This model will form the basis of the extended version, which will be able to set flexible work scenarios and use the full DOS RV network infrastructure. The model is implemented in the SystemC language, which is used to model parallel systems. In the future, we plan to use this model as part of an expert system for preparing and making decisions in the event of an exchange failure in DOS RV.

4. The Concept of Applying the DON Model to Form the Knowledge Base of the Expert System for Preparation and Decision-Making

Currently, an urgent problem is that of designing real-time onboard intelligent systems. Such systems are based on the integration of models capable of adapting, modifying, learning, representing, and operating knowledge, focused on the specifics of the problem area and the corresponding type of uncertainty, which reflects their ability to develop and change their state.

When implementing a real-time decision support and decision-making (RTDSDM), it is necessary to take into account the specifics of such systems. The system should obtain a solution under the time constraints determined by the real controlled process. It should take into account the time factor when describing a problem situation and in the process of finding a solution. Also, it is impossible to obtain all the objective information necessary for the decision, and, in connection with this, the use of subjective, expert information. There is a need to have a multivariate search, and the need to apply methods of plausible inference and active participation in the process of the search and intellectual inference. Case-based decision search methods can be used in many RTDSDM blocks (analyzer, decision search blocks, explanations, modeling, and forecasting) and can improve the efficiency of decision-making in various problem situations.

Precedent-based intelligent inference is an approach that allows a new problem to be solved by using or adapting a solution to an already known problem, i.e., using the already accumulated experience in solving similar problems.

The disadvantages of case-based reasoning include the following:

- When describing precedents, they are usually limited to superficial knowledge about the subject area;
- A large number of cases can lead to a decrease in system performance;
- It is problematic to define criteria for indexing and comparing precedents;
- Difficulties with debugging algorithms for determining similar (similar) precedents;
- The impossibility of obtaining a solution to problems for which there are no precedents or the degree of their similarity (similarity) is less than a given threshold value.

The main purpose of using the apparatus of precedents within the framework of RTDSDM is to issue a ready-made solution to the operator for the current situation based on precedents that have already taken place in the past when managing a given or similar object (system).

It is obvious that in order to structure information and quickly access it from the RTDSDM core, it is necessary to form a specialized knowledge base (KB) containing a set of precedents, each of which describes an emergency situation and certain methods for solving it. One such solution, which allows for parrying the failures that have arisen, is the dynamic reconfiguration of the DOS RV. In addition, information about an emergency situation can be taken both from the real experience of a qualified pilot, and formed using simulation modeling implemented in CAD based on the DOS RV model. The DOS RV model can also be used in RTDSDM to analyze the current state of the network and predict failure situations. This function is implemented by the method of the continuous comparison of the functioning of the network and the DOS RV model implemented in the core of the RTDSDM.

The intelligent inference function is produced by a specialized expert system (SES), which is part of the RTDSDM. To increase flexibility and expand the ability to solve a wider range of problems, SES should rely on both classical search algorithms using standard calculators and specialized parallel algorithms implemented on the basis of coprocessors. One of these coprocessors is a neural network computer or a neurocontroller (NC). On the basis of an NC, it is possible to deploy artificial neural networks with different architectures and different specifics of functioning, which will allow hidden connections and nonlinear patterns to be identified in the data stored in the knowledge base, which, in turn, will significantly increase the flexibility and depth of intellectual inference carried out in the SES. The generalized structure of the RTDSDM is shown in Figure 4a.

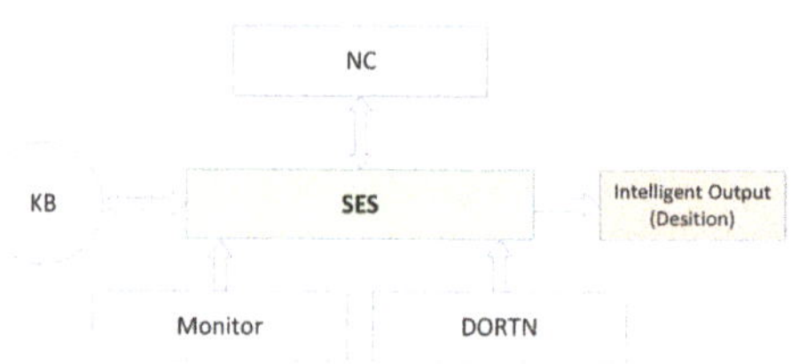

(a)

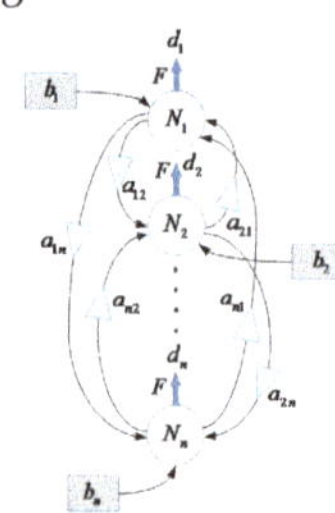

(b)

Figure 4. (**a**) The structure of the RTDSDM based on the DOS RV model; (**b**) NC architecture.

As can be seen from Figure 4a, the RTDSDM includes a tera-network calculator or neurocontroller. This element can be realized using a microprocessor, on the basis of FPGA, and in the form of a special module representing a system-on-chip (SoC). Depending on the variant of realization, the pro-cems in the NC can be sequential (in the case of a microprocessor), parallel (in the case of an FPGA), and combined (in the case of a SoC). The NC architecture is shown in Figure 4b.

As can be seen from Figure 4b, a NC consists of neurons N_j, whose inputs are multiplied by weight coefficients $a_i j$; each neuron transforms the weighted sum of inputs using the activation function F_j and the output is formed d_j, where $i = j = \overline{1,n}$. Taking into account the peculiarity of neuron architecture, namely, the presence of many inputs and only one output, the NC can be described using the neuron interaction matrix M^{nxn} (NIM), whose elements outside the diagonal are the weights of the inputs of the neuron with the number corresponding to the row number, and the diagonal element is the output of this neuron.

In addition to the NIM, it is also necessary to describe the set of activation functions used in the NC with reference to each neuron. For this purpose, we introduce the vector of activation functions F^n (VAF), whose elements f_i indicate the number of the activation function implemented in the neuron N_i. Odds a_{ij} in the NIM affect the slope of the activation function of each neuron; however, to increase the flexibility of the NC, it is also necessary to introduce bias coefficients b_i, which allow one to move the activation function of each neuron to the left or right. To this end, we introduce the vector of bias coefficients B^n, containing b_i. Thus, the NC is mathematically uniquely described by the set

$$\{M^{nxn}, B^n, F^n\}, \tag{1}$$

where n is the number of neurons in the NC. The NC dynamics can be described by the following transformation:

$$d_t = (M \times d_{t-1} + b) \tag{2}$$

Due to the fact that the NIM is used as part of the SES, it is possible to significantly increase the flexibility of decision-making and increase the speed of some (e.g., nonformalizable) computations. This is possible due to the fact that on the NIM, it is possible to implement a large number of different classes of neural networks (multilayer perceptrons, Hopfield network, Boltzmann machine, etc.), both direct propagation and recurrent, without changing the NC architecture. It should also be noted that, in the case of using a parallel implementation of NIM (on FPGA), the response time of the NC is determined by the FPGA production technology and currently reaches units of nanoseconds.

5. Conclusions

The use of a commutation-free optical environment in onboard systems and the construction of a distributed system of support and decision-making on its basis allows one to significantly increase the flexibility and scalability of the onboard equipment complex, simplify and accelerate the process of its reconfiguration (for example, in the case of dynamic reconfiguration), and to achieve high determinism of states of all onboard processes and systems. It should also be noted that the use of the information network model, which is formed and used in the CAD environment during the development of DOS RV, allows one to obtain a dual-purpose model (both at the design stage and at the stage of intelligent control of the information network environment), to create a single, continuous process of development, and the application of DOS RV. In general, the efficiency of work is significantly increased and financial costs are optimized. The considered approach is methodically applicable to the development of other components of real-time technological systems.

Author Contributions: Funding acquisition, V.K.; conceptualization, N.S.; writing—review & editing, E.Z.; writing—original draft preparation, V.N.; software, V.O.; methodology, A.S.; validation, M.S. All authors have read and agreed to the published version of the manuscript.

Funding: The study was financially supported by the Ministry of Science and Higher Education of the Russian Federation under agreement No. 075-15-2022-1024.

Institutional Review Board Statement: Not applicable.

Informed Consent Statement: Informed consent was obtained from all subjects involved in the study. Written informed consent has been obtained from the patient(s) to publish this paper.

Data Availability Statement: Not applicable.

Conflicts of Interest: The authors declare no conflict of interest.

References

1. Novikov, V. Solving the problems of intelligent crew support in terms of reconfiguration during OBE in case of failures. In Proceedings of the Collection of Abstracts of the VI International Scientific and Practical Conference "AVIATOR", Voronezh, Russia, 14–15 February 2019; pp. 189–192.
2. Kopetz, H. *Real-Time Systems. Design Principles for Distributed Embedded Applications*, 2nd ed.; Springer: Berlin/Heidelberg, Germany, 2011.

3. Gorbachev, S.; Koblyakova, L.; Sheynin, Y.; Stepanov, A.; Suvorova, E.; Suess, M. Distributed Interrupt Signalling for SpaceWire Networks. In Proceedings of the 5th International SpaceWire Conference, Gothnburg, Sweden, 10–14 June 2013.
4. Koblyakova, L.; Sheynin, Y.; Suvorova, E. Asynchronous hard real time signals transmission in embedded networks. *Int. J. Embed. Real-Time Commun. Syst.* **2014**, *5*, 24–44. [CrossRef]
5. Koblyakova, L.; Gorbachev, S. *Tasks of Hard Real Time Signals in Built-in Systems and Onboard Networks*; SUAI: Saint Petersburg, Russia, 2016; pp. 79–87.
6. Sheinin, Y.; Solokhina, T.; Petrichkovich, Y. SpaceWire technology for parallel systems and onboard distributed systems. *Electron. Sci. Technol. Bus.* **2006**, *5*, 64–75.
7. Zhu, Q.; Yang, Y.; Scholte, E.; Di Natale, M.; Sangiovanni-Vincentelli, A. Optimizing Extensibility in Hard Real-Time Distributed Systems. In Proceedings of the Real-Time and Embedded Technology and Applications Symposium, San Francisco, CA, USA, 13–16 April 2009.
8. Zhu, X.M.; Lu, P.Z. Multi-Dimensional Scheduling for Real-Time Tasks on Heterogeneous Clusters. *J. Comput. Sci. Technol.* **2009**, *24*, 434–446. [CrossRef]
9. Van Steen, M.; Tanenbaum, A. A brief introduction to distributed systems. *Computing* **2016**, *98*, 967–1009. [CrossRef]
10. Aydin, H.; Melhem, R.; Mossé, D.; Mejía-Alvarez, P. Power-aware scheduling for periodic real-time tasks. *IEEE Trans. Comput.* **2004**, *53*, 584–600. [CrossRef]
11. Novikov, V.; Kosyanchuk, V.; Chuyanov, G.; Platoshin, G.; Stetsyuk, S.; Penkin, S. A Real-Time Information Transmission System Based on a Fully Optical Spectrally Multiplexed on-Board Real-Time Network. Patent RU2694137C1, 22 August 2019.
12. Kosyanchuk, V.; Novikov, V.; Kasatikov, M.; Mishchenko, I. Dual-Loop Real-Time Information Transmission System Based on a Fully Optical Wavelength-Multiplexed Onboard Network. Patent RU2744517C1, 11 March 2021.
13. Olenev, V. Methodology for communication protocols formal design based on Petri nets. *Informaciâ i Kosmos* **2022**, *4*, 37–45.
14. Olenev, V.; Suvorova, E.; Korobkov, I.; Rozhdestvenskaya, K.; Chumakova, N.; Sinev, N.; Novikov, V. Software model of the protocol stack functioning as a means of diagnosing the optical onboard network. In Proceedings of the Collection of Reports of the Sixth International Conference "Perspective Directions for the Development of Airborne Equipment of Civil Aircraft", Moscow, Russia, 22 July 2021; pp. 45–56.

engineering proceedings

MDPI

Proceeding Paper

Research of a Virtual Infrastructure Network with Hybrid Software-Defined Switching †

Yuri Ushakov *, Margarita Ushakova and Leonid Legashev

Mathematical and Information Technology Faculty, Orenburg State University, Pobedy 13, Orenburg 460018, Russia; m.v.ushakova@mail.ru (M.U.); silentgir@gmail.com (L.L.)
* Correspondence: ushakov@unpk.osu.ru; Tel.: +7-353-291-2195
† Presented at the 15th International Conference "Intelligent Systems" (INTELS'22), Moscow, Russia, 14–16 December 2022.

Abstract: Modern trends in the information technology have led to the fact that entire systems of infrastructure are becoming software-defined. Modern hyper-converged solutions use software-defined networking and soft switches for the hypervisor networking subsystem. The paper goal is to study traffic processing in hyperconverged structures with software switching based on OpenFlow versus traditional approaches. The features of the hyperconverged solutions network infrastructure are considered, approaches to the study of software-defined environments are described. A model of the processing traffic internal structure of a converged node, combining the functions of a hypervisor, a storage system and a switch, is proposed. Interface models reproduced traffic switching with the traditional approach and with higher-level OpenFlow processing have been developed. The approaches to the implementation of the developed models based on experimental studies of network equipment are described. The results of an experimental study of a network node and a synthesized model are presented. The possibility of implementing the proposed approaches within the specified accuracy are described.

Keywords: modeling; OpenFlow; switch; traffic; QoS

Citation: Ushakov, Y.; Ushakova, M.; Legashev, L. Research of a Virtual Infrastructure Network with Hybrid Software-Defined Switching. *Eng. Proc.* **2023**, *33*, 52. https://doi.org/10.3390/engproc2023033052

Academic Editors: Askhat Diveev, Ivan Zelinka, Arutun Avetisyan and Alexander Ilin

Published: 19 July 2023

1. Introduction

Today, cloud infrastructures require ever more flexible approaches and technologies that involve the transition to more distributed applications, microservice architecture, and flexible IT models. Modern trends in the development of multi-cloud services lead to the widespread use of software-defined infrastructure, storage, networks, as well as the introduction of convergent and hyper-converged services and equipment [1]. Convergence in networks combines the traffic of heterogeneous services on a single technology stack with a hardware component (traditional switches supporting prioritization, segmentation at levels 2–4, combining several layer 2 protocols), while hyperconvergence is based on a software-oriented architecture. Hyperconvergence does not exclude the use of converged technologies in the network stack, but uniform network management across hypervisors, storage systems, virtual switches and routers, and equipment can greatly improve the quality of flow control, logical segmentation, security and prioritization. The use of virtual dedicated channels from source to destination with guaranteed marginal characteristics based on packet-switched software-defined networks makes it possible to provide traffic quality of service and security with much lower organizational and configuration costs than on the basis of traditional converged networks [2]. The main feature of hyper-converged solutions from different manufacturers is the combination of management of all network devices into a unified software system, and most manufacturers add support for virtual overlay transport technologies (Overlay Transport), such as VXLAN networks (Virtual Extensible LAN), Cisco OTV (Overlay Transport Virtualization), as well as various overlay networks of virtual infrastructure hypervisors (VMWare NSX-T, Microsoft HNV,

OVN, Cisco ACI), and container orchestrators (swarm overlay network, flannel, k-vswitch, OVN) [3]. Open cloud solutions (OpenStack, OpenNebula) and container solutions (Docker Swarm, Rancher, Kubernetes, OpenShift) can use overlay networks through many technologies through a plug-in system, but the most common is the use of VXLAN solutions and transmission control through Open vSwitch. The use of Open vSwitch in conjunction with software-defined networking technologies (both via the OpenFlow protocol and using ovs-ofctl, ovs-vsctl) allows you to combine a virtual switch and equipment into a single managed network, as well as make universal traffic management virtual machines and services running directly on the hardware (bare-metal), for example, software-defined storage systems (CEPH, drbd9, gluster, NFS, VMWare vSAN). At the same time, studying the operation of such networks in conjunction with hyperconverged infrastructure is an open problem, too many factors can affect each element of the infrastructure and, thereby, performance parameters. One of the ways to comprehensively cover all factors for the study of networks can be modeling, but it is complicated by a high level of nesting and mutual dependence of components. The authors of the paper [4] describe the creation of a test bed for experiments with software-defined networking and the cloud using hyperconverged SmartX Boxes distributed over several sites. Each SmartX Box consists of several virtualized functions, which are divided into SDN and cloud functions with the ability to track resources on a distributed cloud platform. The paper [5] discusses an approach to automation for the efficient implementation of a variety of services through distributed hyper-converged blocks towards a converged software-defined infrastructure using the SmartX automation software environment. Within the framework of the paper [6], the principle of integrating software-defined networks and hyperconverged architectures is demonstrated using OpenFlow as a communication protocol and Opendaylight as a controller in the control plane. The authors of [7] note that the hyperconverged architecture has higher availability and performance when evaluating the provision of services in a cloud computing environment. The author of the study [8] deals with the issues of improving the performance of applications in a hyper-converged infrastructure.

2. Theoretical Part

Let us consider one node of hyperconverged infrastructure based on the most common open solutions, such as KVM hypervisor, virtual switching Open vSwitch and Linux Bridge, software defined storage (SDS) (block, object, or file SDS), based on the resources of the same servers on which virtualization is performed (CEPH, Gluster, ScaleIO, SwiftStack, HDFS and the like). Each hyperconverged infrastructure server contains a computing node (processor, memory, disks), a Linux operating system, a network part (network cards, an Open vSwitch, Linux Bridges, veth virtual network cards), a virtualization node controlled by an external orchestrator, a server node SDS with storage on local disks integrated with a common storage subsystem (see Figure 1a). Since there are many options for internal connections (see Figure 1b), it is necessary to foresee and explore all possible options when creating a model [9].

The main problem of modeling networks with a large number of traffic processing technologies and ensuring quality of service is a huge number of possible combinations of various parameters, types and settings of queues, traffic processing policies. Consider the option of creating a network interface with enabled VLAN functionality, a classifier in both directions, incoming and outgoing rate limiting through queues, and traffic filtering in both directions. In this case, the classifier and queues use DSCP; WRED (Weighted Random Early Detection) is selected as the congestion management policy, and prioritization occurs through WRR (Weighted Round Robin) for classes other than ef, for which LLQ (Low Latency Queue) is used. Limiting the amount of incoming traffic also occurs through WRED removal of packets from the queue with different weights of different classes (Class-based policing) [9]. The classifier is configured to mark some packets with the classes af11, af13, af21, af23, ef, and be. The input from the network occurs through a bidirectional mac module, the output and input of the upper levels through filters (ACL). This sequence of actions at the input from the network (decapsulation, classification, bandwidth handling, labeling, filtering) and at the output to the network (filtering, labeling, bandwidth handling, queues, encapsulation) was created based on the description of the sequence of actions Cisco IOS [10] (see Figure 2).

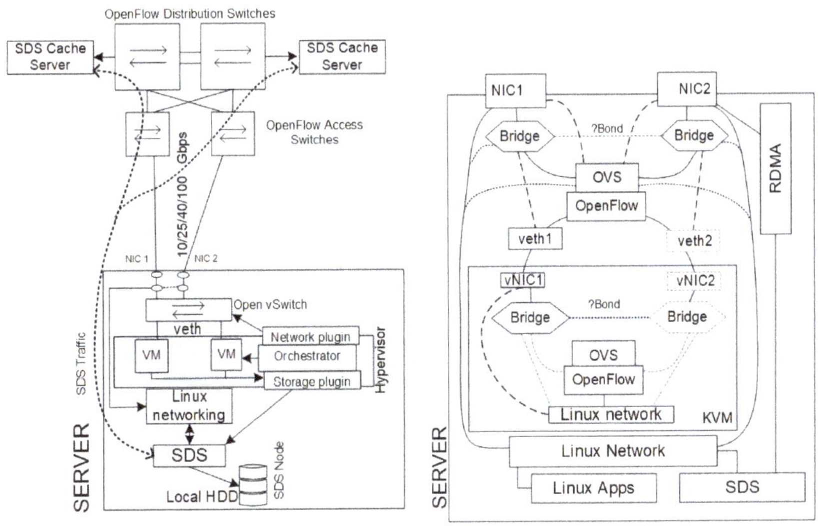

Figure 1. Server internal organization diagram: (**a**) component connection diagram; (**b**) diagram of possible internal network connections.

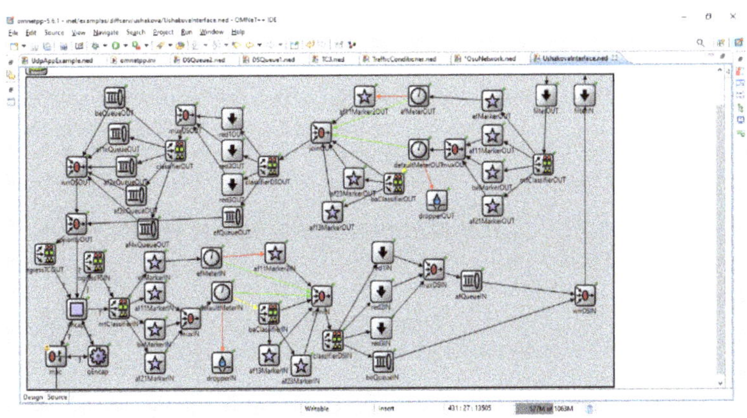

Figure 2. The scheme of the packet passing through the network interface.

To model a realistic delay, the created module of the Traffic-Conditioner type was used to put the ingressTCIN and egressTCOUT components in place. It is used to introduce a delay in the incoming direction, corresponding to the processing time of the packet by the switch on the real network. At the same time, the use of other combinations of queue settings, their marking and connection logic (for example, nested WRR queues) can significantly complicate the interface model. Using OpenFlow for traffic processing generally does not affect the interface scheme in terms of operating with queues, but removes some of the classification and marking, which is now performed at the OpenFlow Data Plane level. In the general case, QoS is implemented in OpenFlow only as a classification, labeling of packets, and measurement of current flow performance [11]. The management of queue parameters, traffic limiters, and prioritization policies in most implementations is left for traditional configuration through the mechanisms of the switch operating system (or the control program, as is the case with Open vSwitch). Some controllers (for example, OpenDaylight, ONOS, Floodlight) have mechanisms for setting up queues on a certain set of switches via NETCONF or directly by commands (for example, for OVSDB). However, this does not solve the problem of creating a model of such equipment. Figure 3 shows the scheme of the interface in OpenFlow mode, all marking and rate limiting takes place in the OpenFlow core, but queues and work with them, including prioritization, remain the same.

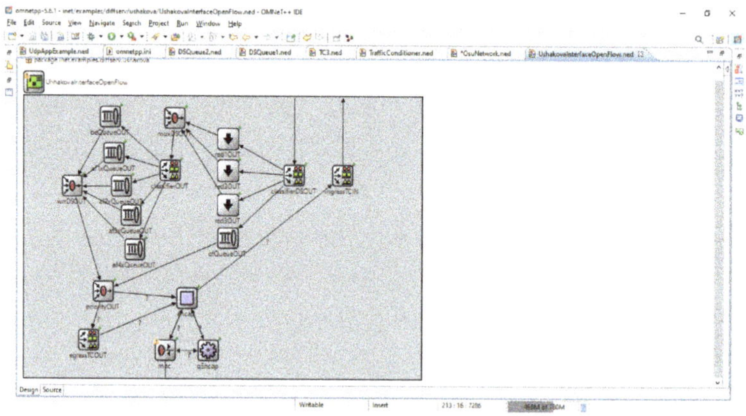

Figure 3. The scheme of the packet passing through the OpenFlow network interface with QoS support.

To study the operation parameters of a software-defined device over a period of time T, we denote the given input parameters that describe incoming traffic and current device settings as $Z(t) = \{Traffic(t), Device(t)\}, \forall t \in T$, and the measured output parameters, which include timing, input and output queue parameters, queue statistics, and a snapshot of device metrics, as $Y(t) = \{Latency(t), Qout_{ij}(t), Qin_{ij}(t), S_{ij}(t), State'_i(t)\}, \forall t \in T$ Then the function of mapping input parameters to output parameters will be defined as:

$$F(t) : Z(t) \to Y(t), \forall t \in T \qquad (1)$$

Let us represent the tree of input parameters Z(t) as a linear sweep

$$Z(t) = \{z_1(t), z_2(t)...z_n(t)\}, \forall t \in T,$$

where is the number of input parameters in the sweep. Similarly, we represent the tree of output parameters $Y(t)$ as a linear sweep

$$Y(t) = \{y_1(t), y_2(t)...y_m(t)\}, \forall t \in T,$$

where $m \in N$ is the number of output parameters in the sweep. The input set of parameters $Z(t_i) at \forall t_i \in [T_{i-1}..T_i) \in T$ should not change, and the duration of the time interval $[T_{i-1}..T_i)$ should be sufficient for the required number of measurements of the output parameters and to achieve stationarity of the probabilistic patterns of the output parameters $y_1(t_i), y_2(t_i)...y_m(t_i)$. In such case mapping (1) can be reduced to the form $F(t) = \{f_1(t), f_2(t)...f_m(t)\}$ in this way:

$$\begin{cases} f_1(t_i) : \{z_1(t_i), z_2(t_i)...zn(t_i)\} \to y_1(t_i) \\ f_2(t_i) : \{z_1(t_i), z_2(t_i)...zn(t_i)\} \to y_2(t_i) \\ ... \\ f_k(t_i) : \{z_1(t_i), z_2(t_i)...zn(t_i)\} \to y_k(t_i) \end{cases}$$

When modeling, it is necessary to ensure the correspondence of the obtained output parameters of the device under study $Y(t_i)$ and the obtained output parameters of the model $Y^*(t)$ with the same sets of input parameters. To do this, we find a mapping $F(t) : Z(t) \to Y^*(t), \forall t \in T$, such that

$$\left| \frac{y_j(t_i) - y_j^*(t_i)}{y_j(t_i)} \right| < \varepsilon, \forall j \in [1..m],$$

where $y_j^*(t_i) \in Y^*(t_i)$ are the measured output parameters of the model; ε is the denoted accuracy. The study of a device with all combinations of input parameters is hampered by the fact that the number of all possible combinations of parameters depends on the number and ranges of parameters of specific devices with specific versions of system software, and, in general, is an NP-complete problem [12]. Therefore, to build a model, the method of determining indicators for boundary values and a certain number of intermediate values of each parameter separately will be used, and the dependencies between the parameters for the maximum possible values will also be studied. This will make it possible to obtain an interpolation estimate of the values of indicators for the model with an accuracy determined by the number of intermediate measurements of parameters and the shape of the dependency curves.

3. Experimental Research

The study of the operation of model devices connected to a network was carried out on the example of a new experimental segment of the data processing center (DPC) of Orenburg State University. Virtual machines in the data center are divided into groups, between which network access must be excluded. There can be more than a hundred such groups, depending on the time and needs, and the groups can have their own routers, DHCP servers, repeating VLANs and VXLAN networks. Therefore, instead of creating overlay networks using the capabilities built into OpenNebula, it was decided to isolate traffic through OpenFlow at the physical and data link levels. The OMNeT++ simulation tool with the INET framework was used as a modeling environment, which is the most suitable tool for modeling networks with SDN support. A feature of OMNeT++ is the decomposition of the model to level 1 of the OSI model with maximum detail of all technologies, protocols, and data formats used. The segment was implemented to provide high-performance data processing and create a private cloud for educational and production tasks, including recording and processing video streams, saving Wi-Fi client traffic dumps, creating training virtual machines, and big data processing tasks. The data center segment includes two switches that play the role of a distribution layer. These switches connect 10 GbE channels to each of the seven access switches. Each server connects to two access switches with two 10 GbE links (storage segment and switching segment) and two 1 GbE links (management). Since the storage system is built on the basis of CEPH, based on the disks of the same servers that are used to run all virtual machines, the requirements for the storage segment lead to the need to provide a peak node speed of at least 10–15 Gb/s when exchanging at the link level and delay. When writing to the storage, duplication occurs on several servers, in addition, regular data rebalancing and sharding occurs, which significantly increases traffic. The switching segment is used for inter-network interaction of virtual machines with external consumers and among themselves. When using distributed neural networks and accessing Hadoop-based storages that physically store data on separate drives of the same servers, the performance requirements for the switching segment also have lower limits. Figure 4 shows the network topology.

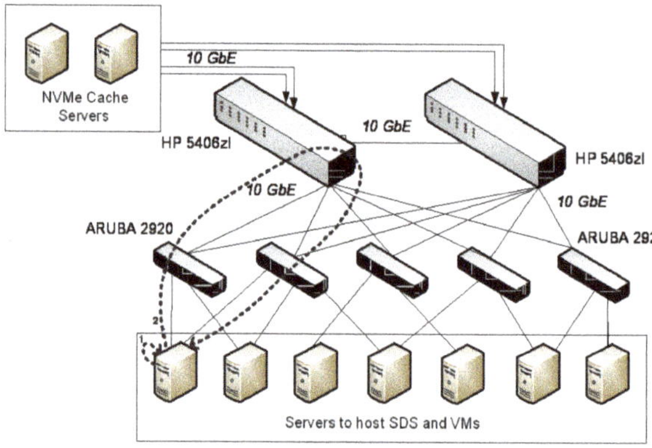

Figure 4. Experimental segment topology.

At the time of research, the experimental segment included seven servers (Xeon Gold, 128 GB of RAM, NVMe SSD, four 10 Gb/s network cards with hardware packet processing) to host storage and virtual machines, two servers to work only as systems storage, five access switches (HPE ARUBA 2920), and two distribution switches (HP ProCurve Switch 5406zl with five HP J9309A ProCurve 4-Port 10GbE SFP+ zl modules). Virtual machines are controlled by the OpenNebula orchestration system, Open vSwitch is used as virtual switches in hybrid switching mode with proactive OpenFlow mode (rules are installed on the local server switch by the orchestrator when performing actions with virtual machines or networks). Using scripts that automate the connection of network components (Figure 1a), delays in the experimental segment of the data center were investigated at traffic generation intensity values Traffic(t) at various time intervals t. To do this, after launching all the necessary components, the script starts intercepting tcpdump traffic with writing to the tmpfs disk, as well as generating traffic with the trafgen utility at various intensities. During the experiment, device state snapshots were taken using the SNMP protocol and OpenFlow metrics. Based on the topology shown in Figure 4, a model of this network was formed in the OMNeT ++ using standard tools and a study of delays in the model network was carried out (see Figure 5).

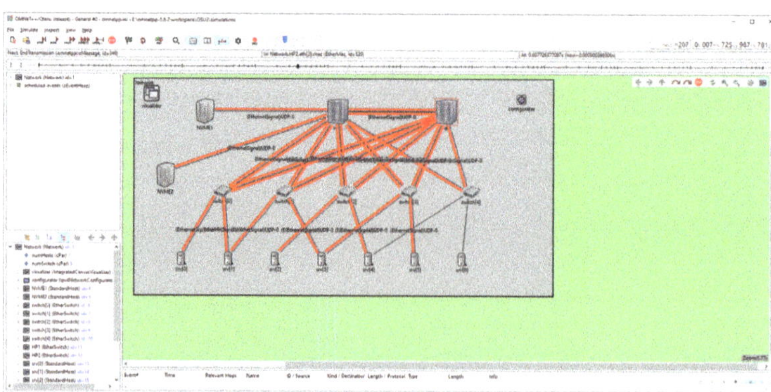

Figure 5. Network model in OMNeT ++.

With the help of the developed scripts for the automation of the experiment, a study of all models of network devices and servers that are part of the network under study was carried out, and models of these devices were synthesized. To study delays on a network model, a traffic generation plan was synthesized, similar to experiments on equipment. Figure 6 shows a graphical interpretation of the results of studying network delay without using queues (on the equipment, queues were minimized to one packet per queue, multiplexing into one channel was excluded), its model obtained using standard tools of the OMNeT ++, and also a model obtained by means of the developed software and hardware complex. The study was carried out under various work scenarios (different paths shown in Figure 1a) for various traffic generation intensities.

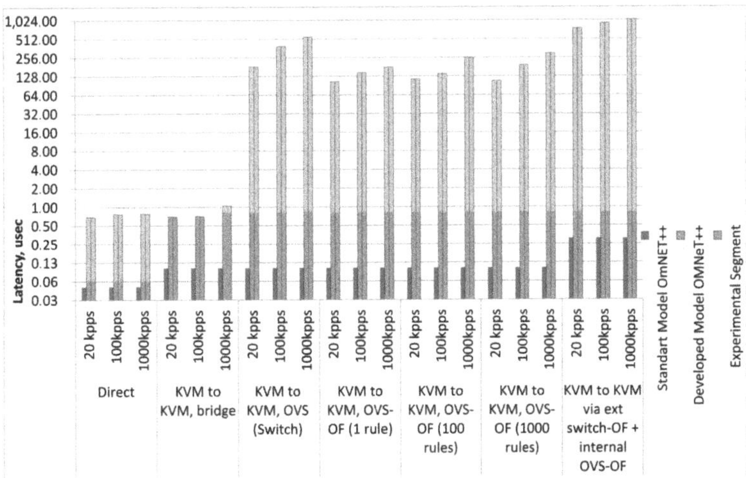

Figure 6. Results of the study of latency in the network and its models.

The delays were introduced by the ingressTCIN module, which interpolated the received packet processing times during an experimental study to the current state of the switch and the available enabled settings (VLANs, queues, classifiers, markers). Figure 6 shows the summary results of measuring delays in a real network, a network model obtained by standard means of the simulation system, and a network model. According to the results obtained, when using the switch models included in the OMNeT++, it is not possible to investigate the influence of the frame processing time by the switch on the total packet transmission delay in the network. Comparison of the results of the study of the model obtained using the tools of the developed software–hardware complex with the delays of the experimental segment of the data center showed a coincidence at similar rates of frame generation within 5%.

4. Conclusions

In this paper, we propose a methodology for studying the processes of traffic flow in hyper-converged structures with software switching based on traditional approaches and OpenFlow tables. Interface models were created for traffic switching in the usual way and with higher-level OpenFlow processing. To obtain the probabilistic-temporal characteristics of packet processing by equipment, a method for the initial synthesis of the equipment model with the developed interface model is proposed, and the possibility of traffic passing along various virtual paths within the virtual infrastructure is also taken into account. The obtained results of research and modeling showed a significant influence of the settings and sizes of the OpenFlow tables of virtual infrastructure elements on the package processing time, including the non-linear form of dependence. The use of interpolation of the results of an experimental study in the model to provide the required form of the dependence of the packet processing time showed high modeling accuracy under similar conditions and traffic generation intensities within 5%.

Author Contributions: Conceptualization, Y.U. and M.U.; methodology, Y.U.; software, Y.U.; validation, M.U. and L.L.; writing—original draft preparation, Y.U.; writing—review and editing, Y.U. and M.U.; visualization, L.L.; supervision, L.L.; All authors have read and agreed to the published version of the manuscript.

Funding: The research was funded by the Russian Science Foundation (project No. 22-71-10124).

Institutional Review Board Statement: Not applicable.

Informed Consent Statement: Not applicable.

Data Availability Statement: Data sharing not applicable due to privacy restrictions.

Acknowledgments: The data for building the models were collected at the Mathematical and Information Technology Faculty of Orenburg State University under the supervision of Canditate of Technical Sciences, D.I. Parfenov.

Conflicts of Interest: The authors declare no conflict of interest.

References

1. Melo, C.; Dantas, J.; Maciel, P.; Oliveira, D.M.; Araujo, J.; Matos, R.; Fé, I. Models for hyper-converged cloud computing infrastructures planning. *Int. J. Grid Util. Comput.* **2020**, *11*, 196–208. [CrossRef]
2. Bhaumik, S.; Chanda, K.; Chakraborty, S. Poster: Controlling Quality of Service of Container Networks in a Hyperconverged Platform. In Proceedings of the IFIP Networking Conference (Networking), Paris, France, 22–26 June 2020; pp. 661–663.
3. Meneses-Narvaez, S.E.; Dominguez-Limaico, P.; Rosero-Montalvo, D.; Narvaez-Pupiales, S.; Vizuete, M.Z.; Peluffo-Ordónez, D.H. Design and Tests to Implement Hyperconvergence into a DataCenter: Preliminary Results. *Adv. Emerg. Trends Technol.* **2019**, *1066*, 54–67.
4. Risdianto, A.C.; Usman, M.; Kim, J.W. SmartX box: Virtualized hyper-converged resources for build-ing an affordable playground. *Electronics* **2019**, *8*, 1242. [CrossRef]
5. Kim, J.W. Realizing Diverse Services Over Hyper-converged Boxes with SmartX Automation Frame-work. In *Conference on Complex, Intelligent, and Software Intensive Systems*; Springer: Cham, Switzerland, 2017; pp. 817–822.
6. Meneses, S.; Maya, E.; Vasquez, C. Network Design Defined by Software on a Hyper-converged Infra-structure. Case Study: Northern Technical University FICA Data Center. In *International Conference on Systems and Information Sciences*; Springer: Cham, Switzerland, 2020; pp. 272–280.
7. Melo, C.; Dantas, J.; Maciel, R.; Silva, P.; Maciel, P. Models to evaluate service provisioning over cloud computing environments-A block-chain-as-A-service case study. *Revista de Informática Teórica e Aplicada* **2019**, *26*, 65–74. [CrossRef]
8. Wen, H. Improving Application Performance in the Emerging Hyper-Converged Infrastructure. Ph.D. Thesis, University of Minnesota, Minneapolis, MN, USA, 2019.
9. Bök, P.B.; Noack, A.; Müller, M.; Behnke, D. *Quality of Service. Computernetze und Internet of Things*; Springer Vieweg: Wiesbaden, Germany, 2020; pp. 251–291.
10. Szigeti, T.; Zacks, D.; Falkner, M.; Arena, S. *Cisco Digital Network Architecture: Intent-based Net-working for the Enterprise*; Cisco Press: Indianapolis, IN, USA, 2018.
11. Karakus, M.; Durresi, A. Quality of service (QoS) in software defined networking (SDN): A survey. *J. Netw. Comput. Appl.* **2017**, *80*, 200–218. [CrossRef]
12. Lei, Y.; Tai, K.C. In-parameter-order: A test generation strategy for pairwise testing. In Proceedings of the Third IEEE International High-Assurance Systems Engineering Symposium, Washington, DC, USA, 13–14 November 1998; pp. 254–261.

engineering proceedings

MDPI

Proceeding Paper

Aircraft Optimal Control for Longitudinal Maneuvers Using Population-Based Algorithm †

Oleg Korsun *, Alexandr Poliyev and Alexandr Stulovskii

State Research Institute of Aviation Systems, 125167 Moscow, Russia; poliyev@gmail.com (A.P.); avstlv2@gmail.com (A.S.)

* Correspondence: marmotto@rambler.ru; Tel.: +7-499-157-9361

† Presented at the 15th International Conference "Intelligent Systems" (INTELS'22), Moscow, Russia, 14–16 December 2022.

Abstract: This report considers optimization of aircraft maneuvers in the vertical plane based on direct control methods. It proposes an object model for longitudinal motion suitable for optimal control, algorithms for control approximation and a numerical solution to the problem via a population-based optimization algorithm. The suggested method proves its applicability by forming the control signals for basic aircraft maneuvers in the vertical plane, including climb, speed increment, descent and speed decrement.

Keywords: optimal control; direct optimization; population-based algorithm; aircraft

Citation: Korsun, O.; Poliyev, A.; Stulovskii, A. Aircraft Optimal Control for Longitudinal Maneuvers Using Population-Based Algorithm. *Eng. Proc.* **2023**, *33*, 53. https://doi.org/10.3390/engproc2023033053

Academic Editors: Askhat Diveev, Ivan Zelinka, Arutun Avetisyan and Alexander Ilin

Published: 19 July 2023

1. Introduction

Optimal control for a maneuverable aircraft is a sophisticated problem, since the control object belongs to a class of multidimensional nonlinear dynamic systems [1]. A common approach to control involves variation methods that require the solution of a two-point boundary value problem [2]. Recently, interest has arisen in direct methods for the formation of optimal control [3,4] due to their scalability, robustness and relative simplicity of implementation.

Direct methods assume that a finite set of parameters could determine the desired control signal on the required time interval with necessary precision. As a result, they reduce the initial problem of finding the optimal control to the unconditional parametric optimization problem.

However, depending on the type of control signal, the number of parameters may be quite large. This requires an approach other than widespread gradient methods. One of the possible solutions to this problem appears to be evolutionary or population-based optimization algorithms known in a variety of applications.

In this paper, we propose using a modification of the direct method developed by the authors [4] for the purpose of optimization of the typical aircraft movement in the longitudinal channel. This method is based on spline approximation of control signals and particle swarm optimization [5,6] (PSO) of the resulting problem.

2. General Formulation of the Problem

The mathematical model of the object could be generally described as

$$\dot{\mathbf{y}}(t) = \mathbf{f}(\mathbf{y}(t), \mathbf{u}(t), t),$$

where $\mathbf{y}(t)$ and $\dot{\mathbf{y}}(t)$—state vector and its time derivative, $\mathbf{u}(t)$—control vector, $\mathbf{f}$—vector-valued function of vector arguments, on time interval $t \in [t_0; T]$, where t_0—is the beginning and T—endpoint.

The initial conditions are assumed to be known

$$\mathbf{y}(t_0) = \mathbf{y}_0$$

and the target functional has the following form

$$J(\mathbf{y}(t), \mathbf{u}(t)) = \int_{t_0}^{T} F(\mathbf{y}(t), \mathbf{u}(t), t)dt. \tag{1}$$

The problem consists of finding control $\mathbf{u}(t)$, which provides a minimum of the target functional (1). The widespread approach which satisfies the necessary optimality conditions leads to the two-point boundary value problem. In the case of direct methods, this problem could be bypassed.

Let us assume that control $\mathbf{u}(t)$ may be presented as projections on some linear basis. Then, it could be uniquely determined by its coefficients in this basis. We denote the vector of these coefficients as $\mathbf{a}$

$$\mathbf{u} = \mathbf{u}(\mathbf{a})$$

Estimation of the coefficients' value allows us to determine the control signal. Replacement of the continuous signal by a set of parameters turns the former task into the problem of multidimensional parametric unconditional optimization, the solution of which provides required coefficients

$$a = \arg\min_{a \in A} J(\mathbf{y}(t), \mathbf{u}(\mathbf{a}(t))),$$

where A is the search domain in the parameter space.

Thus, to solve this problem, it is necessary to form the object model, determine the target functional and select the optimization algorithm.

3. Specification of Problem

3.1. Mathematical Model for Control Object

As a foundation for a control object model, we have chosen the system of equations for the spatial motion of an aircraft as a rigid body [1]. Restriction of the motion to single plane allows considerable simplification of the model. Then, the original model [1] can be reduced to a system of three equations determining the following output signals: angle of attack, velocity and flight altitude. This system takes the form presented below

$$\begin{aligned}
\dot{\alpha} &= \dot{\nu} - \frac{qS}{mV} c_{ye}(\alpha) - \frac{P_x}{mV} \sin(\alpha) + \frac{g}{V}(\sin(\alpha)\sin(\nu) + \cos(\alpha)\cos(\nu)) \\
\dot{V} &= -\frac{qS}{m} c_{xe}(\alpha) + \frac{P_x}{m} \cos(\alpha) + g(-\cos(\alpha)\sin(\nu) + \sin(\alpha)\cos(\nu)) \\
\dot{H} &= V(\cos(\alpha)\sin(\nu) - \sin(\alpha)\cos(\nu))
\end{aligned} \tag{2}$$

where

α—angle of attack, rad;
ν—angle of pitch, rad;
V—airspeed, m/s;
H—flight altitude, m;
$c_{xe}(\alpha)$—drag coefficient;
$c_{ve}(\alpha)$—lift coefficient;
m—aircraft mass, kg;
S—wing surface area, m^2;
P_x—projection of thrust force on axis Ox of body-fixed coordinate system, N;
g—gravity acceleration, m/s^2;
q—dynamic air pressure, Pa.

We assume that at the initial moment state coordinates are known, the beginning and endpoints of the time interval are fixed.

The angle of pitch and projection of thrust become control signals. This choice is opportune for two reasons. Firstly, both of these signals are smooth enough that their approximation does not require introduction of many parameters. Secondly, nowadays the problem of tracking these signals in many cases could be delegated to the inner control loop, where it is solved by the automatic control system [7].

We use model (2) with the assumption that the angle of the engine installation is zero. Values of thrust force, drag $c_{xe}(\alpha)$ and lift $c_{ve}(\alpha)$ coefficients should be estimated based on the results of the identification of the model for motion of the aircraft in question.

3.2. Optimized Functional

Optimization of aircraft motion control in the longitudinal channel could focus on a set of basic maneuvers, including climb, descent, speed increment and decrement. In general terms, the performance of these motions could be thought of as a Lagrange problem with free end.

In this paper, we use as the target functional the sum of the Euclidean norms of the vector of the difference between the outputs of the model and the desired output values. This functional takes the form

$$J(V,\widetilde{V},H,\widetilde{H}) = \int_{t_0}^{T}\left(J_V(V,\widetilde{V}) + J_H(H,\widetilde{H})\right)dt = \int_{t_0}^{T}\left((V-\widetilde{V})^2 + (H-\widetilde{H})^2\right)dt \quad (3)$$

where $\widetilde{V}$, $\widetilde{H}$—desired values of airspeed and altitude.

In cases of maneuvers such as descent, an asymmetric functional can be used. Then, only when values of the variable become less than those specified are they taken into account, which should provide more opportunities for the optimization process to increase speed

$$J_V(V,\widetilde{V}) = \begin{cases} (V-\widetilde{V})^2, & V < \widetilde{V}; \\ 0, & V \geq \widetilde{V}. \end{cases} \quad (4)$$

3.3. Optimization Algorithm

Since the optimization problem requires the estimation of 20–30 parameters depending on the time interval and characteristics of the signal, in this article, we eschew common gradient methods in favor of population-based algorithms [6,8]. Within this class of optimization algorithms, a PSO algorithm was applied.

This algorithm benefits from its highly customizable nature, which allows it to be adjusted to meet the demands of the problem in question, striking a balance between examination of whole search domain and a more thorough investigation of promising solutions [9]. Nonetheless, it should be noted that this approach requires substantial time and computational resources and is less suitable for tasks of online optimization. The suboptimality of the obtained solution, as is commonly the case among population-based algorithms [10], should be pointed-out.

4. Direct Method of Program Control

At the core of direct control methods lies the assumption that the sought control could be projected on the basis of linearly independent functions. In this paper, cubic Hermitian splines [11] are considered on such a basis. Therefore, the approximation requires determining the values of the function and its derivative at the spline nodes. The nodes were evenly distributed throughout the entire time interval. The PSO algorithm requires specification of the boundaries of search spaces for the pitch angle and thrust; for example, $\pm 90^\circ$ in case of pitch angle.

To determine the initial values of parameters, particles were randomly thrown into the search space. After that, the model (2) could be integrated and the value of the targeted functional (3) for each particle obtained.

Then, the PSO algorithm finds those parameter values that provide the minimum for the functional, i.e., ensure that error between the desired signals and model output is minimal.

5. Results Description

Thus, the task consisted of applying the method of finding direct program control described above, optimization of the aircraft's maneuvers in the longitudinal channel. A mathematical model (2) was used to describe the control object's motion; the functional in general case took the form (3). Formula (4) was applied in case of descent problem.

The problem under consideration was solved in two main stages. At first, the problem was solved to determine the solution with the best stabilizing time. As experiments have shown, when we shift the initial time taking the functional into account, the resulting solution changes. With a time shift equal to the stabilization time, it is possible to achieve improvements in fuel consumption and mode of maneuver. Thus, the problem was calculated taking into account this shift.

For the problem of climb and speed increment, results were calculated depending on the initial conditions of altitude, as shown in Figure 1.

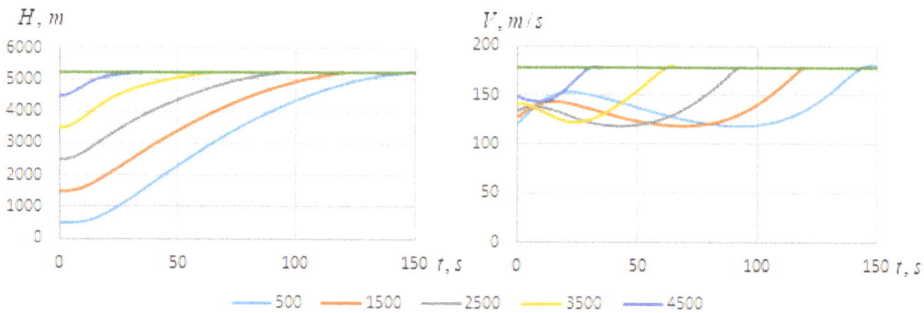

Figure 1. Values of altitude and speed obtained by variation of initial condition for altitude.

Analysis of the descent maneuver shows that taking into account the increase in speed during the maneuver makes it possible to improve the final results. To obtain this, the asymmetrical part (4) is included into the target functional. The problem was considered in a way similar to one described above by variation of initial altitude value. As Figure 2 shows, a family of solutions was found.

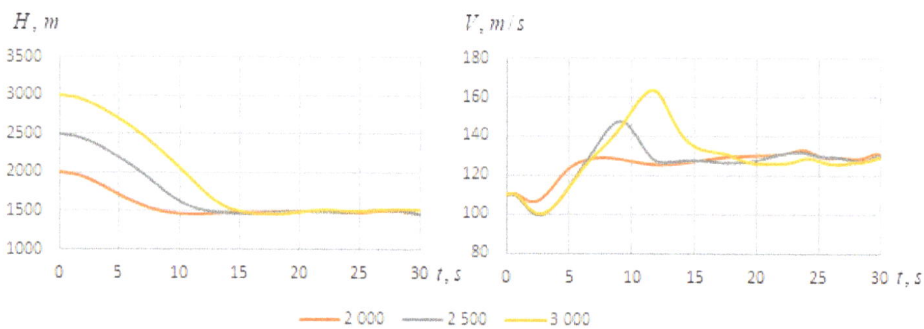

Figure 2. Values of altitude and speed obtained by variation of initial condition for altitude in descent problem.

In all three cases, the maneuver provides a similar picture of performance. The aircraft goes into a dive, losing altitude but gaining speed. When reaching a predetermined altitude, it levels off, trying to keep the pitch angle sufficient to ensure level flight, while gradually losing speed.

Similar results could be observed for the speed decrement problem. The target functional corresponds to Formula (3). The aircraft reduces its speed mainly by increasing its angle of attack, which naturally leads to an increase in altitude. Nonetheless, Figure 3 shows that the requirements for altitude increase have little effect on the overall behavior of the speed on the time interval under consideration. This means that addition of stricter requirements for altitude values could be accommodated without corresponding worsening of the result.

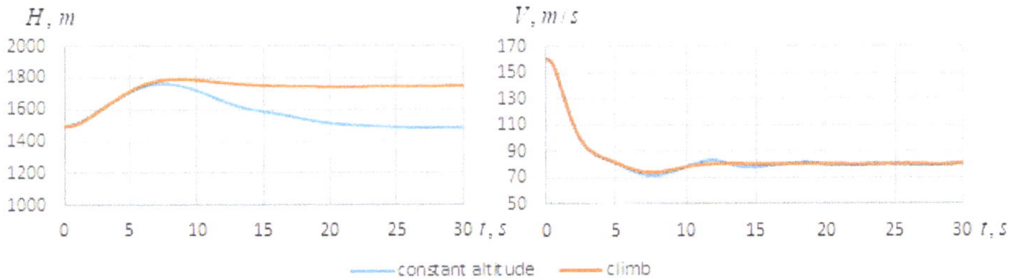

Figure 3. Values of altitude and speed obtained in speed decrement problem.

Above was presented the application of the model for the optimization of relatively simple maneuvers in the longitudinal channel under varying initial conditions. The proposed method could also be used in the case of more dynamic and complex movements, such as, for example, pulling up. Such a problem appears among different applications, including avoiding a collision with the ground.

Within the framework of this task, the idea implemented in the functional (4) is further developed. In this case, the functional becomes asymmetric not only in speed, but also in altitude, including the following expression

$$J_H(H, \widetilde{H}) = \begin{cases} (H - \widetilde{H})^2, & H < \widetilde{H}; \\ 0, & H \geq \widetilde{H}. \end{cases}$$

Pulling up with different initial values of speed was considered. Figure 4 presents the resulting signals for speed and altitude.

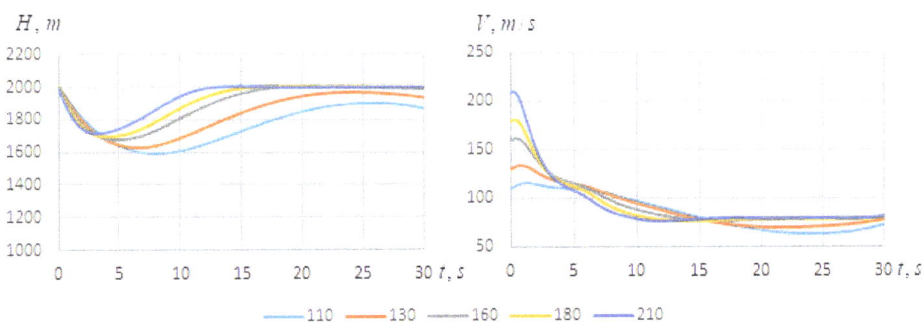

Figure 4. Values of altitude and speed obtained through variation of initial condition for speed in pulling-up problem.

The research showed how an increase in the initial speed leads to a decreasing loss of altitude, but at the same time increases the maximum value of the normal force that accompanies the maneuver. However, it should be noted that extreme values of normal force are not observed over the entire time interval, but are localized during the time of a sharp change in the pitch angle, which means that they last only a few seconds.

6. Conclusions

The paper considers a direct method for optimization of aircraft's maneuvers in the longitudinal channel. The method proposes representation of the desired signals via the cubic Hermitian splines and the solution of the resulting numerical optimization problem via the PSO method. Application of the method to the problems of climb, descent, speed increment and decrement is shown.

The results obtained can be used both for the optimization of individual maneuvers and for determining the reachable areas for the aircraft.

Author Contributions: Conceptualization O.K.; methodology O.K. and A.S.; software A.S. and A.P.; validation A.P. All authors have read and agreed to the published version of the manuscript.

Funding: The work is supported by Russian Foundation for Fundamental Research, project 20-08-00449-a.

Institutional Review Board Statement: Not applicable.

Informed Consent Statement: Not applicable.

Data Availability Statement: The data that support the findings of this study are openly available at https://cloud.mail.ru/public/dS2m/doT5hRp7X.

Conflicts of Interest: The authors declare no conflict of interest.

References

1. Bushgens, G.S. (Ed.) *Aerodinamika, Ustoichivost' i Upravlyaemost' Sverkhzvukovykh Samoletov (Aerodynamics, Stability, and Controllability of Supersonic Airplanes)*; Nauka: Moscow, Russia, 1998.
2. Chernous'ko, F.L.; Ananievski, I.M.; Reshmin, S.A. *Control of Nonlinear Dynamical Systems. Methods and Applications*; Springer: Berlin/Heidelberg, Germany, 2008.
3. Alekhin, D.V.; Yakimenko, O.A. Synthesis of optimization algorithm for planned-route flight trajectory by direct variational method. *J. Comput. Syst. Sci. Int.* **1999**, *38*, 650–666.
4. Korsun, O.N.; Stulovskii, A.V. Direct method for forming the optimal open loop control of aerial vehicles. *J. Comput. Syst. Sci. Int.* **2019**, *58*, 229–243. [CrossRef]
5. Karpenko, A.P. *Sovremennye Algoritmy Poiskovoi Optimizatsii (Contemporary Search Optimization Algorithms)*; BMSTU Press: Moscow, Russia, 2014.
6. Olsson, A.E. *Particle Swarm Optimization: Theory, Techniques and Applications*; Nova Science Publishers: Hauppage, NY, USA, 2011.
7. Beard, R.W.; McLain, T.W. *Small Unmanned Aircraft. Theory and Practice*; Princeton University Press: Princeton, NJ, USA, 2012.
8. Diveev, A.I.; Sofronova, E.A.; Konstantinov, S.V. Approaches to numerical solution of optimal control problem using evolutionary computations. *Appl. Sci.* **2021**, *11*, 7096. [CrossRef]
9. Mirjalili, S.; Dong, J.S.; Lewis, A. (Eds.) *Nature-Inspired Optimizers: Theories, Literature Reviews and Applications*; Springer: Cham, Switzerland, 2020.
10. Clerc, M. *Guided Randomness in Optimization*; John Wiley & Sons, Inc.: Hoboken, NJ, USA, 2015.
11. The MathWorks, Inc. Spline Toolbox // MATLAB.Exponenta. 2021. Available online: http://matlab.exponenta.ru/spline/book1/12.php (accessed on 19 October 2021).

engineering proceedings

MDPI

Proceeding Paper

Providing High-Speed Data Access for Parallel Computing in the HPC Cluster †

Sergey Denisov *, Konstantin Volovich and Alexander Zatsarinny

Federal Research Center "Computer Science and Control" of the Russian Academy of Sciences, 44, Build. 2, Vavilova Str., 119333 Moscow, Russia; kvolovich@frccsc.ru (K.V.); azatsarinny@ipiran.ru (A.Z.)
* Correspondence: sdenisov@frccsc.ru; Tel.: +7-499-135-4320
† Presented at the 15th International Conference "Intelligent Systems" (INTELS'22), Moscow, Russia, 14–16 December 2022.

Abstract: The article discusses approaches to building parallel data storage systems in high-performance clusters. The features of building data structures in parallel file systems for various applied tasks are analyzed. Approaches are proposed to improve the efficiency of access to data by computing nodes of the cluster due to the correct distribution of data in parallel file storage.

Keywords: HPC cluster; parallel file system; Lustre; NVMe

1. Introduction

The growing need of scientific teams, industrial enterprises, and commercial firms in solving problems that require high-performance computing resources demands the creation of computing tools provided to users using cloud and platform technologies. At present, such computing resources are supercomputers and HPC clusters that make it possible to carry out calculations using parallel computing technologies. So, the tasks of mathematical modeling, global optimization, big data analysis, and the training of neural networks cannot be solved using a single server; powerful multi-node clusters united by high-performance computing networks are required. The workflow management in such clusters is a research task aimed at optimizing the computing resources load, minimizing the waiting and computing time, managing job priorities, and providing computing jobs with initial data.

Based on HPC clusters, it is possible to create a research infrastructure that provides tools for miscellaneous scientific calculations to teams of scientists, industry, and developers of high-tech commercial products [1]. The creation of shared research facilities on the basis of scientific centers, enterprises, organizations of the Russian Academy of Sciences and universities makes it possible to increase the efficiency of using computing facilities and expand the circle of consumers of high-performance computing resources [2]. This approach to the use of computational tools involves the parallel execution of miscellaneous types of computational tasks within a single high-performance cluster. In this case, in order to efficiently use computing resources and minimize the loss of time and money for re-configuring the HPC cluster, it is necessary to develop a technology that allows creating individual software environments for various computing tasks and ensuring their parallel execution in a HPC cluster [3], wherein each computational task obtains access to the parallel processing information means and inter-process communication. The means of HPC cluster workflow management create an individual profile for inter-process interaction in a dynamic virtual environment for each computational task.

The optimization of the loading of the shared research HPC cluster is generally aimed at ensuring the maximum loading of computing modules, reducing the waiting time for tasks in the queue, and minimizing equipment downtime.

Citation: Denisov, S.; Volovich, K.; Zatsarinny, A. Providing High-Speed Data Access for Parallel Computing in the HPC Cluster. *Eng. Proc.* **2023**, *33*, 54. https://doi.org/10.3390/engproc2023033054

Academic Editors: Askhat Diveev, Ivan Zelinka, Arutun Avetisyan and Alexander Ilin

Published: 21 July 2023

To implement such requirements for a HPC cluster, it is of particular importance to provide computing tasks with initial data. The performance of the means of providing data to computational jobs should, for example, minimize the processing downtime by waiting for data to be provided. The tasks of providing data to computational tasks in multi-node computing systems are solved by creating a specialized file storage that provides parallel access to cluster computing nodes and data. Thus, the data exchange height speed is ensured in the conditions of parallel operation of virtual software environments for the execution of miscellaneous types of computational tasks.

The classical scheme for organizing parallel file access is the use of a group of data storage nodes connected to computing nodes by a high-speed network. A parallel file system deployed on storage nodes creates a single data space, ensures information consistency, the duplication of information on different nodes, and control of access to files by computing nodes.

Both proprietary and open-source implementations of parallel file systems are available today. For example, proprietary solutions include

- General Parallel File System (GPFS)—developed by IBM;
- Google File System.

IBM GPFS is used by the manufacturer as part of a commercial product, the elastic storage system, which is a scalable hardware and software data storage system. The Google File System is used by Google in the company's high-performance computing clusters.

Among open source parallel file systems, the Lustre project is currently actively developing, originating from Carnegie University, now supervised by Intel Corporation. Another open-source project is the Ceph parallel file system supervised by RedHat (IBM).

The direction of providing parallel access to data is currently an actively developing area of informatics [4–6]. The main efforts of developers and researchers are concentrated in the following areas:

- Maximizing the volume of data storage;
- Ensuring the required performance during data transfer, taking into account the number of computing nodes and data storage nodes;
- Development of methods for scaling the file system while maintaining or increasing performance;
- Development of methods for improving the reliability of parallel file systems.

The architectures and ways of representing data in parallel file systems are discussed below.

2. Parallel File System Architecture

Consider the architecture of a parallel file system using the Lustre project as an example.

Lustre is a high-performance file system composed of servers and storage. The Metadata Server (MDS) keeps track of metadata (such as ownership and access rights to a file or directory). Object storage servers provide file I/O services for object storage targets that host the actual data store. The storage targets are typically a single disk array. Lustre's parallel file system achieves its performance by automatically splitting data into chunks known as "stripes" and writing the stripes in a round-robin fashion across multiple storage objects. This process, called "striping", can significantly increase file I/O speed by eliminating single disk bottlenecks [7].

A parallel storage system consists of many components, including drives, storage controllers, I/O cards, storage servers, SAN switches, and related management software. Combining all of these components together and tuning them for optimal performance comes with significant challenges. Figure 1 shows the Lustre parallel file system architecture.

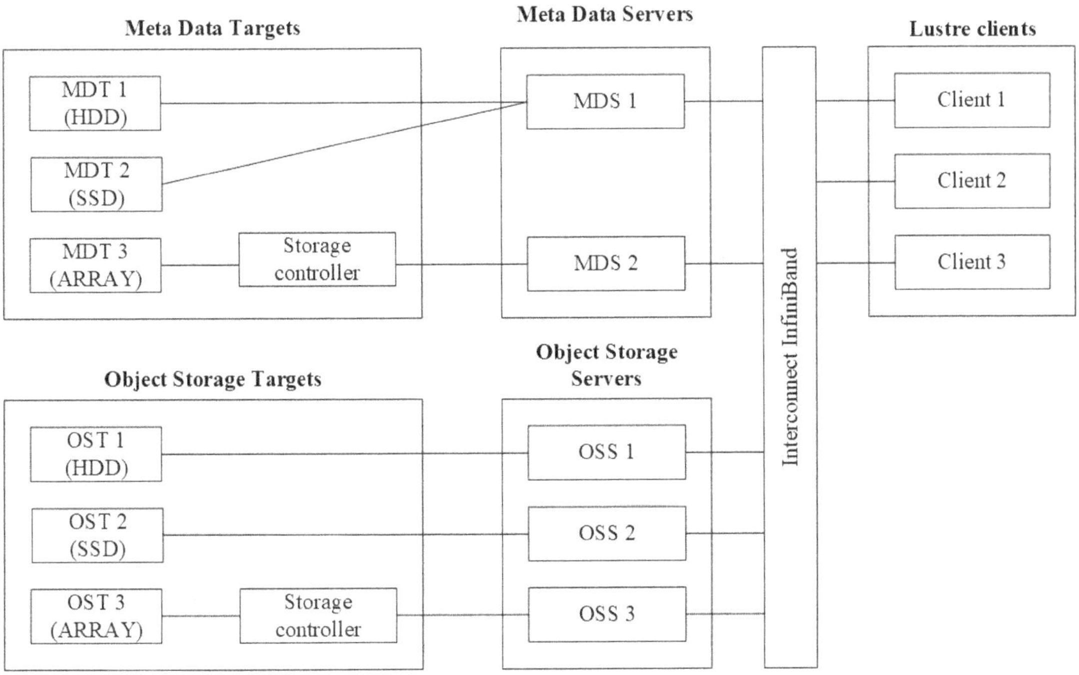

Figure 1. Lustre parallel file system architecture.

The elements of architecture are as follows:

- Control nodes—meta data servers (MDSs);
- Data storage and provision nodes—object storage servers (OSSs);
- Data storage elements—object storage target (OST);
- Metadata storage elements—meta data targets (MDTs);
- High-speed data-processing network.

Metadata servers (MDSs) are the control components of a parallel file system that store information about all the data in the system, as well as serving client requests for access to data. Metadata are stored on information resources called metadata targets (MDTs), implemented either as physical disks or as block devices in a data storage system. To ensure fault tolerance, metadata servers can be duplicated.

Access to parallel file system clients data is carried out through the metadata server. When servicing requests, the metadata server identifies the client and selects one of the nodes for storing and providing data—OSS. The choice is made based on meta-information about the availability of the requested data on storage elements available to this OSS, as well as based on information about OSS loading by requests from other clients. After that, the interaction between the client and the selected OSS is carried out directly through a high-performance data network. This solves the problem of the formation of "bottlenecks" in the access of computing nodes to a single data storage [8].

Data storage elements are disk drives directly connected to the OSS as well as logical drives formed by storage systems based on disk arrays. Data storage and provision nodes provide the caching of information contained on disks and high-speed exchange with the data transmission network.

Thus, the architecture of a parallel file system makes it possible to provide data for a group of computational tasks due to the parallel operation of a group of storage nodes.

3. Data Storage Structure in A Parallel File System

As noted above, the parallel file storage architecture allows you to provide computing nodes with access to different storage nodes, which allows you to avoid "bottlenecks" and prevent performance degradation.

Note that different application tasks may have different requirements for the data structure stored in the file system.

For big data-processing tasks—telemetry streams, the analysis of accumulated information arrays in order to obtain new knowledge (data mining)—where parallel processes of one computing task performed on different computing nodes require access to independent data arrays, it is enough to place one application data copy on each storage node and ensure the interaction of each pair of "computing node–storage node" over a high-speed data transmission network.

For data-intensive tasks (e.g., preparing training sets for neural networks, and training neural networks [9]), a more complex structure data storage is required, allowing different computing nodes to access the same file from several computing nodes. This feature is supported by the parallel file system by fragmenting files into separate blocks (stripes) [10].

The term "number of stripes" refers to the number of fragments into which a file is divided; in other words, the number of OSTs that are used to store the file. So, each stripe of the file will be in a different OST. The "stripe size" refers to the size of a stripe recorded as a single block in the OST.

The benefits of striping include the following:

- Increased I/O throughput due to multiple file areas being read or written in parallel;
- Helping to balance the use of the OST pool.

However, striping has disadvantages if performed incorrectly, such as increased overhead due to internal network operations and contention between servers, and throughput degradation due to inappropriate striping settings.

The default number and sizes of stripes are chosen to balance the I/O performance needs of multiple parallel execution scales and file sizes. Small files should be striped at a value of 1. However, setting the stripe number too low can degrade I/O performance for large files and parallel I/O. Thus, the user must carefully select strip specifications according to application data.

The striping must be compatible with the application's I/O strategy and output size. The increase in the number of stripes and/or the size of the stripes should be proportional to the number of nodes used for I/O. As a general rule, an application should try to use as many OSTs as possible. Thus, when writing a large single file in parallel, the maximum allowable value for the stripe counter is set. Alternatively, when writing a large number of small files in parallel, set the interleave counter to 1. The intermediate number of concurrent output files may work better if the number of stripes is greater than 1. An experimental estimate of the number of stripes is advisable for best performance. Note that for a number of tasks, the file size correlates with the number of computing nodes that perform parallel writing to it. Therefore, adjusting interleaving based on file size is usually sufficient, and a simpler starting point for estimating the number of lanes for subsequent processing optimization.

Figure 2 shows an example of file distribution across six storage elements.

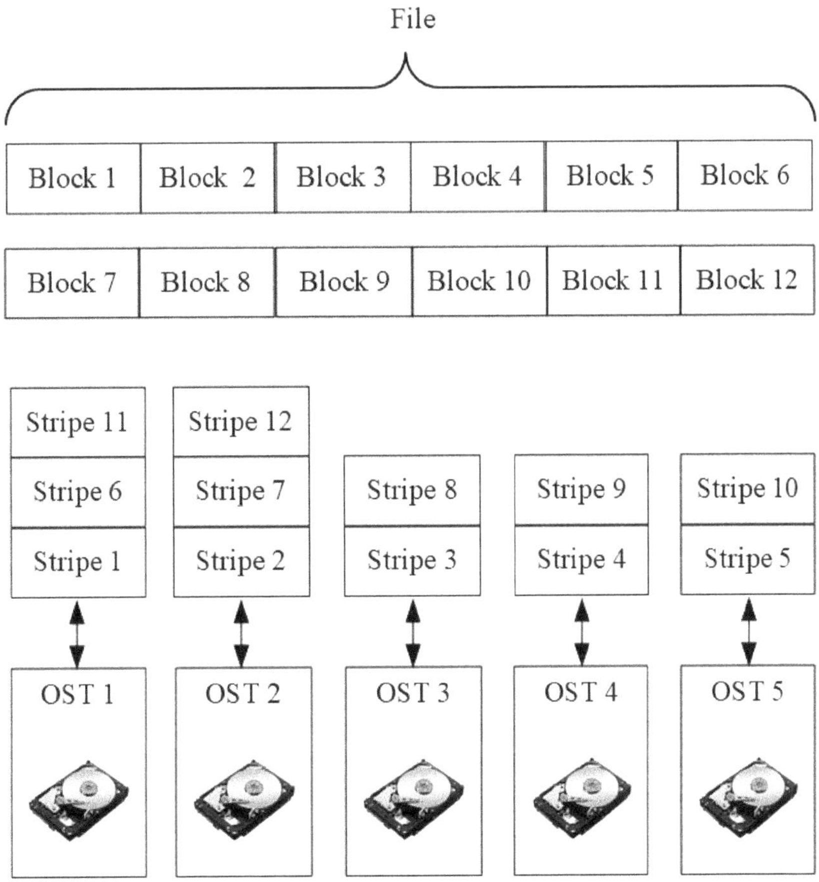

Figure 2. The file distribution example.

4. Experience of Application of Parallel File Systems in The Shared Research Facilities "Informatics"

To conduct research in the field of application of parallel file systems, taking into account the diversity of tasks solved by a HPC cluster, the parallel file storage stand was created using the infrastructure of the Shared Research Facilities "High Performance Computing and Big Data" (CKP "Informatics") of FRC CSC RAS (Moscow, Russia), which allows you to provide the data for different types of computational tasks functioning on the basis of virtual computing environments [11].

During the creation of the stand, the Lustre file system version 2.14.50 was deployed on six servers of the ARM architecture manufactured by Huawei.

During the experiments carried out on the created stand, the following scientific and practical work was carried out:

- Deployment of the Lustre parallel file system by compiling open-source software;
- Assessment of the performance and quality of functioning of the parallel file system in the base case when performing computational tasks of various types;
- Implementation of the developed storage system architecture, data models, access scenarios by setting up various data storage scenarios;
- Evaluation of performance and quality of functioning of a parallel file system for test computing tasks that have different requirements for data storage (size and number of files, multiple access to one file by a group of computing nodes);

- Design of recommendations on "fine tuning" of Lustre file storage in high-performance computing systems of scientific organizations when solving applied problems of various types.

The basic approach for configuration parallel data storage is the approach in which the user creates his own structure of files and directories to store files of various sizes and maps certain storage patterns to these directories that describe the number and size of stripes. Tools for such configuration are provided to the user in the form of executable scripts.

Users have the ability to customize the size and number of stripes for any file they use or create during a computational job. Determining the best settings sometimes requires experimentation, but there are general rules of thumb.

Let us consider the basic process of writing a file to a Lustre parallel file system as an example.

Let us assume that 200 MB is to be written to a file that is created with 10 stripes and a 1 MB stripe. When the file is initially recorded, 10 1 MB blocks will be simultaneously written to 10 different OSTs. Once those 10 blocks are full, Lustre writes another 10 1 MB blocks to those 10 OSTs. This process is repeated a total of 20 times until the entire file is written. When completed, the file will exist as 20 1 MB data blocks in each of 10 separate OSTs.

One of the key factors behind the high performance of Lustre file systems is the ability to stripe data across multiple storage targets (OSTs) in a round robin fashion. Essentially, files can be divided into multiple fragments, which are then stored in different OSTs in the Lustre system.

Larger files benefit from more stripes. By distributing a large file across multiple OSTs, the system's throughput for accessing the file increases, which improves efficiency when multiple processes are working on the same file in parallel. Conversely, a very large file that is interleaved in only one or two OSTs can degrade the performance of the entire Lustre system due to unnecessary OST padding. It is good practice to have dedicated directories with lots of stripes for writing very large files to.

Note that the following method of dividing a file into stripes should be avoided: dividing small files with a large number of stripes. This can negatively impact performance due to the unnecessary overhead of communicating with multiple OSTs. In this case, it is good practice to write small files to a directory with stripes equal to 1, i.e., without striping.

Thus, to configure striping, its main advantages and disadvantages should be considered:

- Advantage 1: increased throughput because multiple processes can access the same file at the same time;
- Advantage 2: the ability to store large files that take up more space than one OST;
- Disadvantage 1: increased overhead due to network operations and server conflicts;
- Disadvantage 2: increased risk of file corruption due to hardware failure.

Methods for improving reliability are the use of RAID of various levels in data storage nodes (OSS). At the same time, RAID can be implemented both by means of storage systems that provide block disk devices to storage nodes, and by means of RAID controllers located directly on storage nodes. Both methods improve the reliability and fault tolerance of the parallel file system. At the same time, as the experience of using such configurations in the CKP "Informatics" shows, the bandwidth for accessing drives increases compared to accessing single disks due to parallel access to drives.

A separate task is to ensure the reliability and fault tolerance of NVMe SSD drives connected directly to the PCI bus. The absence of a RAID controller in this case makes it necessary to take special measures to build fault-tolerant arrays. One such measure is the use of software RAID by means of the operating system. Another method is to use specialized storage systems that include NVMe SSD drives and RAID controllers that combine these drives into arrays of various levels.

Note that due to the use of specialized controllers, creating RAID using storage systems is more productive than using software RAID.

Experiments in the CKP "Informatics" using Huawei OceanStore Dorado 3000 v6 storage systems showed that the speed of access to block devices on RAID5 based on NVMe SSD, provided via the iSCSI interface over a 100 Gb Ethernet network, is about 2 GBps. The access speed to a similar SSD NVMe drive installed in the OSS server is about 1.2–1.4 GBps.

Thus, the use of high-speed storage systems equipped with SSD NVMe drives and RAID controllers as OST allows you to simultaneously increase the throughput of the Lustre parallel file system and improve reliability and fault tolerance.

5. Conclusions

A parallel file system is an integral part of a high-performance computing system designed to solve a wide range of scientific and scientific–practical problems. The parallel file system effectively prevents the formation of "bottlenecks" that reduce the performance of the HPC clusters due to delays in disk operations. Its application in the shared research facilities, providing high-performance computing services, allows to provide sufficient throughput of the disk subsystem for all scientific tasks performed in parallel by the HPC cluster.

The tasks of processing big data, extracting knowledge, and mathematical modeling impose different requirements on the organization of data exchange with disk storage when performed on a group of computing nodes.

At the same time, in order to optimize and increase the efficiency of the functioning of the HPC cluster as a whole, it is necessary to take measures to adapt data storage patterns in a parallel file system to the specifics of the applied problems being solved.

An analysis of computational tasks from various fields of science and technology shows that the requirements for the throughput of access to the file system for various applied tasks differ significantly. So the tasks of optimization, quantum mechanical calculations, aerodynamics do not require significant resources of the file system while loading the computing unit.

The tasks of training neural networks are more demanding on the performance of the file system, which ensures the timely loading of training data on computing nodes, but the ratio between data volumes and calculations is not the maximum.

The greatest requirements for the performance of the file system are observed in the tasks of preparing data for training neural networks and in the tasks of extracting knowledge (data mining). These tasks require a high-speed exchange of large data arrays between a large number of data storage and processing nodes.

Depending on the task type and file storage characteristics, a data storage template is selected. In general, maximizing file system performance requires maximizing the number of storage objects used.

Thus, when deploying projects designed to solve computing problems that require high disk efficiency, you should allocate the maximum number of storage elements, mainly NVMe SSD.

For tasks with medium and low disk intensity, it is enough to allocate a small number of SAS or SATA storage elements.

Ensuring reliability and fault tolerance through the use of a redundant storage architecture is provided on data object storage servers or storage systems.

Author Contributions: Conceptualization, S.D., K.V. and A.Z.; methodology, S.D., K.V. and A.Z.; validation, S.D.; formal analysis, K.V.; investigation, S.D., K.V. and A.Z.; writing—original draft preparation, S.D. and K.V.; writing—review and editing, S.D. and K.V.; supervision, A.Z. All authors have read and agreed to the published version of the manuscript.

Funding: This research received no external funding.

Institutional Review Board Statement: Not applicable.

Informed Consent Statement: Not applicable.

Data Availability Statement: Not applicable.

Acknowledgments: The research was carried out using the infrastructure of the Shared Research Facilities "High Performance Computing and Big Data" (CKP "Informatics") of FRC CSC RAS (Federal Research Center "Computer Science and Control" of the Russian Academy of Sciences, Moscow).

Conflicts of Interest: The authors declare no conflict of interest.

References

1. Zatsarinnyy, A.A.; Abgaryan, K.K. Current problems of creation of research infrastructure for synthesis of new materials in the framework of the digital transformation of society. In Proceedings of the II International Conference Mathematical Modeling in Materials Science of Electronic Components, Online, 19–20 October 2020; pp. 8–13.
2. Zatsarinny, A.A.; Volovich, K.I.; Denisov, S.A.; Ionenkov, Y.S.; Kondrashev, V.A. Methodological approaches to evaluating the effectiveness of the center collective use "Informatics". *Highly Available Syst.* **2020**, *16*, 44–51.
3. Volovich, K.; Zatsarinnyy, A.; Frenkel, S.; Denisov, S. High Performance Computing in a Shared Virtual Infrastructure. In Proceedings of the VI International Conference on Information Technologies and High-Performance Computing (ITHPC 2021), Khabarovsk, Russia, 14–16 September 2021; Volume 2930, pp. 38–46.
4. Kartsev, A.; Malkovsky, S.; Volovich, K.; Sorokin, A. Study of the performance and scalability of the Quantum ESPRESSO package in the study of low-dimensional systems on hybrid computing systems. In Proceedings of the I International Conference Mathematical Modeling in Materials Science of Electronic Components, Moscow, Russia, 21–23 October 2019; pp. 18–21.
5. Abgaryan, K.K. Information technology is the construction of multi-scale models in problems of computational materials science. *Highly Available Syst.* **2018**, *14*, 9–15.
6. Abgaryan, K.K.; Gavrilov, E.S.; Marasanov, A.M. Multiscale modeling for composite materials computer simulation support. *Int. J. Open Inf. Technol.* **2017**, *5*, 24–28.
7. Kokorev, A.; Belyakov, D.; Lyubimova, M. Data storage systems of "hybrilit" heterogeneous computing platform for scientific research carried out in JINR: Filesystems and raids performance research CEUR Workshop Proceedings. In Proceedings of the 9th International Conference "Distributed Computing and Grid Technologies in Science and Education" (GRID'2021), Dubna, Russia, 5–9 July 2021; Volume 3041, pp. 296–303.
8. Seiz, M.; Offenhäuser, P.; Andersson, S.; Hötzer, J.; Hierl, H.; Nestler, B.; Resch, M. Lustre I/O performance investigations on Hazel Hen: Experiments and heuristics. *J. Supercomput.* **2021**, *77*, 12508–12536. [CrossRef]
9. Tipu, A.J.S.; Conbhuí, P.Ó.; Howley, E. Applying neural networks to predict HPC-I/O bandwidth over seismic data on lustre file system for ExSeisDat. *Clust. Comput.* **2022**, *25*, 2661–2682. [CrossRef] [PubMed]
10. Rybintsev, V. Optimizing the parameters of the Lustre-file-system-based HPC system for reverse time migration. *J. Supercomput.* **2020**, *76*, 536–548. [CrossRef]
11. Volovich, K. Estimation of the workload of a hybrid computing cluster in tasks of modeling in materials science. In Proceedings of the II International Conference Mathematical Modeling in Materials Science of Electronic Components, Online, 19–20 October 2020; pp. 30–33.

engineering proceedings

MDPI

Proceeding Paper

Dynamic Job Queue Management for Interactive and Batch Computation on HPC System †

Sergey Denisov *, Vadim Kondrashev and Alexander Zatsarinny

Federal Research Center "Computer Science and Control" of the Russian Academy of Sciences, 44, Build.2, Vavilova Str., Moscow 119333, Russia; vkondrashev@frccsc.ru (V.K.); azatsarinny@ipiran.ru (A.Z.)
* Correspondence: sdenisov@frccsc.ru; Tel.: +7-910-434-08-93

† Presented at the 15th International Conference "Intelligent Systems" (INTELS'22), Moscow, Russia, 14–16 December 2022.

Abstract: The article discusses HPC system computing resources distribution management during execution of interactive and batch jobs. A set of queues for interactive and batch jobs is proposed, and an algorithm for the dynamic resources allocation between the proposed job queues is described.

Keywords: computing resources; hybrid high-performance computing system; hybrid architecture; graphic coprocessor; individual execution environment; interactive mode; computational task management system; job queue

Citation: Denisov, S.; Kondrashev, V.; Zatsarinny, A. Dynamic Job Queue Management for Interactive and Batch Computation on HPC System. *Eng. Proc.* **2023**, *33*, 55. https://doi.org/10.3390/engproc2023033055

Academic Editors: Askhat Diveev, Ivan Zelinka, Arutun Avetisyan and Alexander Ilin

Published: 21 July 2023

1. Introduction

Modern high-performance systems provide solutions to a wide range of tasks and interdisciplinary research, including machine learning, artificial intelligence, big data, digital twins, behavior modeling, and more. Such an extensive set of tasks determines the specifics of the organization of the computational process on such systems. The computing environment is required to be flexible and adaptable to the requirements from both the application and user task sides.

In [1], in addition to the classical batch mode of operation of the computing environment, an interactive mode of operation in an individual execution environment in the computing environment of a high-performance system was substantiated.

The interactive mode of operation is implemented in many software products (graphics editors, spreadsheets, tutorials, computer games, etc.). In other cases, it is used in systems operating in the mode of a computational experiment, including the development and training of artificial intelligence models, when the ongoing processes must remain under the supervision and control of an expert. The interactive mode provides developers, experts, and users with the opportunity to test the functionality of models online (especially during the initial stages of their development), test various hypotheses, and analyze and visualize the obtained data.

The current level of development of microelectronics makes it possible to introduce high-performance computing systems in large corporations as well as in small scientific and scientific-production teams, with the only difference being the "size" of the implemented systems. As a rule, high-performance systems of scientific and scientific-production teams are significantly inferior to the computing complexes of corporations in terms of the amount of computing resources, which means that the issue of efficient allocation of computing resources while taking into account interactive and batch approaches to computing is an urgent task.

Keeping in mind that there are many different families of tools for developing applications for modeling, machine learning, and artificial intelligence, the simultaneous presence of several tools in a single underlying computing environment is inappropriate, as it creates conditions for the occurrence of compatibility conflicts and limits the ability to customize these tools for use by different groups of users. The limitations on the settings are primarily due to the fact that their implementation requires administrator rights to install additional system software and software libraries, taking into account the compatibility requirements of the software components of the entire complex. The performance of such operations by the administrator, which require a change in the computing environment, creates discreteness in the work of users.

In [1], the authors proposed an approach to the placement of computational jobs belonging to the class of batch or interactive jobs while taking into account the organization of an individual execution environment. In this paper, we consider the distribution of computing resources and the allowable time intervals for executing computational jobs for different job queues. Using the example of the "High Performance Computing and Big Data" (CKP "Informatics") Shared Research Facilities of FRC CSC RAS (Moscow) [2], an approach is proposed to dynamically allocate computing resources between job queues serving flows of interactive and batch jobs.

2. Organization of the Computing Process in a High-Performance System

The functions of managing the computing environment of a high-performance system are assigned to the computing task management system (which may be referred to as a "task manager", "workload management system", "job scheduler", etcetera).

As a rule, a computing task management system operates within the "client–server" paradigm and manages the process of executing computational jobs based on the configured service policies in the form of a set of rules (algorithms) for distributing computational resources between computational jobs and managing the progress of their competitive execution.

The process of managing the execution of computational jobs is reduced to the implementation of three key functions of the computing task management system:

1. Providing users with exclusive and/or non-exclusive access to computing resources for a certain period of time;
2. Providing an environment for launching, executing, and monitoring the operation of a parallel task, for example, MPI (Message Passing Interface);
3. Performing resource contention arbitration by managing a queue of pending jobs.

As a rule, computing task management systems have a built-in set of algorithms that implement the following service policies:

- A policy for processing the input stream of computational jobs;
- A policy for managing the behavior of the computational task being performed.

The classical approach to the organization of the computational process in a high-performance system is the batch execution of computational jobs, which ensures the most efficient use of computational resources [3].

However, in a number of cases the established practice of using computing tools dictates the need to implement an interactive mode of operation for users in their individual execution environment together with a batch service mode. An individual execution environment is created through the use of virtual containerization technology based on the Docker system [4]. From the point of view of the computing task management system, there is no difference between the management of a task containing an applied task and that of a container with an applied task. This feature is implemented in the CKP "Informatics", where an individual runtime environment is integrated into the computing environment of a high-performance system, allowing different groups of users to solve their tasks in parallel in different computing environments. In [5], approaches to the formation and maintenance of a functional individual execution environment in the interests of a particular user are substantiated.

Currently, the following types of individual execution environments are implemented in the CKP "Informatics":

- The GROMACS environment [6] for modeling physical and chemical processes in molecular dynamics, with the ability to run several copies and organize MPI interaction between them;
- The Ansys environment [7] for automated engineering calculations;
- A development environment for machine learning and artificial intelligence models based on the Jupyter Notebook solution [8].

3. Queues of the Computing Task Management System

In this paper, we propose the creation of the following queues for simultaneous operation of batch and interactive modes.

The *interactive queue* is designed to perform development and debugging functions, and permits almost-online execution of computational jobs. Service policies for this queue represent a limitation on the execution time of a given job.

The *main queue* is intended for execution of computational jobs in batch mode, and allows computational jobs to be placed for execution, for which the user determines the operation time by different values.

An *additional queue* is intended for executing computational jobs in batch mode, and allows computational jobs to be placed for execution when it is not possible to estimate the time of their execution. This queue has the lowest priority; a limited pool of resources is provided for the queue, and jobs can be suspended.

The listed queues are defined by the following parameters:

- *Interactive queue*: priority is high, availability of resources is based on the "best possible" principle, and the maximum task execution time is very limited;
- *Main queue*: priority is medium, availability of resources is within a limited pool, and the maximum task execution time is limited;
- *Additional queue*: priority is low, resource availability is within a limited pool, and the maximum task execution time is unlimited.

Next, we consider the approach of dynamic distribution of computing resources between job queues serving the flows of interactive and batch computational jobs. The computing task management system (for example, the system based on the Slurm software product [9]) provides users with flexible options for solving their scientific and practical problems by launching computational jobs either

- On one server from the computing system; or
- On a group of servers belonging to the same computing system.

4. Dynamic Distribution of Computing Resources

The parameters that characterize the queue, as indicated in the previous section, are as follows:

- Queue priority;
- Resource pool, i.e., the maximum amount of computing resources available to the queue for distribution between computational jobs;
- Resource portions, i.e., the maximum amount of computing resources allocated to the calculation task in the queue;
- Operation mode;
- Reservation time, i.e., the maximum time for the execution of the calculation task in the queue.

Below, we explain several of these parameters.

4.1. Resource Pool

Usually, in heterogeneous computing systems, all computing resources, which are understood as physical computing servers (nodes), are divided into several classes depending on the type of equipment. For the CKP "Informatics" these are of two classes: hybrid computing nodes (computing servers equipped with graphics co-processor GPUs) and computing nodes (computing servers without graphics co-processors).

4.2. Resource Portions

A resource portion R is described by the set

$$R = \left\{nvCPU, nGPU, nRAM\right\}, \quad (1)$$

where

- $nvCPU$—the requested number of virtual processors (vCPU);
- $nGPU$—the requested number of graphics coprocessors (GPU);
- $nRAM$—the requested amount of RAM.

The $nvCPU$ is the sum of the products

$$nvCPU = \sum_{i=1}^{N}(nCPU * nCore * nThread), \quad (2)$$

where

- N—the number of computing nodes;
- $nCPU$—the number of CPU of the computing node;
- $nCore$—the number of CPU cores of the computing node;
- $nThread$—the number of threads per CPU core.

Typically, resource portions are available to users in discrete sets or user-defined sets. For example, the following types of resource portions are available for users of the CKP "Informatics":

(a) numa node—a resource portion containing n CPUs and n GPUs belonging to the same numa domain [10].
The configuration of numa nodes depends directly on the physical architecture of the hybrid computing nodes.
Possible configuration options are
  - numa.small, containing one GPU and 12 vCPU;
  - numa.medium, containing two GPUs and 24 vCPUs;
  - numa.big, containing four GPUs and 48 vCPUs;
  - numa.large, containing eight GPUs and 96 vCPUs.

(b) user-defined, in which a resource portion is described by the set R and is not more than the size of the "resource pool".

4.3. Operation Mode

The "open" mode means that the computational jobs in the queue have a claim on the allocation of resources, and can begin to be executed. The "closed" mode means that the computational jobs have accumulated in the queue and are waiting for the queue opening time before starting their execution.

The specified parameters are dynamically assigned to the previously designated queues depending on the current mode of operation of the computing task management system. There are two modes of operation for the computing task management system of the CKP "Informatics": "day mode" (daytime operation from 8.31 to 21.30) and "night mode" (night-time operation from 21.31 to 8.30).

It can be seen that there is a time interval of one minute between the "day–night" and "night–day" intervals. During this period, switching between operating modes occurs. The queue parameters are reassigned, while calculation jobs are not executed and remain pending.

The proposed approach for the dynamic distribution of computing resources between job queues serving the flows of interactive and batch jobs is performed twice a day when changing operating modes to day and night modes. The change of modes is carried out automatically by reassigning the parameters of the queues in accordance with the following conditions.

In "day mode", the following parameters are assigned to the *interactive queue*:

1. Priority—high;
2. Resource pool—all hybrid computing nodes;
3. Resource portion—one numa node from the list of available options;
4. Operation mode—open;
5. Reservation time—user-defined, with the following conditions:
    - No more than one hour;
    - If the reservation is made less than an hour before the change of operation mode, then until the change of operation mode.

In "day mode", the following parameters are assigned to the *common queue*:

1. Priority—medium;
2. Resource pool:
    - All computing nodes for computational jobs with reservation times that do not exceed the time limit for the end of "day mode";
    - Three computing nodes for computational jobs with reservation times that exceeds the time limit for the end of "day mode";
3. Resource portion—user-defined;
4. Operation mode—open;
5. Reservation time—user-defined.

In "day mode", only the "operation mode—closed" parameter is assigned to the *additional queue*. The priority, resource pool, resource portion, and reservation time parameters are not assigned due to the mode of operation.

Thus, in the "day mode" the users have access to both interactive and batch computing modes in a high-performance system, while interactive jobs have priority advantage over batch jobs.

In "night mode", only the "operation mode—closed" parameter is assigned to the *interactive queue*. Thepriority, resource pool, resource portion, and reservation time parameters are not assigned due to the mode of operation.

In "night mode", the following parameters are assigned to the *common queue*:

1. Priority—high;
2. Resource pool:
    - All nodes for computational jobs with reservation times that do not exceed the time limit for the end of "night mode";
    - One hybrid computing node for computational jobs with reservation times that exceed the time limit for the end of "night mode";
    - Three computing nodes for computational jobs with reservation times that exceed the time limit for the end of "night mode";
3. Resource portion—user-defined;
4. Operation mode—open;
5. Reservation time—user-defined.

In "night mode", the following parameters are assigned to the *additional queue*:

1. Priority—medium;
2. Resource pool:
    - One hybrid computing node;
    - One computing node;
3. Resource portion—user-defined;
4. Operation mode—open.

Thus, in the "night mode" only the batch mode of calculations is available to users. At the same time, computational jobs for which the resource reservation time is determined have a priority advantage over computational jobs for which it is not possible to estimate the time of execution. When placing a computational job in a computing environment, the user specifies values for the resource portion and reservation time while taking into account the limitation on the maximum execution time of a computational task in the queue.

After that, the computational task management system operates according to the basic algorithm embedded in it:

1. the system places the computational job in the queue in accordance with the policy for processing the input stream of computational job.
   In "day mode":
   If all resource portions are reserved at the current moment, the computing task management system performs the following actions when queuing a computational job:
   - In the interactive queue, the system offers the user either the nearest available time for reservation and available options for numa nodes, or offers the user the option to select a convenient time and available options for numa nodes, including on subsequent calendar days;
   - In the general queue, the system leaves the computational job in the queue in "resources pending" status.

   If the user has specified an invalid reservation time, the computational job is not queued.
   For an interactive queue, it is possible to prolongation the amount of time required by the user 10 min before the expiration of the reservation time; it may be the case that there are no available resource portions, in which case the system offers the user the same options as during the initial reservation of resources.
   When changing the current operating mode to "night mode",
   - For computational jobs pending in queues, the reservation is preserved, while jobs from the general queue are allowed to execute at night;
   - Computational jobs running from the interactive queue are forcibly terminated;
   - Computational jobs running from the general queue continue their execution.

   In "night mode":
   If all resource portions are reserved at the current moment, the computing task management system performs the following actions when queuing a computational job:
   - In the general and additional queues, the system leaves the computational jobs in the queue in "resources pending" status.

   If the user has specified an invalid reservation time, the computational job is not queued.
   When changing the current operating mode to "day mode":
   - For computational jobs pending in queues, the reservation is preserved, while jobs from the general queue are allowed to be executed during the daytime;
   - Computational jobs running from the general and additional queues continue their execution.
2. The system analyzes the parameters of jobs in the queue in a cycle. The system determines which jobs that requires a control action and performs this action in accordance with the policy for managing the behavior of the executed computational jobs;
3. The system removes completed computational jobs from the queue. The job can complete its execution on its own or under the influence of the system in accordance with the policy for managing the behavior of the executed computational jobs;
4. Upon completion of a job, the system checks the correctness of the release of computing resources and returns the resources to the resource pool.

5. Conclusions

This article proposes an approach to the dynamic distribution of computing resources between job queues serving flows of interactive and batch computational jobs. This approach improves efficiency when loading the resources of a high-performance computing system under conditions of joint application of interactive and batch approaches to computing.

The presented mode of operation of the computing task management system with dynamic distribution of computing resources is currently being tested in the computing environment of the CKP "Informatics". The dynamic distribution of computing resources of a high-performance system successfully provides for heterogeneous tasks of interdisciplinary research (including machine learning, artificial intelligence, big data analysis, creation of digital twins, etc.) with the necessary flexibility and adaptability for the requirements of applied applications, including interactive modes.

Taking into account that an interactive mode is used during the initial stages of model creation, while the classic batch mode approach is well suited for operation (i.e., training models in deep learning tasks), the approach to organizing calculations proposed in this article can greatly expand access to computing resources. As a result, the solution proposed in this paper is able to effectively solve the problem of allocating computing resources for interactive and batch computing modes in high-performance computing systems belonging to small scientific and scientific-production teams.

Author Contributions: Conceptualization, S.D., V.K. and A.Z.; methodology, S.D., V.K. and A.Z.; validation, S.D.; formal analysis, V.K.; investigation, S.D., V.K. and A.Z.; writing—original draft preparation, S.D., V.K.; writing—review and editing, S.D., V.K.; supervision, A.Z. All authors have read and agreed to the published version of the manuscript.

Funding: This research received no external funding.

Institutional Review Board Statement: Not applicable.

Informed Consent Statement: Not applicable.

Data Availability Statement: Not applicable.

Acknowledgments: The research was carried out using the infrastructure of the Shared Research Facilities "High Performance Computing and Big Data" (CKP "Informatics") of FRC CSC RAS (Federal Research Center "Computer Science and Control" of the Russian Academy of Sciences, Moscow).

Conflicts of Interest: The authors declare no conflict of interest.

References

1. Volovich, K.; Kondrashev, V.; Posypkin, M.; Denisov S. Some Approaches to Managing Computing Resources of a Hybrid High-Performance Cluster in a Cloud Environment. In Proceedings of the VI International Conference Information Technologies and High-Performance Computing (ITHPC-2021), Khabarovsk, Russia, 14–16 September 2021; Volume 2930, pp. 47–53.
2. CKP "Informatics". Available online: http://www.frccsc.ru/ckp (accessed on 19 July 2022).
3. Volovich, K.I.; Denisov, S.A.; Shabanov, A.P.; Malkovsky S.I. Aspects of the assessment of the quality of loading hybrid high-performance computing cluster. In Proceedings of the V International Conference Information Technologies and High-Performance Computing (ITHPC-2019), Khabarovsk, Russia, 16–19 September 2019; Volume 2426, pp. 7–11.
4. Docker Docs. Available online: https://docs.docker.com/get-started/overview/ (accessed on 19 July 2022).
5. Volovich, K.I.; Denisov, S.A.; Malkovsky, S.I. Formation of an Individual Modeling Environment in a Hybrid High-Performance Computing System, *Russ. Microelectron.* **2020**, *49*, 580–583. [CrossRef]
6. Gromacs. Available online: https://www.gromacs.org/ (accessed on 19 July 2022).
7. Ansys. Available online: https://www.ansys.com/ (accessed on 19 July 2022).
8. Jupyter Notebook. Available online: https://jupyter.org/ (accessed on 19 July 2022).
9. Slurm Documentation. Available online: https://slurm.schedmd.com/documentation.html (accessed on 19 July 2022).
10. The Linux Kernel. Numa. Available online: https://www.kernel.org/doc/html/v4.18/vm/numa.html (accessed on 19 July 2022).

engineering proceedings

MDPI

Proceeding Paper

A Hierarchical Model of a Vector Nash Equilibrium Search in a Control Problem under Conflict and Uncertainty †

Vladimir A. Serov [1,*] and Evgeny M. Voronov [2]

1 Department of Applied Information Technologie, MIREA—Russian Technological University (RTU MIREA), Moscow, 119454, Russia
2 Department of Control Systems, Bauman Moscow State Technical University (BMSTU, Bauman MSTU), Moscow 105005, Russia; emvoronov@mail.ru
* Correspondence: ser_off@inbox.ru

† Presented at the 15th International Conference "Intelligent Systems" (INTELS'22), Moscow, Russia, 14–16 December 2022.

Abstract: A hierarchical model of a vector Nash equilibrium search under uncertainty is developed. The sufficient conditions for a vector Nash equilibrium of a noncooperative game under uncertainty are formulated, which can be used as a criterion to achieve the required degree of nonquilibrium for an acceptable solution to the problem of multi-object multicriteria systems' control optimization under conflict and uncertainty.

Keywords: noncooperative game under uncertainty; vector Nash equilibrium under uncertainty; hierarchical coevolutionary algorithm; hierarchical model; vector minimax

Citation: Serov, V.A.;Voronov, E.M. A Hierarchical Model of a Vector Nash Equilibrium Search in a Control Problem under Conflict and Uncertainty. *Eng. Proc.* **2023**, *33*, 60. https://doi.org/10.3390/engproc2023033060

Academic Editors: Askhat Diveev, Ivan Zelinka, Arutun Avetisyan and Alexander Ilin

Published: 25 July 2025

1. Introduction

The technology of the neuroevolutionary synthesis of algorithms for multi-object multicriteria systems' (MMS) control under conflict and uncertainty is based on the development and widespread use of coevolutionary algorithms for finding a game's stable-effective compromises (STEC) [1–3] and coordinated STEC (COSTEC) [4–6]. However, these coevolutionary algorithms have extremely high computational complexity. Therefore, the practical use of neuroevolutionary technology for solving problems of the class in question requires its implementation on high-performance parallel computing architectures. One of the basic principles of optimality used in the formation of STEC and COSTEC is the Nash equilibrium principle. In this report, a hierarchical algorithm for a vector Nash equilibrium search under uncertainty is developed. The hierarchical structure of the developed algorithm best corresponds to the structure of coevolutionary algorithms and the capabilities of parallel computing technologies.

2. The Problem Statement

The problem statement of multi-object system control optimization under conflict and uncertainty is formalized in the form of a noncooperative game with uncertainty with vector indicators of the effectiveness of subsystems–players

$$\Gamma = \langle \mathbf{N}, \{\mathbf{U}_i, i \in \mathbf{N}\}, \mathbf{Z}, \{\mathbf{J}_i(\mathbf{u},\mathbf{z}), i \in \mathbf{N}\}, \{\boldsymbol{\Omega}_i, i \in \mathbf{N}\}\rangle. \quad (1)$$

In (1), $\mathbf{N} = \{\overline{1,n}\}$—a set of subsystems–players; $\mathbf{U}_i$—a set of admissible strategies of the i-th player $\mathbf{u}_i \in \mathbf{U}_i \subset \mathbf{E}^{r_u}$; $\mathbf{u} = \{\mathbf{u}_1, \ldots, \mathbf{u}_n\} \in \mathbf{U} = \prod_{i \in \mathbf{N}} \mathbf{U}_i \mathbf{E}^{r_u}$—a situation corresponding to the players' choice of their strategies $\mathbf{u}_i \in \mathbf{U}_i, i \in \mathbf{N}; \mathbf{Z} \subset \mathbf{E}^{r_z}$—a set of uncertainty factor admissible values; $\mathbf{z} \in \mathbf{Z}; \mathbf{J}_i(\mathbf{u}) \in \mathbf{E}^{m_i}$—$i$-th player vector indicator of the effectiveness; $\boldsymbol{\Omega}_i = \mathbf{E}^{m_i}_{\leq}$—a convex polyhedral dominance cone defining a binary strict preference relation for the i-th player on the set of achievable vector estimates $\mathbf{J}_i(\mathbf{U},\mathbf{Z}) = \bigcup_{\substack{\mathbf{u}\in\mathbf{U}\\ \mathbf{z}\in\mathbf{Z}}} \mathbf{J}_i(\mathbf{u},\mathbf{z})$ and formalizing the requirement to minimize the components of a vector indicator of effectiveness $\mathbf{J}_i(\mathbf{u},\mathbf{z})$.

To solve problem (1), the vector Nash equilibrium principle under uncertainty is used. When implementing the computational procedure for finding the optimal solution, the principle of vector Nash $\boldsymbol{\varepsilon}$-equilibrium under uncertainty is used.

Definition 1. *Vector estimation $\mathbf{V}_i(\mathbf{u}) \in \mathbf{E}^{m_i}$ is called the point of extreme pessimism for the i-th player in a situation $\mathbf{u} \in \mathbf{U}$ on a set $\mathbf{J}_i(\mathbf{u},\mathbf{Z}) = \bigcup_{\mathbf{z}\in\mathbf{Z}} \mathbf{J}_i(\mathbf{u},\mathbf{Z})$, if it has the following properties:*

$$1)\ \mathbf{J}_i(\mathbf{u},\mathbf{Z}) \subset \mathbf{V}_i(\mathbf{u}) + \boldsymbol{\Omega}_i; \tag{2}$$

2) for any $\widetilde{\mathbf{V}} \neq \mathbf{V}_i(\mathbf{u})$, such that $\mathbf{J}(\mathbf{u},\mathbf{Z}) \subset \widetilde{\mathbf{V}} + \boldsymbol{\Omega}_\mathrm{i}$,

$$\mathbf{V}_i(\mathbf{u}) - \widetilde{\mathbf{V}} \in \boldsymbol{\Omega}_i. \tag{3}$$

Definition 2. *Vector estimation $\mathbf{V}^T(\mathbf{u}) = [\mathbf{V}_1^T(\mathbf{u}),\ldots,\mathbf{V}_n^T(\mathbf{u})] \in \mathbf{E}^m$ is called the point of extreme pessimism for the i-th player in a situation $\mathbf{u} \in \mathbf{U}$ on a set $\mathbf{J}(\mathbf{u},\mathbf{Z}) = \bigcup_{\mathbf{z}\in\mathbf{Z}} \mathbf{J}(\mathbf{u},\mathbf{z})$, where $\mathbf{J}^T(\mathbf{u},\mathbf{z}) = [\mathbf{J}_1^T(\mathbf{u},\mathbf{z}),\ldots,\mathbf{J}_n^T(\mathbf{u},\mathbf{z})] \in \mathbf{E}^m$, $m = \sum_{i\in\mathbf{N}} m_i$.*

Definition 3. *An admissible solution $\mathbf{u}^e \in \mathbf{U}$ is called a vector Nash equilibrium under uncertainty in problem (1) if, for any admissible $\mathbf{u}_i \in \mathbf{U}_i$ $i \in \mathbf{N}$, we have:*

$$(\mathbf{V}_i(\mathbf{u}^e\|\mathbf{u}_i) - \mathbf{V}_i(\mathbf{u}^e)) \notin \boldsymbol{\Omega}_i, \tag{4}$$

where we have the situation $\mathbf{u}^e\|\mathbf{u}_i = \{\mathbf{u}_1^e,\ldots,\mathbf{u}_{i-1}^e,\mathbf{u}_i,\mathbf{u}_{i+1}^e,\ldots,\mathbf{u}_n^e\}$.

Definition 4. *An admissible solution $\mathbf{u}^{e\varepsilon} \in \mathbf{U}$ is called a vector Nash $\boldsymbol{\varepsilon}$-equilibrium under uncertainty in problem (1), where $\boldsymbol{\varepsilon}^T = [\varepsilon_1^T,\ldots,\varepsilon_n^T] \in \mathbf{E}^m, \varepsilon_i \in \mathbf{E}^{m_i}, i \in \mathbf{N}$, if for any admissible $\mathbf{u}_i \in \mathbf{U}_i$, we have:*

$$\mathbf{V}_i(\mathbf{u}^{e\varepsilon}\|\mathbf{u}_i) - (\mathbf{V}_i(\mathbf{u}^{e\varepsilon}) - \varepsilon_i) \notin \boldsymbol{\Omega}_i, i \in \mathbf{N}. \tag{5}$$

3. Representation of the Nash Vector Equilibrium Search Problem under Uncertainty in the Form of a Hierarchical System

We present the problem of searching for a vector Nash equilibrium in a noncooperative game under uncertainty (1) in the form of a hierarchical system, as shown in Figure 1.

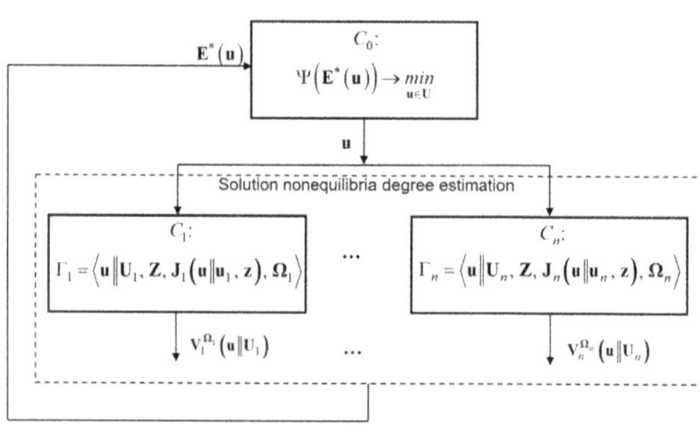

Figure 1. Representation of the Nash vector equilibrium search problem under uncertainty in the form of a hierarchical system.

In relation to the specified hierarchical structure, an algorithm for a vector Nash equilibrium searching under uncertainty can be represented as a sequence of the following stages.

Stage 1. The first move is made by the coordinating subsystem C_0 of the upper level, which forms and communicates its strategy $\mathbf{u} \in \mathbf{U}$ to the subsystems of the lower level. Each subsystem of the lower level C_i estimates the control $\mathbf{u} \in \mathbf{U}$ nonequilibria degree by the corresponding component $\mathbf{u}_i \in \mathbf{U}_i$. To do this, for a fixed $\mathbf{u} \in \mathbf{U}$, the MOU problem Γ_i of the following type is solved:

$$\Gamma_i = \langle \mathbf{u} \| \mathbf{U}_i, \mathbf{Z}, \mathbf{J}_i(\mathbf{u} \| \mathbf{u}_i, \mathbf{z}), \mathbf{\Omega}_i \rangle i \in \mathbf{N}. \tag{6}$$

In (6), it is required to minimize the vector criteria components $\mathbf{J}_i(\mathbf{u}, \mathbf{z})$ on the set $(\mathbf{u} \| \mathbf{U}_i)$ under uncertainty $\mathbf{z} \in \mathbf{Z}$. To solve problem (6), the vector minimax principle and the coevolutionary MOU algorithm are used [4,5]. The result of problem (6) is a vector minimax set $\mathbf{V}_i^{\mathbf{\Omega}_i}(\mathbf{u} \| \mathbf{U}_i) = \mathbf{V}_i(\mathbf{u} \| \mathbf{U}_i^{\mathbf{\Omega}_i}) \subset \mathbf{E}^{m_i}$.

Stage 2. The second step is estimation by the coordinating subsystem C_0 of the coordinating control $\mathbf{u} \in \mathbf{U}$ effectiveness. Based on the results of problem (6), an area is formed for each $i \in \mathbf{N}$:

$$\hat{\mathbf{U}}_i^{\mathbf{\Omega}_i} = \begin{cases} \mathbf{u}_i \in \mathbf{U}_i^{\mathbf{\Omega}_i}, \\ (\mathbf{V}_i(\mathbf{u} \| \mathbf{u}_i) - \mathbf{V}_i(\mathbf{u})) \in \mathbf{\Omega}_i. \end{cases} \tag{7}$$

We construct the objective function:

$$\Phi_i(\mathbf{u}, \mathbf{u}_i) = \| \mathbf{V}_i(\mathbf{u} \| \mathbf{u}_i) - \mathbf{V}_i(\mathbf{u})) \|, \tag{8}$$

and we formulate the optimization problem:

$$\Phi_i(\mathbf{u}, \mathbf{u}_i) \to \max_{\mathbf{u}_i \in \hat{\mathbf{U}}_i^{\mathbf{\Omega}_i}}. \tag{9}$$

Let $\mathbf{u}_i^* \in \hat{\mathbf{U}}_i^*$ be the optimal solution to problem (9). Then, the vector $\boldsymbol{\varepsilon}_i^*(\mathbf{u}) = \mathbf{V}_i(\mathbf{u} \| \mathbf{u}_i^*) - \mathbf{V}_i(\mathbf{u}))$ characterizes the maximum degree of control of the $\mathbf{u} \in \mathbf{U}$ nonequilibrium by component $\mathbf{u}_i \in \mathbf{U}_i$ for the player–subsystem C_i, where $\| \boldsymbol{\varepsilon}_i^*(\mathbf{u}) \| = \Phi_i(\mathbf{u} \| \mathbf{u}_i^*)$.

We construct a vector indicator of the coordinator's effectiveness:

$$\mathbf{F}(\mathbf{u}) = [f_1(\mathbf{u}), \ldots, f_n(\mathbf{u})]^T \tag{10}$$

with components of the form:

$$f_i(\mathbf{u}) = \| \boldsymbol{\varepsilon}_i^*(\mathbf{u}) \|, i \in \mathbf{N}. \tag{11}$$

The efficiency indicator (10) and (11) characterizes the degree of nonequilibrium of permissible control $\mathbf{u} \in \mathbf{U}$ relative to all subsystems–players C_i.

Stage 3. The problem of optimal coordination is solved at the subsystem C_0 level. To do this, we construct an objective function of the form:

$$\Psi(\mathbf{F}(\mathbf{u})) = \max_{i \in \mathbf{N}} \{ f_i(\mathbf{u}) \}, \tag{12}$$

and we formulate the optimization problem:

$$\Psi(\mathbf{F}(\mathbf{u})) \to \min_{\mathbf{u} \in \mathbf{U}}. \tag{13}$$

The following statement is true.

Theorem 1. *Let* $\mathbf{u}^* \in \mathbf{U}$ *be the optimal solution to problem (13), and* $\Psi(\mathbf{u}^*) = \mathbf{e}^*$*. Then, there exists a vector* $\boldsymbol{\varepsilon}^T = [\boldsymbol{\varepsilon}_1^T, \ldots, \boldsymbol{\varepsilon}_n^T] \in \mathbf{E}^m$ *having the properties:*

a) $\|\boldsymbol{\varepsilon}_i\| = e^*, i \in \mathbf{N}$;

b) $\mathbf{u}^*$ *is a vector Nash* $\boldsymbol{\varepsilon}$*-equilibrium of a noncooperative game under uncertainty (1).*

4. Conclusions

A hierarchical model of a vector Nash equilibrium search under uncertainty was developed, which is the basis for constructing parallel hierarchical coevolutionary algorithms for the MMS control optimization under noncooperative conflict interaction and uncertainty.

The sufficient conditions for a vector Nash ε-equilibrium of a noncooperative game under uncertainty were formulated, which can be used as a criterion to achieve the required degree of nonequilibrium of an acceptable solution in hierarchical coevolutionary algorithms of multicriteria conflict optimization.

The proposed hierarchical model is a component of the neuro-evolutionary technology of control algorithms and information processing synthesis in MMS under conflict and uncertainty in real time.

Author Contributions: All authors contributed equally to this work. All authors have read and agreed to the published version of the manuscript.

Funding: This research received no external funding.

Institutional Review Board Statement: Not applicable.

Informed Consent Statement: Not applicable.

Data Availability Statement: Not applicable.

Conflicts of Interest: The authors declare no conflicts of interest.

References

1. Serov, V.A.; Voronov, E.M. Evolutionary Algorithms of Stable-Effective Compromises Search in Multi-object Control Problems. In *Smart Electromechanical Systems*; Gorodetskiy, A.E., Tarasova, I.L., Eds.; Studies in Systems, Decision and Control; Springer: Cham, Switzerland, 2019; Volume 174, pp. 19–29. [CrossRef]
2. Serov, V.A.; Voronov, E.M.; Kozlov, D.A. Hierarchical Population Game Models of Machine Learning in Control Problems Under Conflict and Uncertainty. In *Smart Electromechanical Systems*; Gorodetskiy, A.E., Tarasova, I.L., Eds.; Studies in Systems, Decision and Control; Springer: Cham, Switzerland, 2022; Volume 419, pp. 125–145. [CrossRef]
3. Serov, V.A. Hierarchical Population Game Models of Coevolution in Multi-Criteria Optimization Problems under Uncertainty. *Appl. Sci.* **2021**, *11*, 6563. [CrossRef]
4. Voronov, E.M.; Serov, V.A. A Coordinated Stable-Effective Compromises Based Methodology of Design and Control in Multi-object Systems. In *Smart Electromechanical Systems*; Gorodetskiy, A.E., Tarasova, I.L., Eds.; Studies in Systems, Decision and Control; Springer: Cham, Switzerland, 2019; Volume 174, pp. 147–157. [CrossRef]
5. Serov, V.A.; Voronov, E.M.; Kozlov, D.A. A neuroevolutionary synthesis of coordinated stable-effective compromises in hierarchical systems under conflict and uncertainty. *Procedia Comput. Sci.* **2021**, *186*, 257–268. [CrossRef]
6. Serov, V.A.; Voronov, E.M.; Kozlov, D.A. Hierarchical Neuro-Game Model of the FANET based Remote Monitoring System Resources Balancing. In *Smart Electromechanical Systems*; Gorodetskiy, A.E., Tarasova, I.L., Eds.; Studies in Systems, Decision and Control; Springer: Cham, Switzerland, 2020; Volume 261, pp. 117–130. [CrossRef]

engineering proceedings

MDPI

Proceeding Paper

Neuro-Evolutionary Synthesis of Game Models of Control under Uncertainty Based on Distributed Computing Technology †

Vladimir A. Serov *, Daria L. Popova, Pavel P. Rogalev and Anastasia V. Tararina

Department of Applied Information Technologie, MIREA—Russian Technological University (RTU MIREA), Moscow 119454, Russia; pdl13@yandex.ru (D.L.P.); radugapp@mail.ru (P.P.R.); nharvard2013@gmail.com (A.V.T.)
* Correspondence: ser_off@inbox.ru
† Presented at the 15th International Conference "Intelligent Systems" (INTELS'22), Moscow, Russia, 14–16 December 2022.

Abstract: The methodology basic principles of the neuro-evolutionary synthesis of multi-object multi-criteria systems control models under conflict and uncertainty in real time are discussed. The proposed methodology includes the following main stages: a hierarchical optimization game model under conflict and uncertainty development; a library development of hierarchical coevolutionary algorithms for multi-criteria optimization under conflict and uncertainty; software implementation of hierarchical coevolutionary algorithms library based on distributed computing technology; and game algorithms of control under uncertainty synthesis based on the technology of neural networks ensembles.

Keywords: hierarchical coevolutionary algorithm; multi-criteria optimization under conflict and uncertainty; distributed computing; containerization; orchestration

1. Introduction

The report discusses the problem of multi-object multi-criteria control systems (MMS) optimizing under conflict and uncertainty in real time. In modern concepts of system analysis, a strict description of the MMS should take into account various types of uncertain factors: uncertainty of the goal, conflict uncertainty, uncertainty of environmental conditions. As we know, the above types of uncertainties can be most fully taken into account by gaming approaches based on the integration of various conflict optimality principles [1–4]. In particular, in [1], an approach based on the formation and study of stable-effective gaming compromise properties is being developed. Comparison of different approaches to the formation of stable-effective compromises (STEC) is an important principle of game-theoretic analysis of MMS control models, as well as a source of strict and, at the same time, meaningful reasoning about the motivations of the behavior of conflict participants arising from the structure of conflict models.

When solving applied game problems of STEC search under uncertainty (STECU), a number of problems arise. First, the STECU concept is based on the integration of various game-theoretic principles of optimality. Second, the need for real-time implementation of control algorithms, which often arises in applied problems, requires the representation of control actions in the general case in the form of parameterized program-corrected control laws. Such cases are characterized by a high dimension of criterion space and control parameters space, non-linearity, non-convexity, and the presence of break points of vector effectiveness indicators components of MMS subsystems, which determines the high computational complexity of optimization algorithms. These features, combined with the problem of global optimization, make it difficult or impossible to use well-known optimization methods and algorithms to search for a STECU in real time.

Currently, the Machine Learning Control (MLC) methodology, using machine learning methods to solve control problems of complex technical systems, is actively developing.

Citation: Serov, V.A.; Popova, D.L.; Rogalev, P.P.; Tararina, A.V. Neuro-Evolutionary Synthesis of Game Models of Control under Uncertainty Based on Distributed Computing Technology. *Eng. Proc.* **2023**, *33*, 59. https://doi.org/10.3390/engproc2023033059

Academic Editors: Askhat Diveev, Ivan Zelinka, Arutun Avetisyan and Alexander Ilin

Published: 25 July 2023

One of the main paradigms of MLC is the technology of neuro-evolutionary synthesis, which is considered as a promising means of implementing intelligent control algorithms under uncertainty in real time [5–7].

At the same time, the effectiveness of the neuro-evolutionary approach to the STECU formation is determined primarily by the capabilities of evolutionary algorithms:

- Complex accounting of the influence of the totality of these uncertain factors;
- Integration of the principles of optimality used to solve problems of the specified class;
- The use of advanced architectures, models and methods of distributed computing.

In [8–12], various aspects of improving of the evolutionary computational technology of multi-criteria optimization efficiency, as well as its development in the direction of coevolutionary algorithms constructing and their hybridization, are discussed. In [13,14], the development of coevolutionary technology for a class of multicriteria optimization problems under uncertainty is proposed. Game evolutionary algorithms are considered in [15–18]. In [19–21], an evolutionary computational technology is being developed that provides the possibility of combining various game-theoretic principles of optimality and takes into account various uncertain factors on a single conceptual and algorithmic basis in the task of MMS control optimization under conflict and uncertainty. This technology is implemented in the form of a library of evolutionary algorithms [20] and has been used to solve practical problems of the evolutionary synthesis of neuro-game algorithms of MMS control in real time based on STECU [22–24]. The analysis of the results allows us to draw the following main conclusions: the neuro-evolutionary technology of multicriteria control algorithms under conflict and uncertainty synthesis is effective; at the same time, the algorithms of game MMS control models neuro-evolutionary synthesis have extremely high computational complexity; the practical use of neuro-evolutionary technology for solving problems of the specified class requires its implementation on high-performance distributed computing architectures.

2. Main Methodology Stages of Game MMS Control Models under Uncertainty Neuro-Evolutionary Synthesis

The developed methodology of game MMS control models under uncertainty neuro-evolutionary synthesis is presented in the form of a structural scheme in Figure 1 and includes the following main stages.

Stage 1. Development of a library of hierarchical game models (HGM) of optimization under conflict and uncertainty, implementing the basic principles of conflict optimality for various types of conflict interaction, as well as providing the possibility of their integration in the construction of STECU and coordinated STECU (COSTECU). For this purpose, the problem of the MMS control optimizing under conflict and uncertainty is decomposed into the following problems:

- Local control under uncertainty (LCU);
- Distributed control under uncertainty (DCU);
- Hierarchical control under uncertainty (HCU).

The LCU problem is formalized as a problem of multi-criteria optimization under uncertainty (MCOU). To solve it, the principles of vector minimax and vector minimax regret are used. The hierarchical model of MCOU is considered in [24,25].

The DCU problem integrates game problem statements, that cover a wide range of types of conflict interaction and the corresponding principles of conflict equilibrium:

- Antagonistic interaction (principles of guaranteed result and saddle point);
- Non-coalition interaction (scalar equilibrium and vector Nash equilibrium, Ω—equilibrium);
- Coalition interaction (coalition equilibrium, equilibrium of threats and counter-threats, active equilibrium);
- Cooperative interaction (arbitration schemes, Pareto optimality principle).

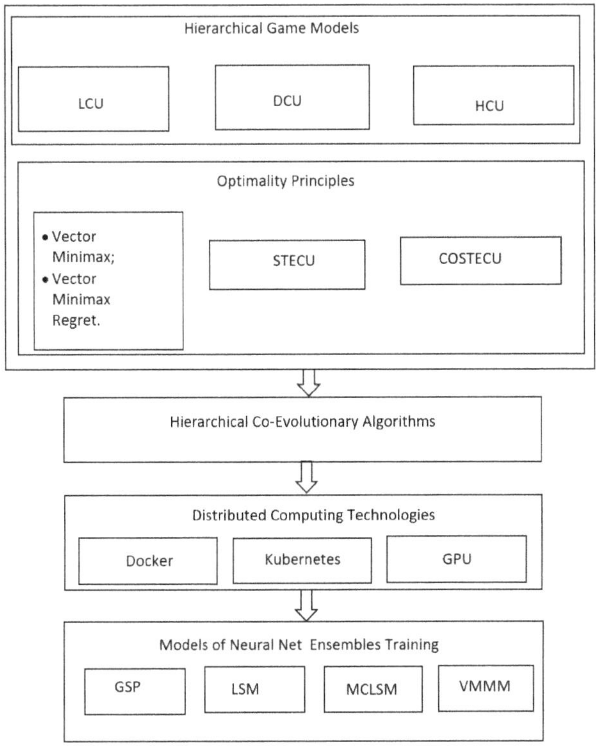

Figure 1. Structural components of the neuro-evolutionary synthesis methodology of game control models under uncertainty.

The hierarchical model of Nash vector equilibrium search under uncertainty is considered in [25].

The hierarchical control problem is formulated as a hierarchical game with the right of the first move. To solve it, the principles of scalar and vector Stackelberg equilibrium are used.

At the same time, all principles of conflict optimality are interpreted, taking into account the uncertainty of environmental conditions.

Stage 2. Development of a library of multi-criteria optimization hierarchical coevolutionary algorithms (HCEA) under conflict and uncertainty. HCEA's structure corresponds to the structure of hierarchical game models of optimization. This provides a flexible algorithm adjustment for the realization of conflict interaction of various types and conflict optimality principles as well as their integration in the formation of STECU and COSTECU. In addition, the hierarchical structure of coevolutionary algorithms best corresponds to the capabilities of distributed computing technologies. All this together provides a significant synergistic effect.

Stage 3. Development of a software package for the synthesis of neuro-gaming control models based on the HCEA library, Docker and Kubernetes [26,27] platforms, graphics processors [28,29], a library of neural network ensemble learning models (NNE). The Docker platform is used for the development, deployment and launch of container applications. The Kubernetes platform is a tool for scaling, managing and coordinating the functioning of containerized applications in a cluster environment. Figure 2 shows the generalized architecture of the neuro-evolutionary synthesis (NES) hierarchical control models (HCM) software, which includes the following structural and functional components.

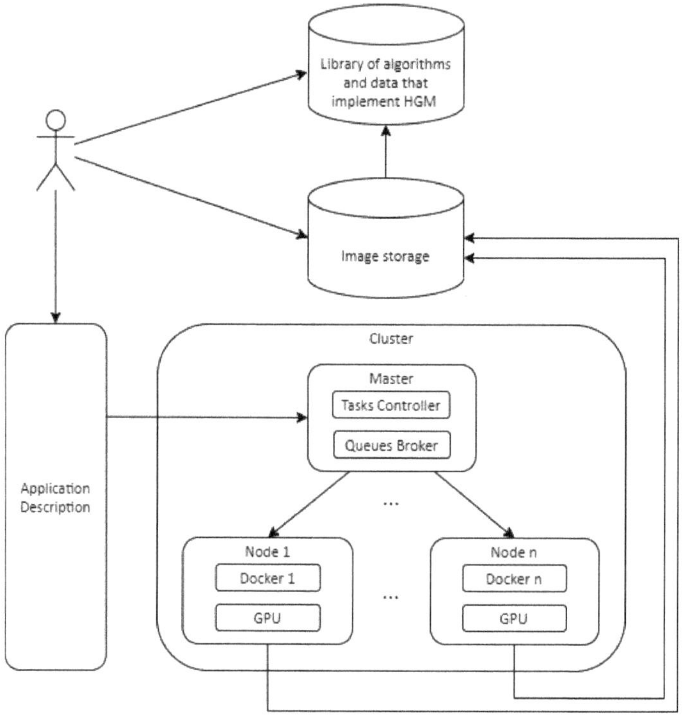

Figure 2. Architecture of the developed software.

Cluster. The Kubernetes cluster includes a master node (master), which provides basic Kubernetes services and orchestrates subordinate (worker) nodes that execute various components of the application. Important components of the master node are the tasks controller, which manages the separation of threads, and the queues broker, which acts as an intermediary between the main and working nodes of the cluster. When solving the problem of integrating the principles of optimality included in the structure of the STECU, the infrastructure of the software environment can be organized in the form of several clusters. To accomplish this, Kubernetes provides a mechanism for organizing a cluster federation.

A Docker container is a set of processes isolated from the main operating system. Applications work only inside containers and do not have access to the main system, except for explicitly connected directories when the container is launched (in this case, GPU—Graphic Processing Unit).

Library of HGM:

$$\Gamma = \{\Gamma_i, i = \overline{1, |\Gamma|}\}. \tag{1}$$

Image Storage:

$$R = \{R_i, i = \overline{1, |R|}\}. \tag{2}$$

Each of HGM Γ_i in the image storage corresponds to an image, R_i, a set of control instructions, according to which a container is formed from the library of algorithms and data that implements the architecture of Γ_i.

Description of applications (STECU):

$$S = \{S_i, i = \overline{1, |S|}\}. \quad (3)$$

The description of each application S_i characterizes the structure of the corresponding container, the composition and the number of replications that must be performed in parallel when implementing the STECU.

Library of Neural Network Ensemble Training Models (NNE). It is used to solve the problem of synthesis of NNE implementing multi-criteria control game algorithms under conflict and uncertainty in real time. The peculiarity of this library is that it presents single-criteria and multi-criteria statements of NNE training tasks: the least squares method (LSM), the multi-criteria least squares method (MLSM), multi-criteria optimization of a general kind, and multi-criteria optimization under uncertainty. Various models of training sets formation oriented on solving game control optimization problems are also presented.

3. Conclusions

The methodology of neuro-evolutionary synthesis of MMS game control models under conflict and uncertainty has been developed. Within the framework of the developed methodology, the following tasks have been solved.

A library of HGM for optimizing the MMS control under conflict and uncertainty has been developed, which allows us to form various schemes for integrating the principles of conflict optimality when a STECU forms on a unified algorithmic basis.

The HCEA library of multi-criteria optimization under conflict and uncertainty has been developed. The hierarchical structure of evolutionary algorithms best corresponds to the capabilities of distributed computing technologies, and in this sense can be considered as a means for structural meta-optimization of coevolutionary parallel algorithms.

A library of NNE training models oriented on solving of multi-criteria conflict control under uncertainty problems has been developed.

Author Contributions: Conceptualization and methodology, V.A.S. and P.P.R.; software architecture, D.L.P. and A.V.T. All authors have read and agreed to the published version of the manuscript.

Funding: This research received no external funding.

Institutional Review Board Statement: Not applicable.

Informed Consent Statement: Not applicable.

Data Availability Statement: Not applicable.

Conflicts of Interest: The authors declare no conflict of interest.

References

1. Voronov, E.M. *Methods of Optimization of Control of Multi-Object Multi-Criteria Systems on the Basis of Stable-Effective Game Solutions*; Egupov, M.D., Ed.; BMSTU: Moscow 2001; 576p.
2. Zhukovsky, V.I.; Zhukovskaya, L.V. *Risk in Multi-Criteria and Conflict Systems under Uncertainty*; Molostvov, M.V.S., Ed.; Editorial URSS: Moscow, Russia, 2004; 272p.
3. Moya, S. The calculation of the Stackelberg–Nash equilibrium as a fixed point problem in static hierarchical games. *Int. J. Dyn. Control.* **2018**, *6*, 907–918. [CrossRef]
4. Ungureanu, V. *Pareto-Nash-Stackelberg Game and Control Theory: Intelligent Paradigms and Applications*; Smart Innovation, Systems and Technologies; Springer: Berlin/Heidelberg, Germany, 2017; 579p, ISBN 13978-3319751504.
5. Kim, E.J.; Perez, R.E. Neuroevolutionary Control for Autonomous Soaring. *Aerospace* **2021**, *8*, 267. [CrossRef]
6. Bernas, M.; Płaczek, B.; Smyła, J. A Neuroevolutionary Approach to Controlling Traffic Signals Based on Data from Sensor Network. *Sensors* **2019**, *19*, 1776. [CrossRef] [PubMed]
7. Salichon M.; Tumer K. A neuro-evolutionary approach to micro aerial vehicle control. In Proceedings of the 12th Annual Genetic and Evolutionary Computation Conference (GECCO'10), Portland, OR, USA, 7–11 July 2010; pp. 1123–1130. [CrossRef]
8. Arias-Montano, A.; Coello Coello, C.; Mezura-Montes, E. Multiobjective evolutionary algorithms in aeronautical and aerospace engineering. *IEEE Trans. Evol. Comput.* **2012**, *16*, 662–694. [CrossRef]

9. Tan, K.C. A Cooperative Coevolutionary Algorithm For Multiobjective Optimization. In Proceedings of the IEEE International Conference on Systems, Man and Cybernetics, The Hague, The Netherlands, 10–13 October 2004.
10. Goh, C.K. Competitive-Cooperation Coevolutionary Paradigm for MultiObjective Optimization. In Proceedings of the 22nd IEEE International Symposium on Intelligent Control, Singapore, 1–3 October 2007; pp. 255–260.
11. Zeng, F. Studies on Pareto Based MultiObjective Competitive Coevolutionary Dynamics. In Proceedings of the IEEE Congress on Evolutionary Computation, New Orleans, LA, USA, 5–8 June 2011.
12. Dirita, V. *Control System Design Applications with Hybrid Genetic Algorithms*; University of Tasmania: Hobart, TAS, Australia, 2002.
13. Serov, V.A. Hierarchical Population Game Models of Coevolution in Multi-Criteria Optimization Problems under Uncertainty. *Appl. Sci.* **2021**, *11*, 6563. [CrossRef]
14. Coelho, R. Co-evolutionary optimization for multi-objective design under uncertainty. *J. Mech Des. ASME* **2013**, *135*, 1–8.
15. Greiner, D.; Periaux, J.; Emperador, J.; Galván, B.; Winter, G. Game Theory Based Evolutionary Algorithms: A Review with Nash Applications in Structural Engineering Optimization Problems. *Arch. Comput. Methods Eng.* **2017**, *24*, 703–750. [CrossRef]
16. Lung, R.I.; Dumitrescu, D. Computing Nash equilibria by means of evolutionary computation. *Int. J. Comput. Commun.* **2008**, *3*, 364–368.
17. D'Amato, E.; Daniele, E.; Mallozzi, L.; Petrone, G. Equilibrium strategies via GA to Stackelberg games under multiple follower's best reply. *Int. J. Intell. Syst.* **2012**, *27*, 74–85. [CrossRef]
18. Periaux, J.; Gonzalez, F.; Lee, D. Multi-Objective EAs and Game Theory. In *Evolutionary Optimization and Game Strategies for Advanced Multi-Disciplinary Design*; Intelligent Systems, Control and Automation: Science and Engineering; Springer: Dordrecht, The Netherlands, 2015; Volume 75, pp. 21–38.
19. Serov, V.A.; Voronov, E.M.; Kozlov, D.A. A neuroevolutionary synthesis of coordinated stable-effective compromises in hierarchical systems under conflict and uncertainty. *Procedia Comput. Sci.* **2021**, *186*, 257–268. [CrossRef]
20. Serov, V.A. Genetic algorithms of conflict equilibriums-based multicriteria systems control optimization under uncertainty. *Vestnik BMSTU. Ser. Instrum. Mak.* **2007**, *4*, 70–80.
21. Serov, V.A.; Voronov, E.M. Evolutionary Algorithms of Stable-Effective Compromises Search in Multi-object Control Problems. In *Studies in Systems, Decision and Control. Smart Electromechanical Systems*; Gorodetskiy, A., Tarasova, I., Eds.; Springer: Berlin/Heidelberg, Germany, 2019; Volume 174, pp. 19–29. [CrossRef]
22. Serov, V.A.; Babintsev, Y.N.; Kondakov, N.S. *Neurocontrol of Multicriteria Conflict Systems*; MosGU: Moscow, Russia, 2011; 136p.
23. Serov, V.A.; Voronov, E.M.; Kozlov, D.A. Hierarchical Neuro-Game Model of the FANET based Remote Monitoring System Resources Balancing. In *Studies in Systems, Decision and Control*; Smart Electromechanical Systems, Situational Control; Gorodetskiy, A., Tarasova, I., Eds.; Springer: Berlin/Heidelberg, Germany, 2020; Volume 261, pp. 117–130. [CrossRef]
24. Serov, V.A.; Voronov, E.M.; Kozlov, D.A. Hierarchical Population Game Models of Machine Learning in Control Problems Under Conflict and Uncertainty. In *Smart Electromechanical Systems*; Studies in Systems, Decision and Control; Gorodetskiy, A.E., Tarasova, I.L., Eds.; Springer: Cham, Switzerland, 2022; Volume 419, pp. 125–145.: 10.1007/978-3-030-97004-810. [CrossRef]
25. Serov, V.A., Voronov E.M. A hierarchical model of vector Nash equilibrium search in a control problem under conflict and uncertainty. *Eng. Proc.* **2023**, *33, in press*.
26. Lukcha, M. *Kubernetes in Action*; DMK Press: Moscow, Russia, 2019; 672p. ISBN 978-5-97060-657-5.
27. Mouat, A. *Using Docker*; DMK Press: Moscow, Russia, 2017; 354p. ISBN 978-5-97060-426-7.
28. Khronos OpenCL Working Group. OpenCL API Specification. Available online: https://www.khronos.org/registry/OpenCL/specs/3.0-unified/html/OpenCL_API.html (accessed on 19 November 2022).
29. Nvidia Corporation. CUDA Toolkit Documentation v11.6.2. Available online: https://docs.nvidia.com/cuda/ (accessed on 24 March 2022).

engineering proceedings

Proceeding Paper

Artificial Neural Networks Multicriteria Training Based on Graphics Processors †

Vladimir A. Serov *, Evgenia L. Dolgacheva, Elizaveta Y. Kosyuk, Daria L. Popova, Pavel P. Rogalev and Anastasia V. Tararina

Department of Applied Information Technologie, MIREA—Russian Technological University (RTU MIREA), Moscow 119454, Russia; evgeniadolgacheva@yandex.ru (E.L.D.); kosyuk.ey@mail.ru (E.Y.K.); pdl13@yandex.ru (D.L.P.); radugapp@mail.ru (P.P.R.); nharvard2013@gmail.com (A.V.T.)

* Correspondence: ser_off@inbox.ru

† Presented at the 15th International Conference "Intelligent Systems" (INTELS'22), Moscow, Russia, 14–16 December 2022.

Abstract: The report considers the task of training a multilayer perceptron, formulated as a problem of multiobjective optimization under uncertainty. To solve this problem, the principle of vector minimax was used. A parallel software implementation of a hierarchical evolutionary algorithm for solving a multicriteria optimization problem under uncertainty based on a GPU is presented.

Keywords: artificial neural network; multicriteria optimization under uncertainty; vector minimax; parallel computing; GPU

Citation: Serov, V.A.; Dolgacheva, E.L.; Kosyuk, E.Y.; Popova, D.L.; Rogalev, P.P.; Tararina, A.V. Artificial Neural Networks Multicriteria Training Based on Graphics Processors . *Eng. Proc.* **2023**, *33*, 57. https://doi.org/10.3390/engproc2023033057

Academic Editors: Askhat Diveev, Ivan Zelinka, Arutun Avetisyan and Alexander Ilin

Published: 25 July 2023

1. Introduction

Currently, the technology of the neuroevolutionary synthesis of management and decision-making models is being intensively developed, which is considered as a promising means of implementing intelligent algorithms for analyzing information and management under conflict and uncertainty in real time [1–5]. The effectiveness of the neuroevolutionary approach for solving this class of problems is determined by the ability to take into account uncertain factors, such as conflict uncertainty, the multicriteria of management goals, and the uncertainty of environmental conditions. In this context, it is expedient to formalize the task of training an artificial neural network (ANN) in the form of a multicriteria optimization problem under uncertainty (MCOU). In [6,7], a coevolutionary technology for solving the MCOU problem was developed, which, as the results of computational experiments show, has an extremely high computational complexity. In [8–12], it was shown that a promising area of research is the parallel implementation of computing technology based on graphics processors (GPUs). In this article, a parallel GPU implementation of a coevolutionary technology for solving the MCOU problem is proposed.

In Section 2, the formulation of the ANN training problem is formulated in the form of an MCOU problem, where the principle of vector minimax is applied for its solution. Section 3 presents a parallel GPU-based implementation of the MCOU hierarchical evolutionary algorithm software. Section 4 presents the results of a computational experiment on a test problem.

2. Problem Statement

The statement of the training problem for a multilayer perceptron (MP) is formulated as an MCOU problem:

$$\langle \mathbf{W}, \mathbf{Z}, \mathbf{F}(\mathbf{w}, \mathbf{z}) \rangle, \tag{1}$$

where $\mathbf{w} \in \mathbf{W} \subset \mathbf{E}^{r_w}$ is a vector of weight coefficients of MP synaptic connections; $\mathbf{z} \in \mathbf{Z} \subset \mathbf{E}^{r_z}$ is a vector of uncertain factors; $\mathbf{Z}$ is a finite set of possible values of the uncertain factor; and $\mathbf{F(w,z)} = [f_1(\mathbf{w,z}), \ldots f_m(\mathbf{w,z})]^T \in \mathbf{E}^m$ is a vector criterion defined on the Cartesian product $\mathbf{W} \times \mathbf{Z}$.

In problem (1), it is required to determine the value of the vector $\mathbf{w} \in \mathbf{W}$, which provides the minimum values for the components of the vector criterion **F(w,z)** under the influence of an uncertain factor $\mathbf{z} \in \mathbf{Z}$, about which it is known only that it can take values from a finite set **Z**.

To solve this problem, it is proposed to use the vector minimax principle. In this case, the original statement of problem (1) is reduced to a deterministic multiobjective optimization problem:

$$\mathbf{V(w)} \rightarrow \min_{\mathbf{w} \in \mathbf{W}}, \quad (2)$$

where **V(w)** is a vector indicator, the components of which are the points of extreme pessimism of the vector criterion **F(w,z)** on the set **Z** with fixed **w**.

To solve problem (1), (2), the hierarchical evolutionary algorithm (HEA) of the MCOU developed in [6,7] is used. As studies [5,6] show, the HEA MCOU, when used in ANN training tasks, shows a high computational complexity. Therefore, it is proposed to implement the HEA software for solving problem (1), (2), based on the GPU architecture and OpenCL technology.

3. GPU-Based Parallel Implementation of the Hierarchical Evolutionary MCOU Algorithm

The architecture of the developed HEA MCOU software reflects the following main stages of the HEA MCOU implementation.

Stage 1. Formation of a set of points of extreme pessimism (Figure 1).

Step 1. The initial population **W**, $|\mathbf{W}| = n$ is formed on the host (CPU).

Step 2. Constant memory is allocated on the GPU, into which arrays **Z** and **W** are entered. A buffer is allocated in the global memory of the GPU for the set of points of extreme pessimism **V(W)**.

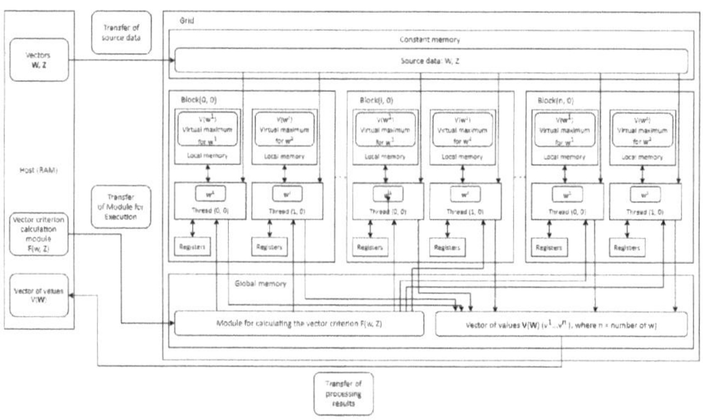

Figure 1. Parallel algorithm for finding a set of points of extreme pessimism on the GPU.

Step 3. A grid is formed on the GPU that determines the number of working blocks and the threads executed in them.

Step 4. The kernel is called with a subsequent transfer from the CPU to each thread of a set of instructions for execution. Threads start to work in parallel. Within each thread, the corresponding set of values of the vector criterion **F(w,Z)** is calculated for each $\mathbf{w} \in \mathbf{W}$ and the extreme pessimism point **V(w)** is calculated on the set **F(w,Z)**, which is stored in the local memory of the thread. Upon completion, each thread transfers its value **V(w)** to the global memory of the GPU. After the GPU has signaled that all threads have terminated, the CPU moves the array **V(W)** from the GPU's global memory to the CPU's RAM.

Stage 2. Assessment of the fitness of each point $\mathbf{w} \in \mathbf{W}$ (Figure 2).

Step 5. Constant memory is allocated on the GPU, into which array **V(W)** is entered. In the GPU's global memory, a buffer is allocated for the set of values of the fitness function $\Phi(\mathbf{V(W)})$.

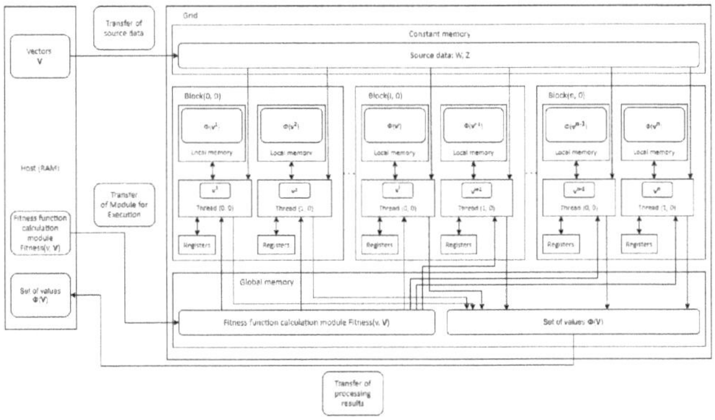

Figure 2. Parallel algorithm for calculating fitness function values on GPU.

Step 6. A new grid is formed on the GPU.

Step 7. The kernel is called and threads are started to work in parallel. Within each thread, the corresponding value of the fitness function Φ**(V)** is calculated for each element of $\mathbf{V} \in \mathbf{V(W)}$, which is stored in the local memory of the thread. Upon completion, each thread transfers its Φ**(V)** value to the GPU's global memory. After the GPU sends a signal to terminate all threads, the CPU transfers array Φ**(V(W))** from the GPU's global memory to the CPU's RAM.

Next, a population of descendants is formed on the CPU and the execution of stages 1, 2 is repeated.

The developed algorithm can be easily modified to solve the MCOU problem (1), (2), where there are many uncertain factors.

The developed software is cross-platform, as CUDA and OpenCL technologies are available on various operating systems, both on Windows and Linux.

4. Computational Experiment

The effectiveness of the developed technology was tested on the following test task MCOU:

$$\Gamma = \langle \mathbf{X}, \mathbf{Z}, \mathbf{F(x,z)} \rangle, \tag{3}$$

where $\mathbf{x} = [x_1, x_2]^T \in \mathbf{X}$ is the vector of control parameters; $\mathbf{z} = [z_1, z_2]^T \in \mathbf{Z}$ is the vector of uncertain factors; and $\mathbf{F(x,z)} = [f_1\mathbf{(x,z)}, f_2\mathbf{(x,z)}]^T$ is the vector performance indicator with components:

$$f_1\mathbf{(x,z)} = x_1^2 + x_2^2 - x_1(z_1^2 - z_2^2), \tag{4}$$

$$f_2\mathbf{(x,z)} = x_1^2 - x_2^2 - x_1(z_1^2 + z_2^2). \tag{5}$$

The restrictions were set in the form:

$$\mathbf{X} = \{0 \le x_1, x_2 \le 2\}, \tag{6}$$

$$\mathbf{Z} = \{\mathbf{z}^i, i = \overline{1, |\mathbf{Z}|} |\ 0 \le z_1, z_2 \le 2\}. \tag{7}$$

It is required to maximize the components of the vector efficiency indicator on the set $\mathbf{X} \times \mathbf{Z}$ based on the vector maximin principle.

Figures 3–5 show the results of searching for a set of vector maximins using the HEA MCOU (elite points in each generation are highlighted in red). Algorithm parameters: population cardinality $|\tilde{\mathbf{X}}| = 1000$; $|\mathbf{Z}| = 1000$; real coding and SBX-crossover were used.

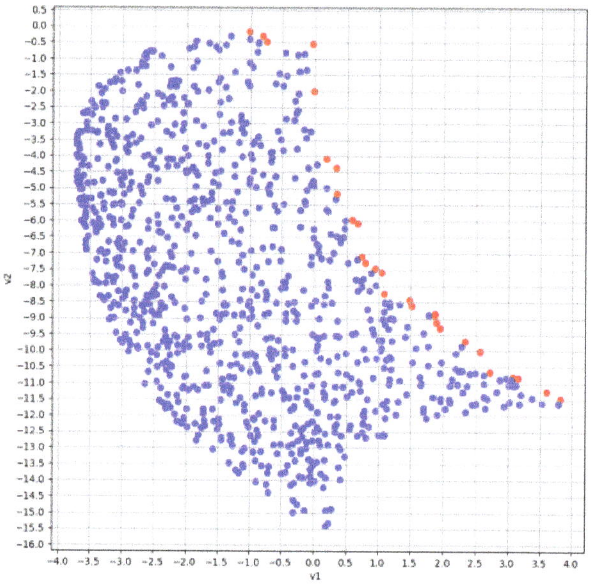

Figure 3. Evolutionary MCOU algorithm, generation No. 1. Elite points in each generation are highlighted in red.

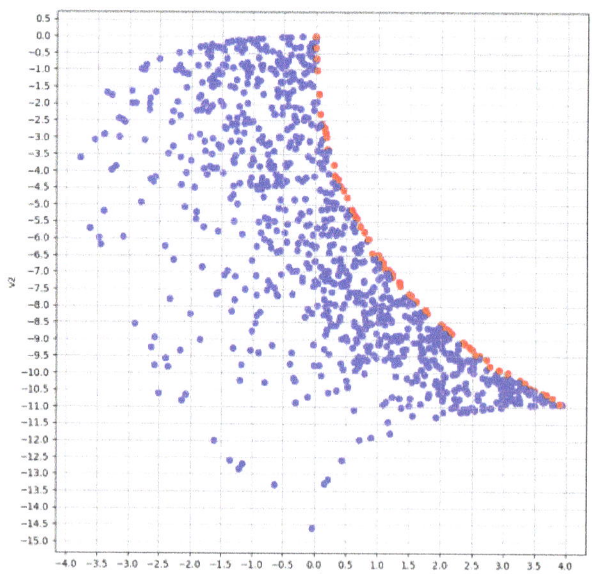

Figure 4. Evolutionary MCOU algorithm: generation No. 5.

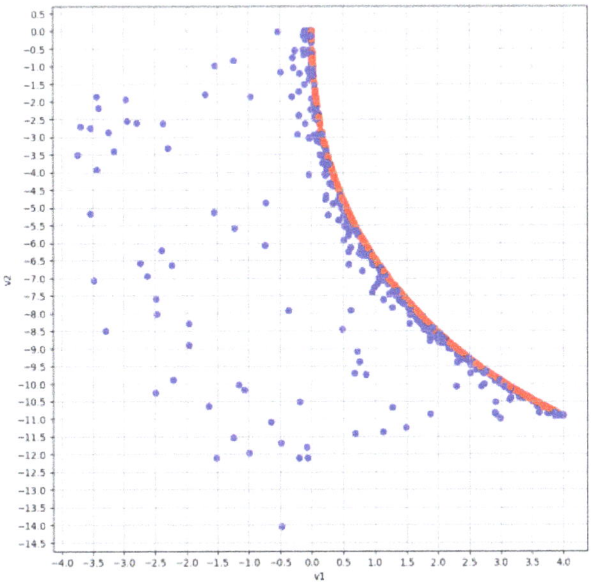

Figure 5. Evolutionary MCOU algorithm: generation No. 10.

Table 1 provides a comparative analysis of the running time of sequential and parallel evolutionary algorithms for solving the considered MCOU test task.

Table 1. Comparative analysis of sequential and parallel MCOU algorithms.

Population Size $\|\tilde{\mathbf{X}}\|$	Running Time of the Parallel Algorithm t_{par}, s	Running Time of the Sequential Algorithm t_{seq}, s
10	0.1353302	0.0003806
50	0.1354179	0.0049311
100	0.1397946	0.0129666
500	0.1476841	0.2834744
1000	0.1611041	1.0414738
5000	0.2373939	26.2836442
10,000	0.3495695	104.6257208
50,000	0.3617121	2623.5825713
100,000	0.6700136	10,134.2378412

A comparative analysis shows that, with a small population size $|\tilde{\mathbf{X}}| \leq 500$, the running time of the parallel evolutionary algorithm MCOU t_{par} is greater than or comparable to the running time of the sequential algorithm t_{seq}. This is due to the fact that the parallel algorithm spends additional time preparing and transferring data to the GPU. However, with a further increase in the size of populations, the advantage of the parallel evolutionary MCOU algorithm in relation to the sequential analog increases. In particular, for $|\tilde{\mathbf{X}}| = 100,000$, the running time of the parallel evolutionary MCOU algorithm is $t_{par} \cong 10^{-4} t_{seq}$.

5. Conclusions

The formulation of the MP training problem was formalized as an MCOU problem, where the vector minimax principle was used for its solution.

A parallel implementation of a hierarchical evolutionary algorithm for searching for a set of vector minimaxes in the MCOU problem based on GPU and OpenCL technology is presented. The developed algorithm can be easily modified to solve the MCOU problem (1), (2), where the set of uncertain factors is infinite.

The results of the computational experiment on the test task show a significant advantage of the parallel GPU implementation of the developed co-evolutionary MCOU algorithm in relation to the sequential analog.

Author Contributions: Conceptualization and methodology, V.A.S.; software and validation, D.L.P. and P.P.R.; computational experiments, E.L.D., E.Y.K. and A.V.T. All authors have read and agreed to the published version of the manuscript.

Funding: This research received no external funding.

Institutional Review Board Statement: Not applicable.

Informed Consent Statement: Not applicable.

Data Availability Statement: Not applicable.

Conflicts of Interest: The authors declare no conflict of interest.

References

1. Kim, E.J.; Perez, R.E. Neuroevolutionary Control for Autonomous Soaring. *Aerospace* **2021**, *8*, 267. [CrossRef]
2. Bernas, M.; Płaczek, B.; Smyła, J. A Neuroevolutionary Approach to Controlling Traffic Signals Based on Data from Sensor Network. *Sensors* **2019**, *19*, 1776. [CrossRef] [PubMed]
3. Salichon, M.; Tumer, K. A neuro-evolutionary approach to micro aerial vehicle control. In Proceedings of the 12th Annual Genetic and Evolutionary Computation Conference (GECCO'10), Portland, OR, USA, 7–11 July 2010; pp. 1123–1130. [CrossRef]
4. Serov, V.A.; Voronov, E.M.; Kozlov, D.A. A neuroevolutionary synthesis of coordinated stable-effective compromises in hierarchical systems under conflict and uncertainty. *Procedia Comput. Sci.* **2021**, *186*, 257–268. [CrossRef]
5. Serov, V.A.; Voronov, E.M.; Kozlov, D.A. Hierarchical Neuro-Game Model of the FANET based Remote Monitoring System Resources Balancing. In *Studies in Systems, Decision and Control. Smart Electromechanical Systems. Situational Control*; Gorodetskiy A., Tarasova I., Eds.; Springer International Publishing: Berlin/Heidelberg, Germany, 2020; Volume 261, pp. 117–130. [CrossRef]
6. Serov, V.A.; Voronov, E.M.; Kozlov, D.A. Hierarchical Population Game Models of Machine Learning in Control Problems Under Conflict and Uncertainty. In *Studies in Systems, Decision and Control. Smart Electromechanical Systems. Recognition, Identification, Modeling, Measurement Systems, Sensors*; Gorodetskiy, A.E., Tarasova, I.L., Eds.; Springer: Cham, Switzerland, 2022; Volume 419, pp. 125–145. [CrossRef]
7. Serov, V.A. Hierarchical Population Game Models of Coevolution in Multi-Criteria Optimization Problems under Uncertainty. *Appl. Sci.* **2021**, *11*, 6563. [CrossRef]
8. Andión, J.M.; Arenaz, M.; Bodin, F.; Rodríguez, G.; Tourino, J. Locality-aware automatic parallelization for GPGPU with OpenHMPP directives. *Int. J. Parallel Program.* **2016**, *44*, 620–643. [CrossRef]
9. Chandrashekhar, B.N.; Sanjay, H.A. Performance Study of OpenMP and Hybrid Programming Models on CPU–GPU Cluster. In *Emerging Research in Computing, Information, Communication and Applications*; Springer: Singapore, 2019; pp. 323–337.
10. Chandrashekhar, B.N.; Sanjay, H.A. Srinivas, T. Performance Analysis of Parallel Programming Paradigms on CPU-GPU Clusters. In Proceedings of the 2021 International Conference on Artificial Intelligence and Smart Systems (ICAIS), Coimbatore, India, 25–27 March 2021; pp. 646–651.
11. Soyata, T. *GPU Parallel Program Development Using CUDA*; CRC Press: Boca Raton, FL, USA, 2018.
12. Karovič, V.; Kaźmierczakb, M.; Pankivb, O.; Górkiewiczb, M.; Zakharchukc, M.; Stolyarchukc, R. OpenCL and CUDA Comparison of MapReduce Performance on Distributed Heterogeneous Platform through Integration with Hadoop Cluster. In Proceedings of the CEUR Workshop Proceedings, IT&AS'2021: Symposium on Information Technologies & Applied Sciences, Bratislava, Slovakia, 5 March 2021; pp. 202–208.

engineering proceedings

MDPI

Proceeding Paper

Convolutional Neural Network Application to Automate the Process of Aliquoting Biosamples †

Sergey Khalapyan [1], Larisa Rybak [2,*], Anna Nozdracheva [3] and Tatyana Semenenko [3]

1 Department of Automated and Information Control Systems, Stary Oskol Technological Institute n.a. A.A. Ugarov NUST MISiS, 42, Makarenko Mkr., 309516 Stary Oskol, Russia; khalapyan@yandex.ru
2 Research Institute Robotics and Control Systems, Belgorod State Technological University n.a. V.G. Shukhov, 46, Kostukova Str., 308012 Belgorod, Russia
3 N.F. Gamaleya Federal Research Center for Epidemiology and Microbiology, Ministry of Health of Russia, 123098 Moscow, Russia; nozdracheva0506@gmail.com (A.N.); semenenko@gamaleya.org (T.S.)
* Correspondence: rlbgtu@gmail.com; Tel.: +7-4722-230530

† Presented at the 15th International Conference "Intelligent Systems" (INTELS'22), Moscow, Russia, 14–16 December 2022.

Abstract: To automate the aliquoting process, it is necessary to determine the required depth of immersion of the pipette into the blood serum. This paper presents the results of a study aimed at creating a vision system that makes it possible to determine the position and nature of the fractional interface based on the use of a convolutional neural network. As a result of training on photographic images of tubes ready for aliquoting, the neural network acquired the ability to determine the visible part of the tube, the upper fraction of its contents, and the fibrin strands with high accuracy, allowing the required pipette immersion depth to be calculated.

Keywords: automation; aliquoting; blood serum; image recognition; convolutional neural network; U-Net

Citation: Khalapyan, S.; Rybak, L.; Nozdracheva, A.; Semenenko, T. Convolutional Neural Network Application to Automate the Process of Aliquoting Biosamples. *Eng. Proc.* **2023**, *20*, 56. https://doi.org/10.3390/engproc2023033056

Academic Editors: Askhat Diveev, Ivan Zelinka, Arutun Avetisyan and Alexander Ilin

Published: 25 July 2023

1. Introduction

Biobanking is a system for collecting, processing, storing, and analyzing samples of biological materials and associated clinical information intended for scientific and biomedical research [1,2].

Currently, this area of biomedicine is actively developing, which requires the introduction of new technological solutions to ensure the quality of biosamples and their compliance with international standards [3]. The types of material stored in the biobank are diverse, but the most common and relatively easily accessible is blood serum.

The technological process of biobanking this type of biomaterial provides for obtaining whole blood samples, separating blood serum, storing it in low-temperature storage, and entering a set of significant information about material donors into a specialized database.

One of the necessary conditions for high-quality biobanking of blood serum is its aliquoting, that is, its division into portions of small volume (200–800 μL), which are then sent to low-temperature long-term storage [4–6].

Blood sampling is carried out in a plastic test tube with a coagulation activator sprayed on its inner wall. After coagulation, the resulting clot is compacted. To speed up retraction, blood tubes are subjected to centrifugation. As a result, the contents of the tube are separated. The lower fraction is a compacted clot and the upper one is blood serum, a material used for aliquoting, biobanking, and further research.

To aliquot the serum, a laboratory assistant uses a pipette to take it from the test tube. This operation requires care; when serum is taken near a clot, erythrocytes may be partially drawn into the pipette from its surface, and the sample will be contaminated. On the other hand, being too careful will result in a significant amount of serum remaining in the tube, resulting in fewer aliquots.

Aliquoting is the process of manipulating an accurately measured fraction of a sample (a volume of solution) taken for analysis that retains the properties of the main sample. The process of aliquoting the blood of various patients involves removing biological material from a tube with a separated fraction (Figure 1a) into smaller-volume tubes (Figure 1b). To do this, it is necessary to move the tube from the rack to the working area, where dosing into aliquots takes place (Figure 1c). The quality of sample preparation is ensured at all stages, including during aliquoting [7].

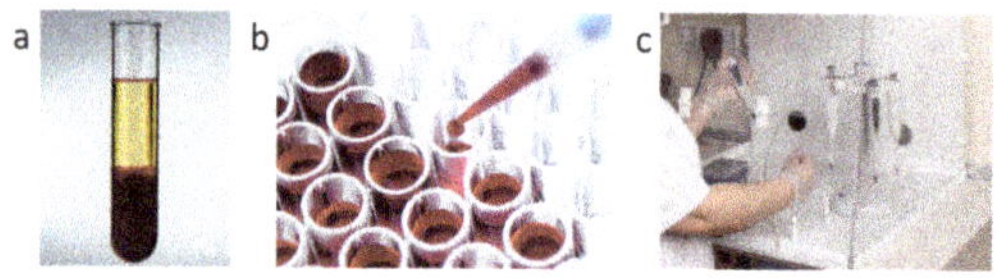

Figure 1. Aliquoting examples.

The results presented in [8] show the efficiency of automated aliquoting in comparison with the use of manual labor. Research in the field of application of robotic systems for liquid dosing and automation of laboratory processes for sample preparation is being carried out by many scientists. Thus, in [9], a technology for determining the level of liquid in a given volume based on a pressure sensor was proposed, making it possible to ensure the stability of the platform for processing liquid and the accuracy of its dosing. In [10,11], the authors proposed the use of a two-armed robot that works together with an operator and is remotely controlled.

There are already optimal programs to prepare for fully automated pipetting and dilution of biological samples in favor of controlled bioassays [12]. To better understand the problem, many typical solutions have been studied. The main developers of robotic equipment for aliquoting biological samples are Dornier-LTF (Germany), TECAN (Switzerland), and Hamilton. The laboratory equipment produced by them includes the PIRO robotic excavation station [13] manufactured by Dornier-LTF (Germany), the Freedom EVO® series robotic laboratory station [14] manufactured by TECAN, and the Microlab STAR automatic dosing robot system [15] by the Hamilton production company. All are designed to perform aliquoting technological processes, and include the same main elements: a robotic module with dosing multichannel pipettes and removable tips, a robotic module with a gripping device for carrying tripods, and auxiliary devices (pumps, barcode readers, thermo-shakers, UV emitters, work surface object position detectors, etc.), the number of which depends on the intended use of the robotic station. The differences between the listed equipment types lies only in the number of operations able to be performed simultaneously; structurally, the systems have no major differences. The structure of the mentioned systems is designed to separate a large number of aliquots of one biological fluid, which is most often used in microbiology and virology laboratories. There is a research gap when searching for new technical solutions in the field of robotic aliquoting and new approaches to organizing a vision system with the possibility of effective image segmentation while taking into account the heterogeneity of biosamples.

The present article proposes a new design of a robotic system (RS) consisting of two robots: a fast robot with a parallel delta structure and a collaborative robot with a sequential structure and a gripper. A dosing device is installed on the delta arm. The dosing device contains a vacuum system and an automatic tip change system. The device must have a small mass, as this significantly affects the speed of the delta robot. The mutual arrangement of the manipulators should ensure the intersection of the working spaces for access of both manipulators to the required objects [4]. The design of the proposed robotic system for aliquoting biological fluid of various patients is shown in Figure 2. The RS (Figure 2a) includes: the body (1), in which there is a parallel delta manipulator (2) that moves the dosing head (3) fixed in the center of the movable platform of the manipulator that performs biomaterial aliquoting. The replaceable tip (4) on the digging head is fixed with a rubber sealing ring. The UR3 manipulator (5) is mounted on a fixed base (6) and ensures the movement of the test tubes (8) within the working area (7) using a gripping device.

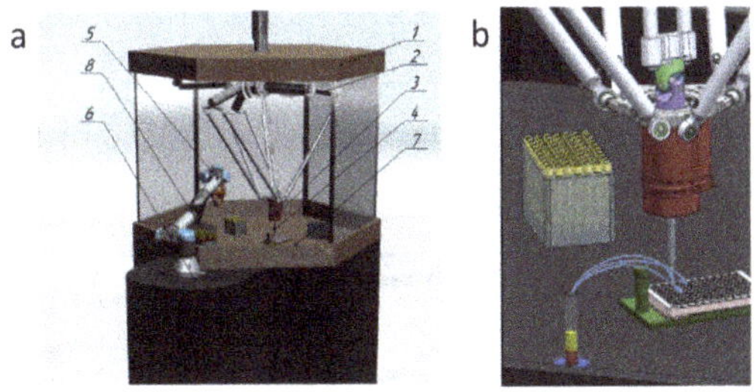

Figure 2. (**a**) 3D model of RS for aliquoting and (**b**) working body with dosing head.

The task of automating the aliquoting process depends on the need to determine the height of the fraction boundary in the test tube. This border can be at different levels, and can either be evenly horizontal (Figure 3a), irregularly shaped (Figure 3b) as a result of improper centrifugation, or even a lump of whitish fibrin filaments (Figure 3c), if the activator did not work well or the tube was shaken during transportation. Because the laboratory assistant currently determines this boundary "by eye", it is advisable to use the technical vision system proposed here instead.

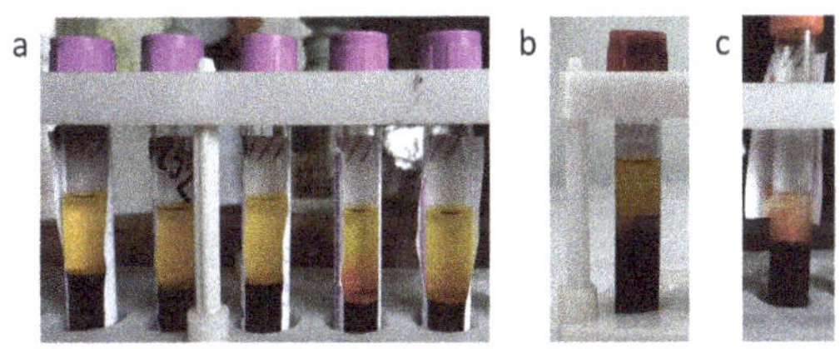

Figure 3. Examples of the boundary between fractions in vitro: (**a**) horizontal, (**b**) irregular, and (**c**) lump of fibrin threads.

Thus, the organization of such a system based on neural network image recognition, which is the goal of this study, is an urgent task.

2. Using U-Net for Image Segmentation

For a set of biomaterial in a pipette, it is necessary to correctly determine the levels at which the upper fraction begins and ends. To do this, it is proposed to photograph the test tube and determine the area corresponding to the serum in the photograph. At the same time, if the clot has an irregular shape or is covered with fibrin threads, determining the lower border of the serum from a photograph can be a nontrivial task. In addition, in order to avoid potentially drawing the threads into the pipette it is necessary to increase the margin on the lower level in the latter case (i.e., to leave more serum in the tube).

Deep learning neural networks developed to solve the problem of image segmentation are currently being used to successfully recognize objects in a photograph.

In this case, segmentation is the process of selecting segments of an image. The initial image is used as the input of a neural network and an image of the same size is formed at its output, with each of its pixels represented as a color that corresponds to the object that this pixel belonged to in the original image (Figure 4).

Figure 4. An example of the result of the operation of a segmenting neural network.

The most effective current neural networks for solving the problem under consideration are networks of the "encoder–decoder" class, such as U-Net [16–18]. Below, we consider the principles of its operation in more detail (Figure 5).

The original picture is subjected to a convolution operation with a 3 × 3 kernel (these are matrices of a specified size which are filled with random numbers when the network is initialized; after the network has been trained for specific pictures, they become means of highlighting essential features in them), indicated by the blue arrows in Figure 5. The number of such nuclei is 32. As a result, 32 feature maps are obtained.

Then, the feature maps, located one above the other, are again convolved with a 3 × 3 × 32 kernel. Again, there are 32 such nuclei. The result is 32 new maps with a clearer selection of features in the image.

Next, a dimensionality reduction operation (max-pooling) is applied with a 2 × 2 kernel, shown by the red down-arrows in Figure 5. This reduces the size of each feature map by half, allowing more abstract features in the picture to be highlighted, which increases the generalization ability of the network. The result is 32 feature maps of half the original size.

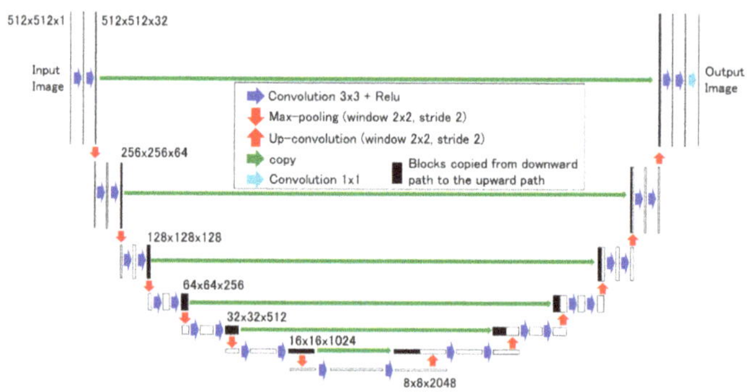

Figure 5. U-Net architecture.

Then, the convolution operation is applied with a kernel of size 3 × 3 × 32. There are 64 such convolution kernels. Therefore, the result is 64 feature maps. The feature maps located one above the other are convolved with a 3 × 3 × 64 kernel. There are 64 such nuclei. The result is 64 new feature maps.

Max-pooling operation and two convolutions are performed five more times in succession. The result is 2048 feature maps with linear dimensions 2^6 times smaller than the dimensions of the original image. Thus, the narrowing network pass (encoder) is completed (the left branch of the structure in Figure 5). Its result is the essential features extracted from the original image, which helps in making subsequent decisions about the presence of objects of any class. The forward pass is the essential part of any convolutional neural network architecture.

Next, the right decoder branch of the network starts working. Its main action is to apply upsampling (the red up-arrows in Figure 4), which has the opposite meaning to max-pooling. Each time this is used, the size of the image is doubled. Pixels that did not exist before are filled using the values of neighboring pixels. After the upsampling operation, the results of the direct network pass from the corresponding level are taken and combined with the resulting image (concatenation), indicated by the green arrows in Figure 5. Then, two additional convolution operations are performed.

As a result of the repetition of the four operations described above, the image is returned to its original size. There are as many nuclei in the last convolution as there are object classes in the source image under consideration.

At the same time, the presence of a group of pixels marked as "1" on a specific output map means that the corresponding object was found by the network in the image.

3. Formation of the Training Sample

To train a neural network, it is necessary to prepare a sample consisting of examples of the "original image–labeled image" type. A labeled image is a picture with the same size as the original image on which each object of interest is painted in its own color.

Image analysis (Figure 3) shows that three classes of objects are of greatest interest:

1. Object No. 1: serum, the upper fraction of the contents of the tube. The greatest difficulty here is the identification of the lower boundary.
2. Object No. 2: test tube. Using its linear dimensions on the image, it is possible to determine the scale of the photograph (Figure 6) and calculate the actual distance from the boundaries of the upper fraction of the contents of the test tube to its lower or upper edge. Selecting this object allows the position of the camera and the focal length of its lens to be changed as if necessary without the need for recalibration.
3. Object No. 3: fibrin threads on a clot (Figure 3c). First, the very fact of its presence/absence in the image is important, as this makes it possible to correctly determine the required distance from the working end of the pipette to the lower boundary of the upper fraction.

The upper fraction can contain erythrocyte hemolysis products. In this case, it turns out to be intensely pink (Figure 7) and is considered unsuitable for further research. With a sufficient number of relevant images, such an object could be recognized by the network; however, tubes showing signs of hemolysis are usually discarded in the early stages of sample handling.

Each photo selected in one of the samples (training, testing, or validation) must be labeled in such a way as to provide a "correct" answer for the neural network that is the goal in the learning process.

Figure 8 shows examples of image markup (cf. Figure 3).

To train the neural network, we used the stochastic optimization algorithm Adam (adaptive moment estimation) [19–21], which is currently one of the most popular methods for training networks of the " encoder–decoder" type. The learning rate α was set to 0.001 and the exponential decay rates $\beta 1$ and $\beta 2$ were set to 0.9 and 0.999, as suggested in [22].

The training itself is traditionally performed as follows: all pictures of the training sample are fed to the neural network, then the accumulated gradient is calculated for each of the parameters of the neural network. However, training on a full sample requires a significant amount of memory, including video memory on a video card.

Therefore, in the course of the present study an alternative method was used: the training sample was divided into minisamples (batches) and the network parameters were corrected after only examples of this sample were fed to the network. Reducing the batch volume leads to a decrease in the required amount of memory, although it slightly reduces the quality of training.

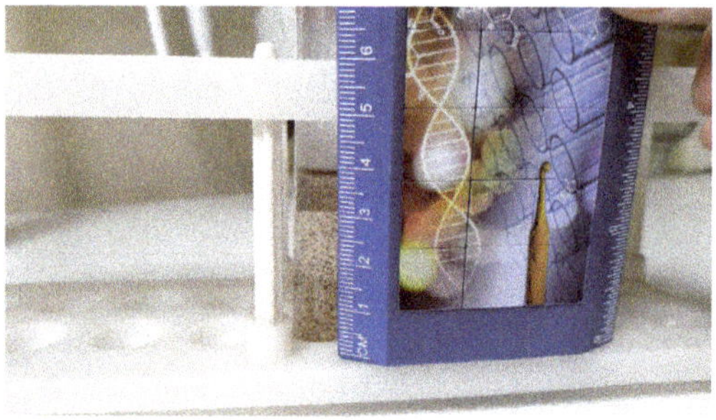

Figure 6. Determining the scale of the photograph by the size of the visible part of the test tube.

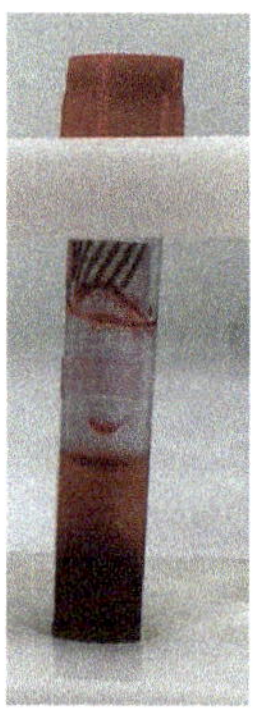

Figure 7. Result of hemolysis.

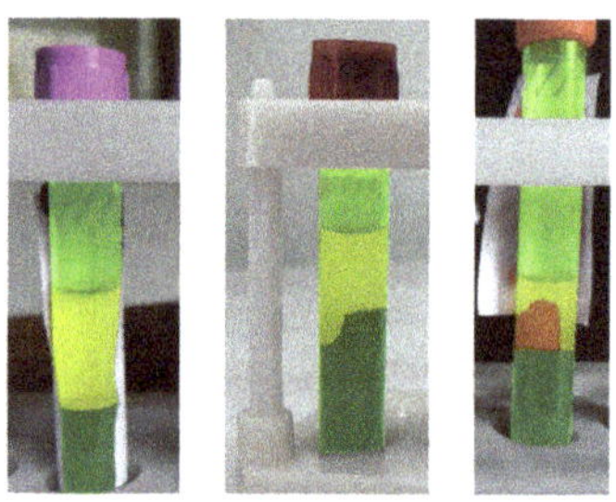

Figure 8. Image markup examples.

Categorical cross-entropy was applied as a loss function (objective learning function). The quality of the network was evaluated using the accuracy metric; that is, the output of the network consisting of three pictures (according to the number of classes of recognizable objects) was combined into a single picture, where in each pixel the color was the class of the object to which the pixel belonged. Then, the picture was compared pixel-by-pixel with the reference markup for the input picture and the percentage of pixels was determined, where the same class was used for both the reference and the output of the network.

4. Training of Neural Networks and Results Obtained

The size of the training sample was increased gradually. The network was first trained on 100 images without fibrin strands. On the test sample, it showed an accuracy of 95%. Figure 9 shows a typical result for one of the images of the test sample combined with the original image.

It can be seen from the figure that there are no serious problems with the recognition of the upper fraction of the contents of the tube, while the tube itself is often not completely recognized. Increasing the number of images in the test set to 200 made it possible to solve this problem, and the quality of recognition on the test set increased to 97% (Figure 10).

In most cases, errors resulted from inaccurate recognition of the vertical boundaries of the test tube, which is associated with glare and reflection in the plastic of paper labels. Such errors do not affect the accuracy when determining the scale of a photograph. In addition, there was an increase in the quality of recognizing the upper fraction. The corresponding figure calculated without taking other objects into account was close to 99%.

Adding 100 images with fibrin threads to the training set did not worsen the quality of recognition of objects No. 1 and No. 2; however, several cases of "phantom" objects being detected in images without fibrin threads (object No. 3) were recorded, and often these were not even inside the test tube (Figure 11). The test tube was successfully recognized in the photo, and objects outside it should not be paid attention to; however the problem of such phantom objects can be solved by simple organizational measures as well, such as by creating a solid background. Testing the network on images with threads showed that they were detected, but often not in full (Figure 12).

Figure 9. The result of the network trained on 100 fibrin-free images.

Figure 10. The result of the network trained on 200 images without fibrin.

An increase in the training set to 500 images (200 without threads and 300 with threads) led to an increase in the quality of recognition for all classes of objects. On the validation set, the recognition accuracy approached 99% (Figure 13), while for objects No. 2 and No. 3 it even exceeded this value.

It should be noted that, despite the detected error, the accuracy of determining the upper and lower edges of the visible part of the tube, which are necessary for calculating the image scale, was 100% based on the results of the network operation. In addition, for all images in the validation set the presence or absence of fibrin strands was correctly determined. As for the lower level of the upper fraction of the contents of the tube, it was accurately determined for all test tubes without filaments. For tubes with filaments the maximum error did not exceed 1 mm, which is sufficient to guarantee safe aliquoting considering the large headroom.

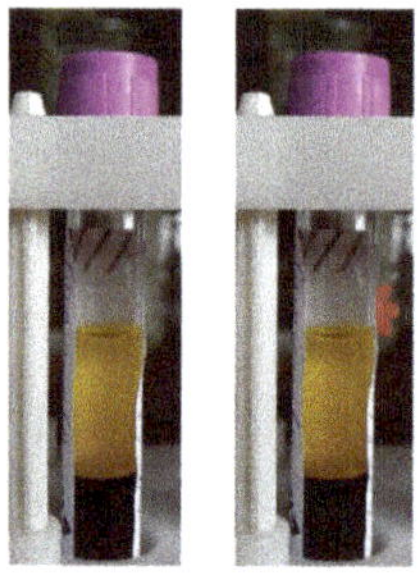

Figure 11. Phantom object outside the test tube.

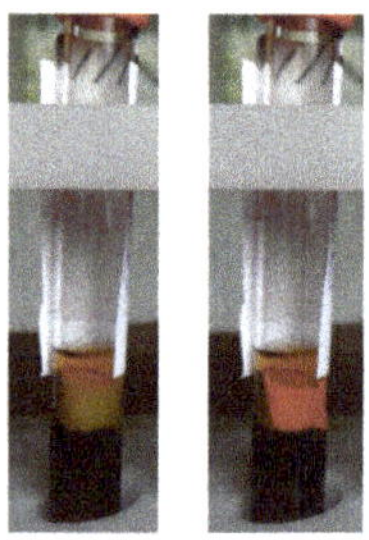

Figure 12. Partial recognition of fibrin strands.

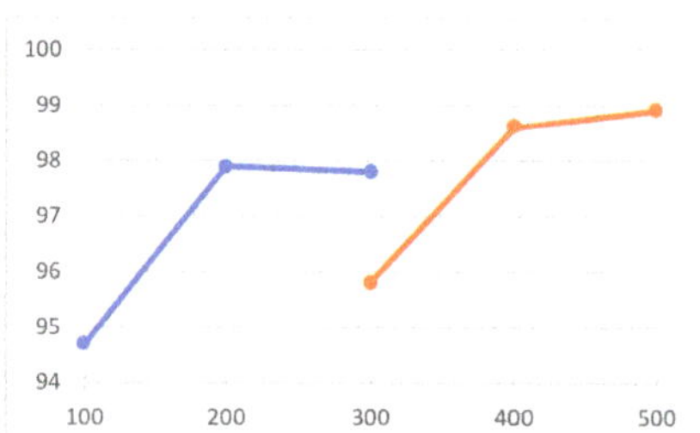

Figure 13. Recognition accuracy on the validation set depending on the volume of the training set: blue graph—validation set containing only test tubes without filaments; orange—validation set containing 50% tubes with filaments.

5. Conclusions

In the course of the study, a neural network was trained to determine the depth to which a pipette must be immersed in order to take serum aliquots based on photographic images of tubes containing blood serum and a clot. In this case, the nature of the fraction interface was determined and taken into account beforehand, making it possible to obtain the maximum number of aliquots while at the same time avoiding (or significantly reducing the likelihood of occurrence) situations such as contamination of an aliquot with erythrocytes drawn from the clot surface, clogging of the pipette tip with fibrin threads, etc. The resulting neural network showed high accuracy, and can be directly used to automate the aliquoting process.

Author Contributions: Conceptualization, S.K., L.R. and T.S.; methodology, S.K. and L.R.; software, S.K.; validation, L.R. and A.N.; formal analysis, A.N.; investigation, S.K. and A.N.; resources, A.N. and T.S.; data curation, T.S. and A.N.; writing—original draft preparation, S.K.; writing—review and editing, L.R. and A.N.; visualization, S.K.; supervision, L.R. and T.S.; project administration, L.R. All authors have read and agreed to the published version of the manuscript.

Funding: This research was supported by a state assignment of the Ministry of Science and Higher Education of the Russian Federation under Grant FZWN -2020-0017.

Institutional Review Board Statement: Not applicable.

Informed Consent Statement: Not applicable.

Data Availability Statement: The investigation was conducted as part of the work approved by the local ethics committee (Protocol No. 31 of 25 November 2022). At the same time, the bio-material used in the work was a "residual material" previously used for laboratory diagnostics from donor patients who gave voluntary informed consent in a medical and preventive organization. For the current investigation, only anonymous blood samples were used (there were no personal data such as full name, residential address, passport details, etc.). Identification of the persons from whom the biosamples were obtained is impossible, therefore their use cannot cause harm to the subjects of the investigation, including a violation of confidentiality. This ensures that the organization and conduct of the investigation comply with the provisions of the Guide for research ethics committee members, Council of Europe, 2010.

Conflicts of Interest: The authors declare no conflict of interest.

References

1. Anisimov, S.V.; Akhmerov, T.M.; Balanovsky, O.P.; Baranich, T.I.; Belyaev, V.E.; Borisova, A.L.; Bryzgalina, E.V.; Voronkova, A.S.; Glinkina, V.V.; Glotov, A.S.; et al. *Biobanking: National Guidelines: Prepared by Experts of the National Association of Biobanks and Biobanking Specialists*; Triumph: Moscow, Russia, 2022.
2. Pokrovskaya, M.S.; Borisova, A.L.; Sivakova, O.V.; Metelskaya, V.A.; Meshkov, A.N.; Shatalova, A.M.; Drapkina, O.M. Quality management in biobank. World tendencies and experience of biobank of FSI NMRC for Preventive Medicine of the Ministry of Healthcare of Russia. *Russ. Clin. Lab. Diagn.* **2019**, *64*, 380–384. [CrossRef] [PubMed]
3. Henderson, M.; Goldring, K.; Simeon-Dubach, D. Advancing professionalization of biobank business opera-tions: A worldwide survey. *Biopreserv. Biobank.* **2019**, *17*, 71–75. [CrossRef] [PubMed]
4. Malyshev, D.; Rybak, L.; Carbone, G.; Semenenko, T.; Nozdracheva, A. Optimal Design of a Parallel Manipulator for Aliquoting of Biomaterials Considering Workspace and Singularity Zones. *Appl. Sci.* **2022**, *12*, 2070. [CrossRef]
5. Voloshkin, A.; Rybak, L.; Cherkasov, V.; Carbone, G. Design of gripping devices based on a globoid transmission for a robotic biomaterial aliquoting system. *Robotica* **2022**, *40*, 4570–4585. [CrossRef]
6. Malyshev, D.; Rybak, L.; Carbone, G.; Semenenko, T.; Nozdracheva, A. Workspace and Singularity Zones Analysis of a Robotic System for biosamples Aliquoting. In *Advances in Service and Industrial Robotics, RAAD 2021 Mechanisms and Machine Science*; Springer: Cham, Switzerland, 2021; pp. 31–38.
7. Malm, J.; Fehniger, T.E.; Danmyr, P.; Végvári, Á.; Welinder, C.; Lindberg, H.; Marko-Varga, G. Developments in biobanking workflow standardization providing sample integrity and stability. *J. Proteom.* **2013**, *16*, 38–45. [CrossRef] [PubMed]
8. Malm, J.; Végvári, A.; Rezeli, M.; Upton, P.; Danmyr, P.; Nilsson, R.; Steinfelder, E.; Marko-Varga, G. Large scale biobanking of blood—The importance of high density sample processing procedures. *J. Proteom.* **2012**, *76*, 116–124. [CrossRef] [PubMed]
9. Zhang, C.; Huang, Y.; Fang, Y.; Liao, P.; Wu, Y.; Chen, H.; Chen, Z.; Deng, Y.; Li, S.; Liu, H.; et al. The Liquid Level Detection System Based on Pressure Sensor. *J. Nanosci. Nanotechnol.* **2019**, *19*, 2049–2053. [CrossRef] [PubMed]
10. Fleischer, H.; Baumann, D.; Joshi, S.; Chu, X.; Roddelkopf, T.; Klos, M.; Thurow, K. Analytical Measurements and Efficient Process Generation Using a Dual-Arm Robot Equipped with Electronic Pipettes. *Energies* **2018**, *11*, 2567. [CrossRef]
11. Fleischer, H.; Drews, R.; Janson, J.; Chinna, P.B.; Chu, X.; Klos, M.; Thurow, K. Application of a Dual-Arm Robot in Complex Sample Preparation and Measurement Processes. *J. Lab. Autom.* **2016**, *21*, 671–681. [CrossRef] [PubMed]
12. Jiang, H.; Ouyang, Z.; Zeng, J.; Yuan, L.; Zheng, N.; Mohammed, J.; Mark, E. A User-Friendly robotic sample preparation program for fully automated biological sample pipetting and dilution to benefit the regulated bioanalysis. *J. Lab. Autom.* **2012**, *17*, 211–221. [CrossRef] [PubMed]
13. INTEGRA Biosciences. Freeling You from Tedious Multichannel Pipetting Tasks. Available online: www.bionity.com/en/products/1128490/freeing-you-from-tedious-multichannel-pipetting-tasks.html (accessed on 11 September 2022).
14. Freedom EVO Platform. Available online: www.lifesciences.tecan.com/freedom-evo-platform (accessed on 11 September 2022).
15. Hamilton Broshure. Life Science Robotics. Available online: www.corefacilities.isbscience.org/wp-content/uploads/sites/5/2015/07/HamiltonBrochure.pdf (accessed on 11 September 2022).

16. Wagner, F.; Ipia, A.; Tarabalka, Y.; Lotte, R.; Ferreira, M.; Aidar, M.; Gloor, M.; Phillips, O.; Aragão, L. Using the U-net convolutional network to map forest types and disturbance in the Atlantic rainforest with very high resolution images. *Remote Sens. Ecol. Conserv.* **2019**, *5*, 360–375. [CrossRef]
17. Ronneberger, O.; Fischer, F.; Brox, T. U-Net: Convolutional Networks for Biomedical Image Segmentation. *arXiv* **2015**, arXiv:1505.04597.
18. Poleshchenko, D.A.; Glushchenko, A.I.; Fomin, A.V. Application of Pre-Trained Deep Neural Networks to Identify Cast Billet End Stamp before Heating. In Proceedings of the 2022 24th International Conference on Digital Signal Processing and Its Applications (DSPA), Moscow, Russia, 30 March–1 April 2022; pp. 1–5.
19. Kida, S.; Nakamoto, T.; Nakano, M.; Nawa, K.; Haga, A.; Kotoku, J.; Yamashita, H.; Nakagawa, K. Cone Beam Computed Tomography Image Quality Improvement Using a Deep Convolutional Neural Network. *Cureus* **2018**, *10*, e2548. [CrossRef] [PubMed]
20. Jais, I.K.M.; Ismail, A.R.; Nisa, S.Q. Adam optimization algorithm for wide and deep neural network. *Knowl. Eng. Data Sci.* **2019**, *2*, 41–46. [CrossRef]
21. Zhang, Z. Improved adam optimizer for deep neural networks. In Proceedings of the 2018 IEEE/ACM 26th International Symposium on Quality of Service (IWQoS), Banff, AB, Canada, 4–6 June 2018; pp.1–2.
22. Kingma, D.; Ba, J. Adam: A method for stochastic optimization. *arXiv* **2014**, arXiv:1412.6980.

engineering proceedings

MDPI

Proceeding Paper

A Hybrid Transdimensional Evolutionary Algorithm for Dynamical System Control Multicriteria Optimization †

Vladimir A. Serov [1,*], Evgeny M. Voronov [2], Evgenia L. Dolgacheva [1] and Elizaveta Y. Kosyuk [1]

1 Department of Applied Information Technologie, MIREA— Russian Technological University (RTU MIREA), Moscow 119454, Russia
2 Department of Control Systems, Bauman Moscow State Technical University (BMSTU, Bauman MSTU), Moscow 105005, Russia
* Correspondence: ser_off@inbox.ru
† Presented at the 15th International Conference "Intelligent Systems" (INTELS'22), Moscow, Russia, 14–16 December 2022.

Abstract: The article develops a new hybrid evolutionary algorithm for the optimal control law multicriteria synthesis of a dynamical system based on transdimensional search models. The transdimensional search model implies the combined usage of finite-dimensional and infinite-dimensional multicriteria optimization evolutionary algorithms, implementing the stages of the global approximate and local clarifying search for optimal solutions. A comparative analysis of the effectiveness of various hybrid transdimensional models of the evolutionary search is carried out for the problem of the bioreactor program control optimal law multicriteria synthesis. It is shown that the transdimensional hybridization of evolutionary algorithms using infinite-dimensional search models provides a higher solving accuracy of the problem.

Keywords: hybrid evolutionary algorithm; multicriteria optimization; transdimensional search model; dynamic system control

Citation: Serov, V. A.; Voronov, E. M.; Dolgacheva, E. L.; Kosyuk, E. Y. A Hybrid Transdimensional Evolutionary Algorithm for Dynamical System Control Multicriteria Optimization. *Eng. Proc.* **2023**, *33*, 58. https://doi.org/10.3390/engproc2023033058

Academic Editors: Askhat Diveev, Ivan Zelinka, Arutun Avetisyan and Alexander Ilin

Published: 25 July 2023

1. Introduction

One of the promising directions of improving the efficiency of solving global search problems is the development of hybrid metaheuristic algorithms (HMHA) and technology of HMHA adaptation to the peculiarities of the applied problem being solved. The main goal of the MHA hybridization is to ensure a balance of global and local search efficiencies and, as a result, to achieve a synergistic effect by combining the positive properties of the combined algorithms.

Currently, a significant number of MHA hybridization methods have been developed. In [1–3], the basic principles of MHA hybridization and classification features that are advisable to use in the construction and comparative analysis of various HMHAs are discussed. In particular, a one-level classification of hybridization models is proposed in [1]. According to those classifications, there are three types of hybrid algorithms: embedded algorithms, preprocessor/postprocessor algorithms and co-algorithms. In [2], a two-level classification is proposed, using the types of hybridized algorithms and decomposition types of the search space as taxonomic features. In [3], the most complete four-level classification of HMHAs is proposed, based on the following taxonomic features: a type of hybridized algorithm; a gibridization level; an execution order; and a control strategy.

In [4], based on a comparative analysis, it is shown that the HMHA classifications discussed above have some intersections, complement each other, and collectively cover almost all currently known HMHA models. Special attention is paid to the study of MHA hybridization possibilities for solving global optimization problems. it is shown that the sequential high-level embedded hybridization model (SHLEH-model) is widely used to solve global optimization problems. This model is usually based on a combination of global search population algorithms and local search algorithms (generally, not populational). A distinctive feature of the SHLEH-model of MHA is that the local search algorithm is an embedded component of the global search populational algorithm. With its help, each agent coordinates in the current population are improved. A new population is formed based on the improved coordinates of each agent in the current population. At the same time, a significant autonomy of the combinable algorithms takes a place. Currently, a large number of high-level embedded hybridization examples of populational algorithms with local optimization algorithmsare known. In particular, memetic algorithms are implemented based on SHLEH-models [5,6]; hybrid evolutionary algorithms [7,8]; hybrid continuous orthogonal ant colony algorithms using orthogonal search methods in the region selected for each ant [9,10]; modifications of a continuously interacting ant colony hybrid algorithm using a combination of global search by an ant algorithm with local search algorithms of a deformable Nelder–Mead polyhedron and dynamic simplex [11], etc.

Thus, hybrid metaheuristic search algorithms implemented on the basis of SHLEH-models generally provide high efficiency in solving global optimization problems. An important advantage of the SHLEH-models of MHAs is also the possibility of their effective parallelization.

However, all the above-mentioned HMHAs are fundamentally finite-dimensional and in their existing form do not allow us to effectively solve the problems of dynamical systems control multicriteria optimization, which in the original formulation are infinite-dimensional. This circumstance determines the relevance of the development of new HMHA-models, a priori focused on solving infinite-dimensional global optimization problems.

2. The Problem Statement of the Dynamic System Program Control Multicriteria Synthesis

We suggest that the a controlled dynamic system mathematical model is described by a system of nonlinear ordinary differential equations

$$\begin{cases} \dot{\mathbf{x}} = \mathbf{f}(\mathbf{x}, \mathbf{u}(t)), \\ \mathbf{x}^0 = \mathbf{x}(t_0). \end{cases} \tag{1}$$

In (1) $\mathbf{x}(t) \in \mathbf{E}^n$ is the vector of the system state; $\mathbf{f}(\mathbf{x}, \mathbf{u}(t))$—given nonlinear; n—dimensional vector function; $\mathbf{u}(t) \in \mathbf{E}^{r_u}$—vector of the system control; $\mathbf{x}^0$—initial conditions.

The control selection is subject to restrictions $\mathbf{u}(\cdot) \in \mathbf{U}$ of the form

$$\mathbf{U} = \{\mathbf{u}(\cdot) \in \mathbf{L}_2^{(r_u)}[t_0, T] | \mathbf{u}_{min} \leq \mathbf{u}(t) \leq \mathbf{u}_{max}, t \in [t_0, T]\}, \tag{2}$$

where $\mathbf{L}_2^{(r_u)}[t_0, T]$ is the Hilbert space of $r_\mathbf{u}$ -dimensional functions, the square of which we integrate on a fixed interval $[t_0, T]$ (hereinafter the entry $\mathbf{u}$ will mean $\mathbf{u}(\cdot)$ as an element of the functional space, and $\mathbf{u}(t)$ is the value at a time t).

The efficiency of the system (1) controlling under constraints (2) is estimated by a vector criteria $\mathbf{J}(\mathbf{u}) \in \mathbf{E}^m$, whose components are Frechet differentiable functionals of the form

$$J_i(\mathbf{u}(\cdot)) = \int_{t_0}^{T} F_i(\mathbf{x}(t), \mathbf{u}(t))dt, i \in \mathbf{M} = \{\overline{1, m}\}. \tag{3}$$

It is necessary to define a program control $\mathbf{u}^* \in \mathbf{U}$ of the system (1) that satisfies the restrictions (2) and minimizes the values of the components of the vector objective functional (3).

The peculiarity of the statement (1)–(3) is that it is an infinite-dimensional nonlinear multicriteria optimization problem with non-convex objective functionals.

3. A Structure of Hybrid Trans-Dimensional Evolutionary Algorithm of Multicriteria Control Optimization

The developed hybrid evolutionary algorithm of multicriteria optimization includes the following stages.

Stage 1. Global finite-dimensional multicriteria search. The initial the problem statement (1)–(3) is formalized as a finite-dimensional multicriteria optimization problem in which the dimension of the vector of variable parameters is equal to p. To solve this problem, an evolutionary algorithm of multicriteria optimization with respect to the polyhedral cone of [12] is used.

Stage 2. Local infinite-dimensional multicriteria search. The initial the problem statement (1)–(3) is presented as an infinite-dimensional multicriteria optimization problem with functional constraints differentiable by Frechet. To solve this problem, an evolutionary multicriteria algorithm of feasible directions has been developed [13]. In this case, the optimal solutions obtained at the stage of global finite-dimensional multicriteria search are transformed into initial approximations for the stage of local infinite-dimensional search.

Thus, the proposed algorithm implements the control optimization process with a changing dimension of the search space. We will call such an optimization model a transdimensional (p/∞)-search model.

4. Multicriteria Optimization of a Bioreactor Program Control Law

Consider a biotechnological process (BTP) occurring in a bioreactor. A bioreactor is a container with reproducing microorganisms, a solution of salts and simple organic compounds, into which a sterile nutrient medium continuously enters, and part of the contents, including living organisms, is constantly removed. It is assumed that the mathematical model of BTP is given in the form of a system of nonlinear ordinary differential equations characterizing the dynamics of the development of a two-age population of bacteria in a dimensionless form

$$\dot{x_1} = \frac{-\alpha x_3 \mu_m x_1}{K_m + x_1} + u_1 - u_3(x_1 + u_1), \quad (4)$$

$$\dot{x_2} = \frac{-\alpha x_4 \mu_m x_2}{K_m + x_2} + u_2 - u_3(x_2 + u_2), \quad (5)$$

$$\dot{x_3} = q\frac{K_m + x_2}{T_2^{min} x_2 + T_2^{max} K_m} x_2 - \frac{K_m + x_1}{T_1^{min} x_1 + T_1^{max} K_m} x_3 - u_3 x_3, \quad (6)$$

$$\dot{x_4} = \frac{K_m + x_1}{T_1^{min} x_1 + T_1^{max} K_m} x_3 - \frac{K_m + x_2}{T_2^{min} x_2 + T_2^{max} K_m} x_4 - u_3 x_4, \quad (7)$$

$$y_1 = x_3, y_2 = x_4, \quad (8)$$

$$\mathbf{x}^0 = \mathbf{x}(t_0). \quad (9)$$

In the considered BTP model, Equations (4)–(7) describe the dynamics of the development of a two-age population of bacteria in a dimensionless form; (8) the equations of the measured output; and (9) initial conditions. The following designations are used: $\mathbf{x} = [x_1, x_2, x_3, x_4]^T$—the state vector of the system; x_1, x_2—the relative concentrations of the substrate that ensures the growth of "young" and division of "old" cells, respectively; x_3, x_4—relative concentrations of "young" and "old" cells, respectively; $\mathbf{y} = [y_1, y_2]^T$—the vector of the measured output of the system; $\mathbf{u} = [u_1, u_2, u_3]^T \in \mathbf{U}$—the control vector; u_1, u_2—the relative rates of entry of the nutrient substrate into the cultivator, ensuring the growth of "young" and division of "old" cells, respectively; u_3—the relative velocity of the toque in the cultivator; and $\mathbf{U}$—a limited closed set of the form

$$\mathbf{U} = \{\mathbf{u}(\cdot) | 0 \le u_i(t) \le 1, i = \overline{1,3}, t \in [t_0; T]; \quad (10)$$

α—quantitative requirements of the microorganism in food; T_1^{max}, T_1^{min}—the maximum and minimum time interval (without feeding with substrates and with maximum feeding with substrate), respectively, necessary for a young cell to change its properties and transition to the phase of development of an old cell; T_2^{max}, T_2^{min}—the maximum and minimum time interval (without feeding with substrates and with maximum feeding with substrate), respectively, required for the old cell to start the division process; q—the number of descendants after division of one cell (it is genetically inherited and does not depend on the conditions of cultivation); μ_m—the maximum rate of absorption of the substrate by microorganisms; K_m—the value of the saturation function is equivalent to the Michaelis–Menten constant, numerically equal to the concentration of the substrate, at which the growth rate of the culture is equal to half the maximum rate.

The efficiency of system control (4)–(9) will be evaluated by a vector criteria $\mathbf{J}(\mathbf{u}(\cdot)) = [J_1(\cdot), J_2(\cdot)]^T$, the components of which are functionals differentiable by Frechet:

$$J_1(\mathbf{u}(\cdot)) = \int_{t_0}^{T} (y_1(\mathbf{u}(t)) - y_1^e(t))^2 dt, \tag{11}$$

$$J_2(\mathbf{u}(\cdot)) = \int_{t_0}^{T} (y_2(\mathbf{u}(t)) - y_2^e(t))^2 dt. \tag{12}$$

In (11), (12) $y_1^e(t) = x_3^e(t), y_2^e(t) = x_4^e(t)$ characterize an etalon BTP mode.

It is required to determine the program control $\mathbf{u}^*(\cdot) \in \mathbf{U}$ of the system (4)–(9) with restrictions on the control vector (10), minimizing the values of the components of the vector functional (11), (12).

Model parameters. The computational experiment was carried out at the following values of the model (4)–(12) parameters: $T_1^{max} = 2$, $T_1^{min} = 1$, $T_2^{max} = 3$, $T_2^{min} = 1$, $q = 2$, $K_m = 0.0495$, $\mu_m = 0.5$, $x_{e_3}(t) = 0.9$, $x_{e_4}(t) = 0.7$, $\mathbf{x}^0 = [0.5; 0.5; 1; 0]$; modeling interval, $t_0 = 0; T = 9$.

To solve the problem (4)–(12), a hybrid evolutionary algorithm of multicriteria optimization based on a transdimensional (n/∞) search model was used, where $n = r_{\mathbf{q}}$.

4.1. Global Evolutionary Multicriteria Search

Perform parameterization of the vector law of program control $\mathbf{u}(t)$ in the form of $\mathbf{u}(\mathbf{q}, t)$, when $r_{\mathbf{u}} = 3$, $r = 3$, $r_{\mathbf{q}} = (r+1)r_{\mathbf{u}} = 12$. To accomplish this this, on the interval $[t_0, T]$ a grid $\{t_j, j = \overline{0, r}\} = \{t_0 = 0, t_1 = 3, t_2 = 6, t_3 = T = 9\}$ is built. The parameterized program control law is represented as natural cubic splines $u_i(\mathbf{q}_i, t) = S_i(\mathbf{q}_i, t), i = \overline{1, 3}$, for which the conditions

$$S_i(\mathbf{q}_i, t_j) = q_{ij}, j = \overline{0, 3} \tag{13}$$

are met at the nodal points.

Thus, the vector of control parameters is represented as $\mathbf{q} = [\mathbf{q}_1^T, \mathbf{q}_2^T, \mathbf{q}_3^T]^T$, where $\mathbf{q}_i = [q_{i0}, q_{i1}, q_{i2}, q_{i3}]^T, i = \overline{1, 3}$. Next, the problem of finite-dimensional multicriteria optimization is solved. The polyhedral cone of dominance is given as a function of uncertainty intervals of the vector criteria components weighting coefficients.

Figure 1 shows the results of the application of the evolutionary algorithm of multicriteria optimization with respect to the polyhedral dominance cone: the points set $\hat{\mathbf{J}}^{\Omega}(\mathbf{Q}) = \{\mathbf{J}^P\} \cup \mathbf{J}^0$ is a discrete approximation of $\boldsymbol{\Omega}$-optimal solutions set in the problem (4)–(12). The polyhedral dominance cone $\boldsymbol{\Omega}$ is given as a function of uncertainty intervals of the vector criteria components weighting coefficients. The uncertainty intervals of the vector criteria components weighting coefficients is given as a set:

$$M = \{^- \in \mathbf{E}^2 | 0.2 \le \mu_1 \le 0.8; 0.2 \le \mu_2 \le 0.8\}.$$

The matrix of the dominance cone

$$\mathbf{B} = \begin{bmatrix} 0.8 & 0.2 \\ 0.2 & 0.8 \end{bmatrix}.$$

The points coordinates of the set $\hat{\mathbf{J}}^{\Omega}(\mathbf{Q})$ in the criterion space are presented in the Table 1. Based on the Hermeyer criterion, point No 3 is chosen as the initial approximation $\mathbf{J}^0$ for the implementation of the local multicriteria search stage. As can be seen from Table 1, the solution $\mathbf{J}^0 = [0.0300; 0.0272]^T$ is characterized by the most balanced component values of the vector criteria.

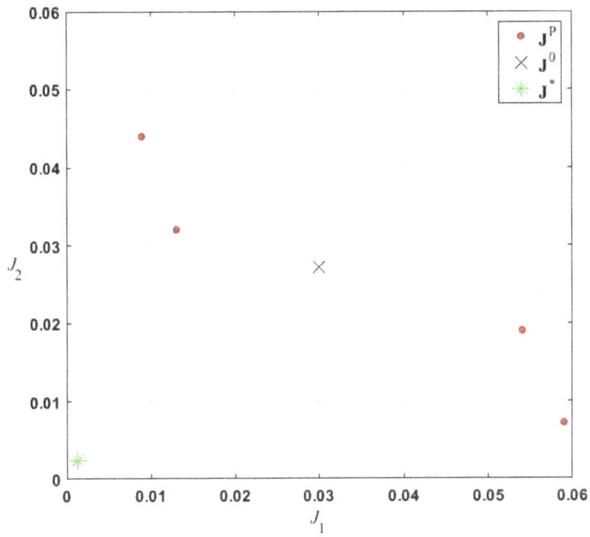

Figure 1. Results of a computational experiment: • = points $\mathbf{J}^P \in \hat{\mathbf{J}}^{\Omega}(\mathbf{Q})$; × = point $\mathbf{J}^0 \in \hat{\mathbf{J}}^{\Omega}(\mathbf{Q})$; * = point $\mathbf{J}^*$.

Table 1. The set $\hat{\mathbf{J}}^{\Omega}(\mathbf{Q})$.

№	J_1	J_2
1	0.0089	0.0440
2	0.0130	0.0320
3	0.0300	0.0272
4	0.0540	0.0190
5	0.0590	0.0072

The point $\mathbf{J}^0 = [0.0300; 0.0272]$ corresponds to the program control $\mathbf{u}^0(t) = \mathbf{u}(\mathbf{q}^0, t)$, where $\mathbf{q}^0$ is the vector of control parameters:

$$\mathbf{q}^0 = [\mathbf{q}_1^{0T}, \mathbf{q}_2^{0T}, \mathbf{q}_3^{0T}]^T = [0.5668 \quad 0.6380 \quad 0.9467 \quad 0.7176;$$
$$0.3114 \quad 0.4255 \quad 0.6911 \quad 0.3112; \quad 0.0000 \quad 0.3094 \quad 0.3846 \quad 0.3700]^T.$$

The components of the subvector $\mathbf{q}_i^0$ are the nodal points of the natural cubic spline $S_i(\mathbf{q}_i^0, t)$, $i = \overline{1,3}$. The graphs of the components of the program control law $\mathbf{u}^0(t) = \{u_i(\mathbf{q}_i^0, t) = S_i(\mathbf{q}_i^0, t), i = \overline{1,3}\}$ are shown in Figure 2. The corresponding trajectory $\mathbf{y}^0(t) = \mathbf{y}(\mathbf{u}^0(t))$ is shown in the figure 3.

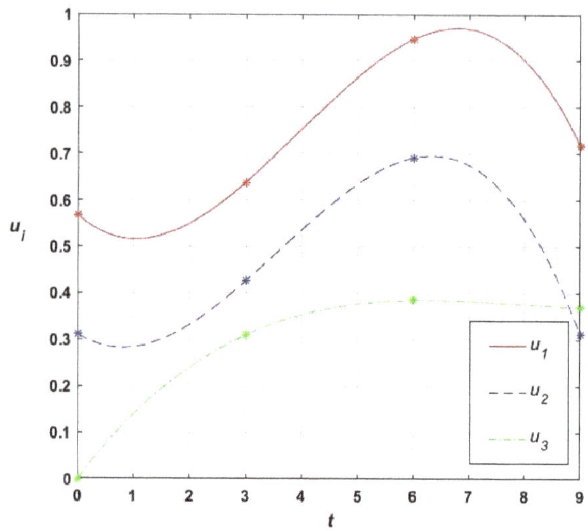

Figure 2. The program control law of BTP $\mathbf{u}^0(t)$.

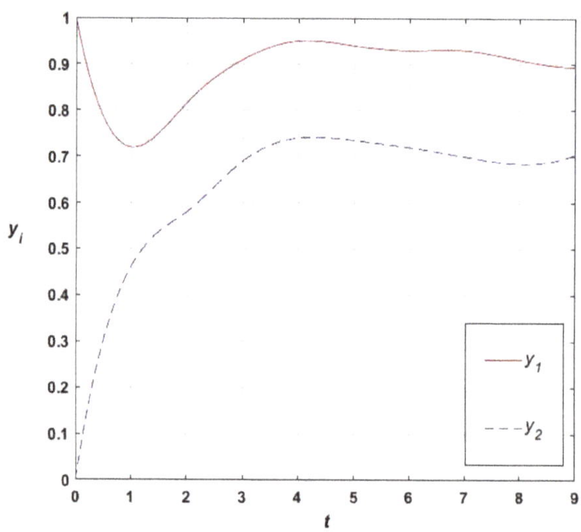

Figure 3. BTP trajectory $\mathbf{y}(\mathbf{u}^0(t))$.

4.2. Local Evolutionary Multicriteria Search

The clarifying search for the optimal law of program control is performed using an evolutionary multicriteria algorithm of feasible directions, in which the degree of the approximating functional of constraints [13] p=3. As a result, the optimal control law of the system $\mathbf{u}^*(\cdot)$ is obtained, the form of which is shown in Figure 4. The corresponding optimal trajectory $\mathbf{y}^*(\mathbf{u}^*(\cdot))$ is shown in Figure 5. The optimal value of the vector criteria $\mathbf{J}(\mathbf{u}^*(\cdot)) = [0.0013; 0.0021]^T$ is also shown in Figure 1.

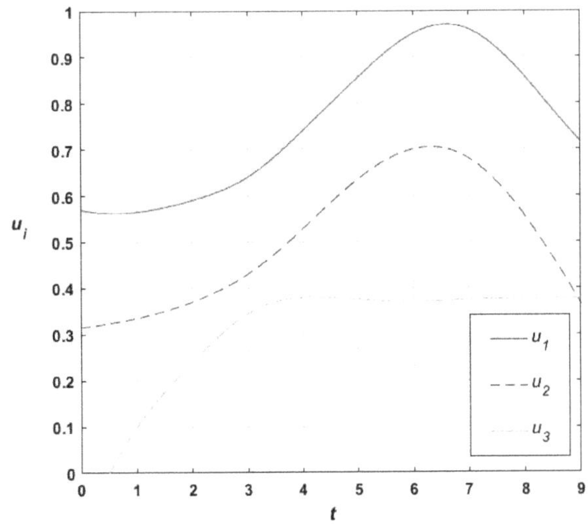

Figure 4. Optimal program control law $\mathbf{u}^*(\cdot)$.

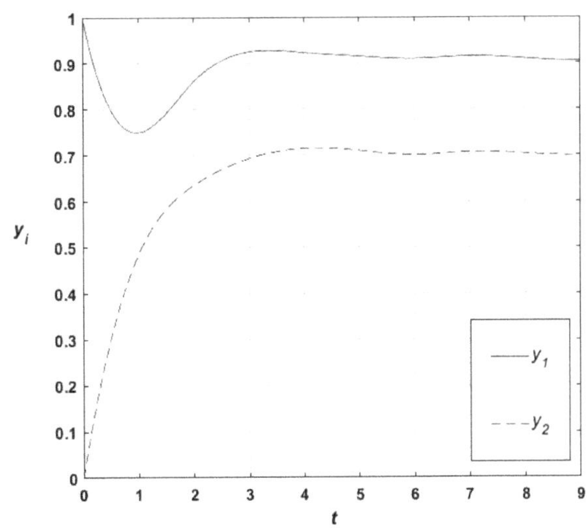

Figure 5. Optimal trajectory $y^*(\mathbf{u}^*(\cdot))$.

4.3. Comparative Analysis of Transdimensional Search Models

Computational experiments were carried out with various transdimensional search (p/s)-models. The results of computational experiments are presented in Table 2.

Table 2. Results of computational experiments with transdimensional search models.

№	p	s	J_1^*	J_2^*
1	12	12	0.0274	0.0205
2	12	21	0.0251	0.0187
3	12	39	0.0164	0.0104
4	12	75	0.0107	0.0072
5	12	147	0.0098	0.0080

In each experiment, the initial approximation $\hat{\mathbf{q}}^0$ for the local multicriteria search stage is formed from the vector $\mathbf{q}^0$ obtained at the global multicriteria search stage based on the transdimensional (p/s)-transition. The transdimensional (p/s)-transition is implemented using the following algorithm.

Transdimensional (p/s)-transition algorithm.

Step 1. Convert the time grid $\{t_j, j = \overline{0,r}\}$ into a time grid $\{t_k, k = \overline{0,\hat{r}}, \hat{r} = 2r$, which is obtained from the original one by dividing each interval $[t_j, t_{j+1}]$ into two equal parts. The values of the transfer transition parameters have the following values: $p = r_{\mathbf{u}}(r+1)$, $s = r_{\mathbf{u}}(\hat{r}+1)$.

Step 2. Form a vector of parameters $\hat{\mathbf{q}}^{0T} = [\hat{\mathbf{q}}_1^{0T}, \ldots, \hat{\mathbf{q}}_{r_{\mathbf{u}}}^{0T}] \in \mathbf{E}^S$, the components of which are determined on the basis of splines (39) with a nodal points vector $\mathbf{q}^0 \in \mathbf{E}^p$

$$\hat{\mathbf{q}}_i^{0T} = [\hat{q}_{ik}^0 = S_i(\mathbf{q}_i^0, t_k), k = \overline{0,\hat{r}}], i = \overline{1, r_{\mathbf{u}}}. \tag{14}$$

Step 3. Construct natural cubic splines $\hat{S}_i = (\hat{\mathbf{q}}_i^0, t)$, $i = \overline{1, r_{\mathbf{u}}}$, with the nodal points vector (14).

Step 4. Form a parameterized program control law

$$\mathbf{u}(\hat{\mathbf{q}}^0, t) = \{u_i(\hat{\mathbf{q}}_i^0, t) = \hat{S}_i(\hat{\mathbf{q}}_i^0, t), i = \overline{1, r_{\mathbf{u}}}\}. \tag{15}$$

The analysis of the computational experiment results leads to the following main conclusion. If the initial the multicriteria control optimization problem statement is infinite-dimensional, then the use of a transdimensional (p/∞)-search model that implements the search for the optimal control law $\mathbf{u}^*(\cdot)$ in space $\mathbf{L}_2^{(r_{\mathbf{u}})}[t_0, T]$ provides a higher efficiency of the optimal program control law $\mathbf{u}^*(\cdot)$ compared to finite-dimensional transdimensional search (p/s)-models.

5. Conclusions

A hybrid transdimensional evolutionary algorithm for dynamical system control multicriteria optimization has been developed. The proposed transdimensional search (p/∞)-model is based on the joint use of a finite-dimensional model for the control law parameters $\mathbf{u}(\mathbf{q}, (\cdot)) \in \mathbf{E}^{r_q}$ global multicriteria optimization and an infinite-dimensional model for the control law $\mathbf{u}(\cdot) \in \mathbf{L}_2^{(r_{\mathbf{u}})}[t_0, T]$ local multicriteria optimization.

An evolutionary algorithm of global finite-dimensional multicriteria optimization with respect to the polyhedral dominance cone has been developed. It is shown that the representation of the polyhedral dominance cone as a function of the uncertainty intervals of the vector criterion components weighting coefficients makes it possible to reduce the uncertainty of choice on the set of Pareto optimal solutions. The resulting subset contains solutions that have a higher degree of balance for various components of the vector criterion. This property is achieved by cutting off solutions located on the "edges" of the Pareto set, which often has properties close to the optimal solutions by Slater and a priori do not meet technical requirements.

An algorithm for calculating the polyhedral dominance cone as a function of uncertainty intervals of the vector criterion components weighting coefficients has been developed.

A multicriteria evolutionary algorithm for feasible directions has been developed to solve the problem of the local multicriteria optimization of dynamic systems program control laws. A comparative analysis of transdimensional evolutionary search (p/s)- and (p/∞)-models' effectiveness is carried out on the example of solving the problem of the optimal law of bioreactor program control multicriteria synthesis. It is shown that the use of a transdimensional search (p/∞)-model that implements the search for the optimal control law $\mathbf{u}^*(\cdot)$ in space $\mathbf{L}_2^{(r_u)}[t_0, T]$ provides higher efficiency of the optimal law of program control $\mathbf{u}^*(\cdot)$ compared with finite-dimensional transdimensional search (p/s)-models.

The general conclusion is that the transdimensional (p/∞)-hybridization of evolutionary algorithms for multicriteria control optimization gives a synergistic effect. This effect is expressed in fundamental increase of the solving accuracy the multicriteria control optimization problem compared to the known hybrid control optimization NIAs by resolving the contradiction between the finite-dimensional global search model and the infinite-dimensional initial problem statement.

Author Contributions: Conceptualization and methodology, V.A.S. and E.M.V.; software, computational experiment, E.L.D. and E.Y.K. All authors have read and agreed to the published version of the manuscript.

Funding: This research received no external funding.

Institutional Review Board Statement: Not applicable.

Informed Consent Statement: Not applicable.

Data Availability Statement: Not applicable.

Conflicts of Interest: The authors declare no conflict of interest.

References

1. Wang, X. Hybrid Nature-Inspired Computation Method for Optimization. Ph.D. Dissertation, Helsinki University of Technology, Espoo, Finland, 2009; 161p. Available online: http://lib.tkk.fi/Diss/2009/isbn9789512298594/ (accessed on 17 September 2022).
2. El-Abd, K.M. A Taxonomy of Cooperative Search Algorithm. In *Hybrid Metaheuristics. Second International Workshop*; Springer: Berlin/Heidelberg, Germany, 2005; Volume 3636, pp. 32–41. Available online: https://www.researchgate.net/publication/221411183 (accessed on 17 September 2022).
3. Raidl, G.R. A Unified View on Hybrid Metaheuristics. In *Hybrid Metaheuristics*; Ser. Lecture Notes in Computer Science; Springer: Berlin/Heidelberg, Germany, 2006; Volume 4030, pp. 1–12. [CrossRef]
4. Karpenko, A.P. Modern search optimization algorithms. In *Algorithms Inspired by Nature*; Publishing House of the Bauman Moscow State Technical University: Moscow, Russia, 2014; 446p.
5. Molina, D.; Lozano, M.; Herrera, F. Memetic algorithm with local search chaining for continuous optimization problems: A scalability test. In Proceedings of the International Conference on Intelligent Systems Design and Applications (ISDA'09), Pisa, Italy, 30 November–2 December 2009; IEEE Publication: Piscataway, NJ, USA, 2009; pp. 1068–1073. [CrossRef]
6. Neri, F.; Cotta, C. Memetic algorithms and memetic computing optimization: A literature review. *Swarm Evol. Comput.* **2012**, *2*, 1–14. [CrossRef]
7. Dirita, V. Control System Design Applications with Hybrid Genetic Algorithms. Ph.D. Thesis, University of Tasmania, Tasmania, Australia, 2002.
8. Bambha, N.K.; Bhattacharyya, S.S.; Teich, J.; Zitzler, E. Systematic integration of parameterized local search into evolutionary algorithms. *IEEE Trans. Evol. Comput.* **2004**, *8*, 137–155. [CrossRef]
9. Hu, X.; Zhang, J.; Li, Y. Orthogonal Methods Based Ant Colony Search for Solving Continuous Optimization Problems. *J. Comput. Sci. Technol.* **2008**, *23*, 2–18. [CrossRef]
10. Zhang, J.; Chen, W.; Tan, X. An Orthogonal Search Embedded Ant Colony Optimization Approach to Continuous Function Optimization. In *Ant Colony Optimization and Swarm Intelligence*; Dorigo M., Gambardella L.M., Birattari M., Martinoli A., Poli R., Stützle T., Eds.; ANTS 2006; Lecture Notes in Computer Science; Springer: Berlin/Heidelberg, Germany, 2006; Volume 4150. [CrossRef]
11. Dreo, J.; Siarry, P. Continuous interacting ant colony algorithm based on dense heterarchy. *Future Gener. Comput. Syst.* **2004**, *20*, 841–856. [CrossRef]

12. Serov, V.A. Genetic algorithms of conflict equilibriums-based multicriteria systems control optimization under uncertainty. *Vestnik BMSTU. Ser. Instrum. Mak.* **2007**, *4*, 70–80.
13. Serov, V.A. Combined evolutionary method of feasible directions in multicriteria synthesis problem of a dynamical system program control. *Procedia Comput. Sci.* **2021**, *186*, 48–58. [CrossRef]

engineering proceedings

MDPI

Proceeding Paper

Hierarchical Cascade Control Systems for Time-Dependent Dynamical Plants as Applied to Magnetic Plasma Control in *D*-Shaped Tokamaks †

Yuri V. Mitrishkin [1,2]

[1] Faculty of Physics, M.V. Lomonosov Moscow State University, Leninskie Gory, Moscow 119991, Russia; mitrishkin.yv19@physics.msu.ru
[2] V.A. Trapeznikov Institute of Control Sciences of Russian Academy of Sciences, 65 Profsoyuznaya Street, Moscow 117997, Russia
† Presented at the 15th International Conference "Intelligent Systems" (INTELS'22), Moscow, Russia, 14–16 December 2022.

Abstract: The systems of poloidal field coils in D-shaped Tokamaks such as ITER, EAST, JET, ASDEX Upgrade, TCV, GLOBUS-M2, DIII-D, SPARC, IGNITOR, JT-60SA, DEMO-9.1, DEMO-1.6, T-15MD, and TRT are analyzed for their efficiency in the application of plasma position, current, and shape control systems in these Tokamaks. The problem of magnetic plasma control in Tokamaks is presented. A methodology for designing hierarchical cascade systems of magnetic plasma control in D-shaped Tokamaks has been developed on the basis of generalizations of existing plasma magnetic control systems. The hierarchical levels are as follows: multivariable robust cascade control level, adaptation level, artificial intelligence level, and decision-making level. To implement these systems in the practice of physical experimentation, it is proposed to use digital twins, the basis of which is a real-time digital testbed created by Lomonosov Moscow State University and the Trapeznikov Institute of Control Sciences of the Russian Academy of Sciences.

Keywords: Tokamaks; poloidal field coils; plasma magnetic control; hierarchical control systems; cascade control; digital twins; digital testbed

Citation: Mitrishkin, Y.V. Hierarchical Cascade Control Systems for Time-Dependent Dynamical Plants as Applied to Magnetic Plasma Control in *D*-Shaped Tokamaks. *Eng. Proc.* **2023**, *33*, 61. https://doi.org/10.3390/engproc2023033061

Academic Editors: Askhat Diveev, Ivan Zelinka, Arutun Avetisyan and Alexander Ilin

Published: 7 August 2023

1. Introduction

1.1. State-of-the-Art

Toroidal vessels with magnetic coils, namely Tokamaks, are currently the leaders in the fusion race [1,2]. In [3–7], a review of the control systems for the plasma position, current, and shape was made both for Tokamaks with a circular cross-section and for Tokamaks with a vertically elongated cross-section (*D*-shaped Tokamaks). As a result, it turned out that the generalized scheme of such systems has the form shown in Figure 1. In general, the magnetic plasma control system consists of an internal circuit for controlling currents in the poloidal field (PF) coils, which also includes a current control circuit in the central solenoid (CS) to control plasma current by the transformer principle, and an external current and plasma shape control circuit (see Figure 1). Plasma shape control requires an algorithm for plasma equilibrium reconstruction in the feedback loop. Two different approaches are used: isoflux control (alignment of magnetic signals at the plasma separatrix) and control of gaps between the plasma boundary and the first wall. The first method is used on DIII-D, TCV, MAST-U, NSTX-U, KSTAR, and EAST Tokamaks with various modifications. The second method is used on JET, ASDEX Upgrade, and ITER Tokamaks. This stabilizes the vertical position of the plasma, or its vertical velocity relative to the zero value, as well as the horizontal position of the plasma by means of the two-dimensional controller C_1. In the development of any Tokamak, it is necessary to solve many different problems simultaneously, which leads to the need to simultaneously meet many different criteria, i.e., it is a multi-criteria problem.

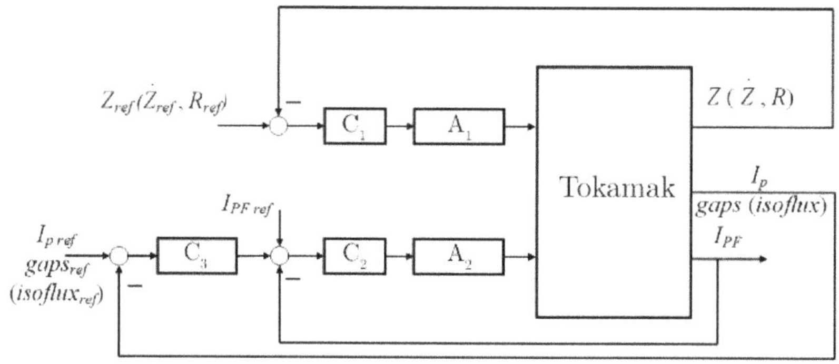

Figure 1. Schematic diagram of the magnetic plasma control system of vertically extended Tokamaks: A_1 and A_2 are actuators, and C_1, C_2, and C_3 are controllers of plasma position, currents in the PF-coils, and plasma shape, respectively. Diagnostics are included in the Tokamak unit.

When designing Tokamaks, it is necessary to consider the following basic provisions related to high-temperature plasma control systems:

- *D*-shaped Tokamaks without plasma position, current, and shape control systems are inoperable (Figure 1);
- The design of poloidal systems for *D*-shaped Tokamaks must be carried out together with the control systems of plasma position, current, and shape (a Poloidal system is a system of poloidal field coils located around the vacuum vessel of the Tokamak (PF coils));
- Efficient plasma control systems lead to improved Tokamak operating modes and a reduced risk of major discharge disruptions;
- Further development of integrated systems of magnetic (position, current, and shape) and kinetic (safety factor q, density, pressure, and temperature profiles) plasma control in Tokamaks should lead primarily to the prevention of major plasma disruptions, which is essential for the operation of fusion power plants based on Tokamaks and the optimal modes of their operation.

On the basis of these provisions in this paper, the following tasks are set:

✓ To analyze the poloidal systems of modern Tokamaks and some of their projects;
✓ To pose the problem of magnetic plasma control in Tokamaks;
✓ To develop a methodology for the design of hierarchical cascade magnetic plasma control systems in *D*-shaped Tokamaks for the purpose of further developing this class of control systems.

1.2. Motivation

The field of plasma magnetic control in modern Tokamaks is very important and complex for their operation because the plasma is gas in a restricted volume with distributed time-varying parameters. In spite of that, during the long history of Tokamaks, 226 Tokamaks were created, and about 50 of them are operating now (http://www.tokamak.info (accessed on 14 December 2022)). Plasma magnetic control systems are of the utmost priority to start and exploit modern Tokamaks, and so far, they require new methods and new ideas to improve their performance and stability, especially for future fusion stations.

1.3. Novelty

In this paper, the new methodology of robust cascade hierarchical plasma control systems is suggested on the basis of some generalizations of plasma magnetic control systems in Tokamaks.

1.4. Paper Structure

After Introduction Section 2 presents a set of poloidal systems for operating *D*-shaped Tokamaks and some of their projects. In Section 3, a comparative analysis of the presented poloidal systems is carried out. Section 4 describes the plasma magnetic control problems in Tokamaks, namely the plant under control, plasma models, and control itself: performance and stability margin estimation. Section 5 shows a new methodology for designing hierarchical cascade systems of magnetic plasma control in *D*-shaped Tokamaks. In Section 6, the real-time test bed for plasma magnetic control in Tokamaks is demonstrated, and the results of real-time simulations on the testbed of the plasma position, current, and shape control system with the plasma reconstruction algorithm in the feedback are shown.

2. Poloidal Systems of *D*-Shaped Tokamaks

First of all, the poloidal system, the system of poloidal field coils around the vacuum vessel, should provide the desired plasma configurations in the discharges at the selected set of coils. This provision is the first priority in the design of Tokamaks. With this, one should start the design of the poloidal system of a *D*-shaped Tokamak, but this is not enough to provide the necessary quality of control of the plasma position, current, and shape in a closed control system, as well as the necessary sufficiently large margins of robust stability. It is also necessary to take into account ensuring the feedback system's proper conditionality and internal stability, as well as the minimization of the control power, which is a very difficult task for which one must correctly design the poloidal system. These problems have not been completely solved, and further professional work in these areas is needed for future fusion reactors and fusion power plants. We present the characteristics of the poloidal systems of a number of *D*-shaped Tokamaks, which allow us to assess the picture of different placements of the coils of the central solenoid and the poloidal field in order to identify patterns that may be useful for the design of new Tokamaks. In the Tokamak labels in Figures 2–8, the first number is the value of the major radius of the Tokamak in meters, and the second number is the value of the aspect ratio (ratio of the major radius to the minor radius of the Tokamak).

ITER (Figure 2a). In the original ITER poloidal system, a special design feature was made: with a major Tokamak radius of 6.2 m, the vertical controllability region for vertical plasma instability was very small, only 3–4 cm. Going beyond this area leads to the loss of stability of the plasma control system, which leads to the accident of the installation. To correct that feature, a new engineering solution was found: horizontal magnetic field coils were introduced inside the ITER vacuum vessel [8], which allowed the vertical control region to be expanded by an order of magnitude with a limited voltage on these coils. This is a well-known position in control theory: when the input effect is limited when controlling an unstable plant, the controllability region of the output signal is also limited [9], which was not taken into account in the design of the ITER poloidal system. In the previous version of ITER with a self-sustaining fusion reaction [7,10], this problem did not arise because the plasma in the Tokamak was slower and the present PF-coils were sufficient to stabilize the shape and position of the plasma simultaneously. Anyway, all modern Tokamaks are working in support of ITER to achieve its main goals: starting ITER, providing its basic scenarios, achieving the plasma with a thermonuclear reaction of $Q > 10$, basic technologies including plasma magnetic and kinetic control, and so on.

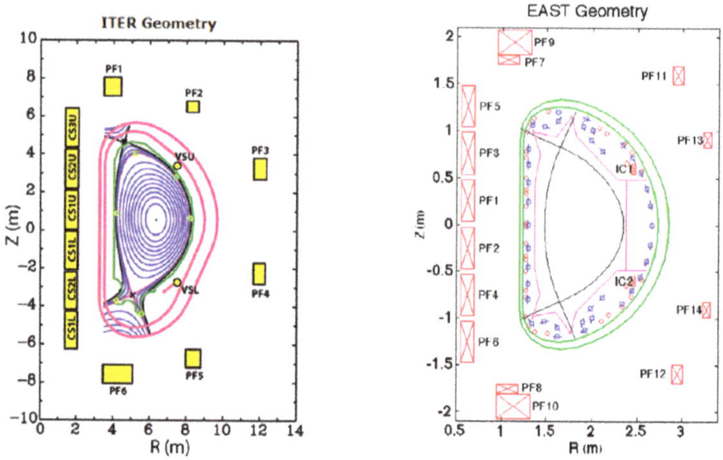

(a) (b)

Figure 2. Tokamaks (**a**) ITER–6.2-3.1 and (**b**) EAST–1.8–4.01. Colors: (**a**) vacuum vessel is red, PF-coils and CS are yellow, (**b**) vacuum vessel is green, PF-coils and CS are red.

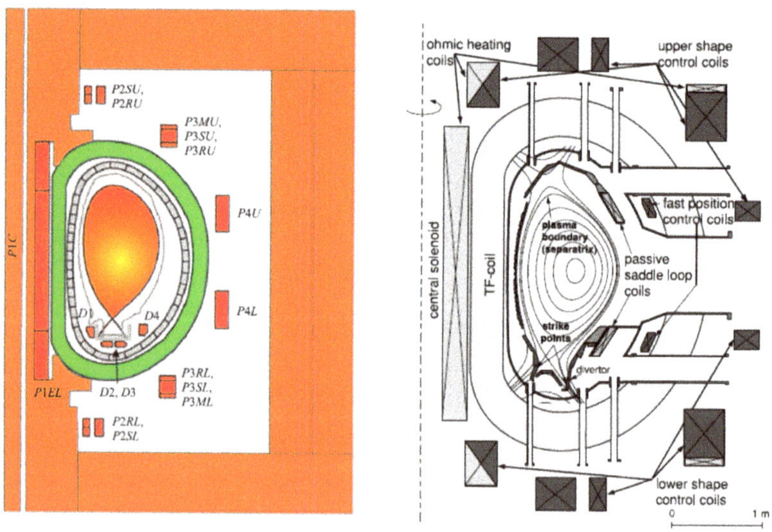

(a) (b)

Figure 3. Tokamaks (**a**) JET-2.96–3.08, the magnet is orange, PF-coils are red, toroidal coil is green, vacuum vessel is gray; (**b**) ASDEX Upgrade-1.65–3.1 Dot line is the axis of the tokamak.

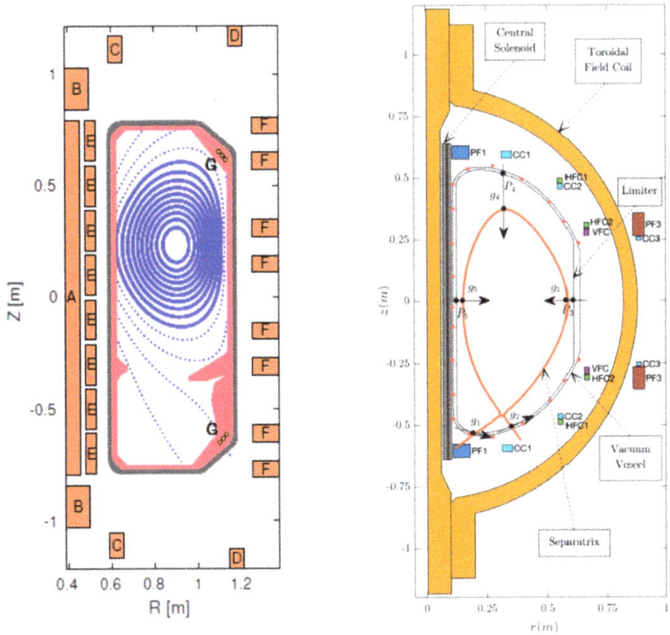

(**a**) (**b**)

Figure 4. Tokamaks (**a**) TCV–0.88–3.5. Level lines are blue, A, B, C, D, E, F are PF-coils, G are horizontal field coils. (**b**) GLOBUS-M2–0.36–1.5. The TF-coil is yellow, separatrix is red, the magnetic loops are red dots, CS is gray, PF is poloidal field, CC is correction coil, HFC is horizon field coil, VFC is vertical field coil, g1–g6 are gaps, the black arrows are directions of measuring of the gaps, *Z* is vertical axis, *R* is horizontal axis.

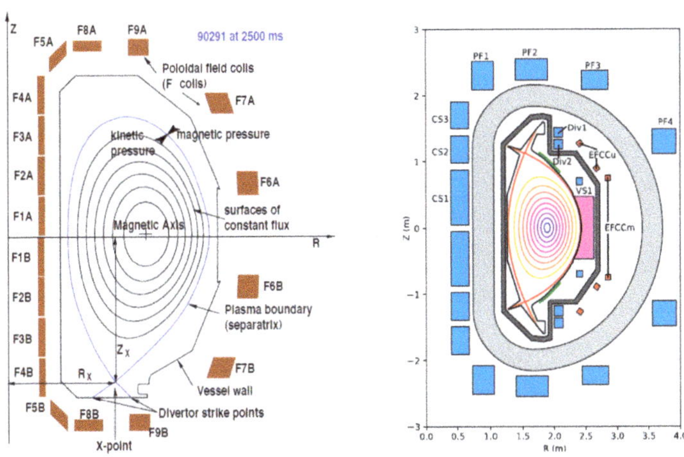

(**a**) (**b**)

Figure 5. Tokamaks (**a**) DIII-D–1.66–2.48; (**b**) SPARC-1.85–3.24. The toroidal field coil is light grey. The central solenoid and poloidal field coils are blue. Error-field correction coils are orange-red. The vacuum vessel is dark grey. The ICRH antenna is pink. The divertor and first limiting surfaces are black. Vertical stability plates are green. The plasma separatrix is red.

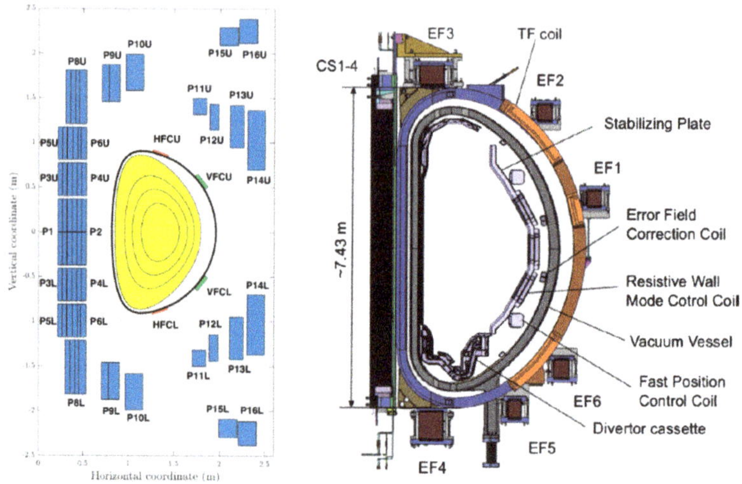

(a) (b)

Figure 6. Tokamaks (**a**) IGNITOR–1.32–2.8. PF-coils are blue, the plasma us yellow, HFCU, HFCL are red, VFCU, VFCL are green; (**b**) JT-60SA–2.96–2.5.

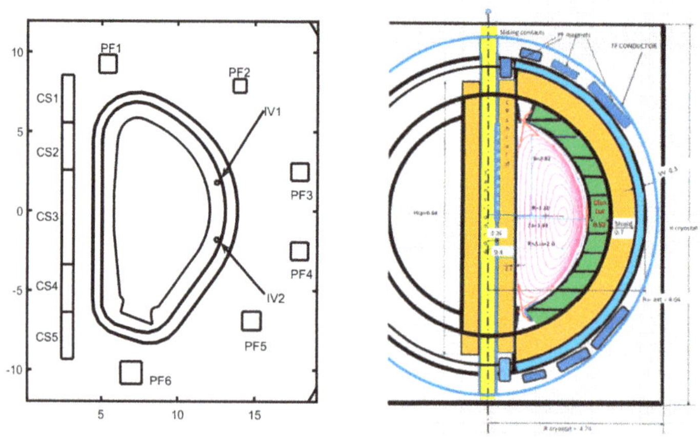

(a) (b)

Figure 7. Tokamaks (**a**) DEMO-9.1–3.1 and (**b**) DEMO-1.6–1.7 module on the spherical Tokamak. The module with aspect ratio A = 1.6, elongation K = 3, triangularity = 0.5, major radius R = 1.6 m, magnetic field B(R) = 3.17 T and plasma current Ipl = 6.7 MA. Blanket is green, TF conductor is light blue, PF coils are dark blue, vacuum vessel is blue, the shield is yellow.

EAST (Figure 2b). Tokamak EAST was designed as a reduced, about 3.4 times smaller in major radius (see Figure 2), copy of ITER with the repetition of the introduction of horizontal field coils in the vacuum vessel [11]. The motivation for such a decision is to simulate in a reduced scale all modes of operation of ITER, although without the fusion reaction, because it is the reduced size of EAST and in this case does not allow to overcome the Lawson criterion to achieve ignition.

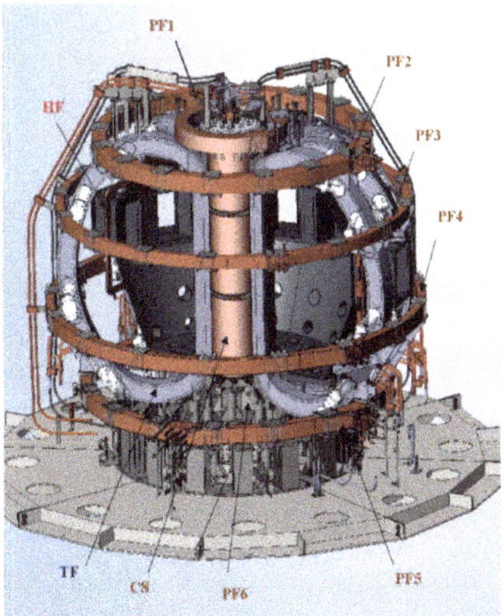

(a)

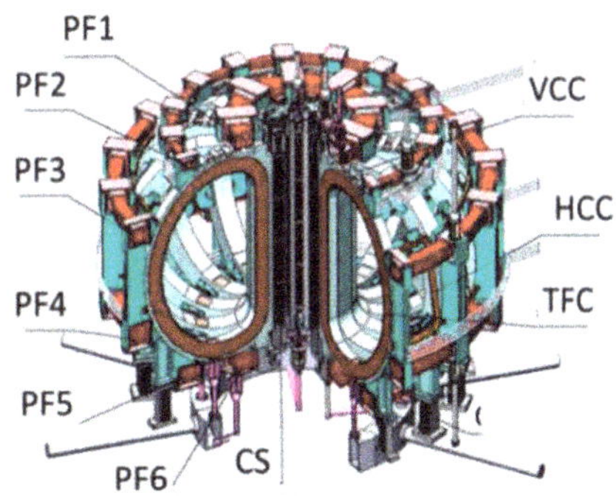

(b)

Figure 8. Tokamaks (**a**) T-15MD-1.48–2.2 и (**b**) TRT-2.15–3.77.

JET (Figure 3a). JET has a major radius of 3 m, and the *D*-shaped vacuum vessel is 2.5 m wide and 4.2 m high [1]. The total plasma volume within it is 100 cubic meters. In 1991, the first experiments including tritium were made, allowing JET to run on the production fuel of a 50–50 mix of tritium and deuterium. In 1997, JET set the record for the closest approach to scientific breakeven. It attained $Q = 0.67$, producing 16 MW of fusion energy while injecting 24 MW of thermal power to heat the fuel, a record that endured until 2021.

ASDEX Upgrade (Figure 3b). This Tokamak has 9 PF-coils, CS in one section, and fast plasma position control coils located between toroidal field (TF) coils and the vacuum vessel [12].

TCV (Figure 4a). Tokamak à Configuration Variable has a set of coils A, B, C, and D that are dedicated to inducing the electric field for driving the plasma current and are commonly referred to as ohmic coils. The G coil located within the vacuum vessel is designed to generate an axisymmetric radial field faster than the unstable growth rate of most elongated plasmas (Figure 4a) [13]. This allows the fast output stage of the G-coil power supply, which is an H-bridge IGBT inverter operating in PWM at a switching frequency of 10 kHz.

GLOBUS-M2 (Figure 4b). The poloidal system of that spherical Tokamak is combined: two PF-coils are located outside the TF-coils, and other PF-coils are inside the TF-coils [14]. There are two control systems with current inverters as actuators for plasma position control and a set of control systems for the control of currents in the PF-coils [15].

DIII-D (Figure 5a). In that Tokamak for plasma shape, the isoflux control is applied to level the magnetic flux on the separatrix. As this takes place, the speed of the plasma's vertical position is stabilized around zero [16].

SPARC (Figure 5b). The Tokamak project has 8 PF coils, 6 CS coils, 6 correction coils, and 2 coils inside the vessel, for a total of 22 coils. SPARC is the most important next step on the way to the commercial use of fusion energy. SPARC is designed with a strong field ($B_0 = 12.2$ T), as a compact ($R_0 = 1.85$ m, a = 0.57 m), superconducting, D-T Tokamak, in order to obtain for the first time the burning gain $Q > 2$ from plasma with magnetic confinement [17].

IGNITOR (Figure 6a). The Tokamak project has 16 PF coils, 14 6 CS coils, 2 HFC coils, and 1 VFC coil, for a total of 33 coils. Two new coils, namely HFC and VFC, were introduced into the poloidal system [18]. IGNITOR is the Tokamak project, which is the closest to ignition.

JT-60SA (Figure 6b). Tokamak has 6 PF coils, 4 CS coils, 4 inner vessel coils, and in whole 14 coils. The JT-60 was planned to be disassembled and then upgraded to the JT-60SA by adding niobium-titanium superconducting coils. It was assumed that the JT-60SA would be able to handle the same plasma shape as ITER. The central solenoid was designed to use niobium-tin because of the higher (9 Tesla) field [19]. The JT-60SA Tokamak will be capable of confining break-even equivalent class high-temperature deuterium plasmas at a plasma current I_p of 5.5 MA and a major radius of $\sim$3 m lasting for a duration longer than the timescales characteristic of plasma processes, pursuing full non-inductive steady-state operation with high plasma beta close to and exceeding no-wall ideal stability limits, and establishing ITER-relevant high density plasma regimes well above the H-mode power threshold.

DEMO-9.1 (Figure 7a). Now there are two roadmaps for creating the first fusion power plant DEMO: on a Tokamak with a large aspect ratio of about 3, and on a modular type on a series of relatively small modules in the form of spherical Tokamaks with a small aspect ratio of about 1.5 [20]. The first roadmap leads to the creation of cyclopean-sized DEMOs with a large radius of the order of 9–10 m and unreasonably expensive power. The leader in the development of such a DEMO is Europe [21], and the terms of commissioning are billed as 2060–2075. A poloidal system copied one-to-one from ITER with horizontal field coils inside the vacuum vessel is proposed for the giant DEMO.

DEMO-1.6 (Figure 7b). The second roadmap says that the DEMO modular type can be created in about 10 years on spherical Tokamaks. In this case, it is proposed to create first one module, on which all the technologies of DEMO will be worked out, and then from a number of modules, assemble a complete fusion power plant. In [20], an estimate of the cost of electricity from such a power plant is made: about 6 US cents per kilowatt-hour of energy.

T-15MD (Figure 8a). The T-15MD poloidal system [22] is copied from ITER and has three CS sections and six PF coils. By placing the PF-coils between the TF coil and the vacuum vessel to stabilize the vertical position of the plasma, it was possible to correct the error of the poloidal system copied from ITER [23–25] and get rid of the internal instability of the plasma vertical position control system.

TRT (Figure 8b). This Tokamak project with reactor technologies also has an ITER-like poloidal system: four CS sections and six PF coils [26]. In TRT, fusion neutrons are flown out of the Tokamak and into uranium or thorium in the blanket, causing nuclear reactions in the blanket. It is the symbiosis of a fusion reactor with a nuclear reactor. Then everything happens as in a conventional nuclear power plant: neutrons are produced, and these neutrons can heat water, obtain steam, and rotate the generator to produce electricity. In principle, the hybrid reactor can not only produce electricity but also nuclear fuel for nuclear power as well as the transmutation of minor actinides, which will not need to be buried.

3. Analysis of Poloidal Systems of *D*-Shaped Tokamaks

From the given data on the Tokamaks poloidal field coils, the tendency to increase the number of these coils is clearly visible. This is explained by the fact that the plasma in Tokamaks is a dynamic plant with distributed parameters, so for greater efficiency in terms of input-output controllability, control quality, and margins of robust stability of magnetic plasma control systems, the poloidal systems should approach the distributed control systems, which leads to an increase in the number of PF-coils surrounding the vacuum vessel.

Comparing and analyzing the poloidal systems of the *D*-shaped Tokamaks, namely ITER, EAST, JET, ASDEX Upgrade, TCV, GLOBUS-M2, DIII-D, SPARC, IGNITOR, JT-60SA, DEMO-9.1, DeMO-1.6, T-15MD, and TRT, shown in Figures 2–8, one can note some regularities:

(1) The initial design of the ITER poloidal system (6 CS sections, 6 PF coils) led to a miserable vertical control area of unstable plasma with a limited control coil voltage, which forced the horizontal field coils to be placed inside the vacuum chamber [8];
(2) Such a poloidal system is repeated in a number of Tokamaks, e.g., EAST [11], DEMO-9.1 [21], JT-60SA [19], etc.;
(3) This poloidal system was not repeated in such current Tokamaks as ASDEX Upgrade, [19] and Globus-M2 [22]. In the ASDEX Upgrade the stabilization of the vertical position of the plasma is achieved by placing the PF coils of the horizontal field between the vessel and the TF-coil, and in the Globus-M2 Tokamak, through 4 HFC coils inside the TF-coil. The same design of plasma poloidal system regarding plasma vertical stabilization as in ASDEX Upgrade has been conducted in T-15MD [24,25];
(4) The most advanced poloidal systems are used in SPARC (Figure 5b), IGNITOR (Figure 6a), and TCV (Figure 4a) because they contain poloidal field coils in sufficiently large numbers and are located fairly evenly around the *D*-shaped vacuum vessel [18]. The fact is that plasma in a Tokamak (gas in a closed volume) is a plant under control with distributed parameters, so it is best to use a controller with distributed parameters to control it most effectively. An example of such a controller are passive structures located around the plasma, in particular the copper shell for round Tokamaks of the past. In passive structures, Foucault currents are induced and inhibit (dampen) the plasma motion. This is the so-called direct-action controller [27]. This controller is energized not from outside but entirely at the expense of the energy of the controlled plant itself, in this case, plasma. To actively influence the plasma in controlling its position and shape, PF coils are used, which are placed around the plasma either completely inside the vessel (Tokamak MAST), outside the vessel completely, or both inside and outside the vessel. The more PF coils used to control the plasma shape, the closer the control system will be to a control system with distributed parameters;
(5) Tokamaks with a large aspect ratio on the order of 3–5 lead to giant-size fusion power plants with a large radius of 8–10 m and an unreasonably high cost of electricity [20,21];
(6) Spherical Tokamaks with aspect ratios on the order of 1.5–1.7 make it possible to create modular fusion power plants with a competitive cost of electricity on the order of 6 US cents per 1 kWh [20];
(7) Many DEMO projects with a large aspect ratio copy the ITER poloidal system, with few exceptions [6].

4. Statement of Plasma Magnetic Control in Tokamaks

The statement of the problem has four basic directions:

- Time-varying parameters plasma models' production on the basis of the first principal equations and experimental data from operating Tokamaks;
- Development of plasma control systems in stabilization, tracking, and disturbance rejection regimes in finite time intervals;
- As in ITER, in RF Tokamaks, namely Globus-M2, T-15MD, and TRT, there are four gaps between the separatrix and the first wall and displacements of two strike points of the separatrix. These output signals are reconstructed on-line by the plasma reconstruction code, namely Improved Moving Filaments [28] or FCDI (Flux-Current Distribution Identification) [29]. These signals are the plasma outputs, which are under the control of plasma magnetic control systems. Simulation of control systems when a plasma equilibrium reconstruction code is included in the feedback for online plasma boundary estimation;
- Implementation of plasma control systems on Tokamaks to conduct physical experiments.

4.1. Controlled Plant

The Globus-M2 Tokamak (Figure 4b) is used in this work as an experimental example for the problem of magnetic control of plasma in Tokamaks (Figure 9). The vertical cross-section of the Globus-M2 Tokamak with the locations of PF coils is presented in Figure 4b.

The main parameters of this spherical Tokamak are as follows: the major radius $R_0 = 0.36$ m, the minor radius $a = 0.24$ m, the maximum plasma current $I_{P\ max} = 0.35$ MA, the maximum toroidal magnetic field $B_{t\ max} = 0.6$ T, aspect ratio $A = 1.5$, and an elongation of about 2.2 [14].

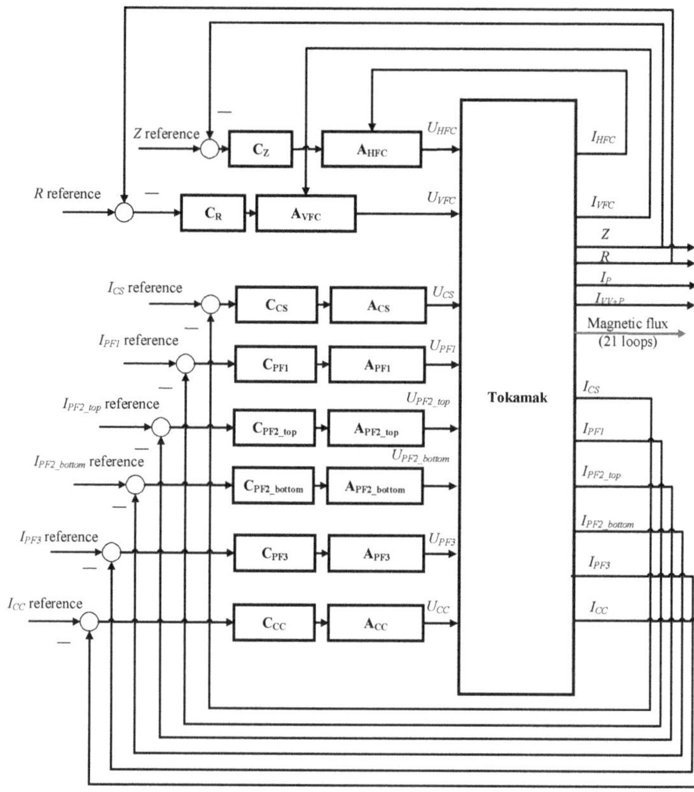

Figure 9. Block diagram of the plasma control system operating in the Globus-M/M2 Tokamak: C_Z, C_R, C_{CS}, C_{PF1}, C_{PF2_top}, C_{PF2_bottom}, C_{PF3}, and C_{CC} are analog controllers; A_{HFC}, A_{VFC}, A_{CS}, A_{PF1}, A_{PF2_top}, A_{PF2_bottom}, A_{PF3}, and A_{CC} are actuators.

The magnetic diagnostics system for the Globus-M2 Tokamak is as follows: Rogowski coils measuring the currents in the PF-coils (I_{HFC}, I_{VFC}, and $I_{PF\text{-}CS}$), VV (I_{VV}), and plasma current (I_P), where HFC is the horizontal field coil and VFC is the vertical field coil (Figure 9); 21 flux loops measuring the magnetic flux Ψ_M outside the VV.

The poloidal system of Globus-M2 (Figure 4b) permits modeling ITER's (http://www.iter.org/, accessed on 14 December 2022) magnetic configurations. In ITER, the plasma configurations have only one lower X-point, but in the Globus-M2 Tokamak experimental data, there are a set of magnetic configurations with an upper X-point, a lower X-point, and two X-points. The plasma boundary is the separatrix, i.e., the magnetic surface containing the most plasma inside the VV. In Figure 4b, the plasma equilibrium configuration is shown with the lower X-point corresponding to the points P_3–P_6 on the first wall for controlling the gaps between the separatrix and the first wall g_3–g_6 and displacements of the strike points g_1–g_2.

To create the gaps' displacements, one tracks the coordinates of six points on the separatrix: strike points g_1–g_2, the rightmost g_3, lowermost g_5, and leftmost g_6 points of the separatrix, and the point g_4 in the middle between the points g_3 and g_5. The displacement vectors are given as the difference between the coordinates of these points and their scenario values. The gaps' displacements δg are calculated as projections of the displacement vectors on the corresponding gaps' vectors presented in Figure 4b.

4.2. Models of Plasma

The Tokamak plasma is the plant under control with distributed parameters having nonlinear time-varying dynamics. Linear controllers in the magnetic feedback stabilize the plasma position, current, and shape in a set of operating Tokamaks with a vertically elongated cross-section and an unstable plasma vertical position [4]. This is possible in reality because the feedback control systems provide relatively small deviations (errors) of dynamical plant outputs from their reference values corresponding to plasma equilibrium in compliance with the Grad-Shafranov equation [1,30]. For this reason, the magnetic plasma linear control systems may be synthesized and tuned on the basis of linear plasma models. Such linear models are usually obtained in the following ways:

- Linearization of plasma nonlinear equations;
- Some numerical linearization procedures of nonlinear plasma-physics codes [7,31,32];
- Using the first principal equations,
- In [33], another technique for obtaining linear plasma models from experimental observations was proposed. In this case, the plasma equilibrium is reconstructed from the recorded experimental signals using reconstruction codes, for example, FCDI [29], and then linear plasma models are obtained from the equilibria reconstructed at given time moments during the discharge. This approach makes it possible to construct linear, time-varying models for the development and investigation of feedback plasma control systems for time-dependent plasmas as controllable dynamical plants.

As a first approximation, mathematical models of plasma dynamics may be represented in the form of a set of ordinary differential equations with time-varying parameters, namely LPV (Linear Parameter-Varying) models, and having additive disturbances [34] for each plasma discharge in Tokamaks:

$$\begin{aligned} \dot{x}_i(t) &= A^{(i)}(t)x_i(t) + B^{(i)}(t)u_i(t) + f_i(t), \\ y_i(t) &= C^{(i)}(t)x_i(t) + w_i(t) \end{aligned}$$

where $x_i(t)$ is the state vector, $u_i(t)$ is the control action vector, $y_i(t)$ is the output vector, $A^{(i)}(t)$, $B^{(i)}(t)$, and $C^{(i)}(t)$ are time-varying matrices of the model, $f_i(t)$ and $w_i(t)$ are additive disturbance vectors at state and output equations, respectively, index i denotes a characteristic plasma discharge, $t \in \left[t_{0i}, T^{(i)}\right]$, $T^{(i)}$ is the duration of the discharge, and t_{0i} is the initial time of discharges when the plasma control system begins to operate.

The evolution of plasma position and currents in the Tokamak is described by Faraday's law [35]:

$$\frac{d}{dt}\Psi(J_\varphi, \psi_p) + RI = U$$

and motion equations

$$m\frac{d^2}{dt^2}\vec{r}_p = \vec{F}(J_\varphi, \psi_p)$$

The plasma shape is described by the gaps between the plasma surface and the VV $g(J_\varphi, \psi_p)$. Here $I \equiv \left[I_c^T, I_v^T, I_p\right]^T$ is the vector of currents in Tokamak coils and the elements of VV and plasma; Ψ, U, and R are the vectors of the magnetic flux through these circuits, the voltage applied to them, and the diagonal matrix of circuits electrical resistance, respectively; $\vec{r}_p \equiv [r_p, z_p]$ is coordinates of the plasma center of a mass; m is the plasma mass. These equations are to be linearized around the reconstructed plasma equilibrium, characterized by distributions of the toroidal current density J_φ and the poloidal flux ψ_p:

$$\begin{aligned}
&M(J_\varphi, \psi_p)\tfrac{d}{dt}\delta I + R\delta I + \tfrac{\partial}{\partial \vec{r}_p}\Psi(J_\varphi, \psi_p)\tfrac{d}{dt}\delta \vec{r}_p = \delta U, \\
&m\tfrac{d^2}{dt^2}\delta \vec{r}_p = \tfrac{\partial}{\partial I}\vec{F}(J_\varphi, \psi_p)\delta I + \tfrac{\partial}{\partial \vec{r}_p}\vec{F}(J_\varphi, \psi_p)\delta \vec{r}_p, \\
&\delta g = \tfrac{\partial}{\partial I}g(J_\varphi, \psi_p)\delta I + \tfrac{\partial}{\partial \vec{r}_p}g(J_\varphi, \psi_p)\delta \vec{r}_p.
\end{aligned}$$

In modern Tokamaks, plasma configurations are usually unstable in regards to vertical displacements and stable in regards to radial displacements. Therefore, small plasma masses may be neglected in the radial motion equation, and radial displacement δr_p can be expressed through vertical displacement δz_p and current disturbances δI. Introducing the state vector $x \equiv [\delta I^T, \delta z_p, \delta \dot{z}_p]^T$, the input vector $u \equiv \delta U$, and the output vector, $y = [\delta r_p, \delta z_p, \delta I_p, \delta I_c^T, \delta g^T]^T$, the model equations take a well-known state space form [34]:

$$\begin{aligned}
\dot{x} &= A(t)x + B(t)u, \\
y &= C(t)x.
\end{aligned}$$

The plasma equilibrium may significantly change during Tokamak discharges, and therefore matrices A, B, and C are dependent on time t. One way to obtain them is to reconstruct a sequence of the plasma equilibria, calculate corresponding time-invariant matrices, and then interpolate them, resulting in LPV models.

4.3. Plasma Control

At present, in the Globus-M2 Tokamak, the controllers C_Z, C_R, C_{CS}, C_{PF1}, C_{PF2_top}, C_{PF2_bottom}, C_{PF3}, and C_{CC} are the analog controllers realizing PD and P control laws. In the future, it is planned to replace all analog controllers of the Globus-M2 Tokamak with digital controllers, which, if necessary, can implement different control laws to optimize the magnetic plasma control system as a whole. Two thyristor current inverters, A_{HFC} and A_{VFC} [15], are used as actuators for plasma vertical and horizon position control, and thyristor multiphase rectifiers, A_{CS}, A_{PF1}, A_{PF3}, and A_{CC} [36], are used as actuators for control of the currents in the CS and PF-coils. At present, the proper values of the plasma current I_P and gaps $g = [g_1, \ldots, g_6]^T$ between the separatrix and the first wall are obtained by pre-programmed currents in the CS and PF-coils without shape feedback.

Therefore, when designing the plasma control system in this paper, the outputs R, Z, I_p, and $g_1, \ldots, g_6$ should be controlled by feedback. Let us introduce the errors in these controlled values:

$$\begin{aligned}
e_R &= r_R - R, \\
e_Z &= r_Z - Z, \\
e_{I_p} &= r_{I_p} - I_p, \\
e_{g_j} &= r_{g_j} - g_j,\ j = 1, \ldots, 6.
\end{aligned}$$

The setpoints of the controlled signals r_R, r_Z, r_{I_p}, r_{g_j}, $j = 1, \ldots, 6$ are, for example, calculated in advance by designing plasma scenarios to solve a direct equilibrium problem or set on the basis of experimental experience. The control problem can be formulated as the need to achieve proper control accuracy at finite time intervals:

$$\begin{aligned}
&\left|e_R^{(i)}(t)\right| < \varepsilon_R,\ t \in \left[\ t_1,\ T^{(i)}\ \right],\ i = 1, \ldots, N, \\
&\left|e_Z^{(i)}(t)\right| < \varepsilon_Z,\ t \in \left[\ t_2,\ T^{(i)}\ \right], \\
&\left|e_{I_p}^{(i)}(t)\right| < \varepsilon_{I_p},\ t \in \left[\ t_3,\ T^{(i)}\ \right], \\
&\left|e_{g_j}^{(i)}(t)\right| < \varepsilon_{g_j},\ t \in \left[\ t_4,\ T^{(i)}\ \right],\ j = 1, \ldots, 6,
\end{aligned}$$

where $t_1, \ldots, t_4$ are the moments of time at which the output values fall into the given tubes, defined by positive constants ε_R, ε_R, ε_{I_p}, and ε_g, after transients caused by the tracking procedure or minor disruptions. The time moments $t_1, \ldots, t_4$ are different due to different plant dynamics on different controlled channels. Index $i = 1, \ldots, N$ is the number of the corresponding plasma discharge.

In order to control the outputs of the plant in accordance with the objectives, it is necessary to make sure that the instability in the vertical direction of the plasma is stabilized and have a sufficiently large controllability region under the input constraints. Also, the separatrix for controlling the shape of the plasma must have a reachability region large enough to position it near the first wall at the desired location. A study of the design of the poloidal system of the Globus-M Tokamak was performed in [37]. The poloidal system in this case has a sufficiently large region of vertical controllability and allows for internally stable regimes of plasma position control systems [28]. The multidimensional reachability region of the separatrix of this poloidal system has lower and upper bounds, which give the possibility to change the location of the separatrix near the first wall accordingly.

For reliable plasma control, it is not only necessary to achieve the desired control accuracy, but it is also important to provide the necessary stability margins in the design of the control system. For multichannel control systems, these margins can be specified and obtained in three ways:

- By means of disconnection of each channel and estimation of phase and amplitude stability margins with the help of the Nyquist hodograph [36,38];
- However, this method may be too optimistic, so in [38], new disk-based stability margins are introduced for SISO and MIMO control systems, which may be closer to reality;
- When the number of channels is large, it is more convenient to use another approach, which makes it possible to characterize stability by only one quantity, namely the robust stability margin. By means of the usage of the small gain theorem [39,40], when the result of the multiplication of H_∞-norms of transfer functions of the uncertainty Δ and the known part of the system Q is less than 1: $\|\Delta\|_\infty \|Q\|_\infty < 1$. In that case, the robust stability analysis gives useful results [41,42], and one can estimate the robust stability margin in the following form:

$$\|\Delta\|_\infty < \frac{1}{\|Q\|_\infty}.$$

5. Methodology for Designing Hierarchical Cascade Systems of Magnetic Plasma Control in *D*-Shaped Tokamaks

This section is devoted to the general notions of further development of the magnetic plasma control systems for *D*-shaped Tokamaks, which arose from the works on the plasma control systems in the Globus-M/M2 Tokamak and reviews on the magnetic plasma control systems in *D*-shaped Tokamaks [3–7] with feedback. A general scheme of the magnetic plasma control systems in them was obtained, which is shown in Figure 1. Most of the magnetic plasma control systems in operating Tokamaks are organized according to this scheme. Figure 10 shows, for the purpose of generalization, a simplified basic scheme of the magnetic plasma control system in *D*-shaped Tokamaks when the actuators are brought into the controlled plant (P).

The resulting basic structural scheme for generality can be represented in operator form by showing the model of the plant P and the controllers C_1, C_2, and C_3 as operators that map the space of input signals into the space of output signals. If the model of the plant P and the controllers is considered linear, then the scheme of Figure 10 connects the input and output signals as follows:

$$\begin{bmatrix} y_1 \\ y_2 \\ y_3 \end{bmatrix} = P \begin{bmatrix} u_1 \\ u_2 \end{bmatrix}, \quad P = \begin{bmatrix} P_{11} & P_{12} \\ P_{21} & P_{22} \\ P_{31} & P_{32} \end{bmatrix}. \qquad (1)$$

From (1) and the scheme of Figure 10, we can obtain a system of equations that describes this cascading hierarchical scheme:

$$\begin{aligned} y_1 &= P_{11}u_1 + P_{12}u_2, \\ y_2 &= P_{21}u_1 + P_{22}u_2, \\ y_3 &= P_{31}u_1 + P_{32}u_2, \\ u_1 &= C_1(r_1 - y_1), \\ u_2 &= C_2(r_2 - y_2), \\ r_2 &= C_3(r_3 - y_3). \end{aligned} \quad (2)$$

Excluding from (2) the quantities u_1, u_2, we obtain y_1 in the form

$$y_1 = -(I + P_{11}C_1)^{-1}P_{12}C_2y_2 - (I + P_{11}C_1)^{-1}O_{12}C_2C_3y_3 + (I + P_{11}C_1)^{-1}(P_{11}C_1r_1 + P_{12}C_2C_3r_3) \quad (3)$$

and arrive at a system of equations of the form:

$$\begin{aligned} a_{11}y_2 + a_{12}y_3 &= R_{21}r_1 + R_{23}r_3, \\ a_{21}y_2 + a_{22}y_3 &= R_{31}r_1 + R_{33}r_3. \end{aligned} \quad (4)$$

Here

$$\begin{aligned} a_{11} &= -P_{21}C_1(I + O_{11}C_1)^{-1}P_{12}C_2 + (I + P_{22}C_2), \\ a_{12} &= -P_{21}C_1(I + P_{11}C_1)^{-1}P_{12}C_2C_3 + P_{22}C_2C_3, \\ a_{21} &= -P_{31}C_1(I + P_{11}P_1)^{-1}P_{12}c_2 + P_{32}C_2, \\ a_{22} &= -P_{31}C_1(I + P_{11}C_1)^{-1}P_{12}C_2C_3 + (I + P_{32}C_2C_3), \end{aligned}$$

$$\begin{aligned} R_{21} &= P_{21}C_1(I - (I + P_{11}C_1)^{-1}P_{11}C_1), \\ R_{23} &= P_{22}C_2C_3 - P_{21}C_1(I + P_{11}C_1)^{-1}P_{12}C_2C_3, \\ R_{31} &= P_{31}C_1(I - (I + P_{11}C_1)^{-1}P_{11}C_1), \\ R_{33} &= P_{32}C_2C_3 - P_{31}C_1(I + P_{11}C_1)^{-1}P_{12}C_2C_3. \end{aligned}$$

Solving the system (4) and substituting the solutions in (3), we obtain the required solutions in the form:

$$\begin{aligned} & y_1 = (I + P_{11}C_1)^{-1}[P_{11}C_1 - P_{12}C_2H_{21} - P_{12}C_2C_3H_{31}]r_1 + \\ & +(I + P_{11}C_1)^{-1}[P_{12}C_2C_3 - P_{12}C_2H_{23} - P_{12}C_2C_3H_{33}]r_3 = H_{11}r_1 + H_{13}r_3, \\ & y_2 = (a_{11}^{-1}R_{21} - a_{11}^{-1}a_{12}H_{31})r_1 + (a_{11}^{-1}R_{23} - a_{11}^{-1}a_{12}H_{33})r_3 = H_{21}r_1 + H_{23}r_3, \\ & y_3 = A_3{}^{-1}(R_{31} - a_{21}a_{11}{}^{-1}R_{21})r_1 + A_3{}^{-1}(R_{33} - a_{21}a_{11}{}^{-1}R_{23})r_3 = H_{31}r_1 + H_{33}r_3, \\ & A_3 = \left(a_{22} - a_{21}a_{11}^{-1}a_{12}\right). \end{aligned} \quad (5)$$

Or, by substituting the expressions for $R_{21}, R_{23}, R_{31}, R_{33}$ in (5), we arrive at the equations of connection between the input actions r_1, r_2 of the closed-loop system and the outputs y_1, y_2, y_3:

$$\begin{bmatrix} y_1 \\ y_2 \\ y_3 \end{bmatrix} = H_0\begin{bmatrix} r_1 \\ r_3 \end{bmatrix}, \; H_0 = \begin{bmatrix} H_{11} & H_{13} \\ H_{21} & H_{23} \\ H_{31} & H_{33} \end{bmatrix}, \quad (6)$$

where

$$\begin{aligned} H_{11} &= (I + P_{11}C_1)^{-1}[P_{11}C_1 - P_{12}C_2H_{21} - P_{12}C_2C_3H_{31}], \\ H_{13} &= (I + P_{11}C_1)^{-1}[P_{12}C_2C_3 - P_{12}C_2H_{23} - P_{12}C_2C_3H_{33}], \\ H_{21} &= a_{11}^{-1}R_{21} - a_{11}^{-1}a_{12}H_{31}, \\ H_{23} &= a_{11}^{-1}R_{23} - a_{11}^{-1}a_{12}H_{33}, \\ H_{31} &= A_3{}^{-1}(R_{31} - a_{21}a_{11}{}^{-1}R_{21}), \\ H_{33} &= A_3{}^{-1}(R_{33} - a_{21}a_{11}{}^{-1}R_{23}). \end{aligned}$$

The obtained solution (6) for (2) is a general solution of the initial system and gives the dependence of the output values of the basic scheme in Figure 10 on the references. To obtain numerical solutions for various structural schemes of hierarchical cascade plasma control systems, the MATLAB/Simulink computer environment is used as the most developed computer tool for solving the problems of automatic control.

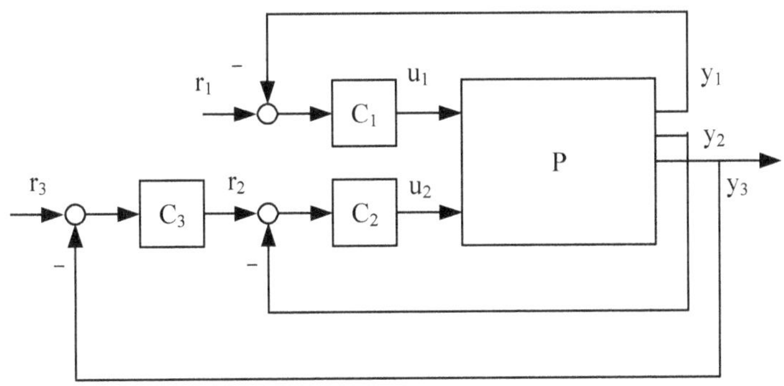

Figure 10. Basic scheme of magnetic plasma control systems in *D*-shaped Tokamaks to obtain generalized schemes.

In Figure 11, a diagnostics unit for the controlled plant is integrated at the lower control level, which actually solves the identification problem, i.e., it reconstructs the signals necessary for control in a closed-loop system based on the measurement signals [29,42,43].

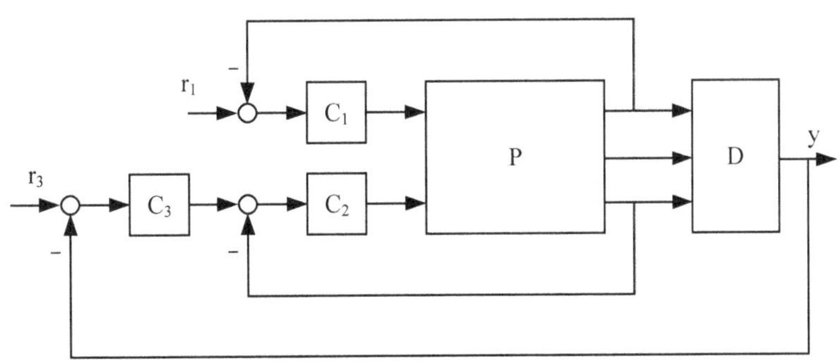

Figure 11. Integration of the diagnostic unit *D* into the control system.

Figure 12 introduces an adaptation level, by means of which the controller C_3 adapts the control system to a preformed control scenario. Adaptation can be carried out by different methods: by switching regulators from a pre-synthesized set or by their interpolation, which leads to a controller with variable parameters [28,44].

When the cascades with controllers C_1 and C_2–C_3 work, a contradiction (conflict) may occur between them since these cascades are connected to each other through the controlled plant *P*. To eliminate this conflict, the matching cascade with the controller C_4 is introduced, which takes into account signals from the controller C_3 and additional signals from the plant *P* by producing matching reference actions on the cascade with the controller C_1 (Figure 13) [42].

All three upgrades in Figures 11–13 of the basic scheme of Figure 10 lead to a generalized hierarchical cascade system (Figure 14), which combines all introductions: integration of diagnostics (Figure 11), adaptation to the scenario (Figure 12), and matching cascade (Figure 13).

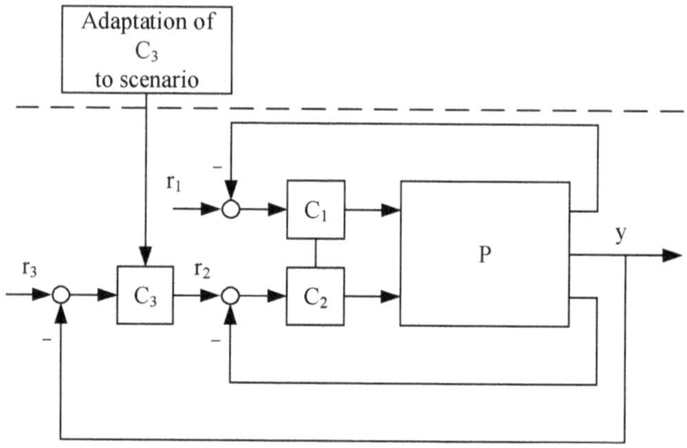

Figure 12. Introduction of the level of adaptation to the scenario in the C_3 controller.

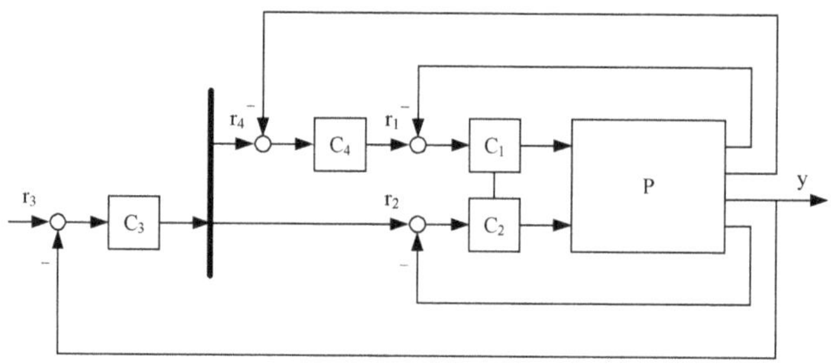

Figure 13. Introduction of the matching cascade with the C_4 controller into the system.

The structural diagram of Figure 14 makes it possible to proceed to further generalization in the form of building a hierarchical system with generalized levels of control (Figure 15). At the lower level, there is a non-stationary, controlled plant with actuators and diagnostics (sensors), exposed to uncontrolled perturbations. The controlled plant is influenced by the level of multivariable cascade robust control, which solves the basic problem of controlling the plant and ensures, due to coordinated control cascades and robustness, its operability in basic control scenarios under the action of uncontrolled perturbations.

To enable the system to operate under changing plant parameters, an adaptation level is introduced, which improves control accuracy and extends stability margins. A higher level of artificial intelligence makes it possible to introduce, in particular, artificial neural networks, which allow, for example, to solve the problem of identification at the rate of observations. The highest level of decision-making is left to humans, who design the hierarchical control system and introduce technical solutions at all its levels, ensuring their interaction. The possibility of using hierarchical systems for plasma control in Tokamaks was indicated earlier in [45].

The carried-out system generalizations arising from the interaction of theory and practice of control made it possible to develop a methodology for designing hierarchical cascade control systems with information feedback, the logic of which is presented in Figure 16. The interaction of theory and practice through processed feedback information flows allows us to eliminate or significantly reduce the gaps between theory and practice, which were revealed in [46]. System analysis of the structures of hierarchical cascade control systems derived from the practice of magnetic plasma control in Tokamaks leads to structures useful for practice, the integration of which provides an overall structure and a generalized structure with a number of hierarchical levels, the interaction of which leads to the achievement of objectives at each level. Control science generates, using synthesis and analysis methods, control algorithms for each level of the hierarchy. And the hierarchical cascade control system has almost unlimited possibilities to combine different control methods at different levels to achieve its objectives at each level, which leads to the achievement of the main objective of the lower-level system. The combination of robust and adaptive methods using identification methods and artificial intelligence is most effective for solving the problems of controlling dynamic plants with variable parameters over a limited time interval.

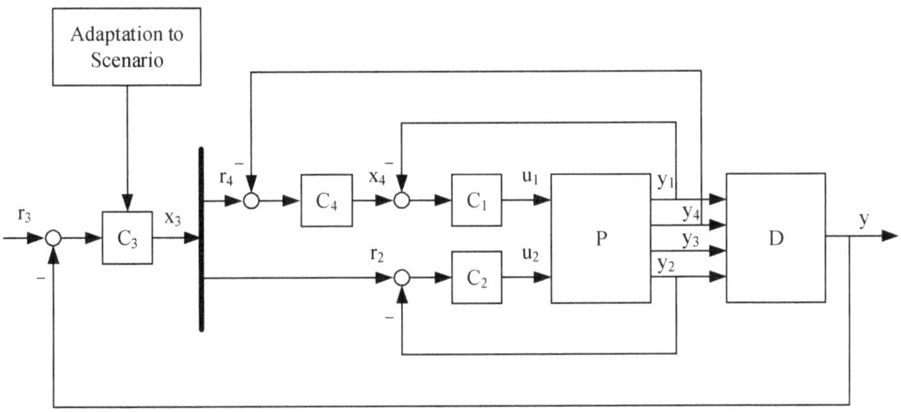

Figure 14. Generalized hierarchical cascade diagram of magnetic plasma control with feedback.

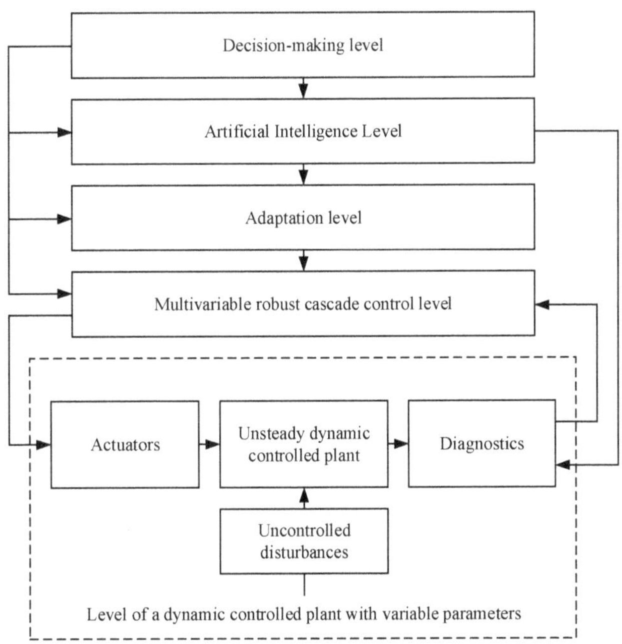

Figure 15. Multilevel hierarchical cascade control system.

There are only three known monographs on the theory and practice of hierarchical automatic control systems [47–49], as well as journal articles, for example [50], and conference papers, in particular [45,51]. It is impossible to construct a unified structure for a hierarchical control system because of the possibility of an almost infinite combination of different control algorithms at different levels of the hierarchy. For this reason, in works on hierarchical control systems, various structural schemes of hierarchical systems are constructed based on applied control tasks and desired control objectives at different levels [45–51]. The same approach is applied in this work to generalize the structural systems of hierarchical magnetic plasma control in Tokamaks. Through this generalization, a general structural diagram of a hierarchical cascade control system (Figure 15) and a methodology for designing systems of this class (Figure 16) were obtained.

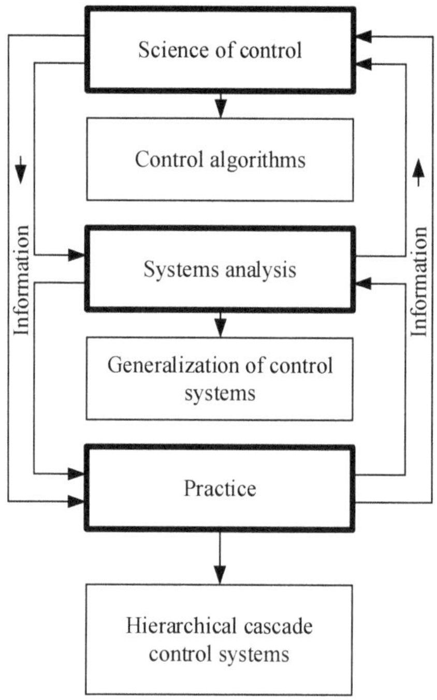

Figure 16. Logical scheme of the design methodology of hierarchical cascade control systems with information feedback.

The above generalization of the structural schemes of hierarchical cascade control systems and the methodology developed for their design is a methodological approach "from above" for the development of magnetic plasma control systems in Tokamaks. At the same time, the methodological approach "from below" was developed in [33], which gives recommendations and an action plan for direct synthesis, analysis, and implementation of magnetic plasma control systems in physical experiments of *D*-shaped Tokamaks, which consists of the following (Figure 17):

- Reconstruction of plasma equilibrium, namely, plasma current distribution and poloidal flux, using measured experimental signals of magnetic diagnostics outside the plasma based on new modifications of known methods;
- Obtaining linear plasma models based on experimental data of Tokamak plasma discharges with respect to the reconstructed plasma equilibrium;
- Application of the model of a thyristor current inverter or voltage inverter in PWM mode as a plasma position control actuator for the design and modeling of plasma control systems;

- Application of an original analysis of the structure of a multivariable plasma control system using a relative gain array (RGA) and μ-analysis procedures to select the optimal system structure in terms of input-output controllability;
- Application of a plasma shape control system with simultaneous adaptive adjustment of the vertical and horizontal plasma positions. This is conducted in order to remove the pole near zero while stabilizing the velocity of the vertical plasma position relative to zero to increase the degree of stability of the system in controlling the plasma shape;
- Application of a new approach to modeling the plasma shape control system with a feedback plasma reconstruction algorithm using experimental scenario data and a linear plasma model with time-varying parameters;
- Implementation of the plasma control system on the testbed to adjust the system in real time;
- Application of the developed control system to a real Tokamak by switching the controller from the plant model to the Tokamak using the real-time testbed.

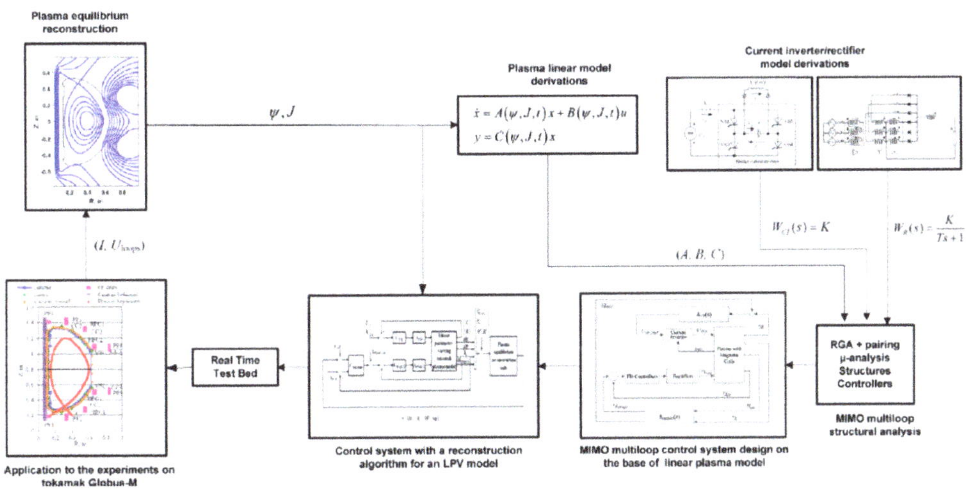

Figure 17. Illustrative diagram of the design methodology of the hierarchical cascade magnetic plasma control system in the Tokamak [33].

6. Real-Time Test Bed for Plasma Magnetic Control in Tokamaks

The real-time test bed is in operation at the Trapeznikov Institute of Control Sciences of the Russian Academy of Sciences (ICS RAS) [52,53]. The real-time test bed was created by a consortium of Lomonosov Moscow State University (MSU) and ICS RAS. It consists of two Speedgoat Performance Real-Time Target Machines (RTTM) (Figure 18): the "Plant Model" (MSU) and the "Controller" (ICS RAS). Both RTTMs run under the SimulinkRT real-time operating system. The real-time algorithms are developed in the MATLAB/Simulink programming environment, and then, using Embedded Coder, the C-code is generated, which is then compiled into the real-time application. The RTTMs are powered by an Intel i7-7700 K CPU running at 4.2 GHz. The RTTM "Controller" contains MIMO switches from an externally controlled plant model to an internally controlled plant model, which is used to synthesize and test the control system under development.

The digital twin [54] is a dynamic/self-developing digital/virtual model or simulation of a real plant that provides the exact state of its physical twin at any given time through real-time data exchange (Figure 19) and data storage. Not only does the digital twin imitate its physical twin, but any changes in the digital twin are reflected by the physical twin.

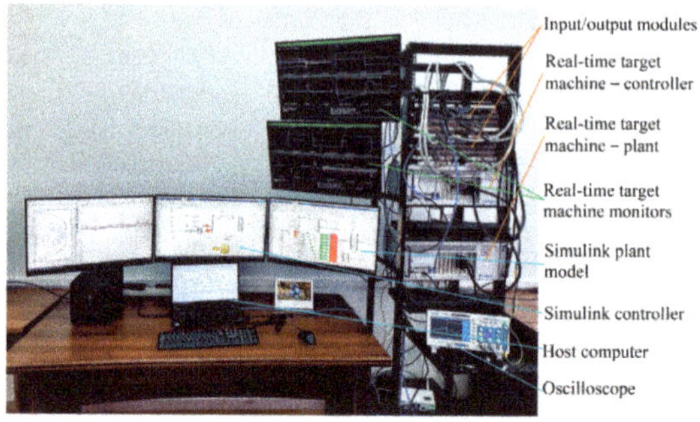

Figure 18. Real-time digital testbed at the V.A. Trapeznikov Institute of Control Sciences of the Russian Academy of Sciences.

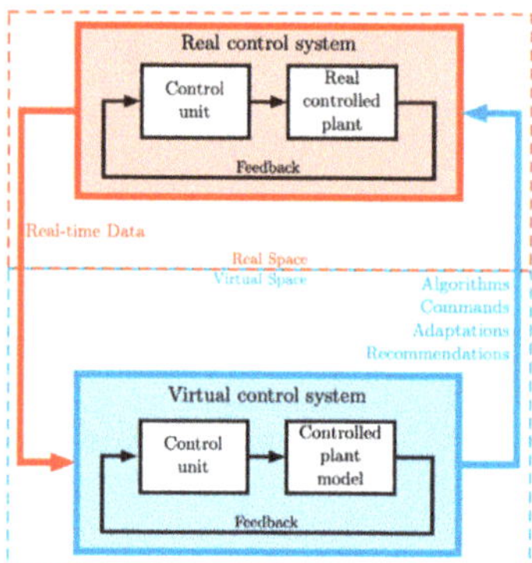

Figure 19. The concept of the digital twin of the control system.

The use of the digital twin provides an increase in the efficiency of the development and implementation of plasma control systems in Tokamaks. Work on the creation of a digital twin of the Globus-M2 Tokamak is being carried out. The virtual part of the digital twin is implemented at the real-time test bed in the ICS RAS, and the physical part will be implemented through the implementation of the third RTTM in the information and control system of the Globus-M2 Tokamak, with both parts remotely exchanging data in real time.

The block diagram of the real-time control system of plasma position, current, and shape for the Globus-M2 Tokamak is shown in Figure 20. The control system is split into two RTTMs. The RTTM "Controller" contains the plasma equilibrium reconstruction algorithm in the feedback [42,43,55] and the external control cascade of the plasma current and shape. The RTTM "Plant Model" contains the internal control cascade of the plasma position and currents in the poloidal field coils, models of the coil power supplies, and the model of the Tokamak plasma and diagnostic system. The control system operates in deviations from the scenario values, so the model outputs are deviations of currents and fluxes from the plasma discharge scenario values. All controllers in that hierarchical control system are synthesized by the method of linear matrix inequalities (LMI) using the approach from [56]. Figure 21 shows real-time simulation results where the control system shapes plasma to provide specified strike point positions and gap values g_1–g_6 (Figure 4b), with the sample time equal to the discretization step of the model $T_s = 100$ µs. The measured average TET (task execution time) value is 67 µs, which satisfies the real-time requirement of TET < T_s.

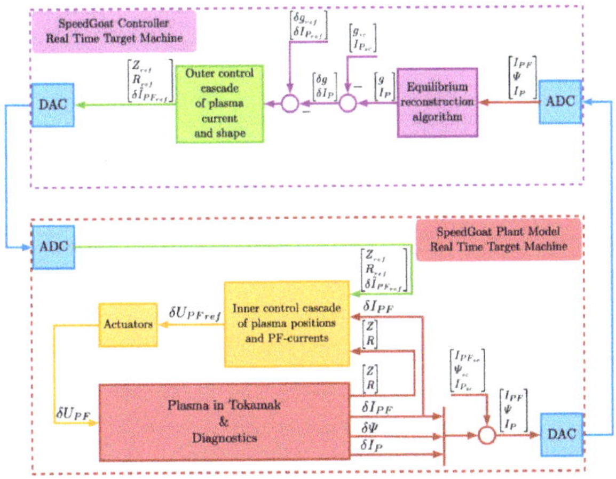

Figure 20. The block diagram of the real-time control system of plasma position, current, and shape for the Globus-M2 Tokamak (meaning of colors: yellow—actuators and controllers for plasma vertical and horizontal positions as well as for currents in PF-coils; blue—DAC/ADC; violet—algorithm of plasma equilibrium reconstruction in the feedback; green—multivariable controller of plasma shape and current).

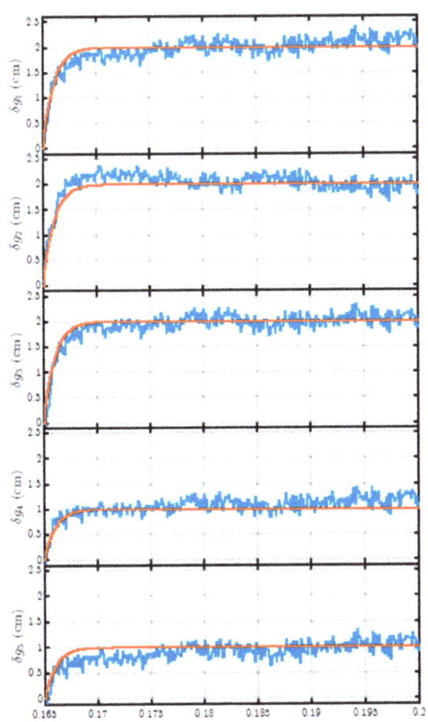

Figure 21. Real-time simulation results of the plasma position, current, and shape control system for the Globus-M2 Tokamak.

7. Conclusions

The paper shows the state-of-the-art in plasma poloidal systems in *D*-shaped Tokamaks. Because plasma magnetic control systems are of the utmost priority for modern Tokamaks, the mathematical statement of the problem on this subject was conducted to see the basic directions to be developed. On the basis of that, a new generalization of plasma magnetic control systems in the area of designing hierarchical cascade control systems was presented. This direction concludes a lot of new possibilities for developing plasma control systems in Tokamaks including developing plasma control systems on a real-time testbed and unifying the testbed with the real plasma control system on the Tokamak creating a new area of digital twin.

Funding: This work was supported by the Russian Science Foundation, grant № 21-79-20180.

Institutional Review Board Statement: Not applicable.

Informed Consent Statement: Not applicable.

Data Availability Statement: All new data obtained in this work are summarized in the contents of the paper.

Conflicts of Interest: The author declares no conflict of interest.

References

1. Wesson, J. *Tokamaks*, 3rd ed.; Clarendon Press: Oxford, UK, 2004.
2. Artsimovich, L. Tokamak devices. *Nucl. Fusion* **1972**, *12*, 215–252. [CrossRef]
3. Mitrishkin, Y.V.; Korenev, P.S.; Prokhorov, A.A.; Kartsev, N.M.; Patrov, M.I. Plasma Control in Tokamaks. Part 1. Controlled thermonuclear fusion problem. Tokamaks: Components of control systems. *Adv. Syst. Sci. Appl.* **2018**, *18*, 26–52. [CrossRef]
4. Mitrishkin, Y.V.; Kartsev, N.M.; Pavlova, E.A.; Prohorov, A.A.; Korenev, P.S.; Patrov, M.I. Plasma Control in Tokamaks—Part 2: Magnetic Plasma Control Systems. *Adv. Syst. Sci. Appl.* **2018**, *18*, 39–78.
5. Mitrishkin, Y.; Kartsev, N.; Konkov, A.; Patrov, M. Plasma Control in Tokamaks—Part 3.1: Plasma Magnetic Control Systems in ITER. *Adv. Syst. Sci. Appl.* **2020**, *20*, 82–97. [CrossRef]
6. Mitrishkin, Y.; Kartsev, N.; Konkov, A.; Patrov, M. Plasma Control in Tokamaks. Part 3.2. Simulation and realization of plasma control systems in ITER and constructions of DEMO. *Adv. Syst. Sci. Appl.* **2020**, *20*, 136–152. [CrossRef]
7. Mitrishkin, Y.V.; Kartsev, N.M.; Kuznetsov, E.A.; Korostelev, A.Y. *Methods and Systems of Plasma Magnetic Control in Tokamaks*; KRASAND: Moscow, Russia, 2020; p. 528. (In Russian)
8. Humphreys, D.A.; Casper, T.A.; Eidietis, N.; Ferrara, M.; Gates, D.A.; Hutchinson, I.H.; Jackson, G.L.; Kolemen, E.; Leuer, J.A.; Lister, J.; et al. Experimental vertical stability studies for ITER performance and design guidance. *Nucl. Fusion* **2009**, *49*, 11. [CrossRef]
9. Stein, G. Respect the Unstable. *IEEE Control Syst. Mag.* **2003**, *23*, 12–25.
10. Mitrishkin, Y.V. *Plasma Control in Experimental Thermonuclear Installations: Adaptive Auto-Oscillatory and Robust Control Systems*; KRASAND: Moscow, Russia, 2016; p. 400. (In Russian)
11. Yuan, Q.P.; Xiao, B.J.; Luo, Z.P.; Walker, M.L.; Welander, A.S.; Hyatt, A.; Mueller, D. Plasma current, position and shape feedback control on EAST. *Nucl. Fusion* **2013**, *53*, 043009. [CrossRef]
12. Mertens, V.; Raupp, G.; Treutterer, W. Chapter 3: Plasma Control in ASDEX Upgrade. *Fusion Sci. Technol.* **2003**, *44*, 593–604.
13. Anand, H.; Coda, S.; Felici, F.; Galperti, C.; Moret, J.-M. A novel plasma position and shape controller for advanced configuration development on the TCV tokamak. *Nucl. Fusion* **2017**, *57*, 126026. [CrossRef]
14. Minaev, V.B.; Gusev, V.K.; Sakharov, N.V.; Varfolomeev, V.I.; Bakharev, N.N.; Belyakov, V.A.; Zhilin, E.G. Spherical tokamak Globus-M2: Design, integration, construction. *Nucl. Fusion* **2017**, *57*, 066047.
15. Kuznetsov, E.A.; Mitrishkin, Y.V.; Kartsev, N.M. Current inverter as self-oscillating actuator in applications for plasma position control systems in the Globus-M/M2 and T-11M tokamaks. *Fusion Eng. Des.* **2019**, *143*, 247–258.
16. Walker, M.L.; De Vries, P.; Felici, F.; Schuster, E. Introduction to Tokamak Plasma Control. In Proceedings of the 2020 American Control Conference, Denver, CO, USA, 1–3 July 2020; pp. 2901–2918.
17. Creely, A.J.; Greenwald, M.J.; Ballinger, S.B.; Brunner, D.; Canik, J.; Doody, J.; Garnier, D.T.; Granetz, R.; Gray, T.K.; Holland, C.; et al. Overview of the SPARC tokamak. *J. Plasma Phys.* **2020**, *86*, 865860502. [CrossRef]
18. Mitrishkin, Y.V.; Korenev, P.S.; Konkov, A.E.; Kartsev, N.M.; Smirnov, I.S. New horizontal and vertical field coils with optimised location for robust decentralized plasma position control in the IGNITOR tokamak. *Fusion Eng. Des.* **2022**, *174*, 112993. [CrossRef]
19. Ishida, S.; Barabaschi, P.; Kamada, Y.; the JT-60SA Team. Status and prospect of the JT-60SA project. *Fusion Eng. Des.* **2010**, *85*, 2070–2079. [CrossRef]
20. Chuyanov, V.; Gryaznevich, M. Modular fusion power plant. *Fusion Eng. Des.* **2017**, *122*, 238–252. [CrossRef]

21. European Research Roadmap to the Realization of Fusion Energy. 2018. Available online: https://www.euro-fusion.org/eurofusion/roadmap/ (accessed on 14 December 2022).
22. Khvostenko, P.P.; Anashkin, I.O.; Bondarchuk, E.N.; Chudnovsky, A.N.; Kavin, A.A.; Khvostenko, A.P.; Sushkov, A.V. Current status of tokamak T-15MD. *Fusion Eng. Des.* **2021**, *164*, 112211. [CrossRef]
23. Mitrishkin, Y.V.; Zenckov, S.M.; Kartsev, N.M.; Efremov, A.A.; Dokuka, V.N.; Khayrutdinov, R.R. Linear and impulse control systems for plasma unstable vertical position in elongated tokamak. In Proceedings of the 51st IEEE Conference on Decision and Control, Maui, HI, USA, 10–13 December 2012; pp. 1697–1702.
24. Mitrishkin, Y.V.; Kartsev, N.V.; Zenkov, S.M. Stabilization of unstable vertical position of plasma in T-15 tokamak: I. *Autom. Remote Control* **2014**, *75*, 281–293. [CrossRef]
25. Mitrishkin, Y.V.; Kartsev, N.V.; Zenkov, S.M. Stabilization of unstable vertical position of plasma in T-15 tokamak: II. *Autom. Remote Control* **2014**, *75*, 1565–1576.
26. Konovalov, S.; Bondarchuk, E.; Krasilnikov, A.; Kuyanov, A.Y.; Khayrutdinov, R.R. Tokamak with Reactor Technologies Concept. In Proceedings of the 28th IAEA Fusion Energy Conference (FEC 2020), Virtual, 10–15 May 2020.
27. Besekerskiy, V.A.; Popov, E.P. *Theory of Automatic Control Systems*; Nauka: Moscow, Russia, 1972; p. 768. (In Russian)
28. Mitrishkin, Y.V.; Prokhorov, A.A.; Korenev, P.S.; Patrov, M.I. Hierarchical robust switching control method with the Improved Moving Filaments equilibrium reconstruction code in the feedback for tokamak plasma shape. *Fusion Eng. Des.* **2019**, *138*, 138–150.
29. Korenev, P.S.; Mitrishkin, Y.V.; Patrov, M.I. Reconstruction of Equilibrium Distribution of Tokamak Plasma Parameters by External Magnetic Measurements and Construction of Linear Plasma Models. *Mekhatronika Avtom. Upr.* **2016**, *17*, 254–266. [CrossRef]
30. Ariola, M.; Pironti, A. *Magnetic Control of Tokamak Plasmas*, 2nd ed.; Springer International Publishing: Cham, Switzerland, 2016. [CrossRef]
31. Dokuka, V.N.; Kadurin, A.V.; Mitrishkin, Y.V.; Khayrutdinov, R.R. Synthesis and modeling of the H_∞-system of magnetic control of the plasma in the tokamak-reactor. *Autom. Remote Control* **2007**, *68*, 1410–1428. [CrossRef]
32. Mitrishkin, Y.; Prokhorov, A.; Korenev, P.; Patrov, M. Tokamak plasma magnetic control system simulation with reconstruction code in feedback based on experimental data. In Proceedings of the 2017 IEEE 56th Annual Conference on Decision and Control (CDC), Melbourne, VIC, Australia, 12–15 December 2017; IEEE: New York, NY, USA, 2017; pp. 2360–2365. [CrossRef]
33. Mitrishkin, Y.V.; Korenev, P.S.; Kartsev, N.M.; Kuznetsov, E.A.; Prokhorov, A.A.; Patrov, M.I. Plasma magnetic cascade multiloop control system design methodology in a tokamak. *Control Eng. Pract.* **2019**, *87*, 97–110. [CrossRef]
34. D'Angelo, H. *Linear Time-Varying Systems: Analysis and Design*; Allyn and Bacon: Boston, MA, USA, 1970.
35. Mitrishkin, Y.V.; Kartsev, N.M.; Prokhorov, A.A.; Pavlova, E.A.; Korenev, P.S.; Konkov, A.E.; Kruzhkov, V.I.; Ivanova, S.L. Tokamak plasma models development for plasma magnetic control systems design by first principle equations and identification approach. *Procedia Comput. Sci.* **2021**, *186*, 466–474. [CrossRef]
36. Mitrishkin, Y.; Pavlova, E.; Kuznetsov, E.; Gaydamaka, K. Continuous, saturation, and discontinuous tokamak plasma vertical position control systems. *Fusion Eng. Des.* **2016**, *108*, 35–47. [CrossRef]
37. Mitrishkin, Y.V.; Kruzkov, V.I.; Korenev, P.S. Methodology of Plasma Shape Reachability Area Estimation in D-Shaped Tokamaks. *Mathematics* **2022**, *10*, 4605. [CrossRef]
38. Seiler, P.; Packard, A.; Gahinet, P. An Introduction to Disk Margins: Lecture Notes. *IEEE Control Syst.* **2020**, *40*, 78–95. [CrossRef]
39. Skogestad, S.; Postlethwaite, I. *Multivariable Feedback Control: Analysis and Design*, 2nd ed.; John Wiley & Sons, Ltd.: New York, NY, USA, 2005.
40. Zhou, K.; Doyle, J. *Essentials of Robust Control*; Prentice Hall: Upper Saddle River, NJ, USA, 1998.
41. Mitrishkin, Y.; Kurachi, K.; Kimura, H. Plasma multivariable robust control system design and simulation for a thermonuclear tokamak-reactor. *Int. J. Control* **2003**, *76*, 1358–1374. [CrossRef]
42. Mitrishkin, Y.V.; Prokhorov, A.A.; Korenev, P.S.; Patrov, M.I. Plasma magnetic time-varying nonlinear robust control system for the Globus-M/M2 tokamak. *Control Eng. Pract.* **2020**, *100*, 104446. [CrossRef]
43. Prokhorov, A.; Mitrishkin, Y.; Korenev, P.; Patrov, M. The plasma shape control system in the tokamak with the neural network as a plasma equilibrium reconstruction algorithm. *IFAC PapersOnLine* **2020**, *53*, 857–862. [CrossRef]
44. Mitrishkin, Y.V.; Korenev, P.S.; Prohorov, A.A.; Patrov, M.I. Robust H_∞ switching MIMO control for a plasma time-varying parameter model with a variable structure in a tokamak. In Proceedings of the IFAC 2017 World Congress, Toulouse, France, 9–14 July 2017. *IFAC PapersOnLine* **2017**, *50*, 11385–11390. [CrossRef]
45. Mitrishkin, Y.V.; Haber, R.E. Intelligent Hierarchical Control System for Complex Processes. In Proceedings of the 6th International Conference on Informatics in Control, Automation and Robotics—ICINCO 2009: Intelligent Control Systems and Optimization, Italy, Milan, 2–5 July 2009; pp. 333–336.
46. Bars, R.; Colaneri, P.; de Souza, C.E.; Allgöwer, F.; Kleimenov, A.; Scherer, C. Trends in Theory of Control System Design: Status report prepared by the IFAC Coordinating committee on Design Methods. In Proceedings of the 17th World Congress the International Federation of Automatic Control, Seoul, Republic of Korea, 6–11 July 2008; pp. 2144–2155.
47. Mesarović, M.D.; Macko, D.; Takahara, Y. Theory of hierarchical, multilevel systems. In *Mathematics in Science and Engineering*; Academic Press: Cambridge, MA, USA, 1970; Volume 68.

48. Findeisen, W.; Bailey, F.N.; Brdys, M.; Malinowski, K.; Taljewski, P.; Wozniak, A. *Control and Coordination in Hierarchical Systems*; International Series on Applied Systems Analysis, 9; A Wiley-Interscience Publication, International Institute for Applied Systems Analysis; John Wiley & Sons: New York, NY, USA, 1980; ISBN 9780471277422.
49. Vasiliev, V.I.; Gusev, Y.M.; Efanov, V.N.; Krymsky, V.G.; Rutkovsky, V.Y.; Semeran, V.A. *Multilevel Control of Dynamic Plants*; Nauka: Moscow, Russia, 1987. (In Russian)
50. Karimoddini, A.; Lin, H.; Chen, B.M.; Lee, T.H. Hierarchical hybrid modelling and control of an unmanned helicopter. *Int. J. Control* **2014**, *87*, 1779–1793. [CrossRef]
51. Liu, J.; Xiaojun, L.; Koo, T.-K.J.; Sinopoli, B.; Sastry, S.; Lee, E.A. A Hierarchical Hybrid System Model and Its Simulation. In Proceedings of the 38th Conference on Decision & Control, Phoenix, AZ, USA, 7–10 December 1999; pp. 3508–3513.
52. Mitrishkin, Y.V. Plasma magnetic control systems in D-shaped tokamaks and imitation digital computer platform in real time for controlling plasma current and shape. *Adv. Syst. Sci. Appl.* **2022**, *22*, 1–14. [CrossRef]
53. Konkov, A.E.; Mitrishkin, Y.V. Comparison Study of Power Supplies in Real-Time Robust Control Systems of Vertical Plasma Position in Tokamak. *IFAC—PapersOnLine* **2022**, *55*, 327–332. [CrossRef]
54. Singh, M.; Fuenmayor, E.; Hinchy, E.P.; Qiao, Y.; Murray, N.; Devine, D. Digital Twin: Origin to Future. *ASI* **2021**, *4*, 36. [CrossRef]
55. Mitrishkin, Y.V.; Prokhorov, A.A.; Korenev, P.S.; Patrov, M.I. Method of Forming a Model of Magnetic Control of Plasma Shape and Current with Feedback in. Tokamak. Patent No. 2702137, 28 April 2017.
56. Konkov, A.E.; Mitrishkin, Y.V.; Korenev, P.S.; Patrov, M.I. Robust Cascade LMI Design of MIMO Control System for Plasma Position, Current, and Shape Model with Time-Varying Parameters in a Tokamak. *IFAC—PapersOnLine* **2022**, *53*, 7344–7349. [CrossRef]

engineering proceedings

Proceeding Paper

Robust System of Algorithms for the Functioning of Biocompatible Artificial Liver Devices †

Alexey Ganshin and Denis Andrikov *

Academy of Engineering, RUDN University, Mikluho-Maklaya str. 6, Moscow 117198, Russia; 1042210064@pfur.ru
* Correspondence: andrikov-da@rudn.ru

† Presented at the 15th International Conference "Intelligent Systems" (INTELS'22), Moscow, Russia, 14–16 December 2022.

Abstract: This article concerns liver transplantation and its associated difficulties and risks, and also describes a more progressive method of saving people in need of a liver transplant—an artificial liver. An analysis of the BioUML software platform for modeling bioartificial liver systems is also presented.

Keywords: re-engineering; energy sector; energy security; cognitive modeling; mathematical modeling

Citation: Ganshin, A.; Andrikov, D. Robust System of Algorithms for the Functioning of Biocompatible Artificial Liver Devices. *Eng. Proc.* **2023**, *33*, 62. https://doi.org/10.3390/engproc2023033062

Academic Editors: Askhat Diveev, Ivan Zelinka, Arutun Avetisyan and Alexander Ilin

Published: 15 August 2023

1. Main Text

The liver is one of the most complex and metabolically active organs in the body, and it performs many functions such as detoxification and protein synthesis, which are necessary for life. Extracorporeal bioartificial liver systems (BAL), consisting of functioning viable hepatocytes, can provide temporary support to patients with acute liver failure and save the lives of patients awaiting orthotopic liver transplantation (OLT).

To date, one of the most dangerous diseases in the world that leads to a fatal outcome is Fulminant hepatic failure (FHF). Few experiments and studies have been conducted in this area, which have brought significant results in therapy, but have had no effect on mortality; rates are close to 80%. Some patients can be saved when provided with short-term liver support, which, in turn, helps theirs regenerate. However, this procedure can only help those with reversible liver failure. It is also worth noting the fact that patients who have gone through the procedure described above, after it was carried out, had liver function completely restored and returned to their usual life. But this was not always the case; for many years, there was only one method of treating FHF—liver transplantation. There are many disadvantages to this method; for example, there are patients with liver failure who absolutely cannot have this operation for medical reasons such as: concomitant infection, cancer with metastases, chronic alcoholism, etc. Another disadvantage of liver transplantation surgery is the lack of donors; as a consequence, the liver, that is, the patient, may die whilst waiting for their turn for transplantation, simply because there was no donor. In this regard, a new method has been invented that will help both patients who cannot undergo liver transplantation for medical reasons and those who are on the waiting list for this operation. For example, patients with acute hepatitis, potentially reversible, may benefit from temporary artificial liver support. For this purpose, special bio-artificial devices have been developed that can replace a number of liver functions: synthetic, metabolic, and detoxification processes. Some of these devices have been evaluated in clinical trials [1–3].

To study both BAL and OLT systems, several articles were analyzed. The primary sources of these articles were also found.

Both adults and children are susceptible to liver diseases. Doctors have noted in recent years that the number of patients suffering from liver diseases has increased sixfold. In addition, mortality in liver diseases remains high. It is also known that it is very difficult to recognize these diseases [4].

Today, transplantation operations are the only method of radical treatment for patients in the late stages of liver disease. More than 25 thousand orthotopic liver transplants are performed annually in the world. Despite this, the need for this operation, according to UNOS, is four times greater and ranges from 20 to 30 per 1 million people. In order to reduce orthotopic liver transplantation, in our work, we will consider replacing an ordinary liver with an artificial one using a robust method. To achieve this, an analysis of BAL systems was carried out. Over the past few years, several clinical studies have been conducted in which the effectiveness of BAL devices has been tested. These preliminary studies have yielded some promising results, although the current generation of liver aid devices has not yet demonstrated sufficient effectiveness for normal use. Some of the problems that the BAL development area continues to struggle with are the following:

1. How to maintain a large cell mass without restrictions on the substrate so that the cells function with maximum efficiency;
2. How to maintain high levels of stable, long-term, liver-specific function in an artificial (and potentially inhospitable) environment;
3. How to minimize the amount of filling (or dead space) that needs to be filled with the patient's blood or plasma.

It is becoming clearer every day that a more fundamental understanding of the influence of environmental parameters on the function of hepatocellular diseases, as well as the interaction of the host and BAL, is needed before the concept of BAL becomes a reality available at a reasonable price [5–7].

Also, a software platform is needed to model bioartificial liver systems. BioUML was taken used as a basis. Figure 1 shows an example of working in the BioUML software environment.

BioUML is an open source software platform for the data analysis of omics scientific research and other advanced computer biology analyses developed by scientists from the Institute of Systems Biology in Novosibirsk, Russia [8].

The Biological Unified modeling language is an integrated extensible environment for the visual modeling of biological systems. The vision of BioUML is to provide a computing platform for creating a virtual cell, virtual human physiology, and a virtual patient. BioUML allows the user to conduct the visual modeling of complex systems, adjust model parameters based on several experiments, analyze data, and conduct joint research, and it also contains tools such as: visual modeling, simulation, parameter selection and analysis, genome browser, scripts (R, JavaScript), and a workflow mechanism. Thanks to its integration with the Galaxy and R/Bioconductor platforms, BioUML provides extensive opportunities for omic data analysis. The plug-in architecture allows the user to add new functions using plug-ins. To help the user focus on a specific task or database, several predefined perspectives have been developed that display only those elements of the web interface that are necessary for a specific task [9–11].

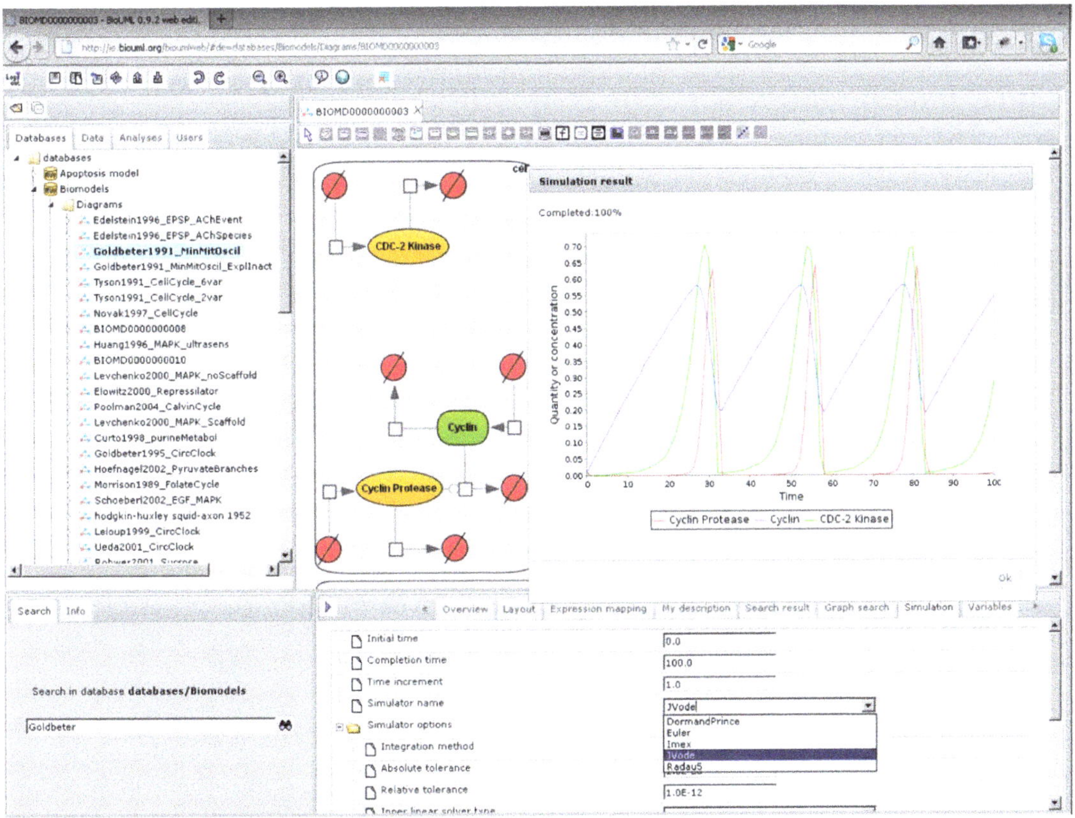

Figure 1. Example of visual modeling.

As mentioned above, one of the advantages of this platform is visual modeling, which greatly simplifies working with complex systems.

2. Considering One of the Languages of Systems Biology

Systems Biology Graphics Notation (SBGN) is a standard graphical representation designed to facilitate the efficient storage, exchange, and reuse of information about signaling pathways, metabolic networks, and gene regulation networks among communities of biochemists, biologists, and theorists (Figure 2). The system was created several years ago by a community of biochemists, model developers, and computer scientists. It is believed that SBGN will contribute to the efficient and accurate representation, visualization, storage, exchange, and reuse of information about all types of biological knowledge, from gene regulation to metabolism and cell signaling.

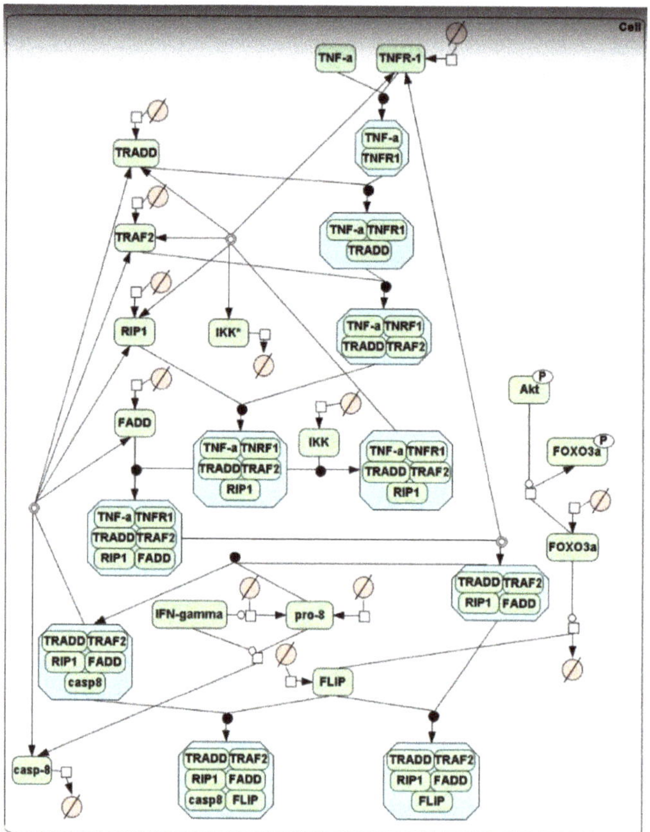

Figure 2. Example TNF-α module (SBGN).

This example demonstrates the convenience of this platform. All elements are reflected and, if necessary, the user can combine some objects together, which is also shown in the diagram [12].

Consider the architecture of a universal platform for analyzing a wide range of biomedical data (Figure 3). It consists of:

1. Computer programs whose main function is data analysis and visualization. For example, Docker is an open source project to automate the deployment of applications in the form of portable standalone containers running in the cloud or on premises. Common Workflow Language (CWL) is a specification for describing the analysis of workflows and tools in such a way as to make them portable and scalable in a variety of software and hardware environments, from workstations to clusters, the cloud, and high-performance computing environments (HPC).
2. The BioUML platform acts as the core [1], which acts as a link with integrated tools (Galaxy, Jupyterhub, R, Docker, noVNC, etc.).
3. IT infrastructure. This means the one on which the platform is installed.
4. External resources are connected via a special API. They are necessary for storing user data.

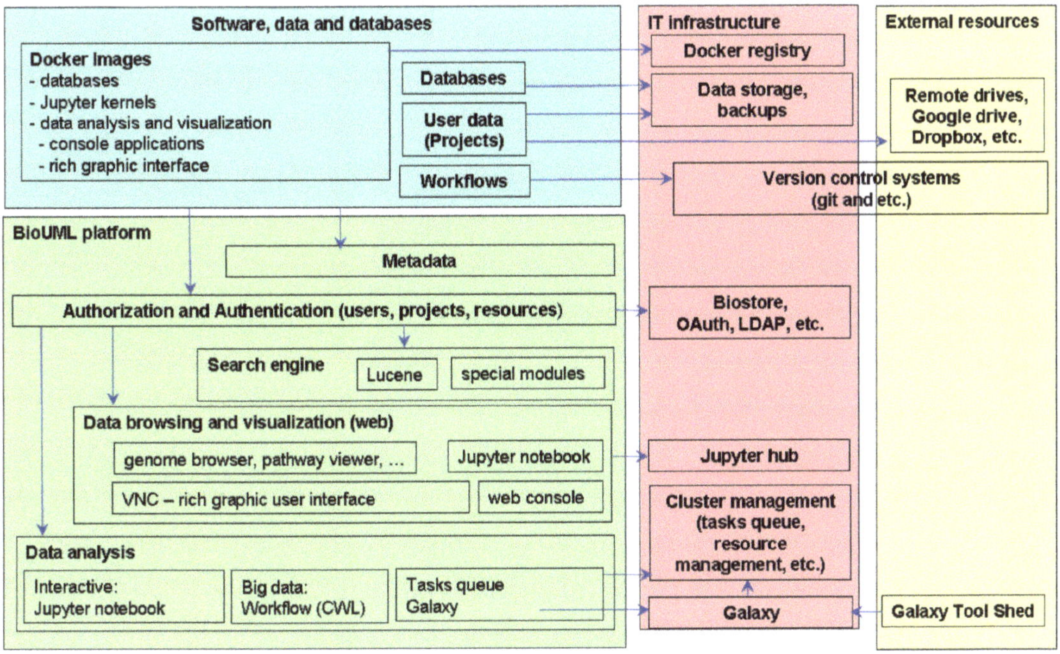

Figure 3. Architecture of a universal platform for the analysis of a wide range of biomedical data.

3. Conclusions

Thus, the shortage of donor organs, in particular, the liver, remains a problem today. This problem is global in nature, and one of the ways to solve it is using a bioartificial liver. In the article, we examined the BAL system, which supports the liver of a person suffering from acute liver failure. The analysis of this system was carried out for the further construction of an artificial liver model using a robust method. The BioUML platform is used to model such complex systems. In the future, this system will be able to save many lives.

Author Contributions: Conceptualization, A.G. and D.A.; methodology, A.G. and D.A.; validation, A.G. and D.A.; formal analysis, A.G. and D.A.; investigation, A.G. and D.A.; resources, A.G. and D.A.; data curation, A.G. and D.A.; writing—original draft preparation, A.G. and D.A.; writing—review and editing, A.G. and D.A.; visualization, A.G. and D.A. All authors have read and agreed to the published version of the manuscript.

Funding: This research received no external funding.

Institutional Review Board Statement: The data was taken from open sources such as ResearchGate.

Informed Consent Statement: Not applicable.

Data Availability Statement: Not applicable.

Conflicts of Interest: The authors declare no conflicts of interest.

References

1. Berthiaume, F.; Chan, C.; Yarmush, M.L. *Liver, Bio-Artificial*; Massachusetts General Hospital, Harvard Medical School, and the Shriners Hospital for Children: Boston, MA, USA, 2008.
2. Tandon, R.; Froghi, S. Artificial liver support systems. *J. Gastroenterol. Hepatol.* **2020**, *36*, 1164–1179. [CrossRef] [PubMed]
3. Bikhchandani, J.; Metcalfe, M.R.; Illouz, F.R.; Puls, F.; Path, F.R.; Dennison, A. Extracorporeal Liver Perfusion System for Artificial Liver Support Across a Membrane. *J. Surg. Res.* **2011**, *171*, e139–e147. [CrossRef] [PubMed]

4. Alves, L.A.; Bonavita, A.; Quaresma, K.; Torres, E. New Strategies for Acute Liver Failure: Focus on Xenotransplantation Therapy. *Cell Med. Part B Cell Transplant.* **2010**, *1*, 47–54. [CrossRef] [PubMed]
5. Abouna, G.M.; Boehmig, H.G.; Serrou, B.; Amemiya, H.; Martineau, G. Long-term hepatic support by intermittent multi-species liver perfusions. *Lancet* **1970**, *2*, 391–396. [CrossRef] [PubMed]
6. Court, F.G.; Wemyss-Holden, S.A.; Dennison, A.R. Bioartificial liver support devices: Historical perspectives. *ANZ J. Surg.* **2003**, *73*, 739. [CrossRef] [PubMed]
7. Kiley, J.; Welch, H. Removal of Blood Ammonia by Hemodialysis. *Proc. Soc. Exp. Biol. Med. Publ.* **1956**, *1*, 57. [CrossRef] [PubMed]
8. Nakao, M.; Nakayama, N.; Uchida, Y. Nationwide survey for acute liver failure and late-onset hepatic failure in Japan. *J. Gastroenterol.* **2017**, *53*, 752–769. [CrossRef] [PubMed]
9. Martínez, J.J.G.; Bendjelid, K. Artifcial liver support systems: What is new over the last decade. *Ann. Intensive Care* **2018**, *8*, 109. [CrossRef] [PubMed]
10. Huang, S.; Hu, D.; Yuan, S. The Serum Metabolomics Study of Liver Failure and Artificial Liver Therapy Intervention. *Med. Sci. Monit. Int. Med. J. Exp. Clin. Res.* **2021**, *27*, e930638. [CrossRef] [PubMed]
11. Bloem, R.; Greimel, K.; Henzinger, T.A. Synthesizing Robust Systems. *Acta Inform.* **2009**, *51*, 193–220. [CrossRef]
12. Kolpakov, F.; Akberdin, I.; Kashapov, T.; Kiselev, L.; Kolmykov, S.; Kondrakhin, Y.; Kutumova, E.; Mandrik, N.; Pintus, S.; Ryabova, A.; et al. BioUML: An integrated environment for systems biology and collaborative analysis of biomedical data. *Nucleic Acids Res.* **2019**, *47*, W225–W233. [CrossRef] [PubMed]

engineering proceedings

MDPI

Proceeding Paper

Natural Data Analysis Method Based on Wavelet Filtering and NARX Neural Networks †

Oksana Mandrikova, Yurii Polozov and Bogdana Mandrikova *

Institute of Cosmophysical Research and Radio Wave Propagation FEB RAS, 684034 Paratunka, Kamchatskiy Krai, Russia; oksanam1@mail.ru (O.M.); polozov@ikir.ru (Y.P.)
* Correspondence: 555bs5@mail.ru; Tel.: +7-41531-33193; Fax: +7-41531-33718
† Presented at the 15th International Conference "Intelligent Systems" (INTELS'22), Moscow, Russia, 14–16 December 2022.

Abstract: A method for analyzing natural data and detecting anomalies is proposed. The method is based on combining wavelet filtering operations with the NARX neural network. The analysis of natural data and the detection of anomalies are of particular relevance in the problems of geophysical monitoring. An important requirement of these methods is their adaptability, accuracy and efficiency. Efficiency makes it possible to detect anomalies timely in order to prevent catastrophic natural phenomena. Wavelet filtering operations include the application of a multi-scale analysis construction and threshold functions. The article proposes a wavelet filtering algorithm and a method for estimating thresholds based on a stochastic approach. The operations of the method implementation are described. It is shown that the use of wavelet filtering allows one to suppress noise, simplifies the data structure and, as a result, allows one to obtain a more accurate NARX neural network model. The effectiveness of the method for detecting ionospheric anomalies during periods of magnetic storms is shown using the data of the critical frequency of the ionosphere as an example.

Keywords: data analysis; wavelets; neural networks; ionosphere

1. Introduction

Investigation of natural data is a meticulous, nontrivial task and is of great concern in the research of different processes and phenomena. Natural data processing and analysis are widely applied in different fields such as technical processes, geophysics, biology, economy, chemistry and so on. In the context of this problem, methods of object statistical analysis and anomaly detection are of special interest. Examples might include the problems of detection of outer space disturbances [1], floods, earthquakes [2], mud flows and landslides, tsunamis [3], changes in the geological environment [4] and other natural emergency situations. The problem of anomaly recognition is also very important in medicine. Timely and highly accurate diagnostics make it possible to detect and identify patient physical state disorders that in some cases may cost someone their life. Such methods should be fast and flexible in order to detect rapidly changing states of systems and objects. These changes characterize anomaly occurrences.

Different methods are used for natural data analysis. A wide spectrum of approaches to this problem is determined by the complexity of the data under investigation and includes deterministic [5] and stochastic [6] tools. At the present time, different combinations of such approaches are being actively developed [7]. As methods for natural data analysis developed, it became evident that classical approaches to modeling and analysis (AR, ARMA models [8], exponential smoothing [9], stochastic approximation [6], etc.) do not allow us to describe properly the complicated nonstationary structure of natural time series.

At the present time, hybrid approaches and the methods [7] allowing one to improve data analysis quality are developing rapidly to solve such problems. Wavelet transform is a flexible instrument and is widely used in the problems of data processing and analysis.

Citation: Mandrikova, O.; Polozov, Y.; Mandrikova, B. Natural Data Analysis Method Based on Wavelet Filtering and NARX Neural Networks. *Eng. Proc.* **2023**, *33*, 63. https://doi.org/10.3390/engproc2023033063

Academic Editors: Askhat Diveev, Ivan Zelinka, Arutun Avetisyan and Alexander Ilin

Published: 16 August 2023

A large library of wavelet filters and a wide choice of constructions for signal expansion provide the possibility to adapt this instrument for the data of different structures. For example, we developed the empirical wavelet filters [10], which are effective in image processing applications.

Now, neural network methods [11] are of great importance. They are known to be able to approximate complex non-linear dependencies in data and automate processes. However, to achieve high efficiency of the neural network apparatus, it is necessary to have high-quality training data and their representativeness. The article proposes a method based on combining the operations of wavelet filtering and NARX neural networks. NARX networks are an evolution of the ARIMA class of models. They approximate non-linear time series and may have different architectures. This allows you to flexibly adapt to the tasks of data analysis and forecasting. Wavelet filtering is used to reduce noise and simplify the data structure. This improves the efficiency of the neural network. Similar hybrid approaches to the analysis of complex data have already been used in other works [7,12]. One such work is the combination of ARIMA and discrete wavelet decomposition together with the LSTM neural network [7]. Another example is the work [12], which proposes a discrete wavelet decomposition with ARIMA and a neural network. Based on this method [12], the authors carry out a forecast of hydrological hazards (high water levels (floods), floods, rain floods, etc.). In this paper, the approach of combined application of wavelet filtering and NARX networks is used to improve the efficiency of autoregressive time series models. With the use of wavelet filtering, it is possible to reduce noise and simplify the structure of the initial data. The article shows that this leads to an increase in the performance of the neural network. The wavelet filtering process includes a combination of multiscale analysis [10] and threshold functions.

The article presents the wavelet filtering algorithm and the threshold estimation method. Thresholds were estimated based on statistical operations. The study of the time series of the critical frequency of the ionospheric layer F2 (foF2) was carried out. The structure of the recorded ionospheric data depends on many temporal, weather, positional, and other factors [13]. Time series of the ionosphere demonstrate a regular course and various anomalies in the form and duration. Anomalies can be observed under conditions of increased solar and geomagnetic activity, as well as during seismic activity [13].

It is known that there are cases of ionospheric anomaly negative impacts on high-frequency radio communication and navigation signals from GPS satellite. One example is an event in January 2005. Due to space weather anomalies, 26 flights of United Airlines were redirected to non-optimal routes in order to avoid catastrophic risks of communication losses. An event in October 2003 is also known when the American Wide Area Augmentation System remained inoperative for more than one day and did not navigate vertically because of ionospehric inhomogeneities [14]. As it was mentioned above, traditional approaches to the analyzed ionospheric parameters do not satisfy the requirements of anomaly detection accuracy. In this paper, we suggest a method that confirmed its efficiency in the task of detection of anomalies (ionospehric inhomogeneities) generated by magnetic storms. It was empirically proved that signal pre-processing based on wavelet filtering improves the accuracy of NARX neural network model for ionospheric parameter time series. The paper presents a comparison of the efficiency of the proposed method with direct application of NARX neural network.

2. Applied Data

In this paper, we applied the ionospheric critical foF2 data [13], which have a one-hour sampling rage. The data have been recorded at the Paratunka site (Kamchatskiy kray, Russia, site coordinates: 53.0 N, 158.7 E) since 1969. Due to instrumentation failures and natural and man-made factors, there are gaps in foF2 data. We used a time series with data gaps no longer than one day. When there were gaps in the data of less than 25 h, they were filled in based on the median method.

3. Method Description

3.1. Application of Wavelet Transform and Threshold Function

Natural time series contain noise formed under a large variety of natural and man-made factors [15]. To improve the quality of natural data analysis based on neural networks, according to the papers [16], we applied noise suppression operations. They included the application of a multiscale analysis construction [17] and threshold function. The algorithm for noise suppression is the following:

1. Signal $f_0(t)$ wavelet decomposition into the components:

$$f_0(t) = \sum_{j=-1}^{-m} g_j(t) + f_{-m}(t),$$

where $f_{-m}(t) = \sum_k c_{-m,k}\phi_{-m,k}(t)$ is the smoothed component, $c_{-m,k} = \langle f_0, \phi_{-m,k}\rangle$, $\phi_{-m,k}(t) = 2^{-m/2}\phi(2^{-m}t - k)$ is the scaling function, $g_j(t) = \sum_k d_{j,k}\Psi_{j,k}(t)$ are detailing components, $d_{j,k} = \langle f_0, \Psi_{j,k}\rangle$, $\Psi_{j,k}(t) = 2^{j/2}\Psi(2^j t - k)$ is the wavelet, j is the decomposition level, the decomposition level $j = 0$ is assumed for the initial signal.

2. Application of the threshold fucntion to the component coefficients $g_j(t)$:

$$T(d_{j,k}) = \begin{cases} 0, \ if \ |d_{j,k}| \le T_j \\ d_{j,k}, \ if \ |d_{j,k}| > T_j \end{cases}, \tag{1}$$

where $T_j = t_{1-\frac{\alpha}{2},N-1}\hat{\sigma}_j$, $t_{\alpha,N}$ are α-quantiles of Student's distribution, $\hat{\sigma}_j$ is the sample standard deviation, the decomposition levels $j = \overline{-1,-m}$.

3. Signal wavelet recovery:

$$\widetilde{f_0}(t) = \sum_{j,k} T(d_{j,k})\Psi_{j,k}(t) + f_{-m}(t).$$

Determination of thresholds T_j in operation (1) was based on the paper results [17] and it was assumed that absolute values of the coefficients $|d_{j,k}|$ of the components $g_j(t)$ were close to zero out of the neighborhood containing signal structural characteristics. In this case, the probability ($\alpha \approx 0.99$) that the coefficients $|d_{j,k}|$ with respect to the argument k are within the interval $(\mu - 3\sigma; \mu + 3\sigma)$ is high. Here, $\mu \approx 0$ is the mathematical expectation of the value $|d_{j,k}|$, σ is the standard deviation (three-sigma rule [18]). Then, for normally distributed $|d_{j,k}|$, we can estimate the thresholds T_j with defined confidence level α as $T_j = t_{1-\frac{\alpha}{2},N-1}\hat{\sigma}_j$, where $t_{\alpha,N}$ are the α-quantiles of the Student's distribution [18].

3.2. Application of NARX Neural Networks

After noise suppression operations, data are analyzed based on NARX neural networks [11]. We applied feedback NARX networks [11], the architectures of which are illustrated in Figure 1.

The value vector applied to the hidden layer neurons consists of the following components:

- Current $\widetilde{f_0}(t)$ and previous $\widetilde{f_0}(t-1), \ldots, \widetilde{f_0}(t-l_x)$ values of the input vector;
- Output values during the previous time intervals $\hat{f_0}(t), \hat{f_0}(t-1), \ldots, \hat{f_0}(t-l_y)$.

The neural network input value $\hat{f_0}(t+1)$ has the form:

$$\hat{f_0}(t+1) = F[\widetilde{f_0}(t), \widetilde{f_0}(t-1), \ldots, \widetilde{f_0}(t-l_x), \hat{f_0}(t), \hat{f_0}(t-1), \ldots, \hat{f_0}(t-l_y)], \tag{2}$$

where $F(\cdot)$ is the neural network display function.

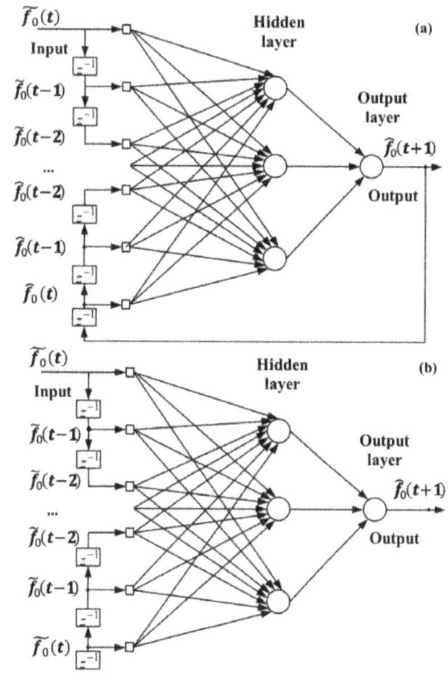

Figure 1. Architecture of a generalized recurrent network. (**a**) NARX PA architecture; (**b**) NARX SPA architecture.

Figure 1 shows the NARX network with Series-Parallel Architecture (NARX SPA) (Figure 1b) and NARX network with Parallel Architecture (NARX PA) (Figure 1a) [11]. Analytical form for the NARX PA architecture is represented in expression (2). In NARX SPA, previous values $\widetilde{f}_0(i)$ are applied to the input instead of the outputs $\hat{f}_0(i), i = \overline{t, t-l_y}$. In this case, the network output value $\hat{f}_0(t+1)$ has the form

$$\hat{f}_0(t+1) = F[\widetilde{f}_0(t), \widetilde{f}_0(t-1), ..., \widetilde{f}_0(t-l_x), \widetilde{f}_0(t), \widetilde{f}_0(t-1), ..., \widetilde{f}_0(t-l_y)].$$

It is known from [11], that NARX SPA are applied when data of a previous value vector are available. NARX PA networks are applied when only modeling results can be used, for example, when making a forecast with a large step ahead.

The network memory parameters were used in the experiments: $l_x = l_y = 2, l_x = l_y = 5$.

The Bayesian Regularization algorithm with Levenberg–Marquardt optimization was used to train the networks [11]. That allows us to minimize the linear combination of error squares and weights. After that, their appropriate combination is formed in such a way that the network has the highest degree of generalization.

3.3. Scheme of Method Realization

The scheme of method realization is illustrated in Figure 2. Anomalies are detected based on estimation of neural network errors

$$\varepsilon_i = \sum_{i=i-s}^{i+s} |\hat{f}_0(i) - \widetilde{f}_0(i)|. \tag{3}$$

There is an anomaly in the data if $\varepsilon_i > 2\sigma + \varepsilon_{mean}$, where σ is standard deviation of errors ε_i during periods without anomalies, and ε_{mean} is mean of errors ε_i in periods without anomalies.

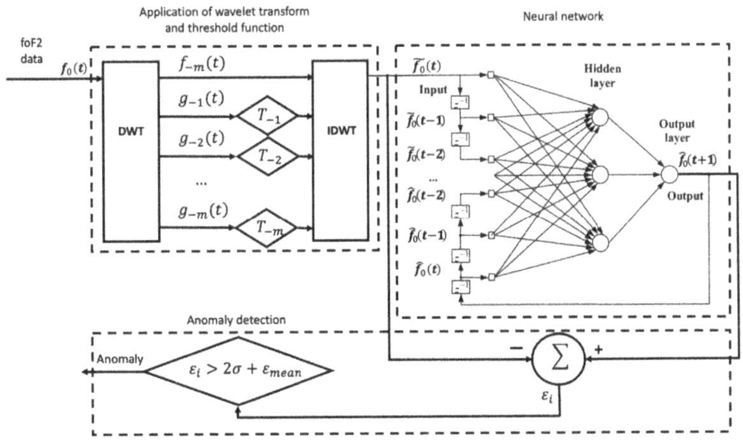

Figure 2. Scheme of method realization.

4. Results of Method Application for Ionospheric Data

Regular foF2 data depend on the time of year and the level of solar activity. Therefore, separate training samples were formed (to build a neural network) for different seasons and levels of solar activity. Two periods of solar activity were studied in the work: high and low. The level of solar activity was estimated based on the average monthly values of the solar index f10.7. We used the ionospheric data recorded during a calm geomagnetic state for the period 1969–2015. When constructing each neural network, training sampling contained about 2000 counts. According to the suggested method, neural networks were trained using the data obtained after wavelet filtering. When making wavelet filtering operations, orthonormal Daubeches third-order wavelets were applied [10]. They were determined by minimizing the data approximation errors. To estimate the method, neural networks were also trained via applying the foF2 initial data. We constructed 16 neural networks using the NARX PA and NARX SPA architectures, which had the memory $l_x = l_y = 2, l_x = l_y = 5$.

Tables 1 and 2 show the results of the obtained neural network operation. Root-mean-square (RMS) deviations of network errors were determined as

$$S = \sqrt{\frac{1}{N-1}\sum_{i=1}^{N}(e_i - \bar{e})^2},$$

where $\bar{e} = \frac{1}{N}\sum_{i=1}^{N} e_i$, $e_i = \hat{f}_0(i) - \tilde{f}_0(i)$.

Table 1. Standard deviations of NARX neural network errors with delays (2) and (5). T_j is the wavelet filtering.

Season	SPA (2)	PA (2)	SPA (2) (T_j)	PA (2) (T_j)	SPA (5)	PA (5)	SPA (5) (T_j)	PA (5) (T_j)
Winter (low solar activity)	0.47	0.57	0.25	0.41	0.42	0.43	0.18	0.30
Summer (low solar activity)	0.42	0.46	0.19	0.29	0.36	0.39	0.16	0.26
Winter (high solar activity)	0.50	0.78	0.32	0.51	0.43	0.61	0.28	0.47
Summer (high solar activity)	0.42	0.45	0.26	0.29	0.36	0.37	0.21	0.22

Table 2. Results of operation of neural networks (Summer, low solar activity). T_j is the wavelet filtering.

Network Memory $l_x = l_y$	Network Architecture	ε_{mean} (T_j)	σ (T_j)	ε_{mean}	σ
2	NARX SPA	0.43	0.37	0.98	0.48
2	NARX PA	0.85	0.48	1.3	0.6
5	NARX SPA	0.33	0.28	0.83	0.37
5	NARX PA	0.80	0.42	1.27	0.56

An analysis of the results from Table 1 shows that NARX SPA architecture describes more adequately the foF2 data time series. When using a large number of delay lines, the results become more comparable in favor of NARX SP (Table 1). A comparison of the results in Table 2 shows that application of wavelet filtering operation allows us to significantly increase the neural network operation quality that confirms the suggested method efficiency.

To test the adequacy of neural network models, the autocorrelation of network errors was evaluated based on the Ljung-Box test $Q = M(M+2)\sum_{s=1}^{L}\frac{p_s^2}{T-s}$, where M is the observation number, p_s is the autocorrelation of the s-th order, and L is the checked lag number. If $Q > \chi^2_{1-\alpha,L}$, where $\chi^2_{1-\alpha,L}$ are the distribution quantiles chi-square with L freedom degree, the presence of autocorrelation of the L-th order in the time series is recognized.

An analysis of the results in Table 3 shows that Ljung–Box test values for the networks constructed without wavelet filtering significantly exceed the critical values $\chi^2_{1-\alpha,L}$. The results indicate a significant correlation of errors of these neural networks and, as a consequence, insufficient quality of foF2 data approximation. When the network memory is increased, data approximation quality improves. The networks, trained applying data wavelet filtering operations, show the best results. For a small memory of networks $l_x = l_y = 2$, in accordance with the Ljung–Box test, network errors are uncorrelated. This confirms the adequacy of the obtained neural network model.

Table 3. Estimate of error autocorrelation of neural networks (Summer, low solar activity). T_j is the wavelet filtering.

Network Memory $l_x = l_y$	Lag Number L	Q (T_j)	Q	$\chi^2_{1-\alpha,L}$
2	1	4.19	3.82	3.84
2	6	13.53	48.18	12.59
2	12	41.41	98.97	21.02
5	1	1.17	0.51	3.84
5	6	11.17	15.53	12.59
5	12	26.38	44.6	21.02

Figure 3 shows the results of neural network operation during a magnetic storm, which occurred on 16 July 2017. The red dashed line in Figure 3 marks the magnetic storm beginning. According to Space Weather site data (http://ipg.geospace.ru, accessed on 1 March 2023), the magnetic storm had a mixed nature, arrival of an accelerated flux from a coronal hole and coronal mass ejection. During strong geomagnetic disturbances, the DST index (DST index of geomagnetic activity) reached the minimum of -72 nTl (Figure 3f,l). Analysis of foF2 initial data (Figure 3a,g) and comparison with the moving median (in Figure 3a,g, the median is a green dashed line) show the distortion of data regular time variation on July 16. That is determined by ionospehric disturbance occurrences. Anomalous changes in foF2 data on July 16 confirm the results of the neural networks, as a significant error increase is observed (Figure 3b–e,h–k). However, as the analysis shows, the application of wavelet filtering improves the network performance quality. The errors of the network with small memory $l_x = l_y = 2$ are close to zero except for the anomalous period.

During the anomalous period, network errors significantly increase and the anomaly is clearly detected.

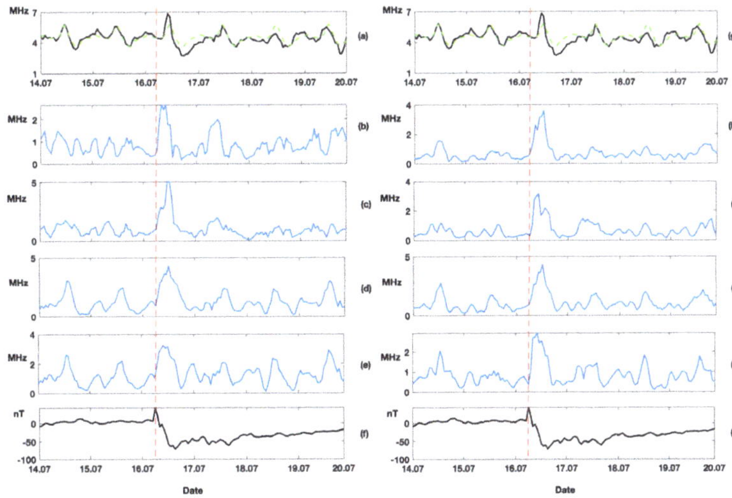

Figure 3. Result of data processing foF2. (**a**,**g**) foF2 data (black) and foF2 median (green); (**b**,**c**) NARX SPA errors with memory 2 and 5, respectively; (**d**,**e**) NARX PA errors with memory 2 and 5, respectively; (**h**,**i**) NARX SPA errors with wavelet filtering and memory 2 and 5, respectively; (**j**,**k**) NARX PA errors with wavelet filtering and memory 2 and 5, respectively; (**f**,**l**) DST. Red dashed line is the magnetic storm beginning.

When comparing the results of NARX PA and NARX SPA architectures, we find that for the delay lines $l_x = l_y = 2, l_x = l_y = 5$, NARX SPA architecture detects anomalous changes more clearly both when applying initial data to the network input and when applying the data after wavelet filtering to the input. A comparison of the results of NARX SPA neural networks with moving median confirms the effectiveness of the method. The moving median has errors in the period after the storm on July 18, 2017, which are absent in the neural network model.

5. Conclusions

The application of wavelet filtering together with a NARX neural network is effective for the task of detection of ionospehric anomalies. Direct application of NARX neural networks does not allow one to approximate ionospheric time variation due to the presence of long-term dependencies. Wavelet filtering operation and application of stochastic thresholds significantly decrease the error level even for small memory networks. Wavelet filtering suppresses noise, simplifies the data structure and, as a consequence, makes it possible to obtain a more accurate NARX neural network model. The application of the Ljung–Box test showed no correlation of network errors and confirmed the adequacy of the obtained neural network model.

Comparison of NARX SPA and NARX PA showed the advantage of the NARX SPA architecture, which makes it possible to obtain a more adequate model for describing the ionospheric time series. Using the example of a strong magnetic storm, the possibility of the method for detecting ionospheric anomalies is shown. Comparison of the NARX SPA network with the moving median, widely used to analyze ionospheric data, showed the efficiency of the proposed approach.

With an extension of the statistics and a more detailed study of ionospheric data during anomalous periods, the authors plan to continue research in this direction.

Author Contributions: Conceptualization, O.M. and Y.P.; methodology, O.M., Y.P. and B.M.; software, Y.P.; validation, O.M. and Y.P.; formal analysis, O.M. and Y.P.; investigation, O.M. and Y.P.; resources, Y.P.; data curation, Y.P.; writing—original draft preparation, Y.P. and B.M.; writing—review and editing, O.M.; visualization, O.M. and Y.P.; project administration, O.M. All authors have read and agreed to the published version of the manuscript.

Funding: The work was carried out as a part of the implementation of the State Task AAAA-A21-121011290003-0. The work was carried out by means of the Common Use Center "North-Eastern Heliogeophysical Center".

Informed Consent Statement: Not applicable.

Data Availability Statement: Not applicable.

Conflicts of Interest: The authors declare no conflict of interest.

References

1. Tang, R.; Zeng, F.; Chen, Z.; Wang, J.-S.; Huang, C.-M.; Wu, Z. The comparison of predicting storm-time ionospheric TEC by three methods: ARIMA, LSTM, and Seq2Seq. *Atmosphere* **2020**, *11*, 316. [CrossRef]
2. Perol, T.; Gharbi, M.; Denolle, M. Convolutional neural network for earthquake detection and location. *Sci. Adv.* **2018**, *4*, e1700578. [CrossRef] [PubMed]
3. Kim, S.-K.; Lee, E.; Park, J.; Shin, S. Feasibility analysis of GNSS-reflectometry for monitoring coastal hazards. *Remote Sens.* **2021**, *13*, 976. [CrossRef]
4. Alperovich, L.; Eppelbaum, L.; Zheludev, V.; Dumoulin, J.; Soldovieri, F.; Proto, M.; Bavusi, M.; Loperte, A. A new combined wavelet methodology: Implementation to GPR and ERT data obtained in the Montagnole experiment. *J. Geophys. Eng.* **2013**, *10*, 025017. [CrossRef]
5. Chen, J.; Heincke, B.; Jegen, M.; Moorkamp, M. Using empirical mode decomposition to process marine magnetotelluric data: Using EMD to process marine MT data. *Geophys. J. Int.* **2012**, *190*, 293–309. [CrossRef]
6. Robbins, H.; Monro, S. A stochastic approximation method. *Ann. Math. Stat.* **1951**, *22*, 400–407. [CrossRef]
7. Wu, X.; Zhou, J.; Yu, H.; Liu, D.; Xie, K.; Chen, Y.; Hu, J.; Sun, H.; Xing, F. The development of a hybrid Wavelet-ARIMA-LSTM model for precipitation amounts and drought analysis. *Atmosphere* **2021**, *12*, 74. [CrossRef]
8. Box, G.E.P.; Jenkins, G.M. *Time Series Analysis: Forecasting and Control*, Rev. ed.; Holden-Day series in time series analysis and digital processing; Holden-Day: San Francisco, CA, USA, 1976.
9. Liu, J.; Kumar, S.; Palomar, D.P. Parameter estimation of heavy-tailed AR model with missing data via stochastic EM. *IEEE Trans. Signal Process.* **2019**, *67*, 2159–2172. [CrossRef]
10. Mallat, S.G. *A Wavelet Tour of Signal Processing: The Sparse Way*; Elsevier: Amsterdam, The Netherlands, 2009.
11. Haykin, S.S. *Neural Networks: A Comprehensive Foundation*, 2nd ed.; Prentice Hall: Upper Saddle River, NJ, USA, 1999.
12. Phan, T.-T.-H.; Nguyen, X.H. Combining statistical machine learning models with ARIMA for water level forecasting: The case of the Red river. *Adv. Water Resour.* **2020**, *142*, 103656. [CrossRef]
13. Danilov, A.D. Ionospheric F-region response to geomagnetic disturbances. *Adv. Space Res.* **2013**, *52*, 343–366. [CrossRef]
14. Kuznetsov, V.D. Space weather and risks of space activity. *Space Eng. Technol. Mag.* **2014**, *3*, 3–13
15. Geppener, V.V.; Mandrikova, B.S. Detecting and identifying anomalous effects in complex signals. *Autom. Remote Control* **2021**, *82*, 1668–1678. [CrossRef]
16. Mandrikova, O.; Mandrikova, B. Hybrid method for detecting anomalies in cosmic ray variations using neural networks autoencoder. *Symmetry* **2022**, *14*, 744. [CrossRef]
17. Chui, C.K. *An Introduction to Wavelets*; Academic Press: Boston, MA, USA, 1992
18. Berger, J.O. *Statistical Decision Theory and Bayesian Analysis*, 2nd ed.; Springer series in statistics; Springer: New York, NY, USA, 1993.

engineering proceedings

MDPI

Proceeding Paper

Optimal Design of a Reliability Experiment Based on the Wiener Degradation Model under Limitations of the Degradation Index †

Evgeniya Osintseva * and Ekaterina Chimitova *

Department of Theoretical and Applied Informatics, Novosibirsk State Technical University, 20 Prospekt K. Marksa, Novosibirsk 630073, Russia
* Correspondence: osinceva.j@gmail.com (E.O.); chimitova@corp.nstu.ru (E.C.)

† Presented at the 15th International Conference "Intelligent Systems" (INTELS'22), Moscow, Russia, 14–16 December 2022.

Abstract: There are many degradation models that are used in reliability analyses. The Wiener degradation model is the most popular across different applications. In this paper, we propose an approach for searching for an optimal design on the basis of the Wiener degradation model. The distinguishing feature of the approach is that the condition that the degradation index cannot exceed a critical level has been taken into account. The elements of the conditional Fisher information matrix have been obtained. This enables us to calculate the optimal stress levels by maximizing the functional of the conditional Fisher information matrix. Moreover, a degradation analysis of light-emitting diodes (LEDs) has been considered.

Keywords: Wiener degradation model; covariates; conditional Fisher information matrix; optimal design; degradation index

Citation: Osintseva, E.; Chimitova, E. Optimal Design of a Reliability Experiment Based on the Wiener Degradation Model under Limitations of the Degradation Index. *Eng. Proc.* **2023**, *33*, 64. https://doi.org/10.3390/engproc2023033064

Academic Editors: Askhat Diveev, Ivan Zelinka, Arutun Avetisyan and Alexander Ilin

Published: 23 August 2023

1. Introduction

A high reliability and long useful lifetime are the most important features of modern equipment. Such requirements make reliability analyses of devices more complicated. Classical methods based on failure data are not suitable for equipment reliability analyses because not all devices fail during experiments. In this case, other characteristics of the equipment can be used, for example, the degradation index or the so-called health index, which demonstrate the condition of the product under study. The Wiener degradation model is the most popular model that can be applied in the case of a non-monotonic degradation path [1–8]. The gamma and inverse Gaussian degradation models are often mentioned in scientific articles [9,10], but can be used for solving tasks where the degradation index only has positive increments.

Despite the fact that observed degradation data allow us to carry out reliability analyses without failure time data, the duration of the experiment can be quite long. Thus, the observation of degradation data during accelerated life testing is widely used in many papers. To conduct an accelerated life test, it is necessary to increase the stress level on a part of the devices, such as voltage, pressure, temperature, etc. [11].

Since in practice it may often be necessary to repeat such an analysis, many scientists have thought about how to optimize the cost and duration of experiment, that is, how to obtain the optimal design for further analysis. The first step for optimal planning was made by R. A. Fisher [12]. Since the 1950s, a lot of scientists have persistently solved the problem of optimal experimental design to obtain both the optimal stress levels and the number of studied units under restrictions on the allowable stress levels, the cost and duration of the experiment [13]. In [14–16], we emphasized that the choice of time points for measuring the degradation index significantly affects the accuracy of the maximum likelihood estimates for the Wiener degradation model parameters. The optimal distribution of measurement time points depends on the model describing the degradation process as well as the experimental conditions, such as the experiment duration, stress levels and the minimum time interval between measurements of the degradation index [14–16]. In [16], we suggested an algorithm for constructing A- and D-optimal designs based on the Wiener degradation model, which includes determining the optimal stress levels, the number of tested devices and the time moments for measuring the degradation index. However, in that paper, we did not consider the fact that in real life the observation of the object is terminated when the degradation path reaches the critical value.

Thus, in this paper, we develop an algorithm for constructing optimal designs based on the Wiener degradation model, which includes determining the optimal stress levels considering the limitation of the degradation index.

2. The Wiener Degradation Model in Reliability Analysis

Let us assume that the observed stochastic process $Z(t)$ is a stochastic process with independent increments and $Z(0) = 0$. For the Wiener degradation model, increments have a normal distribution with the probability density function:

$$f(u, \theta_1, \theta_2) = \frac{1}{\theta_2\sqrt{2\pi}} \exp\left(-\frac{(u-\theta_1)^2}{2\theta_2^2}\right), \tag{1}$$

where $\theta_1 = \mu(\rho(t+\Delta(t)) - \rho(t))$ is the shift parameter, $\theta_2 = \sigma\sqrt{(\rho(t+\Delta(t)) - \rho(t))}$ is the scale parameter, $\sigma > 0$ and $\rho(t)$ is a positive increasing function.

Let us denote the vector of stresses (which are also often referred to as covariates) as $x = (x^1, x^2, \ldots, x^m)^T$. The range of values for each covariate $x^j, j = \overline{1,m}$ is determined by the conditions of the experiment. In this paper, the degradation process $Z(t)$ is supposed to be observed under a constant time stress. Here, we assume that the covariate x influences the degradation paths as in the accelerated failure time model:

$$Z_x(t) = Z\left(\frac{t}{r(x,\beta)}\right),$$

where $r(x,\beta)$ is a the positive covariate function and $\beta = (\beta_1, \beta_2, \ldots, \beta^m)^T$ is the vector of regression parameters. The mathematical expectation of the degradation process $Z_x(t)$ is denoted by

$$E(Z_x(t)) = \mu\rho\left(\frac{t}{r(x,\beta)}, \gamma\right).$$

The time to failure, which depends on covariate x, is defined as:

$$\tau = sup(t : Z_x(t) < z_0),$$

where z_0 is the critical value of the degradation index. Then, the reliability function can be represented as:

$$S(t) = P\{\tau < t\} = P(Z_x(t) < z_0) = \Phi\left(\frac{z_0 - \mu\rho(t/r(x;\beta);\gamma)}{\sigma\rho(t/r(x;\beta)}\right).$$

Suppose the experiment is running over time T. The degradation index values are measured at time points $0 = t_0, t_1, \ldots, t_k = T$.

Let us denote the sample of independent degradation index increments with covariates as the following:

$$\mathbf{X_n} = \{(\Delta Z_{1j}, x_1), \ldots, (\Delta Z_{nj}, x_n), j = \overline{1,k}\},$$

where k is the number of measurements of the degradation index for each object, x_i is the value of the covariate vector for the i-th object and $\Delta Z_{ij} = Z_i(t_j) - Z_i(t_{j-1})$ is the increment of the degradation index during the time from t_{j-1} to t_j.

the unknown parameters of the model can be estimated using the maximum likelihood method:

$$\ln L(\mathbf{X_n}) = -nk(\ln\sqrt{2\pi} + \ln\sigma) - \frac{n}{2}\sum_{j=1}^{k}\ln\left(\rho(t_{j+1}) - \rho(t_j)\right)) -$$

$$\frac{1}{2\sigma^2}\sum_{i=1}^{n}\sum_{j=1}^{k}\frac{\left[\Delta Z_{ij} - \mu(\rho(t_{j+1}) - \rho(t_j)))\right]^2}{(\rho(t_{j+1}) - \rho(t_j)))}.$$

3. Optimal Design of Experiments

We denote the experiment design as a set of values $\xi = \{x_{(1)}, x_{(2)}, \dots x_{(q)}\}$, where $x_{(i)}$ are the reference points of the design. All objects of the sample are divided into q groups corresponding to different values of the covariate vector (reference points of the design). Thus, the problem of direct searching for an optimal design can be written as follows:

$$\begin{cases} M(I(\xi)) \to max, \\ x_{min} \leqslant x_{(i)} \leqslant x_{max}, \end{cases}$$

where $M(.)$ is some functional of the Fisher information matrix and x_{min}, x_{max} are the minimum and maximum values of stress levels determined by the conditions of experiment. The construction of the D-optimal design is based on maximizing the determinant of the Fisher information matrix:

$$M(I(\xi)) = det(I(\xi)).$$

In [16], we did not take into account the fact that in real life the observation of the object is terminated when the degradation path reaches a critical value. To solve this problem, the conditional density function should be considered: $f_{\Delta Z_{ij}}(u|Z_k \leqslant z_0)$, where $Z_k = Z_x(t_k)$.

However, the calculation of the conditional density function requires the calculation of k-fold integrals, which is a composite computational problem. Thus, we propose to use only the last observation Z_k of each degradation path for constructing an optimal design, which enables us to determine the optimal stress levels.

We have obtained the mathematical expectation of the second derivatives with respect to the parameters of the likelihood function to obtain elements of the Fisher information matrix under condition $Z_{t_k} \leqslant z_0$:

$$I(\xi) = -E\left(\frac{\partial^2 \ln f(z_k|z_k \leqslant z_0)}{\partial\theta\partial\theta^T}\right),$$

where

$$\ln f_{Z_k}(u|Z_k \leqslant z_0) = \ln f_{Z_k}(z_k) - \ln F_{Z_k}(z_0).$$

We calculated elements of the Fisher information matrix:

$$I_{11} = \frac{2n}{\sigma^2} - \left[3\frac{(z_0 - \mu\rho(t_k))^2}{2\sigma^4\rho(t_k)} - \frac{(z_0 - \mu\rho(t_k))^4}{4\sigma^6\rho^2(t_k)}c_2\right]c_1 - \frac{c_1}{F_{Z_k}}\frac{(z_0 - \mu\rho(t_k))^4}{4\sigma^6\rho^2(t_k)};$$

$$I_{12} = -\left[2\frac{(z_0 - \mu\rho(t_k))}{\sigma^2} - \frac{(z_0 - \mu\rho(t_k))^3}{2\sigma^5\rho(t_k)}c_2\right]c_1 - \frac{c_1}{F_{Z_k}}\frac{(z_0 - \mu\rho(t_k))^3}{2\sigma^5\rho(t_k)};$$

$$I_{13} = \frac{1}{\sigma}\sum_{i=1}^{q}\frac{1}{\rho(t_{ik})}\frac{\partial\rho(t_{ik})}{\partial\gamma} - \left[\frac{(z_0^2 - \mu^2\rho^2(t_k))}{\sigma^3\rho^2(t_k)}\frac{\partial\rho(t_k)}{\partial\gamma}\frac{(z_0 - \mu\rho(t_k))^2(z_0^2 - \mu^2\rho^2(t_k))}{4\sigma^5\rho^3(t_k)}\frac{\partial\rho}{\partial\gamma}c_2\right]c_1 -$$

$$\frac{c_1}{F_{Z_k}}\frac{(z_0 - \mu\rho(t_k))^2(z_0^2 - \mu^2\rho^2(t_k))}{4\sigma^5\rho^3(t_k)}\frac{\partial\rho}{\partial\gamma};$$

$$I_{14} = \frac{1}{\sigma}\sum_{i=1}^{q}\frac{1}{\rho(t_{ik})}\frac{\partial\rho(t_{ik})}{\partial\beta} - \left[\frac{(z_0^2-\mu^2\rho^2(t_k))}{\sigma^3\rho^2(t_k)}\frac{\partial\rho(t_k)}{\partial\beta} - \frac{(z_0-\mu\rho(t_k))^2(z_0^2-\mu^2\rho^2(t_k))}{4\sigma^5\rho^3(t_k)}\frac{\partial\rho}{\partial\beta}c_2\right]c_1 -$$
$$\frac{c_1}{F_{Z_k}}\frac{(z_0-\mu\rho(t_k))^2(z_0^2-\mu^2\rho^2(t_k))}{4\sigma^5\rho^3(t_k)}\frac{\partial\rho}{\partial\beta};$$

$$I_{22} = \frac{1}{\sigma^2}\sum_{i=1}^{q}\rho(t_{ik}) - \left[\frac{\rho(t_k)}{\sigma^2} - \frac{(z_0-\mu\rho(t_k))^2}{\sigma^4}c_2\right]c_1 - \frac{c_1}{F_{Z_k}}\frac{(z_0-\mu\rho(t_k))^2}{\sigma^4};$$

$$I_{23} = \frac{\mu^2}{\sigma^2}\sum_{i=1}^{q}\frac{\partial\rho t_{ik}}{\partial\gamma} - \left[\frac{\mu}{\sigma^2}\frac{\partial\rho(t_k)}{\partial\gamma} - \frac{(z_0-\mu\rho(t_k))(z_0^2-\mu^2\rho^2(t_k))}{2\sigma^4\rho^2(t_k)}\frac{\partial\rho}{\partial\gamma}c_2\right]c_1 -$$
$$\frac{c_1}{F_{Z_k}}\frac{(z_0-\mu\rho(t_k))(z_0^2-\mu^2\rho^2(t_k))}{2\sigma^4\rho^2(t_k)}\frac{\partial\rho}{\partial\gamma};$$

$$I_{24} = \frac{\mu^2}{\sigma^2}\sum_{i=1}^{q}\frac{\partial\rho t_{ik}}{\partial\beta} - \left[\frac{\mu}{\sigma^2}\frac{\partial\rho(t_k)}{\partial\beta} - \frac{(z_0-\mu\rho(t_k))(z_0^2-\mu^2\rho^2(t_k))}{2\sigma^4\rho^2(t_k)}\frac{\partial\rho}{\partial\beta}c_2\right]c_1 -$$
$$\frac{c_1}{F_{Z_k}}\frac{(z_0-\mu\rho(t_k))(z_0^2-\mu^2\rho^2(t_k))}{2\sigma^4\rho^2(t_k)}\frac{\partial\rho}{\partial\beta};$$

$$I_{33} = \sum_{i=1}^{q}\left(\frac{\partial\rho(t_{ik})}{\partial\gamma}\right)^2\left(\frac{1}{2\rho^2(t_{ik})} + \frac{\mu^2}{\sigma^2\rho(t_{ik})}\right) - \left[\frac{1}{\sigma^2\rho^3(t_k)}\left(z_0^2\left(\frac{\partial\rho(t_k)}{\partial\gamma}\right)^2 -\right.\right.$$
$$\left.\left.\frac{\partial^2\rho(t_k)}{\partial\gamma^2}\frac{(z_0^2\rho-\mu^2\rho^3)}{2}\right) - \frac{(z_0^2-\mu^2\rho^2(t_k))^2}{4\sigma^4\rho^4(t_k)}\frac{\partial\rho}{\partial\gamma}c_2\right]c_1 - \frac{c_1}{F_{Z_k}}\frac{(z_0^2-\mu^2\rho^2(t_k))^2}{4\sigma^4\rho^4(t_k)}\frac{\partial\rho}{\partial\gamma};$$

$$I_{34} = \sum_{i=1}^{q}\left(\frac{\partial\rho(t_{ik})}{\partial\gamma}\frac{\partial\rho(t_{ik})}{\partial\beta}\right)\left(\frac{1}{2\rho^2(t_{ik})} + \frac{\mu^2}{\sigma^2\rho(t_{ik})}\right) - \left[\frac{1}{\sigma^2\rho^3(t_k)}\left(z_0^2\frac{\partial\rho(t_k)}{\partial\beta}\frac{\partial\rho(t_k)}{\partial\gamma} -\right.\right.$$
$$\left.\left.\frac{\partial^2\rho}{\partial\gamma\partial\beta}\frac{(z_0^2\rho-\mu^2\rho^3(t_k))}{2}\right) - -\frac{(z_0^2-\mu^2\rho^2(t_k))^2}{4\sigma^4\rho^4(t_k)}\frac{\partial\rho}{\partial\gamma}\frac{\partial\rho}{\partial\beta}c_2\right]c_1 -$$
$$-\frac{c_1}{F_{Z_k}}\frac{(z_0^2-\mu^2\rho^2(t_k))^2}{4\sigma^4\rho^4(t_k)}\frac{\partial\rho}{\partial\gamma}\frac{\partial\rho}{\partial\beta};$$

$$I_{44} = \sum_{i=1}^{q}\left(\frac{\partial\rho(t_{ik})}{\partial\beta}\right)^2\left(\frac{1}{2\rho^2(t_{ik})} + \frac{\mu^2}{\sigma^2\rho(t_{ik})}\right) - \left[\frac{1}{\sigma^2\rho^3(t_k)}\left(z_0^2\left(\frac{\partial\rho(t_k)}{\partial\beta}\right)^2 -\right.\right.$$
$$\left.\left.\frac{\partial^2\rho(t_k)}{\partial\beta^2}\frac{(z_0^2\rho-\mu^2\rho^3)}{2}\right) - \frac{(z_0^2-\mu^2\rho^2(t_k))^2}{4\sigma^4\rho^4(t_k)}\frac{\partial\rho}{\partial\beta}c_2\right]c_1 -$$
$$-\frac{c_1}{F_{Z_k}}\frac{(z_0^2-\mu^2\rho^2(t_k))^2}{4\sigma^4\rho^4(t_k)}\frac{\partial\rho}{\partial\beta};$$

where

$$c_1 = \frac{1}{2\sqrt{\pi}}\frac{\sigma\sqrt{2\rho(t_{ik})}}{(z_0-\mu\rho(t_{ik}))}exp\left(-\frac{(z_0-\mu\rho(t_{ik}))^2}{2\sigma^2\rho(t_{ik})}\right), c_2 = \frac{\sigma^2\rho(t_{ik})}{(z_0-\mu\rho(t_{ik}))^2} + 1.$$

The function of normal distribution equals:

$$F_{Z_k} = \int_{-\infty}^{z_0}\frac{1}{\sigma\sqrt{2\pi\rho}}\exp\left(-\frac{(z-\mu\rho(t_{ik}))^2}{2\sigma^2\rho(t_{ik})}\right)dz = 0.5 + 0.5\Gamma\left(0.5, \frac{(z_0-\mu\rho(t_{ik}))^2}{2\sigma^2\rho(t_{ik})}\right).$$

4. Optimal Design for a Light-Emitting Diode Reliability Test

In [16], we considered the problem of a reliability analysis for LEDs. Following the proposed algorithm, the D-optimal design for testing the reliability of LEDs has been obtained. It was shown that the determinant of the Fisher information matrix significantly increased for the optimal design in comparison with the initial design. However, the previous algorithm did not take into account the fact that in real life the observation of the object is terminated when the degradation path reaches the critical value. In this paper, we have improved the algorithm on the basis of the conditional Fisher information matrix and have analyzed how the optimal design has been changed.

Let us describe the initial conditions of the experiment. LED lighting offers many benefits for industrial and commercial businesses that are interested in reducing their energy usage and costs. The experiment involves the observation of 24 LEDs for 250 h. The normal stress level of an LED is 25mA. However, the LEDs were tested at two levels of electric current (40 mA and 35 mA) to conduct an accelerated reliability experiment. When the light intensity decreases by 50 percent, the failure of the unit is recorded.

As it was shown in [16], the date can be described by the Wiener degradation model with a log-linear trend function and a power covariate function:

$$f_{\Delta Z_{ij}}(u) = \frac{1}{\sigma\sqrt{2\pi\Delta\rho(t_{ij})}} \exp\left(-\frac{(u-\mu\Delta\rho(t_{ij}))^2}{2\sigma^2\Delta\rho(t_{ij})}\right),$$

where $\Delta\rho(t_{ij}) = \left(\frac{t_{ij}}{e^{x\beta}}\right)^{\gamma} - \left(\frac{t_{i(j-1)}}{e^{x\beta}}\right)^{\gamma}$ with parameters $\sigma = 0.49, \mu = 0.11; \gamma = 0.48$; and $\beta = -0.17$. Let us find the optimal experiment design basing on the Fisher information matrix presented in Section 3:

$$\begin{cases} M(I(\xi)) \to max, \\ 25 \leqslant x_{(i)} \leqslant 50. \end{cases} \quad (2)$$

In Figure 1, the trend functions corresponding to the initial design, the optimal design obtained in [16] and the optimal design obtained on the basis of the conditional Fisher information matrix are presented.

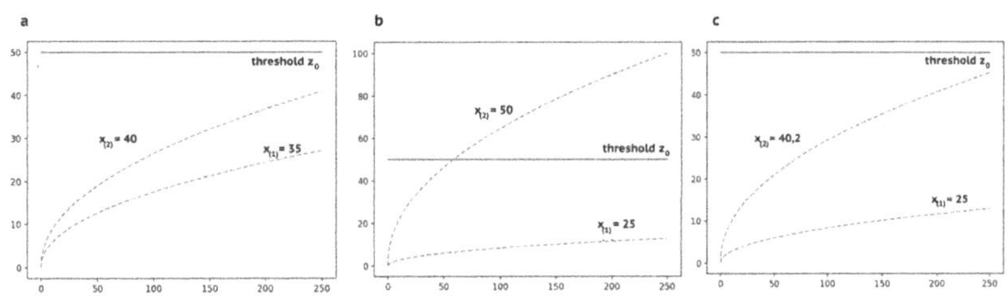

Figure 1. (**a**) The trend functions corresponding to the initial design; (**b**) the trend functions corresponding to the optimal design obtained in [16]; (**c**) the trend functions corresponding to the optimal design.

As can be seen from Figure 1b, the optimal design obtained in [16] includes stress levels equal to the minimum and maximum levels determined by the conditions of the experiment (2). However, in practice, a large amount of information on the degradation path will be lost since the object observation should be terminated when the degradation path reaches the threshold.

Figure 1c clearly illustrates that in the case of using the conditional Fisher information matrix, the reference points of the optimal design correspond to the minimum stress level and the maximum possible stress level under limitations of the degradation index.

Let us check how the determinant of the estimated conditional Fisher information matrix has changed with the new reference points of the design: $M(I(\xi_0)) = det(I(\xi_0)) = 2e+12$, $M(I(\xi^*)) = det(I(\xi^*)) = 9e+13$. On the basis of the obtained result, it is possible to conclude that the accuracy of parameter estimates has increased, as the functional of the conditional Fisher information matrix has significantly increased.

5. Conclusions

In this paper, we have obtained the elements of a conditional Fisher information matrix for the Wiener degradation model under the condition that the degradation path is limited: $Z_x(t) \leqslant z_0$. It is proposed that the optimal stress levels could be obtained by maximizing the functional of the conditional Fisher information matrix. Reliability analyses for LEDs have been considered as an example. Following the proposed approach, the D-optimal design for testing the reliability of LEDs has been obtained. On the basis of the obtained result, it is possible to conclude that the accuracy of parameter estimates has increased, as the functional of the conditional Fisher information matrix has significantly increased.

Author Contributions: Conceptualization, E.C. and E.O.; methodology, E.C.; software, E.O.; validation, E.C.; formal analysis, E.C. and E.O.; investigation, E.O.; writing—original draft preparation, E.O.; writing—review and editing, E.C.; supervision, E.C. All authors have read and agreed to the published version of the manuscript.

Funding: This research received no external funding.

Institutional Review Board Statement: Not applicable.

Informed Consent Statement: Not applicable.

Data Availability Statement: The dataset used in Section 4 is available on https://doi.org/10.17223/19988605/59/3.

Conflicts of Interest: The authors declare no conflict of interest.

References

1. Si, X.-S.; Wang, W.; Hu, C.-H.; Chen, M.-Y.; Zhou, D.-H. A Wiener-process-based degradation model with a recursive filter algorithm for remaining useful life estimation. *Mech. Syst. Signal Process.* **2013**, *35*, 219–237. [CrossRef]
2. Zhou, S.; Tang, Y.; Xu, A. A generalized Wiener process with dependent degradation rate and volatility and time-varying mean-to-variance ratio. *Reliab. Eng. Syst. Saf.* **2021**, *216*. [CrossRef]
3. Wang, Z.; Li, J.; Ma, X.; Zhang, Y.; Fu, H.; Krishnaswamy, S. A Generalized Wiener Process Degradation Model with Two Transformed Time Scales. *Qual. Reliab. Eng. Int.* **2016**, *33*. [CrossRef]
4. Peng, C.Y.; Tseng, S.T. Mis-Specification Analysis of Linear Degradation Models. *IEEE Trans. Reliab.* **2009**, *3*, 444–455. [CrossRef]
5. Tang, J.; Su, T. Estimating failure time distribution and its parameters based on intermediate data from a Wiener degradation model. *Nav. Res. Logist.* **2008**, *3*, 265–276. [CrossRef]
6. Ye, Z.-S.; Chen, N.; Shen, Y. A new class of Wiener process models for degradation analysis. *Reliab. Eng. Syst. Saf.* **2015**, *139*, 58–67. [CrossRef]
7. Hu, C.-H.; Lee, M.-Y.; Tang, J. Optimum step-stress accelerated degradation test for Wiener degradation process under constraints. *Eur. J. Oper. Res.* **2015**, *241*, 412–421. [CrossRef]
8. Chaluvadi, V. Accelerated Life Testing of Electronic Revenue Meters. Master's Thesis, Clemson University, Clemson, SC, USA, 2008.
9. Chetvertakova, E.S.; Chimitova, E.V. Testing significance of random effects for the gamma degradation model. *Anal. Data Process. Syst.* **2021**, *3*, 129–142. [CrossRef]
10. Chimitova, E.V.; Chetvertakova, E.S.; Sergeeva, S.A.; Osinceva, E.A. A comparative analysis of the wiener, gamma and inverse Gaussian degradation models. In Proceedings of the International Workshop Applied Methods of Statistical Analysis, Nonparametric Methods in Cybernetics and System Analysis (AMSA'2017), Krasnoyarsk, Russia, 18–22 September 2017; pp. 160–167.
11. Liao, C.M.; Tseng, S.T. Optimal design for step-stress accelerated degradation tests. *IEEE Trans Reliab.* **2006**, *1*, 59–66. [CrossRef]
12. Fisher, R.A. *The Design of Experiments*, 6th ed.; Oliver and Boyd: London, UK, 1951; 256p.
13. Wu, S.J.; Chang, C.T. Optimal design of degradation tests in presence of cost constraint. *Reliab. Eng. Syst. Saf.* **2002**, *76*, 109–115. [CrossRef]
14. Osintseva, E.A.; Chimitova, E.V. Informacionnaya matrica Fishera dlya vinerovskoj degradacionnoj modeli s uchetom ob'yasnyayushchih peremennyh. [Fisher's information matrix for the Wiener degradation model with covariate]. In Proceedings of the Russian Scientific and Technical Conference on the Information Processing and Mathematical Modeling, Novosibirsk, Russia, 25–26 April 2019; pp. 92–97. (In Russian)

15. Osintseva, E.A.; Chimitova, E.V. Postroenie optimal'nyh planov eksperimenta na osnove vinerovskoj degradacionnoj modeli. [Construction of optimal experimental designs based on the Wiener degradation model]. In Proceedings of the Russian Scientific and Technical Conference on the Information Processing and Mathematical Modeling, Novosibirsk, Russia, 25–26 April 2018; pp. 75–85. (In Russian)
16. Osintseva, E.; Chimitova, E. Optimal Design of Reliability Experiment Based on the Wiener Degradation Model with Covariates. Vestnik Tomskogo Gosudarstvennogo Universiteta. Upravlenie Vychislitelnaja Tehnika i Informatika. 2022; pp. 23–33. Available online: https://elibrary.ru/item.asp?id=49330076 (accessed on 4 July 2022).

engineering proceedings

MDPI

Proceeding Paper

On the Stability of Collinear Libration Points in the Three-Body Problem with Two Radiating Masses †

Abdilda Tureshbaev [1,*], Ulbossyn Omarova [2] and Ramatilla Myrzayev [3]

[1] Faculty of Engineering and Technology, Department of Computer Science, Korkyt Ata Kyzylorda University, Aiteke bi Str. 29A, Kyzylorda 120014, Kazakhstan
[2] Faculty of Economics, Department of Information Technology, Turan University, Satpayev 16A, Almaty 050013, Kazakhstan; ylbosin_kz@mail.ru
[3] Faculty of Engineering and Technology, Department of Computer Science, Turan University, Satpayev 16A, Almaty 050013, Kazakhstan; myrza_ramatilla@mail.ru
* Correspondence: aturesh@mail.ru; Tel.: +7-775-421-5652
† Presented at the 15th International Conference "Intelligent Systems" (INTELS'22), Moscow, Russia, 14–16 December 2022.

Abstract: The stability of cloud accumulations of gas and dust particles in the field of binary star systems is studied. As a dynamic model, we consider a restricted three-body problem in which both main bodies are radiating. We study the stability of collinear libration points (CLL) in a nonlinear formulation. The problem of CLP stability is considered in three-dimensional parametric space. It is shown that at the resonance of the fourth order in the plane problem, the points under study are stable in the sense of Lyapunov. In this case, the invariant normal form and Markeev's theorem are used. The stability of the CLP in the spatial problem is considered. The Birkhoff normal form is used and the Arnold–Moser theorem is used. Results are obtained on stability for most initial conditions (in the Lebesgue measure) and formal stability.

Keywords: collinear libration points; gravity; resonance; stability; field; radiating

Citation: Tureshbaev, A.; Omarova, U.; Myrzayev, R. On the Stability of Collinear Libration Points in the Three-Body Problem with Two Radiating Masses. *Eng. Proc.* **2023**, *33*, 65. https://doi.org/10.3390/engproc2023033065

Academic Editors: Askhat Diveev, Ivan Zelinka, Arutun Avetisyan and Alexander Ilin

Published: 22 August 2023

1. Introduction

In photogravitational celestial mechanics, along with the forces of Newtonian attraction F_g, the light pressure is taken into account F_p, coming from the radiating body (star) [1]. In some cases, the luminous flux is so intensive that the force F_p competes with gravity F_g, and can be even greater than that.

For a particular particle, the magnitude of the light pressure force depends not only on the power of the radiation source (star), but also on the cross-sectional area, the mass and the reflectivity of the particle. To determine the connection between the parameters of the star and the particle, a coefficient Q is introduced, called the particle mass reduction coefficient. For a particular particle, Q has a constant value that characterizes its susceptibility to radiation. The relationship between the parameters of the star [2] and the particle gives the reduction coefficient Q

$$Q = 1 - (1-\varepsilon)A\frac{E}{fM} \tag{1}$$

(f is the gravitational parameter of the star, E and M is the mass and power of the star, A is a windage of the particle, determined by the ratio of the cross–sectional area to its mass, ε is the coefficient of light reflection). Sufficiently large and dense particles with small values of the parameters A and ε are most affected by the gravitational force of the star, therefore, $Q > 0$. For the smallest particles with high windage and reflection coefficient, the action of light is greater than gravity $(Q < 0)$.

The photogravitational three-body problem introduced by V.V. Radzievskiy [3] has become a good dynamic model for studying the motion of microparticles in binary star systems.

In the elliptical version of the problem (when the orbits of a stellar pair are elliptical), we write the equations of motion of a particle in a rectangular system rotating together with the stars in the form

$$\ddot{x} - 2\dot{y} = \frac{\partial W}{\partial x}, \ddot{y} + 2\dot{x} = \frac{\partial W}{\partial y}, \ddot{z} = \frac{\partial W}{\partial z} \quad (2)$$

where x, y, z are dimensionless coordinates related to the distance $r = p/(1 + e\cos v) - 1$ between the stars (p and e are the focal parameter and the eccentricity of the relative orbital motion of the stellar pair) that are the rectangular coordinates of the particle and W is the force function of the system, equal to

$$W = (1 + e\cos v)^{-1}[(x^2 + y^2 - z^2 e\cos v)/2 + Q_1(1-\mu)/R_1 + Q_2\mu/R_2]$$

$$R_1 = [(x+\mu)^2 + y^2 + z^2]^{1/2}, R_2 = [(x - 1 + \mu)^2 + y^2 + z^2]^{1/2} \quad (3)$$

Here, μ and $1 - \mu$ are the dimensionless masses of the stars, and Q_1 and Q_2 are the reduction coefficients of their mass, which represent the ratio of the difference between the gravitational and repulsive forces to the gravitational force. In terms of physical meaning, numerical values Q_1 and Q_2 do not exceed 1. For the classical problem $Q_1 = 1, Q_2 = 1$ (no radiation of bodies) [4].

Libration points—constant solutions of the adopted system of dynamic equations—represent relative equilibria in a circular problem and periodic motions in an elliptic problem. They are found from the system of equations

$$\frac{\partial W}{\partial x} = 0, \frac{\partial W}{\partial y} = 0, \frac{\partial W}{\partial z} = 0, \quad (4)$$

CLP are located on a straight line connecting the main bodies, and for them $y = 0, z = 0$. Their positions on the abscissa axis are determined from the first equation of the system (4). The triangular libration points were carefully studied in [5]. The stability of triangular points in strictly nonlinear formulation were considered in [6,7]. In [8,9], the nonlinear analysis of stable coplanar libration points that are not on the plane of orbital motion of main bodies was completed.

2. Collinear Libration Points and Their Stability in a Plane Problem

The CLP coordinate is determined from the following equation

$$f(x) = x - Q_1\frac{(1-\mu)(x+\mu)}{|x+\mu|^3} - Q_2\frac{\mu(x+\mu-1)}{|x+\mu-1|^3} = 0 \quad (5)$$

Equation (5) contains three mutually independent parameters μ, Q_1, Q_2. Therefore, the problem can consider a three-parameter family of CLPs. Let us consider the case when both components of the binary star radiate and move in circular orbits. From Equation (5), we obtain

$$f'(x) = 1 + 2a(x), a = Q_1\frac{1-\mu}{|x+\mu|^3} + Q_2\frac{\mu}{|x+\mu-1|^3}$$

Thus, there is a parameter a [10], which is included in the characteristic equation

$$\lambda^4 + (2-a)\lambda^2 - (1-a)(1+2a) = 0, \quad (6)$$

its solutions are equal to

$$\lambda_\alpha^2 = \frac{1}{2}(a - 2 \pm \sqrt{(9a-8)a}, \alpha = 1, 2$$

When changing the parameter values a in the intervals $8/9 < a \leq 1$ and $-0.5 < a \leq 0$, roots λ_α are purely imaginary, and at the boundaries there are multiples. Therefore, the parameter values indicated above by the inequalities correspond to the stability region of libration points in the first approximation. In [11], they constructed diagrams of CLP stability.

In the problem, resonances of the third and fourth orders are found; for third-order resonances, the resonant values of the parameter a have the form $a_\pm^* = 41/108 \pm 5\sqrt{145}/108$; for fourth-order resonances, $a_\pm = (68 + 60\sqrt{5})/209$. As expected, the resonance of the third order leads to the instability of the CLP [12].

In [13], it is shown that at the resonance of the fourth order, the CLPs are stable by Lyapunov.

Below, we consider the stability of the CLP in the spatial problem. The Birkhoff normal form is used, and the Arnold–Moser theorem [14] is applied.

3. Equations of Motion and Expansion of the Hamilton Function

Particle motion $P(x, y, z)$ is given by the canonical equations

$$\frac{d\overline{q_i}}{dt} = \frac{\partial H}{\partial \overline{p}_i}, \frac{d\overline{p_i}}{dt} = -\frac{\partial H}{\partial \overline{q}_i}, (i = 1, 2, 3) \tag{7}$$

where q_i are the Cartesian coordinates of the particle $P(x, y, z)$, $\overline{p_i}$ are the corresponding canonical momenta and $H(x, y, z, \overline{p_1}, \overline{p_2}, \overline{p_3})$ is the Hamilton analytic function with respect to coordinates and momenta, which, in our case, has the form

$$\begin{gathered} H = \frac{1}{2}(\overline{p}_1^2 + \overline{p}_2^2 + \overline{p}_3^2) + (\overline{p}_1 y - \overline{p}_2 x) - Q_1(1-\mu)/R_1 - Q_2\mu/R_2 \\ R_\alpha = \sqrt{(x - x_\alpha}^2 + y^2 + z^2, (\alpha = 1, 2) \end{gathered} \tag{8}$$

Here, Q_1 and Q_2 are coefficients of reduction of the masses of the main bodies, which, in the case of CLP, can take both positive and negative values [10].

We study the stability of the CTL under the assumption that the orbit of the main bodies is circular and the particle P of infinitely small mass at the initial moment of time experiences initial perturbations that take it out of the plane of rotation of the main bodies S_1 and S_2.

We introduce perturbations into Equation (1) according to the formulas

$$\begin{gathered} x = x^* + q_1, y = q_2, z = q_3, \overline{p}_1 = \overline{p}_1^* + p_1, \overline{p}_2 = p_2, \overline{p}_3 = p_3 \\ p_1^* = x^*, p_2^* = y^* = p_3^* = z_0^* = 0, \end{gathered} \tag{9}$$

where

$$x^* = 0.5(Q_1^{\frac{2}{3}} - Q_2^{\frac{2}{3}} - 1) - \mu, \overline{p}_1^* = \mp 0.5\sqrt{2(Q_1^{\frac{2}{3}} + Q_2^{\frac{2}{3}}) - (Q_1^{\frac{2}{3}} - Q_2^{\frac{2}{3}}) - 1}. \tag{10}$$

Expanding the Hamilton function in a row according to the degrees of perturbations and in the neighborhood of the considered collinear point, taken as the origin, we obtain

$$H = H_2 + H_3 + H_4 + \ldots \tag{11}$$

Here, H_m are homogeneous polynomials of degree $m(m = 2,3,4,...)$ with respect to generalized coordinates and impulses p_i, so

$$H_m = \sum_{v+l=m} h_{v_1v_2v_3l_1l_2l_3} \cdot q_1^{v_1} q_2^{v_2} q_3^{v_3} p_1^{l_1} p_2^{l_2} p_3^{l_3} \tag{12}$$

Then, in Equation (11), the forms H_2, H_3 and H_4 taking into account (9) will take the following form:

$$\begin{gathered} H_2 = \frac{1}{2}(p_1^2 + p_2^2 + p_3^2) + p_1q_2 - p_2q_1 + h_{200}q_1^2 + h_{020}q2^2 + h_{002}q_3^2 + h_{110}q_1q_2 \\ +h_{101}q_1q_3 + h_{001}q_2q_3, \end{gathered} \tag{13}$$

$$\begin{gathered} H_3 = h_{300}q_1^3 + h_{030}q_2^3 + h_{300}q_3^3 + h_{210}q_1^2q_2 + h_{201}q_1^2q_3 + h_{120}q_1q_2^2 + h_{021q_2^2q_3} \\ +h_{102}q_1q_3^2 + h_{012}q_2q_3^2 + h_{111}q_1q_2q_3, \end{gathered} \tag{14}$$

$$\begin{gathered} H_4 = h_{400}q_1^4 + h_{040}q_2^4 + h_{004}q_3^4 + h_{310}q_1^3q_2 + h_{130}q_1q_2^3 \\ +h_{103}q_1q_3^3 + h_{301}q_1^3q_3 + h_{031}q_2^3q_3 + h_{013}q_3^3q_2 + h_{211}q_1^2q_2q_3 \\ +h_{121}q_1q_2^2q_3 + h_{112}q_1q_2q_3^2 + h_{220}q_1^2q_2^2 + h_{202}q_1^2q_3^2 + h_{022}q_2^2q_3^2, \end{gathered} \tag{15}$$

where

$$\begin{gathered} h_{200} = -8a, h_{020} = 4a, h_{002} = 4a, h_{110} = 0, h_{101} = 0, h_{011} = 0 \\ h_{300} = 16b, h_{120} = -16b, h_{102} = -16b, h_{030} = 0, h_{003} = 0, h_{210} = 0, \\ h_{201} = 0, h_{021} = 0, h_{012} = 0, h_{111} = 0, \\ h_{400} = -32c, h_{040} = -12c, h_{004} = -12c, h_{220} = 32c, h_{202} = 32c, \\ h_{022} = -8c, h_{310} = 0, h_{130} = 0, h_{103} = 0, h_{301} = 0, h_{031} = 0, \\ h_{013=0}, h_{211} = 0, h_{121} = 0, h_{112} = 0. \end{gathered} \tag{16}$$

$$\begin{gathered} a = \frac{Q_1(1-\mu)}{|Q_1^{2/3} - Q_2^{2/3} + 1|^3} + \frac{Q_2\mu}{|Q_1^{2/3} - Q_2^{2/3} - 1|^3}, \\ b = \frac{Q_1(1-\mu)(Q_1^{2/3} - Q_2^{2/3} + 1)}{|Q_1^{2/3} - Q_2^{2/3} + 1|^5} + \frac{Q_2\mu(Q_1^{2/3} - Q_2^{2/3} - 1)}{|Q_1^{2/3} - Q_2^{2/3} - 1|^5} \\ c = \frac{Q_1(1-\mu)}{|Q_1^{2/3} - Q_2^{2/3} + 1|^5} + \frac{Q_2\mu}{|Q_1^{2/3} - Q_2^{2/3} - 1|^5} \end{gathered} \tag{17}$$

4. Stability of CLP in a Spatial Problem

The question of the stability of the investigated spatial CLPs can be considered as a stability problem of equilibrium positions $q_i = p_i = 0 (i = 1,2,3)$ of an autonomous Hamiltonian system with three degrees of freedom. As can be seen from (13), here we have the case when H_2 is not a sign-definite function, and the characteristic equation of the system has no roots with a nonzero real part. Hence, the stability of the complete system does not follow from the stability of a linear system.

Expanding the Hamilton function into a power series q_i, p_i in the vicinity of the considered equilibrium position, first the Hamiltonian H_2 is transformed to the normal form in the form

$$K_2 = \omega_1 r_1 - \omega_2 r_2 + \omega_3 r_3. \tag{18}$$

The structure of the normal form depends on the type of resonance relation

$$\omega_1 r_1 - \omega_2 r_2 + \omega_3 r_3 = 0(|k_1| + k_2 + |k_3| \leq 4), \tag{19}$$

where the frequencies of the principal oscillations for the libration points are equal to

$$\omega_1 = \sqrt{(2-a+\sqrt{(9a-8)a})/2}, \omega_2 = \sqrt{(2-a-\sqrt{(9a-8)a})/2}, \omega_3 = \sqrt{a}. \quad (20)$$

As can be seen from the last expression, for the frequency of spatial oscillations, the parameter a can take only positive values. Therefore, resonances containing the frequency of spatial oscillations can be realized only in a limited part of the region ($8/9 < a \leq 1$ and $-1/2 < a \leq 0$) for necessary stability conditions of the system. Let us investigate the stability of CLP at two-frequency resonances. For CLP, the following two-frequency resonances turned out to be possible:

$$\omega_1 = 2\omega_2, \omega_1 = 3\omega_2, 2\omega_1 = \omega_3, 3\omega_1 = \omega_3, 2\omega_2 = \omega_3, 3\omega_2 = \omega_3$$

Resonances $\omega_1 = 2\omega_2$ and $\omega_1 = 3\omega_2$, discovered in the plane problem, were studied in [12,13]. In the spatial photogravitational problem, resonances of the third and fourth orders turned out to be possible

$$2\omega_1 = \omega_3, 3\omega_1 = \omega_3, 2\omega_2 = \omega_3, 3\omega_2 = \omega_3$$

which respectively correspond to the values of the parameter a defined as

$$a = 4(1+2\sqrt{7})/27, a = 4(-1+\sqrt{10})/9, a = (63+\sqrt{53217})/304, a = (63+\sqrt{53217})/304$$

Note that the last two resonances, $3\omega_1 = \omega_3$ and $3\omega_2 = \omega_3$, match. To construct resonance curves (in the stability region in the linear approximation of the system) for the corresponding specific resonance value of the coefficient a, a curve is constructed, which is determined by the expression

$$\frac{Q_1(1-\mu)}{|Q_1^{2/3} - Q_2^{2/3} + 1|^3} + \frac{Q_2\mu}{|Q_1^{2/3} - Q_2^{2/3} - 1|^3} = a$$

At resonance $2\omega_1 = \omega_3$ (which does not involve the frequency of plane oscillations), which corresponds to the value of the parameter $a = 4(1+2\sqrt{7})/27$, the normalized Hamiltonian takes the form [14]

$$H = 2\omega_1 r_1 - \omega_1 r_3 + A(\omega_1, \omega_3) r_3 \sqrt{r_1} \sin(\varphi_1 + 2\varphi_3) + O(r_1 + r_3)^2, \quad (21)$$

where $A(\omega_1, \omega_3) = -\sqrt{\omega_1(x_{1002}^2 + y_{1002}^2)}$ and the coefficients x_{1002} and y_{1002} look like

$$x_{1002} = -\frac{\omega_1 h_{0111}}{2\omega_1} - \frac{h_{1002}}{2} + \frac{h_{1200}}{2\omega_1^2}, y_{1002} = -\frac{\omega_1 h_{0012}}{2} - \frac{\omega_1 h_{0210}}{2\omega_1^2} + \frac{h_{1101}}{2\omega_1},$$

which, for collinear points, are equal to

$$x_{1002} = -\frac{h_{1200}}{2\omega_1^2}, y_{1002} = 0 \quad (22)$$

hence, the expression

$$A(\omega_1, \omega_3) = -\sqrt{\omega_1(x_{1002}^2 + y_{1002}^2)} = -\sqrt{\omega_1} x_{1002}$$

is not equal to zero anywhere; therefore, according to the Arnold–Moser theorem, at a third-order resonance from the stability region, in the first approximation, we can say that the CTLs are unstable. If there is a fourth-order resonance in the system, corresponding to the value of the parameter $a = (63+\sqrt{53217})/304$, using the Birkhoff transformation in the original Hamiltonian, we annihilate the terms of the third degree. The Hamiltonian normalized in this case in polar coordinates will take the following form [13]:

$$H = 3\omega_1 r_1 - \omega_1 r_3 + c_{20} r_1^2 + c_{11} r_1 r_3 + c_{02} r_3^2 + \\ B(\omega_1, \omega_3) r_3 \sqrt{r_1 r_3} cos(\varphi_1 + 3\varphi_3) + O(r_1 + r_3)^{5/2}, \tag{23}$$

where

$$B(\omega_1, \omega_3) = \frac{1}{3}\omega_3 \sqrt{3(x_{1003}^2 + y_{1003}^2)}$$

It is important to notice that in the classical problem for a fixed value μ, the coefficients $B(\omega_1, \omega_3), c_{200}, c_{100}$ and c_{020} take constant values (which simplifies the investigation of the problem). In this problem, the same coefficients do not remain constant and are functions of arbitrary coefficients of the coefficients Q_1 and Q_2, which makes the task much more difficult. These are denoted by the coefficients of the Hamiltonian (23)

$$N_1 = c_{200} + 3c_{110} + 9c_{020}, N_2 = 3\sqrt{3}B(\omega_1, \omega_3),$$

where

$$B(\omega_1, \omega_3) = \frac{1}{3}\omega_3 \sqrt{3(x_{1003}^2 + y_{1003}^2)}$$

is defined by expressions

$$x_{1003} = \frac{1}{2}\omega_1 h_{0013} + \frac{1}{2\omega_3^3} h_{1300} - \frac{1}{2\omega_3} h_{1102} - \frac{\omega_1}{2\omega_3^2} h_{0211} \\ -\frac{9}{5}(x_{0120} * x_{0012} + y_{0120} * y_{0012}) - \frac{1}{\omega_3}(x_{1002} y_{1011} + x_{1011} y_{1002}) \\ + \frac{4}{\omega_3^2}(x_{1002} x_{0201} + y_{1002} y_{0201}) + \frac{3}{2}(x_{0003} x_{0111} + y_{0003} y_{0111}),$$

$$y_{1003} = -\frac{\omega_1}{2\omega_3} h_{0112} + \frac{1}{2} h_{1003} + \frac{1}{2\omega_3^2} h_{1201} + \frac{\omega_1}{2\omega_3^3} h_{0310} \\ -\frac{9}{5}(x_{0120} * y_{0012} + x_{0012} * y_{0120}) - \frac{1}{\omega_3}(y_{1011} y_{1002} - x_{1011} x_{1002}) \\ + \frac{4}{\omega_3^2}(x_{0201} y_{1002} + x_{1002} y_{0201}) + \frac{3}{2}(x_{0111} x_{0003} + x_{0003} y_{0111}),$$

where coefficients

$$h_{0013}, h_{1300}, h_{1102}, h_{0211}, h_{0112}, h_{1003}, h_{1201}, h_{0310}, x_{0120}, y_{0120}, \\ x_{1011}, y_{1002}, y_{1011}, x_{0012}, y_{0111}, x_{0201}, y_{0201}, x_{0003}, x_{0003}$$

given for CLP above (16) take values equal to zero, therefore, they are identically equal to zero x_{1003} and y_{1003} . Then, the equality takes place

$$N_2 = 3\sqrt{3}B(\omega_1, \omega_3) = 0.$$

Let us now define the value $N_1 = c_{200} + 3c_{110} + 9c_{020}$. Here, the coefficients $c_{200}, c_{110} c_{020}$, which are invariants of the Hamilton function (11) with respect to canonical transformations, depend on the coefficients $h_{v_1 v_2 l_1 l_2}$ that are homogeneous polynomials (12) of degree $m (m = 3, 4)$, which are equal to

$$
\begin{aligned}
c_{200} &= \frac{3}{2\omega_1^2}h_{4000} - \frac{27}{8}\omega_2^2 y_{0030}^2 - \frac{3}{2}x_{1020}^2, \\
c_{110} &= \frac{1}{\omega_1\omega_2}h_{2200} - \frac{2}{3}x_{1002}^2 + \frac{3}{10}\omega_2^2 y_{0012}^2 + 2x_{0111}x_{1020}, \\
c_{020} &= \frac{3}{2\omega_2^2}h_{0400} - \frac{1}{6}x_{0111}^2 - \frac{3}{40}\omega_2^2 y_{0012}^2,
\end{aligned}
\tag{24}
$$

where

$$
\begin{aligned}
y_{0030} &= -\frac{1}{\omega_1^2}h_{3000}, x_{1020} = -\frac{3}{2\omega_1^2}h_{3000}, x_{1002} = \frac{1}{2\omega_2^2}h_{1200}, \\
y_{0012} &= \frac{1}{\omega_1\omega_2^2}h_{1200}, x_{0111} = \frac{1}{\omega_1\omega_2}h_{1200}
\end{aligned}
\tag{25}
$$

Substituting from (16) the values $h_{3000} = 16b$, $h_{1200} = -16b$ in (24), we have

$$
\begin{aligned}
c_{200} &= -\frac{48}{\omega_1^2}c - \frac{864\omega_2^2}{\omega_1^2}(1 + \frac{1}{\omega_1^2})^2 b^2 - 96(1 - \frac{3}{\omega_1^2})^2 b^2, \\
c_{110} &= \frac{32}{\omega_1\omega_2}c - \frac{96}{\omega_2^4}b_2 + \frac{384}{5\omega_1^2\omega_2^2}b^2 - \frac{256}{\omega_1\omega_2}(1 - \frac{3}{\omega_1^2})b^2, \\
c_{020} &= -\frac{18}{\omega_2^2}c - \frac{32}{3\omega_2^4}b^2 - \frac{96}{5\omega_1^2\omega_2^2 b^2}
\end{aligned}
\tag{26}
$$

where a, b and c are parameters that depend on reduction factors Q_1 and Q_2 and dimensionless mass parameter μ. As the calculations showed, the modulus of the expression $N_1 = c_{200} + 3c_{110} + 9c_{020}$ is always different from zero. Consequently, the inequality holds everywhere $|N_1| > N_2 = 0$, which, according to [13], guarantees the existence of Lyapunov stability. In a similar way, it is proved that at a resonance of the third order $2\omega_2 = \omega_3$, CLPs are unstable, and at a fourth-order resonance $3\omega_2 = \omega_3$ they are stable by Lyapunov. Below are the regions of the stability of the linear system (colored), in which the resonance curves of the fourth order are indicated $2\omega_2 = \omega_3$ for two values of the mass parameter μ.

At $\mu = 0.001$, the resonance curve is located closer to the middle of the stability region (Figure 1a). When the mass μ increases up to 0.01 (Figure 1b), the region becomes slightly smaller and the resonance curve becomes closer to the boundary of the stability region and becomes less noticeable than when $\mu = 0.001$. Apparently, this fact confirms the above conclusion that resonances containing the frequency of spatial oscillations can be realized only in a limited part ($8/9 < a$), being the area of necessary conditions for the stability of CLP.

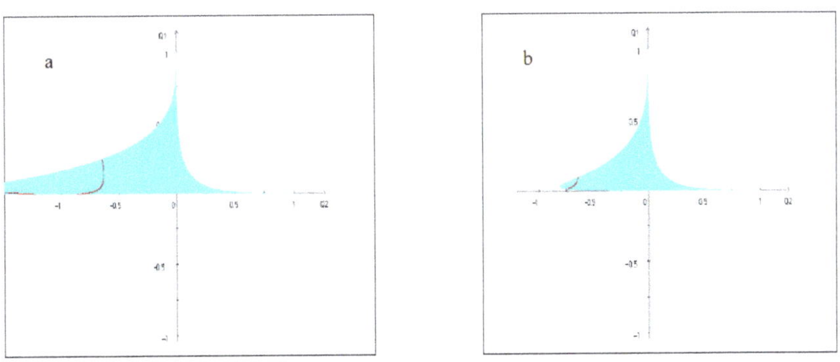

Figure 1. (**a**) Resonance curve of the fourth at $\mu = 0.001$; (**b**) Resonance curve of the fourth at $\mu = 0.01$.

Note that in the classical problem for a fixed value μ, the coefficients $c_{200}, c_{110}, c_{020}$ are constants (which much simplifies the investigation of the problem). However, in this problem, the same coefficients are not constants but functions of the coefficients Q_1 and Q_2, which makes the task much more difficult.

If a ω_i does not satisfy condition (19), then after applying the Birkhoff transformation, the Hamiltonian of the perturbed motion in polar coordinates normalized to the fourth order inclusive has the form

$$H^* = K_2(r_1, r_2, r_3) + K_4(r_1, r_2, r_3) \tag{27}$$

Here, K_4 is defined by the expression

$$K_4 = c_{200}r_1^2 + c_{110}r_1r_3 + c_0 11 r_2r_3 + c_{002}r_3^2 \tag{28}$$

Now, we use Arnold's results on the stability of Hamiltonian systems for most of the initial conditions [13]. It is known that the instability found in the plane problem remains such in the spatial problem.

Assuming that there are no resonances in the system $2\omega_1 = \omega_3, \omega_1 = 2\omega_2, \omega_1 = 3\omega_2, 3\omega_1 = \omega_3, 2\omega_2 = \omega_3, 3\omega_2 = \omega_3$, consider a fourth-order determinant

$$D_4 = det\begin{vmatrix} \frac{\partial^2 K_4}{\partial r_i \partial r_j} & \frac{\partial K_2}{\partial r_i} \\ \frac{\partial K_2}{\partial r_j} & 0 \end{vmatrix} \tag{29}$$

Expanding the determinant (29), we have

$$\begin{gathered} D_4 = \omega_1^2(c_{011}^2 - 4c_{020}c_{002} + \omega_2^2(c_{101^2} - 4c_{200}c_{002}) + \\ +\omega_3^2(c_{110}^2 - 4c_{200}c_{020}) + 2\omega_1\omega_2(c_{101}c_{011} - 2c_{002}c_{110}) - \\ -2\omega_1\omega_3(c_{001}c_{110} - 2c_{020}c_{101}) + 2\omega_2\omega_3(c_{110}c_{101} - 2c_{200}c_{011}) \end{gathered} \tag{30}$$

The balance position $q_i = p_i = 0$ is stable for most initial conditions (by the Lebesgue measure) when the determinant $D_4 \neq 0$. After using numerical analysis, we check the validity of the inequality $D_4 \neq 0$. We see that in the spatial photogravitational three-body problem, the collinear libration points are stable for most initial conditions (by the Lebesgue measure) for all a (except for the values corresponding to internal resonances of the third $2\omega_1 = \omega_3$ and $2\omega_2 = \omega_3$ and fourth $\omega_1 = 3\omega_2, 2_1 = 3\omega_3, 3\omega_2 = \omega_3$ orders) from the stability region in the linear approximation.

The presence of stability in the system for most of the initial conditions means, with a probability close to unity, that the KTLs are stable in the spatial problem.

As shown by numerical calculations, CLPs are formally stable for almost all values of the parameters from the stability region in the linear approximation. The exceptions are, in addition to the values of the parameters corresponding to the studied resonance, perhaps those values μ, Q_1, Q_2 from the stability region, at which resonances above the fourth order are realized.

The presence of formal stability means that Lyapunov instability is not detected over a practically very long time interval. This suggests that the particles will stay near the stable libration points for quite a long time.

Author Contributions: Conceptualization, A.T.; methodology, A.T. and U.O.; software, R.M. and U.O.; alidation, A.T.; writing—rough preparation, R.M. and U.O.; writing—review and editing, A.T.; All authors have read and agreed with the published version of the manuscript.

Funding: This research received no external funding.

Institutional Review Board Statement: Not applicable.

Informed Consent Statement: Not applicable.

Data Availability Statement: Not applicable.

Conflicts of Interest: The authors have declared that there are no conflict of interest.

References

1. Kunitsyn, A.L. and Tureshbaev, A.T. On Collinear Libration Points of the Photogravitational Problem of Three Bodies. *Pis'ma Astron. Zh.* **1983**, *9*, 432–435.
2. Zimovshchikov, A.S.; Tkhai, V.N. Stability diagrams of a heterogeneous ensemble of particles at collinear libration points of the photogravitational three-body problem. *PMM* **2010**, *2*, 221–229.
3. Radzievsky, V.V. Restricted three-body problem with allowance for light pressure. *Astron. Zh.* **1950**, *4*, 249–256.
4. Kunitsyn, A.L.; Tureshbaev, A.T. On the collinear libration points in the photo¬gravitational three-body problem. *Celest. Mech.* **1985**, *35*, 105–112. [CrossRef]
5. Kunitsyn, A.L. On the stability of triangular libration points of the photogravitational three–body problem. *PMM* **2000**, *40*, 788–794. [CrossRef]
6. Tureshbaev, A.T. Stability of libration points of the photogravitational restricted three body problem with two radiating masses. In Proceedings of the 2016 International Conference Stability and Oscillations of Nonlinear Control Systems (Pyatnitskiy's Conference), Moscow, Russia, 1–3 June 2016.
7. Tureshbaev, A.T.; Omarova, U.S. Modeling the dynamics of particles of gas-dust clouds in the photogravitational field of binary stellar systems. *Procedia Comput. Sci.* **2021**, *186*, 521–528. [CrossRef]
8. Perezhogin, A.A.; Tureshbaev, A.T. On the stability of triangular libration points in the photogravitational three-body problem. *Sci. Lett. Astron. J.* **1987**, *18*, 395–408.
9. Perezhogin, A.A.; Tureshbaev, A.T. Stability of coplanar libration points of the photogravitational restricted three-body problem. *Astronomical* **1987**, *33*, 445.
10. Thai, V.N. On stability of the collinear libration points under internal third-order resonance. *Autom. Remote. Control* **2011**, *72*, 1906–1910.
11. Thai, N.V. Stability of collinear libration points at internal resonance of the third order. *AiT* **2011**, *9*, 121–126.
12. Thai, N.V. Stability of collinear libration points of the photogravitational problem of three bodies at internal resonance of the fourth order. *PMM* **2012**, *4*, 610–615.
13. Markeev, A.P. *Libration Points in Celestial Mechanics and Cosmodynamics*; Nauka: Moscow, Russia, 1978; 312p.
14. Sebekhey, V. *Theory of Orbits*; Nauka: Moscow, Russia, 1982; 656p.

engineering proceedings

MDPI

Proceeding Paper

ADRC-Based UAV Control Scheme for Automatic Carrier Landing †

Ruiyang Zhou ‡ and Konstantin A. Neusypin *,‡

Department IU1, Bauman Moscow State Technical University, 105005 Moscow, Russia; zhouruiyang.ray@gmail.com

* Correspondence: neysipin@mail.ru; Tel.: +7-903-613-5859

† Presented at the 15th International Conference "Intelligent Systems" (INTELS'22), Moscow, Russia, 14–16 December 2022.

‡ These authors contributed equally to this work.

Abstract: In this paper the problem of atmospheric disturbances during the UAV carrier landing operation is considered. A UAV dynamics model, and a wind gust and airwake disturbance model are introduced. A LADRC-based cascade control scheme is developed for fixed-wing UAVs. In the control scheme, three ADRC controllers are designed for attitude control, and another two ADRC controllers are designed for course and altitude tracking. Finally, a series of simulations are implemented in Simulink and the results are presented to demonstrate the performance of the proposed control scheme.

Keywords: automatic carrier landing; UAV control; ADRC; atmospheric disturbances; carrier airwake

1. Introduction

Automatic carrier landing for UAVs is one of most critical challenging operations; the marine environment can introduce various disturbances to UAV control when landing on a carrier. Carrier motion, disturbance due to carrier airwake and atmospheric turbulence make it very hard to ensure a safe landing operation. A lot of studies have been conducted on this problem and some control schemes have been proposed accordingly [1–4]. In [5], an MPC controller was designed for a linearized UAV system to handle the landing task with carrier heave motion. In [6], an autoregressive model was used to predict the carrier motion and a preview control scheme was used to reject the disturbances. In [7], an ADRC-based controller was designed for UAV control but only the pitch dynamics were considered. In this paper a detailed UAV model and disturbances model are introduced, and a control scheme with outer loop (navigation control) and inner loop (attitude control) is proposed based on the LADRC method to reject disturbances during the landing operation.

This paper is organized as follows: A typical fixed-wing UAV model and empirical models of atmospheric disturbances are introduced in Section 2. In Section 3, an ADRC double-loop control scheme is proposed and corresponding controllers are designed. Results of simulation experiments are illustrated in Section 4. And some concluding remarks are given in Section 5.

Citation: Zhou, R.; Neusypin, K.A. ADRC-Based UAV Control Scheme for Automatic Carrier Landing. *Eng. Proc.* **2023**, *33*, 66. https://doi.org/10.3390/engproc2023033066

Academic Editors: Askhat Diveev, Ivan Zelinka, Arutun Avetisyan and Alexander Ilin

Published: 9 October 2023

2. Mathematical Model

In this section we briefly describe UAV dynamics, then the considered fixed-wing UAV model is given, followed by a model of atmospheric disturbances.

2.1. Fixed-Wing UAV Model

In the literature, various models are used to describe UAV dynamics, including the linear uncoupled model and the nonlinear cross-coupled model. This section presents a nonlinear 6-DOF UAV model with cross-couplings between yaw, roll and pitch motion.

The body diagram of a fixed-wing UAV is shown in Figure 1. The 6-DOF UAV dynamics can be fully described using 12 states, i.e., $X = [p\_n, p\_e, h, u, v, w, \phi, \theta, \psi, p, q, r]$. In this paper some states are ignored since guidance law design is beyond the scope of this paper (a Dubins path is used as a reference trajectory, see Section 3). Furthermore, we assume that the UAV is not equipped with a rudder and the yaw angle ψ is controlled by the roll maneuver. The equations for describing the necessary controlled variables are given by (1)–(8).

$$\dot{h} = u\sin\theta - v\sin\phi\cos\theta - w\cos\phi\cos\theta; \quad (1)$$

$$\dot{u} = rv - qw - g\sin\theta + \frac{\rho V_a{}^2 S}{2m} C_u + \frac{\rho S_{prop} C_{prop}}{2m}[(K_m otor\delta t)^2 - V_a{}^2]; \quad (2)$$

$$\dot{v} = pw - ru + g\cos\theta\sin\phi + \frac{\rho V_a{}^2 S}{2m} C_v; \quad (3)$$

$$\dot{w} = qu - pv + g\cos\theta\cos\phi + \frac{\rho V_a{}^2 S}{2m} C_w; \quad (4)$$

$$\dot{\phi} = p + q\sin\phi\tan\theta + r\cos\phi\tan\theta; \quad (5)$$

$$\dot{\theta} = q\cos\phi - r\sin\phi; \quad (6)$$

$$\dot{p} = \Gamma_1 pq - \Gamma_2 qr + \frac{1}{2}\rho V_a{}^2 Sb[C_p + C_{p_{\delta_a}}\delta_a]; \quad (7)$$

$$\dot{q} = \Gamma_5 pr - \Gamma_6(p^2 - r^2) + \frac{\rho V_a{}^2 Sc}{2J_y}[C_m + C_{m_{\delta_e}}\delta_e], \quad (8)$$

where u, v, w are roll axis, pitch axis and yaw axis components of airspeed V_a in the UAV body frame, respectively; ϕ, θ, p, q are roll and pitch angles, and their angular velocities; $\delta_t, \delta_a, \delta_e$ are control variables of throttle, aileron and elevator, respectively; C_u, C_v, C_w, C_p, C_m are terms related to UAV aerodynamics; their definitions can be found in [8].

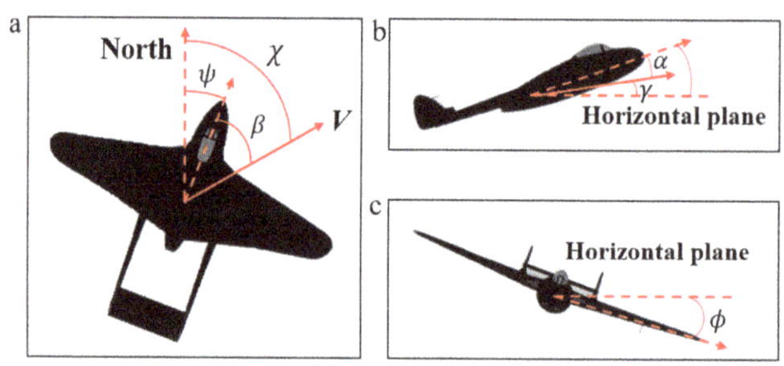

Figure 1. UAV body diagram: (**a**) vertical view; (**b**) end view; (**c**) front view.

Airspeed is calculated using Equations (2)–(4):

$$V_a = \sqrt{u^2 + v^2 + w^2}; \; \dot{V}_a = \frac{u\dot{u} + v\dot{v} + w\dot{w}}{V_a} = \dot{u}\cos\alpha + dV_1,$$

where α is the angle of attack and dV_1 is a disturbance related to sideslip angle β.

Then, Equations (2)–(4) can be replaced by (9) after some substitutions:

$$\dot{V}_a = -g\sin(\theta - \alpha) - \frac{\rho V_a{}^2 S}{2m}[C_D + C_{D_{\delta_e}}\delta_e] + \frac{\rho S_{prop} C_{prop}}{2m}[(k\delta_t)^2 - V_a{}^2] + dV_2, \quad (9)$$

where $C_D, C_{D_{\delta_e}}$ are some coefficients and dV_2 is a disturbance similar to dV_1. Their definitions are given in [8].

2.2. Disturbance Model

In this paper two primary disturbance sources are considered: atmospheric turbulence (wind gust) and carrier airwake. Throughout the whole flight, the UAV is under the influence of wind gust, while disturbance due to airwake only affects the landing process, i.e., when the UAV is close to the carrier. Experiments show that a reliable wind turbulence model should include a steady component and a random component (gust), which can be generated by filtering Gauss white noise with zero mean and unit variance:

$$H_u(s) = \sigma_u \sqrt{\frac{2V_a}{L_u}} \frac{1}{s + \frac{V_a}{L_u}};\ H_v(s) = \sigma_v \sqrt{\frac{3V_a}{L_v}} \frac{s + \frac{V_a}{\sqrt{3}L_v}}{(s + \frac{V_a}{L_v})^2};\ H_w(s) = \sigma_w \sqrt{\frac{3V_a}{L_w}} \frac{s + \frac{V_w}{\sqrt{3}L_w}}{(s + \frac{V_a}{L_w})^2},$$

where H_u, H_v, H_w are the transfer functions; $\sigma_u, \sigma_v, \sigma_w$ are the gust intensities on corresponding axes; and L_u, L_v, L_w are the gust wavelengths [8]. Airwake starts to affect the UAV at about 800 m from the carrier; the resulting disturbance can be divided into four parts: free air turbulence component, steady component, periodic component and random component. The main factor is the steady component, which is usually described by empirical expressions [6]

$$\begin{cases} u_s(t) = 0.0072t^5 - 0.1739t^4 + 1.5532t^3 - 5.9699t^2 + 6.7692t + 10.4033 \\ w_s(t) = 0.0001t^5 - 0.0088t^4 + 0.1769t^3 - 0.8082t^2 - 1.4727t + 7.0433 \end{cases},$$

where u_s denotes the axial airwake disturbance; w_s denotes the normal airwake disturbance; and t denotes time before touchdown.

3. ADRC Control Scheme Design

In this section, a cascade LADRC scheme is designed for UAV control during landing operation. As shown in Figure 2, in the inner loop three single ADRC controllers are designed for attitude control and airspeed control. In the outer loop, navigation control is achieved through two ADRC controllers, which determine the desired attitude. The course, altitude and airspeed command are generated by path planning and a guidance algorithm; in this paper a simple guidance strategy and Dubins path are used; details can be found in [8].

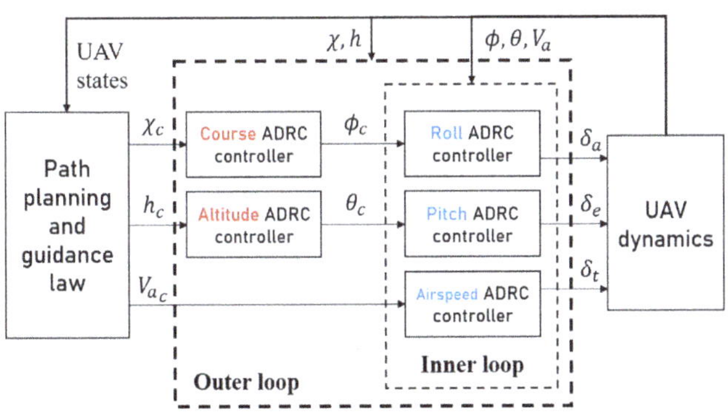

Figure 2. Cascade ADRC control scheme for fixed-wing UAV.

Similar to PID, the ADRC technique can be used only in SISO systems. In Figure 3a the structure of a LADRC controller for roll control is shown. A typical LADRC controller consists of three parts: linear tracking differentiator (LTD), linear state error feedback control law (LSEF) and linear extended state observer (LESO). The LTD is used to generate a smooth transient profile; for simplicity it will not be discussed here; its definition can be found in [8]. In practice, the LSEF is usually a PD controller; its parameter tuning will be discussed later in the end of this section. The LESO is the most important part of the ADRC controller; it estimates the "total disturbance" based on the input and output of the system. Consider an example of a second-order SISO system:

$$\begin{cases} \dot{x_1} = x_2 \\ \dot{x_2} = f(x_1, x_2, d(t), t) + bu, \\ y = x_1 \end{cases} \tag{10}$$

where $d(t)$ is an unknown external disturbance and f is the total disturbance. The LESO for this system can then be constructed as a Luenberger observer. Treating total disturbance f as an additional state x_3 for system (10), the corresponding LESO is then given by

$$\begin{cases} \dot{z_1} = z_2 + l_1(y - \hat{y}) \\ \dot{z_2} = z_3 + bu + l_2(y - \hat{y}), \\ \dot{z_3} = h + l_3(y - \hat{y}) \end{cases} \tag{11}$$

where y is the output measurement, which is state z_1; z_3 is estimation of the total disturbance; h is a derivative of f; and l_1, l_2, l_3 are the observer gains. In [9], a tuning method for these coefficients in a second-order system was proposed: $l_1 = 3\omega_o$; $l_2 = 3\omega_o^2$; $l_3 = \omega_o^3$, ω_o is the LESO bandwidth.

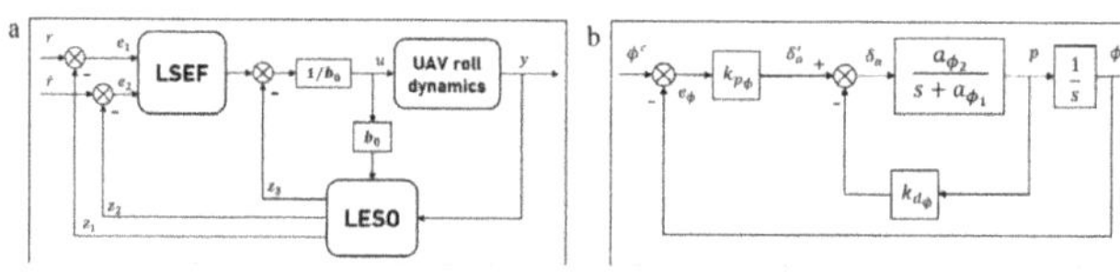

Figure 3. (**a**) ADRC controller structure; (**b**) roll angle control diagram.

The ADRC controllers for different state variables have to be designed individually, as shown in Figure 3a. According to Equations (5)–(9), both of the dynamics of roll and pitch can be simplified as SISO systems:

$$\begin{cases} \dot{\phi} = p \\ \dot{p} = f_\phi + b_\phi \delta_a; \\ y = \phi \end{cases} \begin{cases} \dot{\theta} = q \\ \dot{q} = f_\theta + b_\theta \delta_e, \\ y = \theta \end{cases}$$

where f_ϕ, f_θ are the total disturbances to be estimated and $b_\phi = \frac{1}{2}\rho {V_a}^2 SbC_{p_{\delta a}}$, $b_\theta = \frac{\rho {V_a}^2 Sc}{2J_y} C_{m_{\delta e}}$ are control gains.

Since Equation (9) is nonlinear in control input δ_t, we must first linearize it to design the LESO for the airspeed dynamics. Let $\bar{V_a} = V_a - V_a^*$ be the deviation of V_a from trim, and $\bar{\theta}$ and $\bar{\delta_t}$ be the corresponding deviations. The new airspeed dynamics linearized around the trim condition are described as

$$\dot{\bar{V_a}} = -g\cos(\theta^* - \alpha^*) - \frac{\rho V_a^* S}{m}[C_D^* - \frac{\rho S_{prop} C_{prop}}{m} V_a^*]\bar{V_a} + \frac{\rho S_{prop} C_{prop}}{m} k^2 \delta_t^* \bar{\delta_t} + dV_3,$$

where dV_3 is the disturbance; all the parameter definitions can be found in [8].

Analogously, we can determine a SISO system for airspeed dynamics:

$$\begin{cases} \dot{\bar{V_a}} = f_v + b_v \delta_t \\ y = \bar{V_a} \end{cases}; b_v = \frac{\rho S_{prop} C_{prop}}{m} k^2 \delta_t^*.$$

Then, according to (11) the LESOs for airspeed, pitch and roll systems are designed as follows:

$$\begin{cases} \dot{z_{\phi 1}} = z_{\phi 2} + l_{\phi 1}(\phi - z_{\phi 1}) \\ \dot{z_{\phi 2}} = z_{\phi 3} + b_\phi \delta_a + l_{\phi 2}(\phi - z_{\phi 1}); \\ \dot{z_{\phi 3}} = h_\phi + l_{\phi 3}(\phi - z_{\phi 1}) \end{cases} \begin{cases} \dot{z_{\theta 1}} = z_{\theta 2} + l_{\theta 1}(\theta - z_{\theta 1}) \\ \dot{z_{\theta 2}} = z_{\theta 3} + b_\theta \delta_e + l_{\theta 2}(\theta - z_{\theta 1}); \\ \dot{z_{\theta 3}} = h_\theta + l_{\theta 3}(\theta - z_{\theta 1}) \end{cases} \begin{cases} \dot{z_{v1}} = z_{v2} + b_v \delta_t + l_{v1}(\bar{V_a} - z_{v1}) \\ \dot{z_{v2}} = h_v + l_{v2}(\bar{V_a} - z_{v1}) \end{cases}$$

Altitude dynamics (1) can be simplified as follows and course dynamics in conditions of coordinated turn with zero wind are described in [8].

$$\dot{h} = V_a\theta + d_h,\ d_h = u\sin\theta - V_a\theta - v\sin\phi\cos\theta - w\cos\phi\cos\theta;\ \dot{\chi} = \frac{g}{V_a}\phi + d_\chi,\ d_\chi = \frac{g}{V_a}(\tan\phi - \phi).$$

Then, the LESOs for altitude and course dynamics are given by

$$\begin{cases} \dot{z_{h1}} = d_h + V_a\theta_c + l_{h_1}(h - z_{h1}) \\ \dot{z_{h2}} = h_h + l_{h2}(h - z_{h1}) \end{cases};\ \begin{cases} \dot{z_{\chi 1}} = d_\chi + \frac{g}{V_a}\phi_c + l_{\chi_1}(\chi - z_{\chi 1}) \\ \dot{z_{\chi 2}} = h_\chi + l_{\chi 2}(\chi - z_{\chi 1}) \end{cases}.$$

To ensure adequate performance of LESOs, their bandwidths $\omega_{o_v}, \omega_{o_\phi}, \omega_{o_\theta}, \omega_{o_h}, \omega_{o_\chi}$ must be chosen properly. A large bandwidth allows fast tracking without overshooting but, at the same time, high bandwidth could make the system very sensitive to noise. In this paper, bandwidth choice is made based on the natural frequency of the observer system.

Figure 3b shows a PD control diagram of the roll angle; the natural frequency of this second-order system can be calculated as follows:

$$\omega_\phi^2 = k_{p_\phi} a_{\phi 2};\ k_{p_\phi} = \frac{\delta_a^{max}}{e_\phi^{max}} sign(a_{\phi 2});\ \omega_\phi = \sqrt{|a_{\phi 2}| \frac{\delta_a^{max}}{e_\phi^{max}}},$$

where δ_a^{max} is the maximum elevator deflection and e_ϕ^{max} is the maximum roll angle. All the coefficients in Figure 3b can be found in [8].

Analogously, natural frequency for pitch, airspeed, altitude and course dynamics can be determined. Then, LESO bandwidths can be chosen empirically as $\omega_{o_\phi} = 2\omega_\phi$; $\omega_{o_\theta} = 2\omega_\theta$; $\omega_{o_v} = 2\omega_v$; $\omega_{o_h} = \frac{\omega_{o_\theta}}{2}$; $\omega_{o_\chi} = \frac{\omega_{o_\phi}}{2}$. Since the total disturbance is estimated by LESO, the estimation then can be used to compensate control input to make the control system robust and adaptive. Thus, the LADRC control law is formed as $(u_0 - f)/b$. $u_0 = k_p(r - z_1) + k_d(\dot{r} - z_2)$ is the control law given by LSEF, r is the reference signal, z_1, z_2 are the LESO outputs, $k_p = 2\omega_c, k_d = \omega_c^2$ are the LSEF control gains and ω_c is the LSEF controller bandwidth, in practice it is usually chosen between 4 and 20 [9].

4. Simulation Results

The ADRC control scheme proposed in Section 2 is implemented in the Simulink environment; to verify its advantages over PID controllers in terms of robustness and insensitivity to disturbances, several experiments are conducted to simulate the UAV carrier landing operation with ADRC and PID, respectively. The longitudinal and lateral dynamics of UAVs are usually discussed separately; to validate the effectiveness of the proposed algorithm on both of the dynamics, two stages of landing are considered: the orbit following and the final approach. The orbit is a spiral route over the carrier; it allows the landing officers to prepare the flight deck and ensure a safe landing. In the experiment the orbit is a Dubins path connected by four waypoints, which are shown in Table 1. The Dubins path algorithm can be found in [8]. And, for simplicity, a 3.5-degree glide slope is used as the final approach route.

Table 1. Initial conditions of final approach simulation.

Object	North Coordinate (m)	East Coordinate (m)	Altitude (m)	Course Angle (deg)	Velocity (m/s)
UAV	0	0	500	0	35
Carrier	1000	0	5	0	0

Figure 4 shows wind gust turbulence and airwake disturbance. To better demonstrate the disturbance rejection performance of ADRC, in the experiment the gust intensity is multiplied by 20.

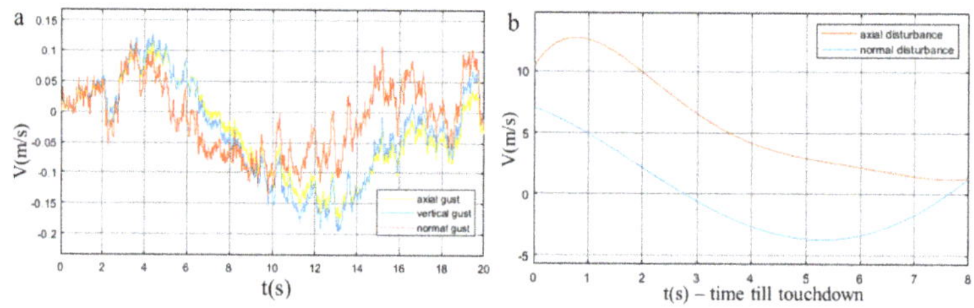

Figure 4. (**a**) Gust velocity in UAV body frame; (**b**) airwake disturbance velocity.

In the experiment a benchmark circle following maneuver that excites all the lateral states is used to illustrate lateral control performance. As shown in Figure 5, even in the existence of intense wind gust ADRC can still follow the circle with high accuracy and from the tracking error result it can be seen that ADRC does suppress high-frequency noise.

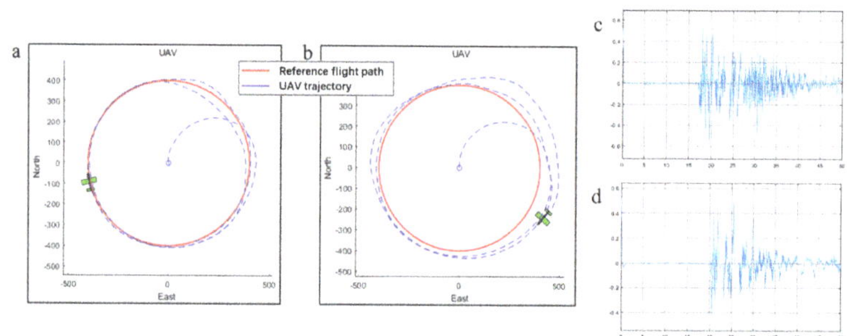

Figure 5. (**a**) Circle following with ADRC; (**b**) circle following with PID; (**c**) PID roll tracking error; (**d**) ADRC roll tracking error.

Figure 6 shows that in terms of path following performance ADRC is far more effective and accurate than PID. The east position error of the ADRC algorithm at the end point is 11.5 m, which is almost two times less than the PID error, 19.4 m.

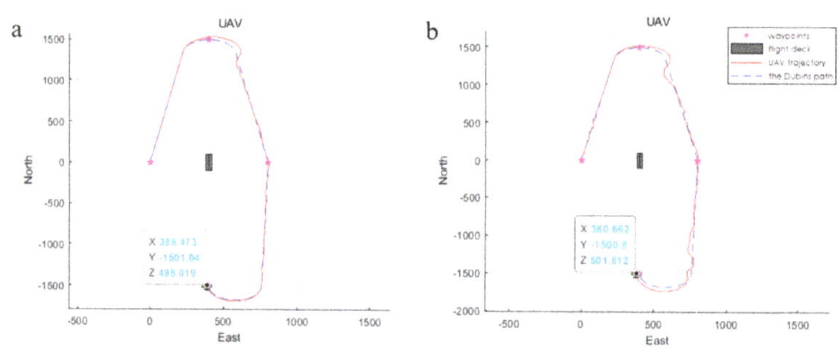

Figure 6. (**a**) ADRC orbit following result; (**b**) PID orbit following result.

Figure 7 shows the results of final approach simulations. It can be seen that ADRC provides a smoother approach route than PID and the ADRC position error at touchdown point is 2.44 m, which is about half the PID error, 4.95 m. The initial conditions of UAV and carrier are shown in Table 1.

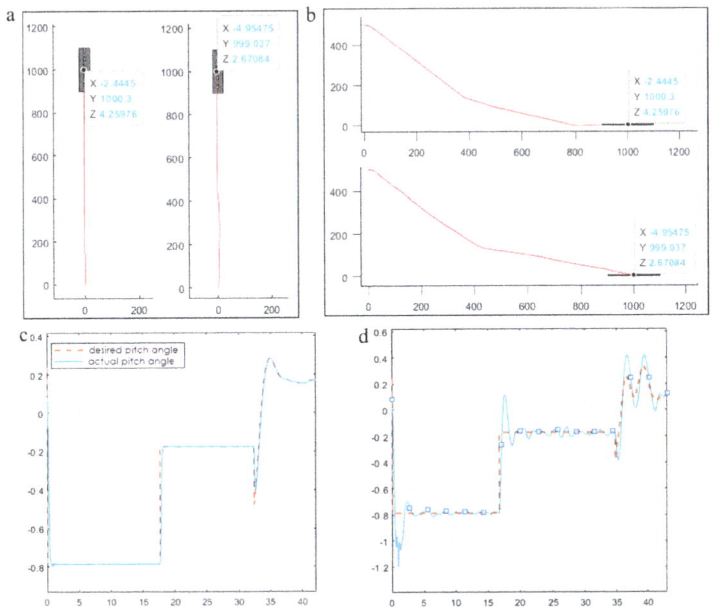

Figure 7. (**a**) Vertical view of final approach results (left: ADRC, right: PID); (**b**) end view of final approach results; (**c**) ADRC pitch tracking result; (**d**) PID pitch tracking result.

5. Conclusions and Future Work

This paper described our work on the topic of automatic UVA carrier landing. In this work a cascade ADRC control scheme was proposed for fixed-wing UAVs. The influence of wind gust and atmospheric disturbance due to carrier airwake was studied. In addition, a UAV dynamics simulation system was developed in the Simulink environment and the proposed control scheme was tested by a series of simulations. The results verified the effectiveness of the ADRC control scheme compared to the PID control:

- The controller design and parameter tuning processes for ADRC are much easier than for PID;
- The ADRC control system has excellent robustness and accuracy even in the presence of intense disturbances.

As further work on this project, we will try to exploit the total disturbance information more comprehensively; an adaptive navigation method based on the LESO will be studied as an extension of the UAV control system proposed in this paper.

Author Contributions: Conceptualization, R.Z.; methodology, K.A.N. and R.Z.; software, R.Z.; validation, R.Z.; formal analysis, K.A.N.; writing—original draft preparation, R.Z.; writing—review and editing, K.A.N.; visualization, R.Z.; supervision, K.A.N. All authors have read and agreed to the published version of the manuscript.

Funding: This research received no external funding.

Institutional Review Board Statement: Not applicable.

Informed Consent Statement: Not applicable.

Data Availability Statement: No new data were created or analyzed in this study. Data sharing is not applicable to this article.

Conflicts of Interest: The authors declare no conflict of interest.

Abbreviations

The following abbreviations are used in this manuscript:

ADRC	Active disturbance rejection control
LADRC	Linear active disturbance rejection control
LTD	Linear tracking differentiator
LSEF	Linear state error feedback control law
LESO	Linear extended state observer

References

1. Hess, R.A. Analysis of the Aircraft Carrier Landing Task, Pilot + Augmentation/Automation. In *2019 IFAC-PapersOnLine*; Elsevier: Amsterdam, The Netherlands, 2019; pp. 359–365.
2. Zhang, L.; Wang, S.; Selezneva, M.S.; Neusypin, K.A. A new adaptive Kalman filter for navigation systems of carrier-based aircraft. *Chin. J. Aeronaut.* **2022**, *35*, 416–425. [CrossRef]
3. Zhang, L. Flight Control Quality Assessment of Carrier-Based Aircraft Using Degree of Controllability Criterion. *Aerosp. Instrum.* **2021**, *12*, 10–18.
4. Shen, K.; Zhang, L.; Selezneva, M.S.; Neusypin, K.A. Modification of federated Kalman filter in the correction scheme of aircraft carrier navigation systems. *Autom. Mod. Technol.* **2020**, *73*, 177–180.
5. Zhou, R.; Neusypin, K.A. Model predictive control for automatic carrier landing considering ship motion. *J. Phys. Conf. Ser.* **2022**, *2235*, 012005. [CrossRef]
6. Zhen, Z.; Miao, P.; Xue, Y.; Jiang, S. Robust preview control and autoregressive prediction for aircraft automatic carrier landing. *IEEE Access* **2019**, *7*, 18273–18283. [CrossRef]
7. Yu, Y.; Wang, H.; Shao, X.; Huang, Y. The attitude control of UAV in carrier landing based on ADRC. In Proceedings of the 2016 IEEE Chinese Guidance, Navigation and Control Conference (CGNCC), Nanjing, China, 12–14 August 2016; pp. 832–837.
8. Randal, W.B.; Timothy, W.M. *Small Unmanned Aircraft: Theory and Practice*; Princeton University Press: Princeton, NI, USA, 2012.
9. Gao, Z. Scaling and Bandwidth-Parameterization Based Controller Tuning. In Proceedings of the 2003 American Control Conference, Denver, CO, USA, 4–6 June 2003; pp. 4989–4996.

engineering proceedings

MDPI

Proceeding Paper

AGI's Hierarchical Component Approach to Unsolvable by Direct Statistical Methods Complex Problems †

Vladimir Smolin * and Sergey Sokolov

Keldysh Institute of Applied Mathematics, Russian Academy of Sciences, Miusskaya Sq., 4, Moscow 125047, Russia; sokolsm@list.ru
* Correspondence: smolin@keldysh.ru; Tel.: +7-916-555-3998; Fax: +7-499-972-7737
† Presented at the 15th International Conference "Intelligent Systems" (INTELS'22), Moscow, Russia, 14–16 December 2022.

Abstract: The amazing deep neural network (DNN) advances over the past 10 years have made it possible, if there is enough data and computing power, to achieve solutions to unexpectedly complex problems. But DNN does not explicitly use decomposition, the main advancement method in complicated task solving. The automatic complex scenes decomposition can be carried out based on mapping by a neural network. The problem is the impossibility to map complex objects and phenomena state spaces. The hierarchical complex scene's division into simple components can be a key for solving the problem. The hierarchically organized structure of simple objects and phenomena maps of different abstraction levels can make it possible to solve problems in a complex environment, in which all properties cannot directly be revealed by statistical methods. Operation modes of such a hierarchical structure can be correlated with terms used in philosophy and psychology.

Keywords: artificial general intelligence (AGI); decomposition; AGI agent

Citation: Smolin, V.; Sokolov, S. AGI's Hierarchical Component Approach to Unsolvable by Direct Statistical Methods Complex Problems. *Eng. Proc.* **2023**, *33*, 67. https://doi.org/10.3390/engproc2023033067

Academic Editors: Askhat Diveev, Ivan Zelinka, Arutun Avetisyan and Alexander Ilin

Published: 10 October 2023

1. Introduction

The neural network revolution in machine learning, based on the use of deep neural networks, has led to tremendous progress in the AI field. Success is usually associated with parallel computing methods development, collecting big data, the neural network structures, and the algorithms embedded in them improving.

The central training algorithm for modern neural networks that solve applied problems is the backpropagation error (BPE) method, which implements the idea of gradient descent. The first ideas of training multilayer networks were expressed by Rosenblatt [1], improved by Rumelhart [2], and were formulated close to modern concepts as early as 1986 in [3]. There were other publications containing similar learning algorithms for neural networks [4].

But it took almost 25 years for BPE methods to demonstrate their effectiveness in solving practical problems [5]. The BPE use without batch-norm [6], dropout, and a number of other algorithms that complement BPE does not allow solving complex applied problems. But even if you use the entire arsenal of modern algorithms and apply it to a neural network with a random structure, then there will be no result either. This begs the question:

1.1. Is BPE Optimization the Best Way to Learn?

Gradient descent implemented by BPE is an optimization algorithm. Although BPE is used to train deep neural networks, the idea of optimizing their parameters by anti-gradient lies on the surface. Optimization means improvement, but it can be small or significant, fast or slow, and has a number of parameters that allow us to compare different optimization methods for solving various problems. Twenty-five years of BPE improvements have made it possible to use this method for solving application problems, and in the past 10 years with commercial financial support, progress has become even more noticeable.

1.2. Are There Other Ideas besides Gradient Descent?

The fact that BPE ideas development has continued for over 35 years shows that optimization is not limited by gradient descent.

Another thing is that BPE does not guarantee the formation of the optimal transformation, such as localization, decomposition, linearization, etc., methods.

Neural networks with tens and hundreds of layers work and successfully solve many applied problems, but require large hardware and computational costs. The article [7] shows that just by analyzing the efficiency of using individual parameters, you can turn off 98% of parameters without losing the transformation accuracy.

The transition from the random formation of advanced optimization methods to their purposeful use allows increasing the efficiency of using (reducing the number of required) parameters of neural networks by thousands of times.

2. Mapping

Neural networks are universal converters that display the input signal vector $\vec{X}$ into the output vector $\vec{Y}$. It is easy to understand that for such a display it is necessary that different $\vec{Y}$ correspond to diverse $\vec{A}$ stated of the internal neural network activity. For this, it is sufficient (but not necessary) that for different $\vec{X}$, diverse $\vec{A}$ are formed.

BPE does not address this condition explicitly. In textbooks [8], it is customary to write that dropout simply improves the performance of BPE. There is reason to believe that the improvement is due to the fact that if different neural network parameter subsets are trained on different $\vec{X}$, then diverse $\vec{A}$ will be formed for different $\vec{X}$. But still, using dropout just makes it more likely for this condition to be met.

Mapping [9] is an algorithm that deterministically ensures the fulfillment of the formation conditions for different $\vec{X}$ diverse $\vec{A}$.

2.1. Low-Dimensional Maps

Neural network mapping algorithms go back to the k-means method [10]:

$$\Delta\vec{M}_i = \eta\left(\vec{X} - \vec{M}_i\right), \eta \ll 1 \quad (1)$$

where $\vec{M}_i$ is i-th element input connection weights vector and $\vec{X}$—input vector.

The i-th element is chosen using the WTA (winner takes all) method, that is, in k-means, the $\vec{M}_i$ connections of the most active elements are changed. Improving the k-means method in neural network mapping algorithms comes down to ensuring that all elements of the neural layer participate in the input vectors $\vec{X}$ mapping.

Low-dimensional (the diapason of dimensions 2–3) neural network mapping is widely used for data visualization. But the limited use of neural network maps in solving complex problems is associated not so much with the difficulty of their visualization, but with the impossibility of mapping high-dimensional X spaces of the input signal X by a dimension of 15–20.

Note that the dimension of X spaces is determined not by the number of sensors, but by the properties of the observed object or phenomenon, because X is a subspace in the space of theoretically possible sensor activities.

2.2. Uneven Mapping

As a result of training, k-means and neural network maps divide the input signal state spaces into subdomains, and each is characterized by approximately the same probability (frequency) of the vector $\vec{X}$ hitting it. Such mapping will be called uniform.

But to solve the transformation problems $\vec{X} \rightarrowtail \vec{Y}$, more accurate mapping (representation of smaller areas) X is needed not so much where $\vec{X}$ appears more often, but where the error

$\Delta\vec{Y}$ is greater than the average. This can be ensured by the dependence of the learning rate on $\Delta\vec{Y}$, for example:

$$\Delta\vec{M}_i = \eta\left(\Delta\vec{Y}\right)^2\left(\vec{X} - \vec{M}_i\right), \eta \ll 1 \tag{2}$$

3. Mapping as Transformation

Neuromapping allows us to transform the input vector $\vec{X}$ into the mapping layer activity $\vec{A}$. The standard WTA rule use causes one element in the mapping layer to have activity equal to 1, and the rest elements activity to zero. If we simply connect each element of the mapping layer with the output layer and assign the weights of the output connections $\vec{L}_i$, equal to the average $\vec{Y}$ value in the area X where $A_i = 1$, then we obtain a zero-order transformation approximation of $\vec{X} \rightarrowtail \vec{Y}$.

If only transformations $\vec{X} \rightarrowtail \vec{Y}$ of dimensions 1 or 2 were considered, everything is simple. But for X dimensionality like 5, 10, and especially 15, increasing the accuracy will require exponentially increasing costs. The transition to at least a piecewise linear approximation will give great savings in the mapping layer elements number.

3.1. Piecewise Linear Transformation

For piecewise linear transformations, it is necessary to go from WTA to kWTA, in which not one, but k of the most active elements of the mapping layer is selected. This is not difficult if you can determine the value of k, the dimensionality of X_i.

In addition, the activities A_i for the (multidimensional) piecewise linear approximation implementation of the transformation $\vec{X} \rightarrowtail \vec{Y}$ must differ from 1 and depend on the $\vec{M}_i$ and $\vec{X}$ ratio. Moreover, it will be necessary to estimate $G_i = max(A_i)$ over all $\vec{X}$. Then both k and all A_i can be found from the equality:

$$S_k = \sum_{j=1}^{k} \frac{A_i}{G_i} = 1 \tag{3}$$

in which $A_i = \vec{M}_i\vec{X} + B_i - T$, where B_i and $\vec{M}_i$ are the parameter learning result of the layer elements by the threshold mapping algorithm, and k and T are chosen as follows: $T = \vec{M}_k\vec{X} + B_k$ is preliminarily taken so that $S_k \geq 1$ and $S_{k-1} < 1$. This determines the k value, and the exact value of T, which provides equality (3), is obtained from the linear equation solution.

3.2. Nonlinear Transformations

The piecewise linear transformations disadvantage is kinks, which significantly reduces the smooth transformations approximation accuracy. This shortcoming can be overcome by non-linear obtained from k neuromaps approximations averaging, moreover, each neuromap can have more than k times fewer elements.

3.3. Mapping Limitations

The transition from the zero-order approximation to the piecewise-linear approximation allows one to move from mapping input signal spaces states X with the diapason of dimensions 2–3 to diapason 5–10, and nonlinear approximation to 12–15. Perhaps more subtle approximation methods will allow to move a little further into the region of high dimensions. Anyway, increasing the X dimension, leads to an exponential rise in costs. Just that advanced methods allow reducing the base and coefficient of the exponential function.

If there is a decomposition possibility and there are no significant time restrictions on using several maps (it is longer than the approximation based on one map), then it is always more efficient to create several low-dimensional maps instead of one high-dimensional. This will require significantly fewer hardware costs and training time.

Conclusion: the main way to use neuromapping is to decompose complex scenes and tasks into simpler (low-dimensional) components, as well as these components' localization and linearization.

4. Localization, Decomposition, Linearization

Decomposition is the most important condition for using neural mapping in applied problems. Localization speeds up the learning process and reduces computational costs. Transformations linearization allows moving further into high dimensions.

4.1. Localization

Modern neural networks operate on a "distributed" memory model. BPE does not provide a special algorithm for the changes localizing. But all changes occur locally on the weights of connections. For efficient operation, a "sparse" representation is needed so that the changes made for different input signals do not spoil each other.

Mapping provides competitive WTA or kWTA activation and input signals recording in the connections weights of various hidden layer elements. Decomposition ensures the complex scenes' various component properties are distributed through different neural network maps, enhancing the data localization.

4.2. Decomposition

The impossibility of complex scenes mapping leads to their decomposition into simple components for which mapping is possible. But for the complex signals perception, it is necessary to compare them with the sum of the simple components memorized earlier. That is, the maps remember not only the transformation $\vec{X}_i \to \vec{Y}_i$, but also $\vec{X}_i \to \tilde{\vec{X}}_i$, where $\vec{Y}_i$ and $\tilde{\vec{X}}_i$ are the output and input vectors stored on the i-th map, corresponding to the i-th component of the input signal $\vec{X}$. Then, we can extract $\vec{X}_i$ from the complex signal $\vec{X}$:

$$\vec{X}_i(t) = \vec{X}(t) - \sum_{j \neq i}^{N} \tilde{\vec{X}}_j(t) \quad (4)$$

Similar splitting input signal ideas were expressed in other works [11,12], but, without the use of mapping, they were not developed. The successful decomposition of complex signals is possible when there are maps of complex signal components. Otherwise, a complex situation cannot be represented as a simple components sum.

4.3. Linearization

Our world in general is described by nonlinear laws. But, for example, most of the laws of physics are reduced to a linear form by taking a logarithm. Similarly, the significant transformations part carried out by neural networks can also be linearized.

The linearization makes it possible to perform transformations of a higher dimension, not only to describe them more compactly. But the linearized transformations' main property is the component's contribution to the independence of the $\vec{X}_i$ to the change $\vec{Y}_i$. This allows selecting the essential variables in $\vec{X}_i$.

5. Hierarchical Structure

The complex signals division into simple components involves the reverse process—the restoration from the components. Partially, such restoration is used at each level when calculating (4), but in a more complete way, it can be implemented in a hierarchical structure. It is even more important that new variables can be distinguished in the maps of the lower levels—the map coordinates, abstractly representing the properties of the lower level signals. And for the upper levels compose complex signals from these more abstract descriptions and also reveal dependencies in them by decomposition and mapping means.

5.1. High-Level Abstraction

Multiple mapping procedure repetitions and signal component map coordinates emerge over the hierarchy levels, opening the way to the high-level abstractions formation directly based on the sensors activity without any human participation.

For the maps' hierarchy successful formation, effectors are also needed to influence the outside world and a block for the action results evaluation, but effectors and evaluation description is beyond the short article scope.

But the central idea for describing the local part of a complex world remains its decomposition into components, indirectly describing by maps, the sensors' activity and more abstract variables changes—the coordinates of maps of different levels.

5.2. Abstract Processes Modeling

Mapping allows identifying relatively simple dependencies between the different hierarchy level signals and building complex signal component models, regardless of the abstraction level.

Maps can describe not only static but also dynamic objects properties. This makes it possible not only to observe the scene's current state, but also to predict the development of this or another scene, not necessarily related to the current observation.

The need to model various (depending on the selected actions) options for the complex scene components development is associated with the inability to collect complex scene statistics and model them directly, without decomposition. But it is not necessary to model complex scenes in all details—first it is enough to use high-level variables for comparing different options and only after choosing the best ones to consider chosen variants in detail, closer to the real-world variables.

6. Screens

Simple components of possible vector states representing complex signals can be stored in the map's structure (this is why simple components differ from complex ones). The complex signals are almost never repeated, and it makes no sense to memorize them. But precisely complex signals are needed to be analyzed and transformed, but mapping cannot cope with them. To work with complex signals inaccessible for mapping, it is necessary to use special structures for displaying complex signals—screens. Screens are not intended for visualizing complex signals (which is impossible for high hierarchy level abstract representations), but for processing complex signals, their formation, and transmission between levels of the map's hierarchical structure.

6.1. What the Seen for the First Time Scenes Could Be Compared to?

The never-repeating complex signal vectors $\vec{X}$ can be compared with the sum of $\tilde{\vec{X}}_i$, stored in simple objects and phenomena maps. The comparison, according to (4), is aimed at identifying individual simple components in the complex signal structure.

At higher levels of the hierarchy, complex signals are formed from the coordinates of the lower levels maps. These coordinates are abstract descriptions of the essential objects' properties. Complex high-level signals are not only constructed from directly observed objects and phenomena. Objects whose state is estimated on the basis of modeling using dynamic properties reflected in their maps can be used too.

The screen also selects which maps to use to form a complex signal. That is similar to the attention not only to observable but also to unobservable objects, affecting the actions' choice. An equally important screen function is to track the set for the lower level goals' achieving process. If the process is going successfully, the upper levels can be switched to modeling actions not related to the lower scene's current state. Otherwise, it is necessary to try to change the current goal.

6.2. How Can Abstract Goals Be Turned into Real Actions?

The AGI agent effectors and muscle actions formation are reduced to control signal creation. Effectors must also have internal sensors and, from the control point of view, represent a part of the world. Simple objects state maps can be formed on the data from both an external and internal sensor basis. For the lowest hierarchy levels associated with effectors, the transformation $\vec{X}_i \to \vec{Y}_i$ consists in generating a direct control signal $\vec{Y}_i$ based on the current state of effectors $\vec{X}_i$. But this is not the only way to determine actions. We need an action goal $\tilde{\vec{X}}_i$, which determines the desirable changes in the effectors' state. This goal comes from higher levels.

If the goal $\tilde{\vec{X}}_i$ achievement from a similar state $\vec{X}_i$ was carried out repeatedly, then (after learning) you can set a goal that is very different from the current state, and the map will cope with achieving such a goal. If the target is being set for the first time, then upper levels should plan small changes to the current effectors' state so that the target can be reached based on the map's available properties.

As we move to higher hierarchy levels, goals become more abstract. First, instead of controlling effectors, changes in the objects in the external world are planned, then the processes organization, and so on. The different goals formations are carried out on the varying abstractness degrees modeling basis, depending on how close the chosen goal is to achieving it and how specific actions are necessary for this.

7. Intuition and Thinking

Considering an objective neural network model (neuromaps multi-level hierarchical structure) from a third-person perspective allows obtaining a new look (not subjectively) at a number of philosophical and psychological concepts. This gives hope for removing the gap between neuronal activity and higher-level concepts such as consciousness, thinking, intuition, etc. The easiest to explain are intuition and thinking.

7.1. Intuition

Intuition corresponds to actions and decisions taken without thinking on the basis of transformations $\vec{X}_i \to \vec{Y}_i$ available in the maps. But this is not just pulling the hand away from the hot (what can be attributed to reflexes), but taking into account the current situation complexity. If there is no time for thinking, then the maps' hierarchical component structure can be based on transformations $\vec{X}_i \to \vec{Y}_i$ that use all levels for changing goals and perform actions to achieve them—an analogue of intuition.

7.2. Thinking

Complex scenes do not repeat themselves, their development prediction (without simulation) is extremely inaccurate. Still, you can act on the basis of previous experience and not think about it. If the situation has not been encountered before, then it is better to simulate the options for its development when performing various actions (based on the available component maps). This will allow us to choose much better action options based on the modeling of consequences—an analogue of thinking.

8. Consciousness, Understanding and Emotions

More difficult to explain are concepts that are not directly aimed at choosing actions. It is not clear why consciousness, understanding, and emotions are needed at all, because there is BPE, RL (reinforcement learning) and it seems that this is enough to solve any problems [13]. For simple tasks—yes, it is fairly.

For complex problems that cannot be solved directly using statistical methods, it is extremely useful to use preliminary action results modeling before they are performed. This requires solving and controlling the neural networks modes operation problem, not directly related to the formation of actions but has analogies with consciousness, understanding, and emotions.

8.1. Consciousness

In a hierarchical component model, all levels (except the bottom one) can be used in two modes: action or simulation (which corresponds to intuition and thinking) [14].

If the lower levels of the hierarchy successfully cope with the achievement of the set goal, then it is known what situation will arise. And there is time for modeling several options and comparing them. This corresponds to proactive behavior when we purposefully achieve goals and prepare new goals and actions in advance for the situation we are striving to reach.

If the set goal is not achieved, then we are forced to return to reactive behavior, which is usually less rational than proactive. Then the upper levels must be used in an intuitive way to set new goals. Within the hierarchical component model framework, as an analog of consciousness, one can consider the mechanism for coordinating and distributing resources between the execution of thinking and intuitive activity. This function can be attributed to screens or to a special part of the system.

8.2. Understanding

Successful actions construction (both on the basis of intuition and thinking) is associated with the correspondence degree of models (maps) to the actually observed situation. And there is an alternative: to use the existing models immediately or try to improve them first in various ways. To solve the mode selection problem, it is necessary to compare the complex input signal (which never repeats) with its available components models. And the considered hierarchical model allows us to carry it out.

Improvement of component models at lower levels is carried out by observing real objects. And at the upper levels—by implementing the simulation of different options based on the existing lower-level models. Separation by the level functions is carried out by consciousness, which also uses the degree of mismatch (understanding) assessment between the observed situation and its component model.

An understanding analog is the ability to assess the correspondence degree of a component's model to the observed state of real objects and phenomena. The assessment of the correspondence (understanding) degree is used by the function of consciousness to select the system operating modes.

8.3. Emotions

In the process of modeling, it is necessary to reduce all estimates to a single indicator and choose the best (forecast) option. But the choice is not only between action options but also between modes of operation, of which there are quite a lot (intuitive and meaningful behavior, achieving goals, avoiding danger, search behavior, and others).

In the hierarchical component model, consciousness controls only the switching between intuition and thinking. Activation degree regulation of other modes can be correlated with emotions. In animals, similar control is carried out due to the hormonal background, which also allows switching the neural system modes.

9. Conclusions

The proposed ways of creating a hierarchical component system are aimed at the model's automatic creation of the world, sufficient for the formation of AGI-level actions. The main idea is to record the surrounding world's simple properties through neural network mapping. The hierarchical component system is essentially aimed at representing complex signals in a form that allows mapping (in the form of low-dimensional components).

The hierarchical structure of maps and screens make it possible to implement the localization, decomposition, and linearization ideas of the performed transformations. Maps work with components of complex signals. Information about the components can be accumulated by statistical methods. Screens process and form complex, non-repetitive signals, which are compared with the components obtained from the created maps.

The complex structure and several operation modes of a hierarchical component system allow correlating some of the properties with philosophical and psychological ideas about intuition, consciousness, etc., by objectively analyzing the properties available for a scientific research formalized structure. The proposed hierarchical component system was created not to argue with philosophers, but to achieve progress in the creation of AGI agents. The development of a hierarchical component system began several decades before the call of the "godfathers" of deep learning [15] to search for new ideas and approaches, but it can serve as a response to this call.

The main ideas and some algorithms of the hierarchical component system are still under development. to create AGI agents based on it, a large additional amount of research is needed. But BPE's path to commercial success was not easy either: about 25 years passed from development to the start of solving practical problems. The modern industry of AI and neural networks allows going this way for a hierarchical component system much faster. The prize for the proposed approaches development can be an increase in the efficiency of using (reducing the number of required) parameters of neural networks by thousands times.

Author Contributions: Conceptualization, V.S. and S.S.; methodology, V.S. and S.S.; software, V.S. and S.S.; validation, V.S. and S.S.; formal analysis, V.S. and S.S.; investigation, V.S. and S.S.; resources,V.S. and S.S.; data curation, V.S. and S.S.; writing—original draft preparation, V.S. and S.S.; writing—review and editing, V.S. and S.S.; visualization, V.S. and S.S. All authors have read and agreed to the published version of the manuscript.

Funding: This research received no external funding.

Institutional Review Board Statement: Not applicable.

Informed Consent Statement: Not applicable.

Data Availability Statement: Not applicable.

Conflicts of Interest: The authors declare no conflict of interest.

References

1. Rosenblatt, F. *Principles of Neurodynamics; Perceptrons and the Theory of Brain Mechanisms;* Spartan Books: Washington, DC, USA, 1962.
2. Rumelhart, D.E.; McClelland, J.L. (Eds.) *Parallel Distributed Processing: Explorations in the Microstructures of Cognition;* MIT Press: Cambridge, MA, USA, 1986.
3. Rumelhart, D.E.; Hinton, G.E.; Williams, R.J. Learning Internal Representations by Error Propagation. In *Parallel Distributed Processing;* MIT Press: Cambridge, MA, USA, 1986; Volume 1, pp. 318–362.
4. Werbos, P.J. Beyond Regression: New Tools for Prediction and Analysis in the Behavioral Sciences. Ph.D. Thesis, Harvard University, Cambridge, MA, USA, 1974.
5. Krizhevsky, A.; Sutskever, I.; Hinton, G.E. ImageNet Classification with Deep Convolutional Neural Networks. *Commun. ACM* **2017**, *60*, 84–90. [CrossRef]
6. Ioffe, S.; Szegedy, C. Batch normalization: Accelerating deep network training by reducing internal covariate shift. In *ICML'15: Proceedings of the 32nd International Conference on International Conference on Machine Learning, Lille, France, 7–9 July 2015;* ML Research Press: Norfolk, MA, USA, 2015; Volume 37, pp. 448–456.
7. Molchanov, D.; Ashukha, A.; Vetrov, D. Variational Dropout Sparsifies Deep Neural Networks. In *ICML'17: Proceedings of the 34th International Conference on Machine Learning, Sydney, Australia, 6–11 August 2017;* ML Research Press: Norfolk, MA, USA, 2017; Volume 70, pp. 2498–2507.
8. Goodfellow, I.; Bengio, Y.; Courville, A. *Deep Learning;* MIT Press: Cambridge, MA, USA, 2016.
9. Kohonen, T. *Self-Organizing Maps (Third Extended Edition);* Springer Berlin: Heidelberg, Germany, 2000.
10. MacQueen, J. Some methods for classification and analysis of multivariate observations. In *Proceedings of the 5th Berkeley Symposium on Mathematical Statistics and Probability, Berkeley, CA, USA, 21 June–18 July 1965 and 27 December 1965–7 January 1966;* University of California Press: Berkeley, CA, USA, 1967; pp. 281–297.
11. Fukushima, K. Neocognitron: A Self-organizing Neural Network Model for a Mechanism of Pattern Recognition Unaffected by Shift in Position. *Biol. Cybern.* **1980**, *36*, 193–202. [CrossRef] [PubMed]
12. Carpenter, G.A.; Grossberg, S. ART 2: Self-organization of stable category recognition codes for analog input patterns. *Appl. Opt.* **1987**, *26*, 4919–4930. [CrossRef] [PubMed]

13. Dickson, B. DeepMind Scientists: Reinforcement Learning Is Enough for General AI. 2021. Available online: https://bdtechtalks.com/2021/06/07/deepmind-artificial-intelligence-reward-maximization (accessed on 10 July 2023).
14. Kahneman, D. *Thinking, Fast and Slow*; Macmillan: New York, NY, USA, 2011.
15. Bengio, Y.; LeCun, Y.; Hinton, G. Deep Learning for AI. *Commun. ACM* **2021**, *64*, 58–65. [CrossRef]

MDPI
St. Alban-Anlage 66
4052 Basel
Switzerland
www.mdpi.com

Engineering Proceedings Editorial Office
E-mail: engproc@mdpi.com
www.mdpi.com/journal/engproc

www.ingramcontent.com/pod-product-compliance
Lightning Source LLC
Chambersburg PA
CBHW061454220926
51850CB00004B/88
9783036592886